THE EVOLVING EARTH

Donald R. Prothero

CALIFORNIA STATE POLYTECHNIC UNIVERSITY POMONA

NEW YORK   OXFORD
OXFORD UNIVERSITY PRESS

Oxford University Press is a department of the University of Oxford.
It furthers the University's objective of excellence in research, scholarship,
and education by publishing worldwide. Oxford is a registered trade mark of
Oxford University Press in the UK and certain other countries.

Published in the United States of America by Oxford University Press
198 Madison Avenue, New York, NY 10016, United States of America.

For titles covered by Section 112 of the US Higher Education
Opportunity Act, please visit www.oup.com/us/he for the latest
information about pricing and alternate formats.

Library of Congress Cataloging-in-Publication Data
Names: Prothero, Donald R., author.
Title: The evolving earth / Donald R. Prothero (California State Polytechnic
   University Pomona).
Description: Oxford ; New York : Oxford University Press, [2021]
Identifiers: LCCN 2019017351 | ISBN 9780190605629 (pbk.)
Subjects: LCSH: Historical geology—Textbooks. | Geology—Textbooks.
Classification: LCC QE28.3 .D678 2020 | DDC 551.7—dc23 LC record available at
https://lccn.loc.gov/2019017351

Printing number: 9 8 7 6 5 4 3 2 1
Printed by Marquis, Canada

This book is dedicated to the memory of my good friend and co-author

Robert H. Dott, Jr.
(1929–2018)

Not only was he a legendary scholar in sedimentary geology and history of geology,
but he was also a pioneer in the effective teaching of earth history to college students

Brief Table of Contents

Contents

PART II   Earth and Life History   140

Preface

The earth is a dynamic planet, always changing and evolving. Its oceans and atmospheres have transformed radically since the earth was formed 4.6 billion years ago. Its crust is in constant motion, pulling continents apart and slamming them together to produce the great earthquakes and volcanoes. The core and mantle beneath are also flowing and moving, causing the changes in the crust above and in the magnetic field that shields us all from cosmic radiation. Finally, life on earth is closely tied to the changes in the earth's atmosphere, oceans, and lithosphere. Major changes in the physical environment of the planet have strongly influenced life (especially during mass extinctions), while life itself changes the earth in ways that occur on no other planet. Life produced the oxygen we breathe, changes the carbon dioxide concentrations of the atmosphere by storing it in crustal reservoirs like coal, oil, and gas, or weathered soil, or releasing it when events remove carbon from the crust. Life controls the flow of nutrients in the ocean, and transformed the surface of the land from a barren wasteland to the lush green surface that it is in most places. Finally, humans are intricately connected to all these cycles and changes, and we would not be here if it were not for some extraordinary accidents, such as the extinction that wiped out the dinosaurs and left the world wide open for our mammalian ancestors. Humans have now become a force of nature ourselves, wiping out entire species and ecosystems, transforming the crust, taking its resources, and now poisoning our own home with our pollution and destructive patterns of growth and consumption.

No subject in college brings these points home better than a survey of earth history. It is the best class to grasp the enormity of outer space and geologic time and the tiny part that humans have at those scales, and to better understand the forces that shape the planet, trigger the evolution of life, and especially inform how we humans view and exploit our home. That is why a course in earth history or historical geology is considered a fundamental background to almost all educations in geoscience, no matter what the specialty. It is also a great course for a non-major in geology to take, since it gives you a perspective that you can get from no other course.

But the nature of how that course is taught, and how textbooks are written for it, has changed radically in my lifetime. For most of the nineteenth and twentieth centuries, historical geology was often taught as a boring, memorization-heavy "march through time", with heavy emphasis on memorization of rock units and the names of index fossils—often at the expense of seeing or understanding why these things are important, or what is behind the events of the earth's past. It was often called the "Roll Call of the Ages". Factual details were often emphasized at the expense of understanding concepts or the context of these facts, or seeing the "big picture" of how they all fit into a broader framework. To paraphrase a famous historian, "history is just one bloody thing after another."

That all changed in the 1960s when geology underwent a scientific revolution known as plate tectonics. At first, historical geology textbooks ignored the topic, or treated it as an interesting but controversial idea, and then resumed teaching the same material in the same old way. All this changed in 1971, when my friends and colleagues Robert H. Dott, Jr., and Roger Batten wrote a groundbreaking textbook, *Evolution of the Earth*. It was the first to incorporate plate tectonics through the entire text, and explore its implications for understanding events in the earth's past in new ways. It was also the first book to take a more inquisitive approach, asking the questions about how and why we know things as much as what we know. Instead of just listing facts and events, Dott and Batten tried to give a background to understanding those events in context, and seeing the causes behind them. This approach worked successfully for over two decades until Roger Batten no longer wanted to work on the book, and I was asked to take it over in its fifth edition in the early 1990s. I was privileged to work on four more editions of the book, the eighth and last edition of which came out in 2010.

The Approach of This Book

But the nature of how earth history is taught, and the interests and skill levels of students, have also changed since the 1960s. Students taking this course today are often less willing to read a long, densely packed, and detailed book like *Evolution of the Earth*, which has become more suitable for an upper-level geology major seminar or graduate student course as taught today. Most instructors, especially in community colleges, now teach this course to a wider spectrum of students, many of whom are not geology majors or not even that familiar with science. With Bob Dott's blessing, I decided to start over again and write an entirely new book with a different publisher to serve the students who have less of a science background, and don't need the long detailed approach found in older books like

Evolution of the Earth. Yet I have tried to preserve some of the best elements of that previous book: exciting, interesting, vibrant writing and descriptions of events and creatures, and striking illustrations to help those who are visual learners. This is true especially the timeline diagrams at the end of many of the chapters, which our book pioneered and are now copied by every other book in the market. Most importantly, I wanted to carry on the Dott and Batten tradition on teaching "How do we know?" or "Why do we know?" and explain the evidence behind our assertions that certain events occurred at certain times and places. This emphasis is implicit throughout the text, where I discuss the evidence for certain facts about the past. However, to highlight important examples, every chapter has one or more separate boxed entries entitled "How do we know?" to explain the background behind a certain line of scientific evidence. Bob Dott taught this subject for over 40 years, and I have also reached my fortieth year of teaching this subject, and we both feel that at the college level, the most important thing a student can take away from a course like this is not a laundry list of factoids that they will promptly forget, but a better understanding about how things work and why things occur, which they will retain for the rest of their lives.

There are several themes that run through the entire book, with detailed discussion in certain key chapters:

1. The enormity of geologic time, and the slow pace of most geologic change, and its implications for humans;
2. The constant irreversible changes in the earth and life upon it, and what forces cause these changes;
3. The importance of the interactions between earth and life, especially in feedback loops and the way that physical environment controls life, but also life changes its own physical environment.
4. In an age when science deniers are more powerful in the United States than they are in any other developed country, I felt it was necessary to take a strong stand for science and be clear about what science knows and why we know it. Thus, the coverage of topics like evolution and climate change states clearly the evidence for why these ideas are accepted by over 99% of research scientists in the country.

Features of This Book

To achieve these goals, the book utilizes some basic pedagogical tools:

1. An emphasis on *how* and *why* we know certain things throughout the text, made explicit in the boxed entries entitled "How Do We Know?"
2. New illustrations and diagrams which emphasize the events in the context of geologic time, especially the time scale diagrams at the ends of the appropriate chapters;

3. A list of "Resources" at the back of each chapter, including great books and journal articles for those interested in further exploring a topic
4. A list of URLs for websites with great animations or videos about key topics within the chapter, found hotlinked in the electronic version of the text, and on the accompanying website for users of the hard copy editions of this book;
5. A short summary of the main themes and facts in each chapter, given in a bulleted list;
6. A list of "Key Terms" for students who need to focus on learning and defining terminology;
7. A series of "Study Questions" to help the student review the material and think about its implications.
8. Detailed chapters on human evolution (Chapter 16) and the Holocene, climatic influences on human history, and the future (Chapter 17), topics often given short shrift in other books or neglected altogether.

Who Is This Book For?

This book is intended for the introductory level non-major earth history course, and assumes only minimal exposure to geology, and only a basic high-school level of understanding of science. I have tried to write with a minimum of jargon, and used the fewest new terms to be learned. Instead I emphasized broad concepts and big themes wherever possible. I have tried to paint "word paintings" that help students visualize the ancient worlds of the past. The writing is guided by the levels of understanding of science that I have seen in teaching this topic for over 40 years to undergraduate students at every level, from Caltech and Columbia University and Occidental, Knox, and Vassar Colleges, to several community colleges where I have had the privilege of teaching. The immediate reactions and feedback of those students has been my best guide to how to explain something, and what topics often give the student trouble or challenges.

Because this course is now often taken in isolation without a previous class in geology, I have included a brief review of chemistry, minerals, and rocks (Chapter 2) and a quick summary of plate tectonics (Chapter 5) so that no student will feel left out. Instructors may choose to skip those chapters if all their students have already come from a previous geology course that covered those topics.

Acknowledgments

I thank the many people who read parts or all of this book, and gave me excellent feedback on how to make it better. I thank Dr. Linda C. Ivany (Syracuse University), Dr. Bruce Lieberman (University of Kansas), and Dr. Raymond Ingersoll (UCLA) for reading nearly all the finished chapters and making great suggestions for improvement. I thank Dr. John Valley (University of Wisconsin Madison) for reading Chapter 7, and Dr. Briana Pobiner (Smithsonian

Institution) for reading Chapter 16 and making sure they were up-to-date and factually correct. I thank the following people for their reviews of the original proposal or the earlier drafts of the manuscript:

William Bartels—Albion College

Paul F. Ciesielski—University of Florida

Peter Copeland—University of Houston

William Garcia—University of North Carolina at Charlotte

Amanda Palmer Julson—Blinn College, Bryan

Christopher Knubley—University of Arkansas, Fort Smith

Julio Leva-Lopez—Lamar University

Ervin G. Otvos—University of Southern Mississippi

Shannon Wells—Old Dominion University

Paul Dolliver—Hill College, Collin College

Veronica Freeman—Marietta College

Keith Mann—Ohio Wesleyan University

John Chadwick—College of Charleston

Kevin Cole—Harper College

Tathagata Dasgupta—Kent State University

David Dobson—Guilford College

Marguerite Moloney—Nicholls State University

Jonathan Sumrall—Sam Houston State University

Peter Voice—Western Michigan University

Cornelia Winguth—University of Texas at Arlington

Thomas Algeo—University of Cincinnati

Takehito Ikejiri—University of Alabama

Jill Lockard—Pierce College

Blaine Schubert—East Tennessee State University

I also thank numerous other anonymous reviewers who gave valuable feedback at earlier stages of the process.

I thank my original editor, Dan Kaveney, for helping me develop this project and giving me great advice and feedback all the way. I thank my new editor Dan Sayre, as well as Megan Carlson, Michele Laseau, and Micheline Frederick at Oxford University Press.

I thank Bob Dott for his guidance and support over many years of teaching this topic, and working together on our own book. I appreciate that he gave me his blessing for this project before he passed away. Last but not least, I thank my wonderful family for their forbearance and support while I worked long hours on this book over several years: my sons, Erik, Zachary, and Gabriel, and my wonderful wife, Dr. Teresa LeVelle. Without them, this would never have been possible.

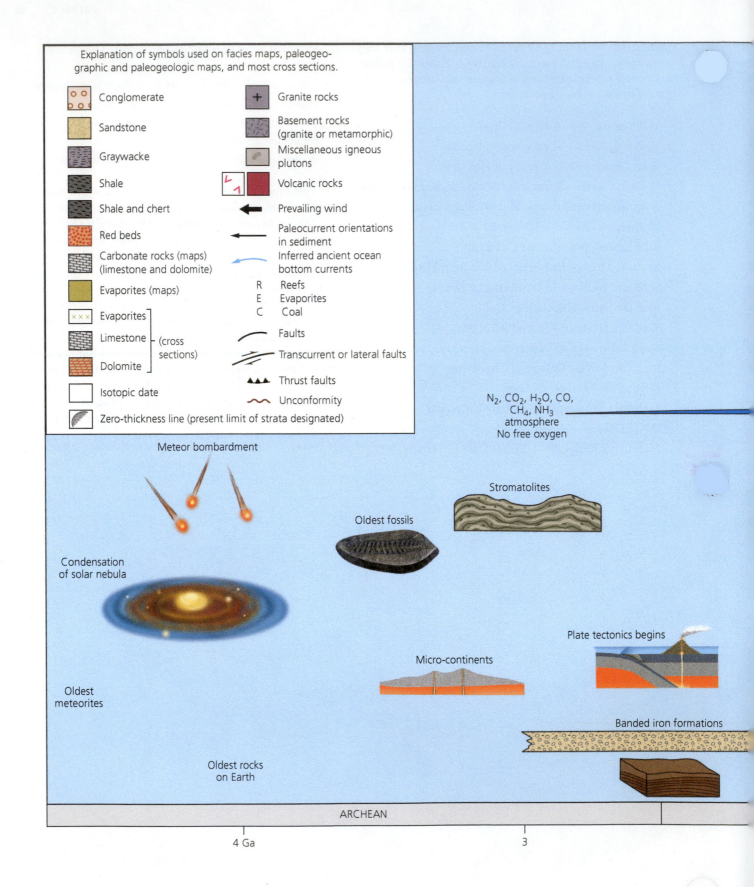

Explanation of symbols used on facies maps, paleogeographic and paleogeologic maps, and most cross sections.

Conglomerate

Sandstone

Graywacke

Shale

Shale and chert

Red beds

Carbonate rocks (maps)
(limestone and dolomite)

Evaporites (maps)

Evaporites

Limestone (cross sections)

Dolomite

Isotopic date

Zero-thickness line (present limit of strata designated)

Granite rocks

Basement rocks
(granite or metamorphic)

Miscellaneous igneous
plutons

Volcanic rocks

Prevailing wind

Paleocurrent orientations
in sediment

Inferred ancient ocean
bottom currents

R   Reefs
E   Evaporites
C   Coal

Faults

Transcurrent or lateral faults

Thrust faults

Unconformity

Meteor bombardment

Condensation
of solar nebula

Oldest
meteorites

Oldest rocks
on Earth

Oldest fossils

Stromatolites

Micro-continents

N_2, CO_2, H_2O, CO,
CH_4, NH_3
atmosphere
No free oxygen

Plate tectonics begins

Banded iron formations

ARCHEAN

4 Ga

3

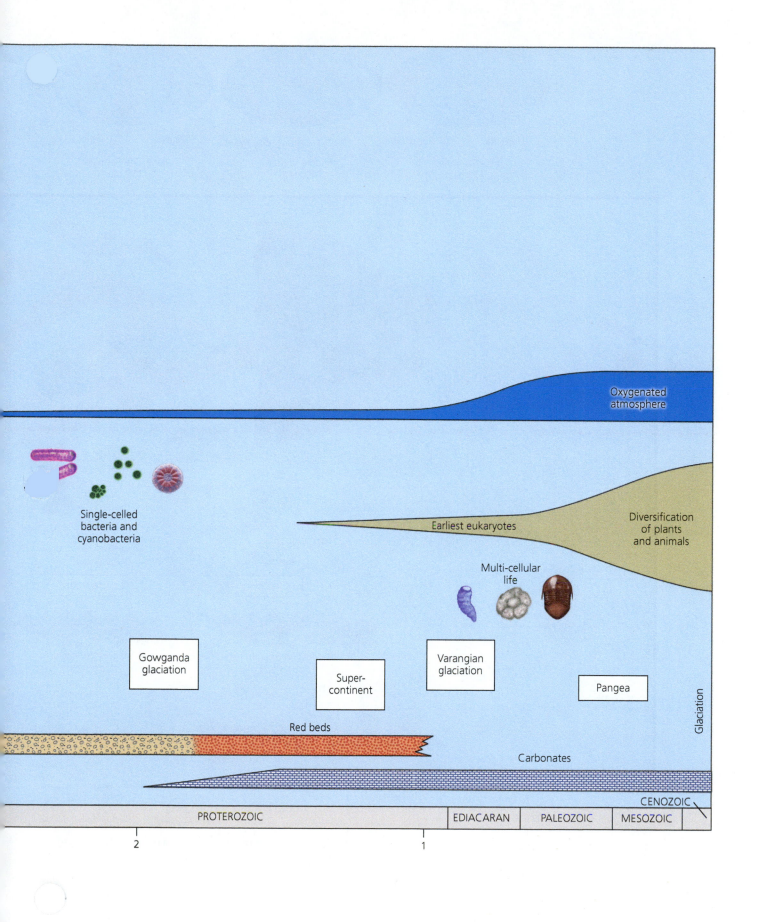

Oxygenated atmosphere

Single-celled
bacteria and
cyanobacteria

Earliest eukaryotes

Diversification
of plants
and animals

Multi-cellular
life

Gowganda
glaciation

Super-
continent

Varangian
glaciation

Pangea

Glaciation

Red beds

Carbonates

| PROTEROZOIC | EDIACARAN | PALEOZOIC | MESOZOIC | CENOZOIC |

2   1

Laurentia   Gondwana   Gondwana   Gondwana   Tethys Sea

Late Cambrian   Middle Silurian   Early Late Permian
520 Ma   430 Ma   260 Ma

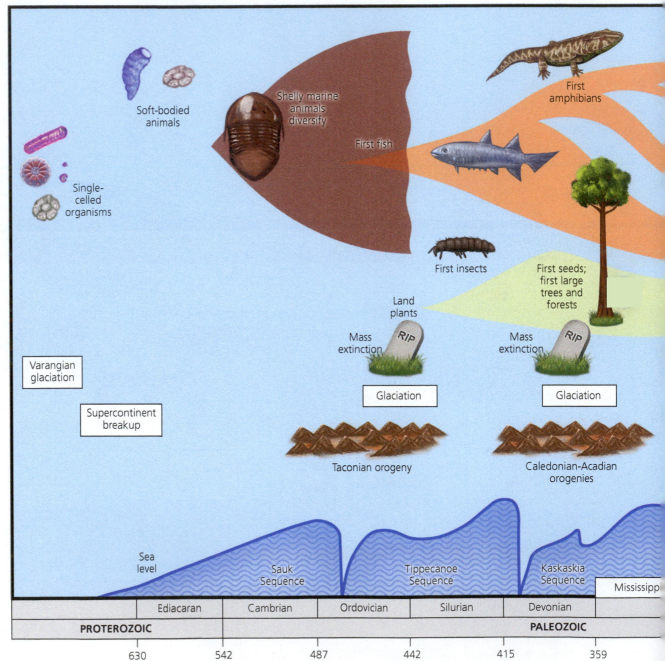

Soft-bodied
animals

Single-
celled
organisms

Shelly marine
animals
diversify

First fish

First
amphibians

First insects

First seeds;
first large
trees and
forests

Land
plants

Mass
extinction   RIP

Mass
extinction   RIP

Varangian
glaciation

Glaciation   Glaciation

Supercontinent
breakup

Taconian orogeny   Caledonian-Acadian
                                   orogenies

Sea
level   Sauk
                          Sequence   Tippecanoe
                                          Sequence   Kaskaskia
                                                          Sequence   Mississipp

	Ediacaran	Cambrian	Ordovician	Silurian	Devonian	
PROTEROZOIC			**PALEOZOIC**			

630   542   487   442   415   359

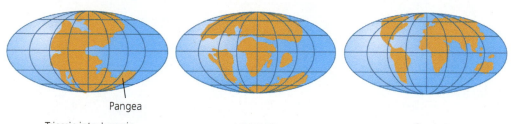

Pangea

Triassic into Jurassic
240-195 Ma

Mid-Tertiary
40-25 Ma

Present

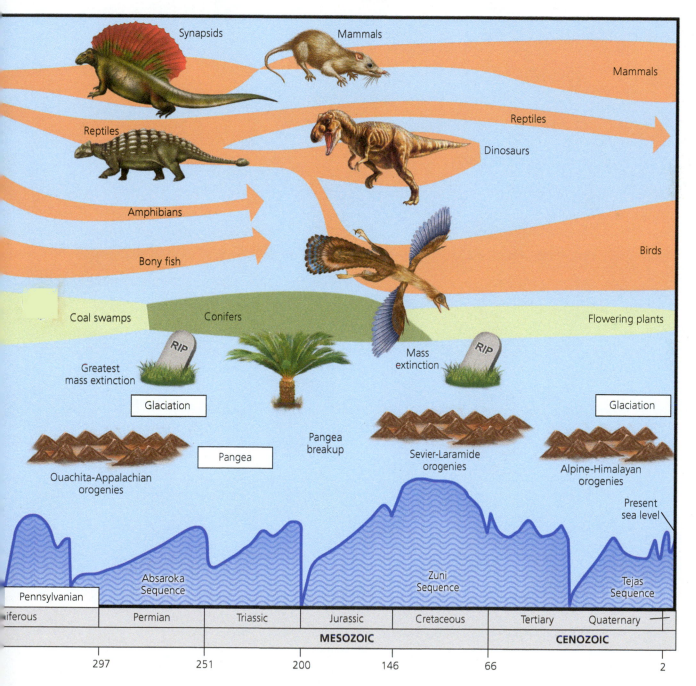

Synapsids

Mammals

Mammals

Reptiles

Reptiles

Dinosaurs

Amphibians

Bony fish

Birds

Coal swamps

Conifers

Flowering plants

RIP

Mass
extinction

RIP

Greatest
mass extinction

Glaciation

Glaciation

Pangea

Pangea
breakup

Sevier-Laramide
orogenies

Alpine-Himalayan
orogenies

Ouachita-Appalachian
orogenies

Present
sea level

Absaroka
Sequence

Zuni
Sequence

Tejas
Sequence

Pennsylvanian

...iferous	Permian	Triassic	Jurassic	Cretaceous	Tertiary	Quaternary	
		MESOZOIC			CENOZOIC		

297   251   200   146   66   2

THE EVOLVING EARTH

PART I
Deciphering The Earth

Time present and time past
Are both perhaps present in time future
And time future contained in time past

T.S. Eliot, Four Quartets

The mind seemed to grow giddy by looking so far into the abyss of time.

John Playfair, 1805

[*The concept of geologic time*] makes you schizophrenic. The two time scales—the one human and emotional, the other geologic—are so disparate. But a sense of geologic time is the important thing to get across to the non-geologist: the slow rate of geologic processes—centimeters per year—with huge effects if continued for enough years. A million years is a small number on the geologic time scale, while human experience is truly fleeting—all human experience, from its beginning, not just one lifetime. Only occasionally do the two time scales coincide.

Eldridge Moores, in Assembling California, *by John McPhee*

The immensity of geologic time, here shown as a spiral of time going back to 4.6 billion years ago.

1.1 Deep Time and Immense Space

Most people think of time in days or hours or minutes or, if we wish to look back, maybe decades or a century at most. Most humans live no more than 70 to 80 years, and only a few live to a century. Human events more than a few thousand years ago are considered "ancient," and we have a hard time comprehending the world of the Middle Ages, let alone the lives of the ancient Egyptians, Greeks, or Romans. Events more than 5000 years ago seem inconceivable to us.

Contrast this with the way geologists see the world. We routinely deal in millions or even billions of years. When looking at events millions of years ago, a few hundred thousands of years either way is considered unimportant. In most cases, we can't resolve events of thousands to hundreds of thousands of years ago. Geologists deal in immense amounts of time, so huge that writer John McPhee called it "**deep time**." The epigraph on page 4 (from one of McPhee's books) captures the essence of the problem of comprehending **geologic time**.

Humans are accustomed to thinking only in the short term and the immediate future and have a hard time even grasping the concept of millions of years. Perhaps an analogy will help. One of the most famous is to squeeze all 4.6 billion years of geologic time into the length of an American football field (Fig. 1.2), 100 yards or 300 feet (91 meters), and one inch is 1.4 million years. On this scale, 1 yard (3 feet) is 50 million years, and 50 yards (half the field) is 2.3 billion years. When you examine the major events of geologic history on this scale, the first thing that impresses you is how long the time before visible fossils (Precambrian time) was and how short the interval of time is for all the events that are familiar to us. If the kick returner caught the ball on one goal line, he would have run 88 yards across the field through all of Precambrian time before the first multicellular animals, such as trilobites, show up—only 12 yards from a touchdown. Just inside the 5-yard line (less than 5 yards from the goal line) is the beginning of the age of dinosaurs (the Mesozoic), and the runner travels to only 1.5 yards from the goal line to reach the end of the age of dinosaurs, when they all vanished (except for their bird descendants). The entire age of mammals takes that final 1.5 yards, the first members of the human lineage occur only 8.3 inches from the goal line, and the Ice Ages begin only 3.6 inches from the goal line. The first member of our own species, *Homo sapiens*, appears about 0.3 inches before the goal line; all of the last 5000 years of human civilization is only 0.08 inches thick—narrower than a blade of grass. If the chalk that marks the goal line is just a tiny bit too wide, it wipes out all of human history.

 See For Yourself: Time Lapse of the Entire Universe

Here is another analogy. Let us squeeze the entire 4.6 billion years of earth history down into a single calendar year, 365 days in length (Fig. 1.3). When you divide up 4.6 billion years into 365 slices, then each day represents 12.3 million years. Each hour in this analogy is equivalent to about half a million years (513,660 years, to be precise), and each minute is 8561 years long. If we start with the origin of the earth as New Year's Day, then the first simple life forms (bacteria) do not appear until February 21. The months roll by, with no life more complicated than single-celled organisms, until we reach November 19th when the first multicellular animals (such as trilobites and sponges) appear. Geologists call this the "Cambrian Period." By November 29, we have reached the Devonian Period, when the seas were full of huge predatory fish

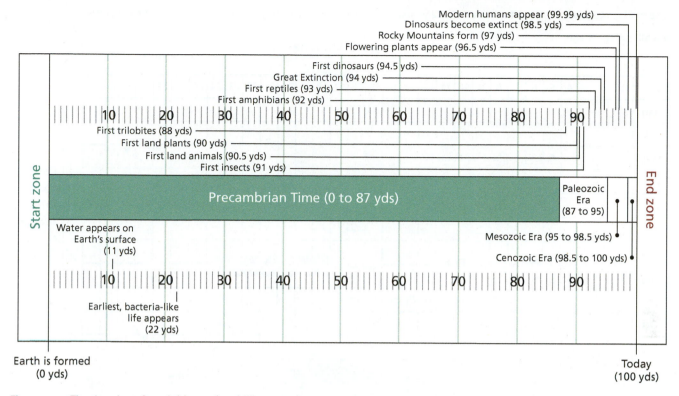

Figure 1.2 The duration of earth history (4.6 billion years) squeezed into the length of an American football field, which is 100 yards (300 feet) long. Major geologic events shown with their time span represented by the distance from the goal lines.

and the first amphibians crawled out on land, cloaked by the first true forests.

See For Yourself: The History of the Earth in 5½ Minutes

See For Yourself: Timeline of Earth

By December 8 we have only reached the Permian Period, about 250 million years ago, when the earth had a single supercontinent called Pangea that stretched from pole to pole and a single ocean that covered almost three-quarters of the globe. The land was dominated by huge amphibians the size of crocodiles, a variety of primitive reptiles, and huge fin-backed relatives of mammals. By December 16, we reach the Jurassic Period, a name familiar from a number of recent hit movies, when huge dinosaurs roamed the planet and we see the earliest mammals, lizards, and birds. The age of dinosaurs ends the day after Christmas, when catastrophic events wiped out not only the huge dinosaurs but also many important groups in the oceans, such as the marine reptiles. The entire last 66 million years of the age of mammals can be squeezed into the final week between Christmas and New Year's. The earliest human relatives do not appear until 7 hours before midnight on New Year's Eve, and the earliest members of our genus *Homo* are found only 1 hour before midnight. All of human civilization flashes by in the last minute before the stroke of New Year's Eve, so if someone starts celebrating a few seconds too early, he or she drowns out all of human history.

See For Yourself: A Brief History of Geologic Time

Putting it this way is very humbling for humans and for our exaggerated sense of self-importance. We are afterthoughts, very late arrivals on the stage of earth history, and have not even been around as long as most species in the fossil record. The human lineage can only be traced back to about 7 million years ago, while dinosaurs dominated the planet for over 130 million years. Think about that the next time you hear someone use the word "dinosaur" to indicate something that is old and obsolete. We should be lucky to last as long as most species on the planet. As legendary author Mark Twain put it (with his caustic wit and sarcasm),

If the Eiffel tower were now representing the world's age, the skin of paint on the pinnacle-knob at its summit would represent man's share of that age; and anybody would perceive that that skin was what the tower was built for. I reckon they would. I dunno.

If the perspective of geologic time was not humbling enough, consider the immensity of space and our position in it. We live on but one tiny planet in our solar system, in a smaller galaxy off in the suburbs of the universe. If the earth were the size of a pea, the sun would be the size of a basketball, and they would be about a mile apart. If we shrunk the entire solar system down to the size of a softball, then our

See For Yourself: Earth Compared to the Rest of the Universe

Milky Way galaxy would be the size of the earth. And there are millions of other galaxies besides ours.

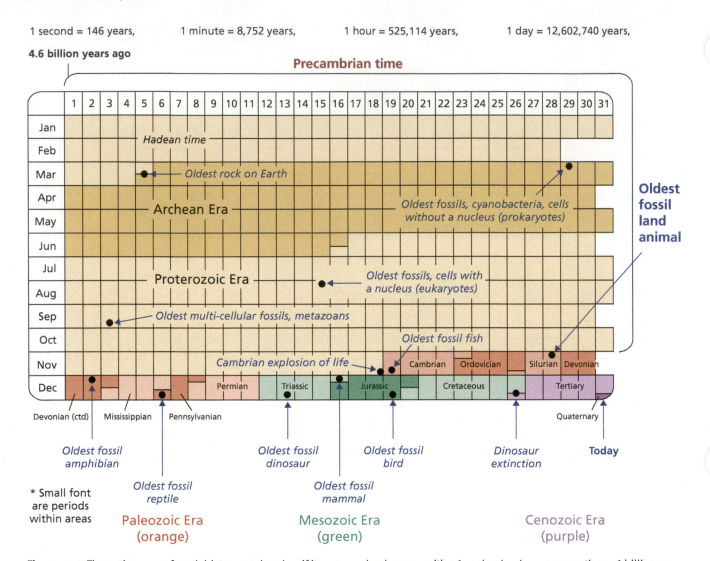

1 second = 146 years,   1 minute = 8,752 years,   1 hour = 525,114 years,   1 day = 12,602,740 years,

Figure 1.3   The entire span of earth history rendered as if it were a calendar year, with 365 calendar days representing 4.6 billion years of earth history and each day representing about 12.3 million years.

As Carl Sagan said,

The size and age of the Cosmos are beyond ordinary human understanding. Lost somewhere between immensity and eternity is our tiny planetary home. In a cosmic perspective, most human concerns seem insignificant, even petty. And yet our species is young and curious and brave and shows much promise. In the last few millennia we have made the most astonishing and unexpected discoveries about the Cosmos and our place within it, explorations that are exhilarating to consider. They remind us that humans have evolved to wonder, that understanding is a joy, that knowledge is prerequisite to survival. I believe our future depends powerfully on how well we understand this Cosmos in which we float like a mote of dust in the morning sky.

We've come a long way since the days when we thought that the earth was a flat disk in the center of the universe and that the planets and the sun moved around us and that the stars were just pinpoints of light in the celestial dome of the

heavens (Fig. 3.1). We now know how tiny and insignificant we are on the scale of space and in the context of geologic time. It's a humbling vision, but this is what science has revealed to us. However, there is a flip side to this coin: we are the only species that has ever been able to see and understand how we got here, and how the earth and universe were formed. As the great anatomist and paleontologist Baron Georges Cuvier put it,

Genius and science have burst the limits of space, and few observations, explained by just reasoning, have unveiled the mechanism of the universe. Would it not also be glorious for man to burst the limits of time, and, by a few observations, to ascertain the history of this world, and the series of events which preceded the birth of the human race?

In this book we will look at some of the events that have happened over the past 4.6 billion years. More importantly, we will examine the scientific evidence for *how* and *why* we know these things.

1.2 "No Vestige of a Beginning"

This conception of the immensity of geologic time is just over 200 years old, and it did not come easily. Although there were some ancient Greeks and Romans who thought the earth was really old, our modern insight about the age of the earth comes from the late 1700s, when the Enlightenment swept across Europe. It was an age of scientific discoveries, and scholarly research was less constrained by the influence of the church or the nobility. Enlightenment thinkers such as Voltaire, Montesquieu, Jean-Jacques Rousseau, and Denis Diderot in France and philosophers such as George Berkeley, Jeremy Bentham, and John Locke as well as scientist Isaac Newton in England were very influential as they focused on evidence and reason and the **scientific method** over supernaturalism and myths handed down for centuries. One of their central rules was **naturalism**: it is better to explain things (especially scientific results) by natural laws rather than by supernatural causes. In fact, science cannot work without naturalism because supernatural explanations for events cannot be tested or evaluated by scientific evidence. An explanation like "the gods just made it that way" is a science stopper. There is no way to evaluate it by scientific evidence since nothing can test that statement.

Surprisingly, one of the hotbeds of intellectual ferment was Edinburgh, Scotland. Even though it was not a large country, Scotland had one of the highest literacy rates in the world at the time. This was because the Presbyterian Church that ruled parts of Scotland believed that everyone should be able to read and interpret the Bible for themselves and not depend on clergy to read it for them. Thus, the churches set up public schools and tried to make sure every Scot, no matter how lowly, was able to read and write. Scots had a thirst for knowledge, with great libraries and many publishing companies churning out books and newspapers. Thanks to the relatively weak influence of the many churches in Scotland, there was no oppression by clergy as there was in England or France or much of Europe. Consequently, in the late 1700s Edinburgh was the capital of the remarkable "Scottish Enlightenment." Brilliant and original thinkers such as the philosopher David Hume (founder of modern skepticism), Adam Smith (who wrote *The Wealth of Nations*, the first great book explaining capitalism), chemist Joseph Black, and inventor James Watt (who built the first practical steam engine that launched the Industrial Revolution) all lived in the same part of Edinburgh. Most of them were drinking buddies in the social clubs, where they debated ideas and argued about science, philosophy, religion, government, and many other topics without fear of persecution. These people were very influential to America's founding fathers, such as Benjamin Franklin and Thomas Jefferson, who visited England, France, and Scotland and met many of the leading thinkers. Much of the Declaration of Independence and Constitution comes directly from the thinking of John Locke and the French philosophers.

Among the geniuses of the Scottish Enlightenment was a gentleman and landowner by the name of James Hutton

Figure 1.4 Portrait of James Hutton, the "father of modern geology."

(Fig. 1.4). He was born in Edinburgh on June 3, 1726, and died there on March 26, 1797. Hutton was a famous chemist and naturalist and is now considered to be the "father of modern geology." Although he was trained in both law and medicine, he was more interested in chemistry and natural history and pursued those hobbies with a very independent way of thinking.

 See For Yourself: James Hutton

As a landowner who had several large farms around Scotland, he used his training in chemistry to fertilize his fields. He also traveled widely, seeking out new methods to improve farming practices. Meanwhile, his curiosity led him to make many observations about the slow process of weathering, how soils form, or how sediments are slowly washed out to sea and then pile up layer by layer. Eventually, he leased his properties out to tenant farmers and returned to live in Edinburgh, mingling with other great thinkers like Adam Smith and Joseph Black, two of his closest friends. He traveled widely around Scotland, adding to his storehouse of observations and seeking answers to his questions about how the earth worked. He finally published his ideas in 1788 in a scientific paper entitled "Theory of the Earth; or an Investigation of the Laws Observable in the Composition, Dissolution, and Restoration of Land Upon the Globe" and then again in a book published in 1795.

Figure 1.5 James Hutton was familiar with the Roman barrier built across the southern border of Scotland known as Hadrian's Wall, built about 122 C.E. He was impressed that it showed little sign of weathering in the over 1600 years since it had been built and from this reasoned that the rocks on the earth's surface must also weather and erode very slowly over hundreds to thousands of years.

Hutton realized that everything he had witnessed demonstrated that earth processes operated very slowly and gradually. Thick soils took years to form; layers of sediment took centuries to build up on the bottom of a lake. He visited Hadrian's Wall (Fig. 1.5), built by the Romans across Scotland over 1500 years earlier, and saw no signs that the stones had changed or even weathered much in all those centuries. In addition, he applied the Enlightenment philosophy of naturalism to geology and reasoned that the natural processes we see operating today—slow weathering, erosion, transportation of sediments—must have operated the same way in the geologic past. Ancient rocks can be explained in terms of observable processes, and those processes now at work on and within the earth have operated with slow, steady uniformity over immensely long periods of time. This came to be known as **uniformitarianism**—the uniformity of natural processes through time. One of his followers, Archibald Geike, summarized this concept as "the present is the key to the past." We must use our understanding of present-day natural laws and processes to understand those that happened in the past.

Although uniformitarianism is central to geology, it is used widely in all the sciences any time we must infer processes that operate in time frames or on scales that we cannot observe directly. Nearly every science uses uniformitarianism to some degree. Until recently, we could not see atoms or molecules and had to infer their properties by their behavior in experiments. The stars and galaxies are hundreds to millions of light years away, meaning that the light we now see from them originated hundreds to millions of years ago and is just now reaching us. We cannot watch these processes in real time but must infer how the universe works using the natural laws of physics. In chemistry and biology, we could not see much below the cellular level until the past few decades, so we had to figure out the processes of biochemistry and molecular biology and the nature of molecules and atoms indirectly through experiments. Geology is just one more science that depends strongly on the uniformitarian approach, which is universal across the sciences. Hutton was focused on the naturalistic approach to the earth and tried to get away from unscientific supernatural explanations, which were often called **catastrophism** (such as the supernatural catastrophe of Noah's Flood).

But these ideas flew in the face of the dogma of the time. Leading church scholars had declared that the earth was only 6000 years old, and few people dared question them. Some scholars thought that all the layered rocks of the earth had been deposited in Noah's Flood, even though there were serious problems with this idea. One school of thought, led by German mineralogist Abraham Gottlob Werner at the Freiburg Mining Academy, declared that all layered rocks, even lava flows, were laid down in water, an idea known as **Neptunism** (after Neptune, the Roman name for the god of the sea).

But Hutton could see that lavas and other igneous rocks were formed from molten rock, or magma, that had come up from the hot interior of the earth, not settled out of water. Where he found igneous magma intruding through older rocks, he could see the evidence that it had forced its way up through cracks as liquid rock, then baked the rocks around it with its intense heat, and cooled into a solid where it came to rest. Salisbury Crags and Arthur's Seat, just south of Edinburgh, are extinct volcanoes with intruded lavas. Hutton used to walk there with his dog Missy and saw clear evidence that the granitic rocks were once a molten mass of magma intruded into older layers of sediment (Fig. 1.6). He saw the same evidence in other places, such as Glen Tilt, in the Cairngorm Mountains north of Edinburgh. For Hutton, this was inescapable evidence that the earth was dynamic and changing, with molten rock rising up from "the earth's great heat engine" deep below our feet, as he put it. By advocating the igneous origin of rocks such as lava flows, Hutton's ideas came to be known as **Plutonism** (after Pluto, the Roman name for the god of the underworld).

How could anyone imagine that lava flows were formed in water? In the twenty-first century, we are used to seeing videos of erupting volcanoes, like Kilauea in Hawaii. But in those days, few people traveled outside their home country, let alone to other countries, and almost no northern Europeans had ever seen a volcano erupt. The closest volcanoes

Figure 1.6 **A.** Hutton's famous outcrop showing magmas intruding into layered sediment, on the southwestern slopes of Salisbury Crags, in southeastern Edinburgh. **B.** Close-up of the right side of the outcrop in the previous image, showing the tilted and cooked layered sediments (bottom) intruded by pink granites (top). **C.** Image from a posthumous edition of Hutton's book by John Clerk, showing magma intruding tilted sedimentary rocks. Notice the scale of the drawing is greatly exaggerated from the real exposure shown in **A.**

were Mt. Vesuvius and Mt. Etna in southern Italy, and they produce mostly ash, not lava flows. In addition, the study of chemistry was still very primitive, so early mineralogists like Werner had no notion about how much heat it would take to melt a rock or crystallize a magma and under what conditions magmas form.

 See For Yourself: Hutton and Siccar Point

Hutton's thoughts were especially stimulated when he saw outcrops of what are known today as **angular unconformities** (see Chapter 3). Here, the rocks on the bottom are tilted up on their sides at an angle, then eroded off the top, and then younger rocks were deposited on that old erosional surface (Fig. 1.7). Hutton reasoned that the lower layers had once been laid down horizontally in the bottom of a river or the ocean. They had then been turned from soft sand and mud into hard sandstone and mudstone, a process that takes millions of years. Sometime later these layers had been tilted on their side by immeasurably strong forces, then uplifted into the air as mountains and eroded away, as we see happening in our mountains today. Finally, those same mountains must have eroded and sunk down so that they could be covered by an even younger layer of sediments. Knowing how slow the modern rates of weathering and erosion were and how long it must take to deposit of thousands of layers of sediment, Hutton realized that an angular unconformity must represent thousands to millions of years of time, not the mere 6000 years that religious scholars believed. As Hutton put it, he saw "no vestige of a beginning, no prospect of an end." The earth was unimaginably old and operated on timescales that humans could barely comprehend. Hutton claimed that the totality of these geologic processes could fully explain the current landforms all over the world and that no biblical explanations were necessary in this regard. Finally, he stated that the processes of erosion, deposition, sedimentation, and upthrusting were cyclical and must have been repeated many

times in the earth's history. Given the enormous spans of time taken by such cycles, Hutton asserted that the age of the earth must be inconceivably great.

Hutton's ideas were radical for his time and hard for most people to comprehend, let alone accept, even today. In addition, Hutton was a not a very clear or lively writer, so few people fully grasped his ideas even if they read his works. In 1802, 5 years after Hutton's death, his friend John Playfair published *Illustrations of the Huttonian Theory of the Earth*, which gave a much clearer explanation of Hutton's ideas.

But it would take another generation for such revolutionary ideas to win acceptance in the geological community. The person who made it possible was a young man named Charles Lyell (Fig. 1.8). Originally trained in the law to become a barrister, he soon tired of the legal profession and instead pursued the young field of geology as a hobby. (This would not be the first time that someone got tired of doing law and moved to a different profession that interested them more.) Lyell traveled widely over Europe, witnessing many different geological phenomena with the uniformitarian eyes of Hutton. Eventually, he wrote his masterpiece, *Principles of Geology*, published in three volumes from 1830 to 1833. Mustering all the observations he had gathered from his travels and his reading, he used his skills as a lawyer to write a decisive case for the uniformitarian view of the earth. Like any good lawyer, he used any arguments and rhetorical tactics necessary to discredit his opponents, the catastrophists, while presenting overwhelming evidence for his own case. As a lawyer, Lyell argued his case very convincingly and made a strong, one-sided case. He tried to exclude any possibility of the old unscientific supernatural Noah's Flood catastrophism ever being taken seriously again. He was so persuasive that within a generation, the last of the old-line catastrophists and Neptunists had all died or given up, and geology became a modern science.

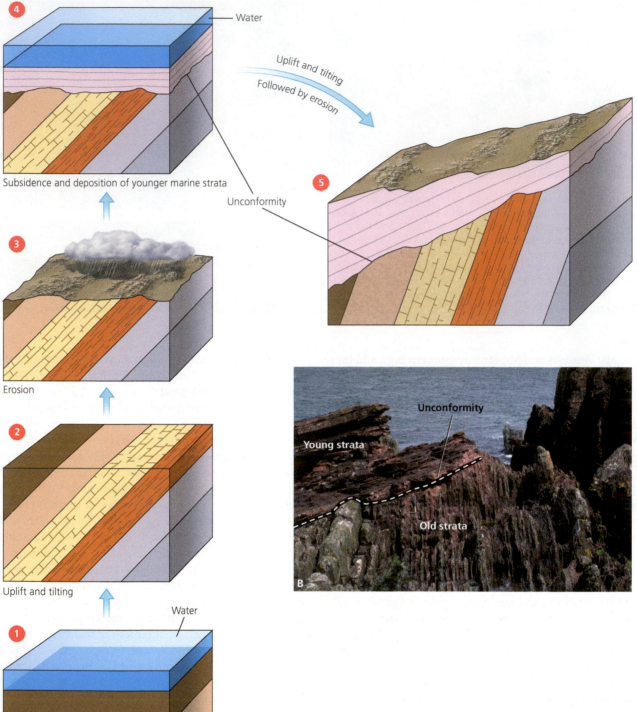

④ Water

Uplift and tilting
Followed by erosion

Subsidence and deposition of younger marine strata

Unconformity

⑤

③

Erosion

②

Uplift and tilting

Water

①

Deposition of marine strata

A

Unconformity

Young strata

Old strata

B

Figure 1.7 A. The famous angular unconformity at Siccar Point, Berwickshire, east coast of Scotland. Hutton first saw this outcrop on his last major field excursion, and it clinched his convictions that the earth was really old. As Hutton realized, the bottom sequence of rocks must have originally formed as a thick stack of sands and muds on the bottom of the ocean, which were compressed into sandstones and shales. Then they were gradually tilted into vertical orientation. Some time later, these rocks were uplifted into a mountain range, where they were eroded off, creating the erosional surface that cuts off the edges of the tilted layers. Then the entire erosional surface sank down again and was buried by younger sands, now turned into sandstones. As Hutton realized, each of these processes takes thousands of years or longer, and together they indicated that the entire unconformity must represent immense amounts of time. **B.** Photograph of the Siccar Point unconformity in outcrop.

Fig. 1.8   Charles Lyell, who took Hutton's ideas and argued them so effectively in his book *Principles of Geology* (1830–1833) that geology never looked back at Noah's Flood, catastrophism, or supernatural causes.

1.3 The Scientific Method

J Harlen Bretz (see Box 1.1) was famous for teaching his students to keep an open mind when looking at the rock outcrops. During field camp, he would force them to study the outcrops on their own and draw their own conclusions, and then only after they were done would they read the other published reports and see if their ideas agreed with those of previous geologists. He didn't want students reading the reports and accepting the prevailing views before they saw the rocks because they would be biased and tend not to question what the earlier scientists had said or see the outcrops from their own perspective.

One of Bretz's mentors, Thomas C. Chamberlain of the University of Chicago, preached that geologists should use the method of **multiple working hypotheses**. Instead of taking one model for how the earth works and trying to fit everything to it, they should keep an open mind with several different models to explain the data, then think critically about what the evidence supports, and rule out the discredited hypotheses. This is still an important aspect of scientific thinking, even though it is not often discussed. Scientists begin with observations or data that lead them to think about several possible explanations for what they have seen and measured. They then construct hypotheses to explain the data.

But the crucial step is the next one: *test your hypotheses* and reject those that are not supported. It's not enough to give a convincing explanation or a "just-so story" to explain why something occurred. A true scientist must go out and try to shoot down, or falsify, his or her hypotheses, rejecting the bad ideas and moving forward with the ones that have survived testing. Only by repeated conjectures and refutations does science move forward.

Why do scientists focus on falsifying their ideas, rather than verifying them? Suppose we make the inductive generalization that "all swans are white." We could observe thousands of swans for many years but never prove that statement true. All it takes is one non-white swan and we can easily falsify this hypothesis. Indeed, there are black swans in Australia and elsewhere, so the statement has been falsified. As philosopher

---

BOX 1.1: HOW DO WE KNOW?

How Do We Apply Uniformitarianism Today?

Lyell argued that to be a good geologist you had to accept not only the uniformity of *natural laws and processes* (**actualism**) but also the slow, gradual uniformity of *rates* (**gradualism**). The burden of fighting against supernaturalistic, unscientific notions had led Lyell to take an extreme opposite position, and most geologists for the rest of the century would not accept any process that seemed to violate gradualism. Yet we do know natural processes do not violate actualism but are sudden events, not gradual. The impact of an asteroid from space is not a gradual event, but it is certainly natural and does happen. So too is an immense landslide or a huge volcanic eruption.

See For Yourself: The Channeled Scablands

During the 1920s, a geologist by the name of J Harlen Bretz was mapping and studying a region in eastern Washington known as the Channeled Scablands (Fig. 1.9). He found areas of hard lava bedrock that were scoured by immensely powerful currents of water, stripping off all the soft glacial soils (the "Palouse") that once covered them and even grinding the sharp edges of the rocks smooth. There were immense canyons, called "coulees," which had almost no water in them capable of carving such gorges, and even old dry waterfalls with almost no water passing over them. To Bretz, this was unmistakable evidence that some immense flood of water had rushed over the area during the last Ice Age.

When Bretz presented his ideas before the meetings of the Geological Society of America in the 1920s, he was rejected because the idea of huge floods violated its gradualistic biases. Generations of geologists had followed Lyell's conflation of actualism with gradualism and refused to accept any natural process that was sudden or large in scale. Bretz continued to stubbornly map the Channeled Scablands, eventually finding more and more evidence, such as huge boulders that could only have been carried by gigantic amounts of water. No matter what he found, however, Bretz was not believed because geologists would not accept anything that violated gradualism.

Part of the problem with Bretz's outrageous hypothesis was that he could point to no source that could produce so much water so quickly. He suggested volcanoes erupting beneath glaciers, but no one could find such an example. Ironically, during the same time as Bretz worked in eastern Washington, US Geological Survey geologist J. T. Pardee was mapping the valleys around Missoula, in western Montana, and found evidence of huge glacial lakes that had flooded the valleys during the Ice Ages. Eventually, Pardee pointed out that these glacial lakes could have released enormous volumes of water if the ice dams that held the water back had melted or broken down.

Finally, in the 1950s the first aerial photographs were taken over the region. From the air, you could see gigantic ripple marks that are 3 to 50 feet (1–15 meters) high, many meters wide, and tens of meters long. They are so huge that you can't detect them on ground level since they are covered in sagebrush, and they are only visible from the air. This, and other evidence, finally proved that Bretz had been right all along. Eventually, the work of other researchers confirmed the conclusions that Bretz had published decades earlier. By the end of Bretz's life, the Scablands floods were used to model the hydraulics of other planets like Mars, which were once scoured by enormous floods of water. For these discoveries, Bretz is recognized as one of the founders of planetary geology. In 1979, at age 96, Bretz finally received the Penrose Medal of the Geological Society of America, the highest award in the profession. In his acceptance speech, Bretz commented that it was no fun to celebrate his award because all his opponents and critics were long dead.

The Scablands floods taught geologists an important lesson: not all processes need to be gradual in order to be natural. A year after Bretz received his Penrose Medal, the discovery of the asteroid impact that ended the age of dinosaurs brought natural but catastrophic events to center stage. Geologists have finally learned that actualism is not the same as gradualism, nor does uniformitarianism require gradual events.

Figure 1.9 Repeated failure of the ice dam in the Idaho panhandle that held back glacial Lake Missoula (as shown by the ancient shorelines) produced gigantic floods that carved the Channeled Scablands.

Karl Popper pointed out, there is an asymmetry between **verification and falsification**. It is easy to falsify something; all we need is one well-supported observation that proves the hypothesis wrong. But we can never prove something true (*verify* it). Additional observations may *support* or *corroborate* the hypothesis but never finally prove it true.

But a scientist must always recognize that even a well-supported idea must remain open to testing and falsification. If it is not, then it is a dogma, not a scientific idea. That is why science is always tentative, subject to testing and refutation, and never about "final truth" in the strictest philosophical sense. Unlike other types of dogmatic human thought, which take their conclusions to be true and then selectively point to data that support them, scientists must be willing to reject their conclusions and throw out all their old ideas if the data demand it.

Scientists aren't inherently negative sourpusses who want to rain on everyone else's parade. They are just extremely skeptical of any idea that is proposed until it has survived the gantlet of repeated testing and possible falsification and risen to the stage of something that is established or acceptable. They have good reason to be skeptical. As we all know, humans are capable of all sorts of mistakes and false ideas and self-deceptions. Scientists cannot afford to blindly accept the ideas of one person or even a group of people who make a significant claim. They are obligated as scientists to criticize and carefully evaluate and test it before it is acceptable as a scientific idea.

Scientists are also human and subject to the same foibles that all mortals are. We love to see our ideas confirmed and to believe we are right. Yet there are all sorts of ways we can misinterpret or overinterpret data to fit our biases. As the Nobel Prize–winning physicist Richard Feynman put it, "The first principle is that you must not fool yourself and you are the easiest person to fool." That is why many scientific experiments are run by the double-blind method: not only do the subjects of the experiments not know what is in sample A or sample B, but neither do the investigators. They arranged to have the samples coded so that no one knows what is in each sample, and only after the experiment is run do they open the key to the code and find out whether the results agree with their expectations or not.

So if scientists are human and can make mistakes, why does science work so well? The answer is testability and another crucial factor, **peer review**. Individuals may be blinded by their own biases, but once they put their ideas forth in a presentation or publications, their work is subject to intense scrutiny by the scientific community. To publish a scientific paper in any legitimate journal, the paper must be reviewed by several other qualified experts (usually anonymous) who scrutinize it carefully for errors or mistakes. These reviewers can be as harsh and critical as they want because they are protected by anonymity. When the authors of the paper get the reviews back, complete with all the (sometimes nasty) criticism and corrections, they must decide whether they can fix the mistakes and resubmit the paper or must redo the entire experiment. The Internet is full of hoaxes, fakes, lies, and false information; the political sphere is full of "fake news"; advertisers and con artists are famous for suckering you into believing something that is not true—but no other field of human endeavor has such a rigorous self-checking mechanism as the peer review system in science. This is why we trust it to get the right answers.

If the results cannot be replicated by another group of scientists, then they have failed the test. As Feynman put it, "It doesn't matter how beautiful your theory is, it doesn't matter how smart you are. If it doesn't agree with experiment, it's wrong." This happens routinely in scientific research, although most people never hear about it because it is so common and never catches the attention of the media. The media loves to report flashy, glamorous ideas that have just been proposed, such as an extraterrestrial impact causing a particular mass extinction, but soon moves on to other glamorous topics and never reports the debunking of the initial claim by other labs a year or two later. Yet ideas that stand the test of time and have been checked and double-checked by many different groups of investigators eventually become part of our common body of scientific knowledge and end up in textbooks like this one.

Eventually, if enough testing has been done and enough experiments support or corroborate (never "prove" an idea—there is no absolute proof in science), the well-supported idea is elevated to a broader category of **theory**. Contrary to the popular meaning of the word "theory" as a wild speculation, to scientists a theory is *a broad idea that successfully explains a whole range of observations*. The very best, most strongly supported ideas are theories. For example, gravity is "just a theory." We still don't even understand how it works; but we know it does, and you can prove it by dropping something on the floor. The germ theory of disease replaced all sorts of false notions about what made people sick and successfully cured illnesses caused by germs (viruses and bacteria). You take real medicine, not go to a witch doctor, when you are sick because you accept the germ theory of disease. So, too, the theory of evolution is one of the best-supported ideas in all of science, yet there are science deniers who misuse the word "theory" to suggest that somehow it is speculative or not accepted by more than 99% of the scientific community. We will look at evolution in greater detail in Chapter 6.

Finally, science is not about asserting that something is true or false. Science is about *testing and supporting your assertions with evidence*, whether they be simple observations, experiments, or other things. In this book, we will try to get beyond the tendency to just cite one fact after another, which is a common problem in earth history courses. Wherever possible, we will discuss the evidence that supports a particular idea about earth history. Whenever you hear a scientist say something, you should not just take it on faith. You should always be asking, "How do you know that? What is your evidence?" Most scientists, if they know their field and how to communicate their ideas clearly, would love to explain it to you. After this course, you will understand the evidence for much of what we know about the last 4.6 billion years of earth history. If you remember the background and explanations for what you learned, you will never forget it because you understand *why* we know it, and you should be able to explain it to others, too. That is one of the great joys of science: not only understanding the world but knowing why things are the way they are and what evidence we have to say that this is how things work.

RESOURCES

BOOKS AND ARTICLES

Baker, Victor. 1973. *Paleohydrology and Sedimentology of Lake Missoula Flooding in Eastern Washington.* Geological Society of America Special Paper 144. Geological Society of America, Boulder, CO.

Bretz, J Harlen. 1969. The Lake Missoula floods and the Channeled Scablands. *Journal of Geology* 77:505–543.

Buchan, J. 2009. *Crowded with Genius: Edinburgh, 1745–1789.* Harper, New York.

Cuvier, G. 1818. *Essay on the theory of the earth.* New York: Kirk & Mercein.

Feynman, R.P. 2015. *The Quotable Feynman,* p.359, Princeton University Press

Geike, A. 2011. *James Hutton—Scottish Geologist.* Shamrock Eden Digital Publishing, Scranton, PA.

Greene, M. T. 1982. *Geology in the Nineteenth Century: Changing Views of a Changing World.* Cornell University Press, Ithaca, NY.

Herman, A. 2007. *How the Scots Invented the Modern World: The True Story of How Western Europe's Poorest Nation Created Our World and Everything in It.* Broadway Books, New York.

Hutton, J. 1788. *Theory of the Earth with Proofs and Illustrations.* Amazon Digital Services, Seattle, WA.

Laudan, R. 1987. *From Mineralogy to Geology: The Foundations of a Science 1650–1830.* Chicago, University of Chicago Press.

Lyell, C. 1830–1833 (3 vols.). *Principles of Geology.* University of Chicago Press, Chicago.

McIntyre, D. B., and A. McKirdy. 2012. *James Hutton: Founder of Modern Geology.* National Museum of Scotland Press, Edinburgh.

McPhee, J. 1993. *Assembling California.* Farrar Straus & Giroux, New York.

Playfair, J. 1802. *Illustration of the Huttonian Theory.* Edinburgh: Cadell & Davies.

Playfair, J. 1805. *The Works of John Playfair, with a Memoir of the Author.* A. Constable & Co, Edinburgh (collected and published in 1822).

Popper, K. 1934. *The Logic of Scientific Discovery.* Basic Books, New York.

Prothero, D. R. 2011. *Catastrophes! Earthquakes, Tsunamis, Tornadoes, and Other Earth-Shattering Disasters.* Johns Hopkins University Press, Baltimore, MD.

Prothero, D. R. 2018. *The Story of the Earth in 25 Rocks.* Columbia University Press, New York.

Repcheck, J. 2003. *The Man Who Found Time: James Hutton and the Discovery of the Earth's Antiquity.* Perseus Books, New York.

Rudwick, M. J. S. 2014. *Earth's Deep History: How It Was Discovered and Why It Matters.* University of Chicago Press, Chicago.

Sagan, C. 1980. *Cosmos.* Random House, New York.

Twain, Mark. 1903. Was the World Made for Man? In *What is Man?: and Other Philosophical Writings* (1973) Franklin Classics, New York p. 106.

SUMMARY

- The earth is immensely old (now estimated at 4.567 billion years old). Geologic time is normally measured on scales of thousands to millions to billions of years.

- For most of human history, people thought the world was only a few thousand years old and that humans had been around since the beginning.

- Before the 1780s, most layered rocks were explained as the product of a supernatural catastrophe, namely Noah's Flood in the Bible (a concept called "catastrophism").

- They even thought that layered lava flows were formed in water. This concept was called "Neputunism."

- The idea of an old earth first emerged in the late 1780s with the work of the "father of geology," James Hutton.

- Hutton came to this insight by using the principle of uniformitarianism and the application of natural laws to geology and rejected supernatural catastrophism as unscientific. He saw how slowly mountains wore down to sand grains in the sea, how gradually erosion worked on the land surface, and how thick the enormous piles of ancient sediment were and concluded that he could see "no vestige of a beginning" of the earth.

- Hutton also recognized that igneous rocks were once fluid molten rock, called "magma," and not formed from water (a concept called "Plutonism").

- A generation later, Charles Lyell masterfully laid out the evidence for the Huttonian view of the earth and finally laid the old notions of catastrophism and Neptunism to rest.

- However, Lyell conflated uniformity of natural processes through time (actualism, a concept all geologists accept) with uniformity of rates of processes through time (gradualism).

- This confusion of gradualism with uniformitarianism meant that many geologists had to struggle to get acceptance of large-scale natural catastrophes, like a meteorite impact or a massive Ice Age flood.

- Geologists, like all scientists, must work within the scientific method: observe and record data, form a hypothesis, gather relevant data, then test that hypothesis and try to falsify it. In particular, we should keep multiple working hypotheses in mind as we try to explain natural phenomena. If enough testing has occurred, a hypothesis that has survived thousands of experimental tests and explained nearly all the evidence becomes a theory, a broader explanation for natural phenomena that explains a wide variety of data.

- For this reason, scientists must adopt naturalism and reject supernatural explanations because they are untestable and cannot lead to any further scientific insights.

KEY TERMS

Deep time (p. 6)
Geologic time (p. 6)
Scientific method (p. 9)
Naturalism (p. 9)
Uniformitarianism (p. 10)

Catastrophism (p. 10)
Neptunism (p. 10)
Plutonism (p. 10)
Angular
    unconformity (p. 11)

Actualism (p. 13)
Gradualism (p. 13)
Multiple working
    hypotheses (p. 13)

Verification versus
    falsification (p. 15)
Peer review (p. 15)
Theory (p. 15)

STUDY QUESTIONS

1. Why is it hard for humans to understand the concept of geologic time?
2. What evidence led Hutton to believe the earth was immensely old?
3. What did Hutton mean when he said that the earth had "no vestige of a beginning, no prospect of an end?"
4. How could people have ever thought that lava flows were formed in water?
5. How did Hutton demonstrate that magmas were formed of molten rock in liquid form, not deposited from water?
6. How is the principle of uniformitarianism applicable to other sciences besides geology?
7. Why must scientists reject supernatural explanations for events that have happened in the past or are still happening?
8. Why was Lyell more successful than Hutton in making the case for our modern understanding of the earth?
9. What is the difference between actualism and gradualism?
10. How is the impact of a giant meteorite or a giant Ice Age flood not a violation of uniformitarian principles?
11. What is the scientific method?
12. What is the method of multiple working hypotheses?
13. Why is it much easier to falsify a hypothesis than verify it?
14. How does peer review work? How does it make science different from any other source of information about the world?
15. What does the word "theory" mean in science? How does this contrast with the popular usage of the word "theory"?
16. Do you think the scientific method is a useful way to investigate natural process? If so, why? If not, why not?

Building Blocks

Minerals and Rocks

The rock I'd seen in my life looked dull because in all ignorance I'd never thought to knock it open. People have cracked ordinary pegmatite—big, coarse granite—and laid bare clusters of red garnets, or topaz crystals, chrysoberyl, spodumene, emerald. They held in their hands crystals that had hung in a hole in the dark for a billion years unseen. I was all for it. I would lay about me right and left with a hammer, and bash the landscape to bits. I would crack the earth's crust like a piñata and spread to the light the vivid prizes in chunks within. Rock collecting was opening the mountains. It was like diving through my own interior blank blackness to remember the startling pieces of a dream: there was a blue lake, a witch, a lighthouse, a yellow path. It was like poking about in a grimy alley and finding an old, old coin. Nothing was as it seemed. The earth was like a shut eye. Mother's not dead, dear—she's only sleeping. Pry open the thin lid and find a crystalline intelligence inside, a rayed and sidereal beauty. Crystals grew inside rock like arithmetical flowers. They lengthened and spread, adding plane to plane in awed and perfect obedience to an absolute geometry that even the stones—maybe only the stones—understood.

—*Annie Dillard, An American Childhood*

Photograph of giant gypsum crystals in the famous Cave of the Crystals, in the Naica Mountains, Chihuahua, Mexico. Under the right conditions of slow, continuous growth, crystals can grow very large.

2.1 Atoms and Elements

Before we can discuss the details of earth history, we must briefly cover the fundamentals. For those who have already taken an introductory course on geology, this chapter may mostly be review or can be skipped altogether. However, this book is often used for college courses that assume no previous exposure to geology. In addition, many non-geologist readers of this book will be baffled if they encounter the names of important rocks and minerals without the proper explanation, so these are introduced here.

First, let's quickly review the most basic principles of physics and chemistry. All matter is made of **atoms**, which are the smaller particle of matter that has the properties of a given **element**. For example, atoms of the element gold have its characteristic properties (such as its high density), but if you break a gold atom into its subatomic particles, they no longer have those same properties. The three main subatomic particles are the **proton** (which has a +1 charge and a mass of 1) and **neutron** (which has no charge and a mass of 1), both of which are found in the **nucleus** of the atom. The nucleus is surrounded by charged clouds of electromagnetic energy known as **electrons**, which have no mass but a –1 charge.

The proton is especially important because the number of protons in the nucleus is the **atomic number**, which determines which element you have. Any nucleus with one proton will be part of a hydrogen atom, a nucleus with two protons is part of a helium atom, and so on. If you change the number of protons in the nucleus, it becomes a different element. Because each proton carries a +1 charge, there should be an equal number of –1 electrons to balance the charge and make the atom electrically neutral. If the number of electrons does not match the number of protons, then the atom becomes a charged **ion**. If it has lost one or more electrons, then there is a net positive charge, and it is a **cation** (KAT-eye-on). If it has gained one or more electrons, then it has a net negative charge, and it becomes an **anion** (ANN-eye-on).

The number of neutrons in the nucleus has no effect on the charge of the atom (since neutrons are uncharged); but each neutron adds a mass of +1 to the nucleus, so they affect the **atomic weight**. For a given element (and thus for a fixed number of protons), there can be different atomic weights depending upon how many additional neutrons are in the nucleus. These different atoms with the same number of protons but different atomic weights are known as **isotopes**. For example, all isotopes of hydrogen have only one proton (they are no longer hydrogen if they don't). But hydrogen has several isotopes. Hydrogen with a mass of 1 (shown as ^1H, with the atomic weight in the left superscript) has only a single proton and no neutrons. This form of hydrogen (sometimes called "hydrogen 1" or "protium") is by far the most common in the universe and makes up 99.98% of the known hydrogen. However, about 0.01% of the hydrogen in the universe has not only a proton but a neutron as well, giving an atomic weight of 2. It is known as "hydrogen 2" or **deuterium** and is shown by the symbol ^2H. A third very rare form of hydrogen is "hydrogen 3", or **tritium**, with one proton and two neutrons. It is shown by the symbol ^3H. It is radioactive and produced only by nuclear reactions, and is a common byproduct of the explosion of nuclear bombs.

Nearly all the elements in nature have several different isotopes, and they are often very useful in geochemistry and in many other fields of science. For example, carbon has 15 known isotopes, only two of which are not radioactive but stable in geological settings. Normal carbon is carbon-12 or ^{12}C (6 protons, 6 neutrons) and makes up 99% of the carbon in your body and in almost anything carrying carbon. But carbon-13 (^{13}C, with 6 protons but 7 neutrons) is also stable. Even though it makes up less than 1% of the carbon on the planet, its occurrence is a powerful tool in geology, geochemistry, and oceanography. Carbon-14 (^{14}C, with 6 protons and 8 neutrons) is produced by bombardment of ^{14}N in the atmosphere that changes it into ^{14}C. It is radioactive and unstable, decaying back to nitrogen at a known rate. This useful property allows it to be the basis of **carbon-14 dating** (or radiocarbon dating), the best tool geologists and archeologists have for measuring the age of human artifacts and anything less than about 60,000 years old (see Chapter 3).

There are over 100 elements in the periodic table, but most of them are extremely rare in geologic settings. Many of them only exist for milliseconds in a high-powered physics lab. In geology, the chemistry is even simpler. It turns out that there are only eight common elements (Table 2.1) in the earth's crust that are worth remembering because they make up over 99% of all rocks on earth.

The first surprise is that so much of most rocks in the earth's crust are made up of oxygen (Table 2.1). An average rock is 46% oxygen by weight percent but 94% oxygen by volume! Why is this? Not only is oxygen a light element, but its ion has a large radius. You pick up a heavy rock, but in reality it's made mostly of oxygen. Why is oxygen by far the most abundant? If you glance at Table 2.1, you'll note that it is the only common anion on the table; and *something* has to balance all the positive charges of the remaining seven cations. Oxygen is very abundant in air and water, so it combines with almost any cation. Next in abundance is silicon, which (like carbon) is an element that readily bonds into long chains and forms complex three-dimensional structures. This is important in making the common rock-forming minerals, which are mostly combinations of silicon and oxygen, or **silicates**. In distant third place is aluminum, which is also an element that readily bonds into complex three-dimensional arrays with silicon, so many silicate minerals are rich in aluminum. These three elements just happened to be the common ones when the solar system

and earth formed, and these three are found in most of the common rock-forming minerals as a result.

The remaining eight elements in Table 2.1 are much less abundant than the "Big Three," with most of them making up only 5% or less in percentage of crustal rocks. All five are metallic cations, which bond with oxygen or with complex silicate structures to make the huge variety of minerals. Note that they have different charges: sodium and potassium are both cations with a +1 charge. They can sometimes replace each other in a mineral since they have the same charge. Likewise, calcium and magnesium and ferrous iron (Fe^{+2}) are all common +2 cations and can switch with each other in many minerals.

Even more surprising is what elements are *not* in the "Big Eight." Hydrogen and helium are the most common elements in much of the solar system, especially the sun and in the outer planets (Jupiter, Saturn, Uranus, Neptune). But they are rare on earth, except where hydrogen is bonded with oxygen to make water. Why? When the earth first formed, it was not massive enough to have strong gravity to hold on to these elements, and they floated off into space. Giant planets like Jupiter and Saturn have much more mass and therefore more gravity, and they held in their hydrogen and helium. Phosphorus and sulfur are also relatively rare and only concentrated in special settings. Perhaps most surprising of all is how rare carbon is in crustal rocks. After all, carbon is the building block of all life! There are a few minerals and rocks that have carbon in them, but they are rare compared to silicates.

2.2 Minerals and Rocks

Now that we have the foundation of the elements available in the earth's crust, let's see how they are combined into more complex molecules (groupings of atoms bonded together) known as **minerals**. The word "mineral" has all sorts of casual and inconsistent meanings in popular culture, but to geologists and chemists, it has a very strict and clear definition. A mineral

- is naturally occurring
- is inorganic
- has a definite crystalline structure
- has a definite chemical composition
- has characteristic physical properties

Let's discuss each of these components.

Naturally occurring: There are lots of complex compounds in the world, but if they are not produced naturally, they are not minerals. Thus, a synthetic diamond produced in a lab has all the properties of a diamond mined from the earth, but technically it's not a mineral. Ice formed as ice crystals or snowflakes is mineral, but not ice in your ice cubes. Most of the stuff sold in a health food store that is called "mineral" was produced synthetically and therefore is not a mineral as a scientist uses the word.

TABLE 2.1   COMMON ELEMENTS IN THE EARTH'S CRUST

Element	Symbol	Percentage of Crust By Weight	Percentage of Crust By Atoms
Oxygen	O	46.6%	62.6%
Silicon	Si	27.7%	21.2%
Aluminum	Al	8.1%	6.5%
Iron	Fe	5.0%	1.9%
Calcium	Ca	3.6%	1.9%
Sodium	Na	2.8%	2.6%
Potassium	K	2.6%	1.4%
Magnesium	Mg	2.1%	1.8%
All others		1.5%	0.1%

See For Yourself: A Brief Introduction to Minerals

Inorganic: Organic chemicals are built of the element carbon, so minerals are usually not made of carbon. However, there are a handful of important minerals with carbon (such as the mineral calcite, or $CaCO_3$), so we'll use "organic" in this context to mean compounds with carbon–hydrogen bonds. Thus, sugar forms beautiful crystals, but it is organic and therefore not mineral. Many of the "minerals" sold in a health food store are organic as well, and thus not minerals.

Definite crystalline structure: Like the word "mineral", the word "crystal" has a different meaning to a scientist from what it has in popular culture. Typically, people use the word "crystal" to describe anything that sparkles. In a scientific definition, a crystal must have a *regular three-dimensional arrangement of atoms* in its internal structure, which repeats over and over again. This three-dimensional array is called a **lattice**. It is analogous to the regular repeated pattern in wallpaper. For example, the atoms of the salt (sodium chloride, or NaCl) crystal (Fig. 2.2A) are arranged in a cubic pattern, with each atom of sodium or chlorine forming 90° angles with the others. The same lattice is found in the mineral galena (Fig. 2.2B), which is made of lead and sulfur in equal amounts (lead sulfide, or PbS). All minerals have a regular three-dimensional lattice of some kind, often very complex and with many other angles between atoms besides the 90° seen in the simple cubic lattice.

Some things in nature may have a three-dimensional arrangement of atoms, but they are not regular and repeating. Take, for example, volcanic **glass** or **obsidian** (Fig. 2.3). At the molecular level, the atoms are not in any kind of repeated pattern but in a random tangle of long chains, like a bowl of spaghetti. Technically speaking, a glass isn't really a solid at all but a supercooled liquid. Over very long spans of time, the glass will slowly flow and change shape. This is clear if you ever see a piece of window glass in a very old house. If it has been in its window frame for about a century, the glass will be thicker at the bottom because it has slowly flowed downhill over the course of decades. Thus, a *glass is by definition not a crystal*. A popular item at many gift shops is "cut-glass crystal" drinking goblets and chandeliers, but this is not the definition of "crystal" that scientists use.

Definite chemical composition: Most minerals have a simple chemical formula, like most other compounds. There is a bit of substitution allowed if you replace one ion with another of a similar charge. For example, the mineral calcite (calcium carbonate, or $CaCO_3$) can have a certain percentage of magnesium replacing calcium sites in its lattice and still be calcite. However, if it gets to be 50:50 Ca/Mg, then it's no longer calcite but a different mineral, dolomite.

Characteristic physical properties: Most of the features of the minerals we have discussed occur at the atomic level. But to identify the mineral, you need to know what physical properties are typical of a small piece of the mineral. These include its color, its **hardness** (from soft minerals like talc and gypsum to the hardest mineral, diamond), whether it fractures with an irregular surface or has a **cleavage** and breaks into many parallel planes, as well as less commonly used properties like density (lead sulfide or galena, for example, is unusually dense because it contains lead), reaction to acid (the mineral calcite fizzes in dilute hydrochloric acid), and magnetism (the mineral magnetite is naturally magnetic).

Many of these properties of minerals can be understood by knowing the

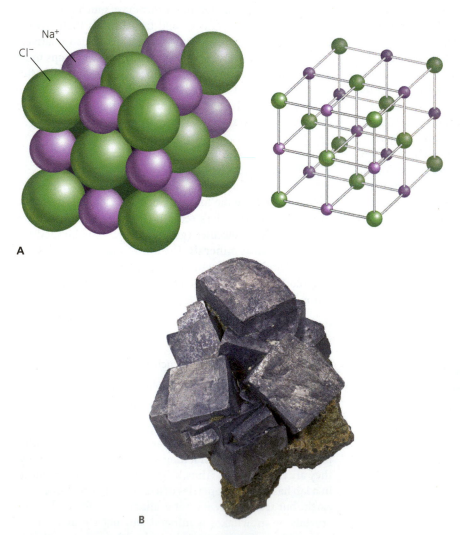

Na⁺

Cl⁻

A

B

Figure 2.2   Minerals with a cubic lattice and cubic cleavage. **A.** A large cubic crystal of salt in front of a ball-and-stick model of the cubic lattice of its atoms. **B.** Galena, or lead sulfide, has the same cubic lattice and breaks into cubic crystals with 90° cleavage angles.

Figure 2.3 Obsidian is a form of volcanic glass, where the magma has cooled so quickly that no minerals had time to crystallize.

crystal lattice. For example, the cubic lattice of minerals like salt or "halite" (NaCl) or galena (PbS) is demonstrated at the hand sample level since any time you hit and break a piece of these minerals, they will naturally cleave to form cubic faces with 90° angles (Fig. 2.2).

The atomic properties and crystal lattice can make a big difference in the behavior of a mineral at the macroscopic level. Let's take as an example the two common minerals formed of pure carbon: **diamond and graphite**. One is the hardest substance in nature and the other one of the softest, yet they are chemically identical. Why are they so different? Diamond has a crystal lattice with all the atoms of carbon tightly bonded together and very short, strong chemical bonds (Fig. 2.4). This structure will survive huge amounts of

pressure, and an expert diamond cutter has to know exactly how to cleave a large stone into several smaller ones. If the cutter misses, the diamond is ruined. Graphite (the "lead" in a pencil), on the other hand, has all its carbon atoms arranged in sheets, with very long, very weak molecular bonds between the sheets. Pushing the graphite tip of a pencil across paper is enough to break those weak bonds, leaving tiny flakes of graphite behind on the paper as pencil markings.

Another example is calcium carbonate, or $CaCO_3$, in two different mineral lattices: calcite (the common mineral in limestones and marbles) and aragonite (also known as "mother of pearl" or "nacre"). They have the same chemistry, but their lattices are very different and lead to very different properties. The most obvious of these is that aragonite is much more soluble than calcite, so in weakly acidic conditions aragonite will dissolve but calcite won't. This is why if you own pearls, it is important to wash your acidic sweat off them once you put them away after wearing them. For some minerals, knowing the chemical composition is not enough; the crystal lattice makes a huge difference in the properties of the mineral as well.

There are literally thousands of different kinds of minerals, but most can be organized into just a few classes based on chemical composition (Table 2.2). There is no room to discuss them all in a brief introduction like this, but their major characteristics are detailed in the table. As we saw in Table 2.1, however, most of the earth's crust is made of silicon and oxygen with minor aluminum, so it's no surprise that the most important rock-forming minerals are the silicates, made of silicon plus oxygen.

[Note: Word endings are very important in chemistry! *Silicon* is the element on the periodic table; a *silicate* is a mineral made of silicon plus atoms of oxygen; *silica* is the

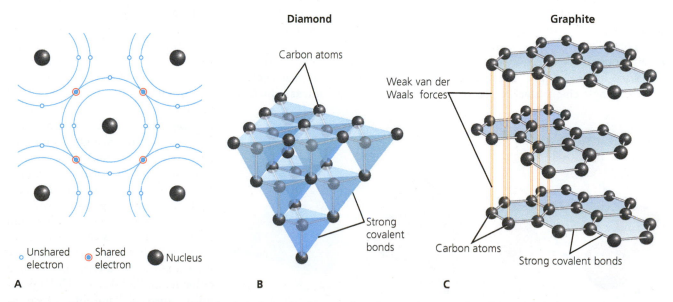

Figure 2.4 Diamond and graphite are both minerals made of pure carbon, but one is very soft and the other is the hardest substance in nature. Graphite is soft because the carbon atoms are arranged in sheets with long, weak intermolecular bonds between the sheets. These bonds break easily when pressure is applied. Diamond has its carbon atoms arranged in a tight cubic lattice, with short, strong bonds in between them, so they are hard to break.

TABLE 2.2   THE MAJOR CHEMICAL CLASSES OF MINERALS

Class	Composition	Examples
Native elements	One element only	Gold, silver, copper, diamond, graphite, sulfur
Oxides	Cation + oxygen	Iron oxides (Fe_2O_3, hematite; Fe_3O_4, magnetite)
Halides	Cation + halogen (Cl, F)	Halite (NaCl), fluorite (CaF_2)
Sulfides	Cation + sulfur	Pyrite (FeS_2), galena (PbS)
Sulfates	Cation + SO_4^{-2}	Gypsum ($CaSO_4 \cdot 2\,H_2O$)
Carbonates	Cation + CO_3^{-2}	Calcite and aragonite ($CaCO_3$), siderite ($FeCO_3$)
Silicates	Silicon + oxygen ± cations	Quartz (SiO_2), olivine (Mg_2SiO_4)

compound silicon dioxide, or silicon plus two atoms of oxygen (SiO_2). But *silicone* is a synthetic compound made from silicon in a lab, used for lubricants, breast implants, and other purposes. Don't confuse them!]

The major classes of silicate minerals are given in Table 2.2. Each has the same basic building block: the silicon–oxygen tetrahedron (Fig. 2.5), or SiO_4 unit. Each class of silicate minerals uses these fundamental building blocks over and over again and links them together in more and more complex structures.

The simplest silicate mineral structures are built of **isolated tetrahedra**, where the silicon–oxygen (SiO_4) building blocks do not bond directly to each other but are held in place by the electrostatic charges of the cations (especially Mg and Fe) between them in the lattice. Common single tetrahedral minerals are the green mineral **olivine**, which is (Mg, Fe)SiO_4, and **garnet**, which is built of SiO_4 tetrahedra with a variety of cations to make the six major types of garnets (from red-brown almandine to green grossular and several others).

The next most complex arrangement is found in the class of minerals known as **pyroxenes** (PEER-ox-eens), which have linked the SiO_4 tetrahedra into long **single chains** (Fig. 2.6). Among the common pyroxenes is the dull greenish black mineral **augite** ($MgSiO_3$) as well as **jadeite**, one of the two minerals that make the beautiful gemstone known as jade. Because of their crystal lattice made of single chains stacked together, pyroxenes have a 90° cleavage in hand samples, the most reliable physical property to recognize them.

The next step in silicate complexity is to link two single chains side by side to make a **double-chain** silicate structure, similar to the way the rails of a train track are linked together by wooden ties. Double-chain silicates are known as **amphiboles**, and their lattice structure gives the hand samples cleavages that are roughly either 60° or 120° (technically, 57° and 124°), their most diagnostic property. The most common amphibole is the shiny jet-black prismatic mineral known as **hornblende**, which is common in many igneous and metamorphic rocks. There are others, such as the greenish amphiboles tremolite and actinolite and the blue amphibole glaucophane, which we'll discuss elsewhere in this book.

Once you have a double-chain structure, the next more complex arrangement is to bond the double chains together side by side to form **sheet silicates**. Sheet silicates are made of two layers of silicon–oxygen tetrahedral ("t") sandwiching a layer of aluminum–oxygen octahedra ("o"). This "t-o-t" structure is like an Oreo cookie, with the "t" layers represented by the chocolate cookie layers and the "o" layer representing the aluminum–oxygen creamy filling. Most sheet silicates are built of stacks of "t-o-t" structures with other materials (different kinds of cations, water molecules) trapped between the "t-o-t" layers. The most familiar sheet silicates are a class of minerals known as **micas**, which are distinctive in that they cleave into large, flat, thin sheets. The silver-white mica is known as **muscovite**, and before it was possible to make glass windows, large sheets of muscovite (called "isinglass") were used as windows and curtains (such as the "isinglass curtains" in the song "Surrey with the Fringe on Top" from the Rodgers and Hammerstein musical *Oklahoma*). There is also a common black mica known as **biotite**, a green mica called **chlorite**, and a lithium-rich lavender mica known as **lepidolite**, which we will see in many different types of rocks. In addition to the micas, all of the clay minerals that make up the

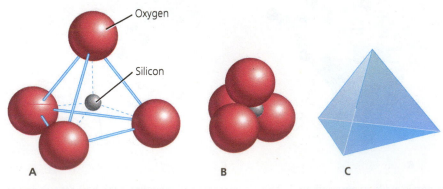

Figure 2.5 The silicate minerals are built from linking together simple building blocks known as silicon–oxygen tetrahedra, here shown with a ball-and-stick model **(A)**, the actual arrangement of the ions drawn to scale **(B)**, and a diagrammatic representation as a tetrahedron.

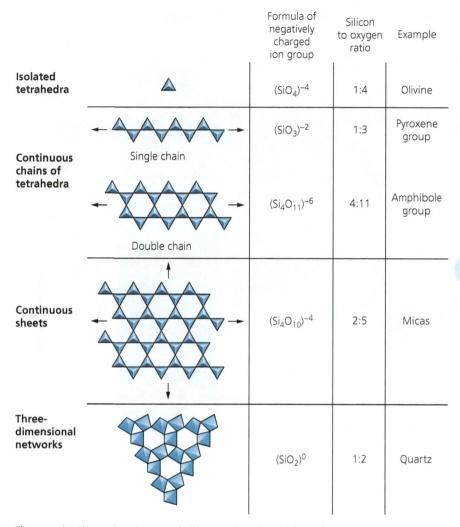

	Formula of negatively charged ion group	Silicon to oxygen ratio	Example
Isolated tetrahedra	$(SiO_4)^{-4}$	1:4	Olivine
Continuous chains of tetrahedra — Single chain	$(SiO_3)^{-2}$	1:3	Pyroxene group
Double chain	$(Si_4O_{11})^{-6}$	4:11	Amphibole group
Continuous sheets	$(Si_4O_{10})^{-4}$	2:5	Micas
Three-dimensional networks	$(SiO_2)^{0}$	1:2	Quartz

Figure 2.6 The major classes of silicate minerals and how they are constructed from linking together silicon–oxygen tetrahedra.

muds of the world are sheet silicates; they are the most common minerals on earth for this reason.

We have progressed from isolated tetrahedra to single chains to double chains to sheets, each structure being more and more complex and linked together (in chemical terms, **polymerized**). The final step is to link the silicon–oxygen tetrahedral into a complex three-dimensional **framework** structure that is almost impossible to render in a two-dimensional illustration. However, there are ball-and-stick models that capture their shape well, and now there are animations online that give some idea of their geometrically complex structure. Some of the most common and important minerals of all are framework silicates. The most important of these is **quartz**, made of pure silica (SiO_2), one of the most abundant sedimentary minerals on earth. The other important class of framework silicates includes the aluminum-rich **feldspars**, the most common minerals in igneous rocks. Two types of feldspar are particularly important: the pink-colored **potassium feldspars** ($KAlSi_3O_8$), which exist in three mineral forms (orthoclase, microcline, and sanidine), and the

plagioclase (PLAJ-o-clase) **feldspars**, which transform continuously from pure calcium-rich plagioclase known as anorthite ($CaAl_2Si_2O_8$, typically bluish-gray in color) to intermediate plagioclases with a mixture of calcium and sodium to pure sodium-rich plagioclase known as albite ($NaAlSi_3O_8$, white in color). Plagioclase crystals typically show lots of fine parallel lines known as **striations** on their cleavage surfaces. Plagioclases are typically the most common minerals in most igneous rocks and are found in all but a few of them.

 See For Yourself: Silicate Structures

These are a lot of mineral names to master, especially for beginning students, yet this is the minimum number of minerals needed to understand the rocks that occur on the earth's surface and in its interior. If you have the opportunity to study hand samples of each of these minerals, test their properties, and compare them to each other, the names make more sense and become easy to remember with lots of experience and practice. Nearly every geology student masters these minerals after the first few classes, so it's not that hard to do. It just takes practice and study! Once you know your minerals, you can understand the rocks that are made from them.

2.3 Igneous Rocks

Igneous rocks are formed by **magma** (molten rock) from deep in the earth that rises up from deep **plutons** (magma chambers) and crystallizes as it cools. The size of the crystals in the rock depends upon how fast it cools and solidifies. If the molten rock is spewed out of a volcano, the magma cools quickly and the crystals have little time to form, so they are **microcrystalline** ("aphanitic"), too small to see with the naked eye. They are only visible in thin polished sections of rocks when viewed under a special microscope. If the crystals cool slowly over years to hundreds of years deep in an underground magma chamber, then they have time to grow larger. Sometimes they are still just barely visible to the naked eye, but still they are **macrocrystalline** ("phaneritic") nonetheless. A few magmas cool extremely slowly over the course of decades or centuries or longer, producing **pegmatites** full of giant crystals.

See For Yourself: Igneous Rocks

Some igneous rocks have a composite texture, with macrocrystalline crystals (**phenocrysts**) floating in a rock that

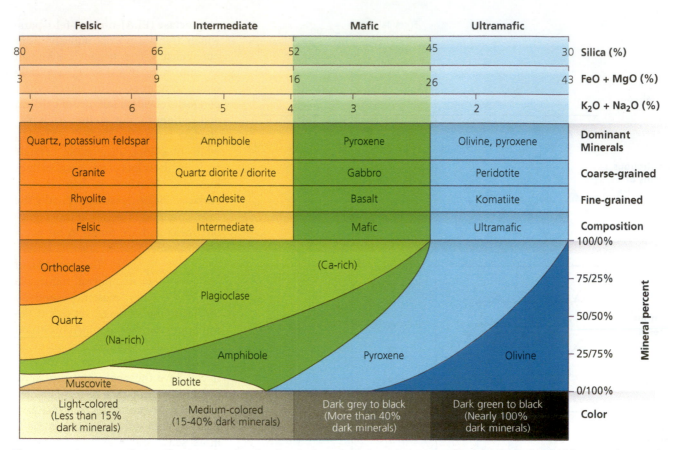

Felsic	Intermediate	Mafic	Ultramafic	

80	66	52	45	30	Silica (%)
3	9	16	26	43	FeO + MgO (%)
7	6	5	4	3	2	K_2O + Na_2O (%)

Quartz, potassium feldspar	Amphibole	Pyroxene	Olivine, pyroxene	**Dominant Minerals**
Granite	Quartz diorite / diorite	Gabbro	Peridotite	**Coarse-grained**
Rhyolite	Andesite	Basalt	Komatiite	**Fine-grained**
Felsic	Intermediate	Mafic	Ultramafic	**Composition**

Figure 2.7   Igneous rock classification, showing the chemical trends and category names (from felsic to ultramafic) across the top, the trends in dominant minerals in the middle, and the color gradient across the bottom.

is mostly microcrystalline (called the **groundmass**). This hybrid texture is called **porphyritic** (the noun form of the word is **porphyry**) and results from a magma that had a two-stage cooling history: the phenocrysts cooled slowly in a large pluton, then the semicrystallized magma was blown out of a volcano, where the rest of the lava cooled quickly to form the groundmass surrounding the phenocrysts. The important thing to remember is that the *crystal size is a result of the time and mode of cooling*, so microcrystalline rocks are volcanic, macrocrystalline rocks cool slowly in plutons (plutonic rocks), and porphyritic rocks have both stages in their history.

The crystal size is one element of a classification of igneous rocks (Fig. 2.7). The other axis is based on their chemical composition. Rocks coming from the mantle are relatively rich in magnesium, iron, and calcium and are known as "**mafic**" rocks for short ("ma" for magnesium plus the chemical abbreviation "Fe" for iron). This magnesium–iron–calcium chemistry produces minerals rich in these elements, such as olivine and pyroxenes (magnesium- and iron-rich silicates) and calcium plagioclase. If a mafic magma cools quickly, it is the familiar black lava called **basalt** (pronounced ba-SALT by American geologists, BASS-alt by British geologists) that erupts out of Kilauea on the Big Island of Hawaii, and other mantle-sourced volcanoes. If it cools slowly, the same magma that makes black basalt instead produces a macrocrystalline rock with visible pyroxene and calcium plagioclase known as a **gabbro**.

Some magmas are so rich in magnesium and iron that they are known as **ultramafics**. These contain almost nothing but olivine derived directly from the mantle or lower crust. A macrocrystalline rock made of pure olivine is called a **peridotite** (peh-RIDD-o-tite) and represents a direct sample of the upper mantle. (The gem name for olivine crystals is "peridot," so a rock made of olivine is a peridotite.) Currently, there are no ultramafic volcanic lavas erupting anywhere on earth, but back about 3 billion years ago, they were common and made an olivine-rich lava called **komatiite** (ko-MATT-tee-ite).

The other extreme of magma chemistry is melts that are rich in silicon, aluminum, potassium, and sodium. These rocks are often called **silicic** or **felsic** (an abbreviation based on combining "feldspar" plus "silica"). With this kind of magma chemistry, you produce minerals such as quartz (pure silica), sodium feldspars (sodium, aluminum, and silica), potassium feldspars, and micas such as biotite and/or muscovite (rich in potassium, aluminum, and silica). If the rock contains quartz and two-thirds of its total feldspar is potassium feldspar, then it is a true **granite**. True granites have so much pink or red potassium feldspar in them that they tend to be pink or red as well. There are volcanic eruptions of magmas with the composition of granites. These are known as **rhyolite**, and they are extremely fine-grained and generally pink or red in color due to the rusting of the iron in them.

The rocks that most people call "granites" do not have enough potassium feldspar in them to be true granites as a geologist defines the term. Many of them have less than 33% of their total feldspar as potassium feldspar, as well as about 20% quartz. Most geologists call these rocks **granodiorite** rather than granite. For example, most of the plutonic rocks in the Sierra Nevada Mountains in California are granodiorites or diorites. However, in this book, we will often use the term "granitics" as a category name to refer to the entire class of felsic plutonic rocks, including granodiorites, diorites, and many other felsic rocks we will not discuss here. The volcanic equivalent of a granodiorite is called a **dacite**. There are more complicated schemes of subdividing felsic igneous rocks, but we will not discuss them here because that requires much more background in geochemistry and mineralogy than is appropriate for this book.

Rocks that are intermediate in composition between basalt-gabbro and granodiorite-dacite are usually made of hornblende, plus an intermediate mixed sodium–calcium plagioclase but no quartz or potassium feldspar. These rocks are known as **diorite** if they are macrocrystalline and usually have a speckled black and white "salt and pepper" appearance. If they cooled quickly, a diorite composition becomes a microcrystalline volcanic rock known as **andesite**. They were named after the Andes Mountains of South America, although it turns out there are few actual andesites in the Andes by modern definition of that term. Most andesites are light to dark gray in color and porphyritic, with tiny black phenocrysts of hornblende floating in a microcrystalline groundmass of intermediate plagioclase plus hornblende.

BOX 2.1: HOW DO WE KNOW?

How Do Magmas Change Chemistry?

Figure 2.7A shows the simplest possible list of names for 10 basic types of igneous rocks. Igneous petrologists recognize hundreds of different types, but that level of detail isn't required here. Now that we have some names and understand some of the basics of magma chemistry, the next question is, *why are there so many different types of igneous rocks?* If their magmas were coming straight from the mantle, they would all be peridotite or komatiite in composition. Something happens to that ultramafic magma as it rises from the mantle and through the crust so that it changes chemistry as it rises, losing magnesium, iron, and calcium and gaining silicon, aluminum, potassium, and sodium in order to make all the different kinds of magmas that produce the hundreds of different kinds of igneous rocks (**magma differentiation**).

This is the central question of all of igneous petrology. It perplexed and puzzled geologists for most of the nineteenth century, when many types of igneous rocks were named and described, but no one could explain how they all formed. The breakthrough came in the early twentieth century, when Norman L. Bowen tried to simulate the process of magma differentiation in the laboratory. He would take a chip of peridotite, melt it in a high-temperature furnace, then drop the molten blob into liquid mercury to instantly quench and chill it. Once the blob had cooled, he would separate out all the material that had crystallized (mostly olivine), then melt the rest and go through the process again and again. (Although he didn't know it, working with mercury is very dangerous, and eventually Bowen lost his sanity due to mercury poisoning. This also happened to the "mad hatters" who used mercury to make felt for hats in the days of *Alice in Wonderland*.) As Bowen repeated his experiments, the remaining material became more and more depleted in magnesium, iron, and calcium as he removed the crystals of olivine, pyroxene, and calcium plagioclase that had formed first. As the residue became depleted in those elements, the magma that remained became richer in silicon, aluminum, potassium, and sodium (Fig. 2.8).

Bowen's experiments simulate what happens in many magma chambers. In a process called **fractional crystallization**, the first mineral to cool and crystallize will settle out to the bottom or sides of the magma chamber and remove the magnesium, iron, and calcium from the remaining magma. Over time, this changes the composition of the magma left behind, so eventually it can no longer crystallize olivine but must make pyroxene instead; and its plagioclase goes from calcium-rich to more sodium-rich. After enough olivine and pyroxene have been removed from the melt, it will no longer have enough magnesium or iron left, and then amphiboles like hornblende would be the next mineral formed. In other words, as each fraction crystallizes out and changes the melt left behind, the magma changes chemistry (hence "fractional crystallization").

As an analogy, imagine a room full of students standing up from their chairs. Let's say that the room is 60% men and 40% women. For this example, the men will represent iron and magnesium atoms and the women represent atoms of silica. Initially, all of the students are standing, as if they were atoms floating around in a melt. Now we'll have three men sit down for every woman who sits down, and do this over and over again. Each student sitting down represents a crystal sinking out of the melt, removing itself from the chemistry of the magma. After enough time, all the men will be sitting, but quite a few women will still be standing. So the ratio of men to women at the start was about three men for every two women, but at the end there will be no men standing and only women left (as if they were still floating in the melt). In the same manner, fractional crystallization pulls out the elements from the minerals that crystallize out first (such as iron, magnesium, and calcium) and deposits them on the walls and floor of the magma chamber. This leaves the remaining melt depleted in those elements and enriched in elements that were rare initially.

Bowen plotted this crystallization sequence of minerals, and today it is known as **Bowen's reaction series** (Fig. 2.8B). This simple arrangement of minerals exactly parallels the mineral sequence of the igneous rock classification scheme

(Fig. 2.7). It is considered one of the greatest discoveries in the science of geology and showed the power of experimental approaches and lab simulation when purely descriptive methods failed.

Bowen's reaction series has two branches (Fig. 2.8A). The **discontinuous path** is the sequence of different classes of dark-colored minerals, from olivine to pyroxene to amphibole to biotite. At the same time, the magma is continuously crystallizing out plagioclase as well, forming a **continuous path**. The first plagioclase to form is calcium-rich, and each later plagioclase is poorer in calcium and richer in sodium. Eventually, the two series converge in felsic magmas, which crystallize out quartz, potassium feldspar, sodium plagioclase, plus biotite and/or muscovite—the minerals found in granites and granodiorites.

Bowen's reaction series was based on a lot of brilliant lab experiments derived from an understanding of the well-documented geochemical patterns of igneous rocks. It has since been confirmed by many field observations. There are many ancient layered magma chambers (now exposed to the surface and eroded) that cooled slowly and built up layers of crystals on their floors as they cooled. The lowest layers are always rich in ultramafic minerals and rocks at the bottom (mostly peridotites and related rocks), then gradually become enriched in gabbros as you go up through the layers into magmas cooled later in the process.

There are other proofs as well. When geologists study a thin section of a gabbro or basalt under the microscope, it is typical to see an early formed olivine crystal that is surrounded by a rim of pyroxene, which crystallized around the olivine core. As the magma chemistry changed, it could no longer make olivine but made pyroxene instead. Also, microscopic examination of plagioclase crystals often shows that they have layers around the outside that are sodium-rich, while the inner core is calcium-rich, also demonstrating that the surrounding magma chemistry changed as the crystal grew.

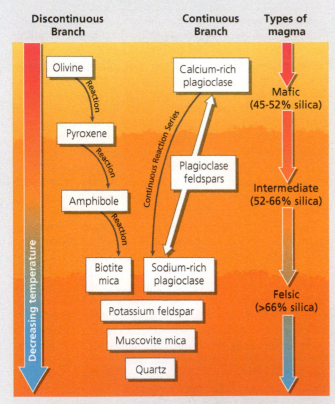

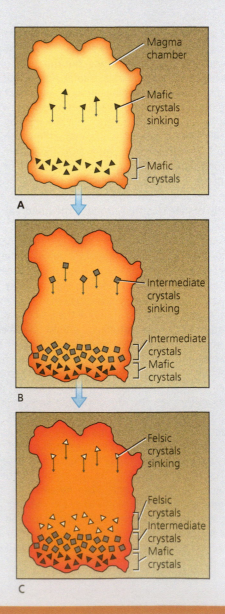

Figure 2.8 Bowen's reaction series displays two simultaneous trends, the continuous series of plagioclases, from calcium-rich to sodium-rich, and the discontinuous series of different dark minerals formed at different stages of melt differentiation. The first melts to crystallize form olivine and calcium plagioclase, taking elements like Fe, Mg, and Ca out of the melt, so each time the magma cools, it has less of those elements and can no longer form olivine, or even pyroxene. Meanwhile, the residue of melt left after those crystals pull out Fe, Mg, and Ca is enriched in Si, Al, K, and Na. This chemistry will tend to form minerals like sodium plagioclase and amphiboles, and eventually quartz, potassium feldspars, biotite, and muscovite.

Fractional crystallization (Box 2.1) is an important way to change an ultramafic magma direct from the mantle into a mafic or even intermediate magma. Another mechanism is called **partial melting**. Imagine that deep in the earth's crust there was a mass of cold gabbro or diorite that was reheated. The first materials to separate away from the original rock would be minerals that melt at the lowest temperatures, such as quartz, potassium feldspar, and sodium plagioclase. If this new low-temperature melt were then cooled, it would become a granodiorite or a granite.

Another mechanism for getting different magmas is to melt a different original rock. Thus, if you start with a rock of felsic or intermediate composition that was already in the crust, melting it will only produce a rock that is just as felsic, if not even more so.

Another important consideration is what causes rocks to melt. It turns out that for rocks that are in subduction zones (Chapter 5) plunging deep in the mantle, there is a lot of water entrapped in the ocean floor basalts that once formed on a mid-ocean ridge. Water and other volatiles (gases) dramatically lower the melting temperature of a rock. Many of the andesite–dacite–rhyolite magmas we will see in later chapters are formed because of different amounts of volatiles in the source material that melted.

Yet another likely mechanism for changing from a gabbro to a diorite to a granodiorite magma is to have it melt its way up through existing crustal rocks, which are typically much richer in silicon, potassium, aluminum, and sodium. As these wall rocks of the magma chamber were melted and digested into the magma, they could change its chemistry from mafic to felsic. If pieces of wall rock are ripped away and melted into the new magma, it is known as **assimilation**. If the wall rock around the hot magma chamber is partially melted into felsic magma and then mixed with the more mafic magma, it is known as **contamination**. Either way, magmas formed deep in the earth's crust that must melt their way through over 100 km (60 miles) of crustal rock above them will end up being very different in chemistry. At one time, assimilation and contamination were thought to be crucial in making magmas more silicic, but now partial melting and fractional crystallization are considered more important.

Finally, there are a few rare instances where a mafic magma chamber from one source melted its way into a felsic magma chamber from a different source (or vice versa), forming a **mixed magma chamber** that is intermediate in composition.

This is the most basic (and highly oversimplified) summary of how the chemistry of magmas changes to form the hundreds of different types of igneous rocks known on the planet. Most of these processes happen deep in the crustal magma chambers to form plutonic rocks. Rising plutons intrude into older rocks ("country rock" or "wall rock") and melt their way to higher levels. They can take many shapes

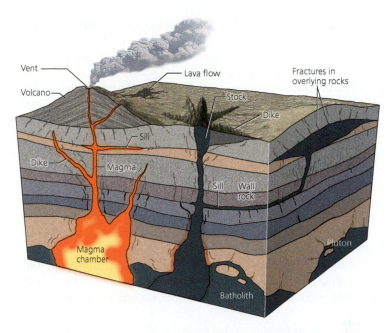

Figure 2.9 Schematic diagram showing different types of igneous bodies. Dikes cut across the structure of the rock, while sills intrude parallel to the bedding. They may be fed but a huge magma chamber called a batholith. Molten magma intrudes upwards, and erupts into extrusive volcanic rocks, which form lava flows, and volcanic cones with a vent that can be a crater or (if the volcano collapses on itself) a caldera.

(Fig. 2.9). Most plutons intrude to form a body of magma cutting through older rock called a **dike** (spelled "dyke" in Britain). If a dike intrudes parallel to the bedding of a sedimentary rock, it is a special kind known as a **sill**. A large chain of plutons which underlies a volcanic mountain range is known as a **batholith**. We will see all of these features, as well as the features typical of volcanoes, in later chapters.

2.4 Sedimentary Rocks

The rocks in the second major class are much more familiar to us since they form at the earth's surface, not in deep magma chambers or dangerous volcanoes. They are known as **sedimentary rocks**. They are of enormous importance since nearly all the economic products we obtain from the earth come from loose sediments or sedimentary rocks. These include energy sources like oil, gas, coal, and uranium; groundwater we rely upon; materials for construction (building stone like sandstone or limestone, concrete made of crushed limestone plus sand and gravel, gypsum for drywall, quartz sand for glass, and so on); and many of our metallic mineral resources (especially uranium, iron, and steel).

 See For Yourself: Sedimentary Rocks Even without these economic incentives, understanding sedimentary rocks is still supremely important. Sedimentary rocks are the source of nearly all our information about earth history and ancient environments and the only source of fossils that demonstrate the history of life and help us tell geologic time. Finally, carbon-rich rocks such as coal and limestone are the thermostats that make our planet livable.

These crustal reservoirs lock up or release carbon to the atmosphere so that earth is neither a hellish supergreenhouse like Venus (where the atmosphere is hot enough to melt lead) nor a frozen iceball like Mars.

All sedimentary rocks undergo some version of a basic pathway. They always start as *weathered material* of a pre-existing rock that can be igneous, metamorphic, or sedimentary. Once that weathered material is picked up by wind and/or water and *transported*, it becomes loose sediment. Eventually, the sediment in motion comes to a stop and is *deposited*. But it's still loose sediment until sometime later in its history (usually after some deep burial) the loose sand grains are cemented together or the mud particles compressed, and the loose sediment is *lithified* into a sedimentary rock. Every step of this history can be detected in clues in the rock, and expert sedimentary geologists act like detectives, gleaning clues about the past from a sandstone or limestone that no one else even notices.

There are two versions of this pathway. Most sedimentary rocks are made of broken particles of pre-existing rocks or minerals, known as **clasts** (Greek *klastos*, meaning "broken fragments"). Therefore, **clastic** (or "detrital") sedimentary rocks are made of different-sized pieces of other rocks, from huge boulders down to fine clay. Eventually these loose grains of gravel, sand, or mud must be lithified into sedimentary rock. The second pathway is much simpler. Instead of fragments of rocks or minerals, the ions from the pre-existing rock weather out and are dissolved in water, where they stay until something causes them to precipitate from the water and crystallize into minerals like halite (forming rock salt), gypsum (hydrous calcium sulfate), calcite or aragonite (calcium carbonate), and many others. Thus, when they crystallize they are already lithified into solid rock. Since this is a purely chemical process, we call these **chemical** sedimentary rocks.

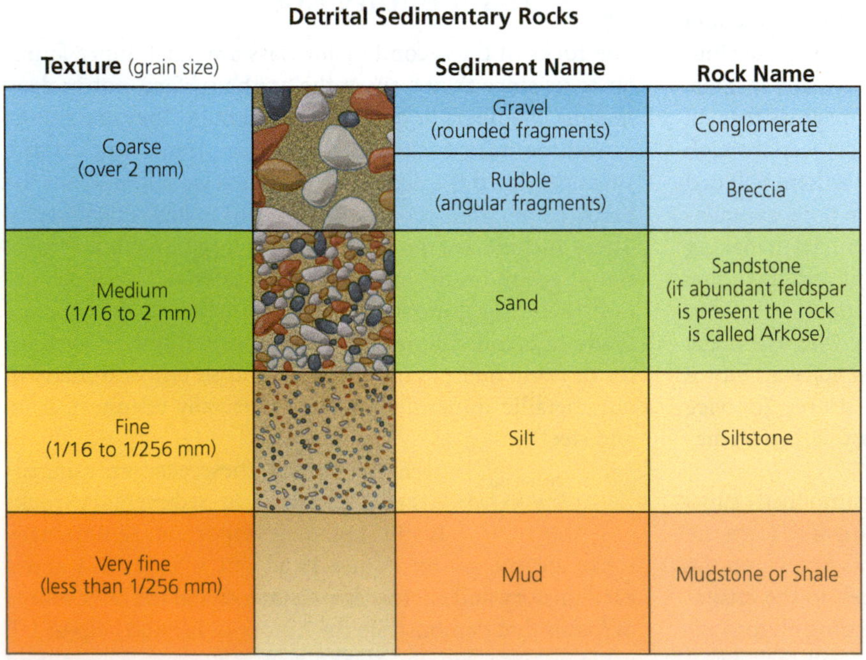

CLASTIC SEDIMENTARY ROCKS

The most important property of rock fragments is their size, so classifications of clastic sediments are based on size. It's also important because the size of grains is a good indicator of the agents of deposition (wind, water, glacial ice), and the grain size decreases downstream from the source, so it helps tell us about transport. Rather than the exotic and obscure names of igneous rocks (try to remember what minerals are in a "lherzolite" or a "jacipirangite"!), most sedimentary rock classifications are based on the practical Anglo-Saxon words that English speakers have used all their lives—"sand," "gravel," "mud"—only those words have a strict definition in classification schemes (Fig. 2.10). Any particle larger than 2 mm in diameter is gravel, which can be further broken down into granules (2–4 mm in diameter), pebbles (4–64 mm), cobbles (64–256 mm), and boulders (larger than 256 mm). A rock made of lithified gravel and finer sand is called a **conglomerate** if it has rounded gravel and a **breccia** (pronounced BRETCH-cha) if its clasts are angular. Conglomerates are the product of flood energies or debris flows and are diagnostic of sedimentary settings near the mountains where such floods occur. Breccias are even more specific since it's hard for large pebbles not to become rounded, even with a few kilometers of washing downhill. Thus, breccias occur in unusual settings (collapsed cave ceilings, impact debris, landslides) and are powerful indicators of these odd conditions.

We all know what sand is and what it feels like (think sandpaper or beach sand), but to a sedimentologist sand is strictly defined as particles between 2 mm and 1/16 mm in diameter. Many sands become cemented to form the rock known as **sandstone**. Most sandstones are rich in quartz, the most stable mineral on the earth's surface, so quartz sandstone is the most common type. However, rare sandstones can be rich in feldspars, which are common in igneous rocks but rapidly break down into clays when weathered on the earth's surface. Thus, a sandstone rich in feldspars (called an **arkose**) is very unusual and typically indicates that the sand grains have not traveled very far or have not been deeply weathered in wet, humid conditions.

Sediment finer than 1/16 mm can be called by the familiar name "mud," and if it is compressed into a rock, it is called a **mudstone**. Mud can be further subdivided into the coarser material known as silt (1/16 mm to 1/256 mm) and clay (finer than 1/256 mm).

Detrital Sedimentary Rocks

Texture (grain size)		Sediment Name	Rock Name
Coarse (over 2 mm)		Gravel (rounded fragments)	Conglomerate
Coarse (over 2 mm)		Rubble (angular fragments)	Breccia
Medium (1/16 to 2 mm)		Sand	Sandstone (if abundant feldspar is present the rock is called Arkose)
Fine (1/16 to 1/256 mm)		Silt	Siltstone
Very fine (less than 1/256 mm)		Mud	Mudstone or Shale

Figure 2.10 A classification scheme for clastic or detrital sedimentary rocks.

A rock made entirely of silt is a **siltstone**, and clay makes a **claystone**. In the field, you can tell silt from clay because silt is still close enough to the sand size range that it is gritty to chew, whereas clay is "creamy" to chew. Almost all mudrocks, however, undergo some burial and pressure. This squeezes down the clay minerals (layered sheet silicates), forces out the water between the mineral grains (often almost 70% of the volume of mud is water), and turns it into a slightly different rock known as **shale**, which breaks into flat sheets along bedding (so it is "fissile"). Shales are by far the most common sedimentary rock on the earth's surface.

CHEMICAL SEDIMENTARY ROCKS

As mentioned, chemical sedimentary rocks form when ions from pre-existing rocks dissolve by weathering and go into solution in the waters of the earth (groundwater, rivers, lakes, or the ocean), then precipitate back out to form new sedimentary minerals. The most common way to do this is for organisms (plants or animals) to pull ions out of seawater (such as calcium and carbonate) and precipitate their shells with calcium carbonate minerals such as calcite or aragonite. When these carbonate shells and coral skeletons of sea creatures accumulate, they build up carbonate sediment that eventually can crystallize into a rock known as **limestone**. Unlike sand and gravel and mud, *limestones are born, not made.* They are nearly always built from fossils (Fig. 2.11A), even if recrystallization might make the fossils invisible. Most limestones forming today are restricted to tropical or subtropical settings, with warm, shallow, clear water and no clastic sand or mud. Places such as Florida, the Bahamas, parts of the Caribbean, the Persian Gulf, and the South Pacific are the main sites forming carbonate sediments today. During the geologic past, huge shallow tropical seas drowned the continents for millions of years, accumulating enormous thicknesses of limestone in much of the world.

Another chemical found in water is silica. It can precipitate to form a rock known as **chert**, which is made of sub-microscopic crystals of quartz. Chert comes in many colors based on impurities, so if it is black we call it flint; if it is red, it is jasper; white chert is novaculite; and so on (Fig. 2.12A). Chert in the form of flint or jasper was once important for arrowheads and spearheads, to start fires, to fire flintlock muskets, and many other purposes.

Chert forms in two main ways. In places where plankton that use silica in their skeletons are extremely abundant, they accumulate to form a silica-rich shale known as **bedded chert** or "ribbon chert" (Fig. 2.12B). This kind of chert is precipitated by organisms, just as limestones are. The other kinds of chert form when silica-rich groundwater percolates through other rocks (usually limestone) and replaces calcite with silica. These are known as **nodular cherts** (Fig. 2.12C).

Figure 2.11 Limestones. **A.** Fossiliferous limestone. **B.** Oolitic limestone. **C.** Cliffs of chalk.

Figure 2.12 Chert. **A.** Photo of different forms of chert, including black chert ("flint"), red chert ("jasper"), and white chert ("novaculite"). **B.** Bedded chert with alternating bands of hard, resistant chert (derived from plankton, which make their shells out of silica) and deep-water shales. These are often called "ribbon cherts" for the narrow ribbon-like appearance of the alternating bands. **C.** Nodular chert is formed by replacing the calcite in limestone. In this case, the black chert (flint) from the Dover Chalk near Birling Gap in England has formed layers and bands of nodules where the silica from sponge spicules and siliceous plankton has been transformed into chert.

Most of the chemical sedimentary rocks just discussed were formed by organisms precipitating ions out of water, so they are often called **biochemical** (or organic) **sedimentary rocks**. But there are also rocks that form by straight chemical precipitation, without the help of organisms to drive precipitation. These are **inorganic chemical sedimentary rocks**. The main mechanism required to make minerals precipitate in this way is to evaporate all the water away and leave the minerals behind to crystallize out of a salty **brine**. Thus, these minerals are known as **evaporites**. Such evaporation is common in dry lake beds but also in tropical lagoons and hot desert seas like the Persian Gulf, where the rate of evaporation is greater than the amount of seawater flowing in to replace the water evaporated. The most common evaporite minerals are salt (sodium chloride), gypsum (hydrous calcium sulfate), and sometimes calcite or aragonite; but there are hundreds of additional evaporite minerals.

2.5 Metamorphic Rocks

The third major class of rocks is metamorphic rocks (*meta* meaning "after" or "changed" and *morphos* meaning "form" in Greek). We are familiar with the word "metamorphosis" to describe many kinds of changes in form, such as the caterpillar changing to a pupa and then a butterfly. Metamorphic rocks are transformed or changed from some original "parent rock" or **protolith** (usually an igneous rock or sedimentary rock) into an entirely new and different kind of rock with new minerals and new textures or fabrics. Some metamorphic rocks are so completely transformed that they are unrecognizable, and we may never know what kind of protolith they started from.

PRESSURE AND TEMPERATURE

Metamorphism occurs when the protolith is buried deep in the crust and experiences extremely *high temperatures* and *directed pressure*. Thanks to the enormous heat flow coming up from the earth's interior, the crustal rock below your feet gets hotter and hotter the deeper you go. In fact, the **geothermal gradient** is about a 30°C increase per kilometer of depth, so at 30 km the crustal rocks would be about 900°C (above the melting temperature of many minerals). Most people can't even imagine this, but if you go down an old abandoned mine shaft with no air-conditioning, you can feel how it gets hot as you descend. South African diamond and gold miners work at depths close to 3.9 km (12,800 feet) and require a continuous supply of refrigerated fresh air to survive shifts of only a few hours because the temperature of the rocks and air down there is 60°C (140°F). The deepest hole ever drilled only reached 12,262 kilometers (7.5 miles), in the Kola region of Siberia, which is only 10% to 15% of the way to the mantle beneath the continental crust.

The increase in pressure is also intense. The pressure gradient is 3 kilobars (kbar) per 10 km depth you descend. One

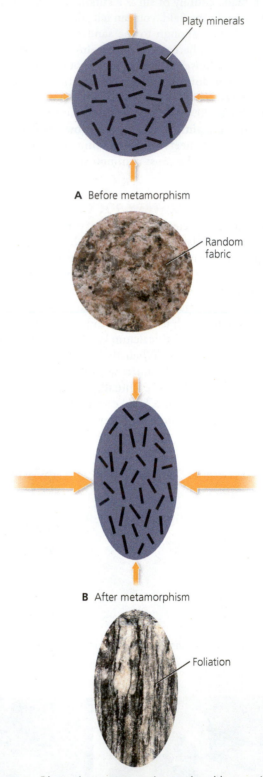

A Before metamorphism

B After metamorphism

Figure 2.13 Directed pressure crushes rocks with a randomly-oriented fabric of crystals, and squeezes and flattens them until those minerals are all aligned, forming foliation.

kilobar (which is 1000 bars in the metric system) is about 14,500 pounds per square inch, so at 30 km down (very shallow crustal levels), pressures are about 9 kbar, or almost

45,000 pounds per square inch. These pressures are far too great for any device to overcome by drilling (no drilling hole has gone much farther than 12 km, or 40,000 feet). You can laugh at any of those science fiction movies that imagine drilling to the earth's interior or a "journey to the center of the earth"! They are complete fantasy because the pressures and heat are so intense even down to depths as shallow as 4 km that no human or machine could survive.

The pressure is also **directed pressure**, or pressure applied in a specific direction, so that one axis of a rock is being crushed in the up–down direction, while there is less pressure in the perpendicular direction, so the rock can be squeezed sideways as it is flattened. This is very different from the uniform pressure that you experience on all sides of your body from the air around you or from water around you when you dive. Directed pressure tends to squash or flatten most rocks, stretching and squeezing them outward

(Fig. 2.13). Old minerals that were present in the rock as it began metamorphism will be flattened out and rotate until they are perpendicular to the direction of maximum pressure. Any new minerals that grow during this metamorphic process (especially platy minerals like the micas muscovite, biotite, and chlorite) will tend to grow perpendicular to the direction of greatest stress as well. As a result, the rock will acquire a strongly planar fabric in the direction of least stress, which is called a **foliation** (after the Latin word *folium*, for "leaf"). Many metamorphic rocks have a planar fabric or layering and will split along this plane of foliation as a result.

TYPES OF METAMORPHIC ROCKS

Let's take a variety of different protoliths and see what happens during metamorphism (Fig. 2.14). If we start with shale, for example, we have lots of chemistry (silicon,

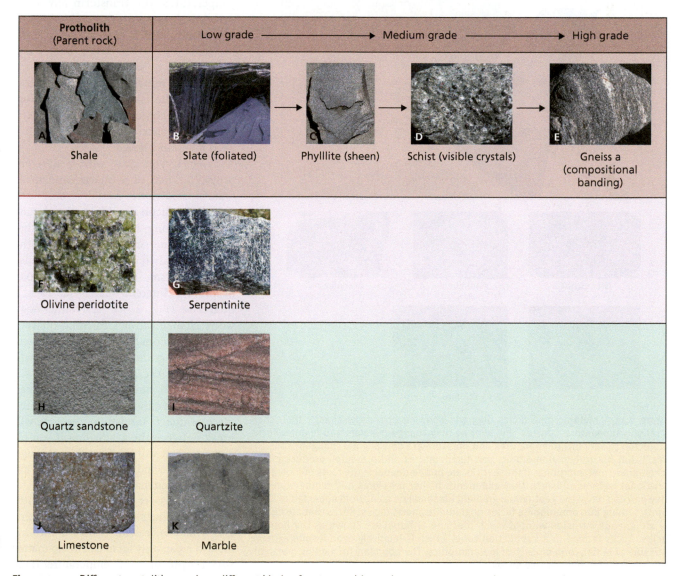

Figure 2.14 Different protoliths produce different kinds of metamorphic products as pressure and temperature increases.

aluminum, potassium, sodium, and other elements) to work with. The first metamorphic product of a shale is a rock known as a **slate**, which is platy and highly foliated so that it readily breaks into things like roofing tiles and (in the old days) slate for blackboards. As the temperature and directed pressure increase, the clay minerals in the shale are transformed into tiny flakes of micas like muscovite or biotite (not yet visible to the naked eye), and the rock acquires a distinctive sheen; this is called a **phyllite** (FILL-ite). Further pressure and temperature allow the

new metamorphic minerals (muscovite, biotite, plus garnet, hornblende, and others) to grow large enough to be visible to the eye; this kind of rock is called a **schist**. Finally, at extremely high pressures and temperatures, some of the minerals begin to melt and segregate into bands of light-colored minerals (typically quartz and plagioclase) and dark-colored minerals (typically biotite and hornblende). This *compositional banding* is the diagnostic feature of a rock known as **gneiss** (pronounced "nice"). Any further increase in pressure and temperature and the gneiss melts completely and can become a magma that could cool to form an igneous rock.

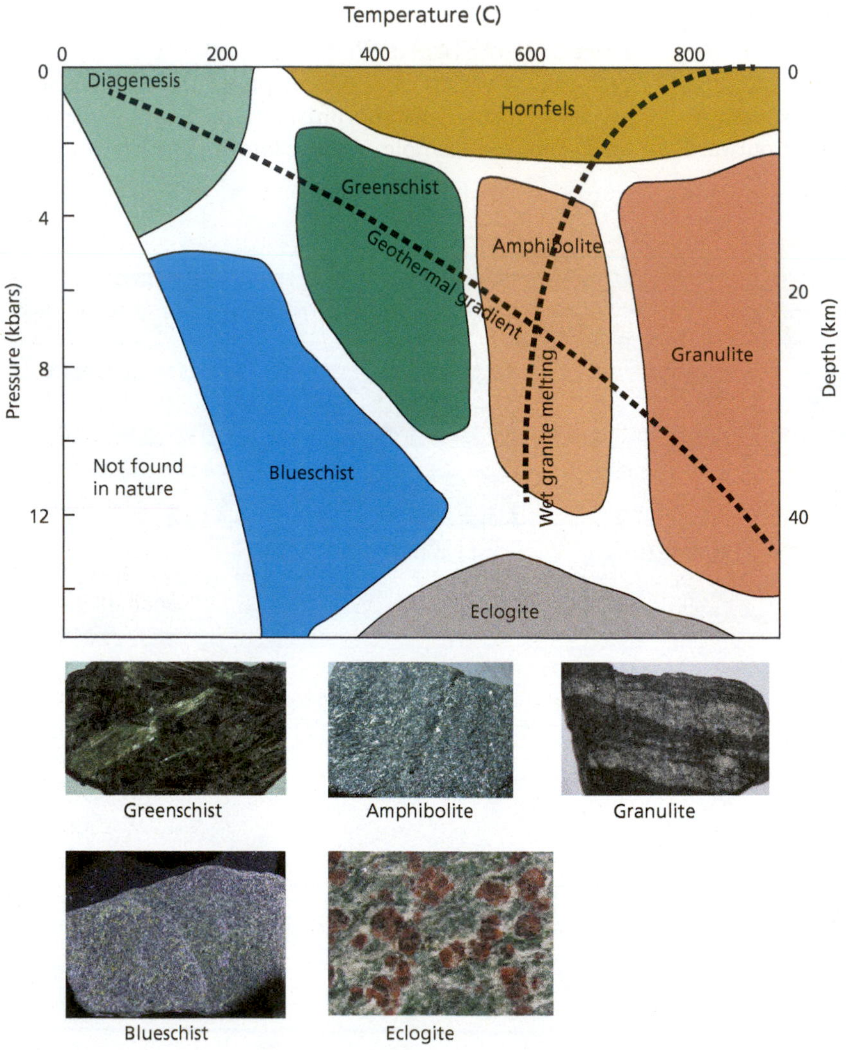

 See For Yourself: Metamorphic Rocks

Let's try a different protolith (Fig. 2.14). If we start with an olivine peridotite or pyroxene-rich gabbro, we have only magnesium, iron, and silica to work with. Under high pressures and temperatures, you transform olivine or pyroxene into a new mineral known as **serpentine**, which gets its name from its snake-like green color and waxy, smooth, scaly, "snakeskin" feel. A rock made of the mineral serpentine is called a **serpentinite** (ser-PENN-ten-ite). Serpentinite is common where ultramafic, olivine-rich slices of oceanic crust have been metamorphosed.

What if you start with a different protolith, such as a quartz sandstone? Quartz sandstone contains only one chemical, silicon dioxide, and you can't make anything but quartz from that chemistry, no matter how high the pressure and temperature. So under metamorphism, the quartz sandstone becomes a different quartz-rich rock, a **quartzite** (Fig. 2.14). The mineralogy won't change (always forming quartz); but the original fabric of spherical quartz sand grains packed together will vanish, and the grain boundaries will fuse together and become interlocked like pieces of a jigsaw puzzle. If you hit a piece of quartz sandstone, the rock will break between the grains; but if you hit a quartzite, it will fracture right through the original sand grains since they are fused into one mass of quartz.

Let's consider one more protolith: limestone. Limestone has only one mineral available, calcite (calcium carbonate), so it can only form calcite no matter what the pressure and temperature. Thus,

Figure 2.15 Metamorphic facies diagram, showing how rocks change their characteristic appearance ("facies" in Latin) as they change minerals and texture. Rocks subjected to very low temperatures and pressure are only slightly changed ("diagenesis"), but not truly metamorphic. Low temperatures and pressures produce greenschist-grade metamorphics. If those rocks are buried deeper in the roots of mountains ("regional metamorphism"), they experience higher pressures and temperatures as they proceed down the geothermal gradient (dashed line going from upper left to lower right), turning into amphibolite facies or granulite gneiss facies. Rocks that are heated by an igneous intrusion without much pressure of burial are known as the hornfels facies. Rocks formed in a subduction zone, which is relatively cool despite extreme pressure of burial, form blueschist metamorphics. The boundary for melting metamorphic rocks into magma is also shown.

a fossil-rich limestone protolith transforms into a **marble** under metamorphism (Fig. 2.14). A true marble is still made entirely of calcite, but the minerals and fossil fragments have completely recrystallized so that no fossils are visible anymore and only large shiny crystals of calcite are left.

So far, most of the metamorphic rocks we have discussed (shale, phyllite, schist, gneiss, serpentinite) are highly foliated. But quartzite and marble never undergo foliation, no matter how high the pressures and temperatures. This is because they are not made of platy minerals like micas or elongate prismatic minerals like hornblende. Minerals like quartz and calcite have no long axis or preferred orientation; no matter how much you cook or squeeze them, they will never foliate.

METAMORPHIC GRADE AND FACIES

We can plot all the different possible regimes of pressure and temperature on the graph shown in Fig. 2.15. Temperature increases to the right on the *x*-axis, and pressure increases downward (as it does in the real world) on the *y*-axis. In the upper left part of the diagram is the region of lowest pressure and temperature (earth surface conditions or very shallow burial), and no metamorphism takes place at this level. If we descend the geothermal gradient line until temperatures are around 330°C–500°C and pressures are about 2–8 kbar, we are in a region of relatively low pressures and temperatures, producing **low-grade metamorphism**. Geologists use a shorthand term for this region, calling it the **greenschist facies**. (*Facies* is the Latin word for "appearance.") Rocks that have undergone low pressures and temperatures *appear* as green schists because they usually grow the green mica chlorite, plus sometimes other green minerals such as amphiboles known as actinolite-tremolite and maybe even the pistachio-green mineral epidote. Moving further down the geothermal gradient, we reach a region of intermediate pressures and temperatures (4–12 kbar and 500°C-700°C). The geologists' term for **intermediate-grade metamorphism** is **amphibolite facies** because these rocks tend to be rich in the black amphibole hornblende. Finally, we go to the very deepest part of the continental crust, where there are extreme pressures (6–14 kbar) and temperatures (at least 700°C), making the highest-grade metamorphics. These rocks have a gneissic texture and are sometimes called **granulite facies**. All three of these facies are produced by **regional metamorphism**, such as when a collision between continents creates a huge uplifted mountain belt (such as the Himalayas today). Rocks that undergo shallow burial will only reach greenschist facies, while those with even deeper burial will become amphibolites or finally granulite gneisses if they are many kilometers down into the crust at the root of the mountains.

Look at Fig. 2.15 again. We can see how the three main facies are found in increasing depths and pressures as they descend deeper into the crust (and further down the geothermal gradient). But what about the peculiar region across the top of the diagram labeled **hornfels**? A glance at the axes of the plot suggests that these would be formed under high temperatures but low pressures. This could not happen by simple burial at greater depths because both temperature and pressure increase together. Where do such peculiar conditions exist? The only possibility is to heat a rock to high temperatures without burying it deeply. This can only happen when a magma body or dike intrudes through the country rock, cooking it without much depth of burial. Such metamorphism caused by the contact of an intruded magma is called **contact metamorphism** and produces hornfels. Such rocks often don't show much in the way of different minerals, but their fabric is welded and baked compared to the unheated rock at some distance from the magma intrusion.

Finally, there is one more peculiar region in Fig. 2.15 to explain. At the lower left is a region marked **blueschist**. This name comes from their often bluish gray or even deep blue color (Fig. 2.15), due to the blue amphibole glaucophane plus another blue or white mineral called lawsonite. The axes in Fig. 2.15 indicate that it is a region of very high pressure but relatively low (less than 400°C) temperature. How could this happen? Most rocks descending into the deep crust get hot when they are under such high pressures. The answer to the mystery of blueschists was discovered in California because they are particularly common in the Coast Ranges and rare elsewhere in the United States. The only place blueschists are found is in the remnants of ancient subduction zones (Chapter 5), where the cold downgoing plate plunges into the hot mantle and reaches depths of 20 km or greater. At this point, the rocks are surrounded by the high pressures of such deep burial, but they are still relatively cold because the old oceanic plate retains a lot of water and heats up slowly. Under these unusual conditions, the pressures can get very high but the temperatures remain low enough to form glaucophane, lawsonite, and some other distinctive minerals. Then, when some of this plate gets scraped off in an accretionary wedge (see Chapter 5), the blueschist can rise to the surface after having been more than 20 km underground in a subduction zone.

2.6 The Rock Cycle

As we have suggested, minerals and rocks can transition from one category to another quite easily. Take the example we used of the sedimentary rock known as shale. It can transition from sedimentary rock to metamorphic rock as it experiences high directed pressures and temperatures. It then goes from shale to slate, phyllite, schist, and gneiss. Eventually, it gets hotter and hotter until it melts. Then it has become a magma and can cool into an igneous rock

Figure 2.16   The rock cycle. Over millions of years, no rock is permanent, but it is part of a continuous slow cycle, changing from one class of rock (e.g., an igneous rock) to another (e.g., weathered sediments from the igneous rocks) to another (e.g., metamorphism of sedimentary rocks into meta-sedimentary rocks). If the metamorphic rock is heated enough it melts and returns to the igneous part of the loop.

(Fig. 2.8). Thus, we have a sequence of sedimentary to metamorphic to igneous rock (Fig. 2.11). Then, if that igneous rock reaches the earth's surface, it will weather and break down into loose sand and mud and return to the sedimentary beginning of this loop all over again.

This is the **rock cycle** (Figure 2.16), and it is a demonstration of the fact that rocks can transform from one category to the next and eventually return to their starting point over millions of years. One of the great lessons you learn from geology is not only that time is immense (millions and billions of years) but that given enough time ocean bottoms can turn into mountains and then weather down into sediment and return to the ocean—or be forced down into the lower crust and transform into metamorphic rocks or even melt into a magma. With enough time, any of these extremely slow processes are inevitable.

RESOURCES

BOOKS

Bowen, N. L. 1956. *The Evolution of Igneous Rocks*. Dover Publications, New York.

Deer, W., and J. Howie. 1996. *An Introduction to the Rock-Forming Minerals*. Prentice-Hall, Englewood Cliffs, NJ.

Dillard, A. 1987. *An American Childhood*. Harper & Row, New York.

Garlick, S. 2014. *National Geographic Pocket Guide to Rocks and Minerals*. National Geographic Society, Washington, DC.

Klein, C., and A. Philpotts. 2012. *Earth Materials: Introduction to Mineralogy and Petrology*. Cambridge University Press, Cambridge, UK.

Pellant, C. 2002. *Smithsonian Handbook of Rocks and Minerals*. Smithsonian Institution Press, Washington, DC.

Philpotts, A. and J. Ague. 2009. *Principles of Igneous and Metamorphic Petrology*. Cambridge University Press, Cambridge, UK.

Prothero, D. R., and F. Schwab. 2013. *Sedimentary Geology*. W. H. Freeman, New York.

Winter, J. D. 2009. *Principles of Igneous and Metamorphic Petrology*. Prentice-Hall, Englewood Cliffs, NJ.

SUMMARY

- Atoms are made of protons (with a +1 charge, mass = 1) and neutrons (with no change, mass = 1), found in the nucleus, and surrounded by clouds of charged particles called electrons (with a –1 charge, no mass).

- The number of protons determines the atomic number, and thus which element is present. The number of neutrons determines the atomic weight, and thus which isotope of the same element is present.

- If the number of protons balances the number of electrons, then the charges balance and there is no net charge. If these numbers do not match, then it is a charged atom or ion. If there are not enough electrons, then there is a net positive charge (cation). If there are too many electrons, then there is a net negative charge (anion).

- Two or more atoms combined together make a compound.

- The most common elements in the earth's crust are oxygen (the only common anion to balance the charges of cations) and silicon (the basic building block element), followed by aluminum and then by five relatively rare metallic elements (iron, potassium, calcium, sodium, and magnesium).

- A mineral is a naturally occurring inorganic crystalline solid with a definite chemical formula and characteristic physical properties.

- Most rock-forming minerals are silicates (silicon plus oxygen plus other elements). Silicates are classed by their structure and linking of silicon–oxygen tetrahedral units, from isolated tetrahedral to single chain to double chain to sheet silicates to the most complex of all, three-dimensional framework silicates.

- Igneous rocks form from a molten liquid (magma) by cooling and crystallization. If the melt cools slowly, the crystals have time to grow large, which usually happens in a magma chamber deep in the earth's crust (plutonic rocks). If they cool quickly, they have microscopic crystals, which usually happens during a volcanic eruption. Some have a hybrid texture of large crystals in a microscopic groundmass (porphyritic texture), formed by crystals that cooled and grew slowly in a magma chamber, then were suddenly erupted so that the rest of the melt cooled quickly in a volcanic eruption.

- Bowen's reaction series showed by experiment how fractional crystallization can turn an ultramafic magma rich in Mg, Fe, and Al into a more silicic (felsic) magma rich in Si, Al, K, and Na. As the crystals rich in mafic components sink to the bottom of the magma chamber, they remove their Mg, Fe, and Ca and the remaining magma becomes richer in more silicic components, producing basaltic and andesitic magmas.

- Partial melting is the primary mechanism whereby reheating a cold igneous rock releases low-temperature minerals like quartz and feldspar, producing granitic magmas. Another mechanism is contamination or assimilation of a mafic magma as felsic components are melted from the wall of a magma chamber as the pluton rises through the crust.

- Sedimentary rocks are formed by the weathering of pre-existing rocks and minerals to form new kinds of rocks at the earth's surface.

- Clastic sedimentary rocks are made of broken fragments of older mineral and rock grains, to produce conglomerates, sandstones, and mudstones.

- Chemical sedimentary rocks form from dissolved ions in solution that crystallize to form new minerals, such as halite, gypsum, and calcite, when the water evaporates or when calcite is precipitated by organisms as they produce their shells.
- Metamorphic rocks are formed from some other parent rock (protolith) after undergoing directed pressures and high temperatures.
- Some protoliths form highly foliated rocks, so a shale protolith will yield slate, phyllite, schist, and gneiss with increasing metamorphic grade. However, a quartz sandstone has only one non-planar mineral (quartz), so it will produce a non-foliated rock like quartzite, while a limestone will yield a rock made of recrystallized calcite known as marble.
- All rock types are constantly changing into other rock types to form a rock cycle.

KEY TERMS

Atom (p. 20)
Element (p. 20)
Proton (p. 20)
Neutron (p. 20)
Nucleus (p. 20)
Electron (p. 20)
Atomic number (p. 20)
Ion (p. 20)
Cation (p. 20)
Anion (p. 20)
Atomic weight (p. 20)
Isotope (p. 20)
Deuterium (p. 20)
Tritium (p. 20)
Carbon-14 dating (p. 21)
Silicates (p. 21)
Mineral (p. 21)
Lattice (p. 22)
Glass (p. 22)
Obsidian (p. 22)
Hardness (p. 22)
Cleavage (p. 22)
Diamond and graphite (p. 23)
Igneous rocks (p. 25)

Magma (p. 25)
Microcrystalline texture (p. 25)
Macrocrystalline texture (p. 25)
Phenocryst (p. 25)
Groundmass (p. 26)
Lava (p. 26)
Plutonic (p. 26)
Volcanic (p. 26)
Porphyritic texture (p. 26)
Mafic versus felsic (p. 26)
Gabbro versus basalt (p. 26)
Peridotite versus komatiite (p. 26)
Ultramafic (p. 26)
Diorite versus andesite (p. 26)
Granite versus rhyolite (p. 26)
Granodiorite versus dacite (p. 27)
Magma differentiation (p. 27)
Bowen's reaction series (p. 27)
Fractional crystallization (p. 27)

Partial melting (p. 29)
Assimilation and contamination (p. 29)
Sill (p. 29)
Dike (p. 29)
Batholith (p. 29)
Sedimentary rocks (p. 29)
Deposition (p. 30)
Clastic sedimentary rocks (p. 30)
Sandstone (p. 30)
Conglomerate (p. 30)
Breccia (p. 30)
Arkose (p. 30)
Mudstone (p. 30)
Shale (p. 31)
Chemical sedimentary rocks (p. 31)
Weathering (p. 31)
Limestone (p. 31)
Chert (p. 31)
Evaporites (p. 32)
Metamorphic rocks (p. 32)
Protolith (p. 32)

Directed pressure (p. 33)
Foliation (p. 33)
Slate (p. 34)
Phyllite (p. 34)
Schist (p. 34)
Serpentinite (p. 34)
Non-foliated metamorphic rocks (p. 00)
Quartzite (p. 34)
Gneiss (p. 34)
Metamorphic facies (p. 34)
Marble (p. 35)
Metamorphic grade (p. 35)
Greenschist (p. 35)
Amphibolite (p. 35)
Granulite (p. 35)
Hornfels (p. 35)
Contact metamorphism (p. 35)
Regional metamorphism (p. p. 35)
Blueschist (p. p. 35)
Rock cycle (p. p. 36)
Transportation (p. 36)

STUDY QUESTIONS

1. Why is oxygen the most common element in the earth's crust?
2. You are at a store which sells "cut-glass crystal." Explain why glass can never be a crystal (as scientists use the term).
3. You are in a store, and you gather up groceries: salt, sugar, organic vitamins, and a bag of ice. As you leave, you see snowflakes falling. Which ones are minerals, and which ones are not? Explain your reasoning.
4. Why do minerals have specific chemical formulas while most rocks do not?
5. How do crystals form, and how do cleavage and hardness reflect the internal atomic structure of a mineral?
6. Diamond is the hardest mineral known, while graphite is one of the softest, yet they are both made of pure carbon. Explain why this is so.
7. Why are silicates the most common minerals in the earth's crust?
8. How do you tell amphiboles, pyroxenes, and micas apart by cleavage?
9. Why are crystals in a lava flow very tiny but crystals formed in a deep magma chamber larger and generally visible to the naked eye?
10. How is a porphyry formed?
11. Explain how crystal settling influences the composition of a cooling mafic magma.
12. Why do magmas which are very felsic form explosive eruptions, while those which are mafic tend to be liquid?
13. Why does magma erupted in a subduction zone have a wider range of compositions than magma erupted in a mid-ocean ridge?

14. Sedimentary rock forms about 75% of the rocks at the earth's surface but only 5% of the rocks inside the crust. Why is this so?

15. Why are sediments produced at the earth's surface?

16. How does a normal conglomerate form? Under what conditions do breccias form?

17. Explain how the grains of sand in a sandstone might become more rounded and spherical as they are transported.

18. What is the difference between a mudstone and a shale?

19. In what setting are most limestones formed? What organisms are often responsible?

20. In what setting are most evaporite minerals formed?

21. Describe the two different origins for cherts.

22. Why do rocks experience higher and higher temperatures and pressures as they go deeper in the crust?

23. Why does a shale protolith produce many different kinds of metamorphic rocks but a sandstone or limestone produce only one?

24. Why are quartzites and marbles non-foliated?

25. Summarize the regional metamorphic setting that produces greenschist, amphibolite, and granulite gneisses.

26. What kinds of settings are high in temperature but low in pressure and produce hornfels?

27. What kind of settings are high in pressure but low in temperature and produce blueschists?

28. Starting with a newly uplifted and eroding granite, describe a loop in the rock cycle from the granite to a sedimentary rock to a metamorphic rock and back again to an igneous rock.

3 It's About Time!

Dating Rocks

The result, therefore, of our present enquiry is, that we find no vestige of a beginning—no prospect of an end.

—*James Hutton, Theory of the Earth, 1788*

A famous engraving from an 1888 book by Nicolas Camille Flammarion, showing the medieval conception of the earth as a flat disk surrounded by the fixed stars on a celestial sphere. The curious explorer pokes his head through the "dome of the sky" to see the sun, moon, and planets moving on great gear wheels as they orbit around us.

3.1 How Old are the Universe and the Earth?

EARLY IDEAS

Since the beginning of recorded history, people have had different notions of when things happened in the distant past. Many cultures, such as some of those in India and other parts of Southeast Asia, thought of time as eternal and cyclic. The earth and life have gone on forever, with no beginning or end. Other cultures had creation myths of how the universe began at some unique point of time in the past. In some versions of the Japanese creation myths, the jumbled mass of elements appeared in the shape of the egg; and later in the story, Izanami gives birth to the gods. In the beginning of the Greek myths, the bird Nyx lays an egg that hatches into Eros, the god of love. The shell pieces become Gaia and Uranus. In Iroquois legend, Sky Woman fell from a floating island in the sky because she was pregnant and her husband pushed her out. After she landed, she gave birth to the physical world. The Australian Aborigines believed in a sun-mother who created all of the animals, plants, and bodies of water at the suggestion of the father of all spirits.

Some of the oldest recorded myths we have come from the ancient cultures of the Sumerians and Akkadians, who lived over 6000 years ago in Mesopotamia, the valley of the Tigris and Euphrates Rivers, in what is now Iraq. These stories come from ancient clay tablets marked with wedge-shaped writing known as cuneiform. One of the oldest recorded myths we have, *The Epic of Gilgamesh*, dates to before 2750 B.C.E. The Sumerians had a hero called Ziusudra, who is warned by the earth goddess Ea to build a boat because the god Ellil was tired of the noise and trouble of humanity and planned to wipe them out with a flood. When the floodwaters receded, the boat was grounded on the mountain of Nisir. After Utnapishtim's boat was stuck for 7 days, he released a dove, which found no resting place and returned. He then released a swallow that also returned, but the raven that was released the next day did not return. Utnapishtim then sacrificed to Ea on the top of Mount Nisir. Another account is the Babylonian myth known as the *Enuma Elish*, whose Sumerian predecessor dates back to over 5000 years ago. It describes a formless void and chaos, with gods dividing the waters from the land and naming the creatures, in language very similar to that of Genesis 1. When the ancient Hebrew peoples lived in ancient Mesopotamia, they were influenced by these more ancient stories when they created their own stories of their origins.

Once the idea of an original beginning or creation event that founded the universe became part of Western culture, the next question was, how long ago did it occur? Most cultures could not image that the origin of the universe was more than a few thousand years old. They viewed the entire universe with the earth at the center and the stars fixed to a great "celestial dome" (Fig. 3.1). They thought that the planets that wandered across the sky (*planetos* means "wanderers" in Greek), plus the sun and the moon, were carried on great wheels around the earth, so they appeared to "move" against the background of the "fixed" stars. The earth itself started out as perfect, created exactly as we see it today without any changes since it was formed. Any evidence that it was changing or eroding or crumbling away was explained away as due to Adam's sin. Until about the early 1700s, nearly all people in the Western world thought of the earth as the center of the universe, only a few thousand years old and unchanged since its formation except for the decay due to the fall of Adam.

But these notions that were carried on for centuries gradually began to crumble. In 1543, Nicholas Copernicus published his book that first proved that the sun, not the earth, was the center of the solar system (the **heliocentric** solar system). By 1632, Galileo made an overwhelming argument for the heliocentric solar system; and by 1687, Isaac Newton had explained the motion of the stars and planets by the laws of gravity, not requiring any supernatural intervention. Thus, the old notion that the earth and humans were the center of creation was abandoned once astronomers truly understood the heavens. Since the mid-twentieth century, we have discovered the evidence that our solar system is small, one of many solar systems on one of the arms of the Milky Way galaxy, which is one of thousands of known galaxies across the entire visible universe.

3.2 Steno's Laws and Unconformities

THE TITULAR BISHOP OF TITIOPOLIS

Over this same period that people changed their notions of the universe, there were discoveries that gradually invalidated the old notion that the earth was young and created exactly as we see it, with erosion caused by Adam's sin. The first of these discoveries was made by a Danish doctor, Niels Steensen (better known today by his Latinized name, Nicholas Steno). Leaving his native Denmark, in 1666 he became the doctor for the Grand Duke of Tuscany in Florence, Italy. When not taking care of the Duke and his family, he was commissioned to improve their natural history collections. From this experience,

he began to think of the earth as much older than a few thousand years, with a long history that could be deciphered.

Steno spent time traveling through the Apennine Mountains in the hills behind Florence. He noticed that the mountains had many sandstone and mudstone beds that were full of fossils, often steeply tilted or folded. Instead of imagining that these rocks were created as we see them, Steno began to think of these layered sandstones and mudstones as products of sands and muds settling out on the bottom of a river or the ocean floor. If so, then the rocks recorded a history, a sequence of events. From these observations, he deduced three basic principles, now known as **"Steno's laws"** (Fig. 3.2):

1. **Superposition**: In any layered sequence of rocks (layers of sedimentary rocks or of lava flows), the oldest rocks are at the bottom of the stack, and they get progressively younger as you move up the pile. This follows from Steno's idea that the layered rocks of the Apennine Mountains were deposited over time, and you can't lay down one layer of rock on top of another without having something on the bottom already. A good analogy is the stack of papers on your messy desk. If you don't clean them up, then you would expect something you worked on last night to be at the top of the pile and stuff from the beginning of the semester to be at the bottom. Another example is a stack of pancakes. The

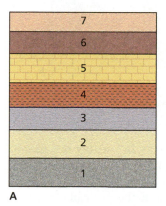
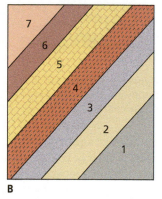
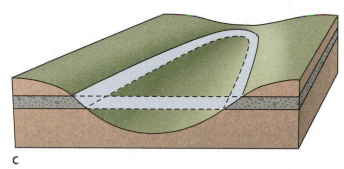

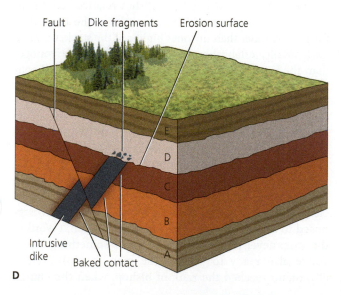

Figure 3.2 Steno's principles. **A.** *Superposition*: the layers at the bottom of the stack are older than the layers placed above them; **B.** *Original horizontality*: Rock layers are normally deposited as horizontal sheets of sediments or horizontal lava flows and volcanic ash layers, so if you find them tilted or folded, their deformation occurred after they were deposited or erupted. **C.** *Original continuity*: If you find a canyon or valley cut through layers which match on both sides, then the canyon is younger than the layers it cuts; **D.** *Cross-cutting relationships*: any event (igneous intrusion, fault, etc.) that cuts through pre-existing rocks must be younger than what it cuts

first ones to be cooked are cooling at the bottom of the stack, while the hot ones fresh off the grill are on top.

2. **Original horizontality**: Steno saw the folded and tilted rocks of the Apennines and reasoned that they had once been sediments lying on the bottom of a seabed or river bottom, which are always deposited nearly horizontally. If you find them tilted or folded today, then they were deformed some time after they were laid down in horizontal beds.

3. **Original continuity**: Today, we can look at the Grand Canyon, see how the layers match up between the North Rim and the South Rim, and imagine how they once connected over that distance but have since been eroded away by the Colorado River. But in Steno's time, people thought the rocks were created exactly as we see them now. Steno realized that matching rock units across a canyon or a landscape must have once been continuous and that later erosion has cut them away so that they no longer connect.

4. **Cross-cutting relationships**: Steno saw many instances where one rock (such as a body of magma) or event (such as a fault) cut through older rocks. Clearly, whatever cuts through a body of rock has to occur after the rock through which it cuts.

While Steno worked in Florence, he spent a lot of time inspecting specimens brought in by fishermen in the port town of Livorno. At that time, there was a big debate over whether the strange objects called "fossils" (literally, "dug up") were actually the remains of ancient organisms or supernaturally created or placed there by the devil to challenge our faith. The strange triangular objects known as *glossopetrae*, or "tongue stones," were thought to have been the petrified tongues of dragons or snakes, capable of curing snakebite. About 40 C.E., the Roman natural historian Pliny the Elder thought they fell from the sky during lunar eclipses. Steno got a chance to see a shark recently caught by fishermen in Livorno and could see that the mysterious "tongue stones" were the triangular teeth (Fig. 3.3) of sharks. Soon, he realized that most fossils he saw in the Apennine Mountains were also the remains of clams and snails that once lived in the ocean, not supernatural creations or pranks of the devil.

But this posed a dilemma: if the rocks were created as we see them, then how did the solid fossils get inside the solid sandstone or mudstone? In 1669, Steno published his book, whose title has been abbreviated as *Prodromus* (its full Latin title, *De solido intra solidum naturaliter contento dissertationis prodromus*, translates to *Prologue to a Dissertation on a Solid Naturally Contained Within a Solid*). What does this strange title mean? Steno realized that the solid rocks had once been soft wet sands or muds and later turned into hard rock around the solid shells and shark teeth that were already on the seafloor. In other words, one solid (the fossil) was older than the solid rock that enclosed it.

Steno's laws were the first attempts to show that *earth had a history* and was not instantly created exactly as we see it, unchanged since its formation. These principles establish

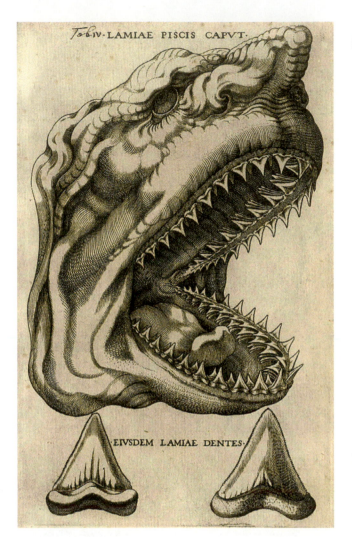

Figure 3.3   Steno's famous 1667 illustration demonstrating that the mysterious "tongue stones" were actually the fossilized teeth of ancient sharks.

the **relative age** of geologic events, or whether one rock or event is older or younger than another. In other words, rock A is older or younger *in relation to* rock B.

 See For Yourself: Geologic Dating

Every one of the principles we have discussed establishes relative age. In superposition, the rocks at the bottom of the pile are older than the ones higher in the stack. Original horizontality establishes that tilting or folding is younger than the horizontal deposition of the rocks. Original continuity establishes that the canyon was cut after the rocks were formed. Cross-cutting relationships show that younger rocks cut through older rocks. All of these essentially modern insights come from realizing that the solid fossils were older than the solid rocks that surrounded them.

After this promising beginning, Steno (raised in Denmark as a Lutheran) converted to Catholicism, and in 1675 he gave up his scientific career to become a Catholic clergyman. The promised "dissertation" which was supposed to follow his *Prologue* was never written. (Word to the wise: never publish the prologue to something unless you're also ready to publish the main book!) Eventually, Steno reached the rank of bishop, when the church

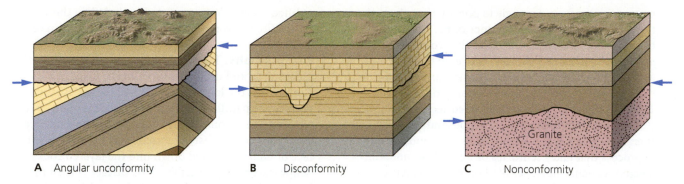

A Angular unconformity   B Disconformity   C Nonconformity

Figure 3.4 The three basic types of unconformities. **A.** Angular unconformities occur when the strata below the erosional surface are tilted or folded, then eroded off and covered by flat-lying strata on top of the erosional surface. **B.** A disconformity is an erosional break between two parallel layers of sediment. It can be tricky to see in the field but is usually indicated by an erosional channel or scouring of the underlying beds filled in by the overlying sediments. **C.** Nonconformities occur when igneous or metamorphic rocks are uplifted and eroded, then the erosional surface is covered by a blanket of sediments.

assigned him to minister to "Titiopolis," the old name for a region of what is now south-central Turkey that had no Catholics. So he returned to northern Europe to live and work, and he was the "titular bishop" ("bishop" in title only, not actually ministering to a region) of Titiopolis. After his death in 1686, he was venerated by the Church and eventually beatified as a saint by Pope John Paul II in 1988. Thus, both the Catholic Church and the field of geology hold him in high regard.

UNCONFORMITIES

We see the thick piles of sandstones, shales, and limestones in the Grand Canyon and imagine them as a continuous recording of all the events of earth history. But over the last century, a number of lines of evidence have shown that this is an illusion. Most thick sequences of layered rocks have big erosional gaps in them, or **unconformities**. An unconformity typically represents an event where the bedrock has been uplifted into mountain ranges, then deeply eroded as it was uplifted, after which it sinks down again so that another layer of sedimentary rock can bury it.

See For Yourself:
Types of
Unconformities

Three types of unconformities can be recognized (Figs. 3.4, 3.5):

1. **Angular unconformity**: This is the easiest to spot in the field since the rocks above and below the erosional

Figure 3.5 Field photos of the three different types of unconformities. **A.** An angular unconformity, here between yellow-orange gravels and conglomerates only 1–2 million years in age lying horizontally across the top and tilted tan sandstones over 10–11 million years in age. There is an 8- to 10-million-year time gap between them. **B.** The nonconformity at the base of the Grand Canyon (the "Great Unconformity"), between the layered Cambrian Tapeats Sandstone on the top (about 550 million years old) and the Proterozoic granite below (about 1700 million years old). I am pointing to the erosional unconformity on top of the granite, which represents 1150 million years of missing time. **C.** A disconformity with the dark orange Devonian Temple Butte Limestone filling a channel carved into the lighter orange Cambrian Muav Limestone in the Grand Canyon. There is a disconformity above the Temple Butte Limestone since most of it (except the small remnant in the channel) is eroded away, then capped by the deposition of the Mississippian Redwall Limestone.

unconformity are at an angle to one another. Beneath the erosional surface is a sequence of rocks that has been tilted, then eroded off. After the erosional event, another sequence of rock layers covered the old erosional surface, or unconformity. Angular unconformities can represent an immense amount of time (Figs. 1.7, 3.4, 3.5). First, the lower stratified sequence must pile up in horizontal layers over thousands of years, then be lithified in hard sandstones and mudstones (a process that takes centuries or longer). Second, the layered sequence must be tilted on its side and the edges eroded off. Finally, another layer of sediment must be deposited on top of the old erosional surface as it slowly erodes down or subsides. A sequence such as this takes millions of years to form.

2. **Nonconformity**: In this kind of unconformity, the rocks beneath the erosional surface are igneous or metamorphic. Such rocks are usually found in magma chambers deep in the crust or are metamorphosed in the roots of mountain belts, so their very presence near the earth's surface implies a huge amount of uplift and erosion. If you hike in many mountains in the world today, you will see igneous and metamorphic rocks that formed many kilometers underground, now

uplifted many kilometers into the air. When those ranges finally erode down, layers of sediment can cover them.

3. **Disconformity**: The third type of unconformity is the most subtle and difficult to spot in the field. In a disconformity, the layers above and below the erosional surface are parallel to one another, so it is often hard to spot the erosional break. Most often, the evidence of erosion is manifested by some sort of scoured surface or channel carved down into the layered rocks below the erosional surface. This surface is then filled in and covered by much younger layered rocks above the erosional surface. In many cases, you can't tell that there's a huge time gap until you find time-diagnostic index fossils indicating that an important chunk of time is missing. (Some geologists use the term "paraconformity" for a virtually invisible unconformity, whose existence is established by fossils but there is no physical evidence of erosion between the two units.)

 See For Yourself: Unconformities

Unconformities are very common in the rock record, and many times it turns out that there is more time missing in the

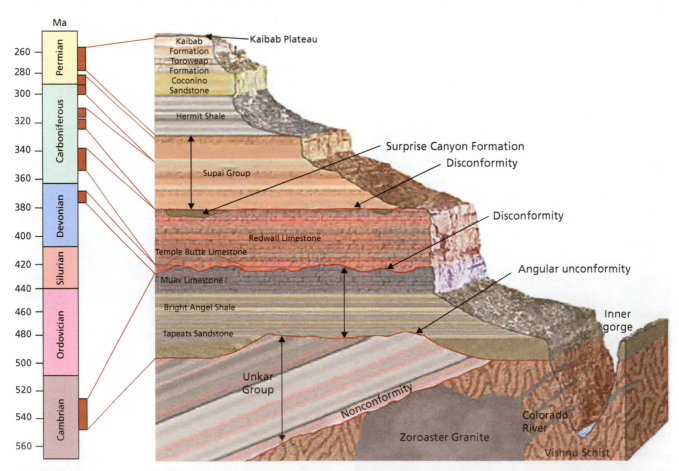

Figure 3.6 The sequence of rock units in the Grand Canyon and the unconformities between them. The age range of each rock unit is plotted on the timescale on the left, showing that only a small portion of the Paleozoic is actually represented by rocks in this apparently "continuous" pile of rocks in the Grand Canyon.

erosional surfaces than is recorded by the rocks that are actually there. For example, several large unconformities in the apparently "continuous" Paleozoic sequence of the Grand Canyon have been documented (Fig. 3.6). Evidence from the fossils and other dating techniques discussed in this chapter indicates that of the 300 million years of the Paleozoic, less than 25% of that time is actually represented by rocks in the Grand Canyon. One unconformity alone (between the Cambrian Muav Limestone and Devonian Temple Butte Formation) spans 150 million years all by itself, wiping out evidence of the Ordovician and Silurian completely. This pattern is typical all over the world so that geologists like Derek Ager describe the sequence of rocks on the planet as "more gaps than record." He uses the analogy of a French schoolchild's definition of a net: "a bunch of holes held together by string."

3.3 Relative Dating and Geologic History

See for Yourself: Relative Dating

Using cross-cutting relationships, superposition, and other principles of relative dating, we can reconstruct the relative sequence of geologic events, or **geologic history**, of any given region. Consider, for example, the cross section of the Grand Canyon shown in Fig. 3.6. Let's see if we can figure out the sequence of events that would explain this diagram.

Following the principle of superposition, let's look for the oldest events at the bottom of the stack. We see two rock units here: the metamorphic Vishnu Schist, which has been intruded by an igneous dike, the Zoroaster Granite. So the oldest unit must be the Vishnu Schist, followed by the granite. Remember from our discussion of metamorphism (Chapter 2) that metamorphic rocks come from a non-metamorphic protolith, or parent rock. So we need to include this as well. Thus, the first three events are 1) protolith 2) metamorphism to Vishnu Schist, 3) intrusion by Zoroaster Granite.

Next, notice that there is an erosional surface cut into the Vishnu Schist and overlain by the Bass Limestone. As we just learned, any unconformity with igneous or metamorphic rocks below the erosional surface is a nonconformity. So event 4 is uplift and erosion to cut the unconformity. Next comes the sedimentary sequence of the Unkar Group (Bass Limestone, Hakatai Shale, Shinumo Quartzite, Dox Formation) and the Chuar Group (Cardenas Lavas, Nankoweap Formation, Galeros Formation, Kwagunt Formation, and Sixty-Mile Formation). But all of these sedimentary units are strongly tilted, so the next event must be tilting and deformation. They are also cut by a fault, which would be younger than the Sixty-Mile Formation that it cuts through but not as old as the erosional surface that cuts it. These tilted sediments plus the underlying schist and granite are all cut by another big erosional surface, which forms an

angular unconformity in some places and a nonconformity in others.

Above this unconformity is the horizontally bedded Paleozoic sequence that makes up the walls of the Grand Canyon. Using superposition, the first three units are the Cambrian Tapeats Sandstone, Bright Angel Shale, and Muav Limestone. You will notice that the top of the Muav has been deeply eroded into channels, which are filled with the Devonian Temple Butte Limestone, so there is a disconformity between those two units. The Temple Butte is in turn overlain by the Mississippian Redwall Limestone, followed by another channel cut into its upper surface, in which the Surprise Canyon Formation was deposited; this is another disconformity. Next comes the Supai Group (Pennsylvanian Watahomigi Formation, Manakacha Formation, Wescogame Formation, and Permian Esplanade Sandstone). Overlying these are the final four Paleozoic rock units of the Grand Canyon: the Permian red shales of the Hermit Shale; the white cross-bedded dune deposits of the Coconino Sandstone; and two cliff-forming limestones at the rim of the Canyon, the Toroweap Formation and the Kaibab Limestone. Finally, we have an erosional event that removed all the rocks younger than the Permian from the top of the canyon (but these can be seen to the north at Zion and Bryce National Parks). The next event would be the cutting of the canyon itself. About a million years ago, a number of basaltic volcanoes erupted in the region, so in the far western Grand Canyon there are very young lavas that flowed down into the deep parts of the newly carved Grand Canyon, forming Lava Falls, one of the biggest rapids on the entire raft trip down the canyon.

Another example of working out a complex series of geologic relationships can be seen in Fig. 3.7.

3.4 Numerical Dating

So far, we have talked about the methods of relative dating, where we establish that a rock unit (or events like faulting, tilting, or uplift and erosion to produce unconformities) is younger than or older than another rock unit or event. The principles of relative dating were worked out over 200 years ago, and the standard relative timescale ("Cambrian," "Ordovician," etc.) was mostly established in the 1820s and 1830s and has been stable ever since. Most geologists use these geologic time terms as everyday language and expect other geologists to know what they mean.

But if we're talking to non-geologists, we need to express these ages in numbers of years before present. Thus, we also need to talk about not only relative age but also **numerical age**, or the age in thousands or millions or even billions of years ago. (Some older books still use the misleading and incorrect term "absolute age" when they mean "numerical age," but use of this term has been discouraged by the North American Commission on Stratigraphic Nomenclature since 1983.) Numerical ages of rocks are not as easy to obtain as relative ages and require special techniques

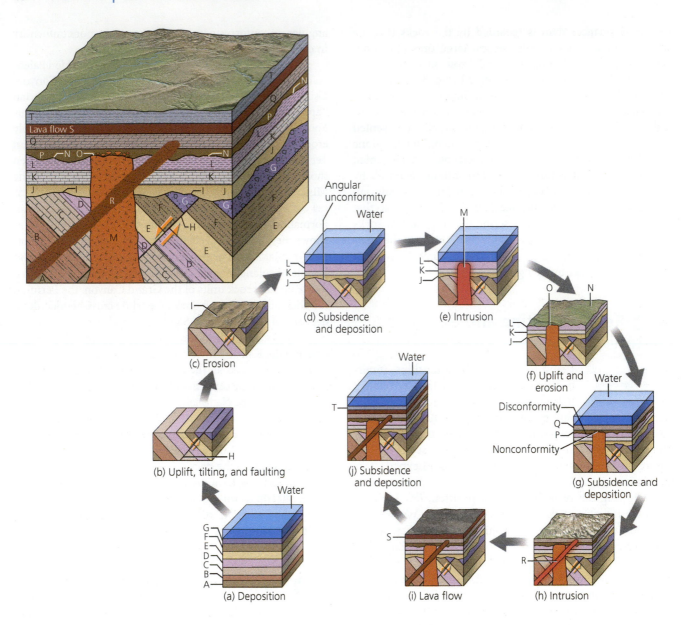

Figure 3.7   A complex diagram like that shown in the upper left of this illustration can be deciphered by careful step-by-step application of the principles of superposition, cross-cutting relationships, and recognizing unconformities and episodes of tilting and faulting.

and machinery, not just field observations. In addition, the modern era of numerical dating is just over a century old (starting in 1915), and the most practical method, potassium–argon dating, has only been in widespread use since the 1960s. Thus, it is often expensive and difficult to get numerical dates, and geologists still use relative ages for most routine age estimates.

Scientists had been trying to estimate the age of the earth for a long time. As we already saw, up until the late 1700s, most scholars regarded the earth as very young and unchanging, with most suggesting it was about 6000 years old. For example, in 1654 Archbishop James Ussher, the Anglican archbishop of Armagh, Ireland, used the ages of the patriarchs in the Bible to calculate that Creation happened on October 23, 4004 B.C.E. Of course, the Bible doesn't give consistent accounts of how much time elapsed between

Creation and Noah's Flood, let alone the time afterward, so a lot of guesswork was involved. Nevertheless, Ussher's estimate was the pinnacle of scholarship for its time, incorporating what was known of the history of the Babylonians, Persians, Greeks, and Romans back then, so we must respect this estimate for the honest attempt that it was—even though we now know that it was a million times too short.

For over a century, the power of the church over European scholarship meant that this estimate was not challenged. However, during the Enlightenment, the hold of the church over scientists began to weaken. Some, like the French Count Buffon in 1779, suggested that the earth was as old as 75,000 years, at least 10 times the estimate based on biblical chronology.

The discoveries of James Hutton in 1788 during the Scottish Enlightenment (see Chapter 1) and their extension by

Charles Lyell in 1830–1833 soon tilted the scales the other way. From a very young earth only 6000 years old, most geologists saw the earth as immensely and immeasurably old. In Hutton's words, there was "no vestige of a beginning, no prospect of an end." Most geologists looked at the enormous thicknesses of strata around the world and knew that the earth must be millions of years old at the very least. A number of them tried to find the maximum thickness of rock known from each period (such as the Cambrian or Ordovician) and, using average rates of sedimentation, calculate the minimum time needed to deposit all the strata of the Cambrian, the Ordovician, the Silurian, and so on. Most such estimates came to about 100 million years, with none less than 50 million years. Of course, we now know that their estimate to the beginning of the Cambrian is off by at least a factor of 5. Why? Because of a faulty assumption: that the sedimentation on earth is continuous. As we just discussed, we now know that there are large unconformities in every sequence of rock, so more time is missing than is included in the rock record.

In the late 1800s, Irish physicist John Joly tried a different approach: calculating how long it would take for the oceans to go from freshwater to their current salinity of about 3.5% salt, using the rate at which salt enters the oceans from the world's rivers. He also came to estimates of 80–100 million years, which we now know to be off by almost a factor of 500. What went wrong? Again, the problem is faulty assumptions: he assumed that the oceans have been constantly adding salt since they formed. It turns out that the salt content of the oceans does not change much through time because when there is too much of it, it is precipitated into salt deposits in the earth's crust. Seawater salinity is in equilibrium and stays very stable over long periods of time.

But the most controversial estimate of the nineteenth century was made by the legendary physicist William Thomson (later known by his title, Lord Kelvin). Kelvin made huge discoveries in the fields of physics, especially thermodynamics (the Kelvin temperature scale is named after him since he pioneered the concept of absolute zero temperature). He was also a great inventor and helped create the transatlantic cable system that enabled telegraph and then telephone communications between Europe and North America. Thus, he was a giant among scientists of his time, and few people dared disagree with him.

In 1862, Kelvin attacked the problem of the age of the earth using thermodynamics. He assumed that the earth had started as a molten ball at the same temperature as the sun and that it had cooled off at rates that we can measure from the heat coming up from the earth's interior. From this method, he estimated that the earth was only 20 million years old, much less than most geologists were willing to accept. It was also a problem for Charles Darwin, who knew that the earth had to be immensely old for his newly proposed concept of evolution to work; Kelvin's estimate didn't seem to offer enough time.

Through the rest of the nineteenth century, physicists and geologists were at an impasse. Neither could comprehend the arguments of the other side or see the flaws in their own

estimates. In the late 1800s, geologists began to back down and fudge their estimates a bit from the original values of 80–100 million years to numbers closer to Kelvin's 20 million years. "Physics envy" was just as powerful then as it is now! But the problem with Kelvin's estimate was just like the others: faulty assumptions. Kelvin made his calculation of the cooling of the earth assuming that the heat was from the original solar system and that no other heat sources were involved.

But in 1896 Henri Becquerel in France discovered radioactivity, and in 1903 Marie and Pierre Curie showed that radioactive materials like radium produced a lot of heat. At the same time, New Zealander Ernst Rutherford was England's foremost authority on this new source of energy. In 1904, he was getting ready to address the Royal Institution of Great Britain about this new discovery, when he suddenly realized that the 80-year-old Lord Kelvin himself was in the audience! The young Rutherford was about to challenge the world's most famous physicist's estimate of the age of the earth! As Rutherford wrote later,

I came into the room which was half-dark and presently spotted Lord Kelvin in the audience, and realised that I was in for trouble at the last part of my speech dealing with the age of the Earth, where my views conflicted with his. . . . To my relief, Kelvin fell fast asleep, but as I came to the important point, I saw the old bird sit up, open an eye and cock a baleful glance at me. Then a sudden inspiration came, and I said Lord Kelvin had limited the age of the Earth, provided no new source [of heat] was discovered. That prophetic utterance referred to what we are now considering tonight, radium! Behold! The old boy beamed upon me.

Kelvin's estimate had been based on the faulty assumption that there were no other sources of heat beyond the earth's original heat when it cooled from a molten mass and that it would cool in no more than 20 million years. But radioactivity provides that additional heat. In fact, radioactivity provides so much heat that it is now the only source of heat that we measure coming from the earth's interior. The original heat from the cooling of the earth Kelvin thought he was measuring dissipated billions of years ago, maybe even during the 20 million years since the earth first formed 4.6 billion years ago.

CLOCKS IN ROCKS

The discovery of radioactivity provided not only a previously unknown source of heat to explain why Kelvin was wrong but also something else: the method to find the true age of the earth. Rutherford and other early nuclear physicists were only concerned for how radioactivity worked. By 1913 (Box 3.1), Arthur Holmes was the first geologist to find a way to use radioactivity for dating, and he obtained ages from samples almost 1.6 billion years old. By the 1920s, Holmes had dated enough rocks that he could say that the earth was over 3 billion years old. Based on these discoveries, not only did Holmes become the "father of radiometric dating" but in the 1920s he was one of the few geologists who supported

Wegener's ideas of continental drift (see Chapter 5), pioneering the concept of convection cells in the mantle driven by radioactive heat from the earth's interior. Before his death in 1965 at age 75, Holmes saw plate tectonics vindicate his ideas about mantle convection and ever-greater refinements to his pioneering efforts at dating rocks. He received all the highest awards that any geologist can receive.

How does radioactivity help us date rocks? It works on a simple principle: there are atomic elements in nature that **decay** (break apart) spontaneously at a known rate. This rate is not affected by temperature or pressure or any other external factor, so it is the most clock-like process that we have in rocks. These radioactive elements are unstable and decay from the unstable **parent atom** to produce a more stable **daughter atom**. If we know how fast the decay from parent to daughter takes place, we can measure their ratio in the rocks and calculate how long that decay has been occurring.

The reaction goes as follows:

Parent atom —> Daughter atom + heat + atomic particles

BOX 3.1: HOW DO WE KNOW?

How Do We Date the Oldest Rocks?

Once Ernest Rutherford and Marie and Pierre Curie had shown how radioactivity works, other scientists soon realized that radioactivity had lots of potential to solve geologic problems. Physicists had been trying to date rocks by measuring the helium given off by uranium decay, but it was almost impossible to capture all the helium gas. Instead, it was Yale chemist Bertram Boltwood who discovered that uranium decayed to lead through radioactive breakdown. Following on a suggestion from Rutherford, Boltwood noticed that the rocks he knew to be older had more lead in them than those that he knew to be younger. Unfortunately, he was using the very primitive understanding of the uranium–lead system that prevailed at the time. He didn't realize, for example, that there are two different radioactive isotopes of uranium, uranium-238 and uranium-235, each with different decay rates and different daughter isotopes of lead. Nonetheless, he analyzed the samples he had, and in 1907 he got samples ranging from 400 million years to as old as 2.2 billion years. This was the first evidence that the earth was indeed billions of years old, as geologists had long suspected, and that Kelvin's estimate was way off. Unfortunately, in his later life Boltwood suffered from severe depression, and his research came to a standstill. He committed suicide in 1927, unable to cope with his personal problems.

Boltwood had done the first analyses and gotten ages on rocks as old as 2.2 billion years but never followed up on his breakthrough. Thus, it fell to a younger British geologist, Arthur Holmes (Fig. 3.8), to take the budding young field of geochronology and develop it into a rigorous science. Born to a family of modest means in 1890 in the tiny town of Gateshead (near Durham and the Scottish border), Holmes was originally planning to be a physics major at the Royal College of Science (now University College London). But in his second year, he took a course in geology, against the advice of his tutors, and found his true calling.

Holmes proved to be a brilliant undergraduate and was soon doing research. He latched on to the hot problem of radioactivity and realized that Boltwood's 1907 paper on uranium–lead dating held immense potential. For his undergraduate research project, he had a granitic rock from the Devonian of Norway to analyze. He cut his Christmas holiday short and spent his "holiday break" in the cold, dark lab in London, working all by himself in the silent building. As his advisor, physicist Robert Strutt, later recalled,

> We are at present largely subsisting on loaned apparatus, some of which belongs to other public bodies, such as the Royal Observatory, the Royal Society, etc., while some has been borrowed from private friends. I need hardly say that it seems rather below the dignity of an institution like the Imperial College that its teachers should have to beg apparatus of their personal friends for the purpose of teaching the students. (cited in Lewis, 2000)

Holmes worked away in the cold and silence in January 1910, crushing the rock into a powder in a mortar made of agate, fusing the mineral grains in a platinum crucible with borax, dissolving it in extremely caustic hydrofluoric acid, then boiling it again and again while measuring the radon emissions (an indirect measure of how much uranium was present). The lead content was measured by fusing the powder in a cake, then boiling it, twice dissolving it in hydrochloric acid, then letting it evaporate all its water. Then, he heated it in ammonium sulfide to make the lead precipitate out as lead sulfide (known as the mineral galena). The precipitate was collected on a filter, dried, ignited, treated with nitric acid, boiled, treated with sulfuric acid, and heated again. Eventually, as Holmes wrote, "A tiny white precipitate then remained. This was collected on a very small filter, washed with alcohol, dried, ignited, and weighed with the greatest possible accuracy (cited in Lewis, 2000)." Often, there were only a few milligrams of material left.

These complicated chemical operations took incredible patience, extraordinary dexterity, and lots of time; and often he used up nearly all his original sample. Then to top it off, the results had to be verified, so the entire analysis was repeated two to five times, depending on how much original sample he had left to work with. Once, he had to discard all his data because radon had been leaking into the room. In other cases, he had to go begging at the British Museum for more sample because he had used up his original allotment. Eventually, however, all this hard work paid off, and he got a

Figure 3.8 Portrait of Arthur Holmes as a young graduate student, just about to publish dates on many rocks that showed the earth was billions of years old.

Holmes was so broke after spending years living on a tiny scholarship of £60 a year that he quit school for a while and took a job prospecting for minerals in Mozambique to earn some real money. He spent 6 months, found nothing, and came down with malaria so bad that his colleagues sent a letter home telling his family he was dead. Eventually, he recovered and managed to get a boat home, where he became a demonstrator (a low-level instructor) at his alma mater, Imperial College London. There, he resumed his studies of the uranium–lead dating technique, discovering that there were two different isotopes of uranium and lead and that this had to be understood in the dating analysis.

By 1913, he had so many new results and so many improvements on the method that he was able to write his groundbreaking book, *The Age of the Earth*, while he was still a graduate student. In it, he not only explained the basic principles of geochronology and discussed the problems with earlier methods of dating the earth but also finally laid Lord Kelvin's mistaken estimate to rest. He had dates on some of the oldest rocks in Britain of 1.6 billion years, although he refused to speculate on the age of the earth. Later editions included results from analyses of older and older samples until by the 1950s he had dates of 4.5 billion years, which is our present estimate. His early research earned him his doctorate in 1917 from University College London. But World War I was raging in Europe, and it was hard to make a living on the paltry salary of a demonstrator in college. To make money for his family, he decided to try a career as an exploration geologist again, this time for an oil company in Burma in 1920. However, the oil company went bankrupt, and Holmes returned to England in 1924 flat broke. This was not his only tragedy: his 3-year-old son had contracted dysentery shortly after arriving in Burma and died there.

Luckily, after his return in 1924, his earlier reputation and research landed him a job as a reader (what we would call a "lecturer") in geology at Durham University, close to his birthplace. There, he spent the next 18 years teaching geology and adding and refining to the database of radiometric dates from around the world. His work was so dominant in the field that he later became known as the "father of geochronology" and the "father of the geologic timescale." In 1943, he moved north of the border to the University of Edinburgh, where he spent the last 13 years of his career until his retirement at age 66 in 1956.

reliable date of 370 million years on his Devonian granite from Norway. He had greatly improved Boltwood's original methods and proven that uranium–lead dating could work and that it was possible to generate dates on rocks. The results were published in 1911, soon after he graduated in 1910.

The heat is a byproduct of this reaction, and it solved the dilemma of Kelvin's age estimate: radioactive decay goes on constantly within the earth and releases huge amounts of heat. Three types of atomic particles can be byproducts of the reaction as well. **Alpha particles** are a helium nucleus (2 protons plus 2 neutrons) and are so heavy and low in energy that wearing a lead vest can stop alpha radiation. **Beta particles** are similar to electrons. **Gamma particles** are a high-energy form of radiation that can pass through the entire earth without stopping.

Although there are hundreds of unstable parent atoms known in nuclear physics, most of them decay in fractions of a second and are of no use to geologists. Only a few have such slow decay rates that they are useful for geologically ancient systems. The most important of these are shown in Table 3.1. Each system has different initial abundance in earth rocks and different decay rates, so each is useful in different types of geologic problems.

The decay of radioactive parent atoms to their stable daughter atoms happens at a known rate, but it is not linear

TABLE 3.1   THE SIX RADIOACTIVE ISOTOPE SYSTEMS MOST SUITABLE FOR RADIOMETRIC DATING

Isotopes		Half-Life of Parent (Years)	Effective Dating Range (Years)	Material That Can Be Dated
Parent	Daughter			
Uranium-238	Lead-206	4.5 billion	10 million to 4.6 billion	Zircon
Uranium-235	Lead-207	704 million		Uraninite
Thorium-232	Lead-208	14 billion		
Rubidium-87	Strontium-87	48.8 billion	10 million to 4.6 billion	Muscovite
				Biotite
				Orthoclase
				Whole metamorphic or igneous rock
Potassium-40	Argon-40	1.25 billion	100,000 to 4.6 billion	Glauconite
				Hornblende
				Muscovite
				Whole volcanic rock
				Biotite
Carbon-14	Nitrogen-14	5730 years	Less than 100,000	Shell, bones, and charcoal

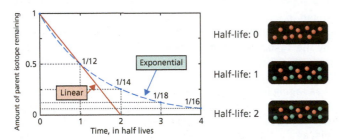

Figure 3.9   Comparison of linear decay versus exponential decay.

decay such as happens when sand pours through the neck of an hourglass or you burn a candle for a fixed number of minutes. These are examples of **linear change**—the same amount of sand passes, or candle burns, in the same amount of time (Fig. 3.9). But radioactive decay is an example of **exponential decay**—the amount of change in the same amount of time varies depending upon how much unstable parent material is still present. When there are a lot of parent atoms (as when the clock starts), decay is very fast. As there are fewer and fewer unstable parent atoms, the decay rate slows down so that after a long period of time the decay rate is almost negligible.

▷ See For Yourself: Radioactive Dating   In atomic physics, the most useful measure of the decay rate is known as the **half-life**. A half-life is *the time it takes for half of the parent atoms present to decay to their daughter atoms.* Most unstable atoms have half-lives of less than a second, but the systems in Table 3.1 have half-lives of thousands to billions of years, so they enable us to date rocks in those age ranges.

Here's a good way to visualize it. Imagine a crystal (Fig. 3.10) cooling out of a magma that has some of the unstable radioactive parent atoms in it. Our radiometric clock is based on the idea that a fixed number of parent atoms

decay into daughter atoms over a predictable amount of time. Thus, we need to lock all the parent atoms into the lattice of a crystal as it cools. Once the lattice is formed and the parent atoms are locked in, we can start to measure the process of decay.

Let's imagine a simple crystal with 16 parent atoms to start with (Fig. 3.9). In one half-life, half of those original 16 parent atoms will have become daughter atoms, so you'll have 8 parents and 8 daughters. In the second half-life, half of the 8 parents decay to 4 daughters, so you'll now have 12 daughters and only 4 parents left. In the third half-life, half of those 4 remaining parents decay, producing only 2 parents and 14 daughters. In the fourth half-life, 1 of the two parents decays, so there are 15 daughters and only 1 parent. Notice how the ratio of parents to daughters changes from 8:0 at the start to 1:1 in one half-life, 1:3 in two half-lives, 1:7 in the third half-life, and 1:15 in the fourth half-life.

In a real sample, we take some fresh, unweathered crystals from the rock we want to date. (We want the material to be fresh and without weathering so that no atoms leak out or soak in from the outside world.) Then we crush the crystal down (or melt it with a laser) and release those atoms into a **mass spectrometer** (an instrument that measures the different isotopes of atoms in a sample). We can then measure the amount of both parent and daughter atoms released and plug them into a formula that calculates their age based on that ratio, the half-life of the system, and a **decay constant** that is known for each parent–daughter atom combination.

The mathematical expression that relates radioactive decay to geologic time is

$$N(t) = N_0 e^{-\lambda t}$$

where t is age of the sample, N_0 is the number of atoms of the parent isotope in the original composition, $N(t)$ is number of atoms of the parent isotope in the sample at time t, and

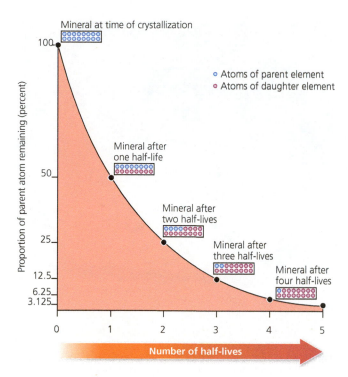

Figure 3.10 Cartoon of radioactive decay in a crystal. In each half-life, have of the original parent atoms (blue dots) decay, so if there were 16 parent atoms to start, in one half-life, there would be 8 parent atoms and 8 daughter atoms (show by the red dots). In a second half-life, half of the parent atoms would decay again, leaving only 4 parent atoms and 12 daughter atoms. In the third half life, only 2 parent atoms remain, creating 14 daughter atoms. In the fourth half life, there is only 1 parent atom left and 15 daughters. The ratio of parent atoms to daughter atoms through each half-life goes from 16:0 to 1:1 to 1:3 to 1:7 to 1:15. By measuring that ratio in a decaying mineral sample, we can determine how far it has moved on age curve, and thus determine its numerical age.

because any kind of weathering releases some of the parent or daughter atoms or might allow extra atoms to diffuse in from the outside world. Plus we need materials that form from cooling in a molten state so that the crystals can form and lock in the parent atoms. This means that *only igneous rocks (and a few high-temperature metamorphic rocks) can be dated directly.*

Notice that this implies that *we cannot date sedimentary rocks directly*! If we took individual sand grains from a sandstone, we could get dates on each one individually—but they would be the dates of the weathered parent material that supplied each grain of sand, *not* the entire sandstone. So how do we date sedimentary rocks, then? After all, our relative timescale is based entirely on sedimentary rocks and the fossils they contain!

The answer is challenging: geologists look all over the world for places where igneous rocks are in contact with sedimentary rocks of known relative age. The best situations are where volcanic ash falls or lava flows are interbedded with fossil-bearing sediments so that their dates give us a numerical age bracket for the relative age determined by the fossils (Fig. 3.11). If we can't have volcanic rocks interbedded with our sedimentary layers, then the next best thing is to have igneous dikes that intrude through fossiliferous sediments. These give us a minimum age since the dike is younger than what it cuts through. Finding places on earth with volcanic ashes or lava flows interbedded with fossiliferous sediments is not an easy task, but for the past few decades geologists have been finding and studying examples from all over the world to produce the standard geologic timescale with numerical ages for the relative time periods. That is also why the numerical dates of the timescale are constantly being adjusted and change slightly every few years. As new dates and better dates come in, the timescale

λ is the decay constant of the parent isotope.

However, we are usually interested in *t*, the age of the sample. Solving the equation to get *t*, we come up with the formula

$$t = \ln (D/P + 1) \times 1/\lambda$$

where D is the number of daughter atoms measured, P is the number of parent atoms measured, and ln is the natural log (log to the base e).

Although it's interesting to see and understand the mathematics of the formula needed to do the calculation, in the real world the computer does the entire calculation for you just as soon as the results come out of the mass spectrometer.

What does this mean in practical terms? First of all, we need to take fresh samples of the rock or mineral

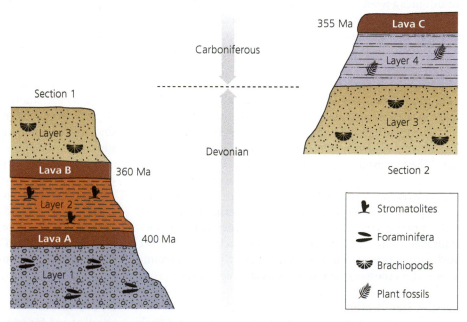

Figure 3.11 The only way to get the numerical ages of sedimentary rocks is to find them interbedded with volcanic ash falls or lava flows. If that's not possible, then a cross-cutting dike will give a minimum age for the beds.

always needs recalibration. For this reason, most geologists don't keep the numerical ages of the timescale in their heads (except in the time interval of their research interest), but all geologists can be expected to know the periods and epochs of the relative geologic timescale by heart since it is stable and hasn't changed much since 1840.

There are a few other things to know as well. Radiometric dates are based on running a number of samples (usually of individual crystals). Each analysis might give a slightly different number, simply because the machine itself cannot perfectly reproduce the same isotope counts every time. Thus, whenever you see a numerical age given, it *should* have a plus or minus error estimate attached to it so that you can judge how precise the date really is. For example, if it quotes a date as "100 ± 5 million years," what that really means is that there's a 95% chance that the true age lies somewhere between 95 and 105 million years. Sometimes the error estimates are omitted or forgotten in popular works about geology, but in a technical journal or professional publication, the error estimates *must* be given—or the reader has good reason to be suspicious about it.

RADIOMETRIC DATING SYSTEMS
Uranium–Lead
Let's look at Table 3.1 again. The very first radiometric dates produced by Arthur Holmes about a century ago were done on the two uranium–lead (U–Pb) systems. Uranium-235 decays to lead-207, and uranium-238 decays to lead-206. Notice that the half-life of the first parent–daughter pair is 0.7 billion years and that of the second is 4.5 billion years. In other words, any uranium-238 that formed when the earth first cooled has just decayed for only a single half-life! Both of these half-lives are very long, so decay is extremely slow, meaning that it takes a long time for measurable daughter products (the two lead isotopes) to accumulate. In addition, uranium is a relatively rare element. It is too large to fit inside the lattice of most of the minerals in a typical magma, so it is left over as a residue after most of the magma has cooled and crystallized out. The lead atoms are concentrated in the crystals that form with the last melt to cool, which commonly include crystals of zircon (zirconium silicate) or apatite (calcium phosphate) and metallic minerals, all of which have large lattice spaces so the huge uranium cation can fit.

Geochronologists using the U–Pb system typically crush a rock down to powder, then dissolve most of it in hydrofluoric acid. The zircons or apatites resist dissolution, so they can be concentrated and then studied. Still, even with this method, the rock has at most about 0.1% uranium in it, so there's not much parent material; and therefore, it takes a *long* time (typically a billion years) for measurable daughter material to accumulate. Thus, the U–Pb system is not practical for younger rocks and is primarily used for very old earth rocks (Precambrian) and especially for moon rocks and meteorites.

Fission-Track Dating
Another dating method related to U–Pb is fission-track dating. Instead of measuring the amount of parent and daughter directly, fission-track dating measures how many

Figure 3.12 An apatite crystal showing the fission tracks caused by nuclear particles flying through and damaging the crystal lattice.

alpha particles have been released during radioactive decay of uranium-238. As the particles fly away from the nucleus, they create a track of damage to the crystal lattice (Fig. 3.12). The more tracks, the longer the crystal has been undergoing decay. Typically, fission-track dating uses weathering-resistant crystals of zircon or apatite in a volcanic ash to date. These crystals are then etched in hydrofluoric acid to make their tracks easy to see. During the lab analysis, several techniques allow us to calibrate the rate of track production, so the more tracks there are, the older the crystal. Fission-track dating works for rocks of almost any age, from 100,000 years to 2 billion years.

Rubidium–Strontium
The rubidium–strontium (Rb–Sr) system (Table 3.1) is very similar to the U–Pb method. Rubidium-87 changes to strontium-87 with a half-life of 48.8 billion years. This process is so incredibly slow that no rubidium-87 has yet gone through even a tenth of its first half-life! Thus, it is hard to find any daughter atoms unless the rock is very old. Rubidium is not a common element, but it has the same charge and ionic radius as potassium, so a tiny percentage of the potassium sites in potassium feldspars and micas are occupied by rubidium. For all these reasons, Rb–Sr dating is only practical for very old rocks: Precambrian earth rocks, moon rocks, and meteorites. Typically, it is used in conjunction with both of the U–Pb systems to get multiple independent dates on the same samples.

Potassium–Argon
In contrast to the U–Pb and Rb–Sr systems that are primarily used for very old rocks, the potassium–argon (K–Ar) system is perfect for nearly every geologic problem. Its half-life is long (1.25 billion years), but unlike the other

systems, potassium is one of the eight most common elements in earth's crustal rocks. Potassium is found in small but significant amounts in potassium feldspars, biotite and muscovite micas, hornblende amphiboles, volcanic glass, and even samples of a whole rock. There is so much parent material initially that (even though it decays slowly) there is nearly always some measurable daughter product as well. Thus, it is used not only for old Precambrian rocks but also especially for rocks from the Phanerozoic (the past 550 million years), when most events that are interesting to the majority of geologists occurred. The only limitation is that the decay rate is slow enough that there is no measurable product yet in just a million years, so it cannot be used to date rocks younger than 100,000 years (Box 3.2).

See for Yourself: Radiocarbon Dating

Radiocarbon Dating

All of the previous radioactive decay reactions work only in igneous rocks (or a few high-temperature metamorphic rocks) since they need to be trapped in a fresh crystal as it forms from a molten state. There is one radiometric dating system that works on a totally different principle: radiocarbon dating, or carbon-14 dating. Instead of measuring crystals of igneous rocks, carbon-14 dating allows you to measure samples of any material that contains carbon: bone, shell, wood, pottery, cloth, peat, lignite coal, and many other materials. Thus, it is the exception to the general rule that you can't get numerical dates directly from sedimentary rocks or fossils.

The carbon-14 system works very differently from the other systems that require a crystal to form from molten

---

BOX 3.2: HOW DO WE KNOW?

How Do We Know the Age of the Earth?

We come back to the question that we posed at the beginning of the chapter: how old is the earth? And, more importantly, how do we know? We have seen the myths of various cultures suggest that the earth is eternal and unchanging and constantly cycling, while other cultures suggested that the earth is only a few thousand years old. Ever since the days of Hutton and Lyell in the early 1800s, most geologists reckoned it was millions to hundreds of millions of years old—"no prospect of a beginning," in Hutton's words. In the mid-1800s, Kelvin's mistaken calculations had geologists shaving their estimate down to 20 million years old, until the discovery of radioactivity and the heat it produces discredited Kelvin's assumptions and gave us a reliable method of getting the real age. In 1907, Bertram Boltwood published the first uranium–lead dates that suggested the earth was billions of years old, and by 1913 Arthur Holmes had worked out the dating method and published many more dates (see Box 3.1).

Over the past century, many more radiometric dates from a variety of different isotopic systems (U–Pb, Rb–Sr, K–Ar, and others) have been obtained. There are several other isotopic methods (such as Pb–Pb, which is derived from U–Pb; samarium–neodymium [Sm–Nd]; and rhenium–osmium [Re–Os]) we have not discussed that work particularly well with the oldest rocks.

There are three different sources of materials that could tell us the age of the solar system, and thus the age of the earth. The first would be rocks and minerals from the earth itself. At the time of this writing, the oldest rock on earth is dated at 4.32 billion years and comes from the east shore of Hudson's Bay in Quebec, Canada. There are individual mineral grains from sandstones in Australia that are even older: 4.4 billion years is the current record holder. But there are good reasons to suspect that no earth rock or mineral will give us a good maximum age of the earth itself. One reason is the fact that the earth has an active, tectonically mobile crust that

is constantly being remelted and recycled, so it is extremely unlikely that any piece of protocrust from the earliest earth still survives today. The other reason is that we have pieces of the original solar system that give older ages.

There are two sources of these rocks from the early solar system: meteorites and moon rocks. A number of different meteorites have been dated. The oldest meteorites are **chondrites,** which formed from the original solar system material before any planets had clumped together out of the disk of solar matter (see Chapter 7) or differentiated into layered planets with a core and mantle. A number of those recently dated meteorites are listed in Table 3.2. Notice that for each meteorite it has been analyzed by multiple different isotopic methods: U–Pb (and its derivative, Pb–Pb), Rb–Sr, K–Ar (and its derivative, Ar–Ar), Re–Os, and Sm–Nd. Every one of these ages from completely different isotopic systems gives answers very close to an average of 4.54 billion years, and not one differs from the others by more than a percent or two (within the normal ± range of measurement error on the machines).

This is the answer to people who doubt that the earth is old and don't believe in numerical dating. All of these isotopic systems are completely independent and have nothing in common except that they are found in the same rock that formed at one time in the past. One could imagine that something might throw one system off, but there is no way that literally *hundreds* of dates using *all* the different isotopic systems in the same rock gave the same answer—*unless* that answer was the true age of the rock. The expert on the K–Ar system, Dr. G. Brent Dalrymple of the US Geological Survey, put it this way: you could imagine a clock shop where one or two clocks don't work right and don't give the right time—but if the vast majority of the clocks all agree, then that is the true time. We do not throw out or ignore every clock in the shop just because one or two runs slow or fast.

TABLE 3.2   SOME OF THE HUNDREDS OF DATES OF THE OLDEST KNOWN METEORITES AND MOON ROCKS, DEMONSTRATING THAT THE SOLAR SYSTEM (AND EARTH) IS ABOUT 4.67 BILLION YEARS OLD, THE OLDEST DATES KNOWN FROM SEVERAL DIFFERENT METEORITES: DATING METHODS INCLUDE LEAD-LEAD ISOTOPES (Pb–Pb), RUBIDIUM-STRONTIUM (Rb–Sr), RHENIUM-OSMIUM (Re–Os), SAMARIUM-NEODYNIUM (Sm–Nd), AND ARGON-ARGON (Ar–Ar)

St. Severin (ordinary chondrite)	
Pb–Pb	4.543 ± 0.019 Ga
Sm–Nd	4.55 ± 0.33 Ga
Rb–Sr	4.51 ± 0.15 Ga
Re–Os	4.68 ± 0.15 Ga
Juvinas (basaltic achondrite)	
Pb–Pb	4.556 ± 0.012 Ga
Pb–Pb	4.540 ± 0.001 Ga
Sm–Nd	4.56 ± 0.08 Ga
Rb–Sr	4.50 ± 0.07 Ga
Allende (carbonaceous chondrite)	
Pb–Pb	4.553 ± 0.004 Ga
Ar–Ar age spectrum	4.52 ± 0.02 Ga
Ar–Ar age spectrum	4.55 ± 0.03 Ga
Ar–Ar age spectrum	4.567 ± 0.05 Ga
Canyon Diablo (Meteor Crater, Arizona, Fe–Ni meteorite)	
Pb–Pb	4.550 ± 0.07 Ga
Acfer 059 chondrules	
Pb–Pb	4.564 ± 0.06 Ga
Efremovka Ca-/Al-rich inclusion	
Pb–Pb	4.567 ± 0.06 Ga
Oldest moon rocks	
Sm–Nd	4.46 ± 0.04 Ma
Sm–Nd	4.54 ± 0.01 Ma
Pb–Pb	4.56 ± 0.02 Ma

Ga = billion years before present; Ma = million years before present.

If you doubt this, then let's imagine an independent test of the system: *another* source of rocks from the early solar system. If they agree with meteorites, then we know the answer is real. Sure enough, we have samples of this third source of dates, the moon, when the Apollo 11 astronauts landed in 1969, up through the last Apollo mission in 1972, Apollo 17 (the only one with a scientist on board, geologist Harrison Schmitt; it collected 111 kg of rocks, the most by any mission). There have been hundreds of dates analyzed on moon rocks, most of which give ages of 3.9 billion to 4.4 billion years. This reflects the age of the different parts of the moon's surface as it went through its earliest history of volcanism and meteorite bombardment. But the oldest dates are the ones that tell us when the moon formed. As can be seen in Table 3.2, there are a number of dates greater than 4.5 billion years in age from the few moon rocks that are this old.

Thus, scientists are confident from so many lines of evidence that the solar system (including the earth, moon, and meteorites) formed around 4.56 billion years ago. Any other conclusion flies in the face of an enormous amount of data from multiple independent sources.

rock (Fig. 3.13). Our atmosphere is composed of over 70% nitrogen gas (N_2), which you breathe constantly. The main isotope of nitrogen is nitrogen-14, which has 7 protons and 7 neutrons. However, cosmic radiation from space is constantly bombarding this atmospheric nitrogen, and it spontaneously changes one of the protons in the nucleus to a neutron so that you have 6 protons and 8 neutrons. Element 6 on the periodic table is carbon, so nitrogen-14 changes the atomic number and element to carbon but does not change the atomic weight, so you get a rare isotope

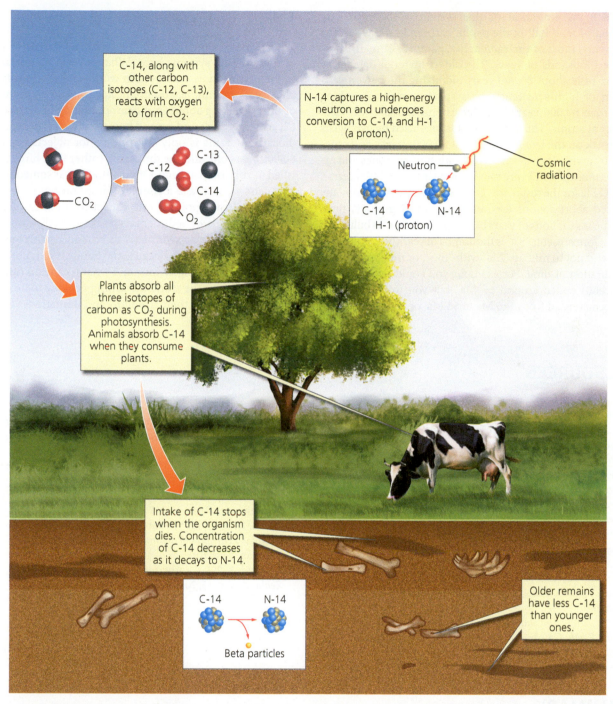

C-14, along with other carbon isotopes (C-12, C-13), reacts with oxygen to form CO_2.

N-14 captures a high-energy neutron and undergoes conversion to C-14 and H-1 (a proton).

Cosmic radiation

Neutron

C-14   N-14

H-1 (proton)

CO_2

C-13

C-12

C-14

O_2

Plants absorb all three isotopes of carbon as CO_2 during photosynthesis. Animals absorb C-14 when they consume plants.

Intake of C-14 stops when the organism dies. Concentration of C-14 decreases as it decays to N-14.

C-14   N-14

Beta particles

Older remains have less C-14 than younger ones.

Figure 3.13 Radiocarbon dating, or ¹⁴C dating, works differently than the other systems, which are based on the decay of atoms within crystals in an igneous body. Carbon-14 is formed in the sky when cosmic rays bombards nitrogen-14 atoms in the air, and converts them to carbon-14 by converting a proton in nitrogen to a neutron, leaving only 6 protons (which makes the element carbon). All organisms take in this carbon-14 while they are alive, either by plants taking in the carbon in the carbon dioxide they absorb for photosynthesis, or animals eating plants and getting a tiny amount of carbon-14 in their diet and thus in their tissues. But when they die, the exchange stops and the carbon-14 clock starts ticking.

of carbon, carbon-14. But carbon-14 is very unstable and breaks down rapidly, with a half-life of only 5730 years.

Most of the time, nitrogen-14 converts to carbon-14 and then decays back to nitrogen-14 and stays in the upper atmosphere. But a tiny percentage of the carbon-14 is taken up in the carbon dioxide that plants absorb by photosynthesis, so it is found in every plant tissue in a minute fraction of a percent. (Over 99% is the common isotope carbon-12.) Animals then consume plants or consume other animals that ate plants, so animals too have a tiny fraction of the carbon in their tissue as carbon-14. These are constantly being exchanged as the plant or animal lives; but when it dies, the carbon-14 is trapped in the wood or bone or other organic tissue, and the radiometric clock begins ticking. Thus, almost any organic material could have carbon-14 in it.

The main limitation of this system is the extremely short half-life. If half of the parent carbon-14 decays in just 5730 years, most of the radioactive parent is gone in a few tens of thousands of years. The oldest material datable by carbon-14 is about 80,000 years old, using the newest generation of high-sensitivity mass spectrometers. Anything older than 80,000 years is radiocarbon-dead and will yield no parent atoms. Thus, carbon-14 is *only* used by geologists who work on the late Ice Ages (last 80,000 years), and especially by archeologists who work with human bones and artifacts (most of which are much younger than 80,000 years).

(NOTE: for the rest of this book, we will use the commonly accepted abbreviations for long intervals of geologic time. "Ma" means millions of years before present; "Ga" is billions of years before present; "Ka" is thousands of years before present. Be careful not to mix up "Ma" with "m.y.", or millions of years as a duration of time. For example, you could say that, "The Cretaceous ended around 66 Ma", but you would say "The Cretaceous lasted from 146 Ma to 66 Ma, or about 80 m.y.")

3.5 Conclusion

We will not go further into the technical details of each of these dating systems here. What is really important is to understand how each system works and where it is applicable. For example, you might hear people on the street or on TV talking about "radiocarbon dating of dinosaurs." Clearly, they don't understand how things work because radiocarbon dating is only appropriate for objects less than 80,000 years old, and the dinosaurs (other than birds) vanished 66 million years ago. Or you might hear someone ask, why can't we just take a chunk of dinosaur bone and date it? As we already pointed out, except for radiocarbon, we can't date fossils or sedimentary rocks directly—we must have those fossils or sediments interbedded with igneous rocks that *can* be dated. Because the public is so ignorant of how numerical dating is done and what it can and can't do, it's important for the student learning geology to be aware of the methods and their strengths and limitations.

RESOURCES

BOOKS

Berry, W. B. N. 1968. *Growth of a Prehistoric Time Scale*. W. H. Freeman, San Francisco.

Burchfield, J. D. 1975. *Lord Kelvin and the Age of the Earth*. Science History, New York.

Dalrymple, G. B. 1991. *The Age of the Earth*. Stanford University Press, Stanford, CA.

Dalrymple, G. B. 2004. *Ancient Earth, Ancient Skies: The Age of the Earth and Its Cosmic Surroundings*. Stanford University Press, Stanford, CA.

Hedman, M. 2007. *The Age of Everything: How Science Explores the Past*. University of Chicago Press, Chicago.

Holmes, A. 1913. *The Age of the Earth*. Harper and Brothers, London.

Lewis, C. 2000. *The Dating Game: One Man's Search for the Age of the Earth*. Cambridge, Cambridge University Press. (Biography of Arthur Holmes)

Macdougall, D. 2008. *Nature's Clocks: How Scientists Measure the Age of Almost Everything*. University of California Press, Berkeley.

Ogg, J. G., G. Ogg, and F. M. Gradstein. 2016. *A Concise Geologic Time Scale 2016*. Elsevier, Amsterdam.

Prothero, D. R. 2018. *The Story of the Earth in 25 Rocks*. Columbia University Press, New York.

Prothero, D. R., and F. Schwab. 2013 (2nd ed.). *Sedimentary Geology: Principles of Sedimentology and Stratigraphy*. W. H. Freeman, New York.

Rutherford, E. 2004. Rutherford's Time Bomb. *The New Zealand*, May 15, 2004.

SUMMARY

- Until the mid-1600s, nearly all cultures thought the earth was very young (only a few thousand years old) and was the center of the universe (geocentrism). They had various creation myths that explained the earth's origins in ways that were meaningful for their time.

- The Copernican system in 1543 first suggested that the sun, not the earth, was the center of our solar system (heliocentrism), a fact confirmed by Galileo Galilei in 1632 and mathematically explained by the physics developed by Isaac Newton in 1687.

- Geologic events can be dated by their relative age (age in relation to something else) or by their numerical age (age in certain number of years).

- Nicholas Steno developed the first principles to determine the relative age of geologic events, including superposition, original horizontality, original continuity, and cross-cutting relationships.

- Unconformities are erosional gaps in the rock record, representing enormous amounts of time for which there is no rock record in a local section. They come in three types: angular unconformities, nonconformities, and disconformities.

- Using the principles of superposition and cross-cutting relationships and looking for evidence of events like unconformities and tilting and folding of rocks, we can reconstruct the geologic history of complex rock sequences and tell the relative age of a long sequence of events.

- Early estimates of the age of the earth were only on the order of tens to hundreds of millions of years, off by a factor of 10 to 500. These estimates were off because of faulty assumptions in the method.

- The only reliable way to get numerical dates of geologic events is by radiometric dating, using the decay of unstable radioactive elements like uranium-235 and uranium-238, rubidium-87, potassium-40, and others to estimate the age since the decay began. The process measures the ratio of radioactive parent atoms to stable daughter atoms to determine the age of a sample.

- Some of the most popular methods of dating use the slow decay of uranium-235 and uranium-238, rubidium-87, and other elements to date very old rocks greater than 500 Ma in age. Potassium–argon dating works for any geologic sample from the oldest known rocks to rocks as young as 1 Ma.

- Carbon-14 only works for very young materials less than about 60,000–80,000 years old, so it is primarily a tool of

archeologists and geologists who study the last Ice Age. However, unlike other numerical dating methods, it can be used to directly date the material, including bones, shells, wood, pottery, and anything else containing carbon.

- In most cases, we do not date sedimentary rocks directly but establish their relative age by their distinctive fossils and then compare the fossils to standard timescales. To numerically date the sediments containing fossils, we need to find igneous bodies (lava flows, ash falls, dikes) that intrude into the sedimentary sequence to calibrate their ages radiometrically.

- The oldest known earth rocks only date to 4.32 Ga, with individual mineral grains from Australia dating to 4.4 Ga. Older earth rocks have long since been destroyed by the processes of erosion and plate tectonics. However, moon rocks and meteorites give dates of the original formation of the solar system, with the oldest known age from meteorites dating to 4.56 Ga.

KEY TERMS

Geocentric system (p. 42)
Heliocentric system (p. 43)
Steno's law of superposition (p. 43)
Steno's law of original horizontality (p. 44)
Steno's law of original continuity (p. 44)

Relative age versus numerical age (p. 44)
Principle of cross-cutting relationships (p. 44)
Unconformities (p. 45)
Angular unconformity (p. 45)
Nonconformity (p. 46)
Disconformity (p. 46)
Radioactive decay (p. 49)

Parent and daughter atoms (p. 50)
Alpha, beta, and gamma particles (p. 51)
Exponential decay curve (p. 52)
Half-life (p. 52)
Mass spectrometer (p. 52)
Decay constant (p. 52)

Uranium–lead dating (p. 54)
Rubidium–strontium dating (p. 54)
Potassium–argon dating (p. 54)
Carbon-14 dating (p. 55)
Meteorites (p. 55)
Age of the earth (p. 55)

STUDY QUESTIONS

1. Summarize some of the early ideas about the age of the earth and universe. What are the assumptions of these ideas?
2. How did Steno's insights about "solids contained within solids" open the door to explaining the principles of relative age dating?
3. Describe the difference between relative dating and numerical dating and how both are used to build the geologic timescale.
4. Describe the sequence of events that produce an outcrop where a package of tilted strata is unconformably overlain by a series of layers of horizontal strata.
5. Most of the major events in the geologic timescale, such as the divisions between the Paleozoic, Mesozoic, and Cenozoic eras, coincide with mass extinctions of fossil life. Why do you think this is so?
6. Lord Kelvin calculated the age of the earth by assuming that it cooled from a solid molten sphere the same temperature as the sun and that no additional heat was added. What was wrong with his assumptions?
7. John Joly calculated the age of the earth by assuming that the oceans started out fresh; then, using the rate of salt supplied to the oceans from the land, he assumed that they reached their current salinity value of 3.5% salt just

recently. According to his calculations, the oceans were not older than 90 million years. What was wrong with his assumptions?
8. What is the current estimate of the age of the earth, and how was it determined?
9. What is radioactivity, and how is it used to determine the numerical age of a sample?
10. A sequence of Paleozoic rocks contains two lava flows, dated at 415 Ma and 420 Ma. The older flow overlies rocks with Silurian fossils, whereas the younger flow overlies rocks with Devonian fossils. What are the ages of the rocks above, between, and below the flows; and how do they constrain the age of the Silurian–Devonian boundary?
11. Why is it impossible to date a sedimentary rock by dating its individual mineral grains? What method must we use to date sedimentary rocks?
12. A science denier says, "This petrified forest in Australia gives an age of 40 Ma when you analyze it by K–Ar dating but only 60,000 years by radiocarbon. Therefore, all dating systems are unreliable, and the earth is only 6000 years old." Analyze what is wrong with this statement, assuming that the K–Ar date is reliable. (Hint: think about the effective age range of carbon-14.)

4 Stratigraphy

Stratigraphy is the backbone of the science. Without sound stratigraphy, structural studies are impossible, and all historical and most economic geology depend upon it.

—*James Gilluly, 1977*

Stratigraphy is the single great unifying agent in geology. Without it the findings of other branches could not be knit into a single historical whole. . . . Stratigraphy makes possible the synthesis of a unified geological science from its component parts. Stratigraphy is the heart of geology.

—*J. Marvin Weller, 1947*

In order to know the anatomy of each mountain range, you have to know the details of sedimentary history. To know the details of sedimentary history, you have to know stratigraphy. Many schools don't teach it any more. To me, that's writing the story without knowing the alphabet. The geological literature is a graveyard of skeletons who worked the mountain ranges without knowing the stratigraphy.

—*J. David Love, Wyoming State Geologist (in McPhee, 1986)*

William Smith's 1815 geologic map of England and Wales, which is still accurate today on the scale on which it was drawn. The colors indicate the different rock units of different ages, with the green in the lower right marking the chalk of the White Cliffs and the Salisbury Plains beneath Stonehenge, and the black showing the Carboniferous coal beds.

Sketch of the Succession of STRATA and their relative Altitudes. Nº 15.

ENGLISH CHANNEL

4.1 The Record in the Rocks

Stratigraphy literally means "the study and interpretation of layered sequences of rocks," or "strata." Normally, this means layers of ancient sedimentary rocks, but it can also apply to stacks of lava flows as well or volcanic ash layers. It was one of the first branches of geology to be developed, when people such as Steno realized that sequences of rocks give geologic history through the principle of superposition (see Chapter 3). In the 1700s, geologists (especially in England) began to search for deposits of coal and other valuable minerals. This led them to carefully study the sequence of strata across the countryside and recognize that it was a record of earth history. Thus, stratigraphy is the fundamental toolbox that geologists use to discover the history of the earth.

 See For Yourself: William Smith and the First Geologic Map

The first person to truly understand and document the sequence of rocks in detail was William Smith (1769–1839). He was not a rich British gentleman, as were most of the founders of geology at that time. They had the education, wealth, and free time needed to study geology, which was considered to be a hobby and not a profession. Instead, Smith was a humble working man, trained in surveying. In the late 1700s, he was given the assignment of planning the route of some of the great canal excavations across England. The Industrial Revolution required a cheap form of transportation for all the coal and iron ore and other commodities that had to be carried to the great industrial cities. As he surveyed and mapped the trenches dug for canals, Smith had the unusual opportunity to see the fresh bedrock exposures through much of England, which is normally covered by vegetation and difficult to map (Fig. 4.1). In the process, he saw the stratigraphic sequence of Britain repeated over and over again (Fig. 4.2), and he got to know not only the formations but also the characteristic fossils found within each unit. More importantly, he realized that similar-looking rock units could best be distinguished by their fossils. Eventually, he could tell where he was in the sequence by the fossils alone, without even seeing the rocks from which they came. This discovery came to be known as the principle of **faunal succession**, or the sequence of fossil faunas through time. The gentlemen collectors and geologists who consulted him were amazed with his ability to accurately guess where the fossils in their collections had come from or to arrange their collections in stratigraphic order.

As Smith himself wrote in 1796 (quoted in Winchester, 2001):,

> *Fossils have been long studied as great curiosities, collected with great pains, treasured up with great care at a great expense and shown and admired with as much pleasure as a child's rattle or his hobby-horse is shown and admired by himself and his playfellows— because it is pretty. And this has been done by thousands who have never paid the least regard to that wonderful order and regularity with which nature has disposed of these singular productions, and assigned to each class its peculiar stratum.*

By 1799, Smith's lists of fossils characteristic of each formation were widely circulated among British geologists, but Smith was too busy working on the first geologic map of England to publish his discovery until 1815. Meanwhile, other geologists were taking his discovery and making their own reputations from it since Smith had not published and established his priority. In addition, Smith suffered from the prejudice of the wealthy, class-conscious gentlemen geologists of the time,

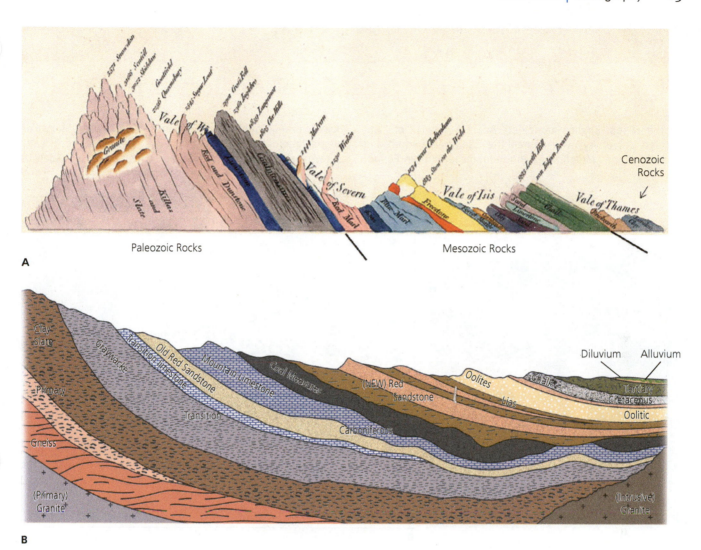

Figure 4.2   William Smith's simplified stratigraphic section across England, **(A)** as originally drafted on his 1815 map of England and **(B)** redrawn and simplified with the modern names and ages of units.

who viewed him as a lowly working man (as engineers and surveyors were then considered). Most of them considered geology a hobby, not a "vulgar" way of making a living. Smith was one of the few who might be considered a "professional" geologist in that he worked on geological problems for an income. He poured most of his earnings into his travel and mapping all over Britain and eventually got into debt when no one would buy his maps and charts because other people had stolen his ideas and published them. He even went to debtor's prison because of his financial problems.

In 1815, however, Smith published the first geologic map of England, the first true geologic map ever published. It is so accurate at the scale it was drawn that it can still be used today (Fig. 4.1). In 1831, toward the end of his life, Smith was finally given credit for all his groundbreaking work and acknowledged by the Geological Society of London as the "father of English geology." More importantly, the development of the geologic map has been a fundamental tool of all geology since then, literally what Simon Winchester calls, "the map that changed the world." Geologic maps are essential tools

for a geologist to understand not only what rocks occur at the surface but also what they might be doing in three dimensions under the surface. Nearly all branches of geology, especially economic fields like the exploration for oil, coal, gas, uranium, and many other resources, depend heavily on geologic maps to decipher the complexities of rocks beneath our feet and find where valuable deposits might occur.

Although Smith's insights about geologic sequences and the succession of fossils were crucial to the birth of modern geology, other people were also realizing their importance. In France, the great anatomist and paleontologist Georges Cuvier and the fossil shell expert Alexandre Brongniart described the sequence of strata in the Paris Basin, independently coming up with the idea of faunal succession. Some say that Brongniart may have heard about Smith's ideas when he visited England in 1806, although the evidence is not clear. Nonetheless, carefully describing and mapping the fossils and rocks and their sequence was becoming the crucial next step in the development of geology.

BOX 4.1: HOW DO WE KNOW?

How Do We Know That "Layer Cake" Geology Is Not Real?

In the 1830s, English geologists Adam Sedgwick of Cambridge University (who named the Cambrian Period) and Roderick Impey Murchison (who named the Silurian Period) were working together to solve a controversial problem. For years, geologists had recognized a distinctive rock unit, known as the "Old Red Sandstone," which crops out widely in

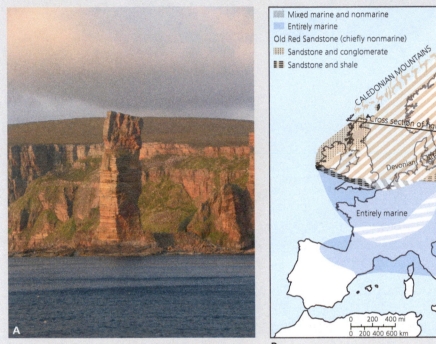

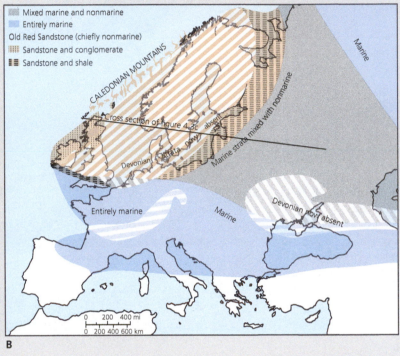

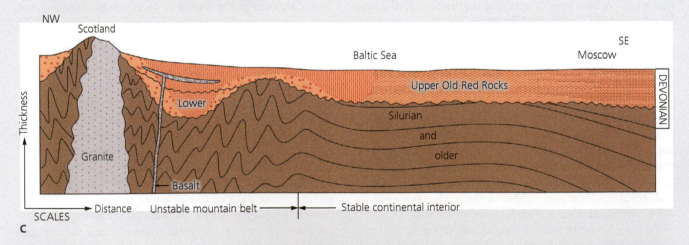

Figure 4.3 Old Red Sandstone and the Devonian. **A.** Outcrop of Old Red Sandstone, showing the distinctive red gravels and sandstones. This pinnacle is known as the "Old Man of Hoy" on the island of Hoy in the Orkney Islands of northern Scotland. **B.** Map of Europe during the Devonian, showing the different depositional environments and facies. Most of the British Isles were covered by Old Red Sandstone, shed from the Caledonian Mountains to the northwest of Norway, Scotland, and Ireland. The Old Red Sandstone outcrops are shown in the stippled pattern, while it has been restored from places where it eroded away in the diagonal line pattern. Most of the British Isles were covered with the floodplains of Old Red Sandstone, except for southwestern England (Devon and Cornwall), which were covered by marine deposits with marine fossils. **C.** Cross section showing how the sandstone-gravel facies in the Old Red Sandstone near the Caledonian Mountains changes to offshore muds to the east and south.

northern England and Scotland (Fig. 4.3A). It was deposited in ancient rivers and streams and was famous for its fossil fish.

However, when Sedgwick and Murchison got to the extreme southwest tip of England in Devon and Cornwall (Fig. 4.3B), there was no Old Red Sandstone, which was normally found between the rocks that Murchison called "Silurian" and below the "coal measures" or Carboniferous. Instead, the rocks between the Silurian and Carboniferous in Devon and Cornwall were marine sandstones and shales, loaded with marine fossils, not freshwater fish fossils. Sedgwick and Murchison correctly guessed that this mystery unit between the Silurian and Carboniferous was formed in marine environments at the same time that ancient freshwater streams were depositing the Old Red Sandstone to the north and east (Fig. 4.3B, C). In 1840, they coined the name "Devonian Period" for any rocks formed at that interval of time, whether they be the Old Red river sandstones or the marine rocks of Devon and Cornwall.

Smith was concerned with mapping units in local areas and establishing their sequence. Most geologists of his time thought that layers of rock were laid down in simple flat sheets, one after another. Some of them even tried to explain these stacks of layered rocks with the Noah's Flood story, as did the "Neptunists" discussed in Chapter 1. But by the late 1700s, it was becoming harder and harder to reconcile the biblical accounts with the real rock record. In the early 1800s, Baron Georges Cuvier proposed that all the recently discovered extinct animals, such as the marine reptiles and mammoths and mastodonts and giant ground sloths, had lived in a dark, dismal, violent "antediluvian" world not mentioned in the Bible but before Noah's Flood. "Cuvier's compromise" was an attempt to salvage the biblical accounts and please the people in power (especially the church), even though the literal interpretation of biblical stories was becoming harder and harder to defend.

A few years later, French paleontologist Alcide D'Orbigny, aware of all the different fossil faunas in each layer, proposed at least 29 separate Creation events not mentioned in the Bible. This was one of the last serious attempts to twist scripture to fit the reality of the rock record. By the 1840s, geologists (all of whom were religious men) no longer even attempted to take the Noah's ark story seriously because the actual rock record was too complex and variable to explain with a single flood.

One of the complexities was that if a single great flood had produced all the rock layers, then there should be uniform sheets of the same type of rock all over the world which had the same thickness and character. This notion of **"layer cake" geology** is simple and intuitive, but by the early 1800s geologists had found that it was also wrong. The details of the rock sequence varied from region to region, and no two areas had exactly the same sequence of rocks and fossils.

Figure 4.4 A 1789 diagram by French naturalist Antoine Lavoisier, showing the coarse nearshore sands and gravels (*Bancs Littoraux*, shown by the pebbly pattern) interfingering with offshore deep-water pelagic muds (*Bancs Pelagiens*). Lavoisier realized that sands and gravels are moved only by rivers and waves near the shore but that finer muds are carried into deeper water offshore. As the diagram shows, he realized that as sea level rose it would flood the land with offshore muddy deposits (*la Mer Montante*), and the deposits would migrate landward. Above that, he shows that as sea level fell (*la Mer descendante*), the nearshore sands and gravels would shift seaward as well.

As the discovery of the Devonian shows (Box 4.1), each formation is typically made of an assemblage of different rock types formed in different sedimentary environments at the same time. Since different rocks (like shales and sandstones and limestones) form in different sedimentary environments at the same time, *rock units are not time units*. The earliest known example of this concept was published by the great French chemist and naturalist Antoine Lavoisier in 1789 (Fig. 4.4), shortly before he lost his head to the guillotine. In 1838, the Swiss geologist Amanz Gressly documented the same phenomenon in the Jurassic of the western Alps and coined the term **facies** (Latin for "appearance") for this concept (pronounced "FACE-eez" in the United States, "FASH-eez" in Great Britain). We can talk about a "sandstone facies" or "shale facies" in some geologic settings or the "river sands facies" and "shallow marine facies" in the case of the Devonian of England.

4.2 Sedimentary Environments and Facies

In retrospect, it should not surprise us that ancient rocks of the same age change facies over distance. If we look at modern sedimentary environments (Fig. 4.5) across the surface of the earth, we can see river sands and floodplain muds being deposited in one place, while shallow marine

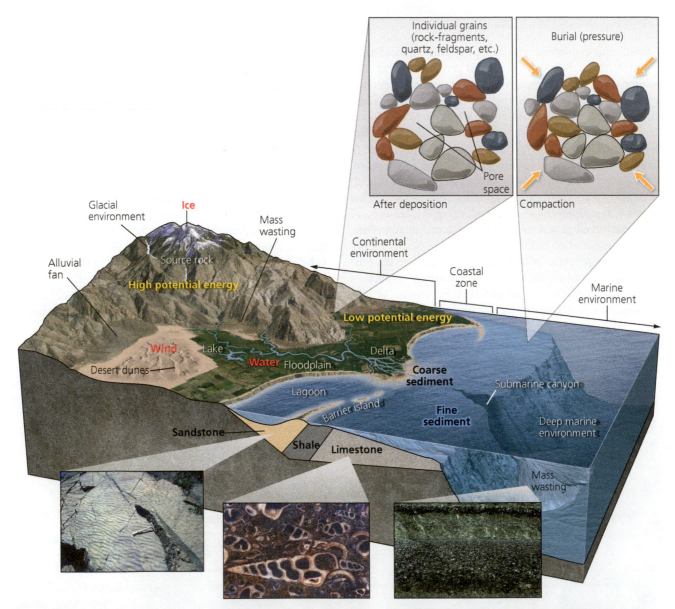

Figure 4.5 Diagram showing different sedimentary environments on the landscape. Sediments erode from the mountains and are first deposited in alluvial fans, then are picked up by rivers and flow toward the ocean. Some sediment accumulates on river floodplains or in swamps and coastal lagoons. Once the sediment reaches the mouth of a river, it can be deposited in the river delta or flow out into the sea, where longshore currents will distribute it along the beach and waves and storms will winnow the nearshore sandstones. Farther offshore, where wave energy is not felt by the sea floor, the finer-grained silts and clays settle out to make mudstones or shales. In shallow regions far offshore, limestones may grow and form reefs and carbonate sand. Some of the sand from the continental shelf is funneled down submarine canyons and ends up on the deep-sea floor by submarine mass wasting events known as turbidity currents.

sands accumulate on the beach and nearshore setting. In the quieter waters farther offshore, the mud settles out of the seawater and piles up as thick layers of mudstone (usually becoming shale when it lithifies into a rock). The sedimentary rock types form a complex mosaic pattern of different environments and facies today, and they always have. There has never been a time when the entire earth was covered by a single sheet of one kind of rock or sediment, as the old "flood geology" model once imagined. We cannot look at rock layers and think of them as representing a single event across a huge area or a single time plane.

See For Yourself: Sedimentary Environments
The study of ancient sedimentary environments and their modern analogues has been one of the most important areas of research in geology over the past several decades. Not only are sedimentary facies crucial to understanding ancient environmental conditions, but knowing what environments you are dealing with is essential to the successful exploration for oil and coal and other natural resources, so there is much economic interest in this field of research. Certain rock associations are particularly diagnostic and easy to recognize as part of a specific sedimentary environment, even for the beginning student. The following is a short and oversimplified list, but there are entire books and courses for the advanced student in sedimentary facies. Like any detective or forensic scientist on crime dramas such as *CSI: Crime Scene Investigation*, sedimentary geologists use a wide range of clues to decipher the forensic history behind a sedimentary rock. These include features only visible under a microscope or mass spectrometer or X-ray diffractometer or ion microprobe to features that can be seen in a hand sample to larger-scale structures that show up in outcrops or even over huge areas. Most people look at a sandstone or limestone and just see a rock, but a skilled geologist can decipher a complex history from the same rock, given the right tools and asking the right questions.

Alluvial fan facies (Fig. 4.6): As they are eroded and transported out of the mountains (Fig. 4.5), the first place that sediments accumulate is in **alluvial fans**, broad aprons of sediment that build up at the base of the mountains. Most of the deposits of alluvial fans arrive in intense rainstorms and flash floods or as gravity flows of liquid mud and rocks with the consistency of wet cement, known as "mudflows" or "**debris flows**." Thus, ancient alluvial fans are easy to recognize. They are dominated by coarse gravel and boulders that could never be carried far from the mountains, building up thick deposits of conglomerate, coarse angular sandstone, and sometimes breccia. The sandstones and conglomerates are often very poorly sorted, with mud mixed with boulders, indicating powerful flood conditions or flows that dumped the sediments abruptly with no winnowing or sorting. In most cases, the sandstones are full of unstable grains, such as feldspars (making an arkose) and rock fragments, which would normally break down if the sands traveled very far. The moment a geologist sees a coarse conglomerate or breccia, he or she begins to think of ancient mountains very close by because that's the only way

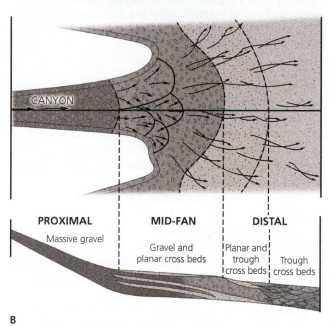

Figure 4.6 The freshly eroded sediments coming out of uplifted mountains form the alluvial fan facies, where flash floods pick up coarse sediment including gravel and even boulders, then dump them at the mouth of the canyon when the floodwaters spread out and their energy weakens. The deposits of alluvial fans are coarse gravels and sands with little or no mud. The sand is angular and full of unstable components like feldspars and rock fragments. **A.** Photograph of the giant alluvial fan coming out of the Black Mountains and Copper Canyon in Death Valley. **B.** The distribution of different sediments in a typical alluvial fan, with the coarsest gravel near the head of the fan.

to get large volumes of such coarse gravel and boulders and sand to accumulate.

Fluvial (river) facies (Fig. 4.7): As the sand and mud are winnowed out of the alluvial fans and carried downhill (Fig. 4.5), they end up in rivers that flow down to the sea. Nearer to the mountains, the rivers tend to be **braided streams**, made of many sandy (and maybe gravelly) channels that crisscross each other and change with every flood. Almost all the mud washes farther downhill, so the ancient deposits of braided streams tend to be thick sandstone bodies, with many channels cutting across one another.

Figure 4.7   Once sediments erode from the mountains and alluvial fans, they are picked up by streams and rivers. **A.** The sands are redeposited in river channels, while the muds spread across the flooplain during major floods and form thick deposits of mudstone or shale, often with terrestrial plant or animal fossils. **B.** Typical river channel sandstone in the middle of a floodplain mudstone, from the Big Badlands of South Dakota.

Farther down the river, the water flow tends to form broad floodplains with a **meandering river** channel down the middle. The meandering river channel fills with fine, well-sorted sand. Meanwhile, the floodplains adjacent to the river channel are covered with a layer of fine mud each time the river floods. As those muds dry up, they form mudcracks and sometimes have little craters on them formed by raindrop impacts on the soft mud. They are also the best place to find fossils of land life, such as dinosaurs and other land animals or land plants. Organisms that lived and died on a floodplain may be buried in the next flood before their remains can be scattered. Thus, an ancient meandering river deposit will be mostly shales (full of mudcracks) with lenses of sandstone from ancient river channels that once shifted across the floodplain.

Lakes (Fig. 4.8): Another depositional setting near a river might be an ancient lakebed. These fossil lakes are usually filled with muds that become mudstones or shales, often with fossils of freshwater organisms that once lived in the lake. Those shales might be very thick and finely bedded in the quiet waters of the lake, and some ancient lakes (such as the Eocene Green River lake shales) are so full of organic matter that they are known as "oil shales." In a few instances, ancient lakes might dry up, in which case you will find not only lots of mudcracks but evaporite minerals (halite, gypsum, and many others) that precipitated out of the briny lake water as it dried up.

Swamps (Fig. 4.9): If you visit a modern swamp, you will find stagnant muddy water filled with large volumes of decaying plant matter. The ancient equivalent would be the major coal deposits of the world, formed in ancient swamps when huge amounts of dead plants fell in the water and accumulated organic material that became coal. When geologists see a coal seam, they almost immediately think of stagnant water and swampy conditions where these rocks form.

Desert dunes (Fig. 4.10): Walk across a modern sand dune, and you will find nothing but extremely well-sorted, well-rounded fine sand grains. As the wind blows the sand, it will flow up over the crest of a dune, then avalanche down the back side where the wind currents are weak. Those tilted

layers of sand that form on the back of the sand dune form cross-bedding when the dune turns into a sandstone (Fig. 4.10B). Between these thick sandstone bodies with prominent cross-beds, you might find small mudflats in the interdune deposits or even ponds and dry lakes with evaporites.

Beach and nearshore sands (Fig. 4.11): Once the sands of the rivers reach the ocean, they are distributed along the coastline and the beaches by longshore currents. This forms a broad zone of fine, well-sorted sands not only on the beach itself but in the shallow water offshore where the strong waves and currents winnow the sand and remove the mud. These sands can be covered with ripple marks formed by tides or waves offshore, as well as cross-bedding from wave and tidal action. In the area beneath the wave action, the sandstone will be full of marine fossils of organisms that burrow in shallow marine sands or live on the sea floor, such as marine snails, clams, heart urchins, and other invertebrates.

Offshore muds: As you get far enough away from land, the water currents become weak enough that the mud carried in suspension from the rivers can finally settle out on the sea floor, making mudstones. Typically, they are far enough offshore that they are too deep to feel the effects of waves, even in the biggest storms. These offshore mudstones (typically turned into shales) will show lots of fine bedding and typically are full of marine fossils of creatures that live in deep, muddy waters (Fig. 4.11C). Such muddy sea floors are found all the way from the outer continental shelf to the deepest part of the oceans, although there will be different fossils in different water depths.

Limestones: As we mentioned in Chapter 2, limestones are the products of organisms that secrete calcite as part of their shells, from certain types of algae to corals to snails and clams and hundreds of other invertebrates with calcite shells. Most limestone forms in water that is very shallow but isolated from the sand and mud flowing out of the rivers on land since sand and mud interfere with these organisms and their shell growth. Today, most limestones are also restricted to clear, warm, agitated, sunny tropical waters, such

Figure 4.8 **A.** Map showing the distribution of the great Middle Eocene lakes of the Green River Formation in Wyoming, Utah, and Colorado. **B.** Diagram of the deposits forming in the Eocene Green River lake system in the Uinta Basin of Utah and the Piceance Basin of Colorado, showing alluvial fans at the edges, finely laminated organic-rich shales in the middle, and even some evaporite minerals when the lake dried out. **C.** The Green River Shale near Hellhole Canyon, Utah, consists of a stack of thousands of feet of finely laminated shale.

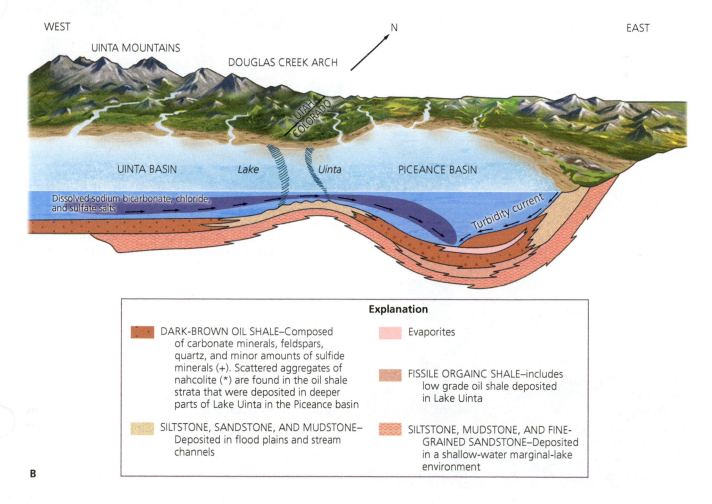

Explanation

DARK-BROWN OIL SHALE—Composed of carbonate minerals, feldspars, quartz, and minor amounts of sulfide minerals (+). Scattered aggregates of nahcolite (*) are found in the oil shale strata that were deposited in deeper parts of Lake Uinta in the Piceance basin

SILTSTONE, SANDSTONE, AND MUDSTONE— Deposited in flood plains and stream channels

Evaporites

FISSILE ORGAINC SHALE—includes low grade oil shale deposited in Lake Uinta

SILTSTONE, MUDSTONE, AND FINE-GRAINED SANDSTONE—Deposited in a shallow-water marginal-lake environment

Figure 4.9 **A.** Marshes and swamps grow huge amounts of plant material that falls in the stagnant water and does not decay but hardens into the rock we know as coal. This is the modern Wolf River swamp near Ashland, Mississippi. **B.** The black coal seams stand out among the tan river sandstones that surround it in the Cretaceous Ferron Sandstone near Price, Utah.

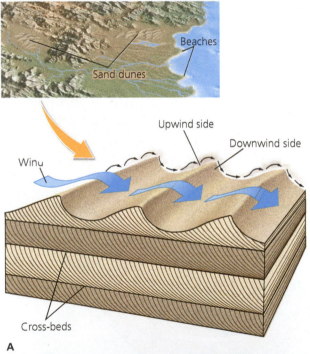

Figure 4.10 The most distinctive deposits of ancient deserts are giant cross-bedded dune sandstones. **A.** Diagram of cross-bedded dune sands, showing how each migrating dune forms a sloping sand deposit on the protected downwind side of the dune crest. These structures are known as cross-beds or cross-stratification. **B.** Photograph of a modern dune field in the Sahara Desert. **C.** Photograph of the huge cross-beds in the Lower Jurassic Navajo Sandstone, Zion National Park, Utah. Each cross-bed set is many meters thick. Notice the small trees near the bottom for scale.

as the Caribbean, the South Pacific, and a few other places, like the Persian Gulf. When we see huge volumes of ancient limestone, we usually interpret it as evidence of shallow carbonate shoals far enough offshore to escape the influx of sand and mud. In other cases, huge volumes of fossiliferous marine limestones covered much of North America, showing that shallow tropical seas covered most of the continent and that there were few mountains or land areas that would shed sands or muds to interfere with them.

Deep marine deposits (Fig. 4.12): In the deeper part of the ocean, the main form of sediment is fine-grained

Figure 4.11 **A.** Low-angle aerial photograph of typical beach depositional setting, with a marshy lagoon behind it and clean well-sorted sands winnowed by waves in the nearshore environment. **B.** These cliffs near Ferron, Utah, show ancient Cretaceous beach sandstones (tan units) alternating back and forth and interbedded with a sequence of gray offshore shales of the Mancos Formation.

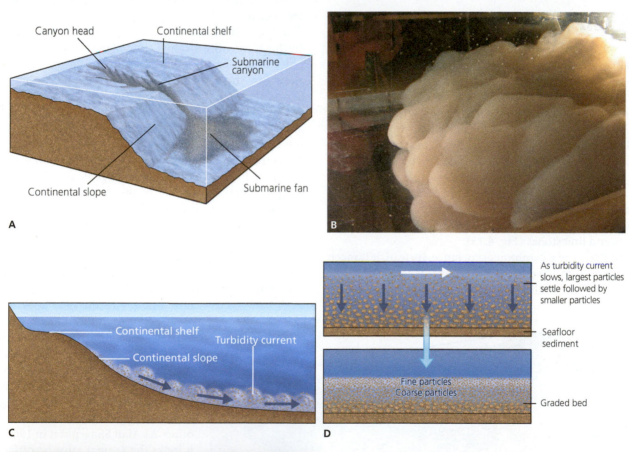

Figure 4.12 **A–E.** The offshore region of the continental shelf–slope–rise (top) is dominated by huge gravity slides of sand and mud **(A–C),** which funnel down through submarine canyons and then reach the submarine fan. There, turbidity currents **(B)** deposit thick sequences of graded beds **(D–E),** which coarsen upward from bottom (sandy) to top (muddy). **(D–E)** The graded beds are formed when the turbidity current slows down and stops, and the heaviest, coarsest sand and gravel settles out first, followed by the finer sands and then muds.

muds from the rivers on land, which settle out of the sea-water very slowly and float to the sea floor. In contrast to mudcracked lake shales, this succession of muds (turned to shales in ancient rocks) can be recognized by the presence of deep-marine fossils and burrows and trackways. In some places, there may be thick layers of sand formed when submarine gravity slides (**turbidity currents**) brought sand and fossils from the shallow marine continental shelf. As these turbidity currents race down the steep continental slope and come to rest on the continental rise and deep-sea floor, the turbulent suspension of sand and mud stops moving, allowing the coarser sand to settle out first and the mud to settle out last. This sequence of coarse-to-fine sorting in a single bed is called "graded bedding," and it is characteristic of **turbidite** deposits in the deep oceans through all of geologic time.

4.3   Transgression and Regression

Let's now focus on a typical association of shallow marine facies: nearshore sands on the beach or just offshore, where wave and current energies are strong. The deeper-water shales form farther offshore, where the water is too deep to feel waves or currents. Farthest offshore, shallow shoals and banks allow limestones to grow in the sunlight (Fig. 4.13). If sea level held constant for a long period of time, the sandstones would pile up continuously in the nearshore area and the shales farther offshore, and the limestones would accumulate offshore where the water remained shallow and clear and tropical. Thus, the sandstone formations would form vertical piles, as would the shales and limestones (Fig. 4.13).

But that's not a common occurrence. Instead, it is much more typical to have a relative change in sea level in any given area over millions of years. When rising seas begin to flood the land and drown the floodplains, this is known as a "**transgression.**" If you look at the coastline in cross section (Fig. 4.14A), you will see the gravels form near the uplands, the sands near the coast, and the muds and limestones farther and farther offshore. As the local sea level rises, the mudstones and limestones will lap over the older sandstones as they flood the former land surface. Eventually, as the sea-level rise reaches its maximum, the limestones and muds will cover most of the former land surface, and their deposits will sit on top of the older gravels and sands.

 See For Yourself: Transgression and Regression

Note something about the diagram in Fig. 4.14A. As the offshore deposits build onto the land, they cover up the older deposits so that in any one vertical sequence, the coarsest sands and gravels will lie at the bottom and the fine muds and limestones will lie on top. This *fining-upward sequence* of sediments is typical of transgressions, no matter where they occur.

Figure 4.14A also demonstrates another key concept. In this figure, the original time planes (representing the old sea surface from the beach to the deep water) are shown by the dashed lines. If you trace these timelines across the rock units (the sandstones, shales, and limestones), you will notice that the rock units cut across the ancient time surfaces. In other words, the rock units are not the same age from one place to another but are **time-transgressive**, changing age as they move landward or seaward. In the past few decades, geologists have come to appreciate how important this is and how difficult it is to recognize in outcrop. If you trace a rock formation across any distance, it is almost certainly different in age. ***Rock units are not time units***. You cannot talk about "the time of the Tapeats Sandstone" or any other rock unit since *rocks are inherently time-transgressive*. No matter how much they look like simple layers laid down all at one time (as in the old "layer cake" view of stratigraphy), that appearance is deceiving. Both theoretical constraints and close examination of real examples show that most rock units (at least in the marine setting) are inherently time-transgressive and differ in age from one place to another. As Alan Shaw put it in 1964, "if it looks the same it must be different in age."

An excellent example occurs with the Cambrian sequence at the bottom of the Grand Canyon. To the casual hiker on the trails or boater in the river the Tapeats Sandstone can be traced for miles along the bottom of the canyon, as can the Bright Angel Shale above it and the Muav Limestone that caps the Bright Angel Shale (Fig. 4.15). But

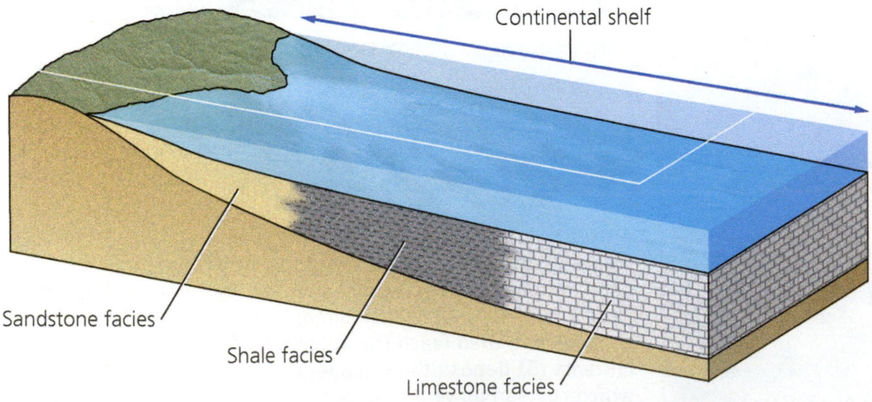

Figure 4.13 The nearshore region is dominated by coarse sandy deposits of the beaches and shallow nearshore region where wave energies are high; this makes the sandstone facies. Farther offshore, the waves are not felt on the bottom, so the muddy deposits settle out, making a shale facies. If the water remains shallow enough but is far from any sources of mud or sand, then a shallow-water carbonate shoal is formed, forming a limestone facies. Facies belts would pile up vertically if relative sea level never changed.

Continental shelf

Sandstone facies

Shale facies

Limestone facies

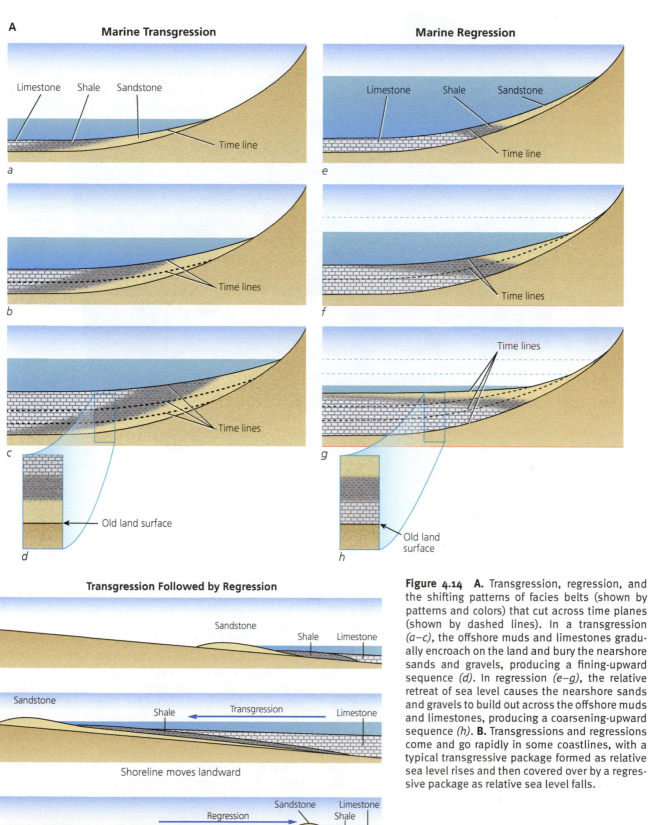

Figure 4.14 **A.** Transgression, regression, and the shifting patterns of facies belts (shown by patterns and colors) that cut across time planes (shown by dashed lines). In a transgression (a–c), the offshore muds and limestones gradually encroach on the land and bury the nearshore sands and gravels, producing a fining-upward sequence (d). In regression (e–g), the relative retreat of sea level causes the nearshore sands and gravels to build out across the offshore muds and limestones, producing a coarsening-upward sequence (h). **B.** Transgressions and regressions come and go rapidly in some coastlines, with a typical transgressive package formed as relative sea level rises and then covered over by a regressive package as relative sea level falls.

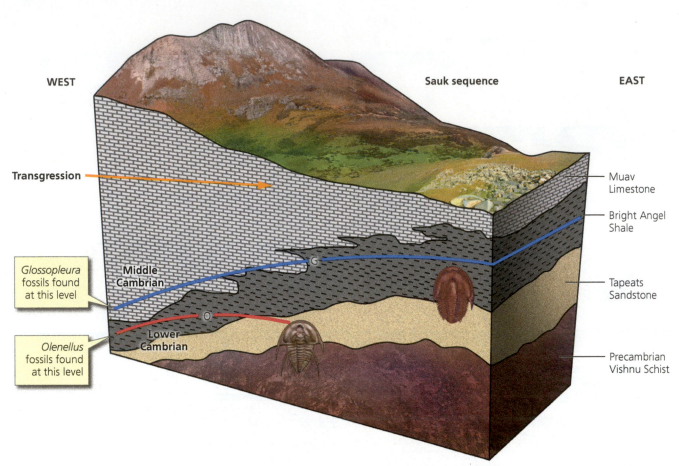

WEST   Sauk sequence   EAST

Transgression

Muav
Limestone

Bright Angel
Shale

Glossopleura
fossils found
at this level

**Middle
Cambrian**

G

Tapeats
Sandstone

Olenellus
fossils found
at this level

O

**Lower
Cambrian**

Precambrian
Vishnu Schist

Figure 4.15   Time-transgressive Cambrian units in the Grand Canyon. The time planes are marked by the first appearance of key trilobites *Olenellus* (level O) and *Glossopleura* (level G).

detailed studies of the fossil trilobites show that each of these units is actually time-transgressive over millions of years as you go from the eastern Grand Canyon (near where the overlooks are) to the western Grand Canyon (far from any paved road). The time planes (established by the fossil trilobites) cut across the formations so that the Tapeats Sandstone in the east is actually millions of years younger than it is in the west; it is the time equivalent of the overlying Bright Angel Shale in the western part of the Grand Canyon (Fig. 4.15).

Or take as another example the sediments off the shores of the Netherlands, formed when sea level rose during the melting of the glaciers of the last Ice Age (Fig. 4.16). The pattern of sands and muds in Fig. 4.16 would become sandstones and shales in the ancient rock record. But we can find time planes in these ancient beds by doing radiocarbon dating of shells at different levels in this deposit. Sure enough, the timeline for 1600 C.E. cuts through deep-water muds offshore, but it cuts through nearshore sands as you approach the land, as do all the other radiocarbon-dated time horizons.

Transgressions, then, are the product of a *relative rise in sea level*. They show a fining-upward sequence of beds in any one vertical section, and we know that each rock

unit within a transgressive wedge also transgresses time (Fig. 4.14A). What could cause such a relative rise of sea level in a local region? The obvious answer would be a global (**eustatic**) rise in sea level, which could be due to a number of factors. The most common force that has changed sea level in the past 33 million years is the melting of glacial ice, especially in the polar ice caps. This occurred rapidly as the Ice Ages ended 18,000 years ago until about 10,000 years ago and is now accelerating due to glaciers melting from global warming. But it is not the only possible factor. Another force that could flood a local coastline would be tectonic movements which cause the coastal crust to sink down (**subsidence**). In the real world, it is not often easy to decide whether a local transgression is due to eustatic changes or local subsidence or both (which can occur at the same time). The best way to separate the two factors is to compare transgression in your local section to other areas at the same time. If they are all undergoing transgression simultaneously, then it is probably due to a global cause like eustatic sea-level change. If they are doing different things, then local tectonics are probably more important.

Let's consider the opposite scenario: a relative fall in sea level, or **regression** (Fig. 4.14B). In this instance, the seas

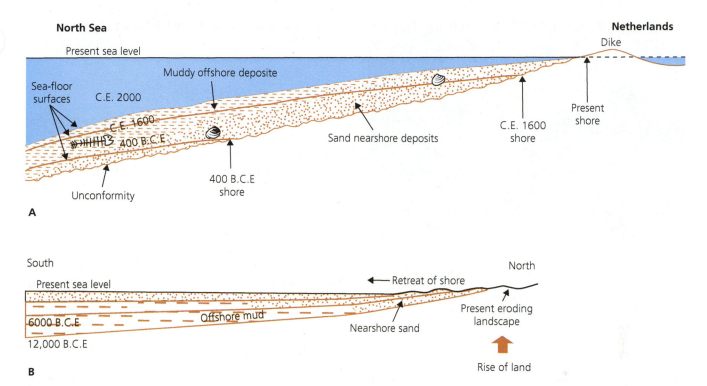

Figure 4.16 A, B. Modern examples of (**A**) transgression and (**B**) regression, showing the time planes cutting through the facies. Figure 4.16A is a cross-section of the post-glacial rise of sea level across the Netherlands. Note the landward shift of the facies, and the shallow-marine molluscs (shell symbol) that move landward with their habitat. Radiocarbon dating establishes time planes (solid orange lines with dates on them) that cut through the facies units of sands and muds, so the rock units are time-transgressive. In Figure 4.16B, the sea level has retreated from the northern Baltic Sea in historic times due to the rapid isostatic uplift of Scandinavia. The facies belts have shifted offshore over time, as shown by the radiocarbon-dated time planes (solid lines with dates) going back over 14,000 years ago.

retreat from the land and even from the shallow shelf so that nearshore and beach sands can build out over areas that were once deep-marine seafloor muds. Once again, we have the rock formations cutting across ancient surfaces and the timelines they represent, *so rock units are not time units; they are time-transgressive.* As is the case of transgression, if we look at a local stratigraphic column, we find a trend; but here the coarse sands and even gravels build up across the top of the offshore muds, so the trend is *coarsening-upward in a regression.*

What could be the causes of a regression? They would again include eustatic sea-level change, but this time it would be a global drop in sea level. Over the past 33 million years, the main cause for global sea-level drops has been the expansion of ice caps, which pull water out of the ocean basins; but there can be other causes as well. The second possible cause would be the opposite of subsidence, which would be local uplift of the coast, pushing the seas back as the land rises out of the water. A third possiblity is the growth of river deltas, with their huge volumes of sandy and muddy sediment pushing the sea back from the land. And as in the case of transgression, we cannot always be sure which factors were most important in any given area, but if the regressions are global and simultaneous, then it is probably due to eustatic sea-level change; if not, it could be due

to more local factors like tectonic uplift or deltas building out from the land.

4.4 Geologically Instantaneous Events

Careful analysis of the factors that control the distribution of shallow marine rocks that we have considered teaches us several important lessons: even though continuous layers of sandstone and shale look like they should be "layer cake" deposits of the same age, they are nearly always time-transgressive over distance. If they look the same, they are *not* the same age! Rock units are not time units, and we cannot talk about rock units as if they represented a time interval.

Are there any exceptions to the general rule that most sedimentary formations (especially those in a marine setting) are time-transgressive? In other words, are there rock types that could form very rapidly (in days or hours or weeks, rapid by geologic standards) across large areas and create "geologically instantaneous deposits" that are all synchronous?

Indeed, there are some, but they are relatively rare and not found in most parts of the world. The best known

Figure 4.17 Exposures of the Deicke bentonite (deeply eroded layer at eye level between the hard vertical faces of resistant limestones), here in a roadcut near Carthage, Tennessee. It was once a huge ashfall that covered much of the Appalachian region during the Ordovician, then the volcanic ash turned into bentonite during underwater weathering.

are volcanic eruptions, especially giant ash clouds that can cover huge areas with a blanket of ash in a matter of hours to days (Fig. 4.17). If the ash falls in the ocean, it sometimes weathers to a distinctive type of clay known as a **bentonite**. In many cases, distinctive volcanic ash layers not only represent time planes but can be "fingerprinted" by their distinctive geochemistry and correlated over large distances, providing excellent time control over an area.

Another possibility is a huge landslide, which takes only minutes or hours to occur, although such deposits are usually only distributed over a limited area. In the deep ocean, the huge submarine gravity slides known as turbidity currents (Fig. 4.12) bring shallow marine sands down to the deep ocean and deposit them in broad sheets in a matter of hours. Again, however, these do not extend over a large area, and often there are so many of them and they look so similar that they are not much use in correlation.

Even rarer are extraordinary events, such as the impact of an asteroid or comet from space. For example, the collision of a large asteroid in the Yucatán Peninsula at the end of the Cretaceous Period produced a layer of the rare element iridium, plus droplets of melted crustal material, that can be traced across the world and marks a unique event in time that only lasted hours to days. But there are only a few such events known in earth history, so they are not found in most geological settings. In places like the center of the North American continent, there are extensive sheets of sandstone, shale, and limestone that formed during the Paleozoic. Yet there were no volcanoes nearby or gravity flows on those shallow sea floors and certainly no impact events. So none of the geological events we just

listed are available. How do we tell time and date rocks in such settings?

4.5 Biostratigraphy

The answer goes back to William Smith, who realized that no matter what formation or facies he was looking at, the fossils provided the best means of correlation and determining age. In fact, in most geological settings (except for the rare instance where we have datable volcanic layers), *fossils are the best means of telling time in geology*. There will always be a need for paleontologists since they are the only ones who can tell other geologists how old most rocks are. For decades, thousands of paleontologists were employed by oil companies and other industries because they could tell the age of a small sample from a drill core just using the biostratigraphy of the microfossils. In marine geology, micropaleontologists are essential to date deep-sea cores, which in turn have unlocked the secrets of the past 100 million years of climate history and oceanographic changes.

Although Smith simply noted the presence of specific fossils in each formation in southern England, today biostratigraphy can be much more precise. The first step is the collection of biostratigraphic data. Every time a geologist or paleontologist picks up a fossil in the field, he or she *must record the exact stratigraphic position in a measured section at the time of collection*. If the collector walks away without recording this information, it may be impossible to reconstruct it later, and the specimen will be useless for biostratigraphy. For a thorough biostratigraphic study, a large collection of fossils is required, each with detailed stratigraphic data as to exactly where it was collected in a local section.

Once the fossils have been collected, they have to be cleaned up and identified. This is not as trivial a task as it sounds. It is often very difficult to correctly identify a fossil to the species level. Sometimes the distinctions between one fossil species and another are so subtle that only a specialist can tell them apart. In other cases, the group may not have been studied in years, so there is no clear understanding of which species are valid or during which geologic time interval they occurred. In many cases, identification is hampered by poor preservation. Incorrect identification can lead to incorrect correlations that may be difficult to detect by non-specialists for years.

Once all the fossils have been identified, they can be plotted on the stratigraphic section, showing the exact level where each fossil came from (Fig. 4.18). As Alan Shaw (1964) pointed out, each **range zone** of a fossil species divides all of geologic time into three parts: the time before it evolved, the time when it existed, and the time since its extinction. We can plot the partial range zones of all the species in a local section (Figs. 4.18, 4.19). It is immediately apparent that the more fossil range zones you have, the finer you can subdivide the time represented by that section. Most often, we use the overlap of two species in time (the **overlapping range zone**) to define the smallest discrete time intervals. This gives the highest possible detail, precision, and resolution (Fig. 4.19). This method of overlapping range zones was pioneered by the Swiss paleontologist Albert Oppel in 1856, using the ranges of Jurassic ammonites in the western Alps to finely subdivide the Jurassic into many small range zones defined by particular ammonite species.

Once we have a partial range zone for our local section, we can compare our biostratigraphy to that from other sections in nearby areas and possibly with other sections around the world. Typically, we will find a good match for most of our range zones, which then allows us to **correlate** sections over long distances. Of course, not every section has an identical assemblage of fossils or an identical sequence of their occurrences. The best method uses overlapping range zones or the rock interval between two biostratigraphic horizons, such as the first appearance of fossil A or the last appearance of fossil B.

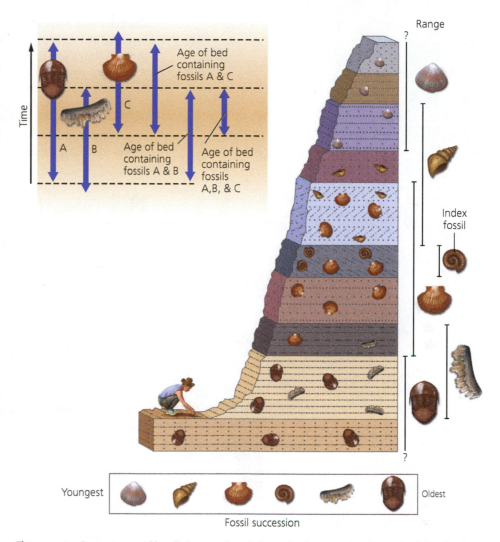

Figure 4.18   Range zones of fossils in a section. Paleontologists can plot the vertical distribution in the rock section of different distinctive fossils, which creates biostratigraphic zones in each section.

Fossils may be missing for many reasons, such as poor preservation or it was the wrong environment for the organisms to live in or an unconformity has wiped out part of the time record in a local section. Nevertheless, biostratigraphers have developed a number of rigorous methods of comparing the ranges in different sections and filtering out the noise and missing data to find the underlying signal of how and when different fossils changed through time. It has worked so well for over 200 years now that all of geology depends upon it, and marine geology, paleoceanography, and oil exploration would not be possible without it.

Not all fossils work equally well for biostratigraphy, of course. Some organisms evolve very slowly or hardly at all over millions of years, so you can't tell much from looking at their fossils. Others are highly restricted to certain environmental settings and conditions and will not occur anywhere else, so when they show up, it's because their favored environment was present, and when they disappear, they may be hanging on somewhere else in the world where their preferred habitat still lingers. The best fossils are known as **index fossils** because they have proven to be the best time indicators. Index fossils tend to have several properties in common: 1) they evolve rapidly so that you can finely subdivide time by the rapid changes in species, 2) they are widespread so that you can reliably find them in many places, and 3) they are abundant so that you can usually expect a few of them any time you have rocks of the right age.

For marine rocks, the best index fossils tend to be planktonic because they float over huge areas of the oceans no matter what kind of sea floor lies beneath them. Thus, they are found in sediments of any water depth because they sink down from the surface when

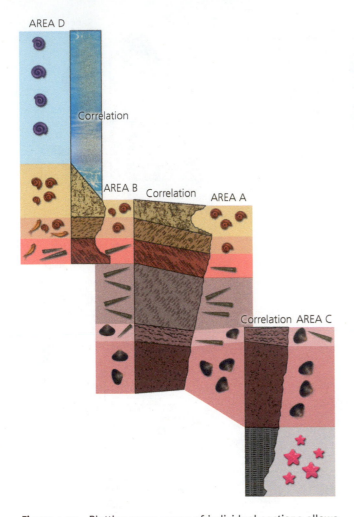

Figure 4.19 Plotting range zones of individual sections allows us to correlate them across distance and piece together a more complete composite section. In this diagram, the sequence of blue oysters, green belemnites, and red snails allows a straightforward match between area **A** and area **B**. In area **C**, however, the green belemnites and the blue oysters are found at the top of the column, so that part of column C is correlated with the bottom of columns **A** and **B**. This establishes that the range zone with the pink sea star fossils must be older than the zone with the blue oysters and gives us a longer composite sequence. Finally, area **D** has a different fossil sequence. The green belemnites and red snails allow us to match the bottom of column **D** with the top of columns **A** and **B**. The presence of the yellow fish fossils in column **D** suggests there may be a range zone missing between the red snails and the green belemnites in columns **A** and **B**. Finally, the blue ammonite zone appears above the red snail zone, giving an even longer composite sequence. Thus, the overall sequence of fossil and time goes from the oldest (red sea star fossils) to the blue oyster zone, the green belemnite zone, the yellow fish fossil zone, the red snail zone, and finally the blue ammonite zone.

they die. Most often these plankton are microscopic algae and amoeba-like creatures that have tiny shells of calcite or silica. These are the basis for most biostratigraphy in marine rocks. The other groups that work well are swimmers (nektonic organisms) such as ammonites, the best index fossils from the Devonian to the end of the Cretaceous. Back in the Cambrian, however, bottom-dwellers

like trilobites were the only common fossils, so they are essential to biostratigraphy in that time interval. On land, the best index fossils are fossil mammals in the Cenozoic, which also evolved rapidly and left an excellent, detailed biostratigraphic record.

The fossil record, then, gives us the best tool for recognizing the relative age of rocks and for correlating them across large regions. Finely divided fossil range zones are the foundation for telling geologic time.

4.6 Time, Time-Rock, and Rock Units

The fact that rock units are time-transgressive and cannot be equated with time units has led to a set of different classification schemes for time units and rock units. Each of them is hierarchical, with smaller units clustered into larger units, which are clustered into still larger units.

The basic foundation of all rock stratigraphy (**lithostratigraphy**) is the **formation**. A formation is a rock unit defined by two main criteria: 1) it must be mappable on some scale and 2) it should have a distinctive rock type or cluster of rock types (**lithology**). Of course, these criteria are very flexible. For example, in places like Minnesota or Illinois, some formations are only a few feet thick, yet they can be mapped over large distances. In others, such as the Canadian Arctic, the formations are many miles thick and are mostly mapped on aerial photos and satellite images since there is too much ground to cover and there are too few days of good weather to walk around and map for long.

A formation must be formally named in the published geologic literature, and the name must be based on a nearby geographical feature. Some formations are just called "formations," but many are given the name of their dominant lithology. For example, the Tapeats Sandstone, Bright Angel Shale, and Muav Limestone are all formations within the Grand Canyon (Fig. 3.6) but take the name of their major rock type. They could just as easily be called the Tapeats, Bright Angel, and Muav formations.

Formations are then subdivided into smaller units called "**members**," which are also formally named. Members can be subdivided into even smaller informal units called "**beds**," although these are not required. Formations can be clustered into larger units called "**groups**," and in a few cases groups are clustered into even larger units called "**supergroups**" (not in the sense of a legendary rock band but of a geologic unit).

For example, in the Big Badlands of South Dakota (Fig 14.20), the White River Group is subdivided into the Chadron and Brule Formations. The Brule Formation has two members, the lower Scenic Member and, above it, the Poleslide Member. Some geologists also recognize informal beds within these members, named after characteristic fossils, such as the *Metamynodon* sandstones in the Scenic Member (named after its fossils of a hippo-like rhinoceros relative) or the *Protoceras* channels in the Poleslide Member (named after the fossils of a weird six-horned animal distantly related to camels). In the Montana Rockies, there are many thick formations, which are clustered into even larger groups, all of which clustered into the Belt Supergroup (see Chapter 8). Rather than list all the formations, it is more convenient to refer to the Belt Supergroup because all of these rocks had a common history.

Time units (**geochronologic units**) have a similar hierarchy of smaller units clustered into larger units (Fig. 4.20). The smallest unit is the **age**, which is clustered into still larger units called **epochs**. Epochs are clustered into **periods**, which are then clumped into **eras** and then into the largest time unit, the **eon**. For example, the Rupelian Age is the first part of the Oligocene Epoch. The Oligocene Epoch is a subdivision of the Tertiary Period (although some geologists now prefer the Paleogene Period), which is part of the Cenozoic Era, which is in turn part of the Phanerozoic Eon.

But geologic time is an abstract concept like the ticking of a clock; there is no physical entity on earth known as "time" that you can touch or see. Yet for practical reasons, we need to recognize geologic time in actual rocks on earth. Thus, stratigraphers use a hybrid unit called the time-rock unit, or **chronostratigraphic** unit, which combines features of both. A time-rock unit is comprised of all the rock units that were formed during a given time interval. The top and bottom of a time-rock unit are marked not by the changes in rock type that define formations but by events that represent timelines. These are typically the boundaries of biostratigraphic zones, which are the closest things we have to time planes in rocks, or occasionally other large-scale geologically instantaneous horizons, such as impact layers or giant volcanic ash layers. The smallest unit in the time-rock hierarchy is the biostratigraphic zone, which is the building block of all time-rock units. Zones are clustered into larger units called "stages." For example, in tropical marine deep-sea cores and rocks, the *Globorotalia opima opima* Range Zone (based on the occurrence of the fossils of this particular subspecies of a planktonic amoeba-like group known as foraminiferans) is one of the subdivisions of the Rupelian Stage, which are all the rocks deposited during the time

interval known as the Rupelian Age. The stage is a subdivision of a larger unit known as the "series," so the Oligocene Series (a time-rock unit) is based on all the rocks deposited during the Oligocene Epoch (a time unit). The next largest unit is the system, so the Paleogene System is based on all the rocks deposited during the Paleogene Period. Moving up one level, the rocks deposited during an era are known as an "erathem," so the Cenozoic Erathem is based on all the rocks formed during the Cenozoic Era. Finally, one could talk about the eonothem as all the rocks formed during an eon, so there is a Phanerozoic Eonothem corresponding to the Phanerozoic Eon. However, such a broad concept is not widely used because it is not very practical or commonly encountered.

So we have a time hierarchy, and a time-rock hierarchy, where the unit names correspond one to one: eon and eonothem, era and erathem, period and system, epoch and series, age and stage. We also have an independent hierarchy of rock units: member, formation, group, and supergroup. But since rock units are not time units, one of the common ways to display these concepts is shown in Fig. 4.20. The time and time-rock units match, line by line. But the hierarchy of rock units is usually drawn perpendicular to the line of time for time and time-rock units so that no one accidentally infers that a formation is equivalent to a stage or a series or a system.

4.7 The Geologic Timescale

Using superposition and careful mapping, the early English geologists like William Smith, Adam Sedgwick, and Roderick Murchison not only were able to work out the sequence of rock units of England but in the process began to name time-rock units based on these rock units. As early as 1795, the great German explorer Alexander von Humboldt named the "Jurassic" for the ammonite-rich marine limestones and shales of the Jura Mountains in the western Alps. Other terms were added bit by bit as they became useful. We saw how Sedgwick and Murchison realized that the Old Red Sandstone was a local rock unit in northern and eastern Great Britain but changed facies to marine rocks in southwestern England, so it required a time-rock term, the "Devonian." Other units were built from characteristic rock units as well, so the "coal measures" of England became the "Carboniferous" (meaning "coal-bearing" in Latin). The characteristic chalk beds of southeastern England led to the naming of a bigger time-rock unit, the "Cretaceous" (from the Latin *creta*, "chalk"). Each of the other units of the standard geologic time scale (Fig. 4.21) has its own interesting history.

When beginning geology students first look at the timescale, it seems daunting to remember the order of all those unfamiliar terms. But anyone who wants to go beyond the first course in geology needs to know the timescale by memory because it is the standard tool of all geologists. Even though geologists must communicate

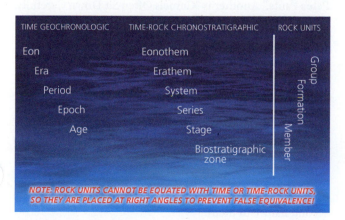

Figure 4.20 The hierarchies of time, time-rock, and rock units.

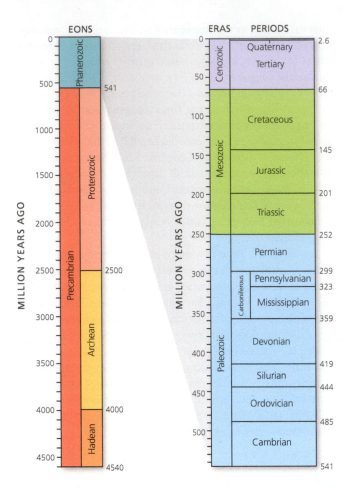

Figure 4.21   The standard geologic timescale.

Fig. 4.22   A cartoon of one of many mnemonics to help remember the sequence of ages in the geologic timescale.

to reporters and non-geologists using "millions of years ago" rather than the proper geologic time term, it's not practical to talk about events as "millions of years ago" when talking to other geologists because the timescale and its numerical calibration are constantly changing as newer and better dates are discovered. Likewise, as we go forward in this book, we must use the standard geologic timescale and assume the reader knows where it fits in the sequence and only rarely mention the numerical age of geologic events.

The geologic timescale grew haphazardly, with different units named by different geologists, so it does not follow a simple logical sequence or allow for simple shortcuts. One way to think about it is to compare the geologic timescale to other chronologies based on events, rather than dates. For example, anyone talking about English history *could* mention the exact year of a historical event (if it is known), but normally it is more practical to refer to longer periods of time by the name of the English monarch at that time. Hence, "Elizabethan," "Georgian," "Victorian," "Edwardian," and so on are more useful to someone discussing British history than an exact year. Nor is it an impossible task to memorize. Most students have taken a foreign language and memorize as many vocabulary words in a single lesson as in the entire geologic timescale.

Some students find it helpful to use a mnemonic device, or memory aid, to remember the order of the sequence of geologic periods. For example, to remember "**C**ambrian **O**rdovician **S**ilurian **D**evonian **M**ississippian **P**ennsylvanian **P**ermian," generations of students have used ancient (and silly) mnemonics like "**C**arl's **O**ld **S**hirt **D**oesn't **M**atch **P**ete's **P**ants." Another is "**C**amels **O**ften **S**it **D**own **C**arefully, **P**erhaps **T**heir **J**oints **C**reak," adding the **T**riassic **J**urassic **C**retaceous to the Paleozoic periods. The famous paleo-artist Ray Troll drew a hilarious cartoon (Fig 4.22) for the mnemonic "**C**rusty **O**ld **S**our **D**oughs **M**ake **P**erfect **P**ancakes, **T**oast, **J**uice & **C**offee" (which covers the sequence of periods of the Paleozoic and Mesozoic). Another of his cartoons is built around the mnemonic "Crying Over Sleeping Dragons May Puzzle People, Terrify, or Joyfully Convert". You are welcome to make up any device that helps you remember the timescale, so long as it helps you keep things in order. However, most geologists use the timescale so much in their daily activities that it becomes second nature to them.

4.8 Conclusion

We have come a long way in the two centuries since the days of William Smith's 1815 map. Stratigraphic theory has developed many important concepts, from the importance of sedimentary environments and facies, to the non-equivalence of time and rock units, to the rigorous application of biostratigraphy to tell time in rocks. Some of these concepts may be challenging at first, but if you study them carefully and try to grasp their meaning, soon you will be able to think like a geologist and see the rocks and fossils for what they really reveal.

RESOURCES

BOOKS

Berry, W. B. N. 1968. *Growth of a Prehistoric Time Scale*. W. H. Freeman, San Francisco.

Cutler, Alan. 2003. *The Seashell on the Mountaintop: A Story of Science, Sainthood, and the Humble Genius Who Discovered a New History of the Earth*. Dutton, New York.

Gilluly, James. 1977. American geology since 1910—a personal appraisal. *Annual Reviews of Earth and Planetary Sciences* 5: 1–12.

Greene, M. T. 1982. *Geology in the Nineteenth Century: Changing Views of a Changing World*. Cornell University Press, Ithaca, NY.

McPhee, John. 1986. *Rising from the Plains*. Farrar, Straus, and Giroux, New York.

Prothero, D. R. 2018. *The Story of the Earth in 25 Rocks*. Columbia University Press, New York.

Prothero, D. R., and F. Schwab. 2013 (2nd ed.). *Sedimentary Geology: Principles of Sedimentology and Stratigraphy*. W. H. Freeman, New York.

Rudwick, M. J. S. 1985. *The Great Devonian Controversy: The Shaping of Scientific Knowledge Among Gentlemanly Specialists*. University of Chicago Press, Chicago.

Rudwick, M. J. S. 2014. *Earth's Deep History: How It Was Discovered and Why It Matters*. University of Chicago Press, Chicago.

Secord, J. A. 1986. *Controversy in Victorian Geology: The Cambrian–Silurian Dispute*. Princeton University Press, Princeton, NJ.

Shaw, Alan. 1964. *Time in Stratigraphy*. McGraw-Hill, New York.

Weller, J. M., 1947, Rhythms in Upper Pennsylvanian cyclothems: Illinois Acad. Sci. Trans., v. 35, p. 145–146.

Winchester, Simon. 2001. *The Map That Changed the World: William Smith and the Birth of Modern Geology*. HarperCollins, New York.

SUMMARY

- About 1795, English canal surveyor William Smith first noticed that there was a sequence of different types of fossils in every formation he encountered across Great Britain. This is now known as the "principle of faunal succession."

- Comparison of rock sequences around Europe using faunal succession showed that rock units are not like a simple "layer cake" but change their composition over distance as they are formed in different sedimentary environments. These are known as "sedimentary facies."

- Sedimentary facies are characteristic of a specific sedimentary environment. For example, the alluvial fan environment produces conglomerates, breccias, and coarse, gravelly sandstones with lots of feldspar and rock fragments. Desert dune deposits are recognized by their huge cross-beds.

- River deposits (fluvial deposits) are characterized by cross-bedded sandstone channels alternating with floodplain mudstones and shales and lots of terrestrial animal and plant fossils.

- Nearshore and delta environments may have nearshore and beach sandstones with features formed by waves, nearshore lagoonal mudstones with features formed in reversing tides, as well as coal formed from swamps.

- Offshore mudstones and shales are recognized by their marine fosssils and great thickness, alternating with nearshore sandstones.

- Limestones form in shallow carbonate shoals just below sea level in tropical or subtropical regions with clear water and good light penetration.

- Deep marine deposits consist mainly of thick shales with deep-sea fossils and burrows, alternating with graded bedding in sandstone layers formed by submarine gravity slides known as "turbidity currents."

- Because rock units form from specific sedimentary environments that migrate onshore (transgression) and offshore (regression) with rising and falling sea levels, most sedimentary rock units are time-transgressive and do not imply a time unit.

- Transgressions can be caused by rising eustatic (global) sea level, subsiding coastlines, or a combination of the two. Regression can be caused by falling eustatic sea level, tectonic uplift along a coast, or the building out of a delta into the ocean.

- Transgressions form a fining-upward local sequence of sedimentary rocks, while regressions coarsen upward.

- A few rare events are geologically instantaneous, such as volcanic eruptions, meteorite impacts, and submarine gravity slides. Volcanic ash layers are the best time markers to establish timelines in rock sequences, especially if the ash can also be dated by radiometric methods.

- Biostratigraphy is the use of the sequence of change in fossils through the rock column to correlate those rocks with other similar sequences bearing the same fossils.

- Biostratigraphy uses the concurrent or overlapping range zone as the best technique for fine-scale subdivision of geologic time.

- Index fossils are groups that tend to give the best results in biostratigraphy. They tend to be abundant, widely distributed, and rapidly evolving groups of fossils, so they can be reliably used to tell fine-scale divisions of geologic time.

- Because time units are not rock units, there are separate hierarchical units for time (eon, era, period, epoch, age) and rocks (group, formation, member, bed) which are not equivalent. To recognize time in real rock units, a hybrid unit known as the "time-rock," or chronostratigraphic, unit is used. It is based on all the rocks that were formed during a given time and has its own hierarchy of terms (eonothem, erathem, system, series, stage, and zone).

- The geological timescale grew haphazardly from about 1795 to 1840 as geologists mapped and discovered more and more units based on their distinctive fossils. The modern geologic timescale is based on the relative sequence of fossils, but it is calibrated by numerical dates from igneous rocks wherever possible. It is a fundamental tool that all geologists use and must have at their fingertips and in their memory.

KEY TERMS

Faunal succession (p. 62)
Geologic map (p. 63)
Sedimentary facies (p. 66)
Facies change (p. 66)
Sedimentary environments (p. 67)
Alluvian fan environments (p. 67)
Desert dune environments (p. 68)
Fluvial environments (p. 68)
Delta and nearshore environments (p. 68)

Offshore sedimentary environments (p. 68)
Limestone sedimentary environments (p. 68)
Deep-sea sedimentary environments (p. 70)
Transgression (p. 72)
Fining upward and coarsening upward (p. 74)
Eustatic sea-level change (p. 74)
Regression (p. 74)
Rock units are not time units (p. 75)

Tectonic uplift versus subsidence (p. 75)
Geologically instantaneous events (p. 75)
Bentonites (p. 76)
Biostratigraphy (p. 76)
Concurrent or overlapping range zones (p. 77)
Index fossils (p. 77)
Lithostratigraphy (rock stratigraphy) (p. 78)
Group, formation, member, bed (p. 78)

Geochronology (geologic time) (p. 79)
Eon, era, period, epoch, age (p. 79)
Chronostratigraphic units (time-rock units) (p. 79)
Eonothem, erathem, system, series, stage, zone (p. 79)
Geologic timescale (p. 79)

STUDY QUESTIONS

1. How did William Smith's careful observations of canal cuts in England revolutionize our understanding of earth history?

2. Why is a geologic map such an important tool to a geologist working on any kind of scientific problem?

3. Why is understanding modern and ancient sedimentary environments so important to finding coal and oil and other resources?

4. Give an environmental interpretation of the following outcrops:
   a. Sandstones with giant cross-beds
   b. Coal seam between black, organic material–rich shales
   c. Thick sequences of black shales alternating with sandstones showing graded bedding

   d. Red oxidized shale deposits with mudcracks and leaf fossils alternating with narrow lenses of sandstone
   e. Coarse conglomerates interbedded with coarse sandstones full of feldspar and rock fragments
   f. Limestone full of shell fragments and pieces of coral
   g. Finely laminated shales with freshwater fish fossils and leaf fossils

5. Why do sedimentary rocks change across horizontal distance?

6. Alan Shaw says, "if a rock unit looks the same, it must be a different age." What does he mean by this?

7. You see a vertical sequence of coarse sandstones fining upward into shales and then limestones, followed by limestones coarsening upward into shales then

sandstones then conglomerates. Describe the changes in relative sea level in this outcrop.

8. Describe some exceptions to the general rule that most rock units are time-transgressive.

9. Why is it important to record the exact vertical position of a fossil as it is found in a stratigraphic section before removing it from the outcrop?

10. What are some of the properties of an ideal index fossil?

11. Why do time units need a different classification scheme from rock units?

12. What is a time-rock, or chronostratigraphic, unit? Why is such a hybrid unit necessary when creating the geologic timescale?

13. Why doesn't the geologic timescale have a simple standardized scheme of units that are easy to learn?

14. Why do geologists need to know the geologic timescale by memory and not just talk about geologic events using their numerical age?

Scientists still do not appear to understand sufficiently that all earth sciences must contribute evidence toward unveiling the state of our planet in earlier times, and that the truth of the matter can only be reached by combing all this evidence. . . . It is only by combing the information furnished by all the earth sciences that we can hope to determine "truth" here, that is to say, to find the picture that sets out all the known facts in the best arrangement and that therefore has the highest degree of probability. Further, we have to be prepared always for the possibility that each new discovery, no matter what science furnishes it, may modify the conclusions we draw.

—*Alfred Wegener, On the Origin of Continents and Oceans, 1915*

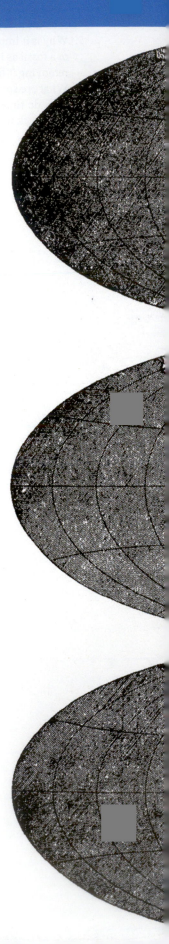

Wegener's original 1915 reconstruction of the Pangea position of the continents 250 million years ago.

5.1 The Way the Earth Works

A SCIENTIFIC REVOLUTION

In his legendary 1962 book *The Structure of Scientific Revolutions*, philosopher of science Thomas Kuhn pointed out that science operates very differently from what most people think. It is not a slow, steady, uninterrupted march toward final truth. Instead, it goes through periods of time where everyone practices "normal science" and accepts certain basic premises and assumptions (a **paradigm** in Kuhn's sense). Eventually, however, anomalies and problems and inconsistencies with the prevailing paradigm start to accumulate. Then someone "thinks outside the box" and comes up with a totally new paradigm for their science, rejecting the assumptions of the old paradigm. This paradigm shift results in a **scientific revolution**.

Kuhn's main example was the Copernican revolution in astronomy, where just one simple change in the basic model (putting the sun, rather than the earth, at the center of the solar system) solved lots of problems with the old Ptolemaic geocentric system and led to a whole new worldview. Newtonian physics transformed the fields of mechanics that were still stuck in the false notions of Aristotle. Einsteinian relativity revolutionized physics once again since Newtonian mechanics does not apply in the realm of things moving near the speed of light.

Likewise, Darwinian evolution overthrew the old creationist notions of life, and biology has never been the same since. It's not clear whether there has been a true scientific revolution in chemistry, although some key ideas have been proposed, like Mendeleev's development of the periodic table.

Unlike other sciences, geology underwent its scientific revolution very recently, in the lifetime of many geologists still alive today. The old paradigm long assumed that continents were fixed and stable, and the first real challenge to this idea came in 1915 when German meteorologist Alfred Wegener published *The Origin of Continents and Oceans*. But the idea was rejected for decades until new data from marine geology and from studying the ancient magnetic fields of rocks on land accumulated in the 1950s to show that continents had indeed moved. In 1962 and 1963, a series of key discoveries launched the new paradigm of geology called **plate tectonics**. Geology hasn't been the same ever since. Most of the key discoveries were made by young scientists in the 1960s, many of whom are still alive and active in research.

WEGENER AND CONTINENTAL DRIFT

People had speculated about the possibility that Africa and South America had once been attached to each other as soon as the first good maps of the South Atlantic were published in the 1500s (Fig. 5.1). But the person most associated with the idea that continents were moving, not fixed, was German meteorologist Alfred Wegener. Born in Berlin in 1880, he was the son of a clergyman, not a scientist, so he was an unlikely revolutionary. He initially studied classical languages (Greek and Latin) before switching to astronomy, meteorology, and climatology; but he never had much formal training in geology. By 1905 he was working at a meteorological observatory. In 1906, at the tender age of 26, he organized and led the first of four expeditions to Greenland to understand climate in polar regions.

During the Christmas holidays in 1910, Wegener happened to glance at a world atlas that was a gift for one of his friends. As he recalled later, he was struck by how well the Atlantic coasts of South America and Africa seemed to fit together. But Wegener didn't stop there, as most people before him had done. He soon began collecting evidence from the distribution of fossils, from the rocks that indicated ancient climates and ancient latitudes of continents, and other data which suggested that all the continents had once been united into one supercontinent he called **Pangaea** (which means "all lands" in Greek). By 1912, he had given a few lectures on his ideas and published three short papers in a German geographical journal. In 1913 he ran his second Greenland expedition and spent the winter on the ice, nearly starving to death before he and his companion were rescued.

When World War I broke out in 1914, Wegener was drafted into the Kaiser's army, as was nearly every able-bodied man in Germany at the time. After being wounded twice (once in the neck) in the infantry at the start of the war, the German high command decided he

Figure 5.2 Alfred Wegener in the polar shelter hut on his final fatal expedition to Greenland, 1930.

was more useful as a meteorologist than as trench fodder. They sent him to work in the army weather service, where he traveled between weather stations all over German-held Europe. Despite all this traveling and army duties, he managed to finish writing his book, *On the Origin of Continents and Oceans*, which was published late in 1915. The book had little immediate impact because of the wartime restrictions, but Wegener remained remarkably productive for a busy active-duty officer, publishing another 20 papers in meteorology and climatology before the war ended.

After the war, he obtained a position at the German Naval Observatory in Hamburg, then at the University of Hamburg, and finally accepted a secure post at the University of Graz. During this time, he wrote an influential book with his father-in-law, the famous meteorologist Wladimir Köppen, on the climates of the geological past. But his ideas about drifting continents were not widely read or accepted yet, especially not in the geological community in Europe and North America. Finally, he presented his ideas at the 1926 meeting of the American Association of Petroleum Geologists in New York City, where everyone rejected his ideas except the chairman of the session who had invited him. Even though his theories continued to be scorned by geologists, Wegener kept working hard, gathering data in Greenland in 1929. In 1930, he led his fourth and final expedition to Greenland (Fig. 5.2), the largest

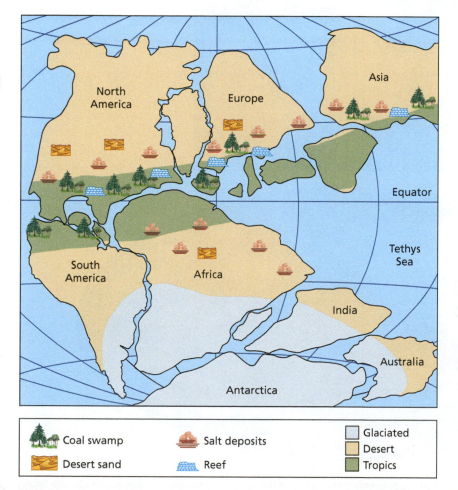

Figure 5.3 The rock types representing the major climatic belts in the Permian (polar ice cap, subtropical deserts, and tropical swampy rainforests forming coal) only make sense in a Pangea configuration. If you move these continents to their modern location, these deposits would be in the wrong latitudes to have formed like they do today.

he had ever mounted, with propeller-driven ice sleds and much innovative equipment. In November 1930, he and a partner were returning from a supply run to their remote camp in the center of the Greenland ice sheet when they ran out of food and got into bad weather. There, Wegener froze to death at the relatively young age of 50. His partner buried his body on the Greenland ice sheet, where he still rests today under many feet of ice. Wegener never lived to see any support for his revolutionary ideas, which would not come for another 30–35 years.

On the Origin of Continents and Oceans was a masterful compilation of all the evidence available at the time. Many

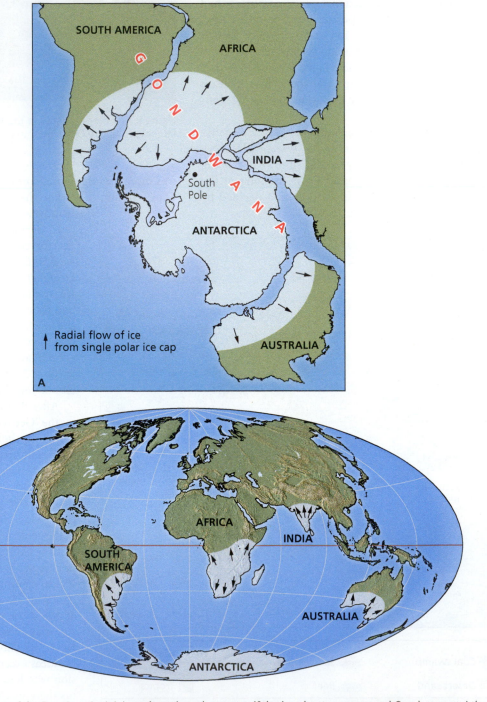

Figure 5.4   A. The occurrence of the Permian glacial deposits only makes sense if the ice sheet once covered Gondwana and the South Pole. **B.** The present location of the continents would require the Permian ice cap to do absurd things, like cross the equator to reach India and cover large areas of the Indian and South Atlantic Oceans. In addition, the glacial scratches and grooves (such as in southwestern Africa) line up with those in Brazil and Argentina, which means the glaciers would have needed to plunge into the Atlantic, cross in a strange curved path, then climbed ashore in South America if the modern Atlantic had existed in the Permian.

people had wondered whether Africa and South America fit together, but Wegener showed that the fit also included India, Madagascar, Australia, and Antarctica (Fig. 5.1). As a climatologist, Wegener was particularly impressed by the way certain deposits are strongly controlled by climate and latitude: ice caps on the poles, rainforests in the tropics, and desert deposits in the mid-latitude high-pressure belt between 10° and 40° north and south of the equator. But when you went back to the Permian Period (250–300 Ma), the location of those ancient deposits made no sense on a modern globe (Fig. 5.3). The south polar ice sheets extended from South America and Africa and apparently stretched across the equator to India, which is a climatological absurdity (Fig. 5.4). Only if you put the continents back into their Permian configuration as part of a single supercontinent, Pangea, did they make sense.

If the continents had not moved, then why did the bedrock scratches carved by the rocks pushed along by ancient Permian glaciers run from Africa to South America (Fig. 5.4)? That would require the glacier to jump into the Atlantic, flow in a straight line across the Atlantic Ocean from Africa to Brazil, and then jump back out of the ocean! Absurd! Likewise, all the ancient Permian desert deposits and coal swamps of tropical Permian rainforests only made sense if they were put back in their Pangea locations, not their modern latitudes. Even the ancient Precambrian bedrock in South America and Africa below the Permian deposits also matched precisely, like pieces of a jigsaw puzzle.

Clinching all this was the evidence from the Permian fossils (Fig. 5.5). South America was loaded with distinctive fossils that were shared with other continents, especially southern Africa. There were distinctive extinct seed ferns known as *Glossopteris* found on all the continents of the southern landmass known as Gondwana. There were small aquatic reptiles like *Mesosaurus*, found in lake beds in Brazil and South Africa, and protomammals (formerly called "mammal-like reptiles") like the bulldog-sized beaked herbivore *Lystrosaurus* and the bear-like predator *Cynognathus* that could never have swum across the modern Atlantic Ocean. To Wegener (and to any modern geologist), this evidence should have been clear and

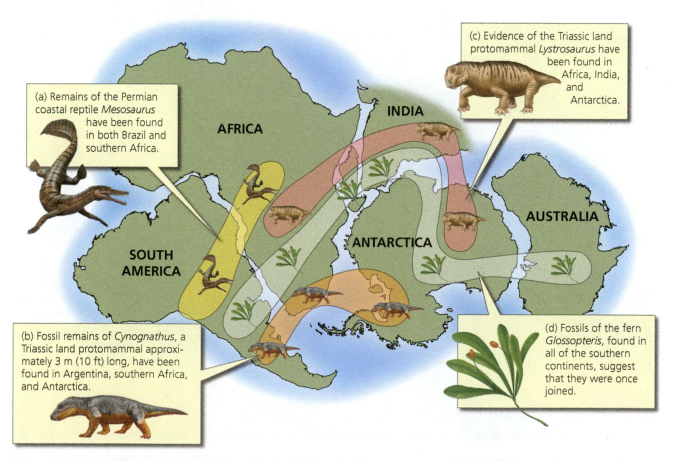

Figure 5.5 The distinctive plants and animals of the Permian link the various Gondwana continents together. The little aquatic reptile *Mesosaurus* and the bear-sized protomammal *Cynognathus* are both found in Brazil and South Africa. The pig-like, beaked protomammal *Lystrosaurus* is found in Africa, Madagascar, India, and Antarctica. The seed fern *Glossopteris* is found on all the Gondwana continents.

conclusive: the continents drifted apart like the ice floes Wegener knew so well.

If the evidence was so solid, why were Wegener's ideas ridiculed and rejected? Why did geologists treat his ideas as crazy and consider him a crackpot for another 45–50 years after his book was published? Not all of it was due to the lack of strength of his scientific evidence. There were also the effects of the sociology of science. First of all, Wegener was not a formally trained geologist, and there is natural reluctance to accept ideas from outside your scientific field that appear to violate your basic assumptions, such as the fixity of continents.

Another problem was that Wegener's evidence came mostly from the Southern Hemisphere (mainly South America and Africa), yet almost all the world's geologists back then lived in Europe or North America. Very few had ever traveled to South Africa or Brazil, which were long, expensive trips by ocean liner in those days. Naturally, the evidence is much more persuasive if you can see it in person, rather than read it in type and tiny black-and-white photos that were normal for journals back then. Indeed, Wegener's biggest boosters, such as South African geologist Alexander du Toit, were mostly in the southern hemisphere and did see the evidence firsthand. They pointed out that the Permian rocks of South Africa and Brazil were virtually identical; the only difference was that the former had Afrikaans names and the latter had names in Portuguese.

Another supporter was British geologist Arthur Holmes, known as the "father of the geological timescale" for his pioneering work in radiometric dating in 1913–1915 (see Box 3.1). He embraced Wegener's ideas—but then he had worked as a geologist in Africa before returning to England. Holmes boldly published diagrams showing drifting continents and mantle currents pushing those continents around for the first time in his widely used geology textbooks in the 1920s, 40 years before the geological community came to accept the notion.

But Wegener's rejection was not all due to sociological biases. Wegener had no mechanism to explain how

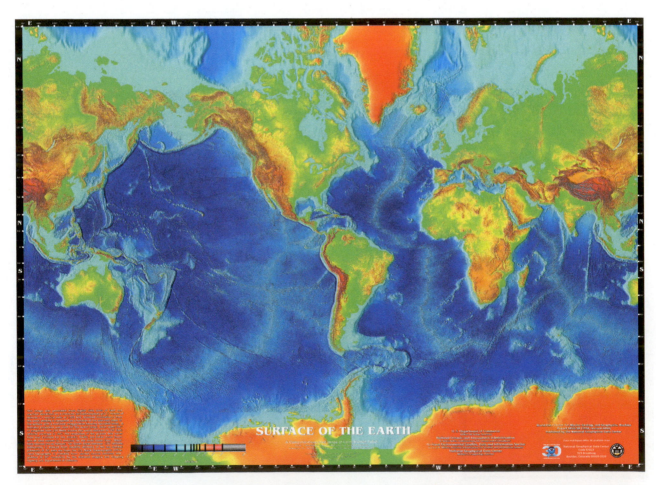

Figure 5.6   A modern version of the Heezen and Tharp map of the world sea floor, showing the Mid-Atlantic Ridge, the East Pacific Rise, the trenches in the western Pacific, and other deep seafloor features.

continents drifted, and the ideas he proposed (such as centrifugal force) weren't very plausible. Wegener postulated that continents were driven away from the poles by the attraction of the earth's equatorial bulge or by tidal forces, but geophysicists showed that this was impossible. Wegener's critics argued that if the continents had plowed across the ocean basins, there should be huge areas of oceanic crust crumpled up on their leading edges like the snow on a snowplow blade—and such deposits had never been found. Meanwhile, they dismissed the matches in rocks across the continents as inconclusive and concocted fantastic land bridges to explain how animals could have walked across the Atlantic Ocean. They continued to denigrate Wegener and hold his ideas up for ridicule for decades. In the 1940s, the leading American paleontologist George Gaylord Simpson published numerous papers arguing that the fossils did not require continental drift. His institution, the American Museum of Natural History, held a big symposium in 1949 dismissing all the evidence of moving continents.

THE BIRTH OF PLATE TECTONICS

Ironically, just as this heap of scorn reached its peak, new evidence was coming from an unexpected direction: the bottom of the ocean. Both Wegener and his critics were completely ignorant of what the oceanic crust was really like. No one really knew anything about the earth's crust beneath the oceans. In fact, the answers did not come until after World War II, when modern marine geology was born. After the war was over, the US Navy was happy to sell or give war-surplus ships and equipment (rather than scrap them) to the newly founded oceanographic institutes such as Woods Hole Oceanographic Institute on Cape Cod, Massachusetts, the Lamont-Doherty Geological Observatory (now Lamont-Doherty Earth Observatory) up the Hudson River from New York City, and Scripps Institution of Oceanography in San Diego, California. In addition, the military and the US government learned from the submarine tactics during the war that they knew too little about the world's oceans. Federal funds to explore the oceans during the Cold War, when American and Soviet submarines stalked each other, would be a good investment.

By the late 1940s and 1950s, several oceanographic institutes sponsored ships that were routinely crisscrossing the world's oceans; getting detailed surveys of sea-floor depth, and the structure of the rocks below the sea bottom; and collecting seismic, gravity, and magnetic data, as well as sediment cores, everywhere they went. By the late 1950s, scientists had the first real image of what 70% of the earth's surface actually looked like, thanks to the first detailed maps of the sea floor by Marie Tharp and her partner Bruce Heezen (Fig. 5.6). Tharp, in particular, was responsible for drawing maps of and discovering more

of the earth's surface than any human ever did or ever will, thanks to data that Heezen brought back from sea. In the mid-1950s, Tharp realized that there was a gigantic range of mountains under the sea known as the Mid-Atlantic Ridge. Not only was it the longest mountain range on the planet (many times as long as mountains on land) but at over 5000 m (18,000 feet) it was also higher than most mountain ranges on earth. In addition, the Mid-Atlantic Ridge had a giant rift valley deeper than the Grand Canyon down the entire length, where it pulled apart and crustal blocks had dropped down along faults. Tharp was a good geologist and realized at once that a rift valley indicated that the sea floor must be pulling apart and spreading. However, her co-authors Heezen and Lamont director Maurice Ewing were too cautious to publish and promote such a heretical idea, so other people got credit for her discovery.

The confirmation of the sea-floor spreading idea (Box 5.1) in 1963 was the crucial piece of evidence to start the plate tectonics revolution. Meanwhile, older discoveries that were once puzzling suddenly began to make sense. Back in 1949, seismologist Hugo Benioff of Caltech (unknowingly rediscovering a 1928 result by Japanese seismologist Kiyoo Wadati) found that there was a zone of earthquakes that plunged beneath the oceanic trenches found around the Pacific Rim (Fig. 5.11). The quakes immediately below the trench were quite shallow, but as you moved inland from the trench, the earthquakes got deeper and deeper. They plunged down in a distinct zone of quakes now called a "**Wadati-Benioff zone.**" In 1949, this zone of quakes was a mystery, a piece of a jigsaw puzzle that was missing the rest of the puzzle. But after sea-floor spreading was confirmed about 25 years later, it became clear what these quakes meant. They were produced as one plate plunged beneath another in a **subduction zone.** The idea of subduction was dramatically confirmed in the 1964 Alaska earthquake, when analysis of the motion of the earthquake showed that the Pacific Plate plunging beneath the Aleutian Islands and Alaska had slipped farther under the continent.

It turned out that Wegener's critics were wrong: the continents *do* drift around the globe, but they don't crumple up the oceanic crust ahead of them because most of the oceanic crust slides beneath other plates and plunges back into the mantle.

5.2 Plate Tectonics

From all these different sources of data, we now know that the earth's surface is covered by **crustal plates** that move around on the fluid mantle beneath. They slide around on the curved surface of the earth like pieces of eggshell on a hard-boiled egg (Fig. 5.12).

BOX 5.1: HOW DO WE KNOW?

How Did Ancient Magnetic Directions Lead to Plate Tectonics?

Ancient magnetic data were gathered from rocks on land on many different continents during the 1950s. Many kinds of rocks record the earth's magnetic field as they form and give not only its direction but also its intensity. As paleomagnetists sampled and measured older and older rocks, their results seemed to show that the north magnetic pole was far from the modern pole and appeared to wander through time. This was called the "polar wander hypothesis" at first. But then they ran into a problem. Each continent had a completely different polar wander curve, which only converged on a common magnetic pole today (Fig. 5.7). These data seemed to suggest that the magnetic field had behaved very strangely in the past, with multiple directions of magnetic north that no longer exist. As outrageous as that idea seemed, the only alternative was just as radical: the continents had moved through time, so it was not the magnetic pole that was changing but the continents that recorded their directions. But when you lined up the polar wander curves for two different continents (Fig. 5.7), like Europe and North America, you found that they matched once you moved the continents back together as Wegener had suggested. In other words, the "polar wander curves" were only **apparent polar wander curves** because it was the continents that moved, not the magnetic poles.

Other paleomagnetists in the 1950s and 1960s found a peculiar phenomenon: some rocks had magnetic directions that pointed in the opposite direction of what it is today (Fig. 5.8). For example, 800,000 years ago, if you held a compass out in the northern hemisphere, its needle would point south, not north. At first they blamed it on some peculiar property of the rock samples. But as more and more examples were found, the idea that the earth's field had reversed direction in the geologic past seemed less outrageous. To test this idea, Allan Cox of Stanford University and Bob Doell of the US Geological Survey (who did the magnetic analysis) and G. Brent Dalrymple of the US Geological Survey (who did the K–Ar dating) performed a crucial analysis. They sampled lava flows all over the world, measured their magnetic directions, and obtained their age. If rocks

the same age all over the world showed the same magnetic direction, then the reversed direction was no longer a quirk of individual self-reversing rocks—it had to be a global phenomenon, such as the earth's magnetic field. By the early

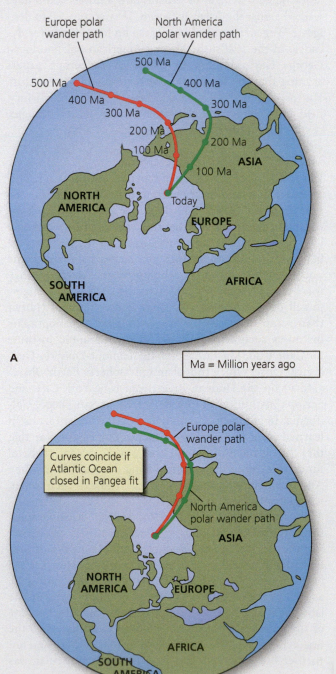

Ma = Million years ago

Figure 5.7 Apparent polar wander curves. If you measure the ancient magnetic directions from North America (green curve), the North Magnetic Pole appears to be a long way from its present position, only reaching its modern location in the magnetic direction of the youngest rocks. If you do the same for the European rocks, you get a completely different location of the North Magnetic Pole in the geologic past (red curve), which just happens to converge on the modern North Magnetic Pole in the youngest rocks. Thus, either there were many different magnetic poles that were widely divergent in the past, and only happened to converge recently, or if you move the continents to their ancient positions, the polar wander curves match.

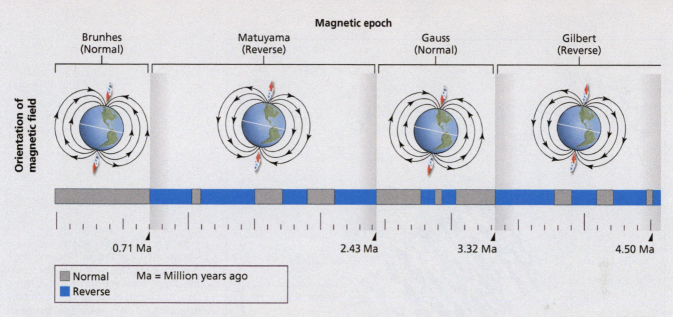

Figure 5.8 The earth's magnetic polarity has flipped from normal (like today) to reversed (180° opposite of today's pole) hundreds of times over the past 50 million years. In just the last 5 million years, there are over a dozen polarity flips, or "magnetic reversals."

1960s, Cox, Doell, and Dalrymple (in friendly competition with Ian McDougall, François Chamalaun, and Don Tarling at the Australian National University) had established that the earth's magnetic field flips back and forth from its modern direction to a direction 180° opposite what we measure today.

Since then, the pattern of magnetic flip-flops has been documented in detail. It has flipped hundreds of times in the past 100 million years, giving a random pattern of normal and reversed polarity that is like the black-and-white stripes of a bar code. Like a bar code, this pattern has a signal imbedded in it, and it is widely used to match the flip-flops in thick sections of rock and date them precisely (Fig. 5.7). This record of the earth's changes in the magnetic field through time is known as the **magnetic polarity timescale**.

This history of magnetic field reversal was the key, the "Rosetta Stone," which solved the final problem. It provided the final crucial piece of data that helped tip the balance of all the evidence and push geologists to take the notion of plate tectonics seriously. Since the 1940s, oceanographic vessels had been towing proton-precession magnetometers behind them, recording the magnetic signal over the sea floor. These devices look like a long torpedo on a cable and were originally developed in World War II to find submarines. Over the years, they collected dense maps of sea-floor magnetic data, but at first marine geophysicists couldn't make sense of what they saw. For example, as they towed the magnetometer over the top of a mid-ocean ridge, the magnetic field they recorded was stronger than the normal earth's magnetic field background that we are always exposed to (Fig. 5.9). This stronger-than-average field direction was called a **positive magnetic anomaly**. But if they towed the magnetometer a few tens of kilometers to either side of the ridge, they got weaker-than-average magnetic field strength, or a **negative magnetic anomaly**. What did these peculiar results mean?

For over a decade this was a great puzzle, mostly because the first magnetic surveys, like those of the eastern Pacific, were very complicated and hard to decipher. The key came in 1963 when Fred Vine and Drummond Matthews of Cambridge University analyzed a much simpler profile over the Mid-Atlantic Ridge. They noticed that the pattern of magnetic anomalies was symmetrical, with positive magnetic anomalies over the center of the ridge but negative anomalies on each side and then a series of symmetrical magnetic "stripes" moving away from the ridge outward (Fig. 5.9). Vine and Matthews realized that the rocks on the sea floor must be magnetized in different directions. The rocks in the center of the mid-ocean ridge had erupted in the last 800,000 years, so their normal magnetic polarity would *add* to the modern background field direction, giving a stronger-than-average field measurement on the magnetometer. But the rocks farther from the center of the ridge had been erupted and magnetized more than 800,000 years ago, where they acquired a reversed magnetic polarity direction. When the reversed directions interact with the modern earth's field, they partially subtract from the background field since they are polarized in the opposite direction. For this reason, the magnetometer records a weaker-than-average measurement, or a negative magnetic anomaly. Thus, the sea floor is like a magnetic tape recorder, picking up a signal from the earth's field as it erupts along the center of the ridge and then passively

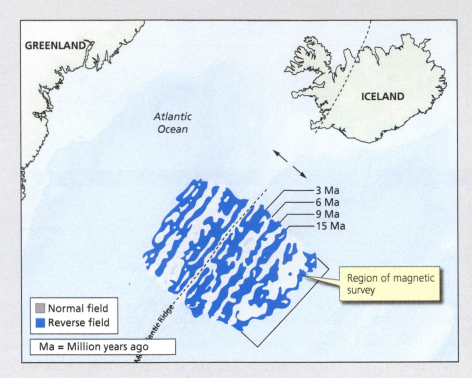

Figure 5.9 Magnetic anomalies were symmetrical over the mid-ocean ridge (here showing the data from the Reykjanes Ridge segment of the Mid-Atlantic Ridge, just south of Iceland) , which only made sense if they were the record of the ancient magnetic field of normal and reversed polarity frozen into the rocks of the spreading sea floor and gradually moving away from the ridge crest like a pair of conveyer belts.

moving away from the "recording head", carrying its magnetic signal with it (Fig. 5.10). The fact that the magnetic signal was symmetrical on both sides of the ridge crest means that the oceanic crust spreads away from the ridge like a pair of conveyor belts moving away from a common center.

See For Yourself: Seafloor Spreading

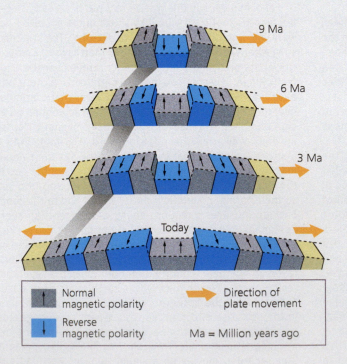

Figure 5.10 Reversal pattern of mid-ocean ridges, which erupted along the mid-ocean ridge, then ripped apart and moved away from the ridge as new seafloor was formed. The magnetic reversal patterns were symmetrical and suggested sea-floor spreading.

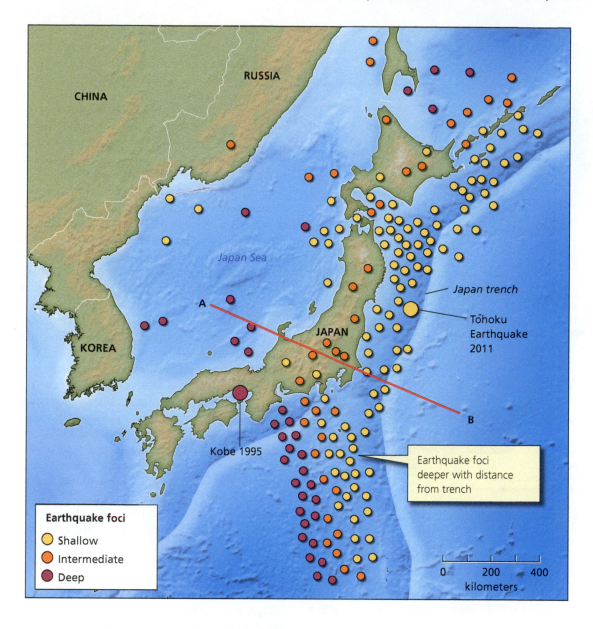

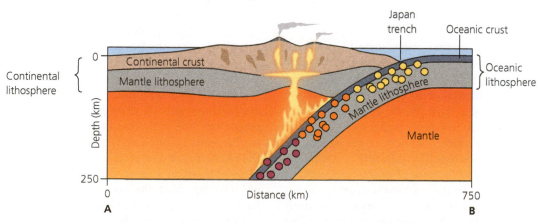

Figure 5.11 Wadati and Benioff first documented that the earthquakes near the trench were very shallow and that they got deeper and deeper in a plunging zone under the continent. These are now called Wadati-Benioff zones.

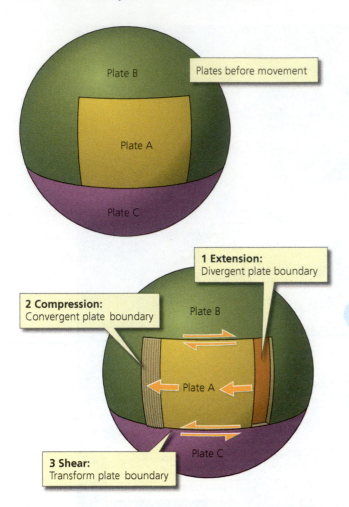

Figure 5.12 The motion of plates is like the sliding of a piece of eggshell on an egg. In some places the crust is pulling apart (divergent), forming extension and seafloor spreading **(1)**. In other places **(2)** plates are convergent and compressing each other, forming subduction zones or collisional mountain belts. But in some places **(3)**, the plate is on its way from a divergent boundary to a convergent boundary, but is neither compression nor extending. These are called transform plate boundaries, with a shearing motion along a strike-slip fault.

See for yourself:
Plate Tectonics

The crustal plates come in two basic types (Fig. 5.13). **Oceanic crust** is relatively thin (only 10 km thick), dense, and made of basalt. **Continental crust** is about three to five times thicker than oceanic crust (30–50 km thick on average) and is made of much less dense rocks, such as granitic rocks and gneisses. The sharp boundary at the base of the crust is called the "**Mohorovicic discontinuity**" ("**Moho**" for short), based on the rapid increase of seismic velocities as earthquake waves pass from the crust to the mantle. The uppermost semi-rigid part of the mantle moves with the overlying crustal plates, and together these two layers are called the "**lithosphere**." Beneath the lithosphere is the next layer of the mantle (from about 50–200 km to about 400 km down), called the "**asthenosphere**," where the mantle is semi-fluid; this is where the main convection currents move with the plates above them. These plates are driven by dense oceanic lithosphere, which sinks down into the mantle. This in turn

produces the huge convection currents in the mantle, which originate from heat rising from the earth's interior bringing up hot mantle plumes and then other areas in the mantle where it cools and sinks down to the core–mantle boundary.

A good analogy for this process would be a pot of hot cocoa with marshmallows on a stove. The stove burners provide the heat (analogous to the heat coming from the earth's interior), and the pot has warm currents of cocoa rising in some areas and cool cocoa sinking in others, like the convection currents in the mantle. At the top surface of the cocoa (especially as it cools) is a thin scum, which is a good analogue for the oceanic crust. The cocoa scum can easily be destroyed or melted or can sink back into the hot cocoa. The floating marshmallows are like continental crust. They can collide with each other and move around, but they never sink down below the surface of the cocoa. If the cocoa scum collides with the marshmallows, the scum sinks down into the cocoa.

 See For Yourself:
Basic Plate
Boundaries

The earth's tectonic plates (made of both oceanic and continental crust) behave in a similar fashion. Three types of plate boundaries are possible (Fig. 5.14). In a **divergent boundary** plates pull apart and produce new oceanic crust by the process known as sea-floor spreading (discussed in the next section). There are relatively quiet plate boundaries, with only small earthquakes and volcanoes on these boundaries, so some parts of divergent boundaries are also called **passive margins**. The second type of plate boundary occurs when two plates come together, so it is called a **convergent boundary**. On most convergent boundaries, one of the plates is made of thin, dense oceanic crust and its underlying dense lithospheric mantle, which sinks down beneath the other plate in a subduction zone. As it does so, the friction of the grinding of the downgoing plate generates most of the world's earthquakes, and the plate also melts as it goes down to produce volcanoes, so this type of plate boundary is called an **active margin**. The third possibility is that two plates are neither converging nor diverging but sliding past one another. This kind of plate boundary is

 See For Yourself:
540 Million Years
of Plate Tectonics

known as a "**transform boundary**." It is typified by huge strike–slip faults, such as the San Andreas fault in California.

DIVERGENT MARGINS

New crustal rock is produced on **mid-ocean ridges**, which are the main part of divergent margins. The mid-ocean ridges were one of the key discoveries of the early days of marine geology in the mid-1950s and the first key piece of the puzzle that led to plate tectonics. They are the longest range of mountains on earth, running for many thousands of kilometers down the middle of the Atlantic, the eastern Pacific, across the Indian Ocean, and several other places, like the seams on a giant baseball. They are also some of the highest ranges of mountains on earth (over 500 km, or over 18,000 feet) with a huge fault valley deeper than the Grand Canyon in some ridges, like the Mid-Atlantic Ridge. Yet no one knew they were there or just how big they were until the late 1950s, when Marie Tharp and Bruce Heezen first mapped the ocean floor and discovered them.

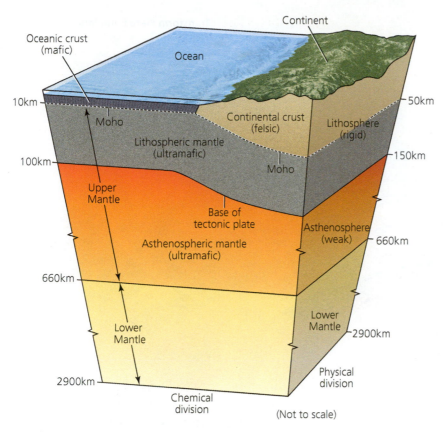

Continent
Oceanic crust (mafic)
Ocean
10km
50km
Moho
Continental crust (felsic)
Lithosphere (rigid)
Lithospheric mantle (ultramafic)
100km
150km
Moho
Upper Mantle
Base of tectonic plate
Asthenosphere (weak)
Asthenospheric mantle (ultramafic)
660km
660km
Lower Mantle
2900km
Lower Mantle
2900km
Physical division
Chemical division
(Not to scale)

Figure 5.13 The layers of the crust and upper mantle, showing oceanic and continental crust, lithosphere and asthenosphere, and the Moho.

dikes, and layered gabbro) are the standard components of oceanic crust all over the world (Fig. 5.15). Since oceanic crust makes up about 70% of the earth's surface, these are the most abundant rock types on the planet. The association of these three distinctive rock types was named the **ophiolite suite** (Greek, *ophis*, meaning "snake," and *lithos*, meaning "rock," since these gabbros and basalts were usually metamorphosed into the mineral serpentine).

The association of pillow lavas, sheeted dikes, and layered gabbros, often capped by oceanic sediments, was found in many other mountain ranges in Europe (especially in the Alps and in the Macedonian region of Greece) and elsewhere (such as the island of Cyprus and in Oman in the Persian Gulf). No one could explain what it meant until plate tectonics came along in the 1960s and studies of active mid-ocean ridges showed that they were producing ophiolites all the time. Today, we recognize ophiolites as slivers of oceanic crust that have been sliced off a downgoing subducting oceanic plate and stuck onto the land, or pieces of oceanic crust trapped between the collision of two continental plates.

Over millions of years, the cooling oceanic crust pulls away from the center of the mid-ocean ridge, and new magma wells up the cracks to replace it (Fig. 5.10). From this point onward, all the oceanic crust gradually spreads and pulls away from the mid-ocean ridge like a gigantic pair of conveyer belts. As the old oceanic crust spreads away, it also sinks because it is cooling and shrinking.

The breaking up of continents to form passive margins and new oceans has a well-known history. We can see it by looking at places on the earth today where each stage is happening.

The first step to breaking up a continent to form an ocean is for upwelling places in the mantle known as "hot spots" to rise up beneath the continental plate. These hot spots cause the crust to bulge upward into a huge blister-like dome. Eventually, the stretching of the "blister" causes it to break open and form a **rift valley** (Fig. 5.17A). These rift valleys can be seen today in places such as the East African Rift, which stretches from the Afar Triangle in Ethiopia down through Kenya and Tanzania to Malawi and South Africa. In modern rift valleys, large volumes of alluvial fan sediment eroded down from the fault scarps on each side, forming deep, narrow basins filled with sands and gravels that become sandstones and conglomerates. In many

In the early 1960s, marine geologists discovered that the ocean floor pulls apart at the mid-ocean ridge, which is why there is a giant fault valley (Figs. 5.6, 5.15) down its entire length. As the plates pull apart, new magma wells up from gabbroic magma chambers below them and chills into new oceanic crust made of basalt. Thanks to many geophysical measurements, as well as observation in small submarines, scientists have witnessed this process over and over again.

 See For Yourself: Formation of Ocean Crust

The magma flowing up through the extensional cracks in the graben valley hits seawater, then instantly chills into blobs of congealed magma called **pillow lavas** (Fig. 5.16). Pillow lavas are produced as hot magma chills in contact with seawater. Additional hot magma from below forces its way through cracks in the older chilled lava, then extrudes from the crack like a blob of toothpaste and quickly chills into a rounded, "pillow"-like shape. This amazing sight has been observed and filmed many times, and you can watch footage of it online with just a search for "pillow lava video."

 See For Yourself: Pillow Lavas

Beneath the ocean floor layer of pillow lavas (Figs. 5.15, 5.16A), the magma that flowed up through the vertical cracks and fissures to feed the pillows eventually congealed in those cracks to form hundreds of vertical slabs of basalt (Fig. 5.16C) called **sheeted dikes**. Beneath the dikes is a gabbroic magma chamber that fed the dikes and pillows. When it finally cools, it forms a layered gabbroic magma chamber (Fig. 5.15). All three of these rock types together (pillow lavas, sheeted

 See For Yourself: Production and Destruction of Oceanic Crust

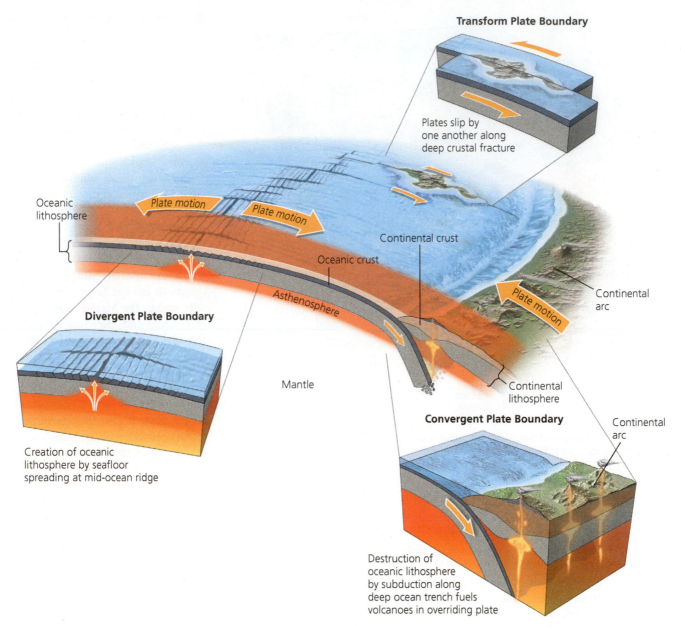

Figure 5.14   The three possible motions of plate boundaries: divergent or spreading (resulting in "passive margins"), convergent or colliding ("active margins"), and sliding horizontally (transform margins).

places, the river drainages pour down into the rift and form lakes, so lake deposits (fine-grained muds and shales and even evaporites) are common in rift valleys, such as those in East Africa. In fact, most of the lakes of Kenya, Ethiopia, and Tanzania lie in the Rift Valley. Finally, the faults and cracks on the edges of the rifts often become conduits for the magma just below the surface, so rift valleys often have abundant volcanoes (such as Mt. Kilimanjaro and Mt. Kenya in Africa).

▷ See For Yourself: The Process of Rifting

The rift valley gets wider and wider as the continents continue to pull apart, until finally one end of the rift is open to the ocean and seawater flows in. At this point, the rift has become a **proto-oceanic gulf** or "linear sea" (Fig. 5.17). Proto-oceanic gulfs form long, narrow, parallel-sided oceanic

troughs, such as the Red Sea or the Gulf of Aden on the south and west of the Arabian Peninsula today. They are typically filled with shallow-marine shales and sands, typically of any shallow-marine basin. If they are in the subtropical desert belt (as is the Persian Gulf and the Red Sea), it is common for large amounts of seawater to evaporate away without rivers to replace it, forming thick deposits of salt and gypsum as evaporites on the bottom. Beneath these sediments, the proto-oceanic gulf is floored by highly stretched and faulted continental crust. Eventually, however, the crust pulls so far apart that magma can flow up the center, generating a mid-ocean ridge and rift system, as are found in every ocean (Fig. 5.15).

From this point on, the slow, steady process of sea-floor spreading pulls the two continents farther and farther apart

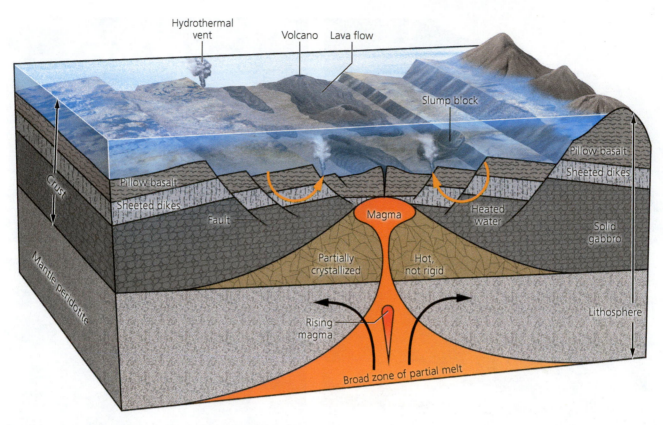

Figure 5.15 The features of a mid-ocean ridge and its rift valley. In the center of the mid-ocean ridge, the crust is extending or pulling apart, forming dropping fault blocks, with magma welling up in the cracks and faults between them. When the magma hits seawater, it erupts underwater and forms pillow lavas. If it never reaches the seawater but chills in the vertical cracks, it forms sheeted dikes. The molten rock that cools in the magma chamber forms a layered gabbro. These three rock layers are typical of oceanic crust around the world, and are known as the ophiolite suite, because in a few places, they have been pushed above sea level and can be studied on land.

as the width of the ocean grows (Fig. 5.17). But the old edges of the continent are not completely without interest. They continue to sink deeper as they pull farther away from the mid-ocean ridge and the oceanic crust next to them shrinks and sinks as it continues to cool. The old buried rift valley deposits, and sometimes the evaporites of the proto-oceanic gulf, become buried under a thick blanket of shallow-marine sediments, forming on what has become the continental shelf (Fig. 5.18). Since the entire complex keeps slowly sinking (fastest on the seaward side), the deposits of shallow-marine shales and sands get thicker and thicker, forming a **passive margin wedge**. The passive margin wedge is a thick prism of sediments forming at the edge of the continental crust where it meets the oceanic crust. The wedge tends to thin toward the onshore direction and thicken rapidly in the offshore direction (Fig. 5.18). It is composed of shallow-marine shelf sandstones and shales on the landward side thousands of meters thick and deep-water shales, turbidites, and cherts on the deep-water edge. It continually sinks and subsides because it is tied to the old oceanic crust far from the mid-ocean ridge, which continues to shrink and sink due to thermal cooling. The entire Atlantic and Gulf coasts of the United States are underlain by thick passive margin wedge deposits, which reach thicknesses of 7000 m (over 25,000 feet) at the far edge of the continental shelf but

thin to nothing as they transition up onto the coastal plans of the southeastern states.

CONVERGENT OR ACTIVE MARGINS

Active, or "convergent" margins, have completely different behavior and characteristics from divergent margins of tectonic plates. They are caused when two plates collide, and the usual result is that oceanic plate slides down beneath the other plate to make a subduction zone (Fig. 5.19). As the downgoing slab scrapes beneath the overlying slab, it causes many huge earthquakes. The subduction zones are the most seismically active regions in the world, with all the biggest quakes.

 See For Yourself: The Process of Subduction

Then, as the subducting plate plunges deeper into the mantle, the mantle rocks above the slab begin to melt due to the abundance of water and volatiles in the weathered ophiolitic basalts, which lowers their melting temperature. This molten rock rises through the overlying plate and creates a chain of volcanoes, making the "Ring of Fire" around the Pacific. Such a volcanic chain is often called an "island arc," or just "arc volcanoes," because many of them form chains of islands and have an arcuate shape because the globe is spherical.

Figure 5.16   The characteristic rock types of the ophiolite suite. **A.** Pillow lavas form whenever magma extrudes and cools under water. These pillows crop out just west of San Luis Pier, Avila Beach, California. **B.** Modern pillows erupting in the deep ocean, making blobs of magma as they are squirted through cracks in the underwater lava flow. **C.** Sheeted dikes from the Troodos ophiolite on the island of Cyprus. **D.** Layered gabbro, here showing the horizontal layers of crystals solidified and turned up on end so that the beds are vertical, from the Smartville ophiolite in the Sierra Nevada foothills.

The boundary between the plates, where one plate plunges beneath the other, is a deep valley on the ocean floor known as an **oceanic trench**. Trenches are found above nearly every subduction zone on the planet, and they are by far the deepest spots in the ocean. The Mariana Trench is the world record holder at 11,034 meters (36,201 feet) below sea level, deep enough to hold Mt. Everest and several other mountains. But there are many other trenches that are 8000–10,000 m (26,000–33,000 feet) deep around the world. Others, like the Cascadia trench off the coast of British Columbia, Oregon, and Washington, are nearly completely filled with the sediment eroding down from the big rivers in the region like the Columbia River.

Between the trench and the volcanic arc are two other important parts of an active margin complex. The boundary zone between the two plates is marked by an **accretionary wedge**, or accretionary prism. Accretionary wedges are amazing geological features. They are formed when slices of the subducting slab are chopped off the plate as it goes down and plastered up against the bottom of the overlying slab (Fig. 5.20). Consequently, the rocks are continuously

added to the *bottom* of the stack, and it gets *older* as you go to the top. This is the reverse of normal superposition, where old rocks are at the bottom and young rocks on the top (see Chapter 3).

These rocks undergo a tremendous amount of shearing and slicing and dicing and being run through the blender, so they no longer have any continuity or bedding or remnants of their original order (Fig. 5.20). They are so mixed up that we use the French word "**mélange**" for them (meaning "mixture"). Several different and unique rock types are found almost exclusively in mélange from accretionary wedges:

1. *Old oceanic sediments*: The most commonly expected rock to get scraped off is the sediments that used to lie on the old plate before it sank down the subduction zone. These are deep-ocean shales with chert layers in them (**ribbon cherts**), along with shreds of sandstones that flowed down in submarine gravity slides (turbidites, discussed in Chapter 4).

2. *Slices of oceanic crust (ophiolite)*: Sometimes not only the sedimentary cover of the downgoing slab is scraped

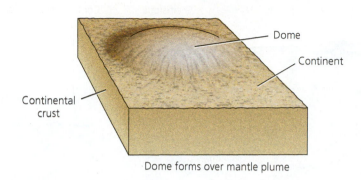

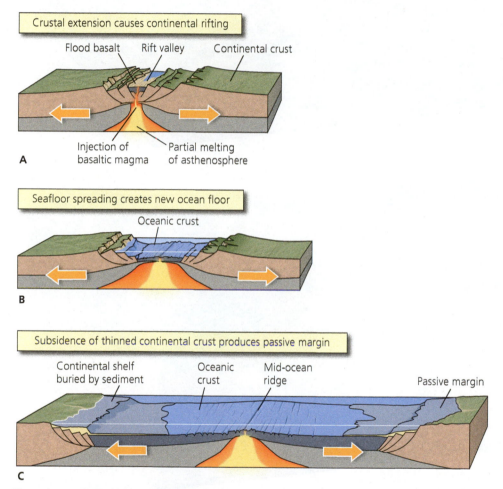

Figure 5.17 Three of the stages of the transition from a solid continent through the rift valley and proto-oceanic gulf to the passive margin.

and pushed up into the accretionary wedge as newer slices are added beneath them.

Finally, between the accretionary wedge and the volcanic arc is a trough-like basin that warps downward. It is usually drowned by seawater, so it fills with marine sandstones and shales, many of which are derived from the volcanic arc above it. This is known as the **forearc basin** (Fig. 5.19). Many of the volcanic arcs have another basin behind them, called a **backarc basin**.

There are three possible configurations for convergent margins (Fig. 5.19):

1. If one oceanic plate slides beneath another oceanic plate (Fig. 5.19A), it is called a **Japan-style volcanic arc**. Not only is this setting well studied and exemplified, but many of the other volcanic chains around the Pacific "Ring of Fire" are Japan-style arcs, including the Aleutians, the Philippines, the Indonesian arc, and the Tonga-Kermadec arc. Japan-style arcs tend to be mostly chains of islands, with only a small forearc basin. The backarc basin, if present, is completely under the ocean and fills with marine sediments.

off but big chunks of the oceanic crust itself. These are the ophiolites we have already discussed.

3. *Blueschist*: Subduction zones are the only places in the world where you find a peculiar metamorphic rock known as blueschist, discussed in Chapter 2. The subducting plate may be 50–100 km down into the mantle, so it experiences extremely high pressures. Yet the cold, old oceanic crust in the slab resists heating up and melting for a very long time as it sinks into the mantle, so the region around it is high-pressure without being the normal high-temperature. Thus, blueschists are made deep in subduction zones and then sliced off

2. If an oceanic plate slides beneath a continental plate, then it is called an **Andean-style volcanic arc**. The Andes (Fig. 5.19) are the most famous example, but the Cascade Range in California, Oregon, Washington, and southern British Columbia is another. Because the overlying plate is continental crust, the volcanic chain erupts on land, and the forearc basin may be filled with seawater and marine sediments or sometimes may be a non-marine basin. The accretionary prism can be uplifted as well, forming a coastal chain of islands (such as the islands off the south coast of Sumatra). The backarc basin is nearly always filled with non-marine

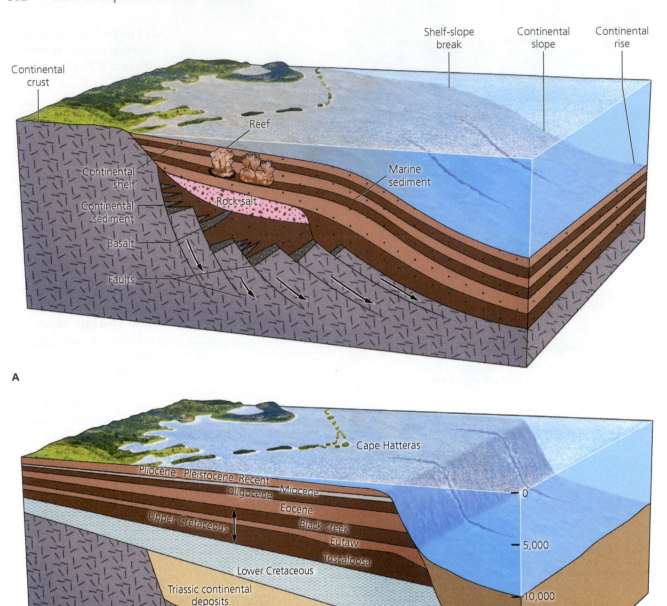

Figure 5.18   A. The passive margin wedge complex, showing the thinning landward and thickening seaward package of shallow-marine shelf sediments overlying earlier deposits of the rift valley fault-block sediments and the salt from the proto-oceanic gulf stage. **B.** Actual cross section of the passive margin wedge beneath the Atlantic coast of the Carolinas.

sediments (rivers and deltas and lakes), and (in the case of the Andes) there is usually a big zone of thrust just behind the arc that pushes out over the backarc basin.

3. Beyond oceanic–oceanic and oceanic–continental convergent margins, the third possible scenario is the collision between two continents. Since continental crust is light and buoyant ("marshmallows in cocoa"), it cannot subduct, so instead there is a huge collision zone

between the plates. This produces enormous folded and thrust-faulted mountains, which are often the highest mountains in the world. Of course, the best modern example is the Himalayas (Fig. 5.21). For this reason, a continental–continental collision is often called a **Himalayan-style margin**, although the Alps and most of the mountain ranges across Europe between them (such as the mountains in the Balkans, Turkey, Iran, and Afghanistan) are part of the same crush zone. The

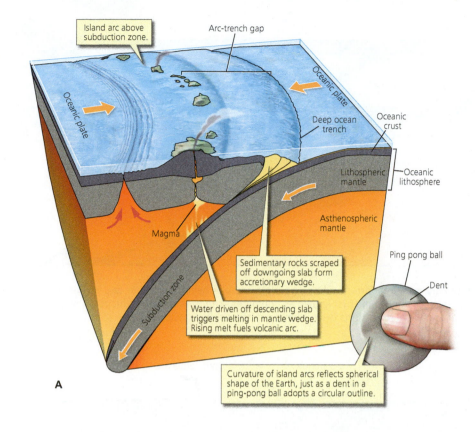

Island arc above subduction zone.

Arc-trench gap

Oceanic plate

Oceanic plate

Deep ocean trench

Oceanic crust

Lithospheric mantle

Oceanic lithosphere

Asthenospheric mantle

Ping pong ball

Dent

Magma

Subduction zone

Sedimentary rocks scraped off downgoing slab form accretionary wedge.

Water driven off descending slab triggers melting in mantle wedge. Rising melt fuels volcanic arc.

Curvature of island arcs reflects spherical shape of the Earth, just as a dent in a ping-pong ball adopts a circular outline.

A

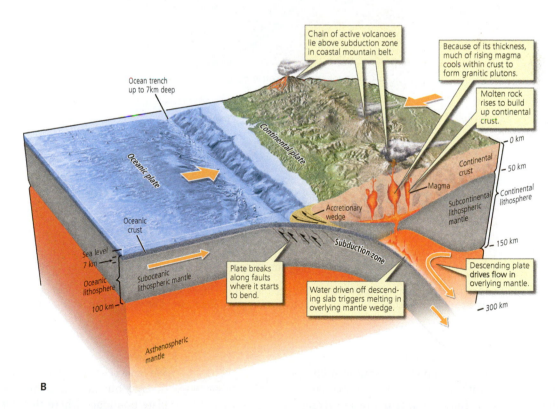

Chain of active volcanoes lie above subduction zone in coastal mountain belt.

Because of its thickness, much of rising magma cools within crust to form granitic plutons.

Molten rock rises to build up continental crust.

Ocean trench up to 7km deep

Continental plate

0 km

Continental crust

50 km

Magma

Continental lithosphere

Oceanic plate

Subcontinental lithospheric mantle

Oceanic crust

Accretionary wedge

150 km

Sea level

7 km

Oceanic lithosphere

Suboceanic lithospheric mantle

Plate breaks along faults where it starts to bend.

Subduction zone

Water driven off descending slab triggers melting in overlying mantle wedge.

Descending plate drives flow in overlying mantle.

300 km

100 km

Asthenospheric mantle

B

Figure 5.19 Two major classes of active subducting margins. Part **A** is a collision between two oceanic plates, forming a "Japan-type" margin, with a volcanic island arc, a forearc basic, and an accretionary wedge just above the trench. The curvature of the island is convex toward the subduction direction, as illustrated by the curvature of a dimple on a ping-pong ball. Part **B** is a collision between an oceanic and continental plate, forming an "Andean-type" margin, with its trench and accretionary wedge. The volcanic edifice is not and island arc, but a volcanic mountain range since it must burn its way up from the mantle through continental crust.

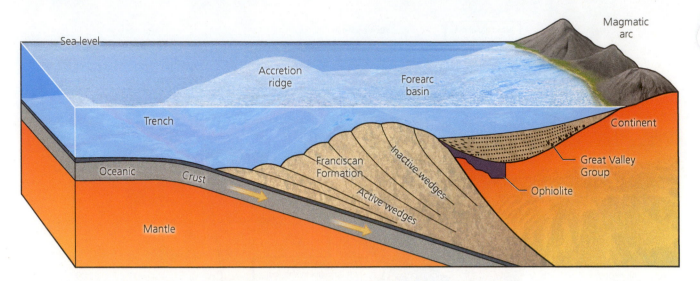

Figure 5.20 The accretionary wedge is formed when one plate scrapes beneath another and accumulates unusual rocks such as deformed ocean sediments, slices of ocean crust known as ophiolites, and even metamorphic rocks such as blueschist. The extreme shearing and shredding of the pile of rocks turns them all into a mixture called mélange.

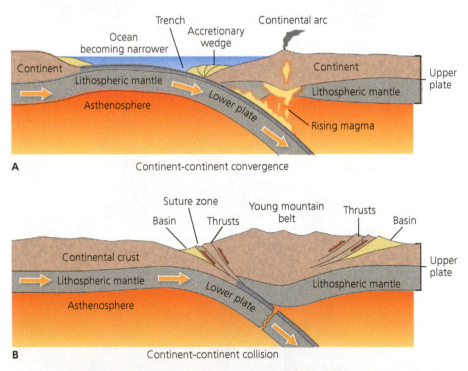

Figure 5.21 In addition to oceanic-oceanic ("Japan-style") and oceanic-continental ("Andean-style") convergent boundaries, the third type of boundary involves a continental–continental collision, to form huge uplifted mountain ranges like the Himalayas.

Alps and Himalayas started forming about 50 Ma, when Africa began to push up against southern Europe and India collided with the belly of Asia, respectively (see Figs. 14.2, 14.3). These mountains are still rising as the continents continue to push against one another after 50 million years, as evidenced by the frequent deadly earthquakes in western China, Nepal, Pakistan, Afghanistan, Iran, Armenia, Turkey, and Greece.

See For Yourself: Development of a Collisional Mountain Belt

TRANSFORM MARGINS

See For Yourself: Transform Faulting

We have seen how plates can converge or diverge, but there is a third possibility: a plate boundary where the plates are neither spreading nor colliding (Figs. 5.14, 5.22). Instead, the plates slide past one another in a gigantic horizontal fault zone, known as a "strike–slip fault." These are **transform plate margins** because the plates are transformed or translated from one place to another without moving together or

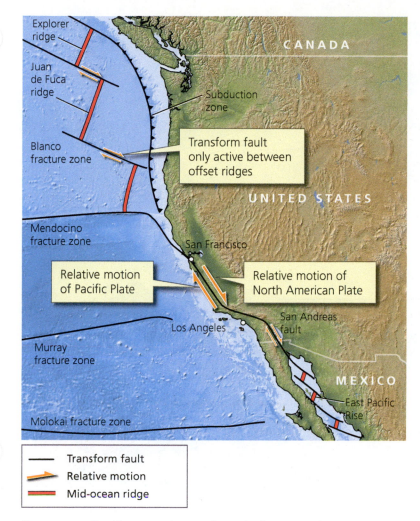

Explorer ridge

Juan de Fuca ridge

Blanco fracture zone

Mendocino fracture zone

Murray fracture zone

Molokai fracture zone

CANADA

Subduction zone

Transform fault only active between offset ridges

UNITED STATES

San Francisco

Relative motion of Pacific Plate

Relative motion of North American Plate

Los Angeles

San Andreas fault

MEXICO

East Pacific Rise

— Transform fault
→ Relative motion
— Mid-ocean ridge

Figure 5.22 Transform margins are formed where one plate grinds past another without either convergence or divergence. Here, the spreading ridge in the Gulf of California turns into the San Andreas transform fault, which runs north through coastal California until it meets the Cascade subduction zone and another transform in the Mendocino triple junction in northern California.

apart. The most familiar and best-studied example of a transform is the San Andreas fault in California. But there are others, including the Queen Charlotte transform on the northern coast of British Columbia and the alpine transform that runs down the spine of the two main islands of New Zealand. There are also huge transform faults that offset the various segments of mid-ocean ridges, allowing each short segment to spread at a different rate in response to its motions across the spherical surface of the earth.

See For Yourself: Where will the Continents be in the Future?

5.3 Sedimentary Basins and Plate Tectonics

As early as the 1850s, pioneering American geologist James Hall noticed that the sedimentary sequence of rocks across New York and the Appalachian region was thinnest toward the center of the continent and thickened rapidly toward the edges. In 1873, Yale geologist James Dwight Dana proposed that this trough on the edge of the continent sank down and accumulated thick piles of sediments over millions of years. He called this concept a "**geosyncline**" because it was a downwarped trough (like the trough-shaped fold known as a "syncline") on an earth-sized scale (hence the prefix "geo"). The part of the geosyncline nearest the edge of the continent was a wedge of shallow-marine sandstones and shales known as a "**miogeosyncline**." Farther offshore from the continent was a thicker wedge of deep-water shales and sandstones, called the "**eugeosyncline**" (Fig. 5.23).

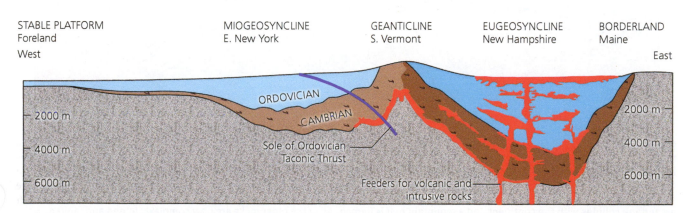

STABLE PLATFORM Foreland · West   MIOGEOSYNCLINE E. New York   GEANTICLINE S. Vermont   EUGEOSYNCLINE New Hampshire   BORDERLAND Maine · East

ORDOVICIAN · CAMBRIAN · 2000 m · 4000 m · 6000 m · Sole of Ordovician Taconic Thrust · Feeders for volcanic and intrusive rocks

Figure 5.23 The classic geosyncline model dating back to the mid-1900s. It consisted of a thick, deep downwarped basin full of deep-water shales, turbidite sandstones, and volcanic ash known as a eugeosyncline and, on the landward side, a shallower downwarped basin filled with nearshore marine sands and shales known as a miogeosyncline.

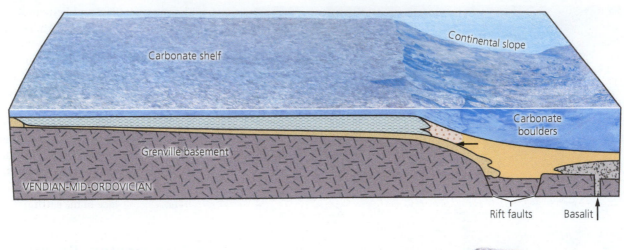

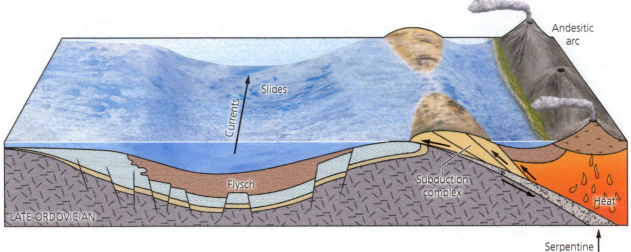

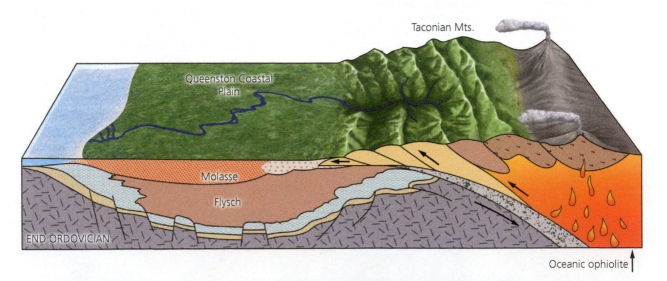

Figure 5.24   Plate tectonics came along in the 1970s and gave an explanation for geosynclines. **A.** First, the sediments are formed by the passive margin wedge, with shallow-marine sediments of the continental shelf (miogeosyncline) and deep-marine shales and tur-bidites of the eugeosyncline. **B.** At some point, the divergent margin turns into a convergent margin, and a subduction zone swallows up oceanic crust and brings a volcanic arc near the old continental margin. This contributes volcanic ash flows and volcanic sediments to the eugeosynclinal basin, eventually crushing and deforming the eugeosyncline and miogeocline. **C.** The orogenic event ends when the arc terrane collides with the continent and a mountain belt is formed. All of the older rocks are now complexly folded and faulted, while the uplifted mountain range sheds river and delta sediments (molasse) across the old continental interior.

In addition to these deep marine rocks, the eugeosyncline contained lots of volcanic rocks and often had sediments derived not from the nearest continent but from some mysterious offshore "tectonic land." All of these rocks of the geosyncline were often intensely folded and crumpled by later mountain-building.

Through the early twentieth century, geosyncline theory got more and more elaborate, with many new terms introduced. The geosyncline model was expanded to describe the Alps in Europe, and more concepts were added. In the 1920s, German geologist Hans Stille coined the word "**craton**" (Greek for "shield") for the stable core of the continent, which doesn't rise or sink much over millions of years. Looking at the Alps, geologists noticed that before these geosynclines formed into crumpled mountain belts, they had been deep-marine troughs that accumulated shales and deep-water sandstones and cherts. This was known as "**flysch**" to the Europeans. Nearly all this pre-orogenic flysch (formed in the crumpling to form a mountain belt, or an **orogeny**) was highly deformed. Then, sitting unconformably on top of crumpled and folded flysch beds would be river and delta deposits known as "**molasse**," eroded off the mountains after they had been folded and uplifted. By definition, molasse was formed after the mountain belt had arisen and started to erode down, so when the term "molasse" is used, it implies that it is post-orogenic.

But there was still no explanation for *why* these basins sank down and were usually crumpled up into mountain belts at a later time. The peak of geosyncline theory was reached in the 1940s and 1950s, just as the discoveries that led to plate tectonics were emerging from the deep sea floor. In 1970, Robert Dietz realized that plate tectonics could explain how and why geosynclines formed, something that had eluded geologists for a century. Most of the classic miogeosyncline was formed by the shallow nearshore continental shelf portion of the passive margin wedge, while the eugeosyncline comprised the deep-water shales and turbidite sandstones that formed in the deep waters of the continental rise and slope.

But the passive margin wedge only explains part of the classic miogeosyncline and eugeosyncline (Fig. 5.24). What about the volcanic rocks in the eugeosyncline and the mysterious "tectonic land" sources of offshore sediments? These are clearly from island arc volcanic chains that formed offshore when the passive margin wedge switched to active margin tectonics after the plates changed their direction of movement from divergence to convergence. As the volcanic arcs approached a continent, not only did they shed volcanic rocks and the mysterious sediments of the "tectonic land" down into the eugeosynclinal trough but their eventual collision with the continent crumpled up the entire geosynclinal package (especially the deep-marine pre-orogenic flysch of the geosyncline). Once the mountain-belt collision was complete, the molasse eroded off the high mountains and formed thick wedges of river and delta sediments.

Thus, the entire field of geosyncline theory became obsolete as plate tectonics explained not only its major features but especially *why* deep-marine troughs sink down, then eventually get crumpled into mountain ranges. Even though the concept of geosynclines is now obsolete, the terminology is still used as a shorthand for these geologic features.

5.4 Conclusion

Plate tectonics revolutionized geology just over 5 decades ago, and nothing has been the same since. Many of the pioneers who made key discoveries that led to plate tectonics are still alive and actively publishing. The entire field of geology was transformed as this whole new way of looking at the world led to many discoveries, as well as reinterpretations of old concepts like geosynclines. Geology is still feeling the effect as new ideas spring up and new connections to plate tectonics become apparent. It is a wonderful time to be a geologist and make such important cutting-edge discoveries every year.

RESOURCES

BOOKS

Cox, A., and R. P. Hart. 1986. *Plate Tectonics: How It Works.* Wiley-Blackwell, New York.

Kearney, P., K. A. Klepeis, and F. J. Vine. 2009. *Global Tectonics.* Wiley-Blackwell, New York.

Kuhn, Thomas. 1962. *The Structure of Scientific Revolutions.* University of Chicago Press, Chicago, IL.

Molnar, P. 2015. *Plate Tectonics: A Very Short Introduction.* Oxford University Press, Oxford.

Oreskes, N. 2003. *Plate Tectonics: An Insider's History of the Modern Theory of the Earth.* Westview Press, San Diego.

Prothero, D. R. 2018. *The Story of the Earth in 25 Rocks.* Columbia University Press, New York.

Roberts, P. 2016. *Tectonic Plates: How the World Changed.* Russet, New York.

Wegener, A. 1915. *On the Origin of Continents and Oceans.* Dover, New York.

SUMMARY

- Geology has undergone a very recent scientific revolution since the late 1950s, with the old notion of fixed continents being replaced by the ideas of mobile continents and a dynamic earth.

- It started with the radical ideas of continental drift by German meteorologist Alfred Wegener in 1912–1915, but his ideas were rejected in his lifetime because he had no mechanism for moving continents, he was not a professional geologist, and professional inertia prevented geologists from seeing his evidence clearly.

- The idea of moving continents was revived in the 1950s and 1960s with the birth of marine geology, which provided a better understanding of how crustal plates moved and much evidence for the motion of plates through time.

- The first evidence of moving crustal plates came from mapping the sea floor and discovering the mid-ocean ridges and their central rift valleys, which Marie Tharp correctly realized meant the sea floor was pulling apart. Other data from seismology, gravity, and especially paleomagnetism eventually led to the discovery of the symmetrical magnetic profiles on each side of the mid-ocean ridge, which was the crucial discovery that led to the widespread acceptance of sea-floor spreading.

- Modern understanding of the earth's upper layers recognizes two kinds of crust: thin but dense basaltic oceanic crust and thick, less dense continental crust. These crustal blocks (plus the uppermost mantle) move around on the semi-fluid mantle, propelled by convection currents in the upper mantle called the "asthenosphere."

- Three types of interaction between plates are possible: divergent, where two plates pull apart at a mid-ocean ridge; convergent, where plates come together in a subduction zone or crumple into mountain belts; and transform boundaries, where plates neither separate nor collide but grind past one another.

- Divergent margins begin with the mid-ocean ridge rift valley, where plates pull apart and new magma comes up from the mantle to add new crust. Typical oceanic crust is composed of several layers, topped by pillow lavas (lavas that erupted under water), sheeted dikes (magma that cooled in the vertical cracks that fed the pillow lava eruptions), and layered magma chambers of gabbro that fed the magma to the opening rifts. A combination of these three igneous rocks in this sequence is known as the "ophiolite suite." These rocks, plus oceanic mantle peridotite, are formed in the process of oceanic rifting and eruption and make up the entire oceanic crust, the most common rock on earth.

- Once sea floor has spread apart, it shrinks due to thermal contraction and sinks deeper in the ocean as it moves away from the mid-ocean ridge.

- Divergent margins that turn into oceans go through a well-known series of stages, starting with hot spots beneath the continent that form continental bulges, which eventually break into rift valleys. One of the three rifts is abandoned and becomes an aulacogen. Two of the three rift valleys pull farther and farther apart until they open up to the ocean and are flooded with seawater, forming a proto-oceanic gulf. Eventually, the proto-oceanic gulf widens enough that it is floored not by thin stretched continental crust but by oceanic crust.

- Where old oceanic crust meets the edge of continents, a passive margin wedge forms. It is composed of a seaward-thickening wedge of sediment thousands of meters in thickness, with mostly shallow-marine sediments near the continent and a thick wedge of deep-marine turbidites and shales on the seaward side.

- Active or convergent margins have very different dynamics since they form by subduction. They are "active" in that most of the world's earthquakes and volcanoes are produced by one plate grinding beneath the other and melting to form subduction zone arc volcanoes. They come in three basic types: oceanic–oceanic plate collisions ("Japan-type"), oceanic–continental plate collisions ("Andean type"), and continental–continental collisions to form huge crumpled and faulted mountain ranges ("Himalayan-type").

- Most subduction zones have a chain of andesitic–rhyolitic volcanoes (volcanic arc), with a back arc basin behind them, a forearc basin in front of them, and in the contact between the two plates an accretionary wedge building up out of material scraped from the downgoing slab and plastered onto the overlying plate. This material can consist of slices of ophiolitic oceanic crust, deep-marine shales and turbidites and cherts, and even blueschists from deeper in the subduction zone, all sheared and shredded into a mixture called a "mélange."

- Transform margins neither subduct nor separate, but they form when two plates grind past one another as they move from one place to another on the earth's surface. They are characterized by huge strike–slip fault zones like the San Andreas fault.

- Geologists had long noticed that ancient mountain belts were often composed of thick wedges of shallow-marine sediments nearer the continent (called the "miogeosyncline") and thicker wedges of deep-marine shales, turbidites, and volcanics formed farther offshore (called the "eugeosyncline"), always crumpled up into a mountain belt during a subsequent orogenic collision. The pre-orogenic deposits were often called "flysch," while those eroding from the newly uplifted mountain belt were known as "molasse."

- Nothing could explain these sedimentary basins until plate tectonics came along and showed that the miogeosynclines are ancient nearshore portions of the passive margin wedge, while the eugeosyncline was the thick wedge of marine sediments in the offshore part of the passive margin wedge. The volcanics, however, come from later active margin arc volcanoes approaching the marine basin as the plates switch from divergence to convergence. The final collision of the approaching crustal block explains why the ancient geosynclinal sediments were always crumpled and deformed.

KEY TERMS

Scientific revolution (p. 86)
Paradigm (p. 86)
Plate tectonics (p. 86)
Continental drift (p. 86)
Wadati-Benioff zone (p. 91)
Subduction zone (p. 91)
Apparent polar wander curve (p. 92)
Magnetic polarity timescale (p. 93)
Oceanic crust (p. 96)
Continental crust (p. 96)

Mohorovicic discontinuity (p. 96)
Lithosphere (p. 96)
Asthenosphere (p. 96)
Mantle convection (p. 96)
Divergent plate boundary (p. 96)
Convergent plate boundary (p. 96)
Transform plate boundary (p. 96)
Active versus passive margin (p. 96)

Mid-ocean ridge (p. 96)
Pillow lavas (p. 97)
Sheeted dikes (p. 97)
Layered gabbros (p. 97)
Ophiolite suite (p. 97)
Hot spot stage of continental breakup (p. 97)
Rift valley (p. 97)
Proto-oceanic gulf (p. 98)
Passive margin wedge (p. 99)
Oceanic trench (p. 100)
Accretionary wedge (p. 100)

Mélange (p. 100)
Volcanic arc (p. 101)
Forearc basin (p. 101)
Japan-type subduction zone (p. 101)
Andean-type subduction zone (p. 101)
Himalayan-type convergent boundary (p. 102)
Miogeosyncline versus eugeosyncline (p. 105)
Craton (p. 107)
Flysch versus molasses (p. 107)

STUDY QUESTIONS

1. In what ways is plate tectonics a scientific revolution? What was the old paradigm, and what changed to cause the new paradigm to emerge?
2. What was Wegener's basic model for continental drift? Why was it rejected by most of the geologists of his time?
3. What evidence did Wegener gather from the southern hemisphere for the existence of a Gondwana continent during the Permian?
4. Why do earthquake foci outline the upper boundary of a descending slab in a subduction zone?
5. Describe how the polar wander curves of different continents are better explained as apparent polar wander curves of moving continents.
6. How do marine magnetic anomalies form? What does their symmetry across the mid-ocean ridge tell us about the motion of the plates beneath the ocean?
7. Describe the differences between oceanic and continental crust. Which one can subduct and which one cannot? Why?
8. What is the difference between the Moho and the lithosphere–asthenosphere boundary?
9. What drives the motions of the plates?

10. Contrast the motions of two plates in divergent, convergent, and transform margins.
11. Describe how magma from the ripping apart of crust along the mid-ocean ridge forms the pillow lavas, sheeted dikes, and layered gabbros of the ophiolite suite.
12. Describe the geometry of the passive margin wedge. How is it formed? Why might there be ancient rift valley sediments buried beneath it?
13. Describe the oceanic trench. Why is it the lowest spot on the ocean floor?
14. Describe the characteristic rock types found in the accretionary wedge and how they got there. Why are they all sheared up into a mélange?
15. Compare and contrast the three basic types of active margins. Which ones involve subduction? Why does no subduction occur in a Himalayan-style margin?
16. Transform margins are called this because they "transform" one plate motion into another. What is meant by this, and how is it achieved?
17. Describe how the old miogeosyncline–eugeosyncline pattern of early geologists is now explained by plate tectonics. What explains the abundant volcanic rocks in the eugeosyncline?

6 Evolution

Everything changes and nothing remains still . . . and . . . you cannot step twice into the same stream

—*Greek philosopher Heraclitus, about 475 B.C.E.*

The discovery of *Archaeopteryx* in 1861, 2 years after Darwin's book was published, was the first strong evidence of transitional forms between major groups (such as birds and dinosaurs) in the fossil record. Although it had feathers and a wishbone like a bird, it still had dinosaurian teeth instead of a toothless bird beak, long bony fingers in its hands (not fused together as in modern birds), and a long bony tail, as well as many other features that showed its dinosaurian ancestry. Darwin could not have asked for a better fossil to show the transition from dinosaurs to birds, although hundreds of additional examples have been found since then.

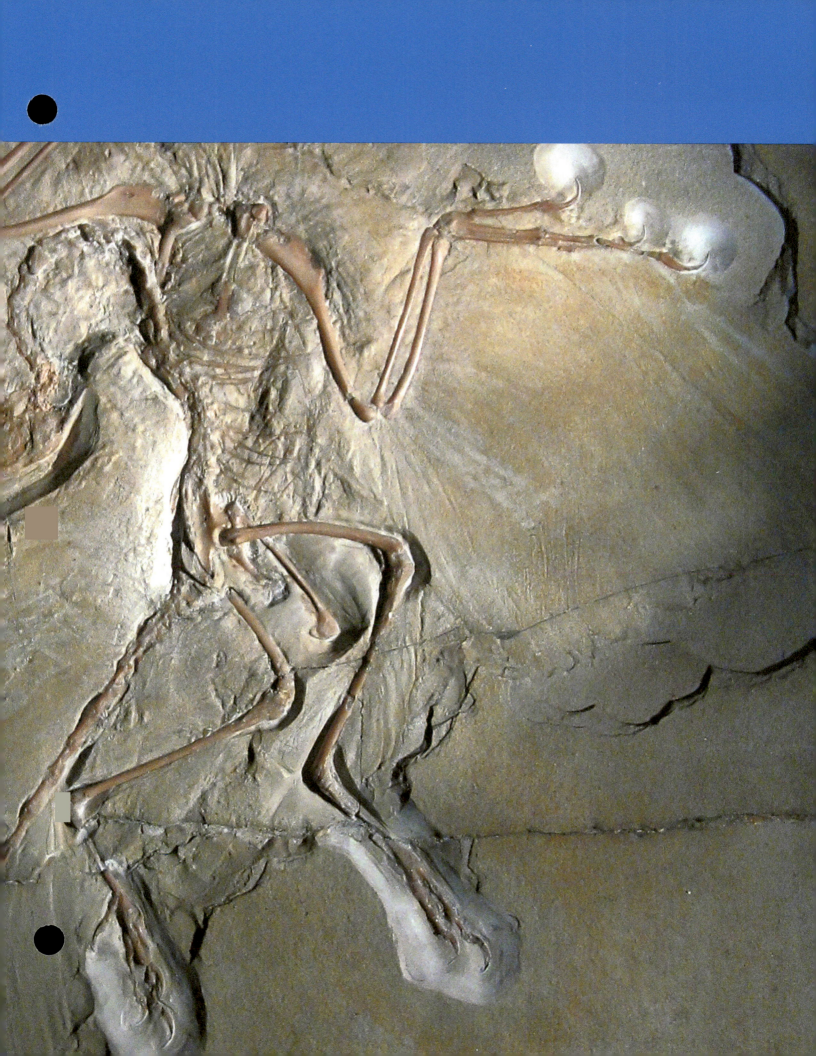

6.1 The Evolving Earth—And Evolving Life

The past 300 years of scientific discoveries, especially in astronomy, geology, and biology, have shown that everything changes through time. Nothing is static or remains unaltered for very long. Galaxies, stars, and planets have changed drastically since they were formed in the Big Bang about 13.7 billion years ago. As discussed in Chapter 2, rocks and minerals are not permanent but erode and weather and melt, are subjected to heat and pressure, and eventually transform to other rocks and minerals in the rock cycle. In Chapters 2 and 5 we saw that even mountains are not permanent features but eventually wear down to sediments. Ocean bottoms can rise out of the depths to form mountains. Every feature of the planet is changing and evolving, including its life.

This notion was not unusual in ancient times. Many Greek philosophers (such as Heraclitus) thought of the earth and life as constantly changing. The Greek philosopher Democritus argued that the world was made of tiny indivisible particles he called "atoms" and that everything was constantly changing. About 50 B.C.E., the Roman poet and philosopher Lucretius proposed that the world was composed of simple matter and followed simple laws of nature, not supernatural gods controlling everything.

But such learning nearly vanished during the Dark Ages, overcome by the church dogma that the earth was unchanging since it was formed only about 6000 years ago. Even the obvious fact that we can see rocks weathering and crumbling and eroding down was attributed to Adam and Eve's sin in the Garden of Eden. According to church dogma, the world would have otherwise been perfect and unchanging as it was created. The prominent natural historian John Woodward was typical of his time when he wrote in 1695 that the "globe is to this day nearly in the same condition that the Universal Deluge left it; being also like to continue so till the time of its final ruin and dissolution, preserved to the same End for which 'twas first formed."

Yet the insights of many Enlightenment scholars gradually challenged this notion. In the mid-1700s, the French naturalist George-Louis Leclerc, the Count of Buffon, suggested the earth might be as old as 70,000 years (thought to be inconceivably old at that time). As we learned in Chapter 1, in the 1780s Scottish naturalist James Hutton's idea that there was "no vestige of a beginning" completely undermined the idea of a young, unchanging earth. By the time Charles Lyell wrote *Principles of Geology* in 1831–1833, the notion that the earth was immensely old and constantly changing was widely accepted.

At the same time, the idea that life could change was also becoming more acceptable. A number of Enlightenment scholars had suggested the idea. The most famous was the pioneering naturalist Jean-Baptiste Pierre Antoine de Monet, the Chevalier ("knight") of Lamarck (1744–1829). (Lamarck was not his name but his noble title, just like we know of William Thomson as "Lord Kelvin" and we know George-Louis Leclerc by his title, "Count Buffon"). Lamarck began as a botanist in the King's Garden, but during and after the French Revolution, he was forced to switch specialties and study the lowly "insects and worms" that no one else cared about. He took advantage of this chance to study a neglected area and ended up revolutionizing our understanding of invertebrate zoology. In the process, he began to see the similarities of plants and animals and realize that zoology and botany were not that different. He even coined the word "biology" to emphasize this unity of all life. In this sense, he might be considered the first "biologist."

But studying the changing sequence of fossil shells of the Eocene rocks of the Paris Basin led Lamarck away from thinking that species were fixed and permanent and toward thinking that they could change through time. In 1809, he published his ideas in a treatise called *Philosophie Zoologique*. Lamarck's idea of animals changing through time was very different from the way we have viewed the evolution of life since Darwin. Lamarck thought that creatures were always climbing the "scale of nature" or the "ladder of life" from their origins in the primordial slime (Fig. 6.2). More advanced creatures had climbed higher up the ladder to become fish or reptiles, while those that had just emerged were still primitive creatures like worms. Instead of the "tree of life" metaphor that all biologists have used since Darwin's time, Lamarck viewed each form of life as hundreds of "blades of grass," each independently emerging from its separate origins and growing to different heights depending upon how long it had been evolving.

This outdated notion, by the way, is still an obstacle for some people trying to understand evolution. Life is not a "ladder" or "scale of nature" or a "great chain of being." Instead, it is a branching, bushy tree of life, with many different lineages that originate side by side and overlap in time (Fig. 6.2). There are no such things as "lower" and "higher" organisms; each is adapted to doing a particular job. In fact, in some respects it could be argued that each of these groups is more successful than mammals or humans. A bacterium or a sponge or a sea jelly is not inferior to a humans in terms of evolutionary success. They do their jobs and play their roles in nature just fine. Bacteria aren't trying to evolve to something else, like sponges, for

instance. Not only that, but bacteria have been around for 3.5 billion years and sponges for at least 600 million years, and they have no reason to become something else. Since life is not a "great chain of being" composed of many different links, the term "missing link" is meaningless and not used by scientists.

Lamarck believed, as did all scientists since the days of Aristotle, that life could spring up from nonliving matter in a form of **spontaneous generation**. Maggots mysteriously appeared and grew in rotting meat, and a broth left out in the open soon got cloudy with mold and bacteria. Tapeworms were thought to have spontaneously grown out of their host organism, and scallops were thought to be formed from the sand of the sea bottom. Although Francesco Redi in 1668 and Lorenzo Spallanzini in 1768 had shown that the idea was probably false, the issue wasn't settled until 1859, when Louis Pasteur showed that a flask of sterilized broth that remained sealed would never go bad, while one that was exposed to the air did. Even more crucially, Pasteur showed that a flask with a long, narrow, curved tubular glass opening would also stay sterile if the tube is long enough and its opening faces down so that almost nothing can float in and begin to grow in the broth.

Lamarck's ideas were daring and revolutionary for his time, even if some of his basic assumptions (such as the "ladder of life" and spontaneous generation, both typical of his time) were wrong. One minor idea in his work was the notion of **inheritance of acquired characters**. In a nutshell, people in Lamarck's time thought that features you acquired during your lifetime (such as the strong muscles of an athlete or a blacksmith) could be inherited by your children. In another example, Lamarck suggested that the long neck of the giraffe was caused by the ancestral short-necked giraffe stretching its neck over and over again to reach leaves high in the tree (Fig. 6.3). Its efforts to strengthen and elongate its neck would be passed on to its descendants, generation after generation.

Nearly every naturalist of that century, including even Charles Darwin, thought that this was true, and it was not until the late 1800s that it was shown to be false. Nonetheless, the word "Lamarckism" today usually refers to this outdated notion that was actually a minor part of Lamarck's broad theories. This distortion trivializes Lamarck's name to an unimportant part of his thinking, and it completely ignores all

Figure 6.2 Life is not a ladder or "chain of being" from "lower" organisms to "higher" organisms, with humans at the top. Life is a branching bush or tree, with many different species radiating out from a common ancestor, so ancestors live alongside their descendants. In this view, "primitive" organisms like bacteria and sponges, which evolved very early, live side by side with very late products of evolution, like humans.

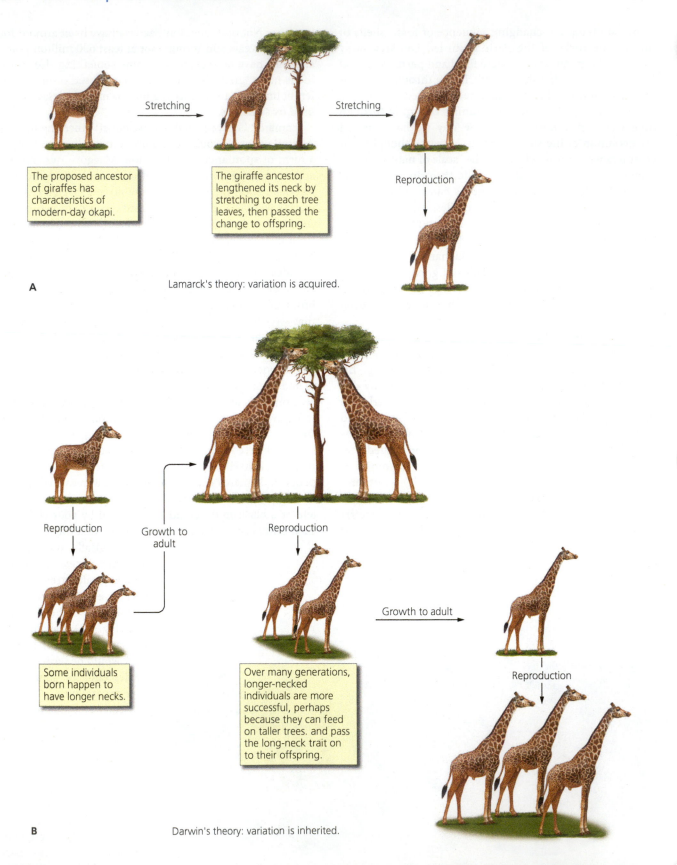

Stretching

Stretching

Reproduction

The proposed ancestor of giraffes has characteristics of modern-day okapi.

The giraffe ancestor lengthened its neck by stretching to reach tree leaves, then passed the change to offspring.

A   Lamarck's theory: variation is acquired.

Reproduction

Growth to adult

Reproduction

Some individuals born happen to have longer necks.

Over many generations, longer-necked individuals are more successful, perhaps because they can feed on taller trees. and pass the long-neck trait on to their offspring.

Growth to adult

Reproduction

B   Darwin's theory: variation is inherited.

Figure 6.3   In Lamarck's concept of evolution, organisms developed special strengths during their lifetimes and passed these features on directly to their descendants. For Lamarck, the giraffe got its long neck by stretching and stretching a little bit each generation, then passing on its longer neck to its offspring. The study of genetics since 1880 shows this is not how animals inherit their features. Darwinian natural selection, argues that in any population of giraffes in nature, there is a range of animals of different sizes and shapes, some with slightly longer necks. During bad times when lower branches are hard to reach, those giraffes with slightly longer necks will have an advantage in reaching higher vegetation and surviving. They then are the parents of the next generation of giraffes, which would inherit the tendency for longer necks. Generation after generation, selection for longer necks would produce the long-necked giraffes we know today.

his pioneering and brilliant insights as one of the founders of biology. In fact, the present-day misunderstanding was due to the influence of Lamarck's rival, Baron Georges Cuvier, who did his best to ridicule and distort Lamarck's ideas, especially after Lamarck died.

6.2 The Evolution of Darwin

The year 1809 was a pivotal one in history. Not only was Lamarck's book published, but on February 12, 1809, both Charles Darwin and Abraham Lincoln were born, just 15 hours apart. Both men ended up being among the most influential people in history as well. They changed the way humans regarded ideas like life and slavery.

Charles Darwin (Fig. 6.4) was born of a wealthy family in Shrewsbury, Shropshire, England, near the border with Wales. His father was a noted doctor, and his grandfather Erasmus Darwin not only was a respected physician but had turned down an offer to be the king's private doctor. Erasmus Darwin was also an important Enlightenment scholar, who helped found the famous "Lunar Society" (because they met once a month in Birmingham during a full moon). These self-described "Lunatics" included such distinguished people as Matthew Boulton and James Watt, inventors of the modern steam engine; chemist Joseph Priestly; and even

Figure 6.4 Charles Darwin in his twenties, shortly after he completed the *Beagle* voyage.

Benjamin Franklin. Erasmus Darwin was also the author of a famous 1794 poem, "Zoonomia," which was full of evolutionary speculations. Because it was a poem, however, it was not taken as a scientific work or a serious threat to religion. Thus, evolutionary ideas were part of young Charles Darwin's inheritance from his own grandfather.

At the age of 16, Charles was sent off to medical school at the University of Edinburgh in Scotland, the best in Britain at the time, to continue the family tradition of doctoring. But Darwin was a failure as a medical student. He could not stomach the dissection of rotting corpses robbed from graves or watching people having their gangrenous limbs sawed off without anesthetics, all common practices of medicine in those days. Instead, he pursued the popular hobby of natural history, a fad in Britain in the early 1800s, learning to collect insects, plants, tidepool animals, and many other natural specimens. One of his professors was Robert Grant, who was the first scientist to prove that sponges were animals and who exposed Darwin to the French evolutionary thought of Lamarck and others.

 See For Yourself: Life of Charles Darwin

After Charles dropped out of medical school, his father sent him off to Cambridge University when he was almost 20, hoping that he could become a country parson and obtain a respectable living while still pursuing his mania for collecting specimens from nature. Here, young Charles learned about plants from the legendary botanist John Stevens Henslow and served as a field assistant in Wales to Adam Sedgwick, the first professor of geology anywhere. By 1831, he had a degree but didn't know what to do with his life.

 See For Yourself: Who Was Charles Darwin?

Then fate stepped in. Through Henslow, he learned of an opportunity to travel on the British naval vessel HMS *Beagle*, charged with the task of surveying the coast of Brazil and Argentina. Although his father objected to such a dangerous voyage at first, eventually he relented. Darwin joined the crew as a gentleman who could share dinner with Captain Robert FitzRoy since a British ship captain was not able to relax or converse with anyone else on the ship except to give orders. It helped keep ship captains sane during the long isolation of the voyage (although FitzRoy did eventually go mad later in life). Originally, Darwin was not the ship's naturalist; that job was officially the responsibility of the ship's doctor, Robert McCormick. But McCormick had so many responsibilities on board, while Darwin was free to collect and do natural history, that McCormick resigned in disgust early in the voyage and was shipped home from Rio de Janeiro. Darwin then became the official naturalist for the rest of the voyage.

 See For Yourself: The Voyage of Charles Darwin

Aboard the *Beagle*, Darwin embarked on what became a 5-year journey around the world, collecting fossils of giant ground sloths, huge armadillo relatives, and other extinct beasts in Brazil and Argentina as well as specimens of many animals and plants in the jungles of South America; climbing the highest peaks of the Andes Mountains; experiencing an earthquake in Chile; and noticing the curious features of animals on many islands. He read Lyell's *Principles of Geology* during the voyage, so he learned to see the world with Lyell's uniformitarian vision of continuous gradual change through time.

 See For Yourself: Darwin's Collections

By the time he returned to England 5 years later, he was so transformed by the experience that his father remarked, "Why, even the shape of his head has changed!" More importantly, what was *inside* his head had changed even more. Darwin quickly settled down to publishing some of his geological discoveries (he correctly explained the origin of the coral atolls in the Pacific) and sending his specimens out to specialists to describe and publish. Eventually, he made his scientific reputation with his book *The Voyage of the Beagle*. By 1842, he had married his first cousin, Emma Wedgwood (of the Wedgwood pottery family), and started a family (only 7 of their 10 children survived past age 11). He settled down on a quiet estate in the village of Down southeast of London and focused full-time on his research.

While on the Galápagos Islands west of Ecuador, Darwin had noticed that each island had its own distinct species of tortoises and mockingbirds, which could easily be told apart. Why was a different species created for each island, and why were they different from those of mainland South America? In Darwin's mind, this suggested that species were not fixed but could change through time. When he returned from the voyage, he sent his Galápagos bird specimens to ornithologist John Gould. Gould told him that what he had identified as wrens, blackbirds, grosbeaks, and nuthatches were in fact all finches that had developed different types of beaks for feeding on different diets on different islands. To Darwin, it made much more sense that these new species had evolved from a common ancestral finch population, blown over from the mainland. They then adapted to different kinds of diets since the Galápagos Islands never had native wrens, blackbirds, grosbeaks, and nuthatches to occupy these ecological niches. Clearly, species were not fixed and unchangeable but could transform under the right conditions (Fig. 6.5).

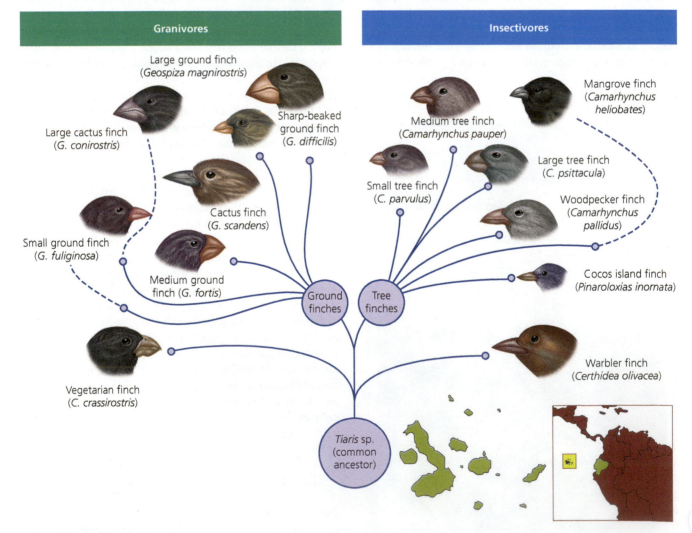

Figure 6.5 The Gálapagos Islands are populated by a number of finch species with different types of beaks for eating different types of food. Their beaks resemble those of grosbeaks, warblers, thrushes, and other birds found on the mainland; but they were all descended from a few finches that blew over from South America, then evolved into a big adaptive radiation of finches occupying the niches that other birds occupied elsewhere in the world.

See For Yourself:
David Attenborough
on Darwin's
Galapagos Discovery

Meanwhile, Darwin read widely about how domesticated animals had changed drastically under the artificial selection of generations of breeders. Darwin himself raised exotic pigeons and saw how much they had changed from the ancestral rock doves. He also read the Reverend Thomas Malthus, who pointed out that human populations are capable of exponential growth and will soon overwhelm their resources unless their numbers are kept in check. Darwin realized that this applied to all populations of living organisms as well.

By 1837 and 1838, Darwin had written down his first short versions of his ideas about how species change. He showed these to his closest friends, including geologist Charles Lyell and botanist Joseph Hooker. He then procrastinated for another 20 years, sitting on the most important idea in all of biology, unwilling and unable to publish his revolutionary ideas in a book-length form. Part of the problem was his continual bad health, which may have been due to the anxiety and stress of worrying about his radical theories or possibly to a disease he may have caught in South America.

He had good reason to worry because the idea of evolution was still controversial and even dangerous to one's reputation. In 1844, an anonymous author published the best-selling *Vestiges of the Natural History of Creation*, which presented French evolutionary ideas and became a scandalous success. (The author was later revealed to be the book's publisher, Robert Chambers, who stayed anonymous for his own safety.) *Vestiges* was very amateurish in its biology (since the author was not a scientist) and easily dismissed at the time, but it showed that the idea of evolution was still in the air. It also demonstrated how dangerous the idea was in England in 1844, especially for someone like Darwin who was trying to build a reputation as a sound scientist. Discouraged, Darwin stalled still more, spending many years doing research on the barnacles he collected on the *Beagle* voyage and eventually publishing four huge volumes on the fossil and living barnacles of the world. This research made him an internationally respected scientist, but he still didn't dare publish his most dangerous and significant ideas. Instead, he sealed his drafts with instructions for his wife to publish them after he died.

See For Yourself:
Darwin and
Wallace

Then another twist of fate occurred. A young naturalist by the name of Alfred Russel Wallace was collecting specimens in what is now Indonesia. Finding so many interesting and exotic animals on different islands, his thinking developed along the same lines as Darwin's thoughts 20 years earlier. While bedridden with malarial fever, Wallace had an inspiration and wrote his inspirations down in a letter that he mailed to (of all people) Charles Darwin. Wallace had independently come up with many of Darwin's ideas, including natural selection. Horrified, Darwin went to his friends Charles Lyell and Joseph Hooker searching for an honorable solution to the dilemma of who got credit for the idea. They arranged to have Darwin's 1842 abstract and Wallace's letter read at an 1858 meeting of the Linnean Society—but hardly anyone noticed. Even the president of the Linnean Society, Thomas Bell (who had studied and published Darwin's Galápagos reptiles, including the marine iguanas and giant tortoises), summarized the events for the year 1858 with the words, "The year which has passed has not, indeed, been marked by any of those striking discoveries which at once revolutionize, so to speak, the department of science on which they bear."

See For Yourself:
Alfred Russel
Wallace

Darwin now had to work quickly, or he would be scooped. He abandoned his planned gigantic work and wrote a shorter (155,000 words, brief by Victorian standards) book summarizing his ideas, *On the Origin of Species by Means of Natural Selection*. It sold out all 1250 copies on the day it was published and eventually went through six editions while Darwin was alive.

See For Yourself:
The Genius of
Charles Darwin

The argument of *On the Origin of Species* is very simple yet powerful (Fig. 6.6). First, Darwin drew an analogy to the artificial selection of domesticated animals practiced by animal breeders. He argued that if they could modify the ancestral wolf into dogs as different as a Chihuahua and a Great Dane, then species were not as fixed and stable as commonly believed. He also borrowed the idea from Malthus that natural populations are capable of exponential growth, yet these same populations remain stable in nature because of high mortality rates. From this he deduced that *more young are born than can survive*. Darwin next described the variability of natural populations and pointed to the evidence from domesticated animals that these variations are highly heritable. He concluded that *organisms that inherit favorable variations are more likely to survive and breed*, and he called this process **natural selection** (later called by others "survival of the fittest").

See For Yourself:
Mechanisms of
Evolution

Natural selection was a very different concept from the ideas of evolution held by Lamarck and earlier naturalists (Fig. 6.3). According to Darwin, the most important thing is that natural populations are highly variable and that no

DARWIN'S ARGUMENT

- Organisms multiply exponentially
- BUT populations stay constant
THEREFORE more organisms are born than can survive
- Organisms vary in nature
- Variation is heritable
THEREFORE, those organisms inheriting favorable variations win the "struggle for survival" and pass those variations on to their descendants

Figure 6.6 Logical structure of the argument for natural selection, based on a few basic observations (bullet points) which lead to deductions or conclusions.

two individuals are truly identical, even if they are siblings. For example, among the ancestral population of giraffes there might be some with slightly longer necks. During periods of drought when vegetation is scarce, they would be able to reach leaves higher on the trees than other giraffes in the population and thus survive. These longer-necked ancestral giraffes then become the parents of the next generation and pass on the genes for longer necks to their descendants. After many generations of doing this, the giraffe's neck would gradually get longer and longer.

As we saw, Darwin was not the originator of the concept of evolution, and at least two others proposed something like natural selection. Why, then, does Darwin deserve most of the credit? For one thing, Darwin was the right man at the right time. In 1844, the idea was still too controversial and Chambers' amateurish efforts only made people scoff at the idea of evolution. But by 1859 the time was right, and many people were thinking along these lines (as Wallace's independent inspiration shows). In addition, Darwin had worked hard to build a sterling scientific reputation and was a member of the Oxford–Cambridge elite, not a radical from the lower-class medical schools of London, which had embraced French ideas of evolution in the 1830s and 1840s. Most importantly, Darwin put all the pieces together in one

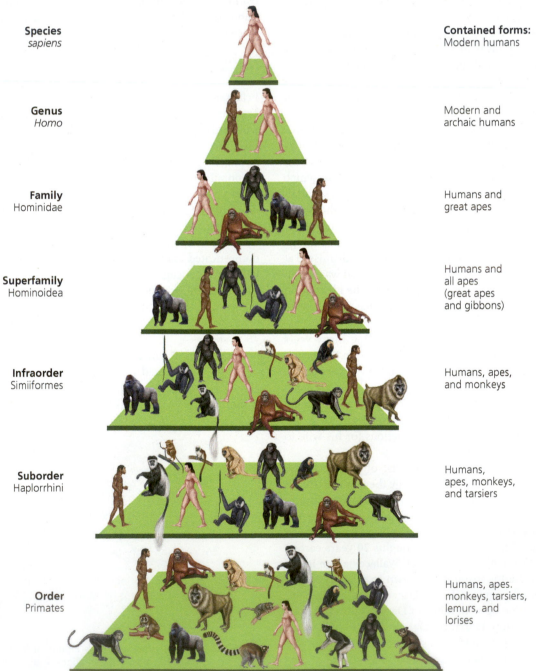

Classification of *Homo sapiens* within the order Primates

Rank	Contained forms:
Species *sapiens*	Modern humans
Genus *Homo*	Modern and archaic humans
Family Hominidae	Humans and great apes
Superfamily Hominoidea	Humans and all apes (great apes and gibbons)
Infraorder Simiiformes	Humans, apes, and monkeys
Suborder Haplorrhini	Humans, apes, monkeys, and tarsiers
Order Primates	Humans, apes. monkeys, tarsiers, lemurs, and lorises

A

(Continued)

book and provided two important concepts: the *evidence* that life had changed through time (the "fact" of evolution) and a *mechanism* for how it occurred, natural selection (the "theory" of evolution). He overwhelms the reader with example after example so that by the end the conclusion is inescapable.

Darwin's ideas were controversial at first, but by the time Darwin died in 1882, the fact that life had evolved was universally accepted in all educated parts of the world (including most of the United States). When he died, Darwin was hailed as one of Britain's greatest scientists. He was buried with honor in Scientist's Corner of Westminster Abbey in

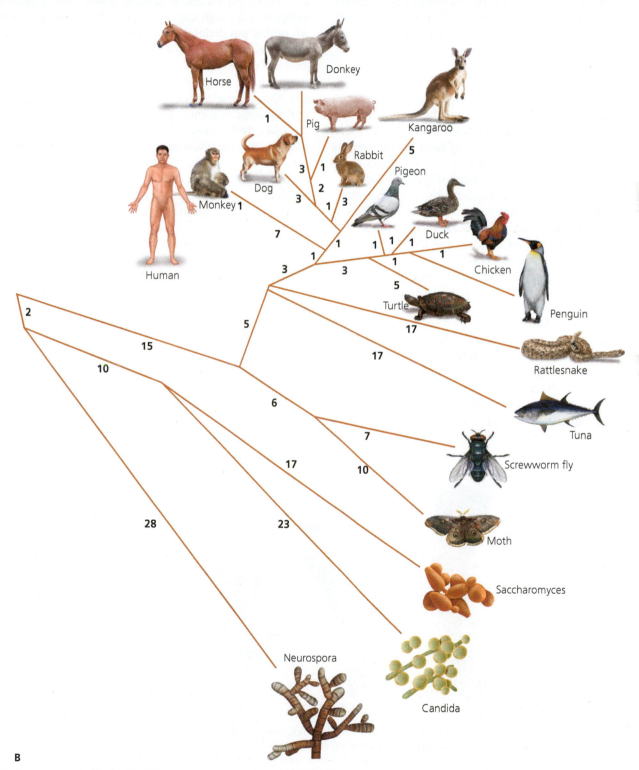

B

Figure 6.7   A. The nested structure of classification as groups nested within large groups strongly suggests that life has a branching history from a common ancestor. **B.** The branching tree of life from molecules closely resembles that produced by drawing a tree of life from external, visible anatomical traits.

London, right next to the grave of Isaac Newton and other famous English scientists.

6.3 Darwin's Evidence of Evolution

What was the evidence that Darwin mustered in the chapter of *On the Origin of Species* that convinced the world of the reality of evolution by 1882? Darwin gave evidence with many examples supporting the idea that life evolved. Among the most powerful (and still important today) are the following:

THE FAMILY TREE OF LIFE

See For Yourself: What Darwin Did Not Know

The first line of evidence had been emerging ever since the days of the Linnaean classification of animals in 1758, a full century before *On the Origin of Species*. The purpose of Linnaeus' classification scheme was to document God's handiwork by discovering the "natural system" of classification that God had used. Inadvertently, Linnaeus stumbled upon the obvious fact of nature: each group (such as a species) of animals and plants clusters with other groups into larger groups (called "taxa" in the plural, "taxon" in the singular), such as a genus or order, and those higher-level supergroups cluster into even larger groups (such as classes or phyla) with additional taxa (Fig. 6.7A). For example, humans are part of a taxon (the family Hominidae) that also includes chimps, gorillas, orangutans, and gibbons. The apes, in turn, cluster together with the Old World monkeys (family Cercopithecidae) as well as New World monkeys, lemurs, and bushbabies into a larger group, the order Primates. The primates are clustered with cows, horses, lions, bats, and whales in the class Mammalia. The mammals are clumped with fishes, birds, reptiles, and amphibians in the subphylum Vertebrata. Together

with sponges, corals, mollusks, and other invertebrates, the vertebrates are part of the kingdom Animalia. The natural system for arranging and classifying life is a hierarchical system of smaller groups clustered into larger groups, which is best represented as a branching tree of life.

By Darwin's time, this branching pattern of life was even more strongly supported and led many people toward the notion that life had undergone a branching pattern of evolution (although not as boldly as Darwin suggested it). All of this was deduced by comparison of features visible to the naked eye or a simple magnifier, primarily in the anatomy of the organism. But not even Darwin could have dreamed that the genetic code of every cell in your body also shows the evidence of evolution. Whether you look at the genetic sequence of mitochondrial DNA or nuclear DNA or cytochrome c or lens alpha-crystallin, or any other biomolecule, the evidence is clear: the molecules show the same pattern of nested hierarchical similarity that the external anatomy reveals (Fig. 6.7B). Our molecules are most similar to those of our close relatives, the great apes, and progressively less similar to those of organisms more distantly related to us.

Teasing out the details of this molecular similarity shows us a simple fact: every molecular system in every cell reveals the fact that life has evolved! If we were not closely related to the apes, why would we share over 98% of our genome with the chimpanzee and progressively less with primates that are less closely related to us?

HOMOLOGY

See For Yourself: David Attenborough on the Tree of Life

As comparative anatomy became a science in the early 1800s, anatomists were struck by how animals were constructed. Organisms with widely differing lifestyles and ecologies used the same basic building blocks of their anatomy but had modified those parts in remarkably different ways. For example, the primitive vertebrate forelimb (Fig. 6.8) has the same basic elements: a single large bone (the humerus), a pair of long bones in the forearm (the radius and ulna), a number of wrist bones (carpals and metacarpals), and multiple bones (phalanges) supporting five digits (fingers). But look at the wide array of ways that some animals use this basic body plan. Whales have modified them into a flipper, while bats have extended the fingers out to support a wing membrane. Birds also developed a wing but in an entirely different way, with most of the hand and wrist bones reduced or fused together and feather shafts providing

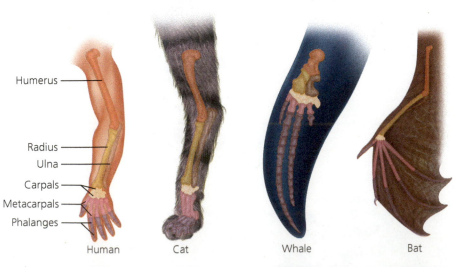

Figure 6.8   Homology in vertebrate forelimbs. Even though the arm of different mammals is constructed of the same basic building blocks (certain bones in certain arrangements), these bones are highly modified for different functions. The fingers of a bat, for example, are hugely elongated to support a wing, while the arm and fingers of a whale are highly modified to form a flipper.

Humerus · Radius · Ulna · Carpals · Metacarpals · Phalanges — Human · Cat · Whale · Bat

the wing support instead of finger bones. Horses have lost their side toes and walk on one large finger, the middle finger. None of this makes any sense unless these animals had a standard body plan in place from their distant ancestors and had to modify it to suit their present-day function and ecology. These common elements (bones, muscles, nerves) that serve different functions despite being built from the same basic parts are known as **homologous structures**. For example, the finger bones of a bat wing are homologous with our finger bones, and so on.

In fact, nature uses a variety of non-homologous ways to build a wing. We have already seen how vertebrates build wings in two completely different ways, even though bats and birds started out with the same bones from a common ancestor—and neither of their solutions resembles the wing of a pterodactyl (which supported its entire wing by the bones of a hugely elongated fourth or "ring" finger). And insects have a completely different structure of their wings, which comes from a different part of the body independent of their arms or legs. The flippers of a whale may perform the same function as the paddles of a marine reptile or the fins of a fish, but all three structures with a common function have completely different bony structure. These different types of wings and flippers found in unrelated organisms are **analogous** organs, which perform the same function but have a fundamentally different structure.

The fact is that organ systems are jury-rigged with whatever bones the animal inherited from its ancestors and not built from scratch in the optimal shape for its current use. This only makes sense if life had evolved to use what anatomy is already available in the ancestral form.

VESTIGIAL STRUCTURE AND OTHER IMPERFECTIONS

As we saw with the way in which homologous structures are reused, most organs are remnants of past structures. Some of them no longer serve a function and are clear evidence that organisms are not perfectly or intelligently designed. The list of such vestigial organs (Fig. 6.9) is overwhelming. They include not only the appendix and tonsils and tail bones of humans (none of which have a significant function now) but the tiny splint bones in the feet of horses, which are remnants of the time when horses had three toes. When these

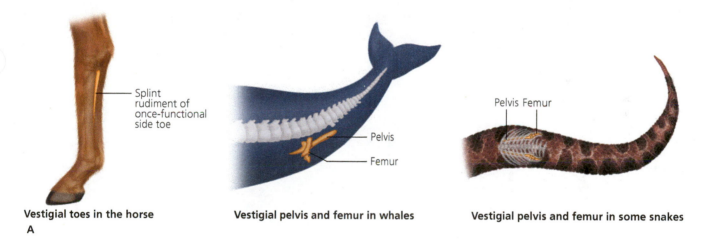

Splint rudiment of once-functional side toe

Vestigial toes in the horse
A

Pelvis
Femur
Vestigial pelvis and femur in whales

Pelvis Femur
Vestigial pelvis and femur in some snakes

Figure 6.9 **A.** Vestigial organs, showing snake hips, whale hips, and horse side toes. **B.** Blue whale skeleton with the vestigial hip and thigh bones still visible behind the rib cage and below the spine (here suspended from the spine by a triangular frame).

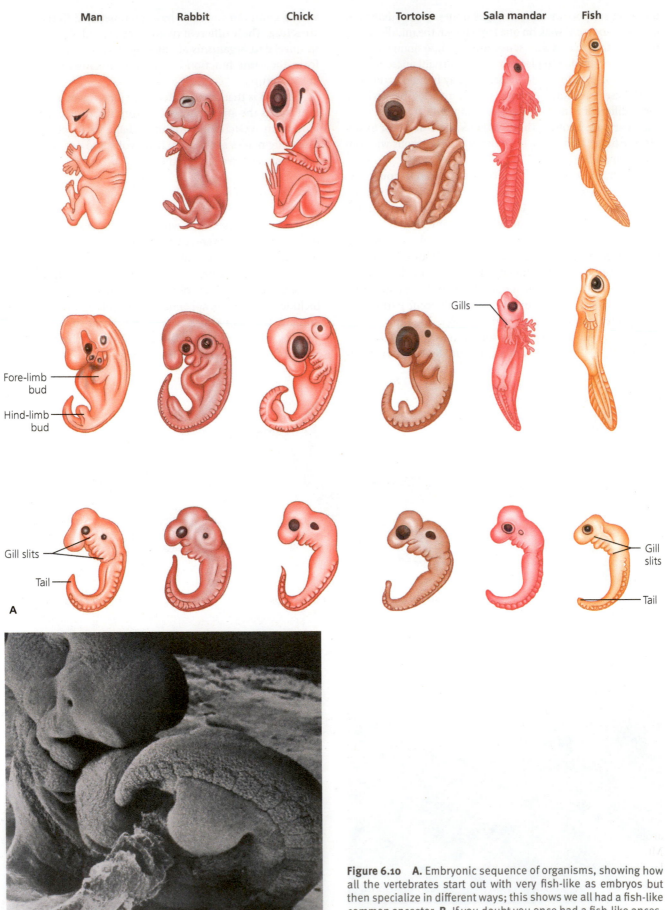

Man **Rabbit** **Chick** **Tortoise** **Sala mandar** **Fish**

Gills

Fore-limb
bud

Hind-limb
bud

Gill slits

Tail

Gill
slits

Tail

A

B

Figure 6.10 **A.** Embryonic sequence of organisms, showing how all the vertebrates start out with very fish-like as embryos but then specialize in different ways; this shows we all had a fish-like common ancestor. **B.** If you doubt you once had a fish-like ancestor, this is how you looked at 5 weeks after conception. You had a long tail, and the embryological precursor of gill slits.

bones break, the horse is crippled for life. Whales and some snakes both have tiny hips and thighbones buried deep in their bodies, with no function whatsoever. Why would they have these features unless they evolved from ancestors that did have hind limbs?

EMBRYOLOGY

Even before Darwin, studies of embryos began to provide important evidence for evolution. In the 1830s, the great German embryologist Karl Ernst von Baer documented that the embryos of all vertebrates show a common pattern (Fig. 6.10). Whether they develop into fish, amphibians, or humans, all vertebrate embryos start out with a long tail, well-developed precursors of gill slits, and many other fish-like features. In adult fish, the tail and gills develop further; but in humans, they are lost during further development. Von Baer was simply trying to document how embryos developed, not provide evidence of Darwin's notion of evolution, which would not be published for many years.

Darwin used this evidence in *On the Origin of Species*, and embryology soon developed into one of the growth fields of evolutionary biology. If you had any doubts that you once had ancestors with fish-like gills and a tail, take a look at a human embryo 5 weeks after fertilization (Fig. 6.10B). Why did you have pharyngeal pouches (predecessors of gills) and a tail if you had not descended from ancestors with those features?

BIOGEOGRAPHY

The great expeditions of discovery in the 1700s and 1800s produced a huge diversity of animals and plants that were unfamiliar to Europeans and not anticipated by the authors of the Noah's ark story, or even Linnaeus in 1758. Not only do these animals and plants render any version of the Noah's ark story impossible, but they created another problem as well. They are not distributed around the earth in a pattern from Mt. Ararat in Turkey (the supposed landing site of the ark) but instead have their own unique distribution patterns that only make sense in light of evolution.

Darwin got a hint of this on the Galápagos Islands, where each island had a slightly different species of giant tortoise or finch. Instead of populating the islands with the same species as occurred on the mainland, there were unique species on each island. This phenomenon is true of islands in general. Later studies of the unusual animals of exotic places further confirmed the fact that they were largely populated with unique animals found nowhere else, and their distribution patterns made no sense in the context of migration from the ark. For example, Australia is

Figure 6.11 Convergent evolution of Australian marsupials (right) and placentals from the northern hemisphere (left). Even though all Australian marsupials are closely related to each other and have a pouch where they raise their joeys, they have evolved to occupy the same body shapes and ecological niches as placentals on other continents. Thus, there is an Australian marsupial equivalent of the "wolf," the "cat," the "flying squirrel," the "ground hog," the "anteater," the "mole," and the "mouse." (Modified from several sources)

home to its own unique fauna of native pouched mammals, or marsupials (Fig. 6.11). These include not only the familiar kangaroo and koala but also many other body forms that apparently evolved to fill the same ecological niches that placental mammals occupy on other continents. There are marsupial equivalents of placental wolves, cats, flying squirrels, groundhogs, anteaters, moles, and mice. If the animals had all migrated away from the ark, why had nothing but marsupials arrived in Australia, then apparently evolved to fill the niches left vacant by the lack of placentals?

Other patterns are equally persuasive. For example, many of the southern continents have at least one large flightless bird species, all members of the primitive group known as ratites. Africa has the ostrich, South America the rhea, Australia the cassowary and emu, and New Zealand the kiwi. This distribution makes no sense in the Noah's ark story but does fit the idea that they were closely related when all these southern landmasses were part of the great Gondwana supercontinent about 100 million years ago. Since the time that these continents have drifted apart, so too have their ratite natives diverged from one another. The same is true of many other groups that evolved in Gondwana over the past 100 million years, such as side-necked turtles and many groups of frogs and fish. Today, they are found on the Gondwana remnants, passengers on the rafts of drifting continents.

See For Yourself:
Evolution: What
the Fossils Say and
Why It Matters

See For Yourself:
Evolution
of Whales

See For Yourself:
Evolution Stated
Clearly

Finally, the fossil record provides the details of how life has evolved and is now the strongest piece of evidence for evolution. The remaining chapters of this book will detail the incredible evolutionary stories revealed by fossils, so we will not discuss them further here.

These were all lines of evidence that Darwin mustered in 1859 and have only grown stronger since with the accumulation of more details and examples. Each alone is strong evidence that life has evolved. Added together, they make the case overwhelming.

6.4 The Origin of Variation

Despite his overwhelming volume of evidence that life had evolved, Darwin's mechanism of natural selection initially did not fare so well. Many of Darwin's critics could not imagine how it was sufficient to shape organisms, and some argued that if favorable variations occurred, they would be blended out of existence in a few generations by breeding with the normal strains of animals. Neither Darwin nor most other scientists of that time knew of a solution to this problem by the time he died in 1882.

However, in 1865 an obscure Czech monk named Gregor Mendel had already found the answer. As he bred strains of pea plants in his garden, he discovered that they had very simple and mathematically predictable inheritance patterns. More importantly, he showed that inheritance does not *blend* the genes of both parents but is *discrete* so that rare genes from one parent can seem to vanish for a generation but then reappear fully functional in the next generation if the genes are crossed in a certain way (Box 6.1).

A good example of how genetic traits can be affected by natural selection is shown by a disease known as sickle cell anemia. If a baby has the sickle cell gene from both parents (homozygous for sickle cell), he or she typically dies very young from the effects of anemia. But the sickle cell gene is a recessive gene that is expressed only when the baby has copies from both parents. If a baby has the normal dominant gene from one parent and a sickle cell gene from the other, he or she does not develop the disease. In fact, in tropical regions where malaria is a major cause of death, these mixed individuals (known as "heterozygotes") are actually protected against malaria and often survive better than people with normal genes from both parents. This means that heterozygotes have an advantage in tropical regions and tend to live and breed more often than homozygotes without the gene at all or homozygotes with copies of the gene from both parents. When these heterozygote malaria survivors become parents, their children have a one in four chance of getting both copies of the sickle cell gene and dying. Thus, the simple Mendelian inheritance pattern means that the sickle cell gene will not blend out and vanish in populations in the tropics, but it is still maintained, even though it is fatal in homozygous recessive individuals.

Mendel's work remained virtually unknown in his own time, and Darwin may have even had a copy of Mendel's 1865 paper—but never read it. In fact, most biologists of Darwin's and Mendel's time would not have noticed the paper, published in an obscure journal. It was full of mathematics, which few biologists knew back then. It was not until three different lab groups independently rediscovered it in 1900 that the time was ripe for appreciating his insights. Genetics made enormous strides over the next 50 years, revolutionized by the discovery of the structure of the DNA molecule and its role in inheritance in 1953. Today, genetics and molecular biology are among the most productive areas of science. We have discovered the complete sequence of the human genome and that of many other organisms. We can identify which genes cause certain traits, especially certain genetic diseases, and know how to treat these conditions. And we find the evidence of evolution in every gene we discover.

6.5 On the Origin of Species

As the title of his book suggests, Darwin thought that by explaining how species could change through time, he could explain the process of speciation. However, it turned out to be far more complicated than simply changing populations through time. When genetic studies of large populations

BOX 6.1: HOW DO WE KNOW?

How Do Genes Work?

In the early 1800s, no one knew how inheritance worked. They could barely see cells under a microscope, let alone resolve the details of tinier structures like chromosomes or DNA that were not discovered until much later. All they could do was conduct breeding experiments and see what kinds of offspring were produced. Many of the early experiments with animal and plant breeding were confusing or misleading because some patterns of inheritance were complex. This "blending inheritance" was a problem for Darwin because his critics thought that any organism with a new favorable variation would have that novel feature blended out when it mated with the other normal individuals in the population.

The breakthrough came when a Czech monk named Gregor Mendel conducted a series of experiments with about 28,000 pea plants in the monastery garden between 1856 and 1863. Luckily for Mendel, pea plants have a very simple pattern of inheritance that would provide insights into the genetics of other more complex inheritance patterns. There were about seven different characteristics that had simple patterns of inheritance: plant size, pod shape and color, seed shape and color, and flower position and color.

As he crossbred pea plants with different characteristics, he found that he got predictable patterns in the features of their offspring. For example, he would cross two plants that produced green seeds, and 75% of their offspring would also produce green seeds—but 25% would produce yellow seeds.

A similar ratio of 75% and 25% showed up in many of the other features of the pea plants.

Mendel came up with an ingenious explanation for this. He suggested that the green seed gene was what he called **dominant** so that if this green gene were present, the seeds would always be green. But every plant has contributions from two parents, so in some cases the plant would carry a **recessive** gene (in this case, the yellow seed is a recessive trait). Recessive genes would not be expressed unless the dominant gene was absent. Thus, if you combine the genes from two different plants, there are four possibilities (Fig. 6.12A). If each parent contributes the dominant gene, the offspring will be **homozygous** (have two copies of the same gene, or GG in this case) and have green seeds. If both parents contribute the recessive gene (g), the offspring will be homozygous for the yellow recessive gene (gg) and produce yellow seeds. However, if one parent contributes the dominant gene and the other a recessive gene, the offspring will be heterozygous (have copies of both genes) and produce green seeds. Thus, only one in four offspring will have yellow seeds (the homozygous recessive occurs 25% of the time), while three-quarters of the offspring will have green seeds (the 25% chance of a homozygous dominant gene plus the 50% odds of either parent combining to form the heterozygous green seeds).

This is just a simple example. Suppose we look at two independent traits: seed color and pod shape (full or constricted). The full pods (coded as F) are dominant, while the

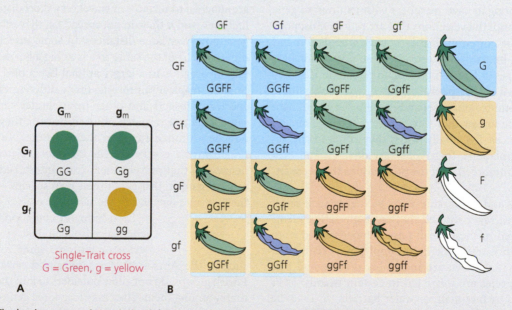

Single-Trait cross
G = Green, g = yellow

A

B

Figure 6.12 The basic pattern of Mendelian inheritance. **A.** Mendel showed that if you cross parent plants with green and yellow genes for seeds, you will get a ratio of 75% green seeds (the G gene for green seeds is dominant), and only if you get the homozygous recessive combination (gg) does a yellow seed develop. **B.** Combining the seed color gene with another gene for normal and constricted seed pods also yields the 75% to 25% ratio for each phenotype but with more possible combinations.

constricted pods are recessive (coded as f). We can make a matrix of all possible results (known as a **Punnett square**) for the likelihood of offspring showing both characters simultaneously (Fig. 6.12B). Of 16 possible combinations, 25% (4 out of 16) will pick up the homozygous recessive gene (gg) and have yellow seeds. Of those yellow-seeded plants, another 25% will have the homozygous recessive gene for pods (ff) and produce constricted pods with yellow seeds (ggff) (Fig. 6.12B). The Punnett square shows that constricted pods will also occur in 25% of the combinations with green seeds, including pure heterozygotes (GGFF); the various heterozygotes (Gg) will produce green seeds, but 25% of them will have constricted pods (Ggff and gGff).

Examples can be multiplied over and over again, but the presence of recessive genes and rare homozygous recessives means that genes can appear to vanish in one generation, only to re-emerge generations later. Thus, important advantageous mutations in the genes can allow a rare recessive gene to hide within the genome and be preserved, invisible to natural selection for a few generations. The "blending problem" that Darwin's critics raised is not really a problem when we know that inheritance is discrete, not blending, and that rare favorable variants can vanish for a while and reappear when the selection forces might have changed in their favor.

were conducted, scientists found that most large natural populations have a lot of interbreeding among their members (known to geneticists as **gene flow**) and tend to be highly resistant to genetic change over time. When a rare mutation arrives, it is very hard for it to establish dominance in the overall larger population because so many individuals are interbreeding that it is swamped.

Instead, the best place for a new mutation or rare variant of a genotype to become established is in a very small population that does not breed with other populations (in other words, experiencing reduced or no gene flow). This kind of genetic isolation means that the population can keep interbreeding among its own members with its unusual gene combinations until the variations that are rare in most populations become common in this isolated population and can be favored by natural selection. These populations are often on the edges or fringes of the main population, so they are often known as **peripheral isolates**. The longer these isolated populations remain separated from the larger main populations, the longer they have time to develop enough genetic differences. Eventually, when they come in contact with other populations, their genes are distinct, and they can no longer interbreed with the populations that were ancestral to them.

In the 1930s and 1940s, studies by biologists such as ornithologist Ernst Mayr showed that most speciation seems to have occurred in these small populations isolated from the mainland ancestral stock. For example, Mayr found that the birds of New Guinea are incredibly diverse and that many of the species are restricted to certain islands, or to areas separated by mountain or water barriers. In 1942, he proposed that most speciation occurs in **allopatric** ("different homeland," or geographically isolated) populations so that the formation of barriers to gene flow (water barriers, mountains, or whatever) fragments the main population into smaller groups that can become genetically isolated and eventually speciate (Fig. 6.13). Once these barriers are

removed and the new peripheral isolate returns to its homeland (becomes **sympatric**, or "same homeland"), the new species can no longer interbreed. This allopatric speciation model was soon supported by much additional research and is still considered to be the most important mechanism by which species diverge and speciate.

Although geographic isolation is the easiest way to restrict gene flow and generate genetically isolated populations that can become new species, it is not required. All that is needed is some mechanism (such as behavior) to prevent interbreeding between two populations, and they can become genetically isolated. For example, many organisms are attached to surfaces or travel very short distances in their lifetimes, and if they do not spread far, they often don't have gene flow over large distances. In some insects, living in a different part of the same tree is as big a barrier as a mountain or ocean is to a larger animal like a bird or mammal. In some tropical rainforests, many different closely related species of insects and other invertebrates can live on the same large tree or patch of forest because they never travel outside a limited range in their lifetimes. In other cases, organisms have very specific behaviors (especially mating behaviors) that prevent them from interbreeding with any other organism that is not their own species—even though these species may live in the same places.

 See For Yourself: Evolution 101

Even human populations can become genetically isolated, just by virtue of behavior or custom. Some religious sects, such as the Amish or Mennonites and certain Orthodox Jewish groups, do not allow outsiders to join their group and marry among them. They may lose some genes when young people leave their community, but they do not receive new genes from the outside; thus, they become highly inbred. As a consequence, they often have high frequencies of unusual genes compared to populations that live among and around them that are not members of their sect. They also often have unusual genetic diseases due to all the inbreeding. In this case, the only thing to restrict outbreeding and gene flow is behavior, especially a strong religious taboo on marrying outsiders. If this cultural isolation continued for thousands of years, conceivably they could be on their way to becoming a new species of humans.

6.6 Darwinism and Neo-Darwinism

Evolution is a change in gene frequencies through time.

—*Theodosius Dobzhansky, 1937, Genetics and Origin of Species*

Evolution is merely a reflection of changed sequence of bases in nucleic acid molecules.

—*John Maynard Smith, 1958, The Theory of Evolution*

Although most of the educated world had accepted the fact that life has evolved by Darwin's death in 1882, not everyone was convinced that natural selection was the most important or effective mechanism for how life had evolved. The newly emerging genetics labs were discovering phenomena such as mutations, which didn't seem to fit the slow and gradual model of Darwinian natural selection. Even though the fossil record provided more and more evidence for the fact that life had evolved, paleontologists were not that influential in proposing mechanisms for how evolution occurred. Some accepted Darwinian natural selection, while others advocated neo-Lamarckian inheritance, others had their own peculiar ideas, and some were completely agnostic as to the mechanism of evolution. Systematic biologists, who study the naming and relationships of organisms, were busy describing new species; but few thought of the evolutionary implications of their work. The early twentieth century was a time of uncertainty and controversy in evolutionary biology, when each area of science (genetics, paleontology, systematics) was going in a different direction. Few scientists saw Darwin's idea of natural selection as the best explanation for how life evolved. How could scientists ever find a way to determine the power of selection in nature?

In the mid-1930s, three scientists with more mathematical training saw a way to test the importance of natural selection. They were the British biologist J. B. S. Haldane, the British mathematician Sir Ronald Fisher, and the American geneticist Sewall Wright. Instead of trying to decipher complex genetic patterns of real organisms like fruit flies, as the geneticists were doing, they simulated the way the frequencies of certain genes changed using mathematical models, known as **population genetics**. These models allowed them to show what happens over hundreds of generations, calculating how much selection pressure or what kinds of mutation rates were necessary to bring about the necessary changes in the gene frequencies. Such models demonstrated that even weak natural selection could explain the rates of evolution seen in genetic experiments and that mutation rates need not be very large to introduce a few new genes that natural selection could work upon.

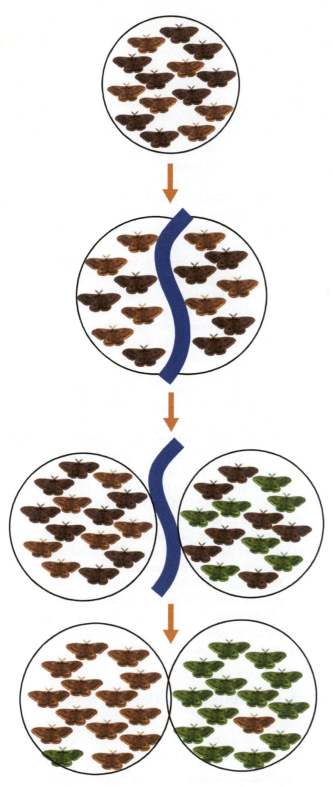

Figure 6.13 Ernst Mayr's allopatric speciation model. From the ancestral population, some kind of barrier to gene flow (whether a mountain, an ocean, or just a changed behavior that prevents interbreeding) separates the original population into smaller isolated populations. Each population can develop its own distinctive gene frequencies, so when they mix again (final stage in diagram), they no longer can interbreed and they have become new species.

By the late 1930s, a number of books tied these new ideas to the current thinking in various branches of biology. This school of thought came to be known as **neo-Darwinism** since it combined the latest developments in genetics, systematics, and paleontology with a revival of the Darwinian emphasis on natural selection. In 1937 the geneticist Theodosius Dobzhansky published a book that synthesized the latest evidence of genetics with the mathematical modeling of Haldane, Fisher, and Wright. In 1942, Ernst Mayr tied the growing consensus about natural selection to the latest ideas in systematics and classification and how species actually formed. Finally, in 1944 paleontologist George Gaylord Simpson documented the connections between evolution in the fossil record and the mathematical models of population genetics and debunked many of the false non-Darwinian notions and explanations of earlier paleontologists. In the 1940s and 1950s, this growing consensus reviving the importance of natural selection as the primary mechanism of evolution came to be known as the "neo-Darwinian synthesis." By the time of the centennial of the publication of *On the Origin of Species* in 1959, neo-Darwinism was the consensus view of nearly all biologists, and most biology textbooks continue to reflect this.

What are the main ideas of the neo-Darwinian synthesis? Its primary insights come from the mathematics of population genetics (with its focus on the **genotype**, or the genetic makeup of the organism), which has shown just how effectively natural selection can change the frequencies of genes in populations. From this, neo-Darwinists often reduced evolution to just the change in gene frequencies through time, without regard to the embryology or development of the organism or the influence of the body (**phenotype**) on the problem. In the view of some neo-Darwinists (like Richard Dawkins), the body is nothing more than a vehicle for genes to make more copies of themselves (this is known as **reductionism**). Mathematical population genetics and lab experiments on organisms such as fruit flies and lab rats showed that most natural variation is due to the recombination of genes from both parents but that additional variation is the result of slight mutations. These random variants are then weeded out by natural selection. The stronger the selection, the more rapid the genetic change.

In some versions of neo-Darwinism, natural selection was treated as an all-powerful, all-pervasive force that, in Darwin's words (from *On the Origin of Species*), "is daily and hourly scrutinizing throughout the world every variation, even the slightest; rejecting all that which is bad, preserving and adding up all that is good; silently and insensibly working." Some biologists think that all changes are adaptive in some way, even if we can't detect how. In this view, no part of an organism is unaffected by natural selection. This idea of all-powerful selection is called **panselectionism**.

The reductionist idea that organisms are nothing more than vehicles to carry their genes led to the extrapolation that the tiny genetic and phenotypic changes observed in fruit flies and lab rats were sufficient to explain all of evolution. This notion defines all evolution as **microevolution**, the gradual and tiny changes that cause a slightly longer tail in a rat or different wing veins in a fruit fly. From this, neo-Darwinism extrapolates all larger evolutionary changes (**macroevolution**) as just microevolution writ large. Finally, most biologists and paleontologists argued that the fossil record should show the gradual transformation of fossils through time (**gradualism**). These central tenets—reductionism, panselectionism, extrapolationism, and gradualism—were the main themes of the neo-Darwinian synthesis of the 1940s and 1950s. Many of them are still followed by evolutionary biologists today.

The evidence for microevolutionary change is abundant throughout nature. We can see small-scale evolution in action all the time. For example, there are a number of different insects that have noticeably changed in response to selection pressure from the environment. The most famous examples include such processes as **industrial melanism** in insects such as the peppered moths of Great Britain. In their natural state, their wings are covered with lots of black and white spots and blotches. These are good disguise when they rest on moss-covered gray tree trunks so that predatory birds cannot see them easily. However, during the Industrial Revolution in the late 1800s and early 1900s, tree trunks became coated in soot and turned black, and the black-and-white speckled moths were no longer camouflaged but conspicuous so that predators selected against them. Instead, a rare mutant with black wings became the most common form of peppered moth because they were well concealed against the dark tree trunks. The frequencies of genes for the dark gray form became much higher as selection worked against the natural phenotype. But since the 1960s and 1970s, environmental laws have cleaned up the smokestacks and reduced the air pollution so much that the tree trunks have returned to their unpolluted color. Consequently, the normal speckled moths recovered their populations, and the black mutant returned to being rare since it is conspicuous again.

Neo-Darwinian evolutionary biology has had many great successes, so there is no reason to doubt that natural selection is the most important engine for evolution. But is it the only factor involved? Is evolution truly reducible to changes in gene frequencies through time?

6.7 Challenges to Neo-Darwinism

The variations detected by electrophoresis may be completely indifferent to the action of natural selection. From the standpoint of natural selection they are neutral mutations.

—*Richard Lewontin, 1974, The Genetic Basis of Evolutionary Change*

When neo-Darwinism swept through the profession in the 1940s and 1950s, it achieved almost a complete consensus. Many evolutionists began to think that the major problems were all solved; only the details needed to be worked out. But it is not necessarily a good thing when an entire field in science seems to have all the answers and is no longer questioning its assumptions. A continuing critical attitude, new unsolved problems, and skepticism and controversy are essential to the health of good science. If a science does not continue to test its ideas but views all essential problems as solved, then it soon stagnates and dies.

Fortunately, the neo-Darwinian synthesis has been continually scrutinized and challenged by legitimate biological and paleontological data, so the field is rife with healthy controversy. Most of these challenges only question some of the more extreme neo-Darwinian tenets or argue that natural selection is not the only mechanism by which life evolves. None of these ideas challenges the well-established fact that *life has evolved* or that we can see life evolving right now or that natural selection is an important (if not exclusive) mechanism for evolution.

NEO-LAMARCKIAN INHERITANCE

As mentioned, most naturalists of the nineteenth century, including Lamarck and Darwin, concluded that features acquired during one's lifetime could be passed directly to descendants (the inheritance of acquired characters). This type of inheritance is unfortunately known as "Lamarckian inheritance," although it was an old notion dating from before Lamarck and only a minor part of Lamarck's ideas and accepted by Darwin as well. It is obvious why this idea is appealing. Instead of the wasteful Darwinian mechanism of the death of many offspring just so a few favorable variants can survive (Fig. 6.3), Lamarckian inheritance allows new variations to be passed on directly in a single generation. By this mechanism, organisms can adapt rapidly on demand.

By the 1880s, however, some geneticists began to doubt whether Lamarckian inheritance was real and to assert the primacy of Darwinian natural selection. The German biologist August Weismann performed a series of experiments that seemed to discredit the idea of the inheritance of acquired characters. He cut off the tails of 20 generations of mice, but each new generation developed a tail, despite this rather extreme form of selection pressure. From this, Weismann concluded that anything that happens to our bodies ("soma" in Weismann's terminology) during our lifetimes does not get back into the genome ("germ line"). This became known as "Weismann's barrier" or the **central dogma** of genetics: the flow of information is one-way, from genotype to phenotype, but not the reverse. When James Watson and Francis Crick discovered DNA in 1953, the central dogma was redefined to mean the one-way flow of information from DNA to RNA to proteins to the phenotype.

For decades, the central dogma seemed to prevail, and even the hint of Lamarckism was considered highly controversial and unorthodox. But as early as the 1950s, embryologist Conrad Waddington showed that repeated environmental stresses could cause abrupt genetic change without direct selection, or what he called "genetic assimilation." The best evidence, however, comes from immunology. When we are born, our immune system is functional but does not yet recognize all the foreign germs and pathogens it must defend against. We acquire immunity through our lifetimes each time we are exposed to a germ and develop an antibody to defend against it. However, a series of experiments showed

that laboratory mice could pass on their immunity directly from the mother to their offspring. It is hard to see how this is explained by anything other than Lamarckian inheritance.

More recently, molecular biologists have found that acquired inheritance is the norm, rather than the exception, in most microbes. Viruses work entirely this way, inserting their DNA into the cell of a host and making more copies of themselves. Many bacteria and some other organisms (including plants such as corn) seem to have "jumping genes" or **lateral gene transfer**, which exchange gene fragments between strains of organisms without sex or even recombination. One group of viruses (the retroviruses that cause AIDS, among other infections) copy their own genetic information from host to host and may be capable of carrying the DNA of one organism into another.

All of these new mechanisms of inheritance suggest that the genome is not so simple or a product of "one-way" inheritance as we thought in the 1950s and 1960s. There is a full range of genetic interactions, starting with the simple "structurally dynamic" genes that respond to a certain environmental stimulus by producing a particular response. At a more sophisticated level are genes that apparently sense their environment and change their response. Automodulating genes change their future responsiveness to stimuli when stimulated. The most Lamarckian of all are "experiential genes," which transmit specific modifications induced during their lifetimes into the genome of their descendants. The example from immunology may fit this, as does bacterial and viral DNA swapping.

Clearly, the simplistic "central dogma" of Weismann no longer applies to microorganisms, which are remarkably promiscuous when it comes to swapping DNA around. It may also not apply to many multicellular organisms either, if the immunology experiments are correctly interpreted.

NEUTRALISM

One of the first challenges to neo-Darwinism came when molecular biology began to understand the details of the genome in the 1960s. Prior to this time, geneticists had assumed that each gene in the chromosome coded for only one protein (and the structures built from them) so that inheritance would be simple (the "one gene, one protein" model). They also asserted that every gene was under the constant scrutiny of natural selection (panselectionism) and that no gene was selectively neutral (even if we can't detect how selection operates). But in the 1960s, a series of discoveries shattered this simplistic idea of the genome. Using a newly developed technique called gel electrophoresis in 1966, Lewontin and Hubby found that organisms had far more genes than they actually use or that can be expressed in the phenotype. Soon, geneticists were discovering that as much as 85%–97% of the DNA in some organisms (including about 97% of human DNA) is not expressed as a phenotypic feature and is either "silent" DNA or "junk" DNA, left over from the distant past when it had some function. If it is not expressed, it cannot be detected by natural selection and is neutral with respect to selective advantages or disadvantages. This new idea of **neutralism** completely shattered the old belief in panselectionism.

At the most basic level, the fundamental structure of the genetic code guarantees that a high percentage of mutations will be invisible to natural selection. The genetic code (Fig. 6.14) consists of a three-letter "triplet" sequence of nucleotides (adenine, cytosine, guanine, and uracil). As the DNA is transcribed, there is a three-letter sequence as the code for one of the 20 amino acids (plus a few codes are used to start and stop the transcription of DNA). Notice in Fig. 6.14 that of the 64 possible combinations of three letters, many of them specify the same amino acid. It is usually the first two letters of the triplet that count, and the third letter makes no difference. For example, if the first two nucleotides are cytosine and uracil, it produces the amino acid leucine, no matter what the third letter is. Clearly, most mutations in the third-letter position (every third nucleotide in the DNA) are invisible to natural selection and must be neutral as a result.

So what is all this useless DNA doing? Some of it may actually have a function in maintaining the spacing between coding regions or be used to help hold the shape of the complex folds of the long DNA strands. Some of these noncoding regions include the following:

1. Introns, chunks of DNA that are initially read but then edited out during final gene splicing

2. Pseudogenes, chunks of DNA that have lost their ability to code for proteins

3. Repetitive DNA: in many parts of the genome, the DNA is made of the same codons repeated over and over again hundreds of time, apparently coding for nothing

4. Transposons, or "jumping genes," which can jump from one part of the DNA to another and yet are not expressed

5. SINEs (short interspersed nucleic elements) and LINEs (long interspersed nucleic elements) that are segments of DNA stuck in the middle of a coding sequence that have no function or ability to code for proteins

6. Highly conserved, noncoding, nonessential DNA which is very consistent in the sequences of many organisms, suggesting that it is important yet can be removed with no effect whatsoever.

Perhaps the most interesting and surprising of all of these junk sequences are **endogenous retroviruses**. These are gene sequences of viruses that once infected us by inserting their DNA into our genome but are no longer active. Instead, every time one of our cells divides we make new copies of this "fossil DNA" from a long-ago viral infection and carry it on through millions of generations. Clearly, these "DNA fossils" hiding in our genome no longer code for the virus or for anything else—they are "junk" that we passively carry around with no ill effects.

First Letter	Second Letter				Third Letter
	U	C	A	G	
U	Phenylalanine	Serine	Tyrosine	Cysteine	U
	Phenylalanine	Serine	Tyrosine	Cysteine	C
	Leucine	Serine	Stop	Stop	A
	Leucine	Serine	Stop	Tryptophan	G
C	Leucine	Proline	Histidine	Arginine	U
	Leucine	Proline	Histidine	Arginine	C
	Leucine	Proline	Glutamine	Arginine	A
	Leucine	Proline	Glutamine	Arginine	G
A	Isoleucine	Threonine	Asparagine	Serine	U
	Isoleucine	Threonine	Asparagine	Serine	C
	Isoleucine	Threonine	Lysine	Arginine	A
	(Start) Methionine	Threonine	Lysine	Arginine	G
G	Valine	Alanine	Aspartic acid	Glycine	U
	Valine	Alanine	Aspartic acid	Glycine	C
	Valine	Alanine	Glutamic acid	Glycine	A
	Valine	Alanine	Glutamic acid	Glycine	G

Figure 6.14 The three-letter sequence of the genetic code, showing how most of the 20 amino acids used by life are specified only by the first two letters of the sequence. The third letter usually does not change the amino acid, so it is redundant and invisible to natural selection.

From these discoveries, geneticists have come to realize that many mutations are adaptively neutral and continue to occur without interference from natural selection. This has led to the hypothesis that there is some kind of **molecular clock**. When molecular biologists began to compare the DNA of closely related organisms, they found that there seemed to be a regular predictable amount of change in their DNA that depended only upon how long ago the two lineages had been separated (Fig. 6.7B). When they calibrated their divergence points on the molecular family tree with the fossil record, they found that they could determine how long ago various lineages branched off, even in the absence of fossil evidence. The only way this can work is if most of the genome is invisible to selection and can constantly change by random mutation without interference. Although the molecular clock has had some great successes, mutation rates can vary unpredictably, so scientists are cautious about putting too much weight on molecular clock estimates for the age of a lineage when all the other evidence disagrees.

STRUCTURAL VERSUS REGULATORY GENES

Early geneticists thought the genome would be very simple, with one gene coding for one protein and nearly all of the genome coding for something (panselectionism). As we have learned, most of the genome is silent or "junk DNA." More importantly, the fact that 80%–90% of the DNA in most organisms codes for nothing at all (so far as we know) says that evolution and selection must work entirely on that remaining few percent of the DNA that *does* code for something. Those remaining genes are known as **regulatory genes**. They are the master switches that control the reading of the rest

of the DNA, some of which is used to make the basic structures of life (**structural genes**) and therefore does not differ between organisms. So from the assertion in the 1950s that *every* gene codes for one protein, we now know that most genes don't code for anything and that only a few regulatory genes exert almost complete control over every other gene in the DNA. By tiny changes in those "switches" or regulatory genes, the organism can make big evolutionary leaps.

 See For Yourself: Evolutionary Genetics

We can see the importance of these regulatory genes when something goes wrong and a bizarre **atavism**, or "evolutionary throwback," occurs. Humans still have the genes for the long tail of their monkey ancestors, and every once in a while, the suppression of those genes fails and a human is born with an external tail (Fig. 16.12). A simple failure in the transcription in the genes of a horse, and you get a horse with three toes. The side toes are poorly developed, but they still resemble the condition of the ancestral horses, which had two functional side toes (Fig. 6.15). This experiment shows that the genes for the ancestral side toes are not lost in modern horses, only suppressed by the regulatory genes, and that when there is a mistake in regulation, these ancient features reappear. Such freakish "horned horses" were thought to have great powers, and Julius Caesar rode one into battle.

The most striking example was an experiment that showed that birds still have the genes for teeth, even though no living bird has teeth. The embryonic mouth tissues of a chick were grafted into the mouth area of a developing mouse. When the mouse grew teeth, they were not normal mouse teeth but conical peg-like teeth similar to those of the earliest toothed birds, or the dinosaurian ancestors of birds. All it took was

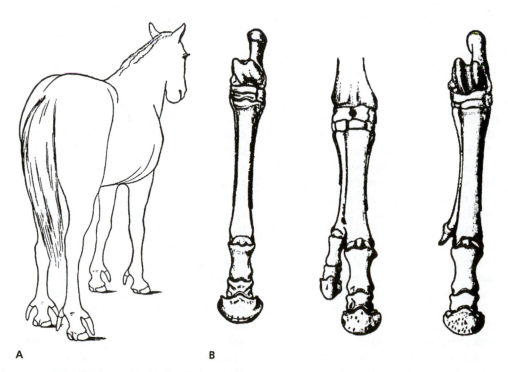

A   B

Figure 6.15 Atavisms, or "evolutionary throwbacks," show us that ancestral genes are not always lost. Humans, for example, still have the gene for a tail but almost never express it. **A.** The "horned horse" which changes from the normal one-toed condition to having its ancestral side toes. **B.** Detail of the bones of the foot of a normal horse (left) and a horse with extra side toes (center and right).

the removal of the regulatory genes that a chick would normally have (by grafting it to a mouse), and the long-suppressed genes for reptilian teeth carried by all birds finally emerged. Other embryonic studies have managed to change the genes that form the birds' short, stumpy, bony tail so that they develop a long bony tail like a dinosaur instead. Another experiment on developing chickens has modified their feet so that they look like dinosaur feet, not bird feet. Yet another experiment turned the beak of a bird into a dinosaurian snout with teeth. Birds have nearly all their old dinosaurian genes still residing in their genome; they are just not expressed.

▷ See For Yourself:
Evolution: Great
Transformations

Figure 6.16 The axolotl is a salamander found in Mexican lakes, which normally spends its life as a breeding adult with larval gills. However, when the lake water goes bad, the axolotl can complete its embryonic history and develop normal salamander lungs; it can then travel to a new body of water. This shows how neoteny allows an organism to make big changes in lifestyle by modifying its existing embryonic history and taking advantage of different developmental stages to survive without major genetic changes.

6.8 Macroevolution and "Evo Devo"

You have loaded yourself with an unnecessary difficulty in adopting Natura non facit saltum ["Nature does not make leaps"] so unreservedly.

—*Thomas Henry Huxley, in an 1859 letter
to Charles Darwin, Nov. 23, 1859*

The importance of regulatory genes goes far beyond neutralism and junk DNA. It raises the question again of whether microevolution, which is so successful at making small changes (such as the number of bristles or wing veins in a fruit fly or the length of the beak of a Galápagos finch), is sufficient to explain macroevolution (the development of large-scale changes in evolution, such as new body plans). If you just keep accumulating tiny microevolutionary changes through time, does this produce wholly novel organisms?

This debate goes back to the earliest days of evolutionary biology. Darwin was a convinced gradualist, but his friend and defender Huxley warned him (in the quotation cited above) that he didn't need to tie his evolutionary ideas to gradualism or rule out evolutionary "leaps" to new body forms. When neo-Darwinism became dominant in the 1940s and 1950s, Richard Goldschmidt, a Jewish German geneticist who fled the Nazis and ended up at the University of California at Berkeley, protested the strict gradualist position. He argued from his studies of gypsy moths that the changes required to build new body plans and new species were not the same as those he found within the normal variation within a species. Goldschmidt argued that some sort of large-scale genetic change was needed (a "systemic mutation" in his words) to jar species out of their normal range of variation and into new body plans. These changes were due to slight modifications in "controlling genes" (what we now call regulatory genes). According to Goldschmidt, speciation was a discontinuous, rapid process that was caused by alterations in controlling genes, not by accumulation of small microevolutionary changes. If a new macromutation appeared that gave the individual a big advantage, it might be a "hopeful monster" that could establish a new species or a new adaptive zone.

Naturally, such opinions appeared highly unorthodox to the gradualistic ideas of the newly dominant neo-Darwinians, and they subjected Goldschmidt to ridicule and scorn. Neo-Darwinists would scoff, "How does the hopeful monster find a mate?" Without more than one "hopeful monster" there is no possibility of breeding or establishing a new population, and thus there would be no chance of a new species forming.

Ironically, the past five or six decades have vindicated Goldschmidt to some degree. With the discovery of the importance of regulatory genes, we realize that he was ahead of his time in focusing on the importance of a few genes controlling big changes in the organism, not small-scale changes in the entire genome as neo-Darwinians thought. In addition, the "hopeful monster" problem is not so insurmountable after all. Embryology has shown that if you affect an entire population of developing embryos with a stress (such as heat shock), it can cause many embryos to go through the same new pathway of embryonic development, and then they all become "hopeful monsters" when they reach reproductive age.

The more we learn about regulatory genes, the more we realize their primary importance to evolution. A common example is the study of **heterochrony**, where organisms change the sequence of their developmental timing. This allows evolution to take advantage of the changes already encoded in our embryology and development. For example, nature frequently makes changes through a particular kind of heterochrony called **neoteny**, where an organism retains its juvenile body form while achieving reproductive maturity. The most famous are the salamanders (such as the Mexican axolotl) that do not complete their metamorphosis into lunged salamanders but hold on to their juvenile gills and body form yet can breed like adults (Fig. 6.16). Whenever these salamanders are exposed to stagnant water conditions, they can complete their metamorphosis into lunged adults and crawl to the next fresh pool of water. Thus, this ability to choose to breed either as the juvenile or adult body form gives them great ecological flexibility, achieved with a few tiny changes in the regulation of their development.

Neoteny is extremely common in nature, especially when the juvenile and adult body forms have radical differences in shape and ecology and allow the organism to "switch-hit" for

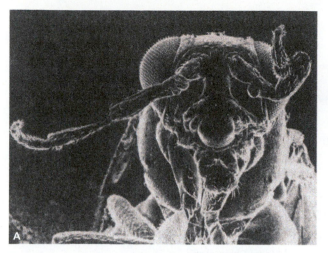 

Figure 6.17 Homeotic mutants show that major changes in body plan are under the control of just a few regulatory genes, called Hox genes. **A.** The antennapedia mutation, where the gene for developing an antenna on the head instead produces a leg on the head of a fruit fly. **B.** The bithorax mutation produces a second thorax and a second pair of wings in flies, where normal flies only have a single pair of wings, plus tiny balancing organs called halteres.

whatever works best. Those pesky aphids that invade your flowers each spring are a classic example. When the food resources are abundant (in the spring and summer), they multiply rapidly, with each immature female giving birth to many daughters as asexual clones (no males are born at all). Those offspring in turn also reproduce asexually as juveniles, so they can make literally hundreds of cloned daughters in a short period of time (which is why they can infest your flowers so quickly). When the fall comes and the food resources dry up and cold weather approaches, they switch to sexual reproduction. A few males are born and mature into adults, then quickly mate with adult females. These lay eggs that can survive the winter and hatch the next spring to start the process all over again. All of this evolutionary flexibility does not require big changes in the genome, just small changes in regulating the normal sequence of embryonic development already encoded in the organism.

The most important recent development, however, has been the discovery of the master regulatory genes, known as the **homeotic genes** (especially the "Hox" genes). These genes are found in nearly all multicellular organisms and regulate the fundamental development of the body plan and how major organ systems develop. They were discovered with experiments on fruit flies that had unusual mutations. Some had legs growing on their heads instead of antennae (Fig. 6.17A); this is known as the "antennipedia" mutation. Some flies developed two pairs of wings instead of the usual single pair (Fig. 6.17B). Normal flies have tiny knoblike balancing organs called "halteres" where the second pair of wings would be, but these mutant flies (known as "bithorax" flies) have apparently changed their regulatory genes so that they develop two thoraxes, each with a pair of wings, instead of the halteres.

From these early discoveries, molecular biologists have identified most of the Hox genes in a number of organisms and found that nearly all animals (including flies, mice, and humans) use a very similar set of Hox genes, with slight variations, deletions, and additions (Fig. 6.18). Each Hox gene is responsible for the development of part of the

organism and all its normal organ systems. Small changes in the Hox genes can put different appendages on a segment of a fly (like the leg where the antenna would go or the wing where a haltere belongs) or even multiply the number of segments. Clearly, then, a tiny change in Hox genes can make a big evolutionary difference. In the arthropods (the "jointed legged" animals, such as insects, spiders, scorpions, and crustaceans), for example, a small change in the Hox genes can multiply the number of segments or reduce them and switch one appendage (for example, a leg) on each segment with another (for example, a crab claw or an antenna or mouthparts). Arthropods (insects, spiders, crustaceans, and their kind) are a classic example of this modular development with interchangeable parts where, with a small change in Hox genes, whole new body plans can evolve easily to exploit new resources.

 See For Yourself: Evolutionary Development

All of these ideas are part of the exciting new research field known as **evolutionary development** (nicknamed "evo devo"), and it is now the hottest research area in evolution. From the neo-Darwinian insistence on every gene gradually changing to make a new species, we now realize that only a few key regulatory genes need to change to make a big difference, often in a single generation. This circumvents many of the earlier problems with ideas about macroevolution and makes it entirely possible that the processes that build new body plans and allow organisms to develop new ecologies are not the small-scale microevolutionary changes extrapolated upward. Some evolutionists still see "evo devo" as just an extension of the neo-Darwinian synthesis, but others argue that it is an entirely different type of process from that envisioned in the 1950s.

Today, many biologists and paleontologists think that the neo-Darwinian synthesis, with its emphasis on tiny changes in the genotype adding up to new species by microevolutionary change, is not sufficient to explain macroevolution but that these new developments showed how macroevolution could occur. Some of the old ideas from the early days

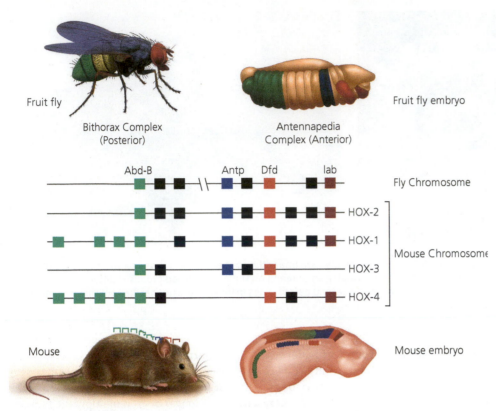

Figure 6.18 The Hox map of the fly and mouse embryo shows that the basic segments of almost all animals are regulated by the same sets of genes, which must have originated at the very beginning of animal evolution.

of neo-Darwinism have been changed or modified as we learn more about the complexities of life.

1. **Central Dogma**: Discoveries suggesting neo-Lamarckian inheritance showed that it is not strictly a one-way flow of information from genes to proteins to phenotype but that there are many ways that genes can be changed after the egg is fertilized and the genome is set.

2. **Reductionism**: The old idea that "evolution is merely change in gene frequencies through time" has been replaced by a realization that changes in structural genes are the most important.

3. **Panselectionism**: The fact that most of the genome is "junk DNA", and is invisible to natural selection and largely unread and not expressed has demolished the old notion that every part of the genome is under the scrutiny of natural selection.

4. **Extrapolationism**: The new discoveries challenge the old notion that the tiny changes visible in the wings of a fruit fly or the beak of a Galápagos finch are sufficient to explain macroevolutionary changes such as the development of new body plans. Instead, we now realize that large-scale changes, especially through evolutionary development, such as the action of Hox genes, and changes in developmental timing (especially heterochrony) are important mechanisms to explain large-scale evolutionary changes.

5. **Gradualism**: The old notion that evolutionary change must be gradual came from the biases of Darwin's time, when the British feared the abrupt revolutionary changes seen in France and tried to make reforms slowly and in small steps. Darwin himself adopted these notions in his thinking, but his friend Thomas Henry Huxley argued that gradualism was not necessary. In fact, the large-scale, rapid changes seen in evolutionary development show that major features, such as body plans, can change quickly (in just a few generations or less).

There are still many hard-core neo-Darwinians who do not agree, of course, so evolutionary biology is in an interesting, controversial time where new ideas are being intensely debated. It may turn out that we understand less about how evolution works than we thought we did back during the heyday of the neo-Darwinian synthesis in the 1950s and 1960s. But the important point is that *this is how normal science operates*. Even if we knew nothing about the mechanisms that drive evolution, it would not change the factual data that show that evolution has occurred and still is occurring. We still don't know exactly how gravity works, but it does not change the fact that objects still fall to the ground. We may never know completely how evolution works, but life keeps evolving. Neo-Darwinism is not all there is to evolution. *Evolution happened in the past and is still happening right now*.

6.9 Evolution Happens All the Time!

Nothing in biology makes sense except in the light of evolution.

—*Theodosius Dobzhansky, 1973*

Biologists finally began to realize that Darwin had been too modest. Evolution by natural selection can happen rapidly enough to watch. Now the field is exploding. More than 250 people around the world are observing and documenting evolution, not only in finches and guppies, but also in aphids, flies, grayling, monkeyflowers, salmon and sticklebacks. Some workers are even documenting pairs of species—symbiotic insects and plants—that have recently found each other, and observing the pairs as they drift off into their own world together like lovers in a novel by D.H. Lawrence.

—*Jonathan Weiner, 2005, "Evolution in Action"*

How can we talk about "the fact that evolution has occurred"? On what evidence can we make that statement? As philosophers of science have pointed out, scientists must use the word "fact" cautiously, as a description of nature or an observation or hypothesis that has accumulated so much overwhelming evidence without falsification that it is a "fact" in common everyday parlance. As Stephen Jay Gould (1981) put it,

> *Moreover, "fact" does not mean "absolute certainty." The final proofs of logic and mathematics flow deductively from stated premises and achieve certainty only because they are not about the empirical world. Evolutionists make no claim for perpetual truth, though creationists often do (and then attack us for a style of argument that they themselves favor). In science, "fact" can only mean "confirmed to such a degree that it would be perverse to withhold provisional assent." I suppose that apples might start to rise tomorrow, but the possibility does not merit equal time in physics classrooms.*

In this sense, the idea that life has evolved and is evolving is "confirmed to such a degree that it would be perverse to withhold provisional assent." We see life evolving all around us, and we have abundant evidence that it has done so in the past. Darwin had some of this evidence in 1859, and it has only become stronger since then.

If the evidence mustered by Darwin and many other scientists since his time were not enough to prove *that* life has evolved, there is an even simpler test: watch life evolve right now! We can see natural selection operating on many different scales and on many different types of organisms. Looking over the shoulders of the hundreds of hard-working, dedicated, self-sacrificing biologists who spend years enduring the harsh conditions in the field to observe evolution in action inspires our admiration.

The classic example, of course, has long been the finches of the Galápagos Islands (Fig. 6.5). Darwin himself collected many of them when he was there in 1835. In the twentieth century, David Lack did a much more detailed study, published in 1947. In recent years, Darwin's finches have been the focus of the research of Peter and Rosemary Grant of Princeton University. The Grants visited the islands year after year, documenting the change in the finch populations. On one island (Daphne Major), the finch population changed wildly from year to year. During a 1977 drought, the finches with strong beaks survived because they could crack the toughest seeds and survive the shortage of food. The next few years, all of the finches on that island were their descendants and had the stronger nut-cracking bills found on other species of Galápagos finches. Since that time, the return of wetter conditions has changed the

finches yet again, so forms that have more normal beaks for eating a wide variety of seeds could also survive. From this, it is easy see how such strong selection pressures could transform the ancestral finches (which still live in South America) into a wide variety of specialized finches that perform the roles that other birds play on the mainland. Instead of nuthatches, there are thick-billed finches; instead of woodpeckers, there are finches with long bills for drilling wood and probing for grubs; instead of warblers, there are finches with similar bills called warbler-finches. One finch has even learned to use a twig as a tool for fishing for insects in the hollows of trees! Recent research has even identified the genes that control beak shape in these finches and artificially duplicated the pattern seen in nature by adding or subtracting those genes.

 See For Yourself: Galapagos Finches Evolving

We don't have to live in the Galápagos to see evolution happen. We can see evolution in action in our own backyards. The common European house sparrow is found all over North America today, but it is an invader, brought from Europe in 1852. These initial populations then escaped and quickly spread all over North America, from the northern boreal forests of Canada down to Costa Rica. We know that the ancestral population was all very similar because they were introduced from a few escaped immigrants, but since they have spread to the many diverse regions of North America, they are rapidly diverging and on the way to becoming many new species. These house sparrows now vary widely in body size, with more northern populations being much larger than those that live in the south. This is a common phenomenon, known as "Bergmann's rule," due to the fact that larger, rounder bodies conserve heat better than smaller bodies. House sparrows from the north are darker in color than their southern cousins, perhaps because dark colors help absorb sunlight and light colors are better at reflecting it in warm climates. Many other changes, in wing length, bill shape, and other features, have been documented.

New species can arise even faster than people once thought. A study by Andre Hendry at McGill University in Montreal analyzed the sockeye salmon near Seattle. These salmon tend to breed either in lakes or in streams and have different shapes dependent upon which environment they breed in. In the 1930s and 1940s, sockeye salmon were introduced to Lake Washington east of Seattle and rapidly became established in the mouth of the Cedar River. By 1957, they had also colonized a beach called Pleasure Point. In less than 40 years, these two populations rapidly diverged. The males of populations that live in the swift-flowing waters of the Cedar River are slender, to fight the strong currents; the females are bigger so that they can dig deeper holes for their eggs, to prevent the river from eroding them away. The populations that live in the warmer, quieter waters of the lakeshore near Pleasure Point have males with deeper, rounder bodies, which are better at fending off rivals for mating privileges, and females with smaller bodies because they do not have to dig deep holes for their eggs. These populations are genetically isolated and already show the differences that would be recognized as separate species in most organisms. Hendry was able to show that this species split started in less than 40 years, and in just a few more generations, they might be genetically isolated and become distinct species.

Another rapidly evolving fish is the three-spined stickleback. Sticklebacks that live in the ocean have heavier body armor than those that live in lakes. In one pond near Bergen, Norway, biologists have been able to document this change in less than 31 years. In Loberg Lake, Alaska, the change took only a dozen years, or just six generations. Sticklebacks also change their spines in response to local conditions. In open water, longer spines are an advantage since they protect against being swallowed by predators. But in shallow water, the long spines are a liability because they make it easier to be captured by dragonfly larvae, which have long pincers. A single Hox gene, *Pitx1*, turns off and on the switches that regulate spine length. Other studies have shown that by artificially modifying captive sticklebacks so that they have unusual new combinations of spines, the females only mate with the males that have new traits, so sexual selection is a driving force in their evolution of novelty. Prior to these studies, ichthyologists would readily assign specimens with different spine counts and different body armor to different species, but these studies show just how easy it is for one stickleback population to transform to another species given the right conditions.

See For Yourself:
Evolution in
Action

See For Yourself:
What Darwin
Never Knew

Examples like these could be multiplied endlessly. In New England, the periwinkles have dramatically changed their shell shape and thickness in less than a century, probably due to predation pressure by newly introduced crabs. In the Bahamas, the anole lizards (the common "chameleon" in the pet shops, which are not true chameleons) have changed the proportions of their hind limbs after people have introduced them to new islands with different vegetation. In Florida, the soapberry bug has evolved a significantly longer beak in response to the invasion of its habitat by a non-native plant with larger fruits. In Hawaii, honeycreeper birds have evolved shorter bills as their favorite food source, the native lobelloids, have disappeared and the birds have switched to another source of nectar. In Nevada, the tiny mosquito fish that live in isolated desert water holes that were once connected by large lakes during the last Ice Age have quickly evolved major differences in less than 20,000 years. And in Australia, the introduced wild rabbits (brought by European settlers less than a century ago) have modified their body weight and ear size in response to the different conditions of the Outback.

Humans are often the strongest agents of selection for many wild animals. In populations of bighorn sheep, trophy hunters have killed off most of the rams with spectacular horns, so the smaller males with reduced horns had a better chance of breeding, and the population no longer has any large-horned rams. Rattlesnakes that are too nervous and buzz when humans approach are quickly killed, so in many regions the rattlers no longer give any warning. Overfishing of the Atlantic cod led to a population crash during the 1980s, and large cod nearly vanished; those that bred quickly while they were small and immature had a better chance of survival.

But the most dramatic and rapid examples of evolution in action occur with microorganisms, especially viruses and bacteria. Every year doctors have new flu strains to battle because last year's flu strain has evolved a new protein coat that makes it unrecognizable to our immune systems and allows it to infect us again. The heavy use of antibiotics has selected for strains of bacteria that are resistant to every drug we throw at them. When sulfoamides were introduced in the 1930s, resistant strains evolved in only a decade. Penicillin was introduced in 1943, and by 1946 there were resistant strains. For this reason, doctors are now much more cautious about issuing antibiotics to sick patients, who want the drug even though it is useless against the viruses that cause cold and flu. Similarly, the heavy use of antiseptic cleansers and wipes has led to strains of bacteria that can resist most antiseptics. Many medical researchers think that our excessive cleanliness in the Western world is to our disadvantage because young people are no longer exposed to many different kinds of germs, and they become vulnerable when a strong strain (that doesn't infect people in the "dirty" underdeveloped parts of the world) invades. Now hospitals are worried—when one of these antibiotic-resistant strains of bacteria appears in a hospital, it can quickly spread to many patients, and nothing can stop it.

See For Yourself:
Debunking Myths
About Evolution

See For Yourself:
How We found Out
Evolution Is True

Likewise, many insects and weeds have evolved resistance to pesticides and herbicides, all within a few decades, causing enormous economic damage to people all over the world. Every modern housefly now carries the genes that make it resistant not only to DDT but also to pyrethroids, dieldrin, organophosphates, and carbamates, so there are few poisons left that can suppress them. The mosquitoes that evolved resistance to DDT and other organophosphate insecticides apparently originated in Africa during the 1960s, spread to Asia, then reached California by 1984, Italy in 1985, and France in 1986. As entomologist Martin Taylor describes it (quoted in Weiner, 1994):,

> It always seems amazing to me that . . . cotton growers are having to deal with these pests in the very states whose legislatures are so hostile to the theory of evolution. Because it is the evolution itself they are struggling against in their fields every season. These people are trying to ban the teaching of evolution while their own cotton crops are failing because of evolution. How can you be a creationist farmer any more?

Evolution is happening all around us. It happens every time a new germ invades your body, a new pest or weed destroys our crops, or a new insecticide-resistant fly or mosquito bites you. We cannot change the fact that life is evolving all around us and threatens our survival if we don't come to terms with that evolution.

RESOURCES

BOOKS AND ARTICLES

Bell, Thomas. 1858. Annual Presidental Report of the Linnean Society. *Journal of the Proceedings of the Linnean Society: Zoology* 3 (1858): 45–62.

Campbell, John. 1982. Autonomy in evolution. In *Perspectives on Evolution*. Ed. R. Milkman. Sinauer, Sunderland, Mass., pp. 190–200.

Carey, Nessa. 2015. *Junk DNA: A Journey Through the Dark Matter of the Genome*. Columbia University Press, New York.

Carroll, Sean. 2005. *Endless Forms Most Beautiful: The New Science of Evo/Devo*. W. W. Norton, New York.

Carroll, Sean. 2007. *The Making of the Fittest: DNA and the Ultimate Forensic Record of Evolution*. W. W. Norton, New York.

Coyne, Jerry. 2009. *Why Evolution Is True*. Viking, New York.

Darwin, Charles. 1859. *On the Origin of Species by Means of Natural Selection, or the Preservation of Favoured Races in the Struggle for Life* (facsimile of the first edition). Harvard University Press, Cambridge, Mass.

Dawkins, Richard. 2004. *The Ancestor's Tale: A Pilgrimage to the Dawn of Evolution*. Houghton Mifflin, New York.

Dawkins, Richard. 2009. *The Greatest Show on Earth: The Evidence for Evolution*. Free Press, London.

Desmond, Adrian, and Jonathan Moore. 1991. *Darwin: The Life of a Tormented Evolutionist*. Warner, New York.

Dobzhansky, T. 1937. *Genetics and the Origin of Species*. Columbia University Press, New York.

Dobzhansky, T. 1973. Nothing in biology makes sense except in the light of evolution. *American Biology Teacher* 35: 125–129.

Eldredge, Niles. 1985. *Unfinished Synthesis*. Oxford University Press, New York.

Gould, Stephen Jay. 1980. Is a new and more general theory of evolution emerging? *Paleobiology* 6:119–130.

Gould, Stephen Jay (1981) "Evolution as Fact and Theory" *Discover, v.* 2 (May): 34–37.

Gould, Stephen Jay. 1982. Darwinism and the expansion of evolutionary theory. *Science* 216:380–387.

Gould, Stephen Jay. 2002. *The Structure of Evolutionary Theory*. Harvard University Press, Cambridge, Mass.

Heraclitus. 2001. *Fragments: The Collected Wisdom of Heraclitus* (translated by J. Haxton). Viking: New York.

Levinton, Jeffrey. 2001 (2nd ed.). *Genetics, Paleontology, and Macroevolution*. Cambridge University Press, New York.

Lewontin, R. 1974. *The Genetic Basis of Evolutionary Change*. Columbia University Press, New York.

Maynard Smith, J. 1958. *The Theory of Evolution*. Penguin, New York.

Mindell, David P. 2006. *The Evolving World: Evolution in Everyday Life*. Harvard University Press, Cambridge, Mass.

Prothero, D. R. 2017 (2nd ed.). *Evolution: What the Fossils Say and Why It Matters*. Columbia University Press, New York.

Prothero, D.R. 2020. *The Story of Evolution in 25 Discoveries: The Evidence for Evolution and the People who Found It*. Columbia University Press, New York.

Ridley, Mark. 1996 (2nd ed.). *Evolution*. Blackwell, Cambridge, Mass.

Schwartz, Jeffrey. 1999. *Sudden Origins: Fossils, Genes, and the Emergence of Species*. John Wiley, New York.

Shubin, Neil. 2008. *Your Inner Fish: A Journey into the 3.5 Billion Year History of the Human Body*. Pantheon, New York.

Stanley, Steven. 1979. *Macroevolution: Patterns and Process*. W. H. Freeman, New York.

Steele, Edward. 1979. *Somatic Selection and Adaptive Evolution: On the Inheritance of Acquired Characters*. University of Chicago Press, Chicago.

Steele, Edward, Robin Lindley, and Robert Blanden. 1998. *Lamarck's Signature: How Retrogenes Are Changing Darwin's Natural Selection Paradigm*. Perseus Books, Reading, Mass.

Weiner, Jonathan. 1994. *The Beak of the Finch: A Story of Evolution in Our Own Time*. Knopf, New York.

Weiner, Jonathan. 2005. Evolution in action. *Natural History* 115(9):47–51.

Wills, Christopher. 1989. *The Wisdom of the Genes: New Pathways in Evolution*. Basic Books, New York.

Woodward, John. 1695. *An Essay Toward a Natural History of the Earth and Terrestrial Bodies*. R. Wilkin, London.

Zimmer, Carl. 2009. *The Tangled Bank: An Introduction to Evolution*. Roberts and Company, New York.

Zimmer, Carl, and Doug Emlen. 2015. *Evolution: Making Sense of Life*. W. H. Freeman, New York.

SUMMARY

- The universe, and everything within it, is constantly evolving. Not even mountains or oceans are permanent, and life is constantly in change as well.

- In 1809 Lamarck published one of the first theories of evolution, which imagined different branches of life climbing up the "ladder of nature" from spontaneous generation (not the branching tree of life with one common ancestor that we know today). Today, the only idea of his that is still remembered is the "inheritance of acquired characters," which was a common notion about inheritance held by almost all naturalists of his time and until the late 1800s when it was discredited by numerous experiments.

- Charles Darwin was the first scientist to combine all the evidence documenting the fact that life was and still is evolving with a mechanism (natural selection) that explained how it evolved. He is given most of the credit for making evolution acceptable because he compiled a huge amount of evidence in support of it, and he was a respected scientist writing the right book at the right time.

- Natural selection combines several lines of evidence to their obvious conclusions. Life is capable of reproducing and multiplying at very fast rates, yet populations in nature remain fairly stable and constant. Therefore, we must conclude that more organisms are born than can survive. Populations of animals and plants vary tremendously in nature, and we know that many of those variations are heritable. From this we conclude that organisms that inherit favorable variations will be more likely to survive and leave offspring to the next generation.

- Darwin mustered many impressive lines of evidence to show that life had evolved. These included evidence from how natural classification schemes show the family tree of life (which has a branching structure showing that we all are interrelated); the evidence of homology (organisms are not perfectly designed but use the same basic building blocks to create very different structures); many organisms have useless structures, or vestigial organs, that demonstrate that they once had fully functional versions of these structures in their evolutionary past; the earliest embryos of vertebrates look virtually identical, with a fish-like body that shows their common ancestry, and only later develop into very different kinds of adults; and organisms are distributed around the world in patterns that only make sense in the light of evolution. Since Darwin's time, these lines of evidence have been strengthened, and many more have been added, to show that life has evolved.

- In his lifetime, Darwin never solved the puzzle of inheritance, which instead was solved by Gregor Mendel in 1865 but not understood or developed until the early twentieth century, when genetics became a major field of research. Mendel did show, however, that inheritance was not blending but discrete, so a new adaptation will not be blended out when the organism interbreeds with the main population but can be expressed in later generations.

- The problem of the origin of new species was solved by Ernst Mayr and other scientists in the 1930s, when it was realized that new species need to be genetically isolated from their ancestral population so that they can develop new gene frequencies, in a model known as "allopatric speciation."

- Although the fact of evolution was indisputable since Darwin's time, natural selection was not the most popular mechanism for many decades. This was true until the 1930s and 1940s, when mathematical population genetics showed how small mutation rates and selection rates could indeed make major changes in gene frequencies. By the late 1940s, natural selection was popular again (combined with modern genetics) and revived as part of the neo-Darwinian synthesis.

- The neo-Darwinian synthesis is continually questioned and challenged, as all good science must be to prevent it becoming a dogma.

- There are now mechanisms by which microorganisms acquire new genomes in a single event ("neo-Lamarckism"), so gene flow is not one way from gene to DNA to protein to organism (the "central dogma" of genetics).

- Most of the genome is junk DNA, unread and unused by most organisms unless the regulatory genes turn on or turn off specific genes that are not being used. For this reason, and many others, much of the DNA is not affected by natural selection and is adaptively neutral ("neutralism").

- The regulatory genes, especially the Hox genes, are more important than any others because they can abruptly change entire organ systems and body plans with a simple mutation and form new species much quicker than the neo-Darwinists imagined. This is the subject of the booming research field of evolutionary development.

- Evolution happens in real time and has been observed by scientists many times. It happens in less than a year in microorganisms like viruses and bacteria. The fact that life evolves is a well-established reality.

KEY TERMS

Scale of nature (p. 113)
Branching tree of life (p. 113)
Spontaneous generation (p. 113)
Inheritance of acquired characters (p. 113)
Lamarckism (p. 113)
Natural selection (p. 117)
Linnaean classification (p. 120)
Evidence for evolution (p. 120)
Homology (p. 120)
Analogy (p. 121)
Vestigial structures (p. 121)
Embryology (p. 123)

Biogeography (p. 123)
Blending inheritance (p. 124)
Discrete inheritance (p. 124)
Dominant gene (p. 125)
Recessive gene (p. 125)
Homozygote (p. 125)
Heterozygote (p. 125)
Gene flow (p. 126)
Punnett square (p. 126)
Peripheral isolate population (p. 126)
Allopatric speciation (p. 126)
Sympatric speciation (p. 126)

Population genetics (p. 127)
Neo-Darwinian synthesis (p. 128)
Genotype (p. 128)
Phenotype (p. 128)
Panselectionism (p. 128)
Reductionism (p. 128)
Microevolution (p. 128)
Macroevolution (p. 128)
Gradualism (p. 128)
Industrial melanism (p. 128)
Central dogma of genetics (p. 129)
Lateral gene transfer (p. 129)

Junk DNA (p. 129)
Neutralism (p. 130)
Genetic code (p. 130)
Endogenous retroviruses (p. 130)
Structural genes (p. 131)
Regulatory genes (p. 131)
Atavism (p. 131)
Heterochrony (p. 132)
Homeotic genes (p. 133)
Evolutionary development (p. 134)

STUDY QUESTIONS

1. How did the idea that the earth was unchanged since its creation only 6000 years ago get replaced with the modern notion that the earth is immensely old and constantly changing?

2. Describe the features of Lamarck's theory of evolution. How does it contrast with our modern concepts?

3. Why is it meaningless to talk about the "missing link"?

4. Evolution is commonly symbolized by a march of monkeys, apes, and primitive humans in a line with modern humans at the end of the line. What is wrong with this icon of evolution?

5. Why are animals like sponges and sea jellies not really inferior to creatures like birds or mammals?

6. What is wrong with the question, *If humans evolved from apes, why are apes still around?*

7. What is spontaneous generation? How was it disproven?

8. Where did Darwin first hear about evolutionary ideas?

9. Why was Darwin invited to be on the *Beagle* voyage? Why was he not the official ship's naturalist at the beginning?

10. How did Lyell influence Darwin's thinking on the *Beagle* voyage?

11. How did Darwin's specimens from the Galápagos Islands change his thinking about the fixity of species as unchanging and permanent?

12. How did the 1844 publication of *Vestiges of the Natural History of Creation* influence Darwin?

13. How did the letter from Alfred Russel Wallace change Darwin's life?

14. How did Darwin's experience with artificial selection by breeders of domesticated animals like pigeons, horses, and dogs influence his thinking about the fixity of species?

15. List the basic observations and the two deductions that make up natural selection.

16. Why was Darwin's explanation of evolution so influential while earlier efforts were not?

17. Distinguish between the fact of evolution and the theory of natural selection.

18. How does classification of animals and plants suggest that evolution from a common ancestor had occurred?

19. How has modern molecular biology further supported the idea of common ancestry and the branching tree of life?

20. Describe the difference between homologous and analogous structures.

21. How do vestigial organs support the idea that life has evolved?

22. What evidence from embryology shows that all vertebrates have a common fish-like ancestor?

23. How are the modern patterns of biogeography best explained by evolution and not the Noah's ark story?

24. Why was the idea of blending inheritance such a problem for Darwin? How did Mendel's work solve it?

25. Why was Mendel's work not appreciated until 35 years after it was first published?

26. How are new species formed? Why is genetic isolation so important to speciation?

27. How does genetic isolation occur even in species that are not allopatric?

28. How was Darwin's mechanism of natural selection regarded among biologists before the 1930s? How did population genetics change this?

29. What fields of biology came together in the 1930s to make the neo-Darwinian synthesis?

30. How is modern evolutionary biology highly reductionist?

31. What is panselectionism? How was it debunked by neutralism and junk DNA?

32. What is the difference between microevolution and macroevolution?

33. What is industrial melanism? How does it demonstrate natural selection in the real world?

34. What is the central dogma of genetics? How was it challenged by the discovery of retroviruses, jumping genes, and other neo-Lamarckian evidence?

35. Why is the third letter in the DNA code often redundant and selectively neutral?

36. What do endogenous retroviruses tell us about the junk content in our DNA?

37. What is the molecular clock? How does it work?

38. What is the difference between structural genes and regulatory genes?

39. What do atavisms like humans with tails tell us about evolution and gene regulation?

40. How does heterochrony allow rapid evolutionary changes without major genetic changes?

41. How do homeotic genes like the Hox genes suggest that macroevolutionary changes are possible?

42. Why do biologists consider evolution to be a fact?

43. How do we see evolution in action every cold and flu season?

PART II
Earth and Life History

There's nothing constant in the Universe,
All ebb and flow, and every shape that's born
Bears in its womb the seeds of change

—*Ovid, Metamorphoses XV (8 CE)*

7

Birth of the Earth

"We live on a hunk of rock and metal that circles a humdrum star that is one of 400 billion other stars that make up the Milky Way Galaxy, which is one of billions of other galaxies which make up a universe which may be one of a very large number, perhaps an infinite number, of other universes. That is a perspective on human life and our culture that is well worth pondering."

—*Carl Sagan, Cosmos*

Telescopic image of HL Tauri, in the constellation known as Taurus, the bull. This image shows a protoplanetary disk forming around the young star HL Tauri, a good analogue for the early stages of the formation of our solar system. This image comes from the Atacama Large Millimeter Array telescope in Chile.

7.1   Origins

Our view of the universe and the solar system has changed dramatically in the last 500 years. Before 1543, most humans thought that the earth was flat and at the center of the universe and that the stars were tiny points of light on the dome of the heavens (Fig. 3.1). In 1543, Copernicus proved that the sun, not the earth, was at the center of our world and that the earth was just one of several planets in orbit around the sun. In 1611, Galileo used one of the first of the newly invented devices called telescopes to discover that the stars were beyond counting and not scattered on a big dome over our heads. He also confirmed that Jupiter has its own moons, showing that it was not surrounded by a perfect celestial sphere if the moons could move completely around it. He debunked the notion that the planetary bodies were perfect and unsullied when his telescope revealed that the earth's moon is covered in craters and not a perfect celestial sphere. Finally, by discovering that Venus has phases like the moon ("full Venus," "half Venus," etc.), he showed that Venus was moving around the sun in an orbit inside our own orbit. Most importantly, he confirmed Copernicus' idea that the earth was just another planet orbiting the sun. By the early 1700s, Isaac Newton had worked out the laws of motion and gravitation and showed how the entire system could be explained by basic physics.

Today, we look at the amazing images from space (Fig. 7.1), coming from both land-based telescopes and the Hubble Space Telescope, and see what no one could have possibly imagined even 30 years ago. We can watch the different stages of how other stars are born and die and how other planets and solar systems have formed. These images, and the astrophysical calculations and models that explain them, give us a new view of the origin of the solar system and allow us to explain much of what was simply guesswork before this century.

As scientists, we must not only describe what we know about the world but always ask the question, *how* do we know this? In a college-level course, it is not enough just to memorize facts, as you might have done in your previous science courses, but you must also *understand how things work*, especially *how we know* about these events that took place in the past. So it's a fair question: how do we know about the formation of universe during the "Big Bang," over 13 billion years ago, and the solar system, when it took place over 4.5 billion years ago?

There are multiple lines of evidence we can examine to get at these seemingly impossible questions.

1. Astronomical observations have shown that almost all the bodies in space—stars, quasars, galaxies, and so on—are rapidly moving away from us, with the most distant objects moving fastest. This observation led to the realization that the universe is expanding and must have started at a single event in the past, now nicknamed the "Big Bang."

2. The accidental discovery of the previously predicted cosmic background radiation confirmed that the Big Bang must have occurred.

3. The motions of all the planets in the solar system are in the same plane, moving in the same direction, suggesting that they all started out together as part of a large disk of matter moving around the sun.

4. Another line of evidence comes from examining the rest of the universe and finding other solar systems that are in the earlier stages of formation. Images from the Hubble Space Telescope and

now from extremely high-resolution telescopic arrays on land, such as the Atacama Large Millimeter Array in Chile, where solar systems such as HL Tauri (Fig. 6.1) and Beta Pictoris can be viewed, give us examples of what our early solar system might have looked like. In 2018, we obtained the first good image of a new planet forming.

5. Numerous space probes have flown by all the planets in our solar system and given us a clear idea of what they are made of, what the inside of the planet is like, and what lies in the space between the planets. Thus, we need to explain the chemical composition of the solar system and why the planets are different.

6. We have actual samples of rocks from space. These include meteorites that came from the earliest stages of solar system formation as well as those that came from planets that formed and then broke up or were blasted off the surface of Mars and reached earth.

7. Last but not least, we have actual pieces of rock brought back from the moon during the Apollo missions from 1969 to 1976, and our Mars rovers have been collecting samples and analyzing the composition of the surface of Mars in detail.

Putting this all together, a remarkably consistent and well-supported model of the origin of the solar system emerges. Throughout our description of how the solar system formed, however, we will also look at the evidence for *why* scientists accept certain aspects of the model.

7.2 The "Big Bang" and the Origin of the Universe

Where did we come from? When and where did it all begin? These questions have fascinated and troubled people for thousands of years. For millennia, the explanations came from a wide variety of religious myths and stories, each different for every culture on earth. But early in the twentieth century, it became possible to go beyond myth and speculation and use the methods of science to find out what really happened.

 See For Yourself: The Big Bang

The first breakthrough came from a number of female astronomers working at Harvard College Observatory under W. C. Pickering. They were known as the "Harvard Computers" because they were talented mathematicians who could make fast calculations in their heads and on paper and do measurements by hand. (Only much later did the word "computer" come to mean the electronic devices we all use.) Pickering hired them because they were not only good at math but also careful and meticulous when studying and analyzing thousands of glass photographic plates of the night sky shot by different telescopes. They were also cheaper than male assistants (25 cents an hour, less than a secretary) and worked hard without complaining, for 6 days a week. Remember, this was a time when most women were barred from scientific careers completely, and those who tried to get an advanced education in science met huge barriers every step of the way.

However, their individual talents soon emerged, and some of them made discoveries that revolutionized astronomy and outshone most other male astronomers of their time. The most famous was Annie Jump Cannon, who catalogued the stars of the night sky and proposed our modern system of star classification, from red giants to white dwarfs. She built upon the first complete star classification system by Antonia Maury.

For our story, however, the key woman was Henrietta Swan Leavitt. She was assigned to study classes of stars known as "variable stars" because their brightness fluctuated from one night to the next. She soon realized that their brightness variations had a regular period of fluctuation, with the brightest stars (most luminous stars) having the longest periods of brightness variation. She found many variables in a cluster in the constellation Cepheus (thus known as "Cepheids") that were all the same distance away. This allowed her to calibrate the brightness spectrum. In 1913, after studying some 1777 variable stars, she worked out the relationship between the period of brightness fluctuation and the luminosity and showed that we could determine how far a star was away from us by measuring its luminosity and its period of fluctuation. Thanks to Leavitt, astronomers now had a reliable tool to tell how far away a star or galaxy was from earth.

The next step was made by a legendary astronomer, Edwin Hubble. In 1919, he was assigned to work at the newly completed Mt. Wilson Observatory (Fig. 7.2) in the mountains above Pasadena, California, and had free use of what was then the world's most powerful telescope, a reflecting telescope with a 100-inch mirror. His first major discovery in 1924 used Leavitt's Cepheid variable stars to show that the spiral nebulae were in fact galaxies outside our own Milky Way galaxy and that the Milky Way was just one of many galaxies. This vastly expanded our understanding of the size of the universe, far beyond what people once thought was possible.

Then he used the telescope to systematically study as many stars and galaxies and other large celestial objects as he could. He not only measured their distance using the Cepheid variable method but also analyzed the spectrum of light from the star. Like taking a prism and splitting sunlight into its major colors, the light from the stars can also be split into a spectrum of colors (Fig. 7.3). The spectrum, however, has white "bands" across the color scale, caused by the absorption of certain elements. We can find these same bands when we analyze the spectrum of burning sodium or other metals in the lab, so each set of bands tells us what elements we are seeing.

His major collaborator in this effort was Milton Humason, who had no education past age 14 but was eager to prove himself. Humason originally drove the mules that hauled the telescope and other materials up that steep mountain. He then became a janitor during the night shift when the astronomers were at work, so Hubble got to know him. He found that Humason had unexpected talents and promoted him to be his assistant. Hubble admired Humason's quiet determination to take the difficult photographs and do the

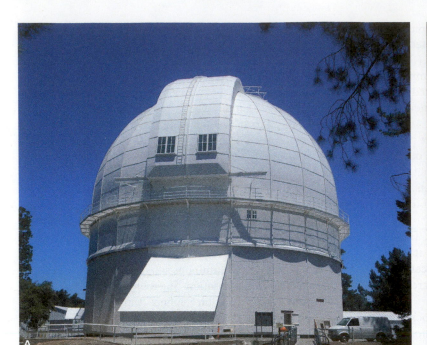

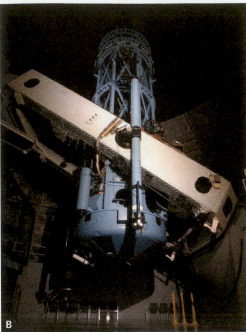

Figure 7.2   A. Mt. Wilson Observatory today, showing the dome of the largest telescope. **B.** Photo of the 100-inch Hooker reflecting telescope on Mt. Wilson Observatory, where Hubble and Humason discovered the evidence for the expanding universe.

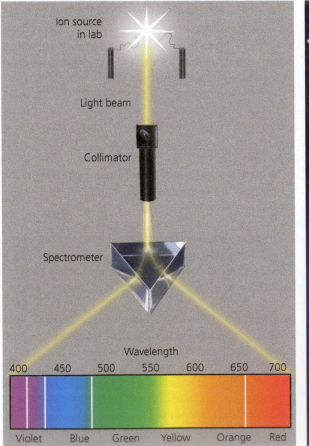

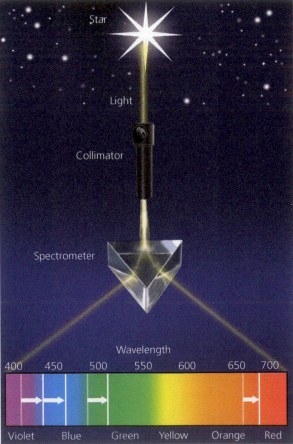

Figure 7.3   When the spectrum of starlight is broken up through a prism, not only are the different colors and wavelengths revealed, but there are also white absorption bands for different elements such as sodium and calcium in the spectrum at different wavelengths. Light from distant stars shows absorption bands that are shifted to the red end of the spectrum compared to their normal positions as determined by a source in the lab.

careful measurements of the spectrum of thousands of photographic plates from the telescope.

After measuring hundreds of different stars and galaxies, Hubble and Humason noticed something very peculiar. The nearest stars had absorption lines in their spectra that resembled the same spectrum for that element on earth. But the farther away the star or galaxy, the more the dark absorption bands were shifted from their original positions toward the red direction of each spectrum (Fig. 7.3).

 See For Yourself: Doppler Effect

Why do the absorption lines move toward the red end of the spectrum? This discovery had first been reported and explained for a few galaxies in 1912 by Vesto Slipher at Lowell Observatory in Flagstaff, Arizona. It is what is known as a Doppler shift, caused by the **Doppler effect**. You have experienced the Doppler effect for sound many times. If you are standing on the street and a car or train rushes toward you blaring its horn, you will notice that the pitch of the sound gets slightly higher as it approaches. Then after it passes you and rushes away from you, you will hear the sound of the horn drop in pitch again (Fig. 7.4). The Doppler effect is caused by the fact that the sound waves approaching you are bunched up because their source is getting closer and closer. If the waves get bunched up, they go higher in pitch. Similarly, when the sound source travels away from you the

waves are stretched out because their source is retreating (Fig. 7.4). Longer, more stretched-out waves are lower in pitch than when they are not stretched out.

The Doppler shift applies not only to sound waves but to light waves as well. If the source is moving very rapidly toward you, then the light waves will get bunched up and have a shorter wavelength (which corresponds to the blue and violet end of the light spectrum). On the other hand, if the light source is rapidly moving away from you, then its waves will be stretched to longer wavelengths, which correspond to the red end of the spectrum.

Slipher's first observations in 1912, and then Hubble and Humason's careful catalogue of over 46 galaxies and many stars, showed not only that galaxies were red-shifted but that almost all of them showed the red shift; there were no blue-shifted objects that might be moving toward us. More importantly, Hubble and Humason found that the farthest objects had the greatest red shifts, so they must be moving away from us the fastest. Hubble realized that this meant the universe must be expanding. It's analogous to making a loaf of raisin bread (Fig. 7.5). When you start with the ball of dough, the raisins are all close together. But as the ball of dough expands, each raisin moves apart from every other raisin, and those raisins on the outer part of the ball of dough move the fastest.

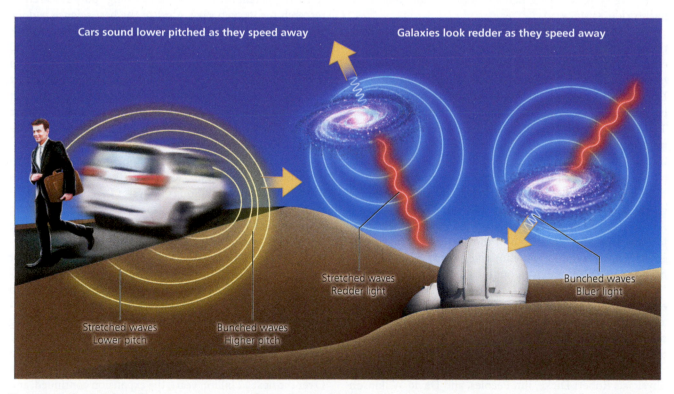

Figure 7.4 The Doppler effect occurs whenever there is motion between a source of waves and the observer. For example, when a moving car honks its horn, the sound waves appear to rise in pitch as they approach you but drop in pitch as they move away. This is caused by the compression of the sound waves when the horn approaches (bunching up the waves and shortening the wavelength, so the pitch is higher). When the sound source is moving away from you, the waves are stretched out and have a longer wavelength, and thus the pitch drops. The same thing applies to the light waves from distant stars. If they were approaching us, their wavelengths would bunch up and shift to the blue-violet end of the spectrum. However, all the stars and galaxies are shifted to the red end of the spectrum, showing that they are moving away from us.

A   B

Figure 7.5 Expanding universe as modeled by a rising loaf of raisin bread. When the universe was smaller, the raisins were closely packed. As it expands outward, the rising loaf of bread separates the raisins so that each is moving away from every other raisin—and those on the edge are moving the fastest.

The universe is expanding. This is a staggering thought, and at first most astronomers were not able to accept it. However, Hubble and Humason's data were solid, and as time went on, more and more objects were analyzed and turned out to be red-shifted. In 1927, Belgian astronomer Georges Lemaître postulated a model where the universe expanded from a single point in the far distant past. Most astronomers did not like the idea that the universe had a beginning but thought that it was in a "steady state" of expansion, with new matter created at the center all the time. One of these steady-state advocates, Fred Hoyle, coined the term "Big Bang" to mock Lemaître's model, and that name has stuck ever since.

The controversy of Big Bang versus steady state continued for about 30 years until the late 1950s without any clear consensus. Then a crucial discovery was made purely by accident, not by astronomers but by two engineers, Arno Penzias and Robert W. Wilson. In 1964, they were employed by Bell Labs, the original research division of AT&T/Bell Telephone, responsible for building the technology of communication for "Ma Bell." They were working on improving the first antennas for receiving and transmitting signals by microwaves, primarily to enable communication with NASA's Project Echo (the first attempts to use satellites for global communication) and later with the TelStar satellite. As the chief electrical engineers on the project, their main job was to get the "bugs" out of the device and improve its efficiency. They found and eliminated many of the sources of "noise" from the antenna, but then they found a source of "background hum" that was 100 times stronger than they expected. It was detected day and night and evenly spread across the sky (so it was not coming from a single point source on earth or in space). It was clearly from outside our own galaxy, and they could not explain it.

Luckily, just 37 miles away in Princeton, New Jersey, physicists Robert Dicke, Jim Peebles, and David Wilkinson just a year earlier had predicted the existence of background "noise" left over from the Big Bang, when everything exploded with a big blast of radiation. The Princeton scientists were just beginning experiments to detect this noise, when a friend told Penzias that he'd seen a preprint of a paper by the Princeton group, predicting the exact same

background noise. The two groups got in touch, and Penzias and Wilson showed them what they had found. Lo and behold, the two Bell Lab engineers had accidentally discovered the proof that the Big Bang had actually happened. For this discovery, Penzias and Wilson eventually received the 1978 Nobel Prize in Physics. And it was discovered entirely by accident!

These stories are classic examples of how "pure" scientific research leads to amazing discoveries. Sometimes discoveries are made by people trying to look for a specific answer to a specific problem. But more often than not, scientists make important breakthroughs by doing "pure research"—research for its own sake. Most of the best science is done by gathering a broad range of data on a particular topic without knowing what scientists might find. Politicians and many other people often scoff at "pure research" without a definite goal in mind and try to prevent it from receiving funding. But pure research is how nearly all the greatest discoveries of science are made, and science would come to an end without it—and so would all the scientific breakthroughs and live-saving discoveries that science makes.

Since this discovery, the Big Bang model has undergone many modifications as physicists use the properties of matter and the equations of physics to figure out how it all happened. The most recent methods date the Big Bang at about 13.8 billion years ago. At the very beginning the universe was in a "singularity"—an extremely small, high-energy region with an infinite density. Ten milliseconds after singularity, the universe was filled with high-energy particles at temperatures over 1 trillion K, expanding rapidly in all directions. It was so hot that it was only radiation, without matter; and space and time did not yet have meaning but were infinitely warped around this superdense region. Over the next billion years, the universe cooled enough to form subatomic particles, then matter in the form of atoms. Over the next 12 billion years, the expansion continued, and random clumps of matter began to coalesce to form stars and galaxies and quasars. Some of these stars have already burned out and exploded, producing the heavier elements like oxygen, silicon, carbon, iron, and so on, which make up most of the matter in the solar system. In that sense, we are all stardust.

7.3 The Solar Nebula Hypothesis

The observation that all the planets are moving in a flat plane in the same direction around the sun emerged as soon as Isaac Newton's laws helped explain planetary motions. In fact, the very idea of a "solar system" wasn't possible until this discovery was made, and the term "solar system" wasn't coined until Newton's time. In addition, astronomers using the latest advances in telescopes had discovered large fuzzy blobs of gases out in space they called "nebulae" (Latin for "clouds"). By the mid-1700s, Swedish scientist Emanuel Swedenborg, French mathematician René Descartes, and German philosopher Immanuel Kant had all independently suggested that our solar system began as a cloud of dust, or nebula, that was spinning around a central axis. This became known as the **solar nebula hypothesis**. The detailed physics of how this system worked was deciphered in the late 1700s and early 1800s by the great French mathematician and astronomer Pierre-Simon Laplace.

See For Yourself: Origin of the Solar System

According to the solar nebula hypothesis, the solar system started out as a ball of gases and matter (Fig. 7.6) that was about 90% hydrogen, 9% helium, and less than 1% all other elements. This nebula was slowly spinning on its axis. In fact, astronomers have found that most of the nebulae in space are spinning at various speeds. Eventually, this loose mass of cosmic dust began to condense, possibly due to the shock of a supernova from a nearby star. As the early solar nebula began to condense, it increased the gravitational attraction of the matter toward its massive center, and it began to spin faster due to the law of **conservation of angular momentum**.

You are familiar with conservation of angular momentum, even if you may not recognize the name. When you watch figure skaters spin, you will notice that they spin slowly if they have their arms and legs far out from the body, but as they pull their arms and legs (and thus their mass) toward the center, they begin to spin faster. You can also experience this by sitting on a spinning stool and holding some small weights in each hand: as you pull in your arms and the weights toward your spin axis, you will spin faster; if your arms are extended and the weight is farther from the axis, you slow down.

Thus, the spinning solar nebula not only goes faster as it condenses but also begins to form a flattened disk. This is comparable to spinning a blob of molten glass around and around on its axis until it flattens out. The simplest way for the mass of the nebula to resist angular momentum is to spread out into a flat disk.

As the disk continued to spin and condense and cool down, the little grains of cosmic dust collided with each other, clumped together by gravitational attraction, and got bigger and bigger. Eventually, their mass was large enough that they had a significant gravitational attraction, which pulled even more tiny bits of matter toward them and enlarged them further, yet again increasing their gravitational pull. Once these growing clumps of matter reach a diameter

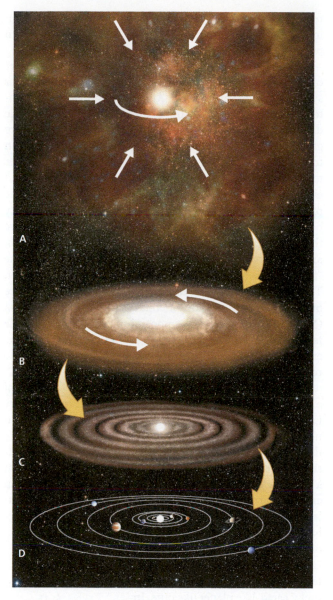

Figure 7.6 Origin of the solar system from the solar nebula. **A.** The solar nebula, a swirling cloud of gas, forms and begins to spin. **B.** Contraction flattens the nebula into a spinning disk. **C.** Matter within the disk coalesces, forming the sun and planetesimals. **D.** The planetesimals coalesce in turn to form protoplanets and eventually planets.

of about 1 km, they have enough gravity to behave like "gravitational vacuum cleaners" and pull in all the cosmic dust in their path as they revolve around the sun. In 1905, geologist T. C. Chamberlain and astronomer F. R. Moulton, called these small bodies **planetesimals** (a mash-up of the words "infinitesimal" and "planet" to describe a body that is a tiny fraction of a planet). As these planetesimals grow larger and larger, their gravitational attraction increases and they pull in more and more loose cosmic debris until they form a clear circular path or trackway within the solar disk (Fig. 7.6). This is particularly true of the gas giant planets like Jupiter and Saturn, which created huge cleared-out tracks as they orbited the central star. The same can be seen in young solar systems like HL Tauri, which have their own protoplanets creating tracks in their disk (Fig. 7.1).

Meanwhile, the energy of the emerging sun in the center of the solar system (99.85% of the total mass of the solar system) changed everything. At first, the gravity of the proto-sun pulled in most of the matter in the original cosmic disk. As the solar nebula cooled, it developed a temperature gradient from the center to the edge of the disk that began to redistribute and shape the growing solar system. Closer to the proto-sun, the temperatures were above 2000°C, and everything was vaporized. About 5 million miles from the proto-sun, the temperatures were cool enough that rocky bodies could solidify. This is known as the "rock line," where the smaller, inner rocky planets (Mercury, Venus, Earth, and Mars) could eventually coalesce. Even farther out is the "frost line," where the temperatures were –375°C or lower. This is cold enough to freeze not only liquid water into ice but also carbon dioxide, methane (CH_4), and ammonia (NH_3). This distinctive composition is the most notable feature of the outer planets like Jupiter, Saturn, Neptune, and Uranus, which are giant frozen gas balls with very little rocky material.

In 2018, scientists recovered the first good image of a new planet in the process of formation, using the Very Large Telescope in the European Space Observatory in northern Chile. It was found in a star named PDS 70, more than 370 light years away. PDS b, as the planet is known, is still condensing in the inner part of the planetary disk at a distance of about 22 times the distance between the sun and the earth. The image clearly shows it cutting a circular track around the star in the center, just as would have happened during the early history of our solar system. Planet PDS b is a gas giant even larger than Jupiter, so it is not going to turn into a rocky earth-like planet like ours. In addition, its surface temperature is still 1200 K (1700°F), so it is a long way from cooling down to become a frozen gas giant like Jupiter.

After 3 million years, the planetesimals clumped together into bigger and bigger bodies until their diameters were about 100 km or larger, and they became **protoplanets**. At this point, their gravity and internal heat are enough to mold them into a roughly spherical shape. Much of the loose cosmic dust of the early solar nebula would have been pulled in by the gravitational attraction of the growing protoplanets.

Then, about 50 million years after the solar nebula first formed, a critical threshold was crossed.

The proto-sun has accumulated enough heat and energy to collapse by its own gravity and trigger **nuclear fusion** of its hydrogen into helium. This is the same reaction that occurs in the hydrogen bomb and powers fusion reactors. Such a reaction allowed the sun to become a full-fledged star that can burn for more than 10 billion years as it continues to power the solar system. The energy of this huge fusion reaction first came out as large bursts of **solar wind**, a huge flux of charged particles (known as a plasma) that poured continuously out from the sun. This first intense burst of solar wind blew away much of the remaining cosmic dust from the inner solar system, almost completely clearing out interplanetary space and preventing the inner planets from gaining much more mass.

▷ See For Yourself: Origin of the Earth

7.4 The Earth Develops Layers

One of these protoplanets eventually became our earth. It is about 149 million km, or 93 million miles, away from the proto-sun, a distance that is sometimes nicknamed the "Goldilocks zone." At this distance it is neither "too hot" (or else it would become a superheated place like Venus, where the atmosphere is hot enough to melt lead) nor "too cold" (so that solar heating is so weak that the planet is frozen solid like Mars).

At the temperatures of the Goldilocks zone, the swirling clumps of matter and planetesimals tended to concentrate (Fig. 7.7), with the large amounts of the common solid elements (silicon, aluminum, iron, nickel, magnesium, calcium, sodium, potassium—see Table 2.1) mixed with lighter gases, like oxygen, nitrogen, carbon, helium, and especially

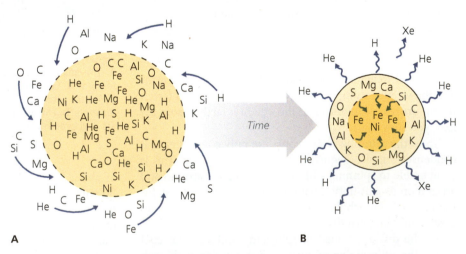

Figure 7.7 Earth's elemental distribution. **A.** When the earth first condensed, it was a random mixture of the common elements in the solar system, especially gases like hydrogen and helium but also rarer materials like oxygen, aluminum, silicon, magnesium, iron, calcium, sodium, carbon, sulfur, and other elements. More material was continually drawn into this swirling mass by gravity. **B.** When the earth melted due to the excess heat from several sources, the elements separated out by density and the earth developed layers. The lightest gases (H, He) were too light to be held in by the earth's weak gravity and escaped. The heaviest elements (Fe, Ni) sank to the center. Those less dense solid elements that that remained (Si, Al, O, Ca, Na, Mg, K, and some Fe) were left behind to form the mantle.

How Do We Know What Is Inside the Earth?

Since the days of Jules Verne and his novel *Journey to the Center of the Earth*, science fiction authors have imagined what the inside of the earth might be like and how people might explore caverns deep in the earth's interior or create machines that could burrow into the earth's center. However, this is purely science fiction and could never actually occur. As we discussed in Chapter 2, the temperatures in just the shallow crust are too hot, and the pressures are too intense, for even hardened drill bits, let alone tunnels or human-carrying machines, to ever survive. Humans have never drilled any deeper than 12,262 meters (7.5 miles) in the Kola region of Siberia, which is only 10%–30% of the way to mantle beneath the continental crust.

If we can't reach or sample the earth's interior ourselves, then how do we know what it is made of? We have some direct evidence of small pieces of rock from the mantle, known as peridotites (which are made of the green silicate mineral olivine—see Chapter 2), which have come up from volcanoes with very deep magma chambers. However, most of the evidence is geophysical in nature, which gives us an indirect picture of what the earth's interior must be like. Plus, there is one additional piece of evidence: meteorites. These lines of evidence include the following.

1. *Seismology*: Every time a large earthquake occurs somewhere on earth, it causes the entire earth to ring like a bell. These vibrations spread out from the earthquake like ripples in a pond, passing through both the interior and the surface of the earth. Those waves that pass through the body of the earth are known as **body waves**. From the same quake, two different types of body waves emerge simultaneously (Fig. 7.8). The fastest waves, known as "primary" or **P-waves**, are the fastest seismic waves of all. They travel with velocities of 5.5 km/second (about 15,000 mph) in the crust and increasing to 8–12 km/second in the mantle and 10 km/second in the core but only about 1.5 km/second in water. Their motion is a pulse of compression and then extension in the direction the wave travels, similar to the pulse of air molecules that transmit sound waves (Fig. 7.8A). After the P-wave, the next wave to arrive at a seismic station is the secondary wave, or **S-wave**. These are much slower than P-waves, with velocities of only about 3.0 km/second in the crust and up to 6 km/second in the mantle. They have an up-and-down shearing motion as they move. However, you cannot shear a fluid, so S-waves die out when they hit water or any other fluid. Think about what happens when you are in a bath or swimming pool. You can slap the water or make a sound underwater, and the sound waves travel easily through the fluid. But if you try to shear the water as you push with your hand, it just flows past your hand.

   ▷ See For Yourself: Seismology and the Earth's Interior

   As these waves spread from the earthquake like the vibrations of a ringing bell, they are picked up in a network of seismic stations that cover the entire globe. Seismologists can then calculate the path that the earthquake waves must have traveled, knowing where and when they started and when they arrived. Those calculations show that for a distance of 103 radial degrees away from the quake, both the P-waves and S-waves have a curved path (Fig. 7.9) caused by **refraction** (bending of the energy waves as they go to a denser and denser layer of the earth). This allows geophysicists to calculate the density of each layer in the mantle below us because the velocity of seismic waves increases with the increasing density of the rocks they travel through.

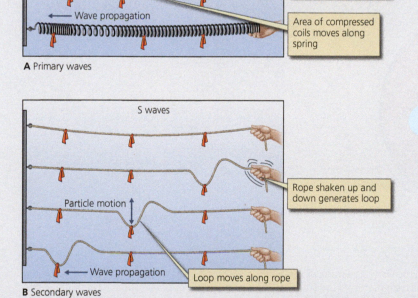

A Primary waves

Spring pushed forward generates area of compressed coils

Area of compressed coils moves along spring

B Secondary waves

Rope shaken up and down generates loop

Loop moves along rope

Figure 7.8 The two main body waves of the earth are the P-waves (with a compressional–extensional motion in the direction of travel, like a sound wave or the wave in a Slinky when it is compressed) and the S-waves (which have an up and down shearing motion).

▷ See For Yourself:
Seismic Shadow
Zones

Beyond 103 radial degrees from the location of the original earthquake, however, strange things happen. P-waves do not show up on seismographs located between 103 and 143 radial degrees, but then they reappear in seismic stations greater than 143° away from the quake. Something is casting a donut-shaped "shadow" for those seismic waves as they travel deeper into the earth so that they never show

up in the **P-wave shadow zone.** This shadow must be caused by a denser body in the center of the earth refracting the seismic waves at angles greater than 143°, which we know as the core.

Even stranger is the behavior of S-waves that travel more than 103° from the source quake. They are completely blocked by the core, so the entire opposite side of the earth greater than 103° away from the quake is the **S-wave**

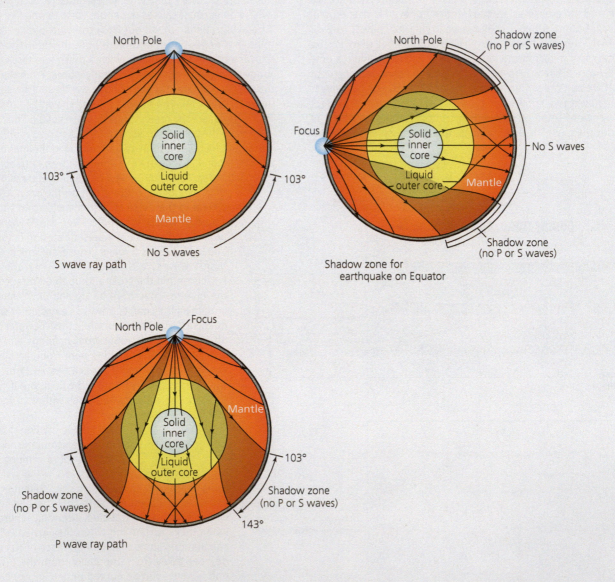

Figure 7.9 If a major earthquake occurs somewhere in the world, the worldwide network of seismographs will record the arrival a number of minutes after it occurred. Those stations that are less than 103° from the source of the quake will pick up waves that have plunged into the mantle, then refracted back up to the surface as they encountered denser and denser layers of the mantle. However, beyond 103° there are no P-waves or S-waves arriving at seismic stations. It's as if the core was creating a big shadow that blocked the arrival of those seismic waves, so these are called "shadow zones." The shadow for P-waves is shaped like a large doughnut on the opposite side of the earth from the source of the seismic waves in the quake zone. No P-waves arrive between 103° and 143° from the source, but P-waves do arrive inside the "doughnut" from 143° to the opposite side of the earth (180°). This is because the even denser core material refracts some of the waves that pass into the core, and they appear in the circular "hole" in the doughnut. However, no S-waves occur past the 103° limit. This is because they are shear waves and are blocked completely by the core, suggesting that the outer core is fluid.

shadow zone (Fig. 7.9). What could block all S-waves from passing through the core? Remember, S-waves do not travel through a fluid, so this shows that the outer core must be at least semi-fluid.

Thus, seismology gives us the depth and density of the major layers of the earth, from the crust to the mantle to the core. Even more sophisticated methods allow us to see detailed structures and flowing currents within the mantle, but that is beyond the scope of this book.

2. *Gravity*: Using Newton's law of gravitation and knowing the mass of the sun, the mass of the earth, and the volume of the earth, we can calculate that the average density of the earth is about 5.5 g/cc, or 5.5 times as dense as water. Crustal rocks are typically no more than 2.7 g/cc (granites and gneisses) to 3.0 g/cc (basalts). But the crust is a tiny percentage of the volume of the earth, so that doesn't matter. The mantle, however, is the biggest part of the earth's volume, and the density of mantle rocks ranges from 3.3 g/cc in the upper mantle to 5.5 g/cc near the core/mantle boundary. If the entire crust and mantle are less dense than the earth average of 5.5 g/cc and more than half of the volume has already been accounted for, then this means that the remaining volume of the core must be incredibly dense to raise the average density for the entire earth. Using seismic wave velocities, we can calculate that the density of the outer core is about 10 g/cc and that the innermost core is as dense as 13 g/cc (13 times as dense as water). This suggests that the pressures in the inner core are about 4 million atmospheres, or 4 million times as strong as the pressure of air on us at sea level, and that the temperatures are over 7000°C! This is why those science fiction movies about traveling to the core are impossible. Any human in a tunneling craft, such as in the sci-fi movies about drilling to the core, would be crushed and melted to a crisp in just the lower continental crust, let alone under the incredible pressures and temperatures of the mantle and core.

3. *Magnetism*: Ever since Michael Faraday's important experiments with electromagnetism and compasses, scientists have known that the earth has a magnetic field that spreads out into space and has many important effects. One of its most important benefits is that the field shields us from the ionized radiation of the solar wind, which is blocked by the magnetic field and flows around us through space. (Astronauts in space, on the other hand, have no protection from the earth's magnetic field and must be shielded from solar wind by their space suits.)

What causes the earth's magnetic field? It can't be a simple bar magnet like you have seen in science class because solid metal magnets lose their magnetization above 650°C and we just demonstrated that the earth's interior is much hotter than this. Instead, it must be some kind of electric **geodynamo**, similar to the dynamos that create electricity in a hydroelectric power plant. There, they use the force of the water to spin coils of conducting metal wires through a magnetic field, and electrical current is generated. Similarly, if the core of the earth was spinning rapidly (as the earth spins) and was made of a conducting metal, it would generate a magnetic field as well. Thus, the earth's magnetic field is only possible if the core of the earth is made of a metal that is a good conductor.</NP>

4. *Meteorites*: The three lines of evidence (seismology, gravity, geomagnetism) we have discussed so far are geophysical. But there is a fourth line of indirect evidence of what the earth's interior is made of, and it comes from outer space: **meteorites**. Meteorites fall to earth on a regular basis and give us samples of other planetary bodies as well as the material of the primordial solar system before there were planets. The three most important types of meteorites (Fig. 7.10) are 1) **chondritic meteorites**, which are from the earliest solar system and supply the data from which we deduced the properties of the early solar system in this chapter (they produce the oldest ages known, about 4.567 billion years, which is thus our estimate of the age of the solar system); 2) **stony meteorites**, which are rich in magnesium silicates, the same composition as rocks that come from our own mantle, and are thought to be remnants of the mantle of another planet that broke up; and 3) **iron–nickel meteorites**, which are a very rare and special kind of meteorite, with only 6% of known meteorites being of this composition since stony meteorites and chondrites are far more abundant, but an iron–nickel meteorites is impressively heavy when you pick one up and heft it.

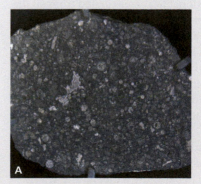  

Figure 7.10   Three types of meteorites. **A.** One piece of a carbonaceous chondritic meteorite that fell near Pueblito de Allende in Mexico in 1969. The little blobs are chondrules, or tiny remnants of the early solar system before differentiation into planets. **B.** A stony meteorite, which fell near Estherville, Iowa, in 1879. **C.** An iron–nickel meteorite, made of almost pure iron and a few other metals.

As their name suggests, iron–nickel meteorites are made mostly of iron, with about 5%–25% nickel and minor amounts of cobalt and other rarer elements. Thus, they are much simpler than the stony meteorites and chondrites, which have many different chemicals and minerals in them. But the most interesting aspect of iron–nickel meteorites is that they provide samples of what the core of many planets (including ours) is made of. When the spectra of certain asteroids (known as M-type) are analyzed, they turn out to be the same composition as iron–nickel meteorites. From the geochemical evidence trapped inside them, we know that iron–nickel meteorites originally formed the core of certain large protoplanets that have since broken up. They also trap isotopes of aluminum-26, which was the radioactive heat source that melted protoplanets and allowed their denser materials (iron and nickel) to sink to the core and separate from the mantle during planetary differentiation. Thus, iron and nickel are the only common metals in the solar system that are dense enough to represent the rocks in our core (when under the pressures and temperatures in our core), and most importantly, iron and nickel are good electrical conductors, so when molten, they could convect and generate the earth's magnetic field.

hydrogen, the element that makes up most of the sun and much of the rest of the solar system. When these elements first got together, they formed a well-mixed mass of uniform composition throughout the proto-earth. Some of the heaviest elements, like iron and nickel, would sink to the center of the protoplanet because of their greater density. Meanwhile, the lightest gases, especially the hugely abundant hydrogen and helium, largely escaped out into space because the gravitational pull of the earth was not strong enough to hold them in. That is why we use them to float balloons today since they are less dense than air and our gravity cannot hold them back.

Simple gravitational settling alone is not enough to explain the earth's **differentiation** into the discrete layers, with an iron–nickel core surrounded by a silicate-rich rocky mantle (see Box 7.1). To completely differentiate the earth into layers of completely different composition (the core and mantle), you need enough heat to melt the entire planet. This would allow almost all the dense iron and nickel to sink to the center of gravitation (nicknamed the "iron catastrophe"), leaving the less dense silicates floating on top.

Where did all this heat come from? There are several possible sources.

1. The early proto-earth was full of unstable elements that were undergoing radioactive decay. Recall from Chapter 3 that during radioactive decay a tremendous amount of heat is released and that 50% of the decay and heat is released in the first half-life. All of the important radioactive elements in the earth today, such as uranium-238, uranium-235, rubidium-87, and potassium-40, were in their first half-lives and were much more abundant as the earth began to form, so their heat production was at its maximum. However, there are also meteorites, such as the Allende carbonaceous chondrite (see Box 7.1), that come from the earliest solar system, just as the earth was forming. They suggest that there might be another element that did most of the heat production. These meteorites had unusual quantities of the rare isotope magnesium-26, which is the daughter product of the radioactive decay of aluminum-26. Aluminum-26 decays very rapidly, with a half-life of only 700,000 years, so none of the earth's original aluminum-26 survives. However, it was apparently hugely abundant in the early condensing proto-earth (judging from how much magnesium-26 is still left behind), so it would have been the most important element of all in melting the earth.

2. In the early condensation of the proto-earth, there was still a lot of cosmic debris left over, so the earth was under constant bombardment from meteorites. The impact of a meteorite represents a huge amount of kinetic energy (energy of motion) that converts to heat energy when it slams into the earth. Dating of the meteorite impact craters on the moon suggests that the early solar system was still under intense bombardment from space debris, which did not slow down until about 3.9 billion years ago. Among these impacts was the one that blasted off a chunk of the mantle to form the moon (discussed below, see "Moonstruck"). Thus, the earth would have had additional heat contributed from all the impacts of rocks from space.

3. As the blobs of iron and nickel sank to the earth's core, they released a lot of potential energy (comparable to the difference in energy between the rock at the top of a cliff and sitting at the bottom). Like any other form of energy, potential energy cannot be destroyed but has to be converted to another form of energy, namely heat.

4. The interior of the earth is under intense gravitational forces (gravitational compression), which increase not only the pressure but also the temperature. This is enough to melt many of the materials that eventually became the mantle. It is also why the outer iron–nickel core of the earth is fluid (see Box 7.1).

5. Finally, as the earth's densest materials began to sink to the center, they also changed the angular momentum of the earth. If the earth were as small as a figure skater, it would cause it to spin faster. However, the earth is too massive to respond to this small change in angular momentum, yet the change of energy must go somewhere—so it is converted to heat.

By the end of this process some 4.5 billion years ago, the earth had its discrete layers of an iron–nickel core and a magnesium–silicate-rich mantle. It was still too hot to allow a crust to cool on the outside, so the earth only had two primary layers.

7.5 Moonstruck

People have stared at the moon for thousands of years and wondered what it is made of and how it formed. There were all sorts of ridiculous or silly ideas, such as the "green cheese" notion, but only a few serious hypotheses were proposed by the scientific community. The ideas fall into three broad categories, which (thanks to the all-male community of astronomers for many years) acquired sexist nicknames that would never be acceptable today.

See For Yourself: Origin of the Moon

1. *The "pickup" or "capture" hypothesis*: For decades, some scientists had suggested that the moon was a foreign body from far outside the earth's orbit that was captured as it flew by the earth and got pulled into orbit by the earth's gravity. But there were numerous problems with this model from the very start. For one thing, the moon's orbit around the earth is almost in the same plane as the earth's orbit around the sun (just 5° out of alignment with our plane around the sun), and it is moving in the same direction as the earth rotates, which would be unlikely if an object coming at any other angle from outer space got captured. Such an orbit would most likely swing around the earth in any plane except the plane of the earth–sun system. In addition, when gravitational capture of a large body occurs, the result either is collision or the object flies back into space with an altered orbit. To allow the moon to slowly be captured by the earth's gravity and stay in orbit without collision or escape, the earth would have to have had a very thick atmosphere back then, which extended much farther out than it does now, to cause friction and drag and slow the object down. No evidence supports the idea that the earth's atmosphere was once that much thicker. Finally, if the moon were an exotic object captured by earth's gravity, its composition would be radically different from that of the earth. Once the Apollo missions brought back samples, scientists studying the moon rocks could test this.

2. *The "daughter" or "fission" hypothesis*: This scenario, first proposed by astronomer George Darwin (son of Charles Darwin) in the late 1800s, argues that the moon is made of the original rapidly spinning undifferentiated earth matter. During this rapid spin, molten earth material flew off into space to form the moon. Some astronomers even suggested that the Pacific Ocean basin is the remnant scar of that event. This scenario seemed plausible for many years, although by the 1960s plate tectonics had shown that the Pacific basin is not an ancient scar but floored by very young lavas, mostly less than 140 million years old. In addition, The "daughter" model also doesn't account for the angular momentum of the earth–moon system. Once again, the crucial test would be the moon rocks brought back by the Apollo missions. If they were the same composition as the primordial earth (before it separated into core and mantle and crust), then it would be plausible.

3. *The "sister" hypothesis*: Similar to the "daughter" hypothesis, this model suggests that the moon started not as a blob spun off from earth but as two separate blobs of matter which got locked into gravitational attraction with each other. Again, there are problems with the angular momentum of the earth–moon system in this model. But like the "daughter" hypothesis, it predicts that moon rocks would have a composition very similar to the primordial earth before it differentiated into core and mantle layers.

These ideas and more were hanging in the balance in 1969 when Apollo 11 and later moon missions brought lunar samples back to labs on earth to study. To everyone's surprise, their composition did not support any of the previous ideas. Instead, it proved a new model that no one had ever thought of.

Figure 7.11 A lunar basalt retrieved from an ancient lava flow on the moon by the Apollo astronauts. It is full of Swiss cheese–like cavities, suggesting that gas bubbles formed and the lava solidified around them while the moon still had some kind of atmosphere.

The lunar rocks brought back by Apollo 11 through Apollo 17 (Fig. 7.11) were not similar to the early earth in composition. Nor were they some exotic composition, as if they had been a body from outside the earth captured by gravity. Instead, they were made of a form of calcium-plagioclase-rich gabbro known as anorthosite and its volcanic equivalent, the familiar black lava known as basalt (Figure 2.8). In other words, their composition was very much like that of the upper mantle, where the lavas that erupt as basalt on the sea floor or in volcanoes like Kilauea on Hawaii have their source.

This was a surprise. If the moon was made almost entirely of material like that from the earth's mantle, with very little iron or nickel such as are found in the earth's core, it must be a piece of the earth's mantle that formed *after* the primordial earth had separated into a core of iron and nickel and the mantle made of silicate minerals. In other words, the moon was formed after the earth had cooled and coalesced and its layers had differentiated and separated.

Even more startling, the only way to get so much mantle material into space was to blast the early earth with a giant impact from another body (Fig. 7.12). Geologists now call this body Theia (the Greek name for the mother of Selene, the moon goddess) and postulated that Theia was a Mars-sized protoplanet that hit the earth obliquely with an impact that blew material sideways off the earth and into orbit. The energy of this collision would have been amazing! Trillions of tons of material would have been vaporized, and the temperature of the earth would have risen to 10,000°C (18,000°F). Once this debris began to orbit the earth (at one-tenth the distance that the moon is today) it would have gradually clumped together and coalesced over the course of about 1000 years to form the proto-moon, a ball of matter about 2000 miles wide. It was only 14,000 miles away back then, not the 250,000 miles it is now. The earth was spinning so fast at that time that each day was only 6 hours long, with 3 hours of nighttime and 3 hours of daylight.

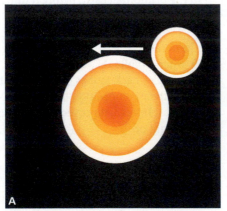

A Mars-sized asteroid nears the planet

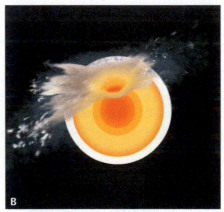

The asteroid collides with Earth

Many fragments are hurtled into space

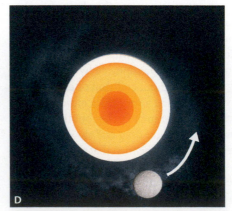

Some particles coalesce to form the Moon

Figure 7.12 Illustration of the impact model of the moon. **A.** A Mars-sized body known as Theia nears the planet. **B.** It collides with the earth, scattering mantle material into space. **C.** The mantle fragments are hurled into space, then captured by the earth's gravity and remain in our orbit. **D.** Some of the particles coalesce to form the moon, while some of the particles burn up as they re-enter the earth's atmosphere.

The heat from its own radioactive minerals would later have remelted parts of the moon completely, and most of the moon would have remained the same composition as the earth's mantle, while the melting also caused huge eruptions of basaltic lava flows that formed the magma oceans that now make the dark "maria" or "seas" on the moon's surface. Meanwhile, the moon has a tiny iron core, only 330–350 km in diameter, thought to be a relict of the core of Theia left behind after the collision; most of Theia's own iron–nickel core must have accreted to the earth's core. By contrast, if the "sister" or "daughter" models (favored before the Apollo missions) were correct, the moon would have a large core, roughly proportional to the size of the earth's core relative to its mantle.

When did this all occur? Once again, moon rocks give the answer. Using the same uranium–lead and lead–lead dating methods discussed in Chapter 3, many labs have dated moon rocks. Most are at least 4 billion years old, suggesting that the moon's surface formed early and has not changed much since. After all, it has none of the forces that change the earth's surface: it has no atmosphere, no water, no weathering, and no plate tectonics. The only major modifications of its surface are huge impacts that left craters, and most of the crater debris has been dated at older than 3.9 billion years, so most of the impacts occurred early and not much has happened since then.

The oldest pre-impact rock dates from the moon are currently 4.44 billion years. This is much younger than the meteorites that date back to the origin of the solar system, so the moon is definitely younger than the events that formed the solar system and the earth and the melting and differentiation episode that separated the earth's core from its mantle.

Since the initial proposal of the giant impact hypothesis, much further evidence has come out of analysis of moon rocks to support the mantle source of the moon. Nearly all the geochemical isotopes (oxygen, titanium, zinc, and many others) that have been studied since the moon rocks were collected have shown that the moon and the earth's mantle have identical chemical compositions. There are also many refinements to the impact model, with some versions having more than one impacting body or positing different-sized impactors or different impact mechanics. But no matter which version is currently favored by scientists, the Apollo samples inescapably

 See For Yourself: A Day on Earth 4 Billion Years Ago

point to the moon as originating from a chunk of the earth's mantle.

7.6 Cooling Down: The Oceans Form

The early earth had a hot core and mantle, but initially the surface of the magma ocean was molten and there was no crust. It is hard to know much about the earliest crust of the earth since none of it survives after weathering and erosion and plate tectonics have destroyed nearly all of it. As we shall see in Chapter 8, when we do find the oldest crustal rocks, they represent very thin crust that has erupted directly from the mantle.

An even bigger question is when did the surface of the earth cool down enough so that it was below the boiling point of water (100°C, or 212°F) so that liquid water could condense out? In other words, when did water collect on the earth's surface to form the first oceans? For decades, most geologists thought that the cooling of the earth's crust was too slow for it to happen much sooner than about 4.0 billion years ago. Most of them thought that the earth would need 500–600 million years after it formed to cool down to below the boiling point of water.

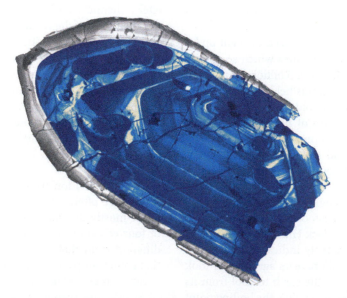

Figure 7.13 One of the Jack Hills zircon grains, which yields dates of 4.3–4.4 billion years in age and contains bubbles of gas and liquid whose chemistry suggests that the earth had liquid water at the time it formed. This specimen is imaged by cathodoluminescence, which highlights the growth rings in the crystal and helps make the tiny bubbles more visible but also makes it appear blue.

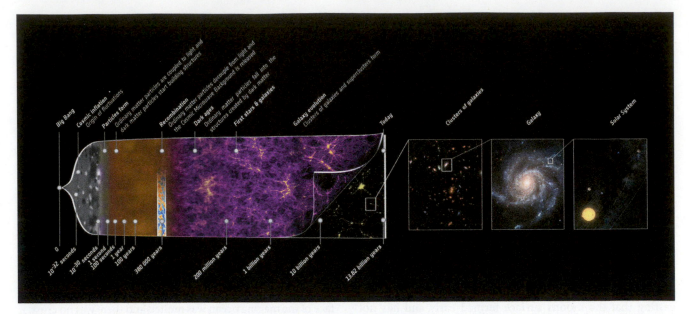

Figure 7.14   Illustrated timeline of the major events since the "Big Bang," about 13.82 billion years ago, with the galaxies forming about 10–12 billion years ago and our solar system forming about 4.567 billion years ago.

Then in 2014, scientists made a startling discovery in a few grains of sand from Australia. To be specific, they were handfuls of zircon (zirconium silicate, or $ZrSiO_4$) sand grains (Fig. 7.13) from a much younger sandstone found in the Jack Hills of Western Australia. Each individual grain can be dated by uranium–lead methods, so they give a scatter of ages. But the oldest grains of all give an age of 4.374 billion years, confirmed and redated by John Valley and his colleagues in 2014. Thus, the current record holder for the oldest crustal material from earth (that is, not a meteorite or moon rock) is 4.4 billion years. These sand grain dates put us closer and closer to the age of moon rocks and meteorites, but we still have a gap of about 160 million years between 4.4 billion and 4.56 billion. And in 2018 scientists announced measurements of chemical isotopes from several samples of mantle rocks that have been brought up by volcanoes. These had chemical signatures which suggested they were part of the original mantle 4.5 billion years ago.

Those same tiny zircon sand grains held even more surprises. Not only did they give the oldest known dates but when scientists analyzed the ratio of the two isotopes of oxygen found within them, they found evidence of the early hydrosphere. These zircons had oxygen isotopes in them that suggested that earth had liquid water on its surface as early as 4.4 billion years ago!

Prior to this discovery geologists had always assumed that the earth took a long time to cool from its molten state at 4.56 billion years ago. But the Jack Hills zircons turn that assumption inside out. If they truly indicate the presence of liquid water on earth 4.4 billion years ago, then it took only about 160 million years for the earth to cool from its molten state to a condition that was below the boiling point of water. A duration of 160 million years seems short when

you consider the billion-year timescales of the early earth, but remember that 160 million years ago the earth was in the Late Jurassic heyday of the Age of Dinosaurs. That was a long time interval in the context of the history of life on earth. This evidence also suggests that there may not have been so many meteorite impacts during this time interval, or the oceans would have been vaporized over and over again. Taken together, these data suggest what is now called the "cool early earth hypothesis."

So where did this early earth water come from? Traditionally, geologists had thought that it was water trapped inside the earth's mantle when it cooled, which gradually escaped through volcanoes in a process called degassing. But lately, chemical analyses of extraterrestrial objects match the chemistry of the earth's oceans (especially carbonaceous chondrite meteorites). This suggests that there was a lot of water trapped in the debris of the early solar system (which the chondrites are remnants of). The same is true of moon rocks, which do not have much water in them today but apparently were wetter when the solar system formed. If this is so, then the earth was born with its water already present as it cooled and condensed. It only required its surface temperature to drop below 100°C for that water to form the first oceans.

One explanation we can rule out is comets. Although comets are often called "dirty snowballs" because they are made mostly of water ice and dust, the chemical analyses of four comets now shows that their geochemistry is very different from that of earth water. Thus, the popular idea that comets impacted the early earth and melted to form its oceans can be dismissed.

So, we have seen how the universe formed from the Big Bang and how the solar nebula condensed to form our solar system. We have looked at the evidence for how the earth formed and how it differentiated into layers. Finally, we described the evidence that shows how the moon formed and what the earliest earth and the earliest oceans were like. Now let us look at the earliest 80% of earth history: the Precambrian.

RESOURCES

BOOKS AND ARTICLES

Bartusiak, M. 2009. *The Day We Found the Universe.* Pantheon, New York.

Bembenek, Scott. 2017. *The Cosmic Machine: The Science that Runs Our Universe and the Story Behind It.* Zoari Press, New York.

Brockman, John (ed.). 2014. *The Universe: Leading Scientists Explore the Origin, Mysteries, and Future of the Cosmos.* Harper, New York.

Canup, R. M., and K. Righter (eds.). 2000. *Origin of the Earth and Moon.* University of Arizona Press, Tucson.

Carroll, Sean. 2016. *The Big Picture: On the Origins of Life, Meaning, and the Universe Itself.* Dutton, New York.

Editors of Chartwell Books. 2017. *How the Universe Works: An Illustrated Guide to the Cosmos and All We Know about It.* Chartwell Books, London.

Hartmann, William K., and Ron Miller. 1991. *The History of Earth: An Illustrated Chronicle of An Evolving Planet.* Workman Publishing, New York.

Hartmann, William K., and Ron Miller. 2005. *The Grand Tour: A Traveler's Guide to the Solar System.* Workman Publishing, New York.

Hawking, Stephen. 1998. *A Brief History of Time.* Bantam, New York.

Hazen, Robert. 2013. *The Story of Earth: The First 4.5 Billion Years from Stardust to Living Planet.* Penguin, New York.

Krauss, Lawrence. 2017. *The Greatest Story Ever Told—So Far: Why Are We Here?* Atria Books, New York.

McKenzie, Dana. 2003. *The Big Splat: Or How Our Moon Came to Be.* Wiley, New York.

Natarajan, P. 2016. *Mapping the Heavens: The Radical Scientific Ideas that Reveal the Cosmos.* Yale University Press, New Haven, CT.

Perlov, Delia, and Alex Velenkin. 2017. *Cosmology for the Curious.* Springer, Berlin.

Ryden, Barbara. 2016. *Introduction to Cosmology.* Cambridge University Press, Cambridge, UK.

Sagan, Carl. 2013. *Cosmos.* Ballantine, New York.

Saraceno, Pablo. 2012. *Beyond the Stars: Our Origins and the Search for Life in the Universe.* World Scientific Publishing, New York.

Silk, Joseph. 2001 (3rd ed.). *The Big Bang.* W. H. Freeman, New York.

Singh, Simon. 2005. *Big Bang: The Origin of the Universe.* Harper, New York.

Sobel, Dava. 2016. *The Glass Universe: How the Ladies of the Harvard Observatory Took the Measure of the Stars.* Viking, New York.

Tyson, Neil de Grasse, and David Goldsmith. 2004. *Origins: Fourteen Billion Years of Cosmic Evolution.* W. W. Norton, New York.

Valley, John W., W. H. Peck, E. M. King, and S. A. Wilde. 2002. A cool early earth. *Geology* 30:351–354.

Valley, John W., and others. 2014. Hadean age for a post-magma-ocean zircon confirmed by atom-probe tomography. *Nature Geoscience* 7:219–223.

SUMMARY

- Early naturalists first showed that the earth rotated around the sun in 1543 and established the laws of gravity and planetary motions in the early 1700s. In the early twentieth century, female astronomers at Harvard University completely catalogued all the objects in the night sky and developed a method of determining the distance to a star or galaxy.

- In the 1920s, Edwin Hubble and Milton Humason at Mt. Wilson Observatory produced evidence that all the stars and galaxies are moving rapidly away from us because all their spectra are Doppler-shifted to the red end of the spectrum. This proved that the universe was expanding.

- The evidence for expansion of the universe implied a singularity at the beginning of the formation of the universe, followed by a "Big Bang" when the universe began to expand. Confirmation of the Big Bang came from the accidental discovery of the cosmic background radiation left over from the Big Bang by two Bell engineers trying to get the noise out of their microwave antenna.

- The fact that the planets are all spinning around the sun in the same plane and same direction suggested that they formed out of a disk-shaped solar nebula. We can see other primitive disk-shaped nebulae in space which have the same shape as our early solar system.

- As the particles of matter in the solar nebula disk began to clump together, they formed larger and larger masses until they became planetesimals. These bodies attracted even more matter to them and pulled in all the matter in their tracks around the sun, until they grew to the size of protoplanets. Eventually, the gravitational attraction of protoplanets pulled in most of the remaining free matter in the solar system until these bodies reached the size of our modern planets.

- Once the early earth had formed, it was made of a random mixture of many gases and solid elements. Hydrogen and helium were the most abundant, but they escaped into space because earth is not large enough to have the gravity to hold them in. The other materials condensed to form the solid earth.

- We know about the interior of the earth from the refractions of seismic waves as they pass through the earth and are detected by seismographs all over the world. P-waves (compressional–extensional waves) disappear about 103° from the earthquake epicenter, then reappear about 143° away, producing a doughnut-shaped "shadow zone" on the opposite side of the earth from the earthquake. This shows the depth down to the core–mantle boundary and how the dense material in the core refracts the P-waves like a lens so that they appear in the circle in the center of the shadow zone. S-waves disappear completely after 103°, suggesting that they have run into a liquid outer core.

- Measurements of the earth's gravity allow us to model the density of the earth's layers, and we find that the mantle ranges from 3.3 to 5.5 g/cc (3.3–5.5 as dense as water) from the upper mantle to the lowest mantle, then the density of the core jumps up abruptly to 10 to 13 g/cc.

- The earth's magnetic field can be explained if the core is made of a spinning conductor, such as iron and nickel, producing a geodynamo.

- The occurrence of meteorites also confirms the composition of the earth and other solar system bodies. Iron–nickel meteorites have the right properties to represent the earth's core, while stony meteorites are very similar to rocks that have come up from the mantle. Both are from some other planet, which had differentiated into layers of core and mantle and then broke up. Chondritic meteorites, on the other hand, come from the primordial solar system before the planets even formed, with the oldest dating to 4.567 billion years ago, the age of the solar system.

- Heat from radioactive decay of aluminum-26 and other elements, plus the heat from the impact of meteorites and from conservation of angular momentum, helped to melt the earth, so the densest elements, iron and nickel, sank to the center to form the core of the earth. The remaining elements were left behind to form the mantle.

- The moon was formed when a Mars-sized body called Theia struck the earth obliquely after the core had separated from the mantle. It blasted mantle material into space, which eventually coalesced to form the moon as we know it (although it was much closer to the earth when it first formed).

- After its formation 4.56 billion years ago, the earth cooled down rapidly so that as early as 4.4 billion years ago its surface was below 100°C and the first oceans could form. The water is thought to have come from the early material that accreted to the earth and not from comets, as once supposed.

KEY TERMS

Nicholas Copernicus (p. 144)
Heliocentric solar system (p. 144)
Galileo Galilei (p. 144)
Isaac Newton (p. 144)
Cepheid variables (p. 145)
Edwin Hubble and Milton Humason (p. 145)
Star spectrum (p. 145)
Doppler effect (p. 147)
Red shift (p. 147)
Expanding universe (p. 148)

Big Bang (p. 148)
Cosmic microwave background radiation (p. 148)
Solar nebula hypothesis (p. 149)
Conservation of angular momentum (p. 149)
Planetesimal (p. 149)
Protoplanet (p. 150)
Nuclear fusion (p. 150)
Solar wind (p. 150)
"Goldilocks zone" (p. 150)

Body waves (p. 151)
P-waves (p. 151)
S-waves (p. 151)
Refraction (p. 151)
P-wave shadow zone (p. 152)
S-wave shadow zone (p. 152)
Geodynamo (p. 153)
Earth's magnetic field (p. 153)
Chondritic meteorites (p. 153)
Stony meteorites (p. 153)
Iron–nickel meteorites (p. 153)

Aluminum-26 (p. 154)
Differentiation of earth's layers (p. 154)
"Pickup" hypothesis (p. 155)
"Daughter" hypothesis (p. 155)
"Sister" hypothesis (p. 155)
Impact hypothesis (p. 156)
Moon rock composition (p. 156)
Theia (p. 156)
Earth's earliest oceans (p. 158)
Jack Hills zircon (p. 158)

STUDY QUESTIONS

1. What are some of the lines of evidence we use to determine what events happened in the distant past when the universe and the solar system formed?

2. Describe some of the accomplishments of the "Harvard Computers." Why do you think so few people have heard about them?

3. How did Hubble and Humason determine that the universe is expanding?

4. What does the discovery of cosmic background radiation tell us about the value of "pure" science versus applied science?

5. What lines of evidence suggest that the solar system started out as a spinning disk of matter?

6. How does the clumping of matter produce "gravitational vacuum cleaners"? How do these explain why there is so little matter in the space between the planets?

7. What are the distinctions between a planetesimal, a protoplanet, and a planet?

8. How does the nuclear fusion of the sun explain why there is so little matter in the solar system outside the planets and the asteroid belt?

9. What is the "Goldilocks zone"? Why might it be important for the existence of life on earth and mean that life is unlikely on other planets?

10. Despite all the science fiction journeys to the earth's interior, why is it impossible to travel more than 12 km into the crust?

11. Describe the lines of evidence that show the internal structure of the earth.

12. Why do we think the earth's layers formed very early in its history? (Hint: The oldest moon rocks date to about 4.5 billion years.)

13. What are the possible heat sources that melted the early earth and allowed it to separate into layers?

14. What did the analysis of the composition of the first moon rocks do to the "pickup," "sister," and "daughter" hypotheses for the origin of the moon? Why?

15. What is the evidence that the earth was cool enough for oceans 4.4 billion years ago?

8

The Early Earth

The Precambrian
4.6–0.6 Ga

In a great number of the cosmogonic myths the world is said to have developed from a great water, which was the prime matter. In many cases, as for instance in an Indian myth, this prime matter is indicated as a solution, out of which the solid earth crystallized out.

—*Svante Arrhenius, Theories of Solutions*

The earliest earth had a molten surface with only limited blocks of cooled crustal rock, floating in a giant lava sea. It was far too hot for liquid water or oceans to form. The surface was subject to a constant rain of meteorite impacts. The moon had recently spun off the earth due to the impact 4.5 billion years ago, so it appeared to be enormous on the horizon.

8.1 The Precambrian or Cryptozoic

The time before abundant fossils are found in the rock record has long been known as the "Precambrian" because it is based on rocks found beneath Cambrian beds full of fossils like trilobites. In more recent years, this immense duration of time, which is almost 4.0 billion years of the earth's 4.567-billion-year history, or about 88% of earth history (Fig. 8.2A), has been given a more detailed chronology (Fig. 8.2B), although the old-fashioned term "Precambrian" is still widely used by geologists. The entire time interval has also been called the **Cryptozoic** (in Greek, *crypto* means "hidden" and *zoic* means "life") because fossils are very scarce and mostly microscopic (see Chapter 9). More recently, however, geological societies have recommended formally dividing the time interval into three eons: the **Hadean Eon** (4.567–4.0 Ga, or billion years ago), the **Archean Eon** (4.0–2.5 Ga); and the **Proterozoic Eon** (2.5–0.542 Ga, or 2500–542 Ma, or million years ago). Each eon is further subdivided into eras, so the Archean consists of the Eoarchean, Paleoarchean, Mesoarchean, and Neoarchean eras and the Proterozoic is broken into three eras: the Paleoproterozoic, Mesoproterozoic, and Neoproterozoic (Fig. 8.2B). Finally, many of the eras are subdivided into periods, so the (for example) Neoproterozoic consists of the Tonian, Cryogenian, and Ediacaran periods. Most geologists don't need to know this level of detail unless they become specialists in Precambrian geology. However, it is important to know the three big subdivisions (Hadean, Archean, and Proterozoic) of this immense span of time because each is very distinct in the geological events and kinds of rocks that were produced.

Part of the reason that we use time units of long duration and have such low resolution of events in the Precambrian is that it is a much more difficult time interval to study. Unlike the abundantly fossiliferous beds of the Cambrian and younger deposits, which are exposed in many places on earth, the study of the Precambrian is handicapped in many ways. First and foremost, these old rocks tend to be buried under much younger deposits, so they are only exposed at the earth's surface in a few places, mostly where mountain ranges have brought deep crustal rocks to the surface. Only 22% of the earth's outcrops are Archean; these occur primarily in Africa, Asia, and North America (Fig. 8.3). Even where they are exposed (such as the "**Canadian Shield**" area of central Canada around Hudson's Bay and parts of Minnesota, Wisconsin, and Michigan), most of the rocks are covered by dense forests and lakes with few roads, so all the early work on their geology had to be done on small outcrops in the middle of lakes or forests with huge areas that were completely covered. Pioneering geologists had to get from one outcrop to the next using canoes since there were no roads and almost no way to walk from one exposure to another. Farther south, Precambrian rocks underlie huge areas of every continent, such as the Great Plains and Midwest of the United States; so the only way to study them except where there are rare uplifts (such as in Minnesota, Wisconsin, and Missouri) is to drill down through thousands of feet of younger rocks.

In addition, most of these ancient rocks have been deeply buried in the crust for a long time, so they are typically metamorphosed. In many cases, it is impossible to tell whether their parent rock (protolith) was even igneous or sedimentary, let alone try to interpret the conditions under which the rock was originally formed. Don't forget—metamorphism destroys all fossils and most sedimentary structures completely. Add to this the fact that the older these rocks are, the more

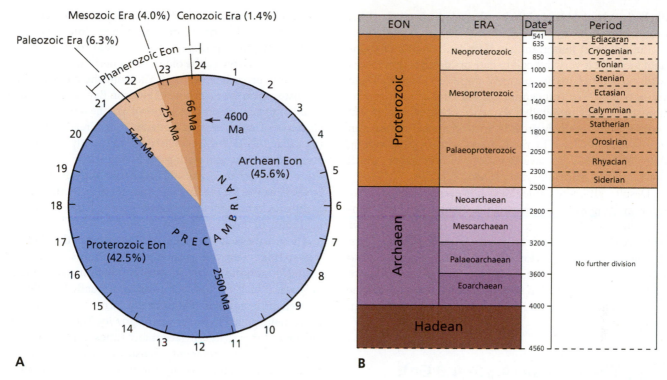

Figure 8.2 Precambrian time makes up about 88% of earth history. **A.** The duration of the Precambrian relative to the Phanerozoic is shown as a 24-hour clock. **B.** The detailed Precambrian timescale.

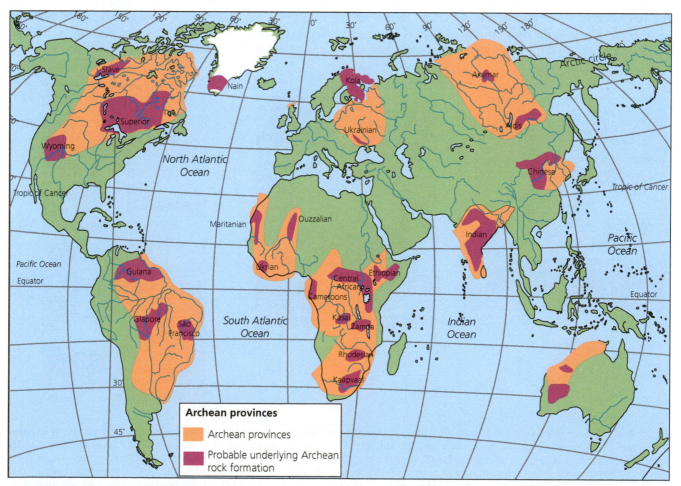

Figure 8.3 Map of Archean and Proterozoic outcrops around the world. Most of the continents have one or more Archean terranes, which were the protocontinents that formed the original nucleus around which the Proterozoic mountain belts and other terranes accreted later.

likely they have been uplifted and eroded, so much of the original record is now missing.

But the most important handicap is that without abundant fossils, it is very difficult to date or correlate these rocks with each other (as is routinely done for Cambrian and younger beds), so you cannot get the kind of high-resolution chronology or detailed sequence of global events that younger beds routinely provide. Instead, most Precambrian events must be dated by radiometric ages on cross-cutting dikes or interbedded lava flows or volcanic ashes, which don't occur in many different places.

Despite all these difficulties, however, geologists have done amazing work deciphering this complex and fragmentary puzzle of the deep geological past. In this chapter, we will race through 88% of earth's history, over 4 billion years. An enormous amount is known about this interval now, and more and more is learned every year; so it is worthwhile to look at it in some detail in this book and to read more in the many excellent books on Precambrian geology coming out every year.

8.2 The Hadean (4.56–4.0 Ga): Hell on Earth

The first half-billion years of earth's history is known as the Hadean eon, after "Hades," the old Greek mythological version of hell or the underworld. Indeed, it was truly a hellish time to be on the earth. Remember, the earth was very hot at the start, and its surface was still molten; so if any crust cooled above the molten surface, it didn't last very long or get very big (Fig. 8.1). No crustal rocks from the earliest earth survive. The oldest dates derived from crustal rocks are currently about 4.32 Ga from the Nuvvuagittuq greenstones on the east shore of Hudson Bay (announced in 2008). The next oldest rocks known are the 4.03 Ga Acasta Gneisses of northwestern Canada, near Great Slave Lake (first dated in 1999). Geologists are constantly looking for even older rocks and dating more and more samples, so there's a good chance they will find a rock with an older date by the time you are reading this. Indeed, in 2018, they found mantle rocks brought up by volcanoes which dated from the beginning of the earth, 4.5 Ga.

Despite the lack of a rock record for most of the Hadean, we can infer many things about the conditions of the earth then. Not only did it start out with a molten surface and not only was it hellishly hot, but it was continually bombarded by meteorites as the debris from the early solar system was continually pulled in by the earth's gravity. These must have pulverized the earth's surface many times, making it even harder to form a permanent solid surface. In fact, the bombardment may have sped up the process of formation of continental crust. Each time a meteorite hit, it sped up the rise of churning and differentiation of lighter crustal rocks from the original komatiites and peridotites. It is analogous to the way shaking up a bottle of carbonated beverage causes the tiny light bubbles of carbon dioxide to combine until they become visible bubbles—and your beverage

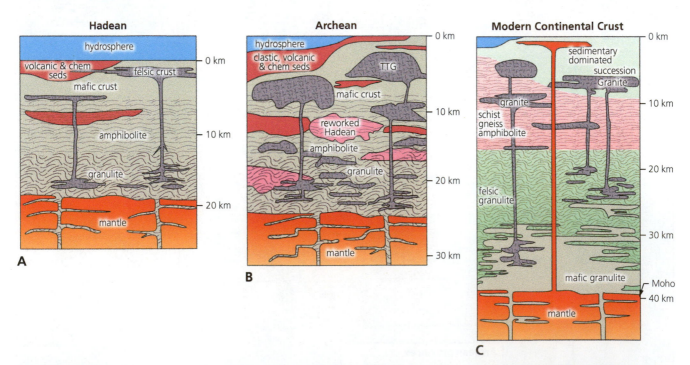

Figure 8.4 Cartoon cross sections of how the crust has changed through time. **A.** Hadean crust was very thin (only 20 km, or 12 miles maximum) and hot and consisted mainly of ultramafic volcanics like komatiite (tan color) made of pure olivine lavas erupting directly from the mantle with little or no differentiation. **B.** Archean crust was slightly thicker and still made of komatiites, but there were more metamorphosed rocks, more differentiation of magmas into granites, and a thicker sedimentary pile on top. **C.** Modern continental crust is 40 km (24 miles) thick or thicker with multiple layers of high-grade metamorphic rocks full of silica intruded by diorites, granodiorites, and granites. The mantle material only rarely erupts at the surface from volcanic vents of deep origin.

becomes foamy. The heavy bombardment apparently went on continuously from 4.5 to about 3.9 Ga, at which time it slowed down tremendously. How do we know this? The Apollo moon samples of impact craters and their melt rock give dates in this age range but almost no dates that are younger.

Geochemical data suggest that the Hadean crust was extremely thin and simple, without the geological complexity we see in modern oceanic or continental crust. Various lines of evidence suggest it was only about 20 km (12 miles) thick, compared to the approximately 30- to 50-km-thick modern continental crust (Fig. 8.4). It was made of material from the mantle that had cooled into lavas near the surface, with the deeper crust undergoing high-temperature, high-pressure metamorphism to become rocks called amphibolites and granulites (Fig. 2.15). These small, thin, hot protocontinental blocks did not yet experience true plate tectonics but simply bumped against each other as they moved around on the highly mobile mantle. We know that there was no plate tectonics in the modern sense because there is no geochemical evidence of volcanoes produced by subduction zones.

However, as we saw in Chapter 7, within about 160 million years after the earth's formation, its surface was cool enough to allow liquid water to form about 4.4 Ga, based on the chemistry of zircon sand grains that date to that time (Fig. 7.13). It now appears that the earth cooled rapidly from temperatures over 1000°C–2000°C at the surface when it initially formed at 4.56 Ga to below 100°C by 4.4 Ga.

Recall also from Chapter 7 that the moon had just pulled away from the earth's mantle and was still condensing and much closer to the earth than it is now. The moon would have looked enormous in the sky (Fig. 8.1). A much closer moon would have immense tidal pull on the earth's surface and caused the earth's crust to rise and fall as the pull of the moon's gravity changed every hour. Its effects were especially strong on the earth's early oceans. The tidal range must have been incredible, with periods of complete exposure and drying when the tide went out, followed by huge walls of water roaring across the surface of the early ocean with the incoming tide—a true "tidal wave." (The immensely destructive waves produced by earthquakes are not "tidal waves" but are more properly known as tsunamis, or seismic sea waves. They are caused by earthquakes, not tides.)

In addition, the earth was spinning much faster then. It rotated on its axis in about 10 hours in the Hadean, so the day was 10 hours long, not 24 hours long, and each day and night cycle was about 5 hours long in the spring or fall. There were almost 600 days in a year, not 365.25 days as we have now. But that rate has been slowing down thanks to the tidal friction caused by the moon, as discussed in Chapter 7.

The early earth also had an unusual atmosphere, made mostly of nitrogen and carbon dioxide, with abundant methane and ammonia, and no free oxygen. This will be discussed in detail in a later section of this chapter, "The Early Atmosphere".

8.3 The Archean (4.0–2.5 Ga): Alien World

After having almost no rock samples representing the Hadean, there are quite a few places that produce rocks dated to the Archean Eon. These occur as the ancient cores of many of the continents (Fig. 8.3), so they were assembled from many different ancient microcontinents that slammed together during the Precambrian, and later rocks have been added to their edges to make our modern continents. Every continent has its own distinctive Archean predecessors, but for simplicity, we will focus on North America.

 See For Yourself: The Archean

The core of North America, for example, is a composite made of several original Archean protocontinents (Fig. 8.5). The largest is known as the **Superior terrane** since part of it underlies Lake Superior beneath Minnesota, Michigan, and Ontario. It also forms much of the basement rock beneath Quebec. It is the oldest part of North America and the core of what has long been called the "Canadian Shield" (since it forms a large gently domed feature that resembles an ancient warrior's shield lying flat). The Nuvvuagittuq greenstones on the east shore of Hudson Bay, dated at 4.32 Ga, are part of this terrane, as are many of the oldest dated Archean rocks in North America. Also forming the core of Canada is a second terrane, the **Slave terrane** (Fig. 8.5A), which underlies Great Slave Lake (hence its name) and large areas of the Yukon and northwestern Canada. Parts of the basement rock to the south of the Slave terrane, known as the Rae and Hearn cratons, were added later in the Archean to the edge of the Slave terrane. The 4.08-Ga Acasta Gneisses come from this crustal block, as well as many younger Archean rocks. The third protocontinent is known as the **Nain terrane**, which crops out mostly in the southern tip of Greenland (originally part of North America until it split away during the Cenozoic). The Isua Supracrustals (including the Amitsôq Gneiss) dated at 3.8–3.7 Ga are the most famous rocks from the Nain block. The fourth and final block to form the core of North America is the **Wyoming terrane**, represented by a number of Archean metamorphic rocks exposed by the uplifted Rocky Mountains in Wyoming, Colorado, and adjacent areas (Fig. 8.5A). All of these original protocontinents were probably formed far apart from one another (Fig. 8.5B), nowhere near their present location in North America. But long after they formed, they collided with one another, crushing younger Proterozoic mountain belts between them as they coalesced into the core of North America.

What kind of rocks do we find in these Archean terranes? Most of them are very different from any of the rocks formed on the earth today. Some have no modern counterparts at all. They tell us that the Archean crust was still very hot and very active, with thin, narrow microcontinents forming and breaking up and colliding with each other on an earth with a much hotter, more active mantle than we have today. Those ancient rocks fall into four classes, which

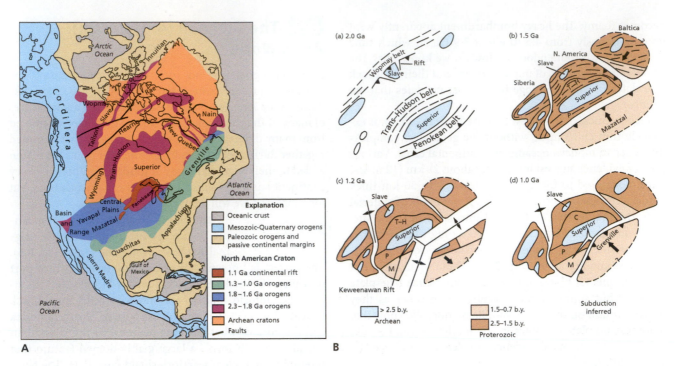

Figure 8.5   **A.** Terrane map of North America. **B.** Cartoon of the stages of assembly of the North American terranes. In (a), the four original protocontinents that would make up North America were the Superior, Slave, Nain, and Wyoming terranes. These floated around in the early earth, and their original position with respect to one another is unknown. By the Proterozoic (2.0 Ga), they were accumulating passive margin wedges on their borders, such as the Wopmay belt, the Trans-Hudson Belt (T-H), and the Penokean (P) Belt. (b) By 1.8 Ga, these early Proterozoic belts had been crushed and fused between the Archean protocontinents. The Siberian and Baltic blocks broke away, while the Yavapai and Mazatzal (M) terranes were accreted to what is now the southern edge of the Superior terrane. (c) By 1.2 Ga, rifting broke a part off the southern region, forming the Keweenawan Rift system. (d) By 1.0 Ga, the Grenville terrane slammed into the southeastern edge of the Superior–Mazatzal blocks to form the basement of the modern Appalachians.

all happen to begin with the letter "G." Thus, the "4 Gs of the Archean" are as follows.

1. **Gneisses** (Fig. 8.6A, B): Gneisses are highly metamorphosed rocks usually formed deep in the crust under high pressures and temperatures (Chapter 2). In many cases they are so transformed that nothing remains of their parent rock textures, and some are in the process of melting to form magmas. They tend to be rich in the lighter silicate minerals, such as quartz, potassium feldspars, and plagioclase, plus darker minerals like biotite and hornblende. Their most distinctive property is that they have compositional banding separating the light- and dark-colored minerals due to their intense heating. When they are found in Archean terranes, they are usually the oldest rocks in the block, as exemplified by the Acasta Gneiss and Amitsôq Gneiss already mentioned. Their composition is very much like modern continental crust, so most geologists consider the Archean gneiss terranes to represent metamorphosed **protocontinental crust** from the earliest microcontinents.

2. **Greenstones** (Fig. 8.6C): Between these gneissic protocontinents were very thin but broad areas of **proto-oceanic crust**. Today, when oceanic crust is formed on the modern sea floor in a mid-ocean ridge (Fig. 5.15), it erupts as blobs of pillow lavas, which ooze from a

crack in a submarine lava flow like red-hot toothpaste. Sure enough, nearly all the Archean proto-oceanic crust shows similar pillow lava features (Fig. 8.6B), proving that it once erupted under water. But these ancient pillow lavas and sea-floor crust are very different from the modern basaltic lavas that erupt every hour from mid-ocean ridges around the world. Instead, they are a weird form of lava known as **komatiite** (Fig. 2.7), which is extremely high in magnesium and thus rich in the mineral olivine (typical of peridotites from the upper mantle), rather than pyroxenes and calcium plagioclases found in modern basalts. In fact, komatiites no longer erupt anywhere on earth but only formed when the Archean sea floor was erupting in the earth's earliest oceans. (There are a few examples from the Proterozoic and later, but they are extremely rare.) Most geologists agree that komatiites suggest a much hotter mantle (hotter than 1600°C, or about 500°C hotter than today), allowing magmas of pure olivine to form. Such a magma would be even more runny and fluid that the basaltic lavas flowing swiftly out of modern volcanoes like Kilauea on the Big Island of Hawaii. The presence of so much komatiite (and no basaltic magmas) also suggests a much more active mantle, pouring out lots of very thin, fluid, hot oceanic crust very quickly, compared to the much slower rates of sea-floor spreading

Figure 8.6 Photos of the four "G" rocks of the Archean. **A.** The Amitsôq Gneiss, from the Isua Supracrustals in western Greenland, dated at 3.8–3.9 Ga. **B.** The Acasta Gneiss, from near the Great Slave Lake in northwestern Canada, is dated over 4.03 Ga. **C.** Pillow lavas from the Yellowknife Group, northern Canada, which have been metamorphosed into greenstone. **D.** A thick sequences of deep-water shales and turbidite graywacke sandstones, from the eastern shore of Great Slave Lake. **E.** Ruby red granites from the Wausau Granite quarry, Wausau, Wisconsin.

and building of oceanic crust today. All of these Archean komatiites that are found today have since been transformed by low-grade metamorphism, so they are made of greenschist minerals such as chlorite as well as serpentine. This is why they are called **greenstones**. The oldest known rocks on earth currently are the greenstones from the east shore of Hudson Bay, which date to 4.28 Ga.

3. **Graywackes** (Fig. 8.6D): The komatiite pillow lavas of the earliest oceans were eventually covered by sediments eroding from the protocontinents. But Archean sediments (especially the sandstones) were very different from typical marine sediments of today, which are normally made of pure quartz sand to form normal sandstones. Instead, they are a mixture of sand and coarser materials, with lots of mud in between the grains, known by the German name **graywacke** (GRAY-wak-ee). Instead of pure quartz sand, like most modern sands, the sands in graywackes are mix of many different components, including unstable minerals like feldspars and even unstable rock fragments (Fig. 8.11A). These suggest that graywackes were freshly eroded out of the protocontinents and dumped right into the ocean basins with little or no weathering and no

recycling of their quartz grains, which typically forms most modern quartz sandstones. Even more interesting, the graywackes are nearly all found interbedded with deep marine shales to form thick sequences of turbidites, which are repeated over and over again and often extend huge distances (Fig. 8.6D). This shows that they were dumped from the continent without much weathering or winnowing or separation of the clays from the sands (as normally happens today) and then deposited in huge submarine gravity flows called "turbidity currents" (see Fig. 4.12). In fact, geologists puzzled over the odd and repetitive occurrence of graded beds made of shale and graywacke (and no quartz sandstones) hundreds of meters thick in the Archean for decades until the mechanism of turbidity currents was finally discovered in the 1940s and 1950s. Archean sediments are also unusual in that there were almost no limestones and few or no evaporite minerals like salt or gypsum. The reasons for this are discussed later in the chapter (see "Earth's Early Atmosphere").

4. **Granites** (Fig. 8.6E): The true granites are often colored deep pink or even red with the abundance of their potassium feldspar. These granites are usually found as dikes intruding into older gneisses, graywackes, and

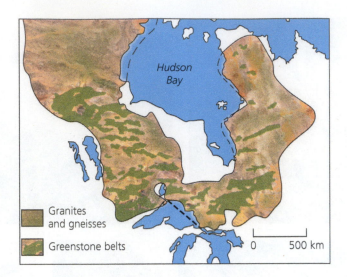

Granites
and gneisses

Greenstone belts

Hudson
Bay

0   500 km

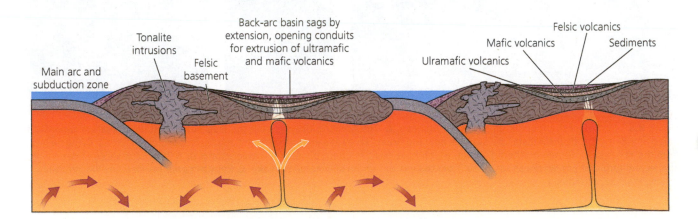

Main arc and
subduction zone

Tonalite
intrusions

Felsic
basement

Back-arc basin sags by
extension, opening conduits
for extrusion of ultramafic
and mafic volcanics

Ulramafic volcanics

Mafic volcanics

Felsic volcanics

Sediments

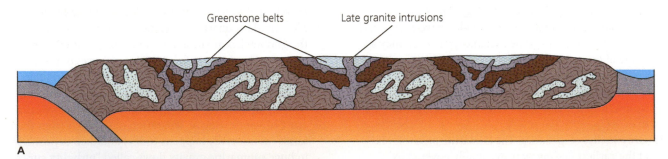

Greenstone belts

Late granite intrusions

A

B

Figure 8.7   A. Assembly of greenstone belts. As the gneissic protocontinents collided with one another, they trapped pieces of proto-oceanic crust and crumpled them between the blocks. These greenstone belts (often covered by marine sediments such as graywackes and shales) are thus found wrapped around and between gneissic protocontinental cores. The final stage occurred when the blocks were all assembled and the lower parts of this crustal material partially melted to form granitic intrusions. **B.** Satellite image of the Pilbara craton in northwestern Australia, showing the Archean blocks (lighter oval-shaped areas) surrounded by greenstone belts.

greenstones, showing that they are the last of the "four Gs" to be formed. They were formed by the remelting of rocks from the bottom of the gneissic protocontinental crust.

Even more distinctive of the Archean than the peculiar assemblage of rocks is how they are formed and distributed. In most of the Archean cores of today's continents, the gneissic protocontinents trapped between them large masses of bedrock known as **greenstone belts** (Fig. 8.7). The ancient proto-oceanic crust of greenstone and the graywacke sands and muds that filled the ocean basin are typically crumpled up and folded and deformed, presumably when the protocontinents collided and smashed the proto-oceanic crust between them. Thus, in map view, the greenstones form long, narrow belts of rock, each of which once was a proto-ocean that has since been squashed into a narrow block by the gneissic protocontinents on either side. Both the crushed greenstone belts and the gneissic protocontinents were intruded by granites from the last stage of these early mountain belt collisions. Presumably, the collision of multiple protocontinents and proto-ocean formed a thicker mass of crust that began to melt at its base, producing magmas rich in silicon, potassium, sodium, and aluminum that intruded the overlying rocks to form the typical Archean granites (Fig. 8.4).

Compared to the rocks forming today, Archean rocks are very strange and indicate a world very different from our own. The protocontinents (Fig. 8.4B) were relatively thin, much smaller, and much hotter than today's true continents. Geochemical evidence suggests that they were only about 25 km (15 miles) thick, compared to modern continental crust that is 30–50 km thick. The rocks at the surface came directly from the mantle (komatiites), and rocks lower down in the crust were under high enough pressures and temperatures that they experienced amphibolite-grade and granulite-grade metamorphism. The deeper parts of the crust were remelted into granites that intruded late in the process. In addition, there are geochemical signatures that suggest that some of the deep crustal material was reworked from even older Hadean crust. These proto-plates did not undergo full-scale plate tectonics in the modern sense, although they moved around on currents in the mantle. There may have been some small-scale versions of subduction zones, but the geochemical evidence suggests that they are not true subduction zones on the scale of those that were formed by modern plate tectonics.

The proto-oceanic crust was much thinner and hotter and formed of magmas erupted directly from the upper mantle, all suggesting that the proto-plates moved much faster and were destroyed much faster than the plates are today. The oceanic sediments were freshly eroded from the continents as angular sands made of all sorts unstable minerals and rock fragments and lots of mud and dumped into the proto-oceanic basins without any winnowing or sorting or recycling into younger sandstones. In addition, there were no limestones and almost no evaporites like gypsum or halite, indicating conditions very unlike our modern sedimentary environments. Finally, the magmas that melted from the bottom of these accreting protocontinents were very rich in silicon, aluminum, potassium, and sodium and did not differentiate into the entire spectrum of igneous rocks that we see in later geologic history.

8.4 Proterozoic Eon (2.5–0.5 Ga): Transition to the Modern World

The Proterozoic by itself lasted about 2 billion years, almost half of earth's total history. Yet by necessity we cannot detail the entire 2-billion-year histories of each continent during the Proterozoic or record every complex event that happened over that immense span of time. Instead, we will try to highlight some of the major trends that happened between the Archean and the Cambrian.

In the Archean, we encountered unusual rocks like greenstones and graywackes formed in proto-oceans and small gneissic continental crustal blocks that formed thin, hot protocontinents. In short, there were some similarities to modern plates and plate tectonics, but everything was hotter and thinner and moved much faster in a world of "proto-plate tectonics." As the transition to the Proterozoic began, however, we encounter rocks that could only be formed by the movement and collision of much thicker, cooler continental blocks and normal basaltic oceanic crust, so the transition to true plate tectonics has begun to occur. By the end of the 2 billion years of the Proterozoic, the outcrops look like any other typical outcrop formed in more recent times, and the earth was the realm of normal plate tectonics.

THE PALEOPROTEROZOIC: PLATE TECTONICS EVOLVES

We can see this transformation in some of the Early Proterozoic rocks of the Canadian Shield (Fig. 8.8). Off to the present-day northwest of the Slave terrane is a region known as the **Wopmay belt** (Fig. 8.5). About 1.9 Ga, the Wopmay belt looked like the classic passive margin wedge that forms when continents rip apart (Fig. 8.8A) during normal plate tectonics (Fig. 5.18). Thick deposits of shallow-marine sandstones and shales were deposited in shallow seas on the edge of the Slave protocontinent. These rocks were not the deep-marine graywackes and shales of the Archean but more modern-looking quartz sandstones and shales such as those that form in the shallow ocean today. There is even evidence of a failed rift, or aulacogen, next to the passive margin wedge, further proof that the Slave terrane must have ripped away from some other Archean protocontinent before 1.9 Ga. By 1.88 Ga, however, a volcanic arc was approaching from offshore, and the shallow-marine sediments of the passive margin wedge were filling with volcanic sediments (Fig. 8.8B), as happens when any passive margin switches to becoming an active margin (Fig. 5.23). By 1.85 Ga (Fig. 8.8C), the collision of this approaching volcanic arc completely smashed up the passive margin of the Wopmay

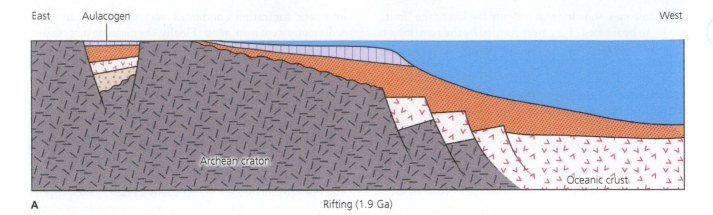

A   Rifting (1.9 Ga)

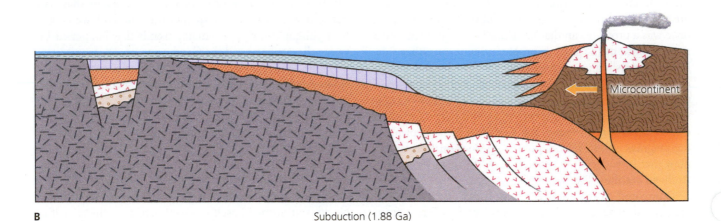

B   Subduction (1.88 Ga)

C   Collision (1.85 Ga)

| Volcanic rocks | Granites | Thrust fault |

Figure 8.8 **A.** Diagram of the Wopmay passive marine wedge forming about 1.9 Ga on what is now the northwestern edge of the Slave terrane. **B.** By 1.88 Ga, the plate directions shifted from divergent to convergent, and the passive margin wedge was crushed by the collision of an arc terrane **C.** By 1.85 Ga, the collision had finished, and the Wopmay passive margin was crumpled into a mountain belt.

belt, intensely deforming the shallow-marine sediments of the passive margin wedge and building a huge mountain range, as is typical of many mountain-building events of the geologic past (Fig. 5.23). In short, the Wopmay belt exhibits all the classic features of modern plate tectonics (such as the plate tectonics that formed the modern-day passive margin wedge beneath the edges of the Atlantic Ocean), suggesting that true plate tectonics was operating on earth by 1.9 Ga.

The Wopmay belt is not the only such example in the Proterozoic. Between the Superior terrane and the terranes to the north was another crumpled passive margin wedge deformed when these protocontinents collided about 1.8 Ga. Known as the **Trans-Hudson belt**, it runs from Hudson Bay all the way down into the northern Plains of the United States (Fig. 8.5).

A third example is found on the southern edge of the Superior terrane, beneath what is now Minnesota, Wisconsin, and northern Michigan. Ancient Archean greenstones (the Keewatin volcanics) were intruded by mid-Archean granites (the Saganagan granites), then uplifted and eroded and covered by Late Archean greenstones and graywackes (the Knife Lake series), and finally intruded by the 2.5-Ga Vermilion or Algoman granite, which marks the end of the Archean events. Together, these rocks form the basement of the Superior terrane in this region. But during the Early Proterozoic, a thick passive margin wedge and ocean basin sediment sequence known as the Animikie series covered up the Early Proterozoic unconformity eroded into the Archean rocks. This entire sequence was then crunched into a mountain belt and deformed during the **Penokean orogeny**, about 1.85 Ga, about the same time as collisions deformed the Wopmay belt and Trans-Hudson belt to the present-day north (Fig. 8.5). Thus, the overwhelming signature of Early and Middle Proterozoic rocks and structure suggests modern plate tectonics much like what has occurred in the past billion years.

THE MESOPROTEROZOIC: SUPERCONTINENTS ASSEMBLE

Once the Trans-Hudson belt had fused the Superior terrane to the Archean terranes to the modern northwest (such as the Slave terrane) and the Wopmay and Penokean belts had formed on its edges, the core of the "Canadian Shield" was fully assembled. It has remained essentially unchanged since the Early Proterozoic (since 1.8 Ga). But there was more to add to North America during the rest of the Paleozoic. About 1.76–1.72 Ga, a huge block fused to the present-day southern edge of the Superior–Wyoming blocks, known as the **Yavapai terrane** (Fig. 8.4), named after the Yavapai region in northern Arizona. It forms the ancient basement rock stretching from Wisconsin down to the Mojave Desert of California. Right after the collision of the Yavapai terrane came another terrane about 1.69–1.65 Ga. Known as the **Mazatzal terrane** (after the Mazatzal Mountains in southern Arizona), it runs southwest to

northeast from southern Arizona to the basement of Indiana and western New York and all the way up into Labrador and Maritime Canada (Fig. 8.5).

The final large terranes to be added to North America in the Proterozoic were the blocks that form the Precambrian basement of the Appalachians. Known as the **Grenville terrane** up in Maritime Canada and down through the northern Appalachians, they accreted to the southeast side of the Mazatzal block about 1.3–1.0 Ga (Fig. 8.5). To their south is the basement of the southern Appalachians, which is known by a variety of names but consists largely of granites and rhyolites that date to 1.55–1.35 Ga.

All of these terranes that gradually accreted to form North America were part of a larger process where the rest of the continents were all growing larger as the Archean protocontinents collided and fused to form their continental cores.

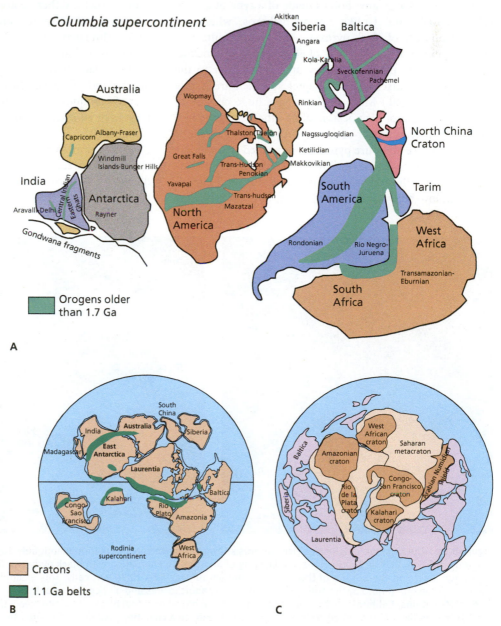

Figure 8.9 Reconstructions of the Proterozoic supercontinents. **A.** Columbia, about 1.3–1.8 Ga. **B.** Rodinia, 0.7–1.1 Ga. **C.** Pannotia, 0.6–0.5 Ga.

All of the other continents also saw additional terranes accreted to their Paleoproterozoic edges during the Mesoproterozoic. Some geologists think they may have fused into a supercontinent called **Valbara** about 3.1–3.6 Ga in the Archean, which then broke apart and fused into another supercontinent called **Kenorland** about 2.7–2.1 Ga in the Late Archean. These ideas are still controversial, but there is widespread consensus that about 2.1–1.8 Ga there was a supercontinent called **Nuna** (also known as the **Columbia supercontinent** or Hudsonland) during the Paleoproterozoic (Fig. 8.9). This supercontinent began to break up after 1.7 Ga, and its fragments moved around the earth's surface until they reassembled as a new supercontinent between 1.0 and 0.7 Ga, known as **Rodinia**. Finally, about 0.5 Ga, the fragments of Rodinia reassembled in a slightly different configuration to form a supercontinent named Pannotia.

Late in the Mesoproterozoic, many of the continents were intruded by magma bodies made of a type of gabbro called **anorthosite**. Unlike normal gabbros, which are an even mixture of black pyroxene and gray calcium plagioclase, anorthosites are made of almost pure plagioclase, with little or no pyroxene. (They are very similar to the moon rocks we discussed in Chapter 7, which are largely anorthosite as well.) Many of them have huge crystals of bluish gray calcium plagioclase, while others form layered gabbros with centimeter-scale layering between layers of pure plagioclase and pure pyroxene, deposited in the floors of ancient magma chambers (Fig. 8.10). Even more intriguing, these anorthosites were all intruded during a narrow time window between 1.3 and 1.0 Ga, and their magma

chambers once lined up across the ancient supercontinent of Rodinia. Today, exposed remnants of the magma chamber are found in the San Gabriel Mountains north of Los Angeles, the Beartooth Plateau of Montana and Wyoming, the Adirondack Mountains in upper New York and Quebec, with additional anorthosites in southern Scandinavia and across eastern Europe.

What could have produced such a large volume of odd intrusions? Geologists have debated this problem for decades, ever since these anorthosite plutons were first described and dated. There is general agreement that some sort of unique event was happening in the upper mantle to generate this peculiar magma, which only intruded in a few places in a limited time window. Somehow as this magma was forming, most of the pyroxenes that make normal gabbros got left behind lower in the crust, and only the plagioclase-rich fraction was allowed to rise upward and intrude into the country rocks. Other geologists have suggested that as the magma chamber was forming at the base of the crust, it may have assimilated a lot of the overlying crust, changing its chemistry and allowing the plagioclase-rich portion to rise to the surface, while leaving the black, pyroxene-rich gabbros behind. Whatever the cause, these fascinating rocks are part of a unique event that happened only once in earth history. Such an event has never been repeated again on such a large scale.

In addition to these major events of the igneous and metamorphic rocks that added to the continental crust during the Proterozoic, there were significant events within the sediments formed on the surface. Back in the Archean,

Figure 8.10 The calcium–plagioclase-rich gabbros known as anorthosites were formed uniquely during a single time interval between 1.0 and 1.3 Ga. **A.** Classic centimeter-scale layering of anorthosite alternating with black pyroxene-rich gabbro, which once formed as horizontal layers on the bottom of the magma chamber but was later tilted vertically. This example is from the Duluth Gabbro near Duluth, Minnesota. (Photo by the author) **B.** The blue-gray anorthosites and black gabbros of the San Gabriel Anorthosite intrusion, San Gabriel Mountains, California. In this case, not only are the once-horizontal black and white layers tilted to vertical, but there are large conical teepee-like mounds of crystals on the floor of the magma chamber (visible as triangular bodies in the picture), which point to the top of the magma chamber (to the right in this photo).

sediments were very simple and monotonous, with mostly graywacke sands formed by submarine turbidity currents dumped into the deep proto-oceanic basins filling with muds to make shales (Fig. 8.6C). No limestones formed at this time, nor did large deposits of gypsum or salt. By the Proterozoic, however, many of these ancient Archean sandstones must have been uplifted and exposed to weathering many times because the sandstones of the Proterozoic are much richer in quartz and show signs of intense weathering, erosion, winnowing, and especially recycling of older quartz-rich sandstones (Fig. 8.11B). In many places across the Proterozoic outcrop belt (especially in Wisconsin, Minnesota, and Michigan), there are thick sequences of Proterozoic **quartz-rich sandstones** (now mostly metamorphosed into quartzite) that show all the classic features of having been deposited in rivers and shallow seas, including many different kinds of ripple marks and cross-bedding. There are even small limestone layers rich in structures called "stromatolites" (Chapter 9) in places like the Proterozoic sequence at the base of the Grand Canyon, along with thick deposits of shale and sandstone. The shales also show mudcracks throughout, proving that they were deposited on drying mudflats, probably in the floodplains of ancient rivers. In short, most of the features of modern sedimentation had appeared by the Late Proterozoic, and the peculiar graywacke–shale sequences that make up the only sediments of the Archean had been replaced by quartz sandstones, floodplain muds and shales, and even small limestone bodies.

In some places, the Mesoproterozoic exposures are spectacular. For example, most of the rocks exposed in the ranges across the western Rocky Mountains in Montana (especially in Glacier National Park) and up into Alberta and British Columbia are part of the **Belt Supergroup**. (A similar sequence called the Purcell Supergroup occurs in the Rockies of British Columbia and Alberta.) Deposited between 1.47 and 1.40 Ga, the total thickness of these shallow lake deposits and river sandstones and shales with minor limestones reaches over 30,000 m (almost 10,000 feet). If you travel over the Rockies in western Montana or southern British Columbia, every range around you is composed of thick deposits of Belt-Purcell sediments, capped by Windermere deposits. They are strikingly displayed in the prominent peaks of Glacier National Park in Montana. If you drive west over the top of these ranges on Going-to-the-Sun Road, you will see amazing outcrops of ripple-marked and cross-bedded sandstone from the belt exposures that make up the entire Rocky Mountain front in this area.

NEOPROTEROZOIC: RODINIA BREAKS UP

By 750 Ma, the supercontinent of Rodinia had begun to break apart (Fig. 8.12). It split in half, and the gap between the fragments to the west of North America (Laurentia) and east of the East Antarctica–Australia block became the predecessor of the Pacific Ocean.

As in the case of breaking any continent apart to form a new ocean, the most obvious remnants of the process are the rift valleys and aulacogens that rip the crust apart, followed by the thick deposits of shallow-marine sediments that accumulate in the passive margin wedge (Fig. 5.18). Numerous aulacogens can be found along the western half of what is now North America from this big event. They include the **Amargosa aulacogen** beneath Death Valley, with over 4000 m (13,000 feet) of upper Proterozoic shales, limestones, and glacial deposits; the **Uinta Mountain Group** in the Uinta Mountains between Utah and Wyoming, which consist of over 7000 m (23,000 feet) of red sandstones and shales deposited in a deep rift valley; and the **Grand Canyon Supergroup** (Unkar and Chuar groups) in the bottom of the Grand Canyon, which consist respectively

Figure 8.11 In the Proterozoic, there was a dramatic change in the sandstones. **A.** Microscopic view of a typical immature Archean sandstone. The large color patches are coarse angular sand grains made of rock fragments and feldspars, surrounded by a muddy matrix (dark speckled area). This sandstone has never been reworked but was dumped into a turbidite after freshly weathering from bedrock. **B.** By the Proterozoic, there were many sandstones, such as this Baraboo Quartzite, that were very mature, so all the rock fragments, feldspars, and mud were gone; all that remains were stable quartz grains (gray patches) of roughly the same size. **C.** Outcrop of the very mature sands and conglomerates of the Upper Proterozoic Hinckley Sandstone, stained yellow and orange by iron oxides ("rust"), located in Sandstone, Minnesota.

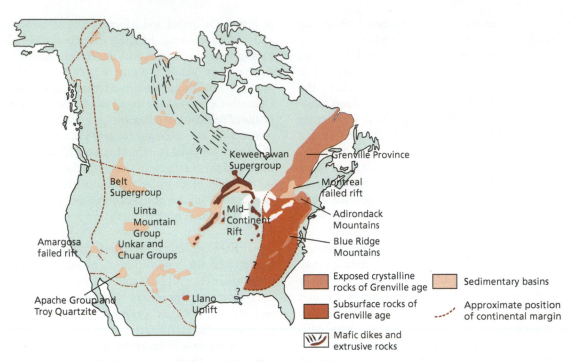

Figure 8.12   The remnants of the Late Proterozoic breakup of Rodinia can be found all over North America. The Uinta Mountain Group in Utah–Wyoming; the Unkar and Chuar groups in the bottom of the Grand Canyon; the Amargosa failed rift in Death Valley, California; and the Apache and Troy quartzites in Arizona are all remnants of the fill of Late Proterozoic aulacogens. The Keweenawan/Mid-Continent Rift valley opened up from Lake Superior down to Kansas, erupting enormous volumes of lavas and filling with sediments as well. Meanwhile, in the east, the Grenville terrane arrived and accreted to form the basement rock of the future Appalachian Mountains and eastern Quebec and Labrador.

of 2200 m (7200 feet) and 1600 m (5250 feet) of red sandstones, shales with mudcracks, and even lava flows now tilted and deeply eroded (Fig. 3.6), dated between 1250 Ga and 700 Ma.

Another place that rifting started and then was abandoned was the center of the North American continent. Known as the **Mid-Continent Rift**, it began to rip apart about 1.1 Ga and continued to about 900 Ma, forming a huge rift valley that today runs through the Minnesota–Wisconsin border region, then plunges into the subsurface beneath Iowa, Nebraska, and as far south as Kansas (Fig. 8.12). Another arm of this Mid-Continent Rift ran beneath Lake Superior and through the Upper and Lower Peninsulas of Michigan, totaling about 1200 miles (2000 km) in length. The enormous rift valley was filled with huge volumes of flood basalts that poured out across the valley and over its rims, known as the **Keweenawan lavas** (Fig. 8.13). They get their name from the Keweenaw Peninsula, which sticks out into Lake Superior from the northern shore of the Upper Peninsula of Michigan. Some of these eruptions were enormous, reaching thicknesses of 500 m (1600 feet), and representing volumes of 1640 km³ (400 miles³) of lava. This is the largest lava eruption ever documented in earth history. Near Duluth, Minnesota, are exposures of the Duluth Gabbro, representing part of the magma chamber that fed these enormous flood basalt eruptions. In other places, such as Palisade Head on the north shore of Lake Superior, there were huge rhyolite eruptions as well. Once the lavas had finished erupting, the basin was filled with at least

7600 m (25,000 feet) of sandstones and shales, typical of any other modern rift valley, making the total thickness of lavas plus sedimentary rocks in the rift sequence up to 17,000 m (56,000 feet) thick.

As incredible as this tectonic event was, it was a failed rift. By 700 Ma, it was all but over, and the deep gash in the continental crust slowly began to heal as it was buried by later Paleozoic sediments starting 500 Ma. Nonetheless, it had effects long after it was extinct. The black Keweenawan lavas crop out in many places in Minnesota, Wisconsin, and Michigan, creating distinctive landforms. Geologists have looked closely at the sedimentary fill of the rift valley, hoping to find oil somewhere. More important, the rift valley may not be entirely extinct. The deeply buried part that plunges beneath the Great Plains may be the source of some of the earthquakes that have occurred in the New Madrid rift zone, which produced the biggest earthquakes in US history in 1811–1812 and are still active today.

Once the rift valleys and aulacogens had formed and filled up, the final stage of opening the Pacific Ocean between western North America and the Australia–East Australia block was the development of a thick passive margin wedge In the Rocky Mountains of Montana and Utah, this non-marine to shallow-marine to deep-marine passive wedge is known as the **Windermere Supergroup**, which sits on top of the Mesoproterozoic Belt-Purcell Supergroup. The Windermere reaches a thickness of 9 km (5.5 miles) in some places and covers an area of 35,000 km² (13,500 miles²).

Figure 8.13 Columnar-jointed rhyolite lava flows from the Keweenawan eruptions on the north shore of Lake Superior just north of Duluth, Minnesota.

Some non-marine river sandstones are interbedded with volcanic lava flows and shallow lake deposits of the late rift valley stage, but most of the Windermere sequence is marine in origin, including quite a few deep-water shales and turbidites. This thick, wedge-shaped sequence of Windermere was the first deposit of the newly opened Pacific Ocean, gradually sinking and accumulating more and more sediments as the ocean rifted open wider and wider.

8.5 The Snowball Earth

One of the most remarkable episodes in our planet's history, postulated to have occurred in the Proterozoic, was a series of events when the planet may have nearly completely frozen over. Known as "snowball earth" episodes, they were first suggested back in the 1950s and 1960s, when geologists found a number of places around the world where limestones were exposed both above and below glacial deposits (Fig. 8.14). As we saw in Chapter 2, today most limestones form in the tropics or subtropics in very shallow water. On the other hand, glaciers never occur in the tropics at sea level; the only modern tropical glaciers occur in the highest mountain peaks, like the Andes of Peru or Mt. Kilimanjaro in Kenya. So how could this "limestone/glacial deposit/limestone" sequence occur at sea level? By the 1980s, it was clear to a number of geologists from the paleomagnetic signatures recorded in these beds that these glacial deposits were also tropical and formed at sea level near the equator. The only way this could occur would be to freeze the entire earth right from the poles to the equator, a condition that came to be known as "snowball earth."

How do you freeze over a planet once the ice sheets start to grow? Geologists discovered a well-known climate effect known as the **albedo feedback loop**. "Albedo" is just a technical word for describing the reflectivity of a surface. If you've ever spent time skiing or snowboarding, you will discover that snow or an ice sheet has high albedo since it reflects most of the sunlight that hits it. That's why you need good dark goggles that reduce glare with tinted lenses when you spend time on the ice—and good sunscreen. By contrast, dark surfaces (like forests or the open ocean) absorb a lot more sunlight and reflect very little.

Figure 8.14 Snowball earth glacial deposits are found in many places around the world. **A.** The Gowganda till, from the Early Proterozoic snowball event (the Huronian glaciation), northern shore of Lake Huron, Blind River, Ontario, Canada. **B.** The Elatina diamictite, an interbedded sequence of limestones and glacial tills from Australia, which was a few degrees from the equator in the later Proterozoic Varangian glaciation. This deposit proves that there had to be sea-level glaciers in the tropics **C.** Upper Proterozoic glacial deposits of the Kingston Peak Formation, near Death Valley, California. It is capped by a carbonate deposit, the Noonday Dolomite, and underlain by a carbonate, the Beck Spring Dolomite, showing that the Death Valley region was warm and subtropical in the Late Proterozoic before the snowball earth event and then went back to tropical conditions. **D.** A thick glacial deposit in Namibia with huge angular boulders, overlain by the "cap carbonate" limestone body precipitated abruptly when the Varangian glaciation ended. Dan Schrag (left) and Paul Hoffman for scale.

The albedo feedback system is very sensitive to small changes, which can change it from frozen to ice-free and back to frozen very quickly (Fig. 8.15). For example, let's say the earth's surface is covered by ice, so it has a high albedo and reflects most of the sun's energy back. But the planet begins to warm slightly, and that ice sheet melts back a bit, exposing dark land and water. This absorbs more sunlight and generates heat, which melts the ice even further. Back and forth these two processes go in a feedback loop that eventually melts the ice in a very short time. Now let's image this dark land and ocean surface has a few really cold winters and the reflective snow and ice layer lasts a bit longer. The increased ice cover reflects more energy back out to space, and the land gets colder, so even more ice sticks around the next few winters, and the ice sheet expands. Before you know it, the entire system has switched back into a complete ice age.

▷ See For Yourself: The Snowball Earth

Albedo is a key feature of the polar regions, and that is why they're so sensitive to small changes in global temperature. Climatic modeling showed that if you had even a small ice sheet in the subtropical or tropical latitudes to start with, the albedo feedback loop would kick into high gear, and the entire planet could freeze over rapidly. The only dilemma with this model was how to thaw the planet once it is completely frozen and has such a high

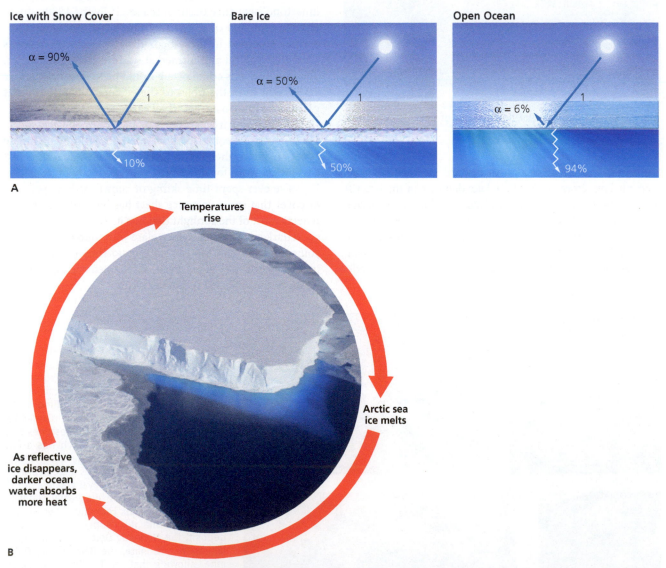

Figure 8.15 Albedo is a measure of the reflectivity of the earth's surface. Ice and snow are highly reflective (have a high albedo) and bounce most of their energy back into space, while dark surfaces like oceans or vegetation have low albedo and tend to absorb heat. Ice with snow cover reflects about 90% of the incoming radiation, and bare ice reflects about 50%, while the dark ocean only reflects 6%. The system works as a big feedback loop. If climate warms slightly in an ice-covered world, more ice melts and exposes dark surfaces, which increases the heat absorption and warms the landscape further. This feedback process causes the ice to melt back faster and faster and further increases heat absorption until the ice is gone. Conversely, a colder summer after a heavy winter will keep the ice on the ground longer, lower the temperature, making more ice stay without melting, increase the albedo, causing more cooling and ice growth, and quickly freeze over the unfrozen world.

albedo that most of its energy is reflected back to space. A completely frozen reflective iceball is a dead end, and the warming part of the feedback loop cannot rescue it. So how does the planet escape becoming a permanently frozen snowball?

The solution was another mechanism: volcanoes. The earth is unlike any other frozen planet in space (such as Mars or many others that have been found) in that it has an active crust with plate tectonics that powers lots of volcanoes. (Mars has some large volcanoes, but they are not produced by plate tectonics.) Volcanic eruptions release lots of gases, especially greenhouse gases like carbon dioxide, water, methane, and sulfur dioxide. If the earth were indeed completely frozen, the volcanic gases would eventually build up and warm the planet through the greenhouse effect so that the ice would finally begin to melt. Once enough dark surface had been exposed, the albedo feedback loop could kick into high gear and quickly melt it from a frozen planet to an ice-free subtropical planet with limestones in the tropics.

Further study of some of the limestone/glacial deposit/limestone sequences yielded interesting clues. One of the unusual features is that the limestones on top of the glacial till are particularly thick and they showed some peculiar geochemical and mineralogical characteristics. These "cap carbonates" on top of the glacial tills may be products of direct precipitation of limestones once the ocean geochemistry, saturated with dissolved carbonate, had been released from the grip of the ice. They are clearly not the normal kind of limestones formed today, precipitated by organic activity. Modern limestones are made largely by the shells of corals, mollusks, and other marine creatures, as well as calcareous algae.

Another suggestive piece of evidence is the brief return of **banded iron formations** or BIFs (see later in this chapter, see "The Oxygen Holocaust") during the peak of the Late Proterozoic snowball conditions. This would make sense if the earth were frozen over because it would shut down the oceans and make them anoxic and saturated with dissolved carbonate so that they become highly acidic oceans (as we are doing to our oceans now thanks to our greenhouse gases). Without runoff from the sediment flowing down rivers (now completely frozen), the sulfate input to the oceans is shut off; and this results in abundant dissolved iron in these acidic, low-oxygen, low-sulfur oceans. Under these conditions, iron could accumulate on the bottom, as it did between 3.7 and 1.7 Ga.

So the main "snowball earth" model runs like this: something causes the planet to begin to cool down dramatically until large ice sheets begin to form. In those days without abundant and complex life (as we have today) to regulate the carbon cycle and keep pumping carbon dioxide into the atmosphere, the planet would begin to experience a runaway albedo feedback loop and eventually freeze over to the equator. Once it was a frozen snowball, it would be stuck in that state for millions of years, just like Mars is completely

frozen now (although it once had liquid surface water with oceans and rivers). Oceanic circulation would shut down, BIFs would accumulate on the anoxic sea floor, and lots of carbon would be frozen into little cages of ice known as "methane hydrates" in the sea-floor sediments. If nothing else happened, the earth would have stayed frozen, and we would not be here.

Unlike Mars or any other planet, however, the earth has plate tectonics and volcanoes, which over a long enough time erupt enough greenhouse gases to finally begin to warm the planet. Once that occurred, another runaway albedo feedback loop kicked in, and the ice melted rapidly until it was almost all gone. The carbon caged in ice in methane hydrates on the sea floor released huge amounts of methane, further accelerating the global warming. The ocean geochemistry was then so rich in dissolved carbonate that huge deposits of calcite precipitated directly out of seawater to form cap carbonates. Finally, the planet was stable again, with warm tropics and cooler poles.

Further research revealed that there were at least two or three separate events in the Late Proterozoic and one in the Early Proterozoic (about 2 Ga), known as the Huronian (Fig. 8.14). It was based on the well-known Gowganda tillite on the shores of Lake Huron and showed that snowball earth conditions are not unique but can happen multiple times if the conditions are right.

Geologists, like all scientists, are naturally skeptical of new ideas, especially those that seem beyond the norm. Over the past several decades, the snowball earth model has piled up an increasing volume of data, so most of the geological community had no choice but to accept the obvious conclusions that something like a snowball earth must have happened at least three or four times.

Still, there are dissenters. A number of geologists accept that there were equatorial sea-level glaciers in the Late Proterozoic but not that the entire tropical region had frozen over so that the earth was a frozen snowball. They prefer a slightly less extreme idea, nicknamed the "slushball." In this model, there was some glaciation on the equator (the data demand it), but much of the tropical region was cold but ice-free. They point to geological evidence of sediments that could only be formed in water, not ice. However, even the original "snowball" model allowed for some ice-free regions in the tropics, so this is not new. Also, many geologists see evidence that the snowball earth episodes had rapid fluctuations of glacial–interglacial cycles like the most recent Ice Ages, so this allows for both glacial sediments as well as sediments formed in running water and unfrozen oceans. Most importantly, the dating of the separate snowball events in the later Proterozoic shows that they were globally synchronous and occurred from pole to equator at the same time. This favors a more extreme snowball earth rather than a slushball because in slushball models the ice lines retreat when the carbon dioxide levels go up—but that is not what we see in the Late Proterozoic snowball model.

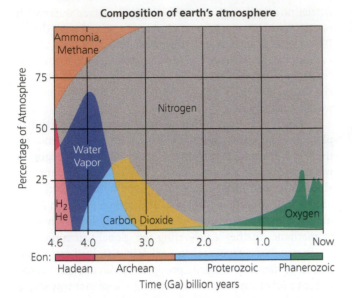

Composition of earth's atmosphere

Figure 8.16 This diagram shows the dominant gases in the earth's atmosphere through the Precambrian and Phanerozoic. Hydrogen and helium would have been abundant at first but escaped to space because earth's gravity was not strong enough to hold in such light elements. Methane and ammonia would have been significant early on but vanished over time as well. Meanwhile, the major gases were nitrogen (still 78% of our modern atmosphere), carbon dioxide (still the dominant gas on Mars and Venus), and water vapor (which gradually rained down to the earth's surface as it cooled). Free oxygen (only 1% of the atmosphere or less) did not appear until 2.4 Ga (the "Great Oxidation Event") and did not become abundant until the Phanerozoic.

8.6   The Precambrian Atmosphere

THE EARLY ATMOSPHERE

See For Yourself: Origin of the Atmosphere

By looking at other giant "gas ball" planets, like Jupiter, Saturn, Uranus, and Neptune, we can get a pretty good idea of what the earliest atmosphere of the earth was like. The most important gases in those giant gaseous planets are hydrogen and helium (as they also are in the sun), which make up about 99% of the total mass of the solar system. However, the earth wasn't large enough to generate a gravitational attraction strong enough to hold such light elements in its atmosphere, and they floated off to space soon after the earth formed (Fig. 8.16). That is why balloons and blimps used to be filled with hydrogen to make them float, until the *Hindenburg* disaster showed how flammable hydrogen gas can be. Today, nearly all balloons, from those at a kids' party to the giant blimps that fly above athletic events, are filled with non-flammable helium.

Once hydrogen and helium escaped, the next most abundant gases would be those that we see around the other smaller planets. Both Venus and Mars have a lot of carbon dioxide in their atmosphere, and it would have been a significant component of earth's early atmosphere

as well. Most important, however, would have been nitrogen gas (N_2), which today makes up 78% of the air you breathe and probably was that dominant in the early atmosphere as well.

Looking at planets like Jupiter and Saturn again, we find that other gases are very important, especially methane ("natural gas" or "swamp gas," CH_4) and ammonia (NH_3). Geochemists have long debated how much of these gases were also in the earth's atmosphere, at least in significant concentrations. A few decades ago, they were thought to be very important; then the thinking shifted, and their importance and abundance were downgraded. However, the pendulum has swung back so that now most scientists think that methane was very important at the start, and probably ammonia (see Box 8.1). This is very important to discussions about how the earliest life formed, as we shall see in Chapter 9.

Thus, the planet started out rich in hydrogen and helium but quickly lost these light gases because earth's gravity was not strong enough. It was left with an atmosphere rich in nitrogen, some carbon dioxide, water vapor, and minor amounts of methane and ammonia. What it did *not* have was any free oxygen, which made it completely inhospitable to life as we know it. We will look at the evidence for this in the next section. First, however, we need to look at an important constraint on the atmosphere and its temperatures.

THE OXYGEN HOLOCAUST

What is the evidence of the earliest oxygen on the planet? The best such evidence comes from certain places, such as the Iron Ranges of Minnesota, the Hamersley Range in Australia, and several other places which are truly unusual. They are the world's main sources of iron that drove the Industrial Revolution. These iron deposits come from BIFs. As their name suggests, these rocks have red or black bands of iron (Fig. 8.18), about a few millimeters to a centimeter thick, alternating with bands that are made of pure silica (in the form of chert or jasper). Sometimes there can be thousands of these alternating bands in a row, extending over huge areas of outcrop. When these were discovered in the mid-1800s their meaning was a mystery. Even more surprising, the rock is made of pure iron plus chert with little or no mud or sand, which you might normally expect to find washing out into the ancient seas when the iron was being deposited.

So how did sediments consisting of dissolved iron and silica settle out on the sea floor without being mixed with sand and mud? The first thing to know is that in modern oceans, iron cannot stay dissolved in seawater because it is rapidly oxidized to various forms of iron oxide ("rust") and clings to other minerals or settles out. The only way to transport and concentrate large amounts of iron in seawater is if the oxygen content is so low that iron cannot rust. This suggests that the ancient sea bottoms must have been completely anoxic when the iron formations were deposited,

BOX 8.1: HOW DO WE KNOW?

How Did the Early Earth Not Freeze Over?

In 1972, the famous astronomer Carl Sagan, along with colleague George Mullen, pointed out something interesting. They showed that our sun is a fairly typical star that follows a well-known sequence of events, which means we know what it would have been like millions of years in the past. In particular, when it was shining on the earth 4 billion years ago, it was not as far along in its evolution as it is now and would have given the earth only about 70% of the energy that it does today. A 30% drop in solar energy on the earth's surface is huge, and if that were the only factor, the planet would have been frozen over from its beginning with a thick layer of ice. Yet we know from the evidence of the zircons discussed in Chapter 7 that there was liquid water and therefore oceans on the earth's surface as early as 4.4 billion years ago, so clearly it was not frozen solid back then. In fact, the sun would not have given the earth enough energy to keep it from freezing from 4.5 until as recently as 2.0 billion years ago. This is the **"faint young sun paradox."** What kept the earth from freezing over when the sun supplied only 70% of the energy it supplies now?

Many ideas have been proposed. Some scientists have suggested that if there is no ice on earth, it would not reflect much sunlight and would absorb much more—but the presence of clouds always reflects some of the sun's energy, 4 billion years ago as it does today, so this is not the answer. Others have argued that the earth's interior was hotter then, so geothermal heat would provide the difference; but calculations show that this is simply not enough heat to compensate for the loss of sunlight. The only other possibility, then, is that the earth's early atmosphere was rich in greenhouse gases, such as carbon dioxide and methane.

We know that carbon dioxide was present in the earth's early atmosphere, but calculations show that it would need to be in concentrations 1000 to 10,000 times as high as today to keep the planet from freezing over—and that is unrealistic. (At its highest, it may have been as high as 5 times present values, but it's not possible that it was 1000–10,000 times as abundant.) The only other reasonable alternative, then, is methane. Methane would have been abundantly produced from the weathering of seafloor lavas, and some may have also reached earth from impacts of comets, which have a lot of frozen methane in them. Later, after life evolved, some of the earliest forms of life were methane-producing bacteria, which break down complex organic matter and release methane gas. Methane is a very powerful greenhouse gas, much more effective at trapping the earth's heat than carbon dioxide, so a little goes a long way. If it was only 1/1000 as common as carbon dioxide, then the earth would need only 15 times its present levels of carbon dioxide to keep the surface waters from freezing. An even richer mix of methane would require even less carbon dioxide.

In fact, there are good analogues for planetary bodies rich in methane. Not only does it occur in Jupiter and Saturn, but Saturn's moon Titan has an atmosphere so rich in methane that it appeared orange when space probes passed by and photographed it (Fig. 8.17). So the old ideas of the earth with its modern clear skies and blue water surface are probably incorrect. Instead, the early earth would have looked orange from space, just as Titan does today.

Figure 8.17   Image of the moon of Saturn known as Titan, which has an orange color because its atmosphere is rich in methane, nitrogen, and ammonia.

and most geologists think the atmosphere was very low in oxygen as well.

Next, you need to have the sea floor far enough from land that almost no sand or mud from the land can mix in the deep-ocean basin with the chemical deposits of iron and silica. Perhaps the iron basins were in the center of ancient seas, while the sands and muds got trapped in basins on the edge of the ancient continents. However, the Hamersley deposits of Australia seem to have formed on a shallow marine shelf, so this model is not true of all BIFs. Finally, it would be a lot easier to deposit huge concentrations of iron if there were some abundant source of dissolved iron entering the ocean. Most geologists think the iron came largely from weathering of the basaltic lavas (which are iron-rich) in the

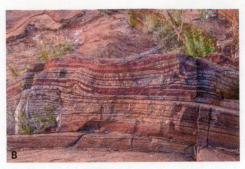

Figure 8.18 Banded iron formations (BIFs) consist of alternating centimeter-scale bands of red iron and black chert, often extending over immense distances and enormous thicknesses. **A.** Contorted BIF from the Mesabi Range, Soudan, Minnesota. **B.** BIF from Fortescue Fall, Karijini National Park, Western Australia. **C.** Thick sequence of BIFs from the Hamersley Range, Australia, the largest deposit of iron in the world.

ancient mid-ocean ridges, plus possibly dissolved iron from weathering of land rocks (which would only be possible if the rivers were completely anoxic as well). Lately, geologists working on BIFs have noticed that some of the biggest deposits occurred when the earth experienced gigantic eruptions of flood basalt, known as "**large igneous provinces**" (LIPs). This excessive eruption of lava would have weathered to produce a lot of iron as long as the atmosphere and ocean were low enough in oxygen that iron could stay in solution and not rust.

BIFs are found in some of the oldest rocks on earth, including the 3.7-billion-year-old rocks of Greenland, the Isua Supracrustals mentioned in Chapters 7 and 9 Most of the world's BIFs (Fig. 8.19) were produced during the

Archean (before 2.5 Ga), when the earth not only had an anoxic atmosphere but also was covered by small proto-continents bashing around in proto-oceans made of the weird lavas known as komatiite. Between 2.6 and 2.4 Ga, the maximum volume of BIFs was deposited, especially the huge mountains of iron in the Hamersley Range of Australia, the Iron Ranges around Lake Superior, as well as similar deposits in Brazil, Russia, Ukraine, and South Africa. This time window was also when the huge eruptions in the LIPs were at their peak.

See For Yourself: Oxygen and BIFs

Then around 2.4–2.3 Ga, something happened. The BIFs began to disappear, although there were still large deposits of iron in granular form rather than banded form, known

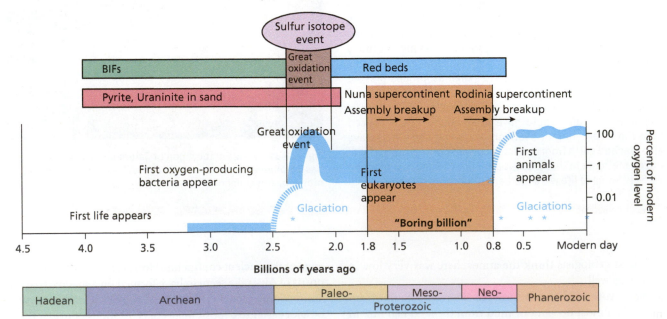

Figure 8.19 Timeline showing the abundance of BIFs, the Great Oxidation Event, and the changes in deposition of pyrite, uraninite, and gypsum.

as "granular iron formations" (GIFs). By 1.9 Ga, BIFs and GIFs vanished completely except for a few freak occurrences (Fig. 8.19) during a snowball earth episode 750–580 Ma. Most geologists regard this as the time when oxygen finally began to reach significant levels in the earth's atmosphere and possibly in the ocean. It has come to be called the **"Great Oxidation Event"** (GOE for short). Oxygen was still nowhere near the 21% that it now is found in the earth's atmosphere. Instead, it went from almost nothing before 2.4 Ga to about 1% of present levels in the oceans, which is enough to begin rusting the dissolved iron in the oceans. Then, about 1.9 Ga, geologists think that the oxygen levels in the oceans were high enough that oxygen was escaping to the atmosphere and possibly weathering rocks on land, although it was still not abundant in the atmosphere. Only in the last 500 million years is it thought that oxygen reached levels like we have today and completely saturated the oceans and atmosphere so that they are now completely oxygenated.

How do we know that oxygen levels were this low? The best evidence comes from the BIFs, which could only form if the oceans were so low in oxygen that iron could stay dissolved, rather than rusting and being left behind. There are also other geochemical clues. Before 1.9–1.8 Ga, we find sand grains and pebbles in the river deposits made of the mineral pyrite or "fool's gold," which is iron sulfide (FeS_2). Today, pyrite only forms in places with very low oxygen levels, such as the bottom of stagnant bodies of water or in deep hot springs and crustal rocks far from the atmosphere. Once pyrite grains weather at the surface, they quickly transform into pebbles of iron oxide (hematite), rather than iron sulfide. As the pyrite breaks down, the iron is released and the sulfur is oxidized into sulfate to make minerals like gypsum (calcium sulfate, or $CaSO_4$). Not surprisingly, we find few significant gypsum deposits older than about 1.8 Ga and no more pyrite pebbles or sand grains after that same time frame. Sand grains made of uranium oxide (uraninite, UO_2) are common before 1.7 Ga but never found after that time. Like pyrite sand and dissolved iron, they are unstable in an oxygen-rich atmosphere.

There are other indicators as well. If we look at the record of isotopes of carbon through time, we no longer see really low values associated with low oxygen after about 2.2 Ga. Sulfur isotope values in Archean rocks are highly variable, fluctuating all over the place. But after 2.4 Ga, the sulfur isotopes are highly stable because they are no longer floating free in minerals like pyrite but stabilized in gypsum and other minerals common in an oxygen-rich world.

Thus, the world went through a dramatic transformation as soon as oxygen became available. The GOE has also been nicknamed the **"Oxygen Holocaust"** because for life on the planet that was used to anoxic conditions, the appearance of such a reactive molecule as O_2 would be poisonous (see Chapter 9). Today, these bacteria and other microbes that are adapted to low-oxygen conditions must live in oxygen-starved places like the bottoms of stagnant lakes and marine basins like the Black Sea. Before 2.3 Ga, however, they ruled the planet. Once the atmosphere became too rich in oxygen, it truly was a holocaust for them, and they lost the world to microbes that can survive oxygen-rich conditions.

The burning question then arises, where did the earth's atmosphere get its free oxygen? The answer is clear: photosynthesis, first from blue-green bacteria or **cyanobacteria** (see Chapter 9) and then eventually when true eukaryotic algae evolved, from plants as well. The big puzzle is that cyanobacterial fossils are known from 3.5 Ga and possibly even 3.8 Ga, yet the GOE didn't start until about 2.3–1.9 Ga. Was their oxygen production so meager that they didn't make a dent on the planet? Did they produce lots of oxygen but it mostly got locked into crustal rocks which were oxidized (such as BIFs) until they finally produced so much oxygen that the crustal reservoirs were saturated and there was free oxygen left over? Or maybe 2.3 Ga is when true eukaryotic algae evolved, with their much larger cells and bigger oxygen production. Maybe only true algae could produce enough oxygen to overwhelm the earth's oxygen sinks, while much smaller cyanobacteria could not. Whatever the reason, the argument remains very controversial and speculative, and there is no consensus on the answer. What is clear is that after 1.7 Ga there were true eukaryotic algae everywhere, and there was an atmosphere with about 1% or maybe even more oxygen, forever changing the earth's oxygen balance.

Here's another thing to think about: without free oxygen, multicellular animals could not have evolved—and we would not be discussing the issue since humans could never have evolved either. In fact, the evolution of all of life as we know it depends on an oxygen-rich planet, which cannot happen without the evolution of photosynthetic microbes and plants. This is a severe restriction on the speculative ideas about extraterrestrials and alien life on other planets as well. It's true that astronomers have found lots of other planets with earth-like properties, including the right size, the right temperatures, and possibly even liquid water oceans on their surface. But so far, not one has evidence of free oxygen in its atmosphere. Without it, there is no multicellular animal life and no alien being like you see in so many science fiction movies (and in the entire culture of people who believe in aliens and UFOs). It's possible that there are anoxic microbes in the deep crustal rocks of other planets, but without abundant free oxygen, the aliens of other planets and our imagination don't exist.

RESOURCES

BOOKS AND ARTICLES

Canfield, Donald. 2015. *Oxygen: A Four Billion Year History.* Princeton University Press, Princeton, NJ.

Dilek, Y., and H. Furnes. 2013. *Evolution of Archean Crust and Early Life.* Springer, Berlin.

Hartmann, J., and R. Miller. 1991. *The History of Earth: An Illustrated Chronicle of an Evolving Planet.* Workman, New York.

Hazen, Robert. 2013. *The Story of Earth: The First 4.5 Billion Years from Stardust to Living Planet.* Penguin, New York.

Hoffman, P. 1988. United plates of America, the birth of a craton: Early Proterozoic assembly and growth of Laurentia. *Annual Reviews of Earth and Planetary Sciences* 16:543–604.

Hoffman, P., and D. Schrag. 2000. Snowball earth. *Scientific American* January:68–75.

Lane, Nick. 2003. *Oxygen: The Molecule that Made the World.* Oxford University Press, Oxford.

Lowe, D. R. 1980. Archean sedimentation. *Annual Reviews of Earth and Planetary Sciences* 8:145–167.

Nisbet, E. G. 1987. *The Young Earth: An Introduction to Archean Geology.* Allen & Unwin, Boston.

Prothero, D. R. 2018. *The Story of the Earth in 25 Rocks.* Columbia University Press, New York.

Shaw, George H. 2015. *Earth's Early Atmosphere and Oceans, and the Origin of Life.* Springer, Berlin.

Walker, Gabrielle. 2004. *Snowball Earth: The Story of a Maverick Scientists and the Global Catastrophe that Spawned Life as We Know It.* Broadway Books, New York.

Ward, Peter, and Joseph Kirschvink. 2015. *A New History of Life: The Radical New Discoveries about the Origin and Evolution of Life on Earth.* Bloomsbury, New York.

Windley, B. F. 1984. *The Evolving Continents.* Wiley, New York.

SUMMARY

- The Precambrian or Cryptozoic spans the interval from 4.5 to 0.5 Ga, or about 88% of geologic time. It is subdivided into the Hadean, Archean, and Proterozoic.

- It is relatively hard to study Precambrian events in detail because the rocks are mostly buried under younger deposits, they are often heavily metamorphosed, they are much more likely to be eroded away at some time in the past, and there are no megascopic fossils to provide fine-scale subdivisions and correlation of time and events.

- The Hadean (4.56–4.0 Ga) was a hellish time on earth, when the crust was very young, thin, hot, and constantly being destroyed and rebuilt. Until 3.9 Ga, the earth underwent heavy bombardment from space debris left over from the early solar system.

- The Archean (4.0–2.5 Ga) was a period of the formation of microcontinents at the core of each continent, including the Superior, Slave, Nain, and Wyoming terranes of North America.

- Typical Archean rocks are dominated by the "4 Gs": protocontinental crust (now turned into **g**neisses), moving between proto-oceanic crust made of komatiite pillow lavas (now turned into **g**reenstones) and covered by deep-marine shales and immature **g**raywacke sandstones. After these blocks crunched into one another in greenstone belts, the lower crust underwent partial melting and produced true **g**ranitic magmas, intruding all the older rocks.

- The Proterozoic (2.5–0.5 Ga) is the longest era of geologic time, spanning 2 billion years. During the Proterozoic, the earth's crust transitioned from proto–plate tectonics of the Archean to true plate tectonics, with volcanic island arcs erupting andesites and rhyolites, passive margin wedges forming on the edges of rifting continents, and sediments typical of modern plate regimes.

- During the Proterozoic, the four Archean terranes of North America were sutured together into one block separated by crushed ocean basins, such as the Wopmay belt, the Trans-Hudson belt, and the Penokean belt. About 1.7 Ga, these Early Proterozoic terranes were enlarged by the addition of the Yavapai and Mazatzal terranes in what is now southwestern and south-central North America and then about 1.3 Ga by the Grenville terrane which underlies Labrador, Quebec, and the Appalachian basement rocks.

- About 1.3–1.0 Ga, weird plagioclase-rich gabbros known as anorthosites intruded many parts of the world, but the reasons for this unique geologic event are still debated.

- By the Late Proterozoic, the simple shales and graywackes of the Archean were replaced by the full suite of modern-looking sediments, including quartz-rich sandstones (representing multiple cycles of erosion and deposition of quartz sand) and the first limestones and abundant gypsum.

- Near the end of the Proterozoic, the Rodinia supercontinent broke up, producing aulacogen deposits in many places in the west (Death Valley, Grand Canyon, Uinta Mountains) and the huge Mid-Continent Rift from Lake Superior to Kansas, filled with enormous volumes of Keweenawan lava flows and quartz sandstones.

- In the Early Proterozoic (about 2.0 Ga) and twice in the Late Proterozoic, the earth was nearly completely frozen over in a "snowball earth" state, with glacial deposits forming in the equatorial regions. The earth escaped being permanently frozen (despite the strong albedo of ice radiating most of the sun's energy back into space) by volcanic eruptions that released greenhouse gases and quickly broke up the ice cover.

- The earth's earliest atmosphere initially consisted of hydrogen and helium (escaped because earth's gravity is not enough to hold such light elements) plus methane and ammonia (also vanished very early on). The main component of atmospheric gas was (and still is) nitrogen, but carbon dioxide and water vapor were abundant early on before they came to be part of the earth's oceans and crust. There was no free oxygen, as shown by deposits like banded iron formations, abundant sedimentary pyrite, and uraninite.

- Between 2.3 and 1.8 Ga, the oxygen-free earth underwent the Great Oxidation Event, or "Oxygen Holocaust," when photosynthesis by cyanobacteria (and eventually algae) finally produced enough free oxygen to overcome the absorption by iron and other crustal reservoir rocks and began to be more abundant in the atmosphere. This is shown by the disappearance of banded iron formations, replaced by rusty red beds, and the end of pyrite sand, with the sulfur going to sulfur-oxygen minerals like gypsum.

KEY TERMS

Precambrian (p. 164)
Cryptozoic (p. 164)
Hadean Eon (p. 164)
Archean Eon (p. 164)
Proterozoic (p. 164)
Canadian Shield (p. 164)
Tidal pull (p. 167)
Superior terrane (p. 167)
Slave terrane (p. 167)
Nain terrane (p. 167)
Wyoming terrane (p. 167)
Protocontinental crust (p. 168)
Proto-oceanic crust (p. 168)

Komatiite (p. 168)
Greenstone (p. 168)
Graywacke (p. 169)
Greenstone belt (p. 171)
Wopmay belt (p. 171)
Trans-Hudson belt (p. 172)
Penokean orogeny (p. 173)
Yavapai terrane (p. 173)
Mazatzal terrane (p. 173)
Grenville terrane (p. 173)
Valbara (p. 174)
Kenorland (p. 174)
Columbia supercontinent (p. 174)

Rodinia (p. 174)
Pannotia (p. 174)
Anorthosite (p. 174)
Quartz-rich sandstones (p. 175)
Belt Supergroup (p. 175)
Amargosa aulacogen (p. 175)
Uinta Mountain Group (p. 175)
Grand Canyon Supergroup (p. 175)
Mid-Continent Rift (p. 176)
Keweenawan lavas (p. 176)
Windermere Supergroup (p. 176)
Snowball earth (p. 177)

Albedo feedback loop (p. 177)
Banded iron formations (p. 179)
Faint young sun paradox (p. 181)
Large igneous provinces (p. 182)
Great Oxidation Event (p. 183)
Pyrite sand (p. 183)
Uraninite sand (p. 183)
Oxygen Holocaust (p. 183)
Cyanobacteria (p. 183)›

STUDY QUESTIONS

1. Why are Precambrian rocks harder to study and get a detailed chronology of their events?
2. What are some of the effects of the moon being much closer in the Early Precambrian?
3. What are the four Archean terranes that were the original protocontinents that made up the nucleus of North America?
4. What is the tectonic significance of Archean gneisses? Archean greenstones? Archean graywackes and shales?
5. What do the pillow lavas in Archean greenstones indicate about their formation?
6. How are the lavas of the Archean different from the basaltic lavas that now erupt on the ocean floor?
7. How are graywacke sandstones different from most of the sandstones formed since the Archean?
8. How were Archean granites formed?
9. How does the proto-crust of the Archean differ from the earth's crust since the Proterozoic?
10. What is the evidence that modern plate tectonics was occurring by the Early Proterozoic?
11. What is the significance of the Wopmay, Trans-Hudson, and Penokean belts?
12. What is anorthosite? How is it different from a typical gabbro? What kind of conditions might have formed it?

13. What does the presence of quartz-rich sandstones in the later Proterozoic indicate?
14. What is the tectonic significance of the Amargosa, Uinta Mountain, and Grand Canyon sedimentary packages?
15. Where is the Mid-Continent Rift located? What might have caused it to form? What is it filled with?
16. How does the occurrence of glacial deposits sandwiched between limestones indicate sea-level glaciers at the equator?
17. How does the albedo feedback effect cause the earth to freeze over into "snowball earth"? What released the earth from being permanently frozen over?
18. What are banded iron formations? When did they form? What do they tell us about the earth's atmosphere at that time?
19. What happened to the earth's original hydrogen and helium? What were the most abundant gases in the earth's earliest history?
20. What is the "faint young sun paradox"? How do methane and carbon dioxide solve the problem?
21. What is the Great Oxidation Event? What three kinds of sedimentary rocks disappear when oxygen becomes common, and what rocks replaced them?
22. How might LIPs help explain the BIFs?

9 The Origin and Early Evolution of Life

It is often said that all the conditions for the first production of a living organism are now present, which could ever have been present. But if (and oh! what a big if!) we could conceive in some warm little pond, with all sorts of ammonia and phosphoric salts, light, heat, electricity, &c., present, that a proteine [sic] compound was chemically formed ready to undergo still more complex changes, at the present day such matter would be instantly absorbed, which would not have been the case before living creatures were found.

—*Charles Darwin, 1871, in a letter to Joseph Hooker*

The early earth for billions of years would have looked much like this, populated mainly by bacteria and blue-green bacteria, making domed mats of bacteria known as stromatolites. These are the famous domed stromatolites in Hamelin Pool in Shark Bay on the northern coast of Western Australia, exposed at low tide. When they are drowned in high tide, the sticky mats trap sediment which continues to build the layered sedimentary structure.

9.1 Describe some of the early notions about the origin of life, and why spontaneous generation cannot happen today but could happen in the early earth.

9.2 List the four basic classes of organic chemicals and their building blocks, and what role they have in primitive microbial life.

9.3 Describe how mud, zeolites, and pyrite might have served as a template to assemble more complex biomolecules.

9.4 Describe the earliest known fossils, and what they tell us about the origins of photosynthesis, and of the eukaryotic cell.

9.5 Describe the transitional fossil assemblages, like the Ediacarans, and the Early Cambrian "little shellies," and why they make the idea of a "Cambrian explosion" misleading and obsolete.

9.6 Discuss why life seemed to evolve so slowly for 3 billion years, and then some of the possible events that triggered the diversification of complex life in the Cambrian.

9.1 How Did Life Begin?

How did life originate? This has long been one of the most fascinating and challenging problems in all of human thought. It was a question that intrigued the ancient Greek and Roman philosophers and scientists. Among the common misconceptions of these people was the notion of **spontaneous generation**. Even as late as the early 1800s, people still thought that life magically arose from non-life. As we saw in Chapter 6, Lamarck and most scientists before the 1860s believed in spontaneous generation. After all, maggots mysteriously appeared on rotting meat, and broth became spoiled when no one had seen another creature touch it. Finally, in 1861 Louis Pasteur conducted a famous series of experiments. He boiled some broth to sterilize it, then put the broth in a sealed flask, and it did not spoil after many months. He put sterilized broth in another flask with a long curved tubular neck opening, and it took a long time before anything grew upon it. Without easy access to microbes from the outside air, the broth remained unspoiled, so microbes were the source of the life you see in spoiled broth. He sealed meat in airtight containers, and no maggots appeared. *Under modern earth conditions*, life only comes from other life.

But if life only comes from previous life under modern earth conditions, how did it originate from non-life in the first place? The key is the phrase "under modern earth conditions." There is good evidence (see Chapter 8) that when life originated around 3.5–3.9 billion years ago (Ga), the conditions were very different (Fig. 9.1). For one thing, the atmosphere back then had no free oxygen, so it was not as reactive and corrosive as it is today. Instead, many chemical reactions could occur because the conditions were not oxygenated. In addition, when life first arose there was no competition from other living organisms, so the world's oceans were a rich soup of organic materials that were easily absorbed. Today, nearly all nutrients in the world's oceans are rapidly consumed by a whole range of living creatures, so there are no spare materials for simple chemical reactions to occur as they did in the early earth. Thus, we need to be careful when applying generalizations based on the biology of today's world to the distant past, when things were very different.

See For Yourself:
Life's Origins

If these experiments ruled out spontaneous generation occurring today, how did life arise in the conditions of the primitive solar system? Clearly, this is a question that can only be solved using experimental biochemistry since such simple molecules do not fossilize. Among the first ideas about the origin of life were suggestions by Charles Darwin in an 1871 letter to his friend, botanist Joseph Hooker (cited in the epigraph at the beginning of the chapter). Darwin speculated that a "warm little pond" with the right combination of chemical compounds (whimsically nicknamed the "primordial soup") and the right sources of energy could produce proteins. Russian biochemist A. I. Oparin and the British geneticist J. B. S. Haldane in the 1920s both came up with the idea that the earth originally had a reducing atmosphere of nitrogen, carbon dioxide, ammonia (NH_3), and methane or natural gas (CH_4) but no free oxygen. Such an atmosphere and corresponding ocean would be the ideal "primordial soup" for producing simple organic compounds.

Then, in 1953, a young graduate student at the University of Chicago named Stanley Miller decided to test Oparin's hypothesis. In collaboration with his advisor, chemist Harold Urey (who later won the Nobel

Prize in Chemistry for his many discoveries in isotope geo-chemistry), Miller wanted to see if such a "primordial soup" could generate simple biochemicals. He built a simple apparatus (Fig. 9.2) out of glass tubing that formed a continuous sealed loop, with all the air removed by vacuum pump. He supplied new "atmosphere" rich in carbon dioxide, nitrogen, methane, ammonia, and water (but *no* free oxygen) in the evacuated tubes. Heating the "ocean" flask at the base to start the steam circulating, he used sparks in another flask to simulate "lightning" as an energy source (Fig. 9.2). Once the methane–ammonia–water-laden steam moved through the "lightning," a condenser cooled the steam back to liquid water, where it flowed back to the "ocean" flask.

This simple experiment produced amazing results. Within days, the clear solution of the "ocean" became brown with new chemicals; and within a week, it was a dark brown organic-rich glop. When Miller analyzed it, he found that he had already produced 4 of the 20 amino acids used to make proteins plus many other simple but crucial organic molecules, such as cyanide (HCN) and formaldehyde (H_2CO). In one remarkable experiment, he launched the whole field of biochemical research into the origin of life.

Although amino acids are much more complex than the chemicals he started with, Miller showed they were remarkably easy to produce. Later, other labs conducted experiments like those performed by Miller and produced 12 of the 20 amino acids found in life. Experimenting with a dilute cyanide mixture produced 7 amino acids. No matter what experiments you try, it does not require supernatural powers or even more than a few days in the lab to make the basic building blocks of life. In fact, Miller's experiment is so simple that anyone with access to a decent chemistry lab can do it, and there are articles online about how to set up your own Miller–Urey experiment. In the years since Miller's original experiments, other scientists have found 74 different amino acids trapped in chondritic meteorites (including all of the 20 found in living systems), so apparently organic compounds

 See For Yourself: The Miller-Urey Experiment

have been produced in many other places in the universe.

9.2 Polymers and Salad Dressing

The Miller–Urey experiments, and all the subsequent work by many different labs, show that amino acids are remarkably easy to produce, so we can assume that they were present in the early earth's oceans (as they were in certain meteorites and apparently throughout space). But we are interested in the more complex biochemicals, of which there are four basic classes:

1. The first are known as **proteins**, which are made of long chains of amino acids. Proteins are the fundamental building blocks of most living systems. For a true living organism, we also need other complex chains composed of simple building blocks.
2. We need **lipids** (common in oils and fats), built out of a combination of a long chain of fatty acids linked to alcohols.
3. We need to assemble **carbohydrates** and starches, which are composed of long chains of simple sugars such as glucose.
4. And we need **nucleic acids**, such as ribonucleic acid (RNA) and deoxyribonucleic acid (DNA), composed of complex chains of sugars, phosphates, and the four bases (adenine, thymine, cystosine, and guanine), which carry the genetic code necessary for making copies of the organism. All of these complex molecules are **polymers**, and they are formed by linking together simpler components (Fig. 9.3), a reaction called **polymerization**. How do we trigger these polymerization reactions?

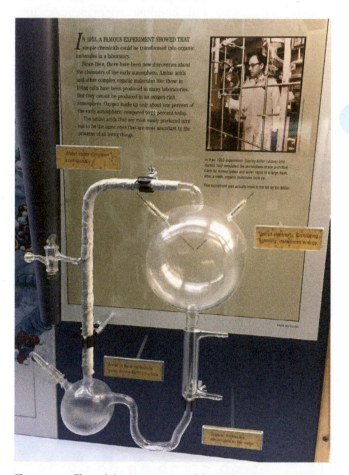

Figure 9.2   The original apparatus of the Miller-Urey experiment, now on display in the Denver Museum of Nature and Science. The sealed set of tubes was evacuated so that no earth atmosphere remained, and a new atmosphere of nitrogen, water, carbon dioxide, methane, and ammonia (but no oxygen) was inserted. The flask in the lower left contained the "ocean," which was continuously heated and boiled. As the steam rose, it passed through the large flask in the upper right, where sparks simulated the energy of lightning. Then the steam passed through the condenser below it, which condensed the steam and returned the liquid water to the starting point. After a week, the clear water of the flask turned brown, and a sample showed that four amino acids had been produced.

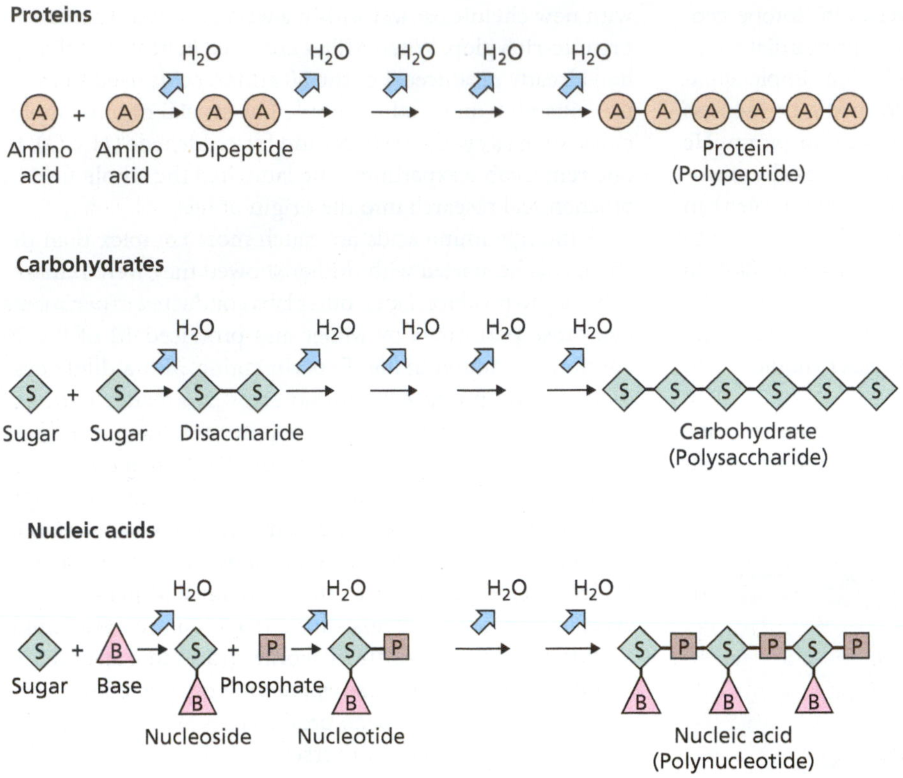

Proteins

Carbohydrates

Nucleic acids

Figure 9.3 Cartoon of the way complex polymers are built from many smaller units (monomers) using polymerization reactions. Many amino acids polymerize into proteins, simple sugars polymerize into complex carbohydrates, and nucleic acid form from sugars plus bases plus phosphates.

It turns out that many of these reactions are easy to produce and readily make at least short-chain polymers. In the 1950s, Sidney Fox splashed a solution of amino acids on hot, dry volcanic rocks and formed many of the proteins found in life. In the presence of formaldehyde, certain sugars readily form complex carbohydrates. Miller's early experiments produced the components of nucleic acids, such as the nucleotide base adenine (by heating aqueous solutions of cyanide) and adenine plus guanine (by bombarding dilute hydrogen cyanide with ultraviolet radiation).

Lipids are even easier to produce. Their basic building blocks are an alcohol, glycerol (Fig. 9.4), at one end and a long fatty acid chain sticking out at the other end. The alcohol is **hydrophilic** (soluble in water), while the fatty acid is **hydrophobic** (does not dissolve in water). When you put some lipids in water, the hydrophilic end of the molecule orients toward the water, while the hydrophobic fatty acid points away from the water. With enough lipids and water, you can generate a **lipid bilayer** (Fig. 9.4), with all the lipids lined up and closely packed together in the same orientation, allowing them to link together. You have seen such reactions any time you have a droplet of oil in water or a droplet of water in oil. As everyone knows, oil and water don't mix but form discrete fluid masses separated by lipid bilayers. It turns out that the simple membranes of most primitive organisms are also lipid bilayers. So generating a membrane like our most primitive living organism is simpler than making oil and

vinegar plus water droplets, like in salad dressing.

When these lipid droplets are dried and then wetted again, they form spherical balls that also concentrate any DNA present up to 100 times. Thus, little lipid bilayer droplets with nucleic acids trapped inside have all the properties of "proto-life." Oparin produced droplets he called "coacervates," and Sidney Fox produced just such structures, which he called "proteinoids." These droplets behave much like living cells, holding together when conditions change, growing, and budding spontaneously into daughter droplets. They selectively absorb and release certain compounds in a process similar to bacterial feeding and excretion of waste products. Some even metabolize starch! Even though these "protocells" are not living, they have most of the properties of living cells—all without much more than simple chemical reactions plus heat.

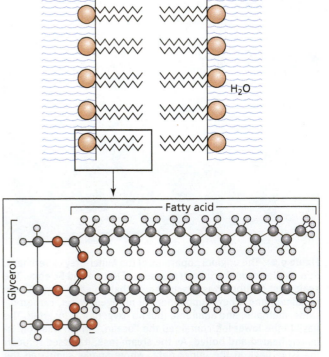

Figure 9.4 Lipids naturally form bilayers when mixed in water since the water-soluble alcohol end faces the water and the water-insoluble fatty acid chain turns away from the water. When many lipids are in a solution of water, they form a natural bilayer or membrane, very similar to the membrane of the simplest bacteria. The same process happens when you have a droplet of oil in water or a droplet of water in oil. The two cannot really mix but form a sharp boundary between them, made of a lipid bilayer.

9.3 Mud and Mosh Pits, Kitty Litter and Fool's Gold

▷ See for Yourself: Chemical Evolution

So far, simple organic chemistry in a lab has demonstrated that it is relatively easy under natural conditions to form the basic building blocks of life: amino acids and short-chain proteins, simple sugars and starches, fatty acids plus alcohol to make lipid bilayers and cell membranes, and short nucleic acids to pass on the genetic information. But most living systems are built of molecules that are many hundreds to thousands of units long. These are hard to assemble when the individual building blocks are randomly bumping into each other in a solution. But there is a more efficient, natural way to bring these molecules closer together in the right arrangement to link up into complex molecules. Organic chemists use a **catalyst**, some sort of substance added to the solution to speed up the reaction. In nature, there are many such catalysts that could serve to line up the building blocks of life until they formed a tightly packed framework of molecules. These catalysts can be thought of as "templates" or "scaffolds," an external inorganic framework which holds the smaller organic molecules in position until they are all lined up in the right direction and jostling against one another. It is analogous to what you see in a mosh pit, with all the people packed in shoulder to shoulder and oriented the same way facing the stage. If they are packed closely enough, their large earrings and other piercings might link together, and they would be assembled into a tightly linked chain of people. Likewise, if you pack organic molecules in the proper orientation and very closely together, their "earrings" (OH– and H+, or hydroxyls and hydrogens sticking out at the end of each chain; see Fig. 9.3) connect, leaving two larger molecules linked together.

What kinds of scaffolds or templates might be able to catalyze such reactions? Several candidates have been proposed. They include:

1. *Mud*, specifically, clay minerals. Clays are silicates with a very complex, highly repeated, sheet-like structure that propagates and copies its internal arrangement as the crystal grows. Many types of clay minerals naturally attract organic molecules in the water-filled spaces between the aluminosilicate sheet layers and, in doing so, arrange them across the clay scaffold so that they can link up. In addition, clay minerals have "mutations" in the form of atomic substitutions or lattice defects within the crystal. When new crystals grow from that parent crystal, they inherit that "mutation" and pass it on to their "descendants."

2. *Kitty litter*, or, more precisely, the minerals known as **zeolites**. Zeolites are a class of more than 40 kinds of aluminosilicate minerals that naturally crystallize in places like gas bubbles in volcanic lava flows. They have a complex porous silicate crystal structure that makes them excellent catalysts for organic reactions, and they are widely used in industry for that purpose. They also are helpful in pulling out organic molecules from fluids, which is why they are heavily used in water purification, laundry detergents, and kitty litter. Their porous structure means that when water moves through them, they act as "molecular sieves" and separate large molecules from water. These larger organics can then align along the charged surfaces of the zeolite crystals and then polymerize to form larger organic molecules.

3. *Fool's gold*, or, more precisely, the mineral known as **pyrite** (iron sulfide, FeS_2). Pyrite is a common mineral that has a metallic golden appearance, hence the name "fool's gold." But pyrite is an important mineral in certain settings (especially in fluids which are highly reducing and poor in oxygen so that the iron combines with sulfur rather than oxygen). Iron sulfides are also common in deep anoxic oceanic waters where the reducing conditions form FeS_2 rather than iron oxides, such as the bottom of the Black Sea, so they are typical of black shales. Pyrite is especially abundant in the mid-ocean ridge volcanic vents, or "black smokers" (Fig. 9.5), where dissolved sulfides from the oceanic crust rise with the superheated water from the heat of the magma chamber of the mid-ocean ridge. This is interesting because a number of scientists have argued that the chemistry of these sulfide vents is ideal for the production of earliest life. Not only is there plenty of energy in the form of volcanic heat, but as several chemists have argued, pyrite is also a good scaffold or template. Its crystal surfaces are electrically charged, so organic molecules are naturally attracted and attached by their oppositely charged ends. Once again, when they become a densely packed "mosh pit," they will link together by the condensation reaction and form long-chain biochemicals on the pyrite "scaffold." The "mid-ocean ridge vent" theory for life's origins has another advantage: that environmental setting is isolated from most events at the earth's surface and very stable. Even when meteorites pounded the earth between 4.6 and 3.9 Ga, vaporizing the shallow oceans over and over again, the deep ocean vents were protected. And the final interesting convergence of lines of evidence: the most primitive organisms on earth, the "Archaebacteria" or **Archaea**, are found in these same extreme settings, where there are boiling waters and an abundance of sulfur.

Thus, there is no shortage of good mechanisms to naturally assemble small organic molecules into the long-chain biochemicals that life requires. Some of these proposed templates also fit with an increasing body of evidence suggesting that life originated in the deep-sea volcanic vents, not in Darwin's "warm little pond" as was long supposed.

But there is one other issue that has been widely debated among scientists working on the origins of life: what was the first genetic material? Today, the information for reproduction and making more copies of living organisms is encoded in the nucleic acids, RNA or DNA, of each cell. The nucleic acids then code for certain strings of proteins, which are the stuff of life. But nucleic acids are far more

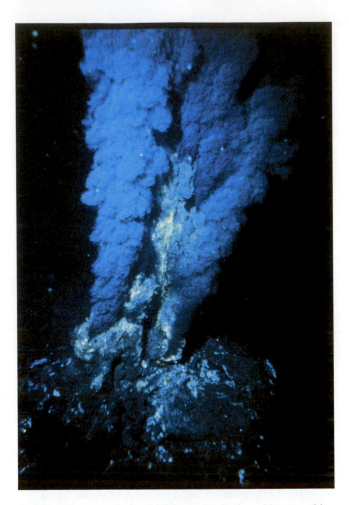

Figure 9.5 On the bottom of the ocean in the mid-ocean ridge rift valleys are areas where hot magma is just below the surface. As seawater percolates down through cracks in the rock, it is quickly boiled, then flows upward in superheated "hot springs" on the sea floor that accumulate a lot of dissolved metals. The dissolved metal particles in the plume of boiling water make it look a bit like "smoke," so these structures have been dubbed "black smokers." These structures are also full of iron pyrite and dissolved sulfur in the form of H_2S, which support an ecosystem based on sulfur-reducing bacteria. Such bacteria are among the most primitive forms of life known, and life may have originated when organic molecules were aligned along a template or scaffold of the charged surfaces of pyrite crystals.

complex and difficult to produce than are proteins, which we saw are the long-chain biomolecules that are easiest to generate. Protein biochemists, like Sidney Fox, long argued that it would be easier for the first self-replicating organism to make its genetic code out of readily available protein chains (which still execute the commands of the nucleic acids today). At some later point in time, more complex nucleic acids were produced, which eventually hijacked the system of replication from one protein to its descendants.

On the other hand, many scientists have suggested that this is an overly complicated and implausible hypothesis: build a genetic code of proteins first, then replace it with another, more complex one. Instead, they argue, it makes

more sense to evolve the genetic code in nucleic acids from the very beginning, even if nucleic acids are harder to produce in chemical reactions than are proteins. Thus, we have a classic "chicken or egg" problem. Which came first: the protein replication system or the nucleic acid replication system?

Fortunately, there is a way to resolve this conundrum. In the early 1980s scientists discovered certain types of RNA, known as **ribozymes**, that perform multiple functions. The RNA in these molecules not only acts as a genetic code but also catalyzes reactions and binds together proteins. In fact, the functional part of the ribosome in the cell, which translates the cellular RNA into proteins, is a ribozyme. Thus, ribozymes perform not only their familiar role as replicators but also the role that proteins play. Further research led to the idea that the simplest scenario for the origin of living, self-replicating systems would be an "**RNA world**." The very first self-replicating form of life would be a single-stranded RNA, perhaps enclosed in a lipid bilayer membrane and perhaps using simple carbohydrates for food storage. Using both its replication powers and enzymatic powers, it would make more copies of itself and perform the role of the proteins as well until later, more complex reactions involving many different proteins could evolve.

Every year, more discoveries are made that add details to our understanding of the origin of life and the RNA world. For example, according to Lehmann and colleagues (2009), coding sequences of amino acids are easily built on small RNA templates in normal prebiotic conditions. Experiments by Kun and others (2005) show that the first ribozymes in the RNA world were much longer and more stable. According to Costanza and colleagues (2009), experiments have shown that nucleotides easily merge in water to form RNA over 100 nucleotides long. Pino and colleagues (2008) demonstrated that RNA molecules link up into long chains easily under normal earth conditions. And finally, a range of experiments by Long and colleagues (2003) and by Patthy (2003) showed that new genes have been produced repeatedly by evolution.

The RNA world hypothesis is now accepted as the most likely scenario for the origin of the first self-replicating system that can be truly called "life," although there are still additional conundrums that are being worked on: How did the RNA world get replaced by the DNA world of today? And what preceded the RNA world? Could it have been (some suggest) a PNA world (peptide-nucleic acid) system that had amino acids in the nucleic acid chains instead of the sugar ribose? Or something else? Like any good scientific problem, the solution of one mystery leads to new and more interesting problems to solve.

▷ See For Yourself: RNA World

9.4 Planet of the Scum

So far, we have examined natural mechanisms that could produce simple organic molecules, simple life, simple prokaryotic cells, and simple eukaryotic cells. What does the

How Did Complex Eukaryotic Cells Evolve?

So far, we have shown that the origins of the basic biochemicals (amino acids and proteins, sugars and carbohydrates, lipids and cell membranes, plus nucleic acids) are easily produced by simple natural chemical reactions, and some even occur in space. We have seen that these simple short-chain polymers are naturally and easily linked together into the long-chain polymers we have now by catalysis on scaffolds of some non-organic matrix, whether it be clay minerals, zeolites, or pyrite. We have found that the evidence from ribozymes argues that the earliest self-replicating form of life was an RNA strand that could both copy itself and act as a protein. All of these steps give us a nucleic acid wrapped by a lipid bilayer coat, with some other metabolic functions. In short, this hypothetical earliest life form is not too different from the most primitive bacterial cells, the simplest known organisms, which are a nucleic acid wrapped in a cell membrane with additional other functions.

Bacteria and other very simple organisms are **prokaryotes**, organisms that have their nucleic acid genes (either RNA or DNA) floating in the cell without a nucleus. These tend to be very small (only a few microns in diameter) (Fig. 9.6) and very simple. The earliest fossils known (see "Planet of the Scum") are of the size and shape that we can confidently attribute them to prokaryotes, including blue-green bacteria (cyanobacteria, once known incorrectly as "blue-green algae") and other types of bacteria. But the more complex type of cell (found in animals, plants, and fungi) is about 10 times larger. All of the nucleic acid genes (DNA) are enclosed in a nucleus, so these organisms are known as **eukaryotes**. As well as the nucleus, eukaryotic cells almost always have additional structures (**organelles**) within the cell wall (Fig. 9.6). These might include **chloroplasts**, which are the sites of photosynthesis in plant cells; **mitochondria**, the

"power plants" of the cell where energy is exchanged using adenosine triphosphate and adenosine diphosphate; **Golgi bodies**, which process and package proteins; **endoplasmic reticulum**, which synthesizes proteins, lipids, steroids, and other chemicals and regulates the concentration of calcium and other steroids; and external structures, such as the hairlike cilia used in propulsion and the whip-like **flagellum** used to power the cell rapidly through a fluid. All of these are complex structures, and for a long time it was a great puzzle how they had evolved from scratch.

In 1967, the biologist Lynn Margulis proposed a radical solution to this problem (although an early version of the idea was proposed by the Russian botanist Konstantin Mereschowski in 1905 but then forgotten). Instead of the difficult process of evolving organelles out of nothing, Margulis argued that each of the organelles found in the eukaryotic cell was once a free-living prokaryote that had come to live symbiotically within another cell and eventually became part of it (Fig. 9.7). This idea is known as the **endosymbiosis** theory. Chloroplasts apparently started out as cyanobacteria, which are photosynthetic even though they are prokaryotes without organelles. Purple nonsulfur bacteria have much the same structure and function as mitochondria, and apparently that's where these organelles came from. The flagellum has the identical 9 + 2 fiber structure (9 sets of microtubule doublets surrounding a pair of single microtubules in the center) as the prokaryotes known as spirochetes, which also cause syphilis. As these smaller prokaryotes came to live within a larger cell, they sublimated their functions to that of their host so that the cyanobacteria became chloroplasts that are now homes for photosynthesis, and the purple nonsulfur bacteria became mitochondria and performed the role of the energy converter for the cell.

In addition to the detailed similarities of these prokaryotes to the organelles, Margulis pointed to many other suggestive lines of evidence. Usually, organelles are not floating within the eukaryotic cell membrane but separated from the rest of the cell by their own membranes, strongly suggesting that they are foreign bodies that have been partially incorporated within a larger cell. Mitochondria and chloroplasts also make proteins with their own set of biochemical pathways, which are different from those used by the rest of the cell. Chloroplasts and mitochondria are also susceptible to antibiotics

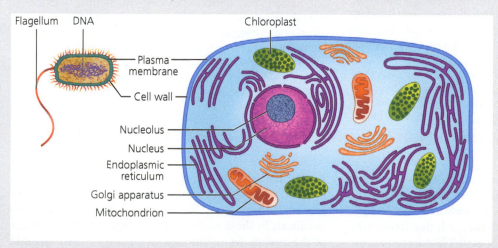

Figure 9.6 The differences between prokaryotic and eukaryotic cells. Prokaryotic cells are very simple, with no organelles and no nucleus surrounding their genetic material. Eukaryotic cells tend to be about 10 times as large as prokaryotic cells and have many different organelles inside them, as well as a membrane surrounding the DNA in the nucleus.

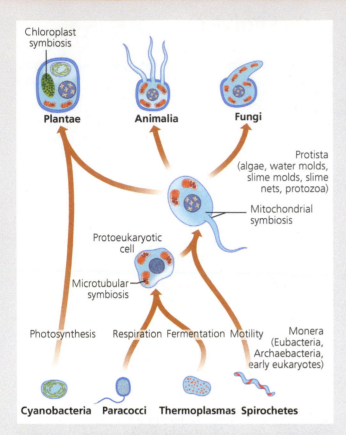

Chloroplast
symbiosis

Plantae   **Animalia**   **Fungi**

Protista
(algae, water molds,
slime molds, slime
nets, protozoa)

Mitochondrial
symbiosis

Protoeukaryotic
cell

Microtubular
symbiosis

Photosynthesis   Respiration Fermentation Motility

Monera
(Eubacteria,
Archaebacteria,
early eukaryotes)

Cyanobacteria **Paracocci** **Thermoplasmas Spirochetes**

Figure 9.7   A cartoon sketch of Lynn Margulis' idea of symbiotic origins of eukaryotes from prokaryotes.

of endosymbionts (mostly bacteria) on our skin and inside us. Our intestines are full of the bacterium *Escherichia coli* (*E. coli* for short), familiar from Petri dishes and news alerts about sewage spills or contaminated kitchens. These bacteria actually do most of our digestion for us, breaking down food into nutrients in exchange for a home in our guts. Most of our fecal matter is actually made of the dead bacterial tissues after digestion, plus indigestible fiber and other material that we cannot metabolize. There are many other examples of endosymbiosis in nature. Termites, sea turtles, cattle, deer, goats, and many other organisms have specialized gut bacteria that help break down indigestible cellulose, so these animals can eat plant matter efficiently. Tropical corals, large foraminiferans, and giant clams all house symbiotic algae in their tissues, which produce oxygen, remove carbon dioxide, and help secrete the minerals for their large skeletons.

The strongest evidence came when people started studying the organelles more closely and found that not only did they have the right structure to have once been independent prokaryotic cells but they also have *their own genetic code!* Mitochondria and chloroplasts both have their own DNA, which has a different sequence from the DNA found in the cell nucleus. In fact, the mitochondrial DNA is different enough and evolves at a such a different rate from nuclear DNA that it can be used to solve problems of evolution that the nuclear DNA cannot. This would make no sense if eukaryotic cells had tried to generate the organelles from scratch. They would not have their own genetic code if that were true.

 See For Yourself:
The Endosymbiotic
Theory

The final clincher is that there are many living endosymbiotic cells which show that this process is occurring right now. The simpler eukaryotes, such as the freshwater amoebas *Pelomyxa* and the *Giardia* (famous for causing dysentery from contaminated water) lack mitochondria but contain symbiotic bacteria that perform the same respiratory function. In the laboratory, scientists have observed amoebas that have incorporated certain bacteria in their tissues as endosymbionts. The parabasalids, which live in the guts of termites, use spirochetes for a motility organ instead of a flagellum. Thus, from the wild speculation of 1967, Margulis' idea is now accepted as the best possible explanation of the origin of eukaryotes and organelles.

like streptomycin and tetracycline, which are good at killing bacteria and other prokaryotes but have no effect on the rest of the cell. Even more surprising, mitochondria and chloroplasts can multiply only by dividing into daughter cells like prokaryotes and, thus, have their own independent reproductive mechanisms; they are not made by the cytoplasm of the cell. If a cell loses its mitochondria or chloroplasts, it cannot make more.

When Margulis' startling ideas were first proposed, they were met with much resistance. But as biologists began to see more and more examples of symbiosis in nature, the notion became more plausible. We humans have millions

fossil record tell us about the plausibility of these chemical mechanisms and the timing of how they occurred?

Finding these early fossils is no easy task. First of all, there are just a handful of places on earth that have rocks from the first half-billion years of earth's history; and most of those are highly metamorphosed, and all evidence of fossils has been destroyed. As we saw in Chapter 8, the older the rock, the greater the chance it was eroded away long ago, is still buried under younger rocks, or is highly metamorphosed, so much of the key evidence is missing.

As discussed in Chapter 8, the oldest rocks on earth are just over 4.32 billion years old (from the Quebec shore of eastern Hudson Bay), and there are structures and biochemicals in these rocks that are suggestive of life. Some of the zircons we mentioned in Chapter 7 have organic carbon in them, and they are 4.1 billion years old. There are lower-grade meta-sedimentary rocks in Greenland (the Isua supracrustals) that have yielded organic molecules suggestive of life at 3.9 Ga and the oldest structures that might be megascopic fossils. There is still a lot of argument as to whether these structures are truly fossils or just pseudofossils and whether the carbon in these rocks is truly ancient or just a later contaminant, but it seems likely that some form of life was established by at least 4.1 Ga.

In addition, the dates from meteorite impacts on the moon show that the early solar system was heavily bombarded by meteoritic junk left over from the coalescence of the planets, with large impacts pounding both the earth and moon from 4.6 to 3.9 Ga. This period of "impact frustration" prior to 3.9 Ga probably vaporized the planet's surface waters over and over again, preventing life from becoming established in the world's shallow oceans or lakes (but not affecting life if it arose in deep-sea volcanic vents, as discussed already).

Finally, there are several localities in Australia and South Africa that yield specimens that are undoubted fossils of ancient prokaryotes, from about 3.4–3.5 Ga (Fig. 9.8). These fossils are of blue-green bacteria, or cyanobacteria, prokaryotic bacteria with the ability to photosynthesize and produce their own food. These were called "blue-green algae" in the past, but they are not true algae; algae are eukaryotic members of the plant kingdom, while cyanobacteria are prokaryotes. These fossils show that long strings of filamentous cyanobacteria were well established in the shallow oceans all over the world, forming sticky mats that became covered with sediment. If the sediment-filled mats build up, they form mound-shaped and dome-shaped structures known as stromatolites (Figs. 9.1, 9.9), which have a distinctive layering in their sediment formed by the daily growth and entrapment of sediment of new cyanobacterial mats. In cross section, they resemble sliced cabbages; and in some places, like Shark Bay in Australia, they grow into large knob-like pillars that stick up over a meter above the sea floor.

See For Yourself: Stromatolites

As it was in the beginning (3.5 Ga), so it was for almost another 2 billion years. There are hundreds of Precambrian microfossil localities around the world in rocks dated between 3.5 and 1.75 Ga, and they yield plenty of good examples of prokaryotes (and occasionally stromatolites). This was an extraordinarily slow rate of evolution; in fact, it appears that cyanobacteria evolve slower than anything we know.

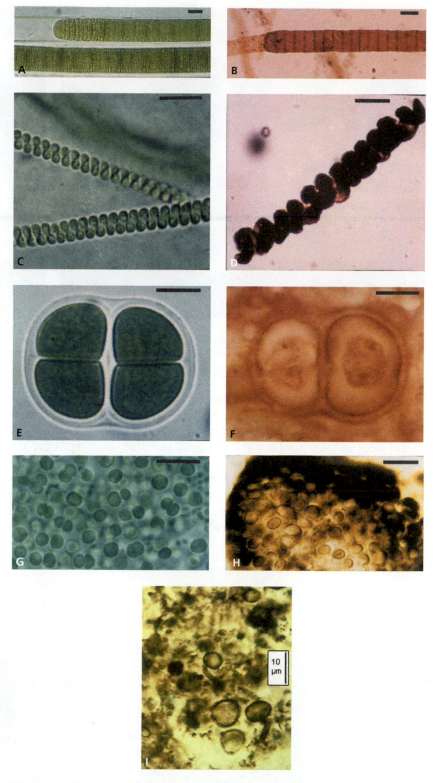

Figure 9.8 Photographs of the earliest known fossils compared with their modern equivalents. **A.** The living filamentous cyanobacterium *Lyngbya*, with its distinctive string of cells and the rounded terminal cell. **B.** *Palaeolyngbya* from 950 million–year-old rocks in Siberia. **C.** The living cyanobacterium *Spirulina* with the corkscrew shape. **D.** *Heliconema* from 850 million–year-old rocks in Siberia. **E.** The four-celled stage of *Gloecapsa*, another cyanobacterium. **F.** The very similar *Gloeodiniopsis* from 1.55 billion–year-old rocks in Bashkiria. **G.** *Entophysalis*, a colonial cyanobacterium. **H.** *Eoentophysalis* from the 2.1 billion–year-old Belcher Supergroup of Canada. **I.** Photomicrograph of the dense clumps of eukaryotic cells from the 1.9 billion–year-old Gunflint Chert of Canada.

Figure 9.9 Stromatolites are sedimentary structures formed by domed cyanobacterial mats that accumulate sediment on their sticky top surfaces as they grow. **A.** In vertical cross section, they show clear layering and a wrinkled, bumpy top surface. **B.** In horizontal cross section, they look like sliced cabbages made of stone)

They show almost no visible change in 3.5 billion years. Everywhere we look in rocks between 3.5 billion and about 1.8 billion years old, we see nothing more complicated than prokaryotes and stromatolites. The first fossil cells that are large enough to have been eukaryotes do not appear until 1.8 Ga, and multicellular life does not appear until 600 Ma. For almost 2 billion years, or about 60%, of life's history, there was nothing on the planet more complicated than a bacterium or a microbial mat; and for almost 3 billion years, or 85% of earth's history, there was nothing more complicated than single-celled organisms. As Bill Schopf of the University of California, Los Angeles, puts it, the earth was truly the "planet of the scum." If aliens existed and had visited the planet long ago, odds are they would have come at a time when there was nothing more interesting to see (Fig. 9.1) than mats of cyanobacteria—and they would have probably left immediately because life on this planet was so primitive and boring!

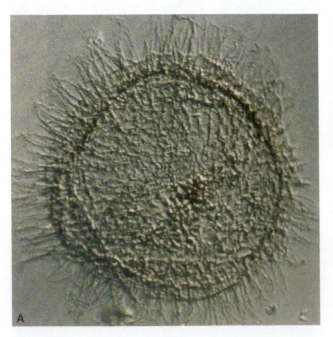

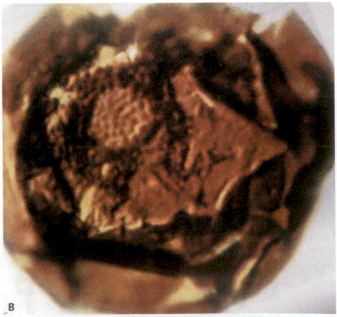

Figure 9.10 Acritarchs are common in Late Proterozoic sediments. Some are believed to be the resting spores of eukaryotic algae and are shaped like tiny balls, often with spikes sticking out of them.

For whatever reason, the first undoubted eukaryotic cell fossils (based on their large size because they are too large for prokaryotes and because they appear to have structures inside them resembling a nucleus) are found in rocks about 1.85–1.6 billion years old. However, there are distinctive organic chemicals (**biomarkers**) known to form only in eukaryotes in rocks up to 2.7 billion years old, so eukarotes may be this old. The most common early eukaryotic fossils are called **acritarchs** (Fig. 9.10). They are organic-walled structures that resemble the resting spores of eukaryotic algae, so they were probably the first fossils of eukaryotic groups that eventually gave rise to plants. For the later part of the Proterozoic, acritarchs increased in abundance and diversity and became the only common microfossils in the generally unfossiliferous rocks of the Late Proterozoic (so much so that they are the main biostratigraphic index species). Then their numbers and diversity crashed in the latest Proterozoic during the "snowball earth" events (Cryogenian, or Varangian glaciations), when the planet froze over even at the equator (see Chapter 8). Acritarchs did survive into the Cambrian but always in reduced numbers until their final extinction in the early Paleozoic.

9.5 The Cambrian "Explosion"—Or "Slow Fuse"?

For the pioneering European geologists in the 1800s, very little was known of the fossil record, and the lack of suitable fossil-bearing sedimentary rocks below the oldest fossiliferous strata (then called "Silurian" but now recognized as Cambrian) was a puzzle. To the early geologists, the apparent absence of Precambrian fossils and then the apparently rapid appearance of diverse trilobites in the Lower Cambrian strata seemed abrupt, so the mistaken term "Cambrian explosion" was born. Of course, Precambrian strata are *not* unfossiliferous—they are full of microfossils (in rocks like cherts that preserve them well) but lack any megascopic fossils except stromatolites, whose biological origin was not confirmed until the 1950s. Thus, the absence of Precambrian fossils was an illusion; the geologists were expecting to find megafossils in strata that had only abundant microfossils.

▷ See For Yourself: The Ediacaran Biota

But elsewhere in the world, there are strata that preserve latest Proterozoic marine conditions well, and they *do* produce megascopic fossils. The first to be described were the deposits of the Ediacara (pronounced "Ee-dee-AKK-ara") Hills in the Flinders Ranges of southeast Australia. These date to 600 Ma, and they are known as the **Ediacara fauna.** This period of time from 600 Ma to the beginning of the Cambrian 545 Ma is known as the Ediacaran Period of the Proterozoic Eon. The Ediacara fauna is now known from a wide variety of localities around the world, including many spectacular localities in China, Russia, Siberia, Namibia, England, Scandinavia, the Yukon,

and Newfoundland. Most of these fossils (Fig. 9.11) are the impressions of soft-bodied organisms without shells, so they had no hard parts that make up the bulk of the later fossil record. Instead, these impressions have reminded some paleontologists of the impressions made by jellyfish, worms, soft corals, and other simple unshelled organisms. Over 2000 specimens are known, usually placed in about 30–40 genera and about 50–70 species, so they were relatively diverse.

Although the Ediacara fauna clearly represents fossils of multicellular organisms (some reach almost a meter in length), paleontologists have a wide spectrum of opinions about what type of creatures made these impressions. The older, more conventional interpretation (Fig. 9.11E) is that they are related to groups we know today: sea jellies, sea pens, and worms of various sorts. Some do look a bit like a sea jelly (Fig. 9.11A), but if so, they have symmetry and structure unlike any living sea jelly. Others vaguely resemble some of the known marine worms (Fig. 9.11B, C), although their symmetry and segmentation do not match any groups of worms alive in the ocean today. Nor do the "worms" have evidence of eyes, mouth, an anus, locomotory appendages, or even a digestive tract.

For this reason, some paleontologists have suggested that Ediacaran fossils were made by organisms unlike any that are alive today. They point to the lack of modern patterns of symmetry and the apparent large size of many of the fossils (some *Dickinsonia* are over a foot across) and argue that they are an early experiment in multicellularity. Some paleontologists think they are not true animals at all but unique organisms that are constructed in a quilted or "water-filled air mattress" fashion, which maximizes surface area relative to the volume of the organism. This is a necessity if they lacked structures like a mouth and internal digestive, respiratory, and circulatory systems. These simple organisms had no internal organs but instead received all their nutrients and oxygen, and got rid of waste, through the huge surface area of their outer membranes. Others have suggested the "Garden of Ediacara" hypothesis: the Ediacarans lived by incorporating symbiotic algae in their tissues (as do many large invertebrates, such as reef corals and giant clams today). Their large surface area maximizes the area of exposure of sunlight for these internal algae, which then helps such large organisms metabolize. Unfortunately, since the Ediacarans are known entirely from the impressions on the soft sea bottom and not from any hard-shelled fossils with internal organs or other important features, it is very difficult to resolve this controversy. Although the possibility that they were passive "water-filled air mattresses" holding symbiotic algae is plausible for the shallow-marine specimens of Australia, the oldest Ediacarans are the branching, frond-shaped fossils from Mistaken Point, Newfoundland, which lived in deep water without light. Whatever the biological affinities of the Ediacara fauna, it is very clear that they are some kind of multicellular organisms, whether animals, plants,

Figure 9.11   Fossils from the Ediacara fauna, composed of impressions of large soft-bodied organisms found all over the world in the latest Proterozoic. **A.** *Mawsonites*, originally interpreted as a sea jelly. **B.** *Dickinsonia*, a flattened worm-like creature with many ridges and a central axis. **C.** *Spriggina*, an elongated, segmented, arthropod-like creature. **D.** *Tribrachidium*, with three "arms" on a central disk. **E.** Conventional interpretations treat Ediacarans as worms and sea jellies, although they do not have the same structure and symmetry as any modern group of animals.

fungi, or some early experimental organisms not related to any living group.

So, the old concept of the "Cambrian explosion" of no fossils followed abruptly by trilobites is now completely out of date. We now have a record of 3 billion years of microfossils, then at 600 Ma the appearance of the first megascopic soft-bodied multicellular organisms. Even though some of them were large, they still didn't have modern body plans, nor did they have any hard parts. The next evolutionary step was to develop some sort of shelly framework. This also means that these organisms must leave behind the first hard-part fossils. As you would expect, the first shelled organisms acquired a shell in small steps, forming tiny (less than 1 mm) hard parts nicknamed "the **little shellies**" or "small shellies" (Fig. 9.12). Many of these "little shellies"

appear to be primitive cap-shaped mollusks or even clam-like shells, others appear to be bits of "armor" that made up a "chain-link" covering on some animals, and still others (like the tiny spiky "maces" known as *Chancelloria*) are spicules of creatures that may be related to sponges. The affinities of most of the "small shellies" are still unclear. They appeared in the first series of the Cambrian Period (Nemakit-Daldynian Stage, beginning at 545 Ma). The megascopic skeletons of trilobites finally appeared in the second Cambrian Stage, the Atdabanian, about 520 Ma (Fig. 9.13).

 See For Yourself: Cambrian "Explosion"

Thus, the term "Cambrian explosion" is obsolete and inaccurate. It was proposed when we knew almost nothing about the many different steps (from single-celled to multicelluar soft-bodied Ediacarans to "little shellies") in the origin of large skeletonized fossils like trilobites. If you look at it from the perspective of 520 million years later, then the 80 m.y. between 600 and 520 Ma, or the 25 million years of the first stage of the Cambrian, looks a little bit like an "explosion." But 80 million years of time from the Ediacaran to the Middle Cambrian (longer than the entire Cenozoic) or even 25 million years of the Early Cambrian is hardly an "explosion" by any objective standard. Geologists are gradually abandoning the misleading and obsolete term "Cambrian explosion" when it was clearly a "Cambrian slow fuse."

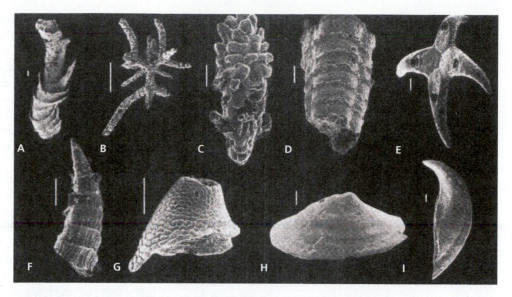

Figure 9.12 The tiny earliest Cambrian fossils from the Terreneuvian stage of the Cambrian, nicknamed the "small shellies." These were the first skeletonized fossils on earth, showing that the process of building hard parts had begun but on a limited scale. Some were tiny elements of bigger organisms, like sponges (B, C, E); others may have been mollusk shells (F, G, H, I); most do not correspond to any living organism.

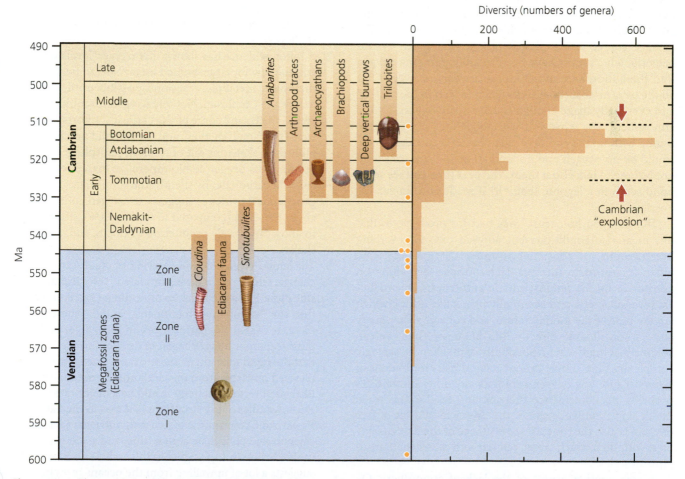

Figure 9.13 The multiple gradual stages of the "Cambrian slow fuse." Diversity and complexity gradually increase from single-celled eukaryotic algae over 1.8 Ga to the soft-bodied Ediacarans around 600 Ma to the "small shellies" of the first stage of the Cambrian and finally to the diversification of hard-shelled trilobites in the second stage of the Cambrian. Depending upon where you define the events (600–520 Ma or 80 million years from Ediacarans to trilobites, or 545–520 Ma, or 25 million years from small shellies to trilobites), the event took millions to tens of millions of years, hardly an "explosion"!

9.6 Why Did Life Change so Slowly Before the Cambrian?

So why did life linger in a state of incredibly slow evolution for almost 3 billion years, then in about 80 million years finally evolve toward multicellularity, then to the "little shellies," and finally to the multicellular skeletons in trilobites? This is one of the great puzzles of geology and paleontology and remains a fertile field for debate and research, with exciting discoveries and hypotheses being published every few months. Still, there are some considerations that are important.

First of all, there is a great variety of geologic evidence about the earth's early atmosphere. We know from the banded iron formations (BIFs, laminated deposits of chert and pure iron) plus the existence of detrital uraninite (uranium oxide) and pyrite (both unstable in the modern atmosphere), the absence of abundant gypsum (which requires oxygen to form), and many other geochemical clues that the earth's atmosphere and surface oceans had no free oxygen from the very beginning (Chapter 8). Then about 1.8 Ga, the BIFs disappear, along with the other geochemical indicators, indicating that there was at least a low percentage of free oxygen in the oceans and probably the atmosphere. Where did all this oxygen come from? The likeliest source is the photosynthetic blue-green cyanobacteria, which must have been steadily pumping oxygen into the atmosphere over billions of years to overwhelm all the oxygen-absorbing sinks (such as iron in the oceans) and eventually make enough to have spare oxygen for the atmosphere. Say what you will about the "planet of the scum," without the efforts of cyanobacteria and algae there would be no oxygen and no land animals—and we would not be here to talk about them. The next time you see some pond scum in a lagoon, thank it! It made all complex life possible.

This "oxygen holocaust" was bad for prokaryotes that live in anoxic conditions but allowed oxygen-absorbing **heterotrophs**, or consumers ("animals" which consume other organisms to survive). Heterotrophs feed either on detritus or on the photosynthetic **autotrophs**, or producers (cyanobacteria and eventually plants like algae), which are the base of the food pyramid, and convert solar energy and carbon dioxide directly into biomass. Thus, we can assume that there was no possibility of much more ecologic complexity than simple cyanobacterial mats for billions of years because low oxygen prohibits any other lifestyle. It's probably no coincidence that about 1.85–1.6 Ga, when the geologic evidence shows the presence of a bit of free oxygen, is also the time we see evidence of the first eukaryotes, with their much larger cell size and mostly heterotrophic lifestyles.

There's another aspect of the lack of atmospheric O_2. Without it, there is no ozone layer, which is produced in the lower stratosphere when ultraviolet (UV) radiation strikes O_2 molecules and converts them into O_3 (ozone) molecules. When you get a sunburn, you have been exposed to too much UV radiation. It could be much worse, but the ozone layer protects us from the damaging effects of UV radiation on living cells, especially in its ability to cause harmful mutations (including skin cancer). Water is a good filter for UV, so marine organisms are protected; but there could be no significant life on land until the ozone layer was thick enough.

Lack of free oxygen and the ozone layer probably explain a lot of the slow evolution in a world with only cyanobacteria for over 1.5 billion years, and low oxygen might also explain why eukaryotes remained single-celled for over a billion years. What kinds of events might have led to the diversification of the "Cambrian slow fuse" after 3 billion years of almost no change? Again, there are many ideas out there, some of which are more plausible than others.

1. The Late Proterozoic "snowball earth" conditions just before 600 Ma were certainly important. The acritarchs were nearly wiped out by the freezing conditions, but when the freezing ended, the planet warmed rapidly, which melted the ice caps, raised sea level, and drowned the continents to form large areas of shallow seas.

2. A lot of evidence has shown that oxygen levels were still too low for large-bodied metazoans to perform their basic metabolic reactions, possibly until shortly before the Cambrian. Notably oxygen is required to create shells, and some of the earliest Cambrian organisms (the "small shellies" and the lingulid brachiopods) built their skeletons out of calcium phosphate, which requires less oxygen to biomineralize than does calcium carbonate. By the Middle Cambrian, geochemical evidence suggests an oxygen content of about 6%–10% (compared to today's 21%), sufficient to allow large-scale production of calcite and aragonite shells.

3. The latest Proterozoic was a period of rapid tectonic change, which might have stimulated biological diversification. From the breakup of Rodinia (Fig. 8.9) and then the Pannotia supercontinents (and the subsequent increase in shallow marine areas on the fringes as their passive margin wedges began to develop) to the idea that the rapid movement and erosion of new continents released nutrients from the land into the ocean, it seems likely that tectonic activity would be a major factor. Recent studies of the strontium isotopes in the Late Proterozoic confirm the increase in nutrients such as calcium and phosphate at that time, and a sudden pulse in the carbon isotope signal in the earliest Cambrian suggests a lot of upwelling from the oceans bringing deep-water carbon back into the organic system.

4. The end of the Late Paleozoic "snowball earth" marked the beginning of a new "greenhouse planet" phase of earth history in the Early Paleozoic. During the Cambrian global sea level rose and caused a transgression that drowned nearly all the continents under shallow epicontinental seas, much like those of the Bahamas but thousands of times bigger (see Chapter 10). These huge areas of shallow seas were ideal for shallow-marine organisms to diversify, and we see this not only in the radiation of trilobites around the world but also in the diversification of many other lineages.

5. All of the previous factors are largely environmental, rather than biological, in origin. However, there may be some ecological and biological factors as well. Consider the earth's shallow-marine realm through most of life's first 3 billion years. Near the shoreline the sea floor was covered by thick cyanobacterial mats and stromatolitic domes, which coated the sea floor and prevented many other organisms from breaking through the scum, establishing burrows or roots, or otherwise colonizing the scarce open sea floor. So far as we can tell, no grazer had evolved who could eat this vast biomass of cyanobacteria and algae until the latest Precambrian and earliest Cambrian. But many of those "small shellies" may have been analogous to simple limpet-like mollusks, some of which must surely have been good at grazing cyanobacteria and algae (as limpets and chitons do today in tide pools). They would have been able to mow down a huge untapped resource of food to create a whole new world with open areas uncoated by bacteria or algae. This is known as the "**cropping hypothesis**." With all this newly exposed sea floor, there was room for many of the quilted Edicarans to live and create a diverse array of bottom-dwellers and even those which apparently stuck up from the sea floor. In the Late Precambrian, we also see the first explosion of trace fossils indicating that burrowing worms were present, creating whole new niches beneath the sea floor as well. By the Early Cambrian, there was a much more diverse world of burrowing inarticulate brachiopods and worms and large colonial archaeocyathids forming reefs. By the Middle Cambrian, there were not only mud-grubbers like trilobites in great abundance but also the first large predators, like the meter-long *Anomalocaris* from the Burgess Shale. Thus, the ecological complexity of the shallow-marine biosphere began to escalate from the simple two-dimensional world of cyanobacterial mats in the Late Proterozoic to the more complex world of the Early Cambrian, with shallow burrowers and tall filter-feeding reef builders, and finally by the Ordovician to a food web almost as complex as the one that exists today (Fig. 10.17).

Of course, these different explanations and causes are not mutually exclusive, and most are plausible enough to assume that they all had some effect on the "Cambrian slow fuse." As new evidence emerges, the relative importance of these possible causes is being reassessed and revised. Altogether, the wealth of new data and intriguing new explanations make the study of early life an exciting field of research!

RESOURCES

BOOKS AND ARTICLES

Cairns-Smith, A. G. 1985. *Seven Clues to the Origin of Life.* Cambridge University Press, New York.

Cone, Joseph. 1991. *Fire Under the Sea: The Discovery of the Most Extraordinary Environment on Earth—Volcanic Hot Springs on the Ocean Floor.* Morrow, New York.

Costanza, G., S. Pino, F. Ciciriello, and E. Di Mauro. 2009. Generation of long RNA chains in water. *Journal of Biological Chemistry* 284:33206–33216.

Erwin, D. H., and James M. Valentine. 2013. *The Cambrian Explosion: The Construction of Animal Biodiversity.* W. H. Freeman, New York.

Fry, Iris. 2000. *The Emergence of Life on Earth: A Historical and Scientific Overview.* Rutgers University Press, Piscataway, NJ.

Hazen, Robert M. 2005. *Gen-e-sis: The Scientific Quest for Life's Origins.* Joseph Henry Press, Washington, DC.

Knoll, Andrew H. 2003. *Life on a Young Planet: The First Three Billion Years of Evolution on Earth.* Princeton University Press, Princeton, NJ.

Knoll, Andrew H., and S. B. Carroll. 1999. Early animal evolution: emerging views from comparative biology and geology. *Science* 284:2129–2137.

Kun, A., M. Santos, and E. Szathmary. 2005. Real ribozymes suggest a relaxed error threshold. *Nature Genetics* 37:1008–1011.

Lehmann, J., M. Cibils, and A. Libchaber. 2009. Emergence of a code in the polymerization of amino acids along RNA templates. *PLoS One* 4(6):e5773.

Lipps, H. H., and P. W. Signor. 1992. *Origin and Early Evolution of the Metazoa.* Plenum, New York.

Long, M. 2001. Evolution of novel genes. *Current Opinions in Genetics and Development* 11(6):673–680.

Long, M., E. Betran, K. Thornton, and W. Wang. 2003. The origin of new genes: glimpses from the young and old. *Nature Review of Genetics* 4:865–875.

Margulis, Lynn. 1981. *Symbiosis in Cell Evolution.* W. H. Freeman, San Francisco.

Margulis, Lynn. 2000. *Symbiotic Planet: A New Look at Evolution*. Basic Books, New York.

Miller, Stanley L. 1953. A production of amino acids under possible primitive earth conditions. *Science* 117:528–529.

Nutman, A. P., V. C. Bennett, C. R. L. Friend, M. J. van Kranendonk, and A. R. Chivas. 2016. Rapid emergence of life shown by 3700-million-year-old microbial structures. *Nature* 537:575–538.

Patthy, L. 2003. Modular assembly of genes and the evolution of new functions. *Genetica* 118:217–231

Pino, S., F. Ciciriello, G. Costanzo, and E. Di Mauro. 2008. Nonenzymatic RNA ligation in water. *Journal of Biological Chemistry* 283:36494–36503.

Schidlowski, M., P. W. U. Appel, R. Eichmann, and C. E. Junge. 1979. Carbon isotope geochemistry of the 3.7 × 109 yr old Isua sediments, West Greenland; implications for the Archaean carbon and oxygen cycles. *Geochimica Cosmochimica Acta* 43:189–200.

Schopf, J. W. 1999. *Cradle of Life*. Princeton University Press, Princeton, NJ.

TIMESCALE OF PRECAMBRIAN EVENTS

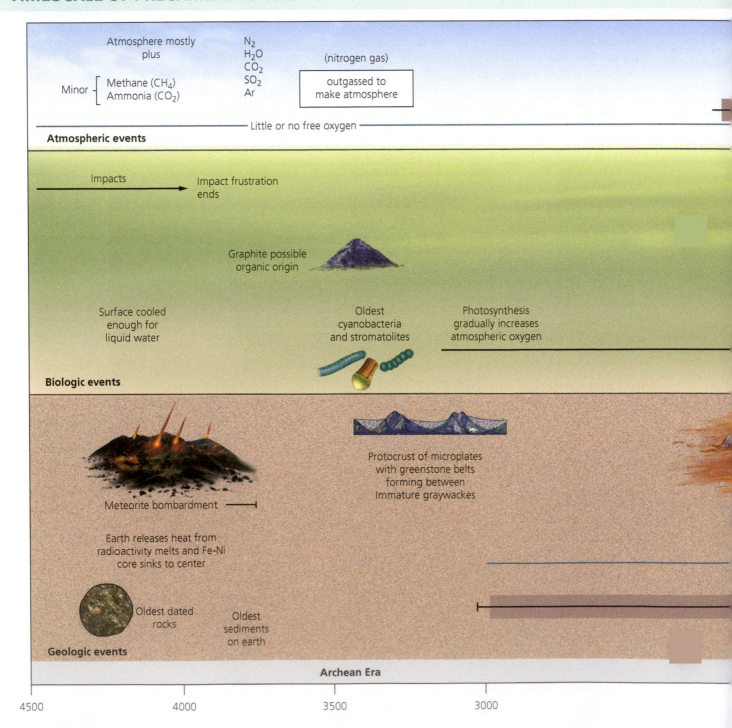

Atmosphere mostly plus — N₂, H₂O, CO₂, SO₂, Ar

(nitrogen gas)

Minor — Methane (CH₄), Ammonia (CO₂)

outgassed to make atmosphere

Little or no free oxygen

Atmospheric events

Impacts → Impact frustration ends

Graphite possible organic origin

Surface cooled enough for liquid water

Oldest cyanobacteria and stromatolites

Photosynthesis gradually increases atmospheric oxygen

Biologic events

Protocrust of microplates with greenstone belts forming between Immature graywackes

Meteorite bombardment

Earth releases heat from radioactivity melts and Fe-Ni core sinks to center

Oldest dated rocks

Oldest sediments on earth

Geologic events

Archean Era

4500   4000   3500   3000

Schopf, J. W. 2002. *Life's Origin: The Beginnings of Biological Evolution*. University of California Press, Berkeley.

Shapiro, R. 1986. *Origins, A Skeptic's Guide to the Creation of Life on Earth*. Summit, New York.

Wächtershäuser, G. 2006. From volcanic origins of chemoautotrophic life to Bacteria, Archaea, and Eukarya. *Philosophical Transactions of the Royal Society of London B* 361:1787–1806.

Wächtershäuser, G. 2008. Origin of life: life as we don't know it. *Science* 289:1307–1308.

Ward, P. D., and D. Brownlee. 2000. *Rare Earth: Why Complex Life Is Uncommon in the Universe*. Copernicus, New York.

Ward, Peter, and Joseph Kirschvink. 2015. *A New History of Life: The Radical New Discoveries About the Origin and Evolution of Life on Earth*. Bloomsbury, New York.

Wills, C., and J. Bada. 2000. *The Spark of Life: Darwin and the Primeval Soup*. Perseus, New York.

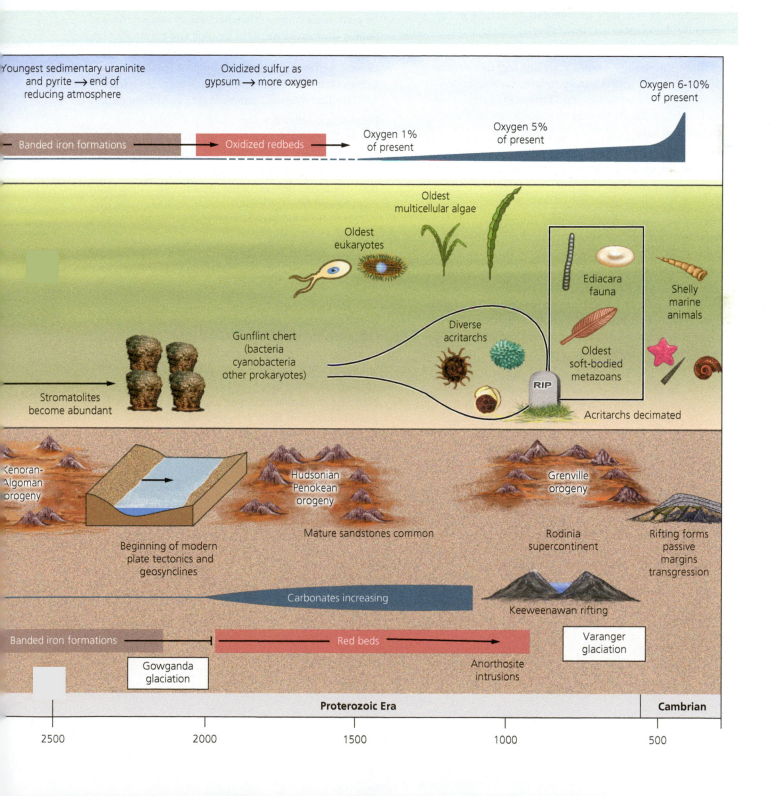

SUMMARY

- Spontaneous generation of life cannot occur under modern earth conditions, but in the Early Precambrian the conditions were actually favorable to life forming from non-life.

- The earliest biochemicals do not fossilize, so the Miller–Urey experiment began the field of modern origin-of-life research by showing that simple biochemicals, like amino acids, are relatively easy to produce under early earth conditions. In fact, they are found in meteorites and apparently abundant across the universe.

- Complex biomolecules are made by polymerizing chains of simpler molecules: amino acids link together to form proteins, simple sugars join up to form carbohydrates, lipids are produced by linking a fatty acid to an alcohol, and nucleic acids are formed by linking sugars, phosphates, and bases.

- To get cell membranes, lipid bilayers spontaneously form whenever lipids mix with water, making cell-like droplets with properties of life.

- To make very long biomolecules, the organic compounds need to be lined up by an inorganic template or scaffold, such as clay minerals, zeolites, or pyrite. Pyrite is especially likely because it is abundant in deep-sea vents, where the simplest forms of life are found.

- Which was the first genetic code, a protein-based system or the nucleic acid–based code we have now? The discovery of ribozymes, RNA with protein properties, suggests that the earliest life was a strand of RNA capable of copying itself and performing the most basic functions within a lipid bilayer membrane.

- To change the simple prokaryotic cell into a eukaryote, the organelles began as symbiotic prokaryotes living inside a larger cell. Mitochondria came from purple non-sulfur bacteria, chloroplasts from cyanobacteria, and flagella from spirochetes.

- The earliest fossils are known from rocks 3.5 to 3.4 Ga in Australia and South Africa, but there are possible fossils and ancient biomolecules in rocks 4.32 Ga and younger. However, life probably didn't get a good foothold until the end of the bombardment from leftover space debris about 3.9 Ga.

- For most of life's history (3.5–0.6 Ga), there was nothing on earth more complex than bacteria, cyanobacteria, and domed stromatolitic mats of bacteria and eventually algae. This "planet of the scum" demonstrates extraordinarily slow rates of evolution, possibly because environmental factors prohibited life from becoming more complex. By 1.8 Ga, eukaryotic cells known as acritarchs were common.

- From single-celled life, the next step was the larger soft-bodied Ediacara fauna about 600 Ma, followed by simple tiny-shelled animals in the earliest Cambrian.

- The misconception of the "Cambrian explosion" of life is due to outdated notions of the fossil record. Life actually transformed through many progressive steps, from single-celled life 3.5 Ga to multicellular soft-bodied Ediacarans 600 Ma to simple shelled animals 545 Ma and finally to larger, more complex animals like trilobites about 525 Ma. The process took tens of millions of years, so it was not an "explosion" but a "slow fuse."

- The reasons that life took so long to reach complexity then began to diversify in the Cambrian are still controversial. The end of the snowball earth, the breakup of Rodinia and the flooding of continents, the increased abundance of oxygen, and the opening of new space for animals when the first grazing snails broke through the blanket of cyanobacterial and algal mats are all likely possibilities.

KEY TERMS

Spontaneous generation (p. 188)
Primordial soup (p. 188)
Miller–Urey experiment (p. 189)
Amino acids (p. 189)
Polymerization (p. 189)
Lipids (p. 189)
Carbohydrates (p. 189)
Nucleic acids (p. 189)

Hydrophilic (p. 190)
Hydrophobic (p. 190)
Lipid bilayer membrane (p. 190)
Catalyst (p. 191)
Zeolites (p. 191)
Black smokers (p. 191)
Archaea (p. 191)
Ribozymes (p. 192)

RNA world (p. 192)
Prokaryotes (p. 193)
Eukaryotes (p. 193)
Organelles (p. 193)
Mitochondria (p. 193)
Chloroplasts (p. 193)
Flagellum (p. 193)
Endosymbiosis theory (p. 193)

Biomarkers (p. 197)
Acritarchs (p. 197)
"Cambrian explosion" (p. 197)
Ediacara fauna (p. 197)
"Little shellies" (p. 198)
heterotrophs (p. 200)
autotrophs (p. 200)

STUDY QUESTIONS

1. Why is it impossible for life to form from non-life today? Why was it extremely easy in the conditions of the early earth?
2. How did the Miller–Urey experiment open the door to the study of the origin of life?
3. What are polymers? What kinds of building blocks are used to make polymers like proteins, carbohydrates, lipids, and nucleic acids?
4. How are lipids (and the behavior in water) crucial to the formation of the simplest cells?
5. Why is it difficult to get very long-chain biomolecules in a random mixture in a test tube solution? Why are templates or scaffolds essential to the next step in producing life?
6. How are templates like a mosh pit?
7. Summarize the three main inorganic materials that would make good organic templates. Why is pyrite a very good candidate for the template that formed complex biomolecules?
8. Which came first, a protein genetic code or the modern nucleic acid code? How do ribozymes solve that chicken-and-egg problem?
9. What is the RNA world? What does it suggest the earliest forms of life were built like?
10. What evidence suggests that mitochondria, chloroplasts, and flagella were once separate prokaryotes that came to live symbiotically inside the eukaryotic cell?
11. How long from the oldest known signs of life did it take before complex multicellular animals appeared? What might explain this long delay?
12. What is "impact frustration"? When did it end?
13. What were the earliest known eukaryotes like? When did they appear?
14. How does the Ediacara fauna represent a step up from single-celled life?
15. Why is the Ediacara fauna not easy to shoehorn into modern classifications of "worms," "sea jellies," and other known life forms?
16. How did the Early Cambrian "little shellies" represent an important next step from the Ediacarans to trilobites? Why do you think scientists missed finding them for so long?
17. Why is the term "Cambrian explosion" outdated and misleading?
18. What are some of the possible factors that explain the diversification of multicellular life in the Cambrian?

Cambrian–Ordovician
541–444 Ma

A mountain's giddy height I sought,
Because I could not find
Sufficient vague and mighty thought
To fill my mighty mind;
And as I wandered ill at ease,
There chanced upon my sight
A native of Silurian seas,
An ancient Trilobite.
So calm, so peacefully he lay,
I watched him even with tears:
I thought of Monads far away
In the forgotten years.
How wonderful it seemed and right,
The providential plan,
That he should be a Trilobite,
And I should be a Man!

—*May Kendall, The Lay of the Trilobite, 1885*

The Lower Cambrian Potsdam Sandstone in Ausable Chasm, New York, was formed on beaches and shallow nearshore settings. It represents the initial phase of the Sauk transgression in the New York region.

10.1 Transgressing Seas in a Greenhouse World

The Early Paleozoic marked an important transition between the world of the Proterozoic and the beginning of the Phanerozoic. The presence of abundant fossils has several important implications. The immediate result is that it is possible to tell time very precisely in the Cambrian and Ordovician (compared to the Precambrian) because the biostratigraphy of rapidly evolving fossils like trilobites allows for very fine time resolution. Thus, we know what events occurred when in much greater detail than we do for any part of the Precambrian.

More importantly, the abundance and diversification of life transformed the environment itself. Gone forever was the almost two-dimensional world where the shallow sea floor was covered by a scum of cyanobacteria and algae, and no organisms could burrow beneath it or rise above it. Once complex organisms began to diversify, they soon developed the ability to burrow deep into the sea-floor sediments or rise above them and live in previously uninhabited parts of the ocean, feeding on underexploited food resources. Eventually, more and more complex food chains and food webs developed, and the world went from extreme ecological simplicity to the complexity of ecosystems we take for granted today. Finally, during the Proterozoic and Cambrian, the land was barren and uninhabited except for a few lichens and mosses that managed to cling to the moister places. By the end of the Ordovician, there is evidence for the first significant land plants and the first land animals, primarily millipedes.

The Early Paleozoic was also a time of dramatic transition in the types of rocks deposited and the changing atmospheric composition. Emerging from the last of the "snowball earth" events in the Late Proterozoic, during the Early Paleozoic the earth was typified by warm climates and high sea levels. The seas eventually drowned nearly all the continents and left little dry land exposed. The sedimentary record is dominated by shallow-marine sandstones and mudstones or shales (Figs. 10.1, 10.2), especially by the first abundant production of limestones as animals that secreted calcite shells dominated the seas for the first time.

 See For Yourself: The Cambrian

Finally, there is good evidence that the atmosphere continued to evolve. Not only was it rich in carbon dioxide, but the low levels of atmospheric oxygen that may have hampered the development of life in the latest Proterozoic increased so that the atmosphere had significantly more oxygen, although present-day atmospheric levels of oxygen (21% of the modern atmosphere is free oxygen) may not have occurred until after the end of the Ordovician. The planet switched from the "snowball earth" **icehouse planet** of the Late Proterozoic (with lots of polar ice pulling sea level down and low carbon dioxide warming the planet) to a **greenhouse planet** (with lots of carbon dioxide in the atmosphere melting the ice and causing sea level to rise and drown the continents). As we shall see throughout this book, the earth has alternated between greenhouse and icehouse states repeatedly through its history. The Early to Middle Paleozoic greenhouse world we will discuss in this chapter and the next was replaced by an icehouse world with polar glaciers in the Late Paleozoic and Triassic. By the Middle Jurassic, the greenhouse world returned to flood the continents again in the Cretaceous, and these warm conditions did not end until the late Eocene, when the world slipped into our modern-day icehouse as the Antarctic glaciers returned 33 Ma.

Let us look at each of these events in detail.

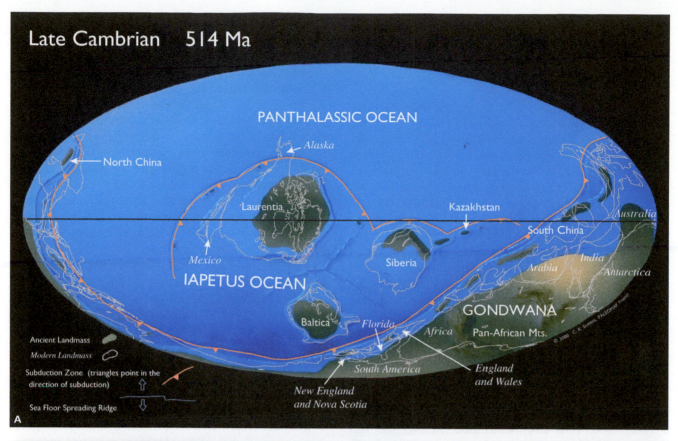

Late Cambrian   514 Ma

PANTHALASSIC OCEAN

North China

Alaska

Laurentia

Kazakhstan

South China

Australia

Mexico

Siberia

Arabia

India

Antarctica

IAPETUS OCEAN

Baltica

Florida

GONDWANA

Africa

Pan-African Mts.

South America

England and Wales

New England and Nova Scotia

Ancient Landmass
Modern Landmass
Subduction Zone (triangles point in the direction of subduction)
Sea Floor Spreading Ridge

A

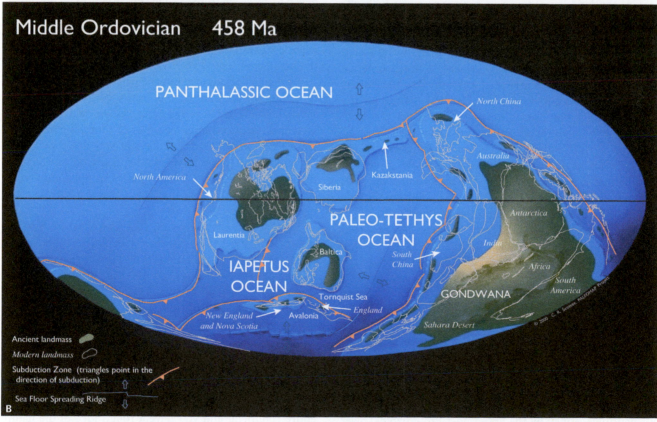

Middle Ordovician   458 Ma

PANTHALASSIC OCEAN

North China

North America

Siberia

Kazakstania

Australia

PALEO-TETHYS OCEAN

Antarctica

Laurentia

Baltica

South China

India

Africa

IAPETUS OCEAN

South America

Tornquist Sea

England

GONDWANA

New England and Nova Scotia

Avalonia

Sahara Desert

Ancient landmass
Modern landmass
Subduction Zone (triangles point in the direction of subduction)
Sea Floor Spreading Ridge

B

Figure 10.2 A. Global paleogeographic map of the Late Cambrian. North America straddled the equator, with the modern east coast facing south and the modern west coast facing north. By the Late Cambrian, nearly all of the United States and much of Canada was drowned by the shallow seas of the Sauk transgression. Gondwana remained as the last piece of the Late Proterozoic Pannotia and Rodinia supercontinents. The Baltic block and Siberian block were floating away from the rest of these continents. The future New England and Nova Scotia (Avalon terrane) was still close to Gondwana at this time. **B.** By the Middle Ordovician, North America was still tropical but had begun to rotate counterclockwise; it was still drowned by huge, shallow Tippecanoe seas. Both Baltica and Siberia were approaching collision with North America. The Avalonia microcontinent had pulled away from Gondwana on its way to collide with North America by the Devonian. And a volcanic arc of the Piedmont terrane was beginning to rise in the future region of the Appalachians. The rest of Gondwana was still one supercontinent but was shifting from the South Pole to a more equatorial position.

10.2 The Sauk Transgression (Latest Proterozoic–Early Ordovician)

If you travel around the world looking at lower Paleozoic rocks, you will see a characteristic pattern (Fig. 10.3). The lowest Paleozoic unit is nearly always a sandstone with sedimentary features (such as ripple marks and bedding formed by tides) that prove it was formed in a shallow-marine beach or nearshore environment. As you go up the vertical section of rock, the sandstones are overlain by shales that have marine fossils (such as trilobites) in them, showing that they were deposited in a quieter, more offshore setting. Similar mudstones and shales are formed today on the continental shelf far enough offshore, deep enough that they seldom feel the effects of waves or storms. At the top of this sequence is typically a limestone, crammed full of marine fossils, which represents a shallow shoal far offshore, where limestone-producing marine life could grow. The same kinds of rocks are being produced in shallow offshore marine settings like the Bahamas and the Great Barrier Reef today.

This sequence of progressively finer-grained rocks (sandstone, then shale, then limestone, also known as a "fining-upward sequence") is the classic pattern formed by a transgression (Fig. 4.14), when local sea level rises and drowns the landscape, covering nearshore sands with offshore muds and finally shallow-marine limestones. This transgressive pattern is nearly universal around the world in the early Paleozoic, including the margins of North America. This major transgressive event in North America is known as the **Sauk sequence**. It was named by the famous stratigrapher Larry Sloss in 1963. He chose names for the sequences to honor Native American tribes (in this case, the Sauk tribe of Illinois) or locations from the American Midwest. Sloss was trying to avoid any confusion with existing formation names or geologic terms or the geological periods that are based on European rock sequences. (For example, the Cambrian, Ordovician, and Silurian periods were based on rocks in Wales and western England.)

The oldest evidence of the Sauk transgression occurred on the outer edges of the continent (Fig. 10.3). Moving toward the continental interior, the transgression began later and later. For example, in the Appalachians there was a rift valley sequence in the Upper Proterozoic Ocoee Group, followed by transgressive sandstones, shales, and limestones of the Lower Cambrian Chilhowee Group (the river deposits of the Weverton Formation, capped by the Harpers Formation Shale, and the Antietam Sandstone, a beach sand) (Fig. 10.3A). At the opposite edge of the continent, the Paleozoic succession in Death Valley also began with Upper Proterozoic nearshore sandstones (Johnnie and Stirling Formations), capped by the uppermost Proterozoic–Lower Cambrian Wood Canyon Formation, which is mostly shales representing the more offshore portion of the transgression (Fig. 10.3G). Moving from the center to the edge of the continent, the Sauk transgressive sequence began later and later. Thus, in Death Valley the transgression began in the Late Proterozoic, but in the Grand Canyon (Figs. 4.15,

10.3F) the transgression began with the Lower Cambrian Tapeats Sandstone, followed by the Bright Angel Shale and finally by the Muav Limestone. In the northern Rockies, the Proterozoic Belt and Windermere Groups are overlain by the Lower Cambrian Gog Group, composed of the classic sandstone–shale–limestone transgressive package. In the northern Appalachians (upstate New York–Vermont–Quebec), the sequence begins with the Middle Cambrian Potsdam Sandstone (Fig. 10.1), followed by the Teresa Shale and then a series of Ordovician limestones. Similarly, there is a Middle Cambrian transgressive nearshore sandstone in the Colorado Rockies (the Sawatch Formation, capped by the Ordovician Manitou Limestone) (Fig. 10.3D), as well as a similar sequence starting with the Middle Cambrian Deadwood Sandstone in the Black Hills of South Dakota (Fig. 10.3E). Finally, by the time the transgression reached the middle of the continent, the basal transgressive sandstone started forming in the Late Cambrian, some 30 million years after the transgression had begun at the edge of the continent. Thus, in places like Wisconsin, the basal marine transgression begins with a series of Upper Cambrian sandstones known as the Mt. Simon, Galesville, and Eau Claire Formations (Fig. 10.3B). Putting all the pieces of this puzzle together, it is clear that the transgressing seas began to encroach from the edge of the continent during the Late Proterozoic and Early Cambrian. They slowly drowned the continent from the edge toward the middle during the Middle Cambrian, then reached their maximum extent in the Late Cambrian.

By the end of this event in the Late Cambrian–Early Ordovician, North America was nearly completely drowned by shallow-marine seas full of carbonate-secreting organisms that produced huge volumes of limestone across the entire continent. The only area that was not drowned by this global rise in sea level was the Transcontinental Arch in North America (Fig. 10.3), which was apparently a low ridge or chain of islands that stood high enough to avoid being immersed.

What would a sea creature, such as a trilobite, have experienced living in this enormous shallow-marine seaway of the Sauk transgression? The best modern analogue is a place like the Bahamas (Box 10.1) in the Atlantic or the Great Barrier Reef off the east coast of Australia, where an enormous area of shallow-marine shoals is covered with carbonate sediment made of broken pieces of shells, corals, and other lime-secreting marine animals (plus plants such as algae that secrete calcite as well). But the Bahamas Platform is only about 500 km (300 miles) across, while North America was at least 10 times that large. So imagine the shallow carbonate shoals of the Bahamas on a scale of an entire continent, and you have a rough idea of the nature and scale of the Sauk seas. Like the Great Barrier Reef or the Bahamas, the region would have been tropical since North America straddled the equator at that time (Figs. 10.2, 10.3), with its modern-day west coast on the north and its modern-day east coast on the south. Added to that, the climate was unusually warm with greenhouse climate conditions,

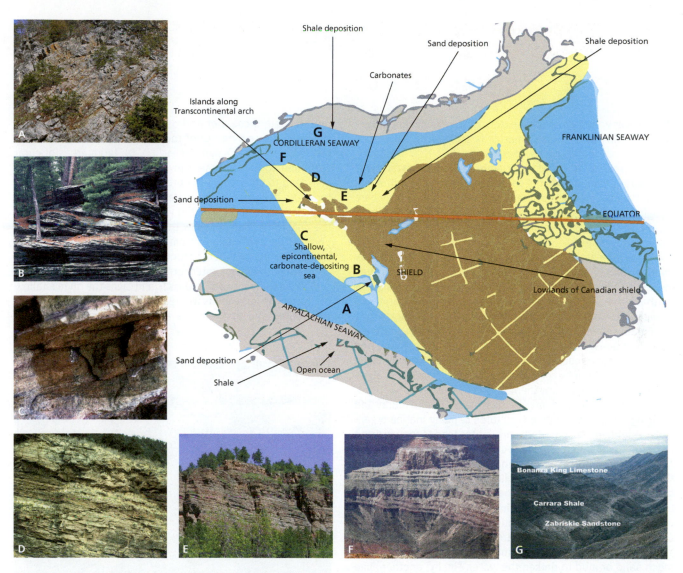

Figure 10.3 Paleogeographic map of North America during the Late Cambrian. North America straddled the equator, with the modern west coast facing north and the modern east coast facing south. Nearly all the United States was drowned by the maximum transgression of the Sauk sequence, except for the Canadian Shield and a low string of islands called the Transcontinental Arch that ran from Minnesota to Texas. Inset are photos of typical outcrops of the Cambrian Sauk transgression. **A.** Sequence of shallow marine sandstones of the Chihowee Group, including the Weverton Sandstone, near Thoroughfare Gap, Virginia. **B.** The cross-bedded sandstones of the Upper Cambrian Galesville Formation, Wisconsin Dells, Wisconsin. **C.** Cambrian Lamotte Sandstone nonconformably overlying Precambrian granites, St. Francois Mountains, Missouri. **D.** The shallow-marine Sawatch Sandstone overlain by the Manitou Limestone, in the eastern Rockies near Colorado Springs, Colorado. **E.** The thick beds of shallow-marine Deadwood Sandstone, near Deadwood, South Dakota. **F.** Apollo Temple in the Grand Canyon. The tilted Precambrian rocks are overlain unconformably by the Lower–Middle Cambrian Tapeats Sandstone (lower horizontally bedded cliff), Bright Angel Shale (greenish slopes), and Muav Limestone (lower part of upper vertical cliff) in the Grand Canyon. **G.** The enormously thick Cambrian sequence of Zabriskie Sandstone, Carrara Shales, and Bonanza King Limestone, Aguereberry Point, looking east and down into Death Valley, California.

so it was warm and tropical not only in equatorial North America but also across much of the world.

In these shallow seas was an enormous diversity of marine life, as we shall see in the next section. Recall also that the moon was much closer then, so the tides sweeping across these shallow-marine shoals would have been huge. They would have flowed across some areas as huge tidal waves coming in, then drained out until only a few areas were immersed (somewhat like the tidal waves that come in and out of the Bay of Fundy in Canada every

12 hours). In many places (such as the Upper Cambrian rocks in Wisconsin) we find enormous storm deposits that dumped huge boulders into the sea, so the storms crossing this shallow seaway could be fierce. Not only was the moon closer, but both the earth and moon were still spinning much faster and the moon orbited the earth faster, so a month had only 18 days. Likewise, the year was over 430 days long, so the rapid spin of the earth packed more days into a single revolution around the sun than the 365 days we have now.

BOX 10.1: HOW DO WE KNOW?

What Do Limestones Tell Us?

The sheer volume and areal extent of Paleozoic limestones over nearly all of North America is staggering. Enormous sheets of carbonate rocks covered most of North America starting in the Middle and Late Cambrian, and limestone was the predominant rock on the continent through the entire rest of the Paleozoic except at the edges of the continent, where collisions with other continents and microcontinents built mountain ranges and warped down deep basins full of sandstones and shales.

Yet we look at the earth today, and we notice there are only relatively small amounts of limestone being produced. Even more critically, modern limestones form in only a few places on earth which have just the right conditions: shallow, warm, and tropical because most of the carbonate-secreting organisms like corals and mollusks prefer warm waters or cannot tolerate cold conditions. In addition, the water must be clear and free of mud because much of the carbonate is secreted by algae that require light or by coral reefs, which have symbiotic algae in their tissues that allow them to build large reef structures. Mud, on the other hand, makes the water dark and murky and can clog the gills or feeding structures of many marine organisms. This means that modern carbonates tend to form in areas that are shallow and tropical but far from a source of mud. Thus, many islands in the Caribbean, the Gulf of Mexico, and the tropical Atlantic are rich in limestones, as is the Atlantic coast of Florida—but the Gulf coast from Florida west toward Louisiana is relatively poor in limestones because the water in the Gulf of Mexico is full of mud from the Mississippi River. Carbonates also grow in the warm shallow waters of the Persian Gulf because the local bedrock is also limestone, without much sand or mud to darken the water. Finally, another huge area of carbonate growth is the tropical islands of the southern and western Pacific, which are warm and shallow but a long distance from any mud coming out of rivers on land.

One of the most studied examples of a modern shallow-marine platform covered by carbonates is the Bahamas (Fig. 10.4A). Located to the east of the Atlantic coast of North America just east of Florida and just north of Cuba, the Bahamas Platform rises out of the ocean to form a large submerged plateau, with only a tiny portion above water to form islands. It is bathed in the warm waters of the Gulf Stream heading north from around the tip of Florida, so it is always tropical and warm in climate; and the entire area has no large rivers carrying mud from the land (the bedrock in Florida is mostly limestone), so the water is extremely clear. The entire platform is very shallow, with a water depth of less than 12 m (40 feet) covering a vast area of the submarine plateau and only a tiny portion (such as Andros Island, the largest island in the Bahamas) above water (Fig. 10.4B). The edges of the plateau are quite steep, dropping off from the shallow platform itself to deep water that surrounds the platform (Fig. 10.4C).

For decades, geologists have visited the Bahamas to dive in the warm clear water and study carbonate sediments being formed and deposited on a large scale. It turns out that there is a well-defined pattern of what kinds of carbonate sediments grow in different parts of the platform. For example,

on the eastern coast of Andros Island, which faces the hurricane winds coming out of the east and the deep water of the Tongue of the Ocean (Fig. 10.4B, D), large areas of coral reefs and coralline algae grow along the shore. Reef corals also tend to grow along the break between the shallow platform and the steep slope to deeper water (Fig. 10.4D). Many of these corals are tolerant of waves and storms, so they are the only thing that can grow there. They are also dependent upon nutrients brought up from deeper water by upwelling, so it is not surprising that they form on the edges of the platform adjacent to the deep-water currents.

The western shore of Andros Island faces the huge shallow platform with large areas where the tidal flats are alternatively submerged and then emerge as the tides come in and out

Figure 10.4   Carbonate facies of the Bahamas. **A.** Satellite image of the Bahamas, Florida, Cuba, and adjacent areas. The light blue areas are shallow water, and the darker blues are deep water. **B.** Index map of the Bahamas, showing the water depths of the carbonate platform. Areas in brown are land, and those in gray are deep water. **C.** Simplified cross section of the Bahamas platform, showing the characteristic sediment types in each area. **D.** Map of the platform around Andros Island, showing the typical sediment types. **E.** Close-up of ooids. **F.** Microscopic section through an oolitic limestone, showing the concentric layering "snowball" structure.

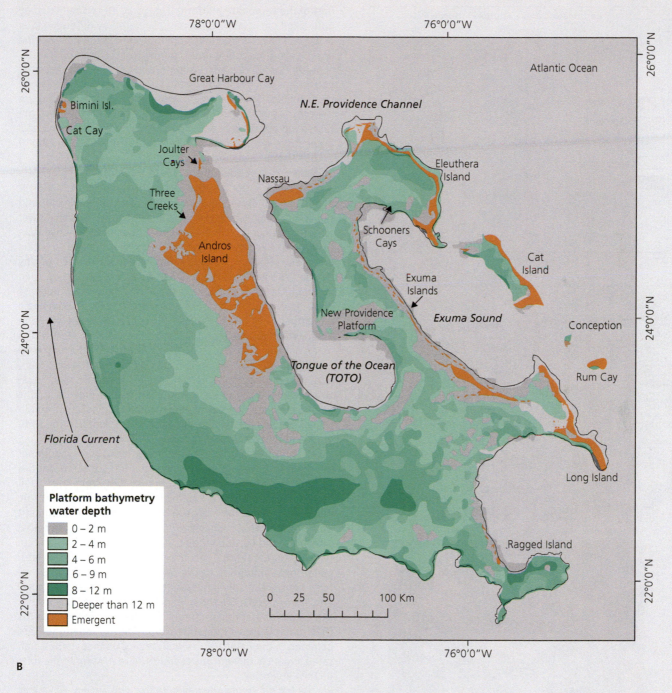

B

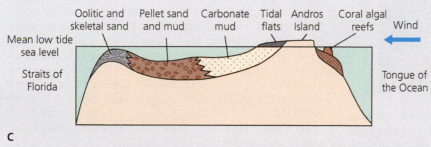

C

Figure 10.4   (Continued)

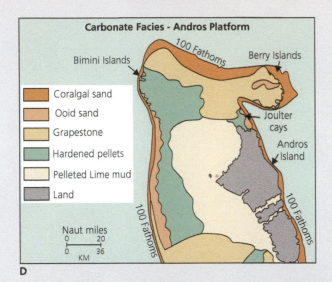

Carbonate Facies - Andros Platform

- Coralgal sand
- Ooid sand
- Grapestone
- Hardened pellets
- Pelleted Lime mud
- Land

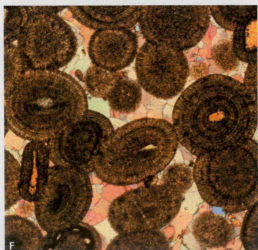

(Fig. 10.4C). These conditions favor animals that tolerate the dry conditions during low tides and thick mats of algae and bacteria making stromatolites. The central, deeper, quieter

part of the carbonate platform (Fig. 10.4C,D) is dominated by mud-sized particles of carbonate, secreted by different kinds of algae. These quiet, muddy lagoons are excellent for marine organisms that require quiet, undisturbed conditions, especially delicate bryozoans, crinoids, and certain delicate corals and mollusks. The carbonate muds also accumulate the small fecal pellets of marine organisms, so most of the lime mud of the central platform is full of such pellets. In certain areas, the tiny round pellets are cemented together with calcite to make grapestone, a mass of tiny pellets that resembles a cluster of grapes (Fig. 10.4D).

Near the west edge of the platform before the coral reefs that mark the dropoff (Fig. 10.4C, D) is an area of shallower shoal that is buffeted by the currents of water moving around the lagoon, especially when the tides change. This shallower agitated area of the shoal tends to produce a distinctive type of carbonate sediment known as **ooids**, which make an oolitic limestone (Fig. 10.4E). Ooids look like large masses of tiny BBs or fish eggs (*oos* is the Greek word for "egg"), but when you look closer, you find that these tiny sand-sized spheres only a few millimeters across are concentrically layered like a miniature snowball (Fig. 10.4F). Indeed, a snowball is a good analogy for how they form. Like a snowball, they start with a much smaller nucleus (usually accreting around a tiny fragment of shell) that then rolls back and forth in the lime mud, adding layer after layer until it reaches a certain size. Thus, whenever a geologist finds an ancient oolitic limestone, it is a good indicator not only of very shallow tropical shoal conditions but also of rapid currents which have agitated the sediment and rolled it around thousands of times.

All of these modern examples of carbonate sediment have ancient analogues in Paleozoic limestones. Tidal flat sediments with stromatolites are known from the fossil record even before the Paleozoic. We often find Paleozoic limestones made of mud-sized carbonate with abundant delicate animals, such as branching bryozoans, crinoids, blastoids, and certain kinds of corals, along with many fecal pellets; these almost certainly formed in the quiet-water conditions analogous to the central platform of the Bahamas. Many Paleozoic limestones show well-developed coral and sponge reefs like those in modern settings, except that the corals and sponges are all members of extinct groups that didn't survive the Paleozoic. Oolitic limestones are common in the Paleozoic as well, telling us that shallow agitated shoals with vigorous currents were present in the past as they are today.

The big difference, however, is that the ancient Paleozoic limestones covered almost the entire North American continent for most of the Paleozoic. Thus, the small-scale Bahamas Platform doesn't even begin to approach the immensity of the enormous carbonate platforms that covered whole continents for millions of years. When you see Paleozoic limestone exposures in the American Midwest or Rockies, full of fossil shells, ooids, pellets, and even coral reefs, start with thinking of the Bahamas—and then multiply the scale by 10 to a 100 to get a sense of the enormous scale of these shallow, warm, tropical continental seas of the past.

These Cambrian and Ordovician sedimentary rocks also showed differences from sedimentary rocks of the past. Recall that during the Archean, the only sandstones we encountered were muddy, angular graywackes full of unstable minerals and rock fragments, which had clearly gone from a source area and been dumped in a deep-marine basin with no sorting or winnowing of their components (Fig. 8.11A). By the later Proterozoic, on the other hand, we find the first evidence of sandstones that are nearly pure quartz, with few or no other minerals or rock fragments and no mud in the pores between the sand grains (Fig. 8.11B). Such sandstones are clear evidence that ancient sandstones and other sources of quartz grains had been weathered out and reworked into younger sands, again and again. Such sandstones are called **mature**, in that they are reworked and weathered until only the most stable minerals, such as quartz, remain. The sandstones of the basal Sauk transgression are also highly reworked and mature.

Thus, the Sauk transgression marks not only the first great drowning of the continent by shallow seas but also a major change in the sedimentary regime. The sandstones that covered its beaches were all pure quartz and mature in their textures. The shales are full of many different types of fossils and deeply churned and burrowed, something that did not occur in the Proterozoic when there were no worms or other burrowing organisms yet. Finally, the Early Paleozoic marks the first time that huge bodies of limestone were forming across entire continents by a wide variety of shelly animals in the ocean, whereas limestones were rare in the Proterozoic and mostly made of local patches of stromatolites.

What caused the Sauk transgression? Recall that at the end of the Proterozoic, the world (Fig. 10.2) was experiencing a time of rapid breakup of the supercontinent Pannotia (Fig. 8.9). Some geologists suggest that the continents were going through extraordinarily fast plate movements, producing rapid sea-floor spreading, and possibly even shifting the position of the rotational pole of the earth with respect to the continents. Such intense activities have many consequences.

The huge volumes of lava produced by high rates of sea-floor spreading pumped lots of greenhouse gases into the atmosphere, contributing to the warming of climates around the globe after the last "snowball earth" event. Whatever ice was present in the Late Proterozoic melted, so there were no icecaps or glaciers left and all the meltwater caused sea level to rise.

High rates of sea-floor spreading have other effects as well. On modern mid-ocean ridges with a slow rate of

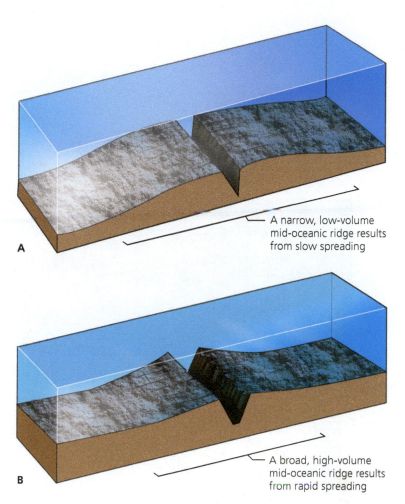

A narrow, low-volume mid-oceanic ridge results from slow spreading

A broad, high-volume mid-oceanic ridge results from rapid spreading

Figure 10.5 Mid-ocean ridges which are spreading slowly have a long time for the recently erupted, hot, expanded oceanic crust to cool and contract, so they tend to slope steeply down from the central rift valley. However, fast-spreading ridges slide their hot, expanded oceanic crust away from the central rift quicker, so they tend to be higher in elevation, gentler in slope, and thicker in volume. When fast spreading occurs, this increase in the volume of sea-floor rocks decreases the volume for ocean water and forces the seas onto the land.

spreading, their cross section (Fig. 10.5) is relatively small. A good example is the modern Mid-Atlantic Ridge, which spreads only 6 cm/year, about as fast as your fingernails grow. It has a steep, relatively small profile because the hot rocks that erupted from the center of the ridge have plenty of time to cool and contract as they slowly move away from the ridge axis. But a much faster spreading ridge (like the modern East Pacific Rise) spreads apart about three times as fast (18 cm/year). Its cross-sectional profile (Fig. 10.5) is much thicker and more gently sloped because the hot rocks move a long way from their origin in the rift valley where they erupted before they have much chance to cool and shrink down. Thus, the profile of a fast-spreading ridge is much broader and has a much greater volume than that of a slow-spreading ridge. If you increase the volume of rocks at the bottom of the sea floor, the water in the ocean basins has only one place to go: up and out of the ocean. It spills out of the oceans and begins to flood the continents.

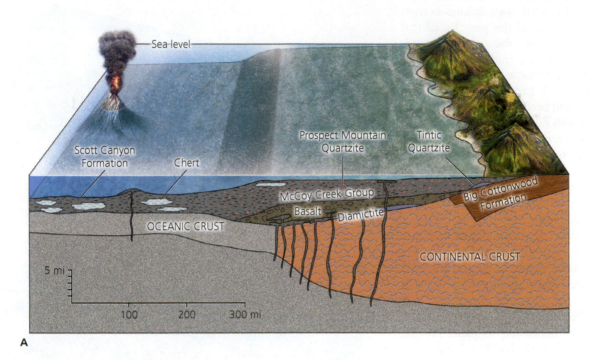

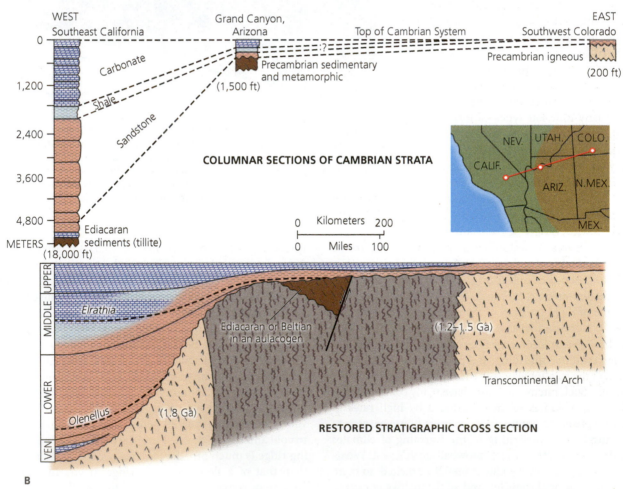

Figure 10.6 **A.** The thick Upper Proterozoic–Lower Cambrian passive margin wedge sequence as seen in Nevada (left) and Utah (right), formed as Pannotia broke up and left the western edge of North America as a passive margin. **B.** Diagrammatic cross section showing how the passive margin wedge sequences are incredibly thick due to continual subsidence. Meanwhile, the deposits on the craton at the same time (Grand Canyon to the center of the continent) are much thinner and full of unconformities.

Not only these events contribute to the flooding of the landmasses, but other factors contribute as well. At the edges of the continents where they pulled apart, the huge passive margin wedges of sediment accumulated, producing a thick sequence of shallow-marine sediments (Fig. 10.6). This passive margin wedge can be seen in places like Death Valley, where the Cambrian sequence is over 2500 m (8200 feet) thick (Fig. 10.3G). But moving inland from the steadily subsiding wedge, with its thick piles of sediment, there is a rapid transition to the central core of the continent. This area has long been known as the "shield," or **craton** (the Greek word for "shield"), since it resembles the gently domed surface of an ancient warrior's shield. Unlike the passive margin, the craton is very stable tectonically and sinks or rises very little during the course of millions of years. In North America, for example, the craton consists of the most ancient Archean and Proterozoic rocks of the "Canadian shield" (Fig. 8.5A) plus the additional terranes that accumulated during the middle and later parts of Proterozoic, such as the Yavapai, Mazatzal, and Grenville terranes (Fig. 8.5A).

The differences between the craton and the passive margin wedge can be striking. Instead of 2500 m of Cambrian sediment in Death Valley, the Cambrian deposits on the craton are only about 10–100 m (33–330 feet) in most places such as Wisconsin, Colorado, and New York. The Tapeats Sandstone in the Grand Canyon is on the craton and only about 70 m (230 feet) thick, and the Cambrian formations in Minnesota are barely a few meters in thickness; but the rocks of the same age in Death Valley (on the passive margin wedge) are over 1000 m (3300 feet) thick. Thus, the passive margin sinks steadily through

millions of years, continuously accumulating thousands of meters of shallow-marine sediment. Meanwhile, the craton barely sank at all during the same span of time and only accumulated thin deposits of shallow-marine sediment when there were extraordinarily high sea levels, such as occurred

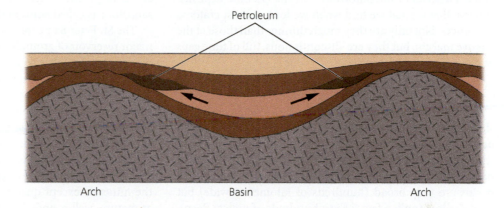
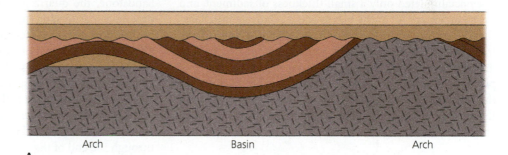
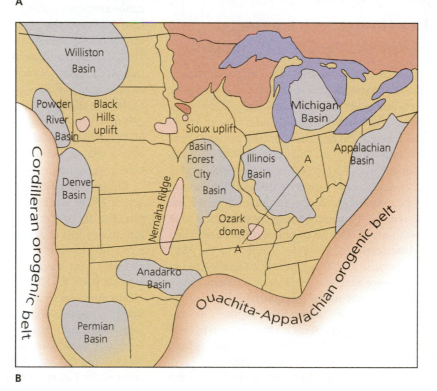

Figure 10.7 A. Cartoon cross section of cratonic basins and arches, showing how the basins tend to be small and lens-shaped, thinning at the edges, with many unconformities. **B.** Map showing the location and size of some major Paleozoic cratonic basins, which tend to be large, bowl-shaped depressions filled with sediment.

in the Sauk transgression. The sedimentary deposits of the craton are thus not only very thin (compared to the passive margin) but also discontinuous because every lowering of sea level (short-term local regression) during the overall transgression would have exposed the old marine beds and created a series of unconformities in the cratonic deposits. Indeed, that is what we find when we look at most cratonic sequences. Not only are they much thinner than those of the passive margin, but they are discontinuous, full of time gaps at the major unconformities.

See For Yourself: Ordovician

In fact, cratonic sedimentary deposits operate very differently from those of the passive margin wedge. When the craton sinks or rises at all, it does so in broad, shallow, bowl-shaped basins and low-domed arches (Fig. 10.7A), rather than thick, deep sedimentary basins such as the fault-bounded basins of the rift valley sequence (Fig. 5.17). Cratonic basins are very broad (hundreds of kilometers wide) but very shallow (only a few tens to hundreds of meters deep). They accumulated only a small thickness of sediment, and the sedimentary fill of the basins was a relatively thin, lens-shaped wedge of sediment, thickest in the center of the basin and thinning toward the edges. Such bowl-shaped or dish-shaped cratonic basins are known from a number of places in North America, such as the Michigan Basin (over the center of the lower peninsula of Michigan), the Illinois Basin, the Williston Basin in Montana and North Dakota, and a few others (Fig. 10.7B).

10.3 The Tippecanoe Sequence (Middle Ordovician–Early Devonian)

The Sauk sequence continued into the Early Ordovician, when huge areas of limestone covered the shallow seas across most of the continents. At the end of the Early Ordovician, however, the seas regressed from nearly all of North America, leaving an erosional unconformity carved into the top of the uppermost Sauk limestones. During the Middle Ordovician, a new transgression flooded the continent and covered the deeply eroded surface of Lower Ordovician limestones. This second great epicontinental transgression across the craton is known as the **Tippecanoe sequence**. Like the Sauk sequence, it was named after an important place associated with Native Americans. In this case, it refers to the 1811 Battle of Tippecanoe in Indiana, where the Shawnee forces under Tecumseh were defeated by General William Henry Harrison and American troops. (You may have heard of the word "Tippecanoe" from the old political slogan "Tippecanoe and Tyler Too." Harrison got his nickname from his victory, so when he ran for president with Vice-President John Tyler in 1840, this became his campaign slogan.)

Like the Sauk sequence, the first unit at the base of the Tippecanoe transgression is a beach and nearshore sandstone, which covered the erosional surface in the Lower Ordovician limestones as the sea level rose. In the Midwestern states, this sandstone is known as the St. Peter Sandstone

(Fig. 10.8A). It covers over 20,000 km², from the north over much of Wisconsin, Illinois, Iowa, and the upper Midwest to as far south as Oklahoma. The St. Peter Sandstone is an extremely pure sandstone (over 99% quartz), with very well-rounded grains that are all the same size (well sorted) and excellent porosity between the grains (Fig. 10.8B). Such a sandstone is called **supermature**.

The St. Peter Sandstone is so porous, in fact, that it is the major reservoir of groundwater in the Midwest. In addition, the quartz content is so pure that it has been mined using high-pressure water hoses to produce clean quartz sand to melt to make glass. Sedimentary geologists have argued for a long time about what processes might produce such a high concentration of pure quartz, as well as the excellent rounding and sorting of the grains. They concluded that just reworking old quartz sandstones is not enough, nor is intense chemical weathering of tropical soils that breaks down all the minerals except quartz. It appears that to make sands as mature, well rounded, and well sorted as in the St. Peter Sandstone, the quartz grains must have also spent some time in a coastal sand dune. There, they were abraded and bashed and their surfaces frosted by the wind before they were reworked at least one more time and dumped into the nearshore setting of the shallow Sauk seas.

Once the nearshore sands were deposited at the base of the Tippecanoe sequence, they were followed by the classic transgressive package of shale (in a few places) but then enormous volumes of limestone that persisted not only through the end of the Ordovician but for the entire Tippecanoe sequence ending in the Early Devonian. In the American Midwest, the St. Peter Sandstone is capped by thick sequences of the Platteville and Galena limestones (Fig. 10.8C), which are legendary for their fossils. In the Cincinnati Arch region of Indiana, Kentucky, and Ohio, the thick sequence of Middle–Upper Ordovician limestones with shales was the basis for the standard Late Ordovician timescale in North America because of their abundant fossils (Fig. 10.8D). These fossiliferous beds have been so productive for over a century (and still are very rich) that they were the training ground for a veritable "who's who" in paleontology.

Ordovician rocks of the Tippecanoe sequence are found across North America, so wherever the Ordovician is exposed at the surface, there are usually good exposures of limestones full of fossils. Our previous examples demonstrate how they crop out widely in the Midwest, but you can also find them in the Criner Hills of Oklahoma, a famous place for Ordovician trilobites as well. The Ordovician Simpson Group, Viola Limestone, and Fernvale and Sylvan Formations are exposed in the Arbuckle Mountains of Oklahoma and reach up to 2100 m (about 7000 feet) in thickness. In many of the desert ranges of Utah and Nevada and in Death Valley, California, are excellent exposures of the Lower Ordovician Sauk sequence limestones of the Pogonip Group, over 3500 m (11,500 feet) thick (Fig. 10.9A), unconformably overlain by the transgressive sandstone (now Eureka quartzite) and Ely Springs limestone of the Tippecanoe sequence. And on the tilted flanks of the

Figure 10.8 **A.** The base of the Tippecanoe sequence in the upper Midwest is represented by the supermature quartz sandstones of the St. Peter Sandstone. This waterfall in St. Louis Canyon at Starved Rock State Park, Illinois, cuts through a thick sequence of the finely bedded St. Peter Sandstone. **B.** Microphotograph of a thin section of St. Peter Sandstone, showing the well-sorted, well-rounded pure quartz sand grains with almost no other components. This is a classic example of a supermature sandstone. **C.** Above the St. Peter Sandstone, the Tippecanoe sequence continues with the fossiliferous limestones of the Platteville and Galena Formations. This roadcut near Dickeyville, Wisconsin, is loaded with Ordovician fossils. **D.** The Upper Ordovician shallow-marine shales and sandstones of the Cincinnati Arch are legendary for their fossils.

Figure 10.9 Examples of the Sauk and Tippecanoe sequence in the western United States. **A.** Arrow Canyon Range in Nevada, showing a thick sequence of Ordovician rocks. The lower half of the cliff is the Lower Ordovician Pogonip Group, part of the end of the Sauk sequence. The light tan band midway up the cliffs is the nearshore marine sandstone, now turned into the Eureka Quartzite, that represents the nearshore deposits of the Tippecanoe transgression in the Middle Ordovician. The dark gray layer above the light tan Eureka is the Ely Springs Dolomite (Silurian), followed by Devonian rocks on the peaks, all part of the Tippecanoe sequence. **B.** Photo of the Ordovician Bighorn Formation, formerly a limestone and now a dolostone, Shell Canyon in the Bighorn Mountains, Wyoming.

uplifted Rockies of Wyoming and Montana, there are spectacular cliffs of the Bighorn Formation (Fig. 10.9B). It is another Tippecanoe limestone which has been transformed into a magnesium-rich carbonate rock known as dolostone, made of the mineral dolomite ($CaMg[CO_3]_2$).

Limestone deposition of the Tippecanoe sequence continued through the Silurian and Early Devonian. We will look at some of their other features in Chapter 11.

10.4 The Mountains Rise: The Taconic Orogeny (Middle–Late Ordovician)

From the Late Proterozoic (about 580 Ma) until the Middle Ordovician (about 460 Ma), an interval of over 120 million years, the entire North American continent was the realm of passive margins and shallow cratonic seaways. All of the edges of North America were divergent margins, with thick passive margin wedges. There was no sign of convergent tectonic margins, mountain building, or subduction or compressional tectonics and no significant volcanism. But during the Middle–Late Ordovician, this began to change on the present-day Appalachian margin of North America. In many places, we see signs that the passive margin that had persisted along the Appalachian seaboard was changing tectonically. We find Middle and Late Ordovician volcanic ashes as far from the coast as central Tennessee (Fig. 4.17). Not only that but the rock sequences are no longer the shallow-marine sandstones, shales, and limestones that dominated the rest of the continent during the Tippecanoe transgression. Instead, they are very different.

The most obvious signs of change occur in Ordovician rocks in Newfoundland, eastern New York, eastern Pennsylvania, and the Carolinas, where we see evidence of intense folding and deformation. These tightly folded rocks are not shallow-marine limestones so typical of the rest of the Upper Ordovician but deep-marine shales and turbidite sandstones (Fig. 10.10A). For example, in the Hudson Valley near Poughkeepsie, New York, is a thick sequence of deep-water turbidites and shales known as the Normanskill Formation. In eastern Pennsylvania, a similar-looking unit is called the Martinsburg Shale (Fig. 10.10B, C).

When these rocks were first studied, they were a puzzle to early American geologists. What had caused the edge of the continent to change so abruptly from shallow-water limestones of the Lower Ordovician to deep-water shales and turbidites by the Middle Ordovician? An even closer examination yielded further puzzles. The deep-water flysch deposits (see Fig. 5.23) had abundant volcanic ash layers throughout them and grains of volcanic rocks in the sandstones. But there were no volcanoes of any kind in North America to the west, so they must have come from some mysterious volcanic source offshore to the east. In addition, many of the ancient current directions showed that the turbidites flowed downslope from an uplifted area to the east of this region, where only deep ocean lies today. What was this mysterious uplift that provided all these gravity slides and shed so much volcanic ash and debris into the deep-water marine basin?

In some places, such as in Newfoundland, there were even more puzzling features associated with the Late Ordovician deformation. These included large outcrops of weathered pillow lavas as well as other rocks (sheeted dikes, layered gabbros, even mantle peridotite) that we now recognize as slices of ancient oceanic crust (Fig. 10.10D). At the time these were first described, no one knew what they meant, but now we know that they are part of the ophiolite suite, typical of oceanic crust the world over (Fig. 5.16).

Finally, the strangest features were large slices of shallow-water limestone sitting right in the midst of the deep-water shales in the Hudson Valley and Taconic Mountains to the east of the Hudson River (Fig. 10.10E). When these out-of-place rocks were discovered, they were thought

Section without vertical exaggeration

Figure 10.10 The Taconic orogeny produced dramatic changes in what is now the Appalachian region of North America. **A.** Cross section across the region, showing the deeply downwarped sedimentary basin on the continent to the east (Michigan, western Pennsylvania) and the highly deformed, faulted, and folded Taconic mountain belt or "Taconian land" to the west (eastern Pennsylvania, New Jersey, eastern New York). **B.** Before the collision, huge volumes of deep-water shales and turbidites formed flysch deposits in many of the basins. This is the thick stack of Middle Ordovician Normanskill Formation along Highway 44 just west of Poughkeepsie, New York. **C.** Similar thick sequences of shales are found in Pennsylvania, where they are known as the Martinsburg Formation. Here in Jackson State Quarry, Pennsylvania, the horizontal lines are actually cleavage along the axial plane of a huge overturned fold, which can be seen by the curved color patterns on the back wall of the quarry. **D.** Pillow lavas and pillow breccias from Cape St. Francis, Newfoundland. Ordovician ophiolites are found in many places in the Taconic mountain belt as subduction sliced off pieces of ocean crust and emplaced them in the accretionary wedge. **E.** Exposure of the Cow Head Breccia on western Newfoundland. These blocks of

(Continues)

Molasse facies	Irregularly stratified coarse sandstone (with cross-stratification), conglomerate, minor shale and coal, may be red-colored	Nonmarine
	Coal	
Flysch facies	Evenly stratified graywackes (with graded bedding) alternating with dark shale	Shallow-marine
		Deeper marine
Ophiolite suite	Black shale	Deepest marine
	Bedded chert	
	Basalts with pillow structure	Oceanic crust
	Swarms of dikes	
	Gabbro	
	Ultramafic rocks (peridotite or serpentine)	Oceanic mantle

G

(*Continued*)   shallow-marine limestone, some over 3 m (10 feet) in diameter, were rolled into the deep-marine trough of shales and turbidites by submarine landslides. This is one of many examples of Taconic shallow-water limestones that were once interpreted as klippen, or fault slices, but are now known to be deposited by gravity slides. **F.** Exposure of the red Queenston Shale, a typical deposit of the post-orogenic Taconic molasse, formed by the wearing down of the Taconic Mountains. **G.** Typical sequence in many Taconic basins: the base is a slice of ophiolite from oceanic crust. Then as this crustal rock sinks, it is covered by a thick pile of deep-water turbidites and shales, which precede the orogenic collision, so they are flysch deposits. Finally, the deformed flysch sequence is capped by post-orogenic river sands, deltas, and lagoons of the molasse sediment that eroded off the mountains after they formed. A similar sequence is found in many mountain-building events.

to be crustal blocks of limestone from other formations nearby that had been pushed by thrust faults into the shale. For this reason, they were called the **Taconic klippen**. (A klippe is an isolated eroded fault block pushed horizontally on a thrust fault.) But on closer examination, it was clear that these blocks had not been faulted at all. Instead, they appeared to be gigantic gravity slides of limestone from shallow-water environments from some offshore block to the east that had slid down into the deep-water marine trough.

Nobody could explain all these puzzling features until 1970, when plate tectonics was first applied to the region (Fig. 5.23). This series of events is a classic transition from the quiet shallow-water deposition of a passive margin (Fig. 5.18) to the compressional tectonics and sediments of an active margin subduction zone. Suddenly, all the odd pieces of the puzzle fell into place. The deep-water trough was formed as an exotic terrane came out of the east and began to approach North America, causing the crust to buckle downward and then rapidly fill with turbidites and shales. This approaching terrane was apparently some sort of volcanic arc complex formed over a subduction zone, explaining all the volcanic ashes in the marine section and volcanic sand grains in the graywackes of the turbidites. The ancient current directions that came out of the east must have been influenced by the steep slope down the front of the approaching volcanic arc. Likewise, the presence of ophiolites made sense if the collision had sliced off slabs of oceanic crust and stuck them into the accretionary wedge ahead of the subduction zone (Fig. 5.20). The strange limestone blocks known as the Taconic klippen must have been gravity slides of limestones formed on the shorelines of these approaching tropical volcanic islands. Finally, the fact that all of these rocks are now intensely deformed tells us that the subduction zone and island arc must have finally collided with North America in the Late Ordovician, crumpling up all the rocks of the deep-marine trough that formed between them before the collision.

Immediately above these intensely deformed deep-marine rocks of the Taconic flysch lies an unconformity in most of the region. This unconformity is capped by another distinctive sequence of rocks known as the Queenston (in New York) or Juniata (in Pennsylvania) sequence (Fig. 10.10F). Unlike the contorted pre-Taconic collision rocks, these rocks are undeformed and composed mostly of alluvial fan, river, and delta deposits that eroded off the Taconic uplifts after the collision was finished. The volume of these post-orogenic molasse (Fig. 5.23) deposits is enormous— over 600,000 km³ in the Appalachian region alone. If you calculate the size of the Late Ordovician Taconic Mountains that must have shed this much sediment, it suggests a Himalayan-scale mountain range over 12 km (40,000 feet) high and 200 km (124 miles) wide. Thus, the collision of the volcanic arc during the Taconic orogeny must have produced an enormous collisional mountain belt that uplifted 12 km into the sky, or 3100 m (10,300 feet) higher than the modern Himalayas (which are no higher than 8848 m, for 29,029 feet high)! As this immense Taconic mountain range wore down during the Late Ordovician and into the Silurian, it continually poured enormous amounts of sand, gravel, and mud into the Queenston–Juniata molasse.

What was the source of this huge collision? Where does it reside today? Careful tectonic analysis shows that this ancient volcanic arc complex that caused the Taconic orogeny is now lodged in the basement rocks of the eastern Appalachians, from the eastern edge of Newfoundland, New Brunswick, and through Vermont, western Massachusetts,

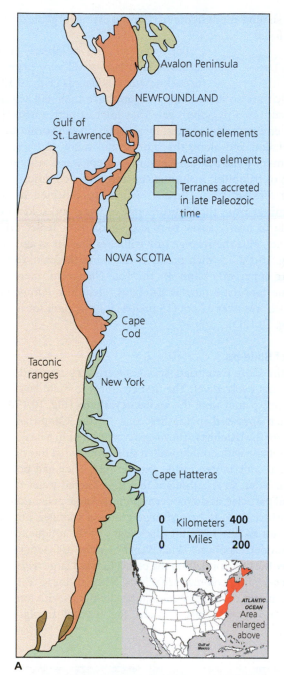

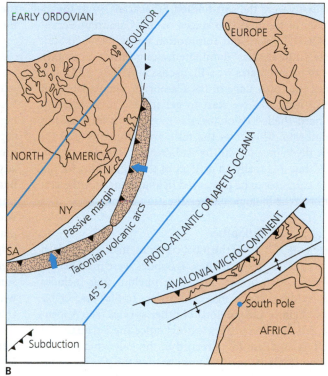

Figure 10.11 Map showing the location of the Piedmont terrane and the Avalon terrane basement in the Appalachian region and eastern North America. **A.** The Piedmont terrane (here labeled "Taconic elements") makes up the basement rock of much of the Appalachians west of the Piedmont (hence the name) and was accreted during the Late Ordovician Taconic orogeny. To its east is the Avalon terrane, accreted in the Late Devonian during the Acadian orogeny (see Chapter 11). **B.** Paleotectonic map of the Iapetus (Proto-Atlantic) Ocean during the Early Ordovician, with the Piedmont volcanic arcs just about to collide with North America during the Taconic orogeny. Farther offshore was the Avalonia microcontinent, which would arrive during the Acadian orogeny in the Late Devonian, followed by the African portion of Gondwana in the Pennsylvanian.

eastern New York, Connecticut, and the Piedmont region of the Appalachian Mountains (Fig. 10.11). For this reason, the entire tectonic block is known as the **Piedmont terrane**. During the Cambrian, it sat out in the Iapetus Ocean, which was the predecessor of the modern Atlantic, gradually approaching North America as the subducting plate consumed more and more of the oceanic crust that lay between it and the continent. Finally, in the Middle Ordovician it was close enough that its effects (volcanic deposits, downwarping of the deep-marine trough) were being felt. When the collision of the Piedmont terrane finally occurred in the Late Ordovician, it slammed into North America with such force that it built a mountain range 3 km higher than the Himalayas. Then during the Late Ordovician and Early

Silurian, this enormous range was eroded down, shedding the Queenston–Juniata molasse across most of eastern North America (Fig. 10.8).

10.5 Early Paleozoic Life

CAMBRIAN INVERTEBRATES

As we saw in Chapter 9, the Cambrian began with a huge diversification of hard-shelled marine animals, long mislabeled the "Cambrian explosion." It was not an "explosion" but a "slow fuse," a steady diversification of multicellular animal life, lasting through the 55 m.y. (600–545 Ma) of the Ediacaran Period in the Neoproterozoic (when most of the fossils were the soft-bodied Ediacarans), and the 25 m.y. years of the first two stages of the Cambrian Period (Nemakit-Daldynian and Tommotian stages, 545–520 Ma), when the "little shellies" dominated, along with several other groups of invertebrates we will discuss here (Fig. 9.13). It is not until the third stage of the Cambrian, the Atdabanian (520–515 Ma), that we start to get the big diversification of the hard-shelled trilobites, the most common fossils in the Cambrian. This gives the false appearance (Fig. 9.13) that the diversification in the Atdabanian was also an "explosion," but it is really an artifact of the conditions for the shells of trilobites to be preserved in abundance for the first time, not a true biological diversification event.

Trilobites

By far the most common fossils are indeed the trilobites. Trilobites were members of the phylum Arthropoda, the same phylum that includes insects, spiders, scorpions, centipedes, millipedes, crabs, lobsters, shrimp, and millions of other species (see Appendix A). Trilobites have been extinct since the Permian, but they left an excellent fossil record in the Paleozoic because their outer shells are not just composed of the protein chitin (which is found in insects, spiders, and most other arthropods) but also mineralized with calcite, so the shell fossilizes easily. Trilobites have a discrete head shield (the cephalon), many different segments in the middle of the body (the thorax), and a tail plate called a pygidium. They get their name "trilobite" because they have three lobes from left to right: a central lobe down the middle (axial lobe) and two lobes on the sides (pleural lobes). Although most trilobites follow this basic body plan, they evolved a huge diversity of shapes and sizes that make many of them easy to tell apart—and very popular with fossil collectors. Some trilobites have been preserved in such a way as to show their softer parts, like the antennae in front and the multiple pairs of legs and gills on the underside of the hard shell. Most trilobites were apparently mud-grubbers, burrowing through

the shallow muds of the sea floor to extract organic detritus for food.

For example, the trilobites of the earliest Cambrian are all very primitive forms known as olenellids (Fig. 10.12A), which are so archaic they do not have their tail segment fused into a pygidium yet. They are distinctive with their big crescent-shaped eyes, long spines off the corner of the cephalon and along the thorax, and many thoracic segments. By the Middle Cambrian, there was a wide diversity of trilobites, including the amazingly common forms like *Elrathia* (Fig. 10.12B), which are collected in huge numbers from the House Range in Utah and sold worldwide in nearly every rock shop and fossil dealer on the planet, plus the strange tiny trilobites known as agnostids (Fig. 10.12C), which probably floated in the plankton and were apparently blind. By the Late Cambrian, there were some more spectacular trilobites, including *Paradoxides*, which was over 37 cm (15 inches) long, a giant for its time (Fig. 10.12D).

Reef Builders

In the Early Cambrian, sponges were very rare and corals had not evolved yet. The first creatures to build large reefs on this planet were the **archaeocyathans** (Fig. 10.13A, B), which appeared in the first stage of the Cambrian, long before the trilobites. They were constructed in a basic cone-like shape with double-walled sides, somewhat like one ice-cream cone nested inside another one. They had pores on their sides, so it is thought that they filtered water for tiny food particles, as do modern sponges. Some paleontologists consider them to be a strange group of sponges, but most are not so sure since they lack many of the unique features of modern sponges. Whatever they were, archaeocyathans were the first animals on the planet to build large reefs full of filter-feeding colonial organisms. Some of these reefs were enormous, covering many square kilometers with hundreds of individual archaeocyathans, found today in places like

Figure 10.12 Typical Cambrian trilobites. **A.** *Olenellus*. **B.** *Elrathia*. **C.** Agnostids known as *Peronopsis*. **D.** *Paradoxides*.

Figure 10.13 Typical Cambrian fossils. **A.** The structure of an archaeocyathan, with the double-walled construction and porous exterior. **B.** Cross section of a branching archaeocyathan from Rowland's Reef near Lida, Nevada. **C.** The coiled mollusk fossil *Oelandiella*. **D.** The long conical shell of the possible mollusk *Hyolithes*. **E.** The living inarticulate brachiopod *Lingula*, with the long fleshy pedicle for digging and the tongue-shaped phosphatic shells. **F.** The primitive echinoderms known as eocrinoids were predecessors of the abundant crinoids of the later Paleozoic. This is *Gogia* from the Middle Cambrian Wheeler Shale, showing part of the stalk, the head, and the arms flaring out. **G.** The peculiar spindle-shaped fossils known as helicoplacoids apparently lived with their points stuck in the mud, and their food grooves ran in a spiral around them.

Siberia and Nevada. Then, mysteriously they vanished at the end of the Cambrian, to be replaced by sponges and eventually by corals, although there were no equally large reefs until the Silurian.

Shelly Fossils

Although simple limpet-like mollusks evolved in the "little shellies" (Fig. 9.12) and there are a handful of archaic mollusks known in the Cambrian (Fig. 10.13C, D), the most abundant shelled invertebrates of the Paleozoic were not mollusks but the **brachiopods** (Fig. 10.13E). As discussed in

Appendix A, the brachiopods (or "lamp shells") are nearly extinct today, but they were the most common fossils of the entire Paleozoic. The earliest brachiopods appeared in the first stage of the Cambrian (long before the trilobites), and they are the most primitive ones in the group. Known as the **lingulids**, these brachiopods have very simple cap-shaped or dish-shaped shells that are held together with bands of muscle, rather than the teeth-and-socket hinges of more advanced brachiopods. They have a long fleshy stalk (called a pedicle) for probing and digging down deep in the mud, and they still live today in muddy bottoms where they burrow and filter-feed. Finally, they are unusual among marine invertebrates, most of which make their shells out of calcite (calcium carbonate), which is easily extracted from seawater. Instead, lingulid brachiopods are one of the few invertebrates that make their shells out of calcium phosphate, the same material of which vertebrate bones and teeth are made. The reason for this odd chemistry is not known, but some scientists think that the earliest Cambrian oceans did not yet have enough oxygen or the right seawater chemistry to form large calcite shells, but it was a good chemistry for phosphatic shells like the "little shellies" and the earliest shelled brachiopods of all, the lingulids. This might also explain why the first appearance of trilobites and most other shelly fossils was delayed until partway through the Cambrian.

The lingulid brachiopods were not the only shelled invertebrates in the Cambrian. As mentioned, there were cap-shaped, limpet-like mollusks and some of the earliest relatives of clams (Fig. 10.13C, D). Another group with hard shells is the earliest echinoderms, the phylum that today includes sea stars, sea urchins, and their relatives (see Appendix A). Cambrian echinoderms, however, are members of extremely primitive groups that mostly vanished before the end of the Cambrian. One of the more common forms are the eocrinoids, which have a hard outer shell made of calcite plates and a number of long arms for filter-feeding like the more advanced crinoids that evolved later but are otherwise extremely primitive (Fig. 10.13F). Another group is the strange spindle-shaped echinoderms known as helicoplacoids, which were built in a spiral arrangement and apparently lived with their pointed ends stuck in the mud (Fig. 10.13G). The main trend that summarizes the hard-shelled marine life of the Cambrian is that most of them are *archaic, experimental forms within each phylum*, most of which vanished at the end of the Cambrian and were replaced by more advanced members of the same phylum.

These are the Cambrian animals that we have long known about because they had hard shells and left a good fossil record of their evolution. But recent research has emphasized the amazing discoveries of a handful of fossil localities that also preserve animals with soft bodies that left no fossilizable shells behind. The first of these to be discovered and studied was the famous Middle Cambrian (508 Ma) **Burgess Shale** locality, in the Canadian Rockies near Field, British Columbia. Since the rediscovery and reinterpretation of the Burgess Shale fauna, more of these soft-bodied faunas have been found, including the late Early Cambrian (520 Ma) Chengjiang fauna of China and the latest Early Cambrian (518 Ma) Sirius Passet fauna of Greenland. All of these amazingly preserved fossils from these faunas have been a revelation about just how diverse in body form Cambrian life really was. We would never know this fact if we had only the animals that left easily fossilized hard shells, which give a poor representation of the true diversity of forms that once evolved. Not only do these faunas produce abundant shrimp-like arthropod crustaceans, hard-shelled trilobites, sponges, and other

familiar forms but there are some that are truly bizarre. *Opabinia* (Fig. 10.14) from the Burgess Shale, for example, had a long, segmented body; five large eyes in the middle; and a nozzle-like protuberance in front. The aptly named *Hallucigenia* looked something like a worm but with paired spikes on its back (Fig. 10.14). When it was first found, it was reconstructed upside down, but now there are enough similar fossils that we know it is related to a living group, the "velvet worms" (phylum Onychophora). The dominant predator was the alien-looking creature *Anomalocaris*, which was almost 1 m long and the largest animal known in the Cambrian (Fig. 10.14). Its long body had multiple flap-like lobes to propel it while swimming, and it had two large mouthparts for grabbing prey and a circular mouth with sharp teeth that closed like a camera diaphragm and chopped pieces off its prey. In fact, there are trilobites with bite marks that match *Anomalocaris*, suggesting that it was the most feared creature of the Cambrian.

In recent years, paleontologists have noted that these rare windows into life, especially the organisms that are soft-bodied and rarely preserved, force us to rethink the entire radiation of animal life in the Cambrian. Instead of the slow, steady, progressive diversification of life from the beginning, paleontologists like the late Stephen Jay Gould have argued that the Burgess Shale shows just how many bizarre and experimental body plans first appeared in the Cambrian, many of which represent whole new phyla that are no longer around. Thus, the Cambrian represents a broad burst of diversification into many experimental body forms, most of which are pruned down by evolution and never even survived into the later Paleozoic. According to Gould, if you were to time-travel back to the Cambrian, you could

Figure 10.14 Reconstruction of the Burgess Shale animals. At the top is the predator *Anomalocaris*, with the long, jointed appendages in front of the circular mouth. Tall sponges are visible on the left, along with numerous trilobites. The strange creature in the center with the five eyes and the nozzle is *Opabinia*. The spiky little "worm" to the left is *Hallucigenia*, and to the left of it is the spiky armored *Wiwaxia*. The large worm emerging from the burrow in the right foreground is *Ottoia*. To the left of it with the antennae is the tiny fish relative *Pikaia*.

not anticipate which phyla would be the dominant groups 500 million years later and which would be extinct. Most of the Burgess Shale creatures that were common and successful back then are no longer around, while rare unimpressive forms like the earliest fish and their relatives are also found there—and they are the ancestors of all backboned animals alive today, including us. Gould's point is that evolution is not a steady, progressive march to greater and greater success and diversity but a game of chance ("contingency") where the winners and losers of the extinction game cannot be predicted and random events like the impact of an asteroid from space could potentially wipe out groups (like the dinosaurs) that had been hugely successful for a very long time during normal earth conditions. As Gould puts it, if we were to rewind the history of life as a long videotape and play it back again, there would be a different result every time because life is not predictable and random accidents and unexpected events change the ending of the story every time.

The last part of the Cambrian was marked by a series of extinction events that hit many trilobite groups particularly hard. The causes are still debated, but there is strong geochemical evidence that the oceans were becoming more and more stagnant and depleted in oxygen, which may have triggered the Late Cambrian extinctions. Then the geochemical data show a sudden change about 500 Ma, when the conditions reversed and the oceans were flooded with oxygen. Some geochemists estimate that the atmospheric level of oxygen was as high as 30%, compared to the 21% oxygen we breathe today. (The rest of the air is nitrogen gas.) On top of that, carbon dioxide levels reached very high values as well, making the entire planet warmer and more tropical than any "greenhouse planet" conditions that had ever been seen before. As we already saw, most of the Ordovician was dominated by very high sea levels, creating enormous areas of shallow carbonate shoals across most of the continent, providing a lot of shallow sea-bottom habitat for life to diversify. This rapidly changing world triggered not only extinctions among the archaic Cambrian groups but a whole new world of animals in the Ordovician.

ORDOVICIAN INVERTEBRATES

The Cambrian fauna (dominated by trilobites and archaic, experimental groups among the mollusks, brachiopods, echinoderms, and the bizarre experiments in the soft-bodied fauna) ruled the earth for at least 40 million years. But in the latest Cambrian and especially in the Early Ordovician, archaic Cambrian creatures were replaced by a huge evolutionary radiation of more advanced invertebrates that pushed most of their predecessors into the background or completely into extinction. This event is known as the "**Great Ordovician Biodiversification Event**" (GOBE), when hundreds of new genera and species appeared, the number of orders of marine invertebrates doubled, and the number of families tripled. There was a small diversity increase in the Early Ordovician, but the big radiation of new forms didn't really get going until the Middle and Late Ordovician.

 See For Yourself: Ordovician Life

The most important feature of the GOBE is that invertebrates evolved many new forms that exploited niches that had not yet been occupied in the Cambrian. Whereas the Cambrian was a time of experimental forms that first tried lots of unique body plans, the Ordovician marine life "filled out" all the underutilized resources and habitats in the sea floor. For example, the Cambrian fauna is dominated by mud-grubbing trilobites and lots of soft-bodied worm-like forms that burrowed but not many other ecological types. What kinds of animals evolved in the Ordovician to replace them?

After the archaeocyathan reefs vanished in the Middle Cambrian, there were no great reefs through the rest of the Cambrian and few large reef-building organisms. Into this void, a wide variety of reef-building organisms evolved by the Late Ordovician. The most important of these were the corals (see Appendix A). There were two main groups of corals that dominated the Paleozoic, both of which died out at the end of the Permian. One of them was the **tabulate corals**, whose individual tubes held little sea anemone–like creatures known as polyps. Some of the tabulates had their tubes holding the polyps packed closely together like honeycomb, so their nickname is the "honeycomb coral" (formally known as *Favosites*) (Fig. 10.15A). The other main group of corals was shaped like a long conical shell, on the top of which once sat a large anemone-like polyp. Because of their shape, they are known as **horn corals**, or rugosans. By far the most common horn coral in the Late Ordovician was a distinctive fossil known as *Streptelasma* (Fig. 10.15B).

But the reefs were built of not just corals but sponges as well. The most common were the layered sponges known as **stromatoporoids** (Fig. 10.15C, D). Their fossils look somewhat like a stack or mound of layered calcite, often with distinctive features on the top surface that looked like little bumps or pimples, called mamelons. The most common reefs were built not by corals but by a distinctive fossil known as *Ischadites* or *Receptaculites* (Fig. 10.15E). In some Late Ordovician localities, these fossils form dense reefs and massive structures that mimic coral reefs. For years, the collectors nicknamed them "sunflower corals" because their top surface resembles the spiral patterns found in a sunflower head. But recent research shows they are not corals or even sponges but a structure formed by a group of algae.

Living in and around these coral–stromatoporoid–receptaculitid reefs were many smaller groups of marine invertebrates. Most important of these were the two groups of filter-feeders that use a feathery structure called a lophophore to trap tiny food particles in the water (see Appendix A). The most abundant and important lophophorates were the brachiopods, or "lamp shells." We saw the very primitive lingulid brachiopods of the Cambrian (Fig. 10.13), which had simple cap-shaped phosphatic shells and still survive in muddy bays around the world. By the Ordovician, much more advanced brachiopods had evolved. These all had durable calcite shells, which had tooth-and-socket connections in the hinge area between their two shells (rather than the simple strands of muscle that lingulid brachiopods have

Figure 10.15   Typical Ordovician fossils. **A.** The "honeycomb" coral *Favosites*. **B.** The horn coral *Streptelasma*. **C, D.** Stromatoporoid sponges in top view showing the bumps known as mamelons (**C**) and the layered structure in side view (**D**). **E.** The "sunflower coral" *Receptaculites*, a common reef former in the Ordovician, was actually formed by a large colonial alga. **F, G.** The "D"-shaped brachiopods known as strophomenides, here exemplified by *Rafinesquina* shown in dorsal and ventral views and in life position. **H.** *Hebertella*, a typical orthide brachiopod, with the fine ribs and straight hinge. **I.** The massive bryozoan *Constellaria*, with the hundreds of tiny pinprick-sized holes that housed individual animals. **J.** A long-stalked modern crinoid, showing the umbrella-like fan of arms curled back into the current for feeding. **K.** The huge Ordovician snail *Maclurites*, with its oddly-shaped shell. **L.** The giant "snowplow" trilobite *Isotelus rex*. **M.** The tiny "lace collar" trilobite *Cryptolithus*. **N.** The multisegmented trilobite *Flexicalymene*.

in their hinge). Brachiopods exploded not only in diversity but also in abundance, so in many Ordovician localities the ground is literally paved with hundreds of their shells and you cannot avoid stepping on them. In the Upper Ordovician rocks in the Cincinnati Arch, you commonly find the "D-shaped" brachiopods known as strophomenides (Fig. 10.15F, G). These creatures had one shell that was concave and one that was convex, so they nested inside one another like a pair of saucers. They laid on the sea bottom with the concave side of the shell facing down and opened their shells just enough to allow the current bearing food and oxygen to filter through the inside. The second characteristic Ordovician group of brachiopods is the orthids, typified by fossils such as *Hebertella* (Fig. 10.15H) and *Resserella*. Orthids typically had a long straight hinge, a convex shell with lots of fine ribs radiating away from the hinge area, and a fold in the center of the shell.

The other major phylum that feeds with a lophophore is the **bryozoans**, or "moss animals" (see Appendix A). They are the only new phylum to appear in the Ordovician since they are unknown in the Cambrian. Individual bryozoan animals are extremely tiny and never fossilized, so they leave behind fossils that are covered by hundreds of tiny pinprick-sized holes where the animals lived (Fig. 10.15I). Most Ordovician bryozoans formed massive colonies that supported hundreds of little animals, although many formed interesting branching shapes that raised the little bryozoan animals high above the sea floor and let them feed in currents well above the bottom. They formed dense masses not only in isolated patches on the sea floor but especially in the crevices of the coral–stromatoporoid reefs.

But the tallest creatures of the Ordovician sea floor were the **crinoids**, or "sea lilies" (Fig. 10.15J). These are one of the five major living groups of the phylum Echinodermata, related to sea stars, brittle stars, sea urchins, and sea cucumbers (see Appendix A). Although they look superficially like a flower, they are an animal that attaches to the sea bottom

with a long, rooted stalk up to 3 m (10 feet) long. The "head" of the crinoid has a mouth in the center, surrounded by arms that formed an umbrella-shaped fan and trapped food particles as they flowed through the fan. With their long stalks, crinoids could feed well above the sea bottom and exploit a source of food that was unavailable to any animal that had ever lived up until that time.

The mollusks also diversified in the Ordovician. Archaic snails were particularly common, including the huge snails known as *Maclurites* (Fig. 10.15K), which carried their heavy coiled shells with the point over the top of their heads, rather than pointed backward and sideways as in modern snails. Clams were also found in the Ordovician, although they were rare compared to the hugely abundant brachiopods.

But the most impressive example of mollusk evolution in the Ordovician was the radiation of the cephalopods, the group that today includes octopus, squid, cuttlefish, and the chambered *Nautilus*. Recall that the largest predator known from the Cambrian was the meter-long, soft-bodied *Anomalocaris*, not something that would frighten anything larger than a small trilobite. By the Ordovician there were much bigger, more formidable predators, some of which resembled gigantic squid-like creatures living in conical shells up to 10 meters (33 feet) long (Fig. 10.16). These giant creatures were called orthocone nautiloids, distantly related to modern *Nautilus*, only their shells grew as a long, straight cone that was counterweighted on the inside to allow it to float or rest horizontally or lying on its side. The giant ones must have had huge parrot-like beaks in their mouths (as modern squids and octopus do) and long tentacles that could reach many meters from their bodies, so no trilobite or any other prey animal was safe from their suckered grasp or crushing beak.

 See For Yourself: Giant Orthocones

These are all the new groups of marine invertebrates that make up the bulk of the GOBE. Meanwhile, what happened

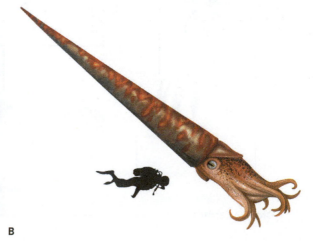

Figure 10.16 **A.** Diorama of the Ordovician seafloor, featuring large nautiloid. (Photo by the author) **B.** The size of the largest orthocone nautiloids, the largest predators the earth had seen up to this time.

to the survivors of the Late Cambrian extinctions? Many of those typically archaic or experimental Cambrian groups (especially the soft-bodied kinds from the Burgess Shale and similar places) vanished or were replaced by more advanced members of their phylum. Thus, the primitive lingulids were replaced by advanced brachiopods, primitive echinoderms like eocrinoids were replaced by the crinoids, and simple mollusks were replaced by more advanced creatures like coiled snails and giant nautiloids. So too with the trilobites. Many of the archaic Cambrian trilobites vanished during the Late Cambrian extinctions. In their place were much more advanced and specialized trilobites that look much more different from each other than any Cambrian trilobites did. For example, there are the "snowplow" trilobites, such as *Isotelus*, which had a front cephalon that formed a smooth plow, and the pygidium was the same size and shape; their bodies had only a few thoracic segments (Fig. 10.15L). The biggest of these trilobites was *Isotelus rex*, almost 70 cm (28 inches) long and 40 cm (16 inches) wide, the largest trilobites ever found as a complete fossil. Another huge species, *Isotelus maximus*, is the official Ohio State Fossil. Their smooth front and back ends suggest that

they were mostly shallow burrowers that lived just below the surface of the sea floor, covered with a light dusting of sediment to hide themselves from huge nautiloid predators. Other distinctive Ordovician trilobites had similar anti-predatory devices. Some were tiny, like the little "lace collar" trilobites known as *Cryptolithus* and *Trinucleus* (Fig. 10.15M), which were about the size of a thumbnail and had a huge "lace collar" around the rim of their broad cephalon with a bulging "nose-like" glabella but an extremely tiny thorax and pygidium behind. Still others could roll up into an armored ball like the "pill bugs" or "roly-polies" that you might find in modern leaf litter today. The most famous of these were the calymenid trilobites, such as *Flexicalymene* (Fig. 10.15N), which had many thoracic segments to make them flexible and eyes that could allow them to see in any position, and some even had a little hook on their pygidium that allowed them to lock their defensive "ball" shut. In summary, the trilobites of the Ordovician all look very different from one another, even to the untrained eye, because each had different specializations that allowed it to hide from the newer, more efficient predators, especially the giant nautiloids.

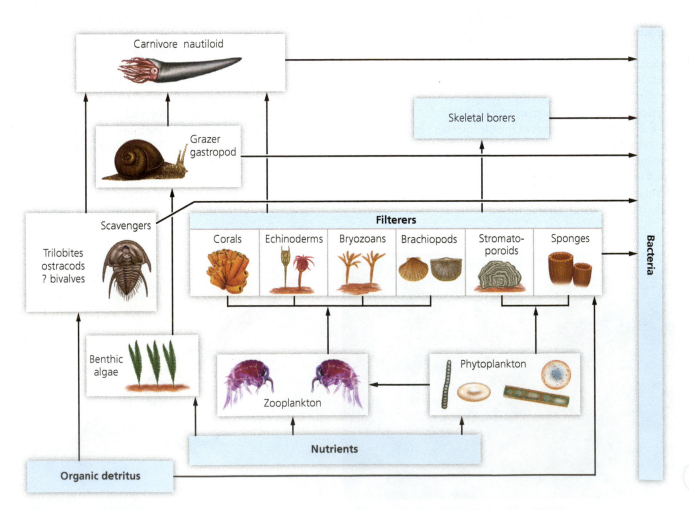

Figure 10.17 During the Ordovician, complex food webs emerged in the oceans for the first time. A large food base of algae and phytoplankton supported a diverse community of filter-feeders, including sponges (especially stromatoporoids), corals, brachiopods, bryozoans, and echinoderms. They also supported grazing snails like *Maclurites* and trilobites grubbing in the mud. At the top was the earth's first large predator, the orthocone nautiloids.

If you collect fossils in classic Ordovician localities such as those around Cincinnati, Ohio, and adjacent parts of Indiana and Kentucky, you will pick up dozens of brachiopods, lumpy massive bryozoans and branching bryozoans, corals, stromatoporoid sponges, and lots of broken fragments of crinoids (especially the pieces of their stems, which look like Life Savers candies). Now and then you get the huge snail *Maclurites* and some clams. The top predators, the nautiloids, are rare; but it's not unusual to find a broken segment of their long conical shells. In some localities, like the famous roadcut near Graf, Iowa, nautiloids occur in huge numbers packed like jackstraws. Trilobites are not nearly as common as they were in the Cambrian, but every once in a while one can get a good *Isotelus* or *Cryptolithus* or *Flexicalymene*.

This enormous diversification of different new groups implies a whole new ecology for the shallow sea floor. Instead of the relatively simple world of trilobites and worms that dominated the Cambrian, we see a **complex food web** in the oceans for the first time (Fig. 10.17). The enormous areas of shallow tropical seas, warm conditions, and high levels of atmospheric oxygen and carbon dioxide meant a huge bloom of plankton at the base of the food chain. Indeed, this is one of the highest times of diversification of the planktonic algae whose resting spores are known as acritarchs (Fig. 9.10), suggesting that the oceans were loaded with microscopic food, primarily planktonic organisms. The richness of the food supply in the water was exploited by a huge diversity of filter-feeders, from the abundant brachiopods to the reefs crowded with corals, stromatoporoid sponges, bryozoans, crinoids, and many others. Trilobites were still doing their jobs filtering the mud for food particles, and the algae-eating snails like *Maclurites* were abundant as well. Topping it all was the superpredator that ruled the Ordovician world, the huge nautiloids.

The Ordovician marine community was different not only in its ecological complexity, filling many previously unfilled niches, but also in the level at which they fed. Back in the Cambrian, there were no really tall marine creatures but simply mud-grubbing trilobites living just on the surface and some shallow burrowing worms (Fig. 10.18). By the Late Ordovician, however, animals had evolved that could feed at multiple levels above and below the sea floor. Not only were there deeper burrowers, but there were tall corals, sponges, and bryozoans. Tallest of all, however, were the crinoids, whose long stalks allowed them to reach half a meter or more above the sea floor, harvesting food in the upper part of the water column that no bottom-hugger could reach. This evolution of more complex, multilevel feeding is known as **tiering** and represents another dimension in the complex exploitation of the ecospace that Ordovician animals pioneered.

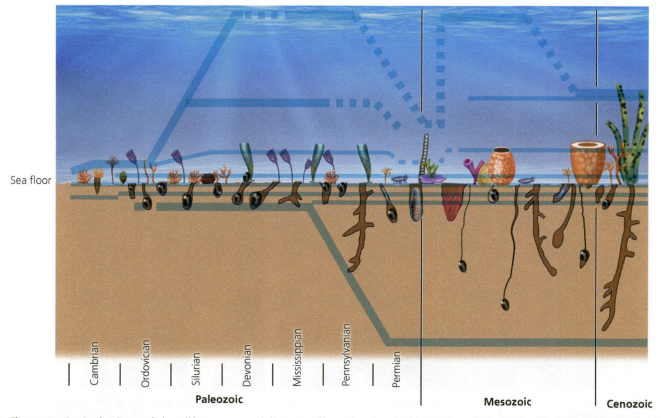

Figure 10.18 In the Precambrian, life was essentially a two-dimensional mat of bacteria and algae. By the Ordovician, there were numerous tall filter-feeders like crinoids, sponges, and corals that could reach water currents far above the sea floor and numerous burrowers which reached food sources deeper and deeper in the sediment. This exploitation of resources far above and below the sea floor is called "tiering" and increased steadily after the Ordovician.

Early Vertebrates and Their Relatives

So far, we have focused on just the invertebrates, which mostly lived in and around the sea bottom. But what about our own relatives, the vertebrates? When did the first fish appear? For the longest time, there were no teeth or scales or bone fragments of any fish known before the Silurian, when the first good fish fossils appeared. Then, bit by bit, little pieces of bone that represented fish body armor began to appear in the fossil record, suggesting they were around in the Ordovician. In the late 1990s, Chinese scientists looking at the Early Cambrian soft-bodied faunas like that at Chengjiang found faint impressions of **jawless fish** that had no bony scales or jaws (Fig. 10.19A). Most were the size of a goldfish, with a simple slit-like mouth and well-developed gill slits; and the only skeletons they had were made of cartilage, just like you have between your joints. These simple jawless fish became more and more diverse so that by the Late Ordovician there are quite a variety of shapes. By that time and especially during the Silurian, we see the first fish with bony armor (Fig. 10.19B) covering most of their bodies. They still had simple slit-like mouths without jaws, but they had bony plates that covered nearly their entire bodies (even the tail in some), with simple fin-like protuberances on the side. These primitive jawless fish were not very good swimmers, so they could probably swim about as efficiently as a tadpole. But their armor may have been necessary in a world with predators like the huge nautiloids.

Finally, we mentioned the numerous algae and invertebrate larvae that formed most of the plankton. But there was another group that floated across the open ocean, which for over 200 years have been known from Ordovician black shales, called the **graptolites**, whose name literally means "writing on stone" (Fig. 10.20). These fossils look like miniature saw blades made of flattened smears of graphite, almost as if they had been drawn with a soft pencil. But they are extremely abundant in Ordovician and Upper Cambrian rocks, and they appeared to evolve so rapidly that as early as the 1830s geologists were using them as the principal fossil to tell the precise age of Ordovician beds around the world. Still, no one knew what animal made them until the 1940s when uncrushed three-dimensional specimens were found in limestones and cherts. These were carefully extracted or sliced into serial sections to recover their detailed structure. Lo and behold, the mystery of the graptolites was solved! Their detailed structure proved to be identical to that of a group of animals floating in the oceans today, known as the pterobranchs. Along with acorn worms, pterobranchs are the primary members of the phylum Hemichordata, which are closely related to vertebrates (see Appendix A). So these mysterious graphite markings on shales turned out to be

Figure 10.19   A. One of the earliest fish relatives, a small, soft-bodied, jawless fish from the Cambrian of China known as *Haikouichthys*. (Courtesy Wikimedia Commons) **B.** Reconstruction of the earliest fish fossils such as *Haikouichthys* and *Haikouella*. (Courtesy N. Tamura). C. By the Ordovician and Silurian, some jawless fish were completely encased in armor from head to tail. This is *Athenaegis* from northern Canada.

Figure 10.20   A. Typical Ordovician graptolites, such as this *Diplograptus*, had two rows of cups on each side of the main stalk. (Courtesy Wikimedia Commons). **B.** Graptolites apparently lived in large colonies which hung down from floats or driftwood on the ocean surface and included double-sided *Diplograptus* (right) and the more primitive bushy *Dictyonema* (left).

among our closest relatives in the Ordovician (other than true vertebrates like the jawless fish). They were apparently made of large colonies with tiny filter-feeding animals like modern pterobranchs, which floated across the surface of the ocean attached to driftwood or seaweed or to their own little raft of bubbles (Fig. 10.20B). As they floated across the surface waters around the world, they evolved very quickly, so their worldwide distribution and rapid change made them the best index fossil of the Ordovician. Most importantly, when they died and sank to the bottom, they were fossilized in every marine setting, from shallow-water limestones to the deepest anoxic black shales where no shelly animals lived on the sea bottom, so they are powerful tools for correlations of rocks representing every environment.

Life on Land

Finally, we've focused entirely on the life in the ocean because that's where 99% of the fossil record of the early Paleozoic has been found. What about life on land? So far, there is little evidence that life on land in the Cambrian or the early part of the Ordovician was any more complex than existed on land through most of the Precambrian. The fossil record only provides evidence of simple lichens, fungi, and algae forming crusts on top of the soil (as seen in many barren regions on earth today). On the moist, well-watered shores of oceans and lakes, dense mats of algae were certainly thriving as well. But in the Middle Ordovician, we get the first glimmerings of more complex life. We see the evidence of spores of more complex plants (Fig. 10.21A), as well as fossils of sheets of plant tissue with complex cell structure much more advanced than simple algae (Fig. 10.21B). So there were some kinds of land plants by the later Ordovician, although we still don't have complete fossils of such plants until the Silurian (see Fig. 11.18).

But what about land animals? Again, the fossil record is very sparse, but it is apparent that no animals had managed to come out of the ocean and colonize the dry land through most of the Cambrian and Ordovician. But in Upper Ordovician rocks we find ancient soil horizons that have distinct burrows in them, very similar to those made by millipedes today. If this is true, then millipedes (one of the most primitive of all arthropods on the planet) were the first creatures to ever crawl out on and feed on its unexploited food resources of simple plants and decaying matter, just as they do today. And they did it more than 100 million years before the first amphibian left the water.

Late Ordovician Mass Extinction

The huge radiation of complex invertebrate communities that arose during the GOBE came to an abrupt crash near the end of the Ordovician, about 445 Ma. Almost the entire marine realm was hit with the second largest extinction event in earth history, bigger than the extinction that wiped out the dinosaurs at the end of the Mesozoic. About 50%–60% of the genera of marine invertebrates died out, and about 85% of the marine species on the planet vanished. The brachiopods and bryozoans were hit particularly hard, so about one-third of the families of both groups were wiped out, for a total of over 100 families of invertebrates. These include the characteristic Ordovician brachiopods (like the D-shaped strophomenides and the orthids), as well as nearly all the massive and branching bryozoans (Fig. 10.12). Many of the typical Ordovician trilobite groups were affected as well, along with the corals, echinoderms, and many of the characteristic Ordovician graptolites.

See For Yourself:
Ordovician Ice Age

Lots of explanations have been proposed for this tremendous catastrophe, including gamma ray bursts in space, metal poisoning, meteorite impacts, and huge volcanic events. The

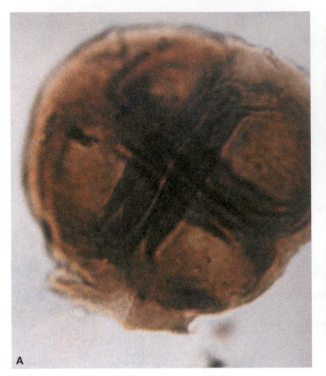
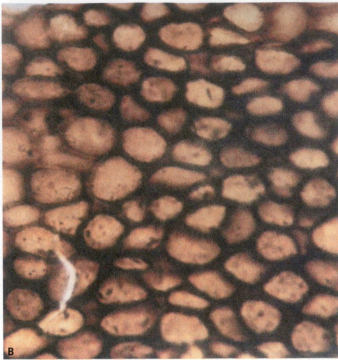

Figure 10.21   Fossils of earliest land plants. **A.** A four-part spore from the Late Ordovician of Libya. **B.** The cellular structure of the vascular tissues from an Ordovician plant stem.

best evidence, however, comes from Africa and South America, which were sitting over the South Pole at the time. Both continents have extensive Upper Ordovician glacial deposits, which suggest at least five separate pulses of glaciation and deglaciation. This was a culmination of a long-term cooling trend near the end of the Ordovician. This was the first time the planet had experienced glaciation since the end of the Late Proterozoic "snowball earth" event, and it brought the "greenhouse world" conditions of the Early Paleozoic to an abrupt close. Not only did the oceans become dramatically cooler but the carbon dioxide levels dropped, and the expansion of glaciers caused sea level to lower and exposed huge areas of shallow-marine shelf habitat to erosion, so the habitable area of shallow sea floor was greatly reduced. In many places, black shales and geochemical evidence suggest that the ocean bottom became anoxic and poisoned by hydrogen sulfide, which killed off any animals that lived then.

These conditions were particularly hard on the shallow-marine species and the plankton, but the species that lived in deep, cold water did fine, with many spreading widely over the Early Silurian seas that were colder and depleted in most of their species. Indeed, most of the Early Silurian marine animals are opportunistic "weedy" species, generalists that could live in any habitat. As in many other cases, such generalists tend to quickly spread around the world in the aftermath of a mass extinction after all the highly specialized species have been wiped out by the changing environment, leaving lots of open habitats on the sea floor. Throughout the Early Silurian, the marine communities remained less complex, with broader ecological niches and worldwide distributions. The highly distinctive local biogeographic provinces of the Ordovician virtually disappeared as all their hyperspecialized localized species vanished.

RESOURCES

BOOKS AND ARTICLES

Bird, J. M., and J. F. Dewey. 1970. Lithospheric plate-continental margin tectonics and the evolution of the Appalachian orogeny. *Bulletin of the Geological Society of America* 81:1031–1060.

Bjerrum, C. J. 2018. Sea level, climate, and ocean poisoning by sulfide all implicated in the first animal mass extinction. *Geology* 46:575–576.

Brannen, Peter. 2017. *The Ends of the World: Volcanic Apocalypses, Lethal Oceans, and Our Quest to Understand Earth's Past Mass Extinctions.* Ecco, New York.

Bruton, D. L. (ed.). 1984. *Aspects of the Ordovician System: A Handbook.* Universitets-Forlaget, Oslo, Norway.

Erwin, D. H., and James M. Valentine. 2013. *The Cambrian Explosion: The Construction of Animal Biodiversity.* W. H. Freeman, New York.

TIMELINE OF EARLY PALEOZOIC EVENTS

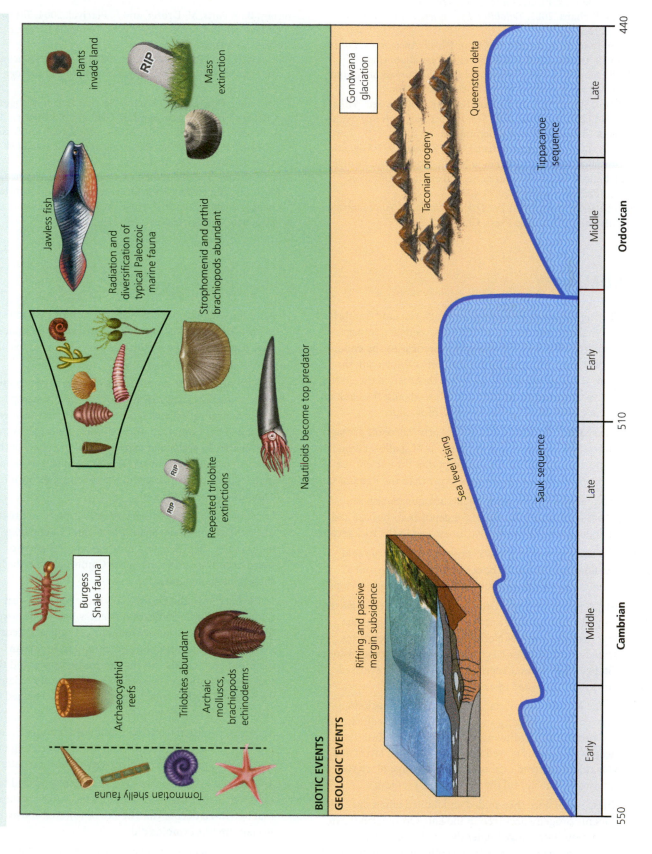

BIOTIC EVENTS

Tommotian shelly fauna

Archaeocyathid reefs

Trilobites abundant

Archaic molluscs, brachiopods echinoderms

Burgess Shale fauna

Repeated trilobite extinctions

RIP   RIP

Nautiloids become top predator

Strophomenid and orthid brachiopods abundant

Radiation and diversification of typical Paleozoic marine fauna

Jawless fish

Plants invade land

Mass extinction

RIP

GEOLOGIC EVENTS

Rifting and passive margin subsidence

Sea level rising

Sauk sequence

Taconian orogeny

Queenston delta

Tippacanoe sequence

Gondwana glaciation

550					510			440
Early	Middle	Late		Early	Late	Middle	Late	
	Cambrian				**Ordovician**			

Fisher, Donald W., and Stephen L. Nightingale. 2006. *The Rise and Fall of the Taconic Mountains: A Geologic History of Eastern New York*. Black Dome Press, New York.

Fortey, Richard. 2001. *Trilobites: Eyewitness to Evolution*. Vintage, New York.

Foster, John. 2014. *Cambrian Ocean World: Ancient Sea Life in North America*. Indiana University Press, Bloomington.

Gould, Stephen J. 1989. *Wonderful Life: The Burgess Shale and the Nature of History*. W. W. Norton, New York.

Holland, Steven M., and Richard R. Davis. 2009. *A Sea Without Fish: Life in the Ordovician Seas of the Cincinnati Region*. Indiana University Press, Bloomington.

Huff, W. D., S. M. Bergström, and D. R. Kolata. 1992. Gigantic Ordovician volcanic ash fall in North America and Europe: biological, tectonostratigraphic, and event-stratigraphic significance. *Geology* 20:875–878.

Levi-Setti, Ricardo. 2014. *The Trilobite Book: A Visual Journey*. University of Chicago Press, Chicago.

MacLeod, Norman. 2015. *The Great Extinctions: What Causes Them and How They Shape Life*. Firefly Books, London.

Webby, B., F. Paris, M. Droser, and E. Percival. 2004. *The Great Ordovician Biodiversification Event*. Columbia University Press, New York.

Williams, H., and R. D. Hatcher, Jr. 1982. Suspect terranes and accretionary history of the Appalachian orogen. *Geology* 10:530–536.

Xian-Guang, Hou, David J. Siveter, Derek J. Siveter, Richard J. Aldridge, Cong Pei-yun, Sarah E. Gabbott, Ma Xiao-ya, Mark A. Purnell, and Mark Williams. 2017. *The Cambrian Fossils of Chengjiang, China: The Flowering of Early Animal Life*. Wiley-Blackwell, New York.

Zou, C., Z. Qiu, S. W. Poulton, D. Dong, H. Wang, C. Chen, B. Lu, Z. Shi, and H. Tao. 2018. Oceanic euxinia and climate change "double whammy" drove the Late Ordovician mass extinction. *Geology* 46:535–538.

SUMMARY

- During the latest Proterozoic through the Late Cambrian, the shallow-marine seas of the Sauk sequence transgression gradually inundated North America. At the end of the Cambrian and during the Early Ordovician, the Sauk seas covered nearly all of North America except the Trans-Continental Arch in the Midwest, and the Sauk sequence continued into the Early Ordovician.

- The shallow limestone seas of the Early Paleozoic are best understood by looking at the deposits of shallow carbonate platforms like the Bahamas, except that the ancient seas were tens to hundreds of times larger than the Bahamas.

- Sauk seas were dominated by mature, quartz-rich sandstones in the lower transgressive deposits at the base and introduced the first extensive limestones in the rock record.

- The Sauk transgression was probably a result of rapid sea-floor spreading as the continents broke up and a greenhouse world without any ice on the poles developed.

- At the end of the Early Ordovician, the Sauk seas regressed, only to be followed by another extensive transgression that drowned most of North America, the Tippecanoe sequence. Its basal unit is a supermature quartz sandstone (St. Peter Sandstone), representing extremely reworked dune sand deposited in a nearshore setting.

- In the Middle Ordovician, the Appalachian region underwent downwarping and filled with thick sequences of pre-orogenic turbidites and shales, or flysch. Into these deep-water basins flowed many volcanic ash layers and volcanic sand from offshore, as well as huge blocks of limestone that slid off the shallow shelf region.

- These rocks were then folded and faulted when the Taconic orogeny occurred in the Late Ordovician, forming a mountain range higher than the Himalayas. The Taconic orogeny was due to the collision of the Piedmont terrane across most of the Appalachians and Quebec.

- After the Taconic collision ended, enormous volumes of post-orogenic river and delta sandstones and shales were deposited as the mountains wore down, a typical post-orogenic molasse deposit.

- Cambrian seas were dominated by a huge diversification of trilobites. The major reef-formers were conical sponge-like creatures called archaeocyathans, which vanished by the end of the Cambrian. Most of the other phyla were represented by archaic or experimental members of their groups, such as cap-shaped mollusks and simple clams, inarticulate brachiopods, and weird echinoderms like eocrinoids and helicoplacoids.

- Localities with extraordinary preservation of soft tissues like the Burgess Shale, Sirius Passet, and Chengjiang faunas show that there was much greater diversity of weird experimental forms in the Cambrian than the hard-shelled fossils suggest.

- In the Early–Middle Ordovician, there was the Great Ordovician Biodiversification Event, when the seas were filled with an evolutionary radiation of typical Paleozoic invertebrates, like articulate brachiopods, bryozoans, tabulate and horn corals, stromatoporoid sponges, and the first diversification of crinoids. The dominant predators were huge straight-shelled nautiloids, which reached up to 10 m (33 feet) in shell length.

- Trilobites persisted from the Cambrian, but they became highly specialized for burrowing, hiding, or rolling up into a ball to protect themselves against new predators.

- The incredible Ordovician diversification produced the earth's first complex food webs and tiering of organisms that burrowed deeper than before or towered above the sea bottom like crinoids did.

- The Cambrian and Ordovician seas saw the first fish, although they were simple, small, jawless forms with only a small amount of bony armor. The surface waters were dominated by planktonic graptolites, which floated around the world's oceans and are good biostratigraphic index fossils in any facies, shallow-water limestones, or deep-water shales.

- Not until the later Ordovician do we see the first evidence of simple land plants, and the only animals

on land at the end of the Ordovician appear to be millipedes.

- The Late Ordovician was marked by a huge mass extinction, which wiped out much of the diversity of typical Ordovician marine invertebrates. The likely causes appear related to a brief series of Gondwana glaciations and stagnant oceans which were low in oxygen.

KEY TERMS

Sauk sequence (p.210)
Limestone environments (p.212)
Grapestone (p.214)
Ooids (p.214)
Mature sandstones (p.215)
Ridge volume and sea level (p.215)
Cratonic deposition (p.217)

Tippecanoe sequence (p.218)
Supermature sandstone (p.218)
Taconic orogeny (p.220)
Taconic klippen (p.222)
Queenston-Juniata molasse (p.222)
Piedmont terrane (p.223)
Trilobites (p.224)
Archaeocyathans (p.224)

Lingulid brachiopods (p.225)
Burgess Shale (p.226)
Great Ordovician Biodiversification Event (p.227)
Tabulate corals (p.227)
Horn corals (p.227)
Stromatoporoid sponges (p.227)

Articulate brachiopods (p.227)
Bryozoans (p.229)
Nautiloids (p.229)
Food web (p.231)
Tiering (p.231)
Jawless fish (p.232)
Graptolites (p.232)

STUDY QUESTIONS

1. What is the typical sequence of sedimentary rocks during the Sauk transgression?
2. Why are the transgressive deposits of the Sauk sequence oldest on the edge of the continent (Late Proterozoic in Death Valley and in the Appalachians) but much younger in the center of the continent (Late Cambrian in the northern Midwest)?
3. Describe the distribution of the typical limestone facies on the Bahamas Platform? Where do hardy corals grow? Where do the quiet waters full of lime mud and fecal pellets occur?
4. What are ooids? What kind of sedimentary conditions do they suggest?
5. What is the significance of mature sandstones in the Cambrian deposits?
6. What might have been the causes of the Sauk transgression?
7. How are cratonic deposits different in terms of thickness and continuity from the passive margin wedge deposits?
8. What is the Tippecanoe sequence? When did it begin to form?
9. How did sandstones like the St. Peter Sandstone become supermature, with 99% pure quartz and extraordinary rounding and sorting?
10. What do the deep-water deposits of the Middle Ordovician, such as the Normanskill and Martinsburg Shales, tell us about the tectonic events about to occur in the Appalachians?

11. What are the Taconic klippen? How were they first explained? What is their current interpretation?
12. What did the presence of volcanic ashes and volcanic sands from a mysterious land offshore of North America tell us about the approaching Piedmont terrane?
13. Compare the Late Ordovician Taconic Mountains to the Himalayas in terms of size and amount of deposits weathered off them.
14. What are trilobites? What did they do to feed?
15. How are the brachiopods of the Cambrian different from most later brachiopods in shell hinges and in shell chemistry?
16. What are the common trends seen among Cambrian mollusks, echinoderms, and brachiopods?
17. According to Stephen Jay Gould, what is the significance of the unexpected diversity and weird shapes of the animals from the Burgess Shale?
18. Which groups of marine animals dominated the Great Ordovician Biodiversification Event?
19. What was the most terrifying predator of Ordovician seas? What did the trilobites do in response to new predators?
20. How does the food web of the Ordovician compare to that of the Late Proterozoic or the Cambrian?
21. What is tiering? How did it increase in the Ordovician?
22. If you took a time machine to the Cambrian or Ordovician, what would you see on the land?
23. What are the suggested causes of the Late Ordovician extinctions?

Silurian and Devonian
444–355 Ma

We were fish in a wide and humming day,
No limbs on land, no left-hand brains,
The world was prehistoric:

With silvery springs and shimmery shoals,
Billowing reefs, rumbling stones,
Our unfolding jaws with flaps and bones.

No rings to wear, no words to own,
No fur, no quills, no left-hand poems,
No limbs on land, no right-side brains,

Just watery weeds and whirling winds,
Our shimmering scales and fluttering fins,
The shifting land, the rumbling stones,
Our unfolding jaws with flaps and bones.

—*Paula Weld-Cary, "Devonian Dream"*

Taughannock Falls in the Cayuga Valley near Ithaca, New York, was formed by hundreds of feet of Silurian and Devonian shales and sandstones shed from the rising Acadian Mountains.

11.1   Reefs, Limestones, and Evaporites

As discussed in Chapter 10, by the end of the Ordovician, all of North America (except in the Taconic mountain belt) was drowned by huge areas of shallow-marine limestones from the Tippecanoe transgression (Fig. 11.1). These conditions persisted throughout the Silurian and even into the Early Devonian (Fig. 11.2). North America still lay across the equator, although it had begun to rotate counterclockwise so that the modern west coast faced more northwest and the modern east coast faced southeast (Fig. 11.2). The rest of the world was also in the grips of a global transgression and greenhouse conditions; so once the brief glaciations of the Late Ordovician were over, the climate swung to the other extreme. The only areas not immersed by shallow tropical seas were the eroding areas of the old Taconic mountain belt and a few other mountainous areas on other continents (Fig. 11.2).

REEFS

The conditions in the shallow Tippecanoe seas of the Silurian were not identical to those of the Ordovician. There were many new features added to the seascape at this time, especially during the Silurian. The most striking change was the extensive growth of huge reef complexes, much bigger than the small patches of corals found in the Ordovician (Fig. 11.1). By the Middle Silurian and especially in the Devonian, there were some truly impressive areas of reef growth, as big as or bigger than the Great Barrier Reef of Australia today. Most of these reefs towered as high as 10 m (33 feet) above the sea floor, and individual reefs stretched for 3 km (2 miles) or more, although in a few places they spanned tens of kilometers and were as tall as 100 m (328 feet).

Reef complexes can be found in many places across the Tippecanoe seaway, just about any place where Silurian or Devonian limestones crop out at the surface. If you visit the Silurian rocks around the Midwest, you can find many ancient reefs that are now exposed in the walls of limestone quarries, such as the huge Thornton Reef in a quarry just outside Chicago. There are also many big quarries in reefs in Indiana, Michigan, Ohio, Iowa, and several other places. In fact, the reefs formed a ring-like barrier complex around the Michigan Basin (Fig. 11.3). If you walk through one of these limestone quarries, you can see the detailed structure of the reef in the quarry walls (Fig. 11.4). On the flanks of the reef, the beds dip gently away from the crest of the reef. The reef core, on the other hand, often is made of nothing but massive corals and sponges, with little or no visible bedding.

Even more interesting is the fact that reefs demonstrate a classic biological principle known as **ecological succession** (Fig. 11.5). You might remember succession from a previous science class. Succession is a predictable pattern of change from one ecological community to another. The most familiar example might be a freshwater pond in a forest, which is the **pioneer stage** of succession. In many cases, the pond fills with vegetation and so much sediment that it dries up, and then the **intermediate stage** would be low shrubs and grasses growing on the filled-in pond. These are eventually replaced by trees that get denser and taller until the **climax stage** is reached, which is represented by a fully mature forest. If the forest is cut down or burned down and the region flooded with ponds, the sequence of different communities can start all over again.

Normally, a rapidly changing ecological phenomenon like succession is hard to find in the fossil record, which typically represents

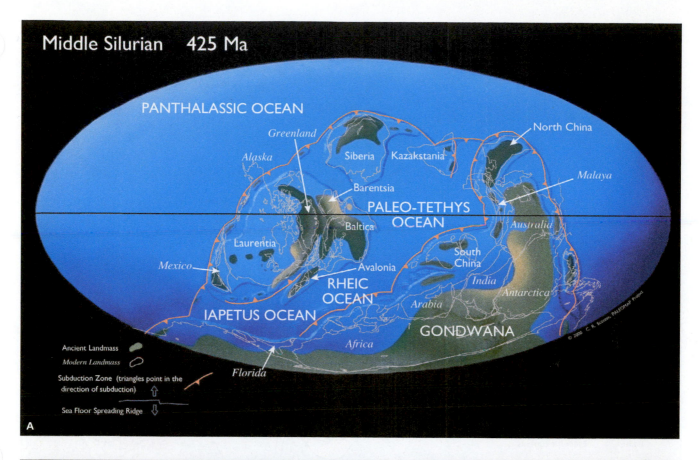

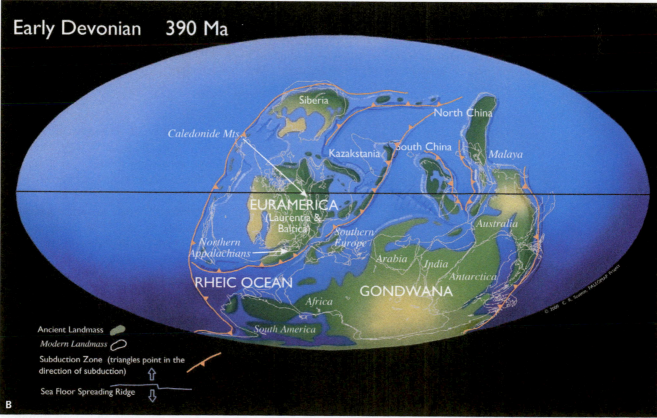

Figure 11.2   A. Paleogeographic map of the Middle Silurian, showing the Gondwana continent approaching North America and the collision of Baltica with northern Canada and Greenland to create the Caledonian orogeny. **B.** By the Devonian, the Caledonian Mountains were rising up to the sky, while the Avalon terrane that contained much of Europe plus the coastal part of the Appalachians was about to collide with North America.

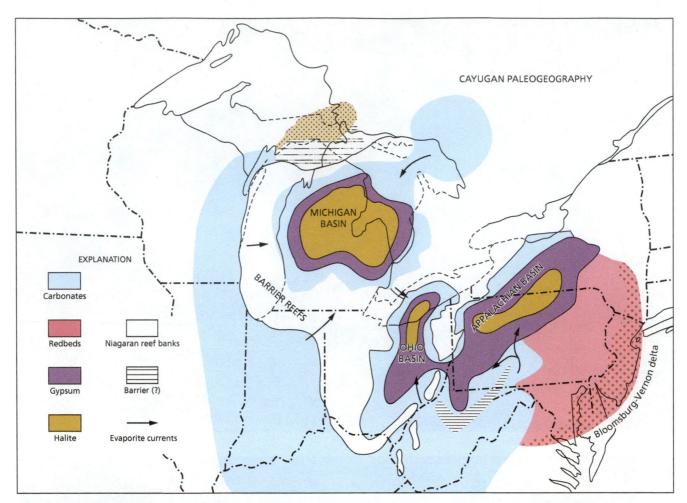

Figure 11.3 Paleogeographic map of reefs and evaporites in the Midwest during the Late Silurian (Cayugan Stage). To the east, the Bloomsburg-Vernon redbeds were molasse deposits shed off the slowly eroding Taconic Mountains.

Figure 11.4 Cross-sectional view of Mount Hawk Formation Devonian reef in Canadian Rockies, right on the Alberta–British Columbia border.

enormous amounts of time between each bedding surface. But reefs change in such a way that they grow over and preserve their earlier stages in the core of the reef, so the entire succession can be seen in the reef cross section. The pioneer stage (Fig. 11.5) is dominated by organisms that can stabilize the loose sea-floor sediments, such as crinoids that bind the sediment with their root-like holdfasts and small corals. As they continue to grow denser and flourish, they establish a hard surface, and larger corals and sponges can attach and grow. The climax of the reef community occurs when the masses of corals and stromatoporoid sponges are so dense that they can grow up into the surf zone and survive the pounding of waves and storms (Fig. 11.5). These walls of hard, dense, resistant reef-builders then protect and shelter the areas behind the wall, providing a quiet lagoon that can be colonized by crinoids, delicate bryozoans, brachiopods, trilobites, and other inhabitants of quiet sea bottoms with no wave disturbance.

You can see this succession preserved in the cross sections of ancient Silurian and Devonian reefs found in quarry walls (Fig. 11.5). The bottom of the reef core will be dominated by crinoids, small corals, brachiopods, and other pioneering animals that first stabilized the sea bottom. As the reef core grows on top of the fossils of its pioneers, the next layer might be dominated by corals and stromatoporoid sponges, until the uppermost layer of the core of the reef is entirely massive corals and sponges that represent the climax community. Meanwhile the rocks on the flank of the

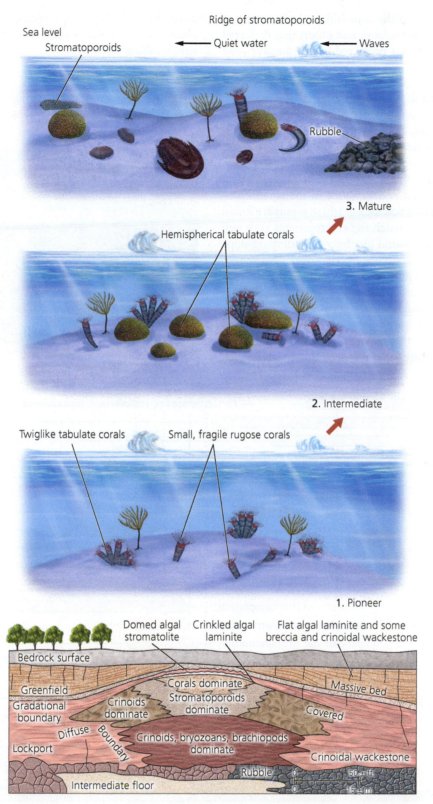

Figure 11.5   Cartoon of reef succession. At the top are three cartoons showing the pioneer stage, when sediment-stabilizing organisms such as crinoids, brachiopods, bryozoans, and small corals lived in the area and helped bind the loose sediment and make a hard surface. In the intermediate stage, additional colonizers including bryozoans and corals helped build a large marine community. In the climax stage (top), dense reef organisms like stromatoporoid sponges and certain corals lived on the top of the reef mound, where they could tolerate wave pounding, while the quiet flanks of the reef were colonized by crinoids, mollusks, and other types of corals. At the bottom is a sketch of the face of a quarry wall of an actual Silurian reef in Indiana, showing the shape of the reef and the different fossil communities found from the bottom to the top. The pioneer organisms are found in the bottom central part of the reef, followed above by stromatoporoid sponges and corals that form the wave-resistant crest of the reef in the climax stage. Meanwhile, the gently dipping flanks are made largely of crinoids and other organisms that lived in the quiet water below the reef crest.

reef dip gently away from the core and are built mostly of crinoids, brachiopods, bryozoans, and others that preferred the quiet waters of lagoons protected by the massive crest of the reef.

An even more impressive example is the huge Devonian reef complex exposed in Windjana Gorge in the Australian outback (Fig. 11.6). At one end, the wall of the gorge beautifully preserves the sloping beds on the flank of the reef. In the center are massive limestone beds dominated by the reef core corals and stromatoporoid sponges that built up the reef. And on the other side of the reef core are horizontal beds representing the quiet backwaters of the lagoon protected by the reef crest. If you look at which fossils are found in these exposures, they match the environments very well (Fig. 11.6B). The fossils found just below the dipping slope beds of the reef are mostly those of open-ocean swimmers, like fish, crustaceans, nautiloids, and ammonoids. The dipping reef-flank beds were inhabited by animals that could live on this steep sloping habitat, such as crinoids and brachiopods. The massive core of the reef is built entirely of corals and stromatoporoid sponges. Finally, the back-reef lagoon is full of creatures that required quiet waters, such

as brachiopods, bryozoans, snails, and clams, and in some places the lagoon was shallow enough for bacterial mats and stromatolites.

EVAPORITES

The ring of reefs around the Michigan Basin and parts of the Appalachian Basin in western New York (Fig. 11.3) created

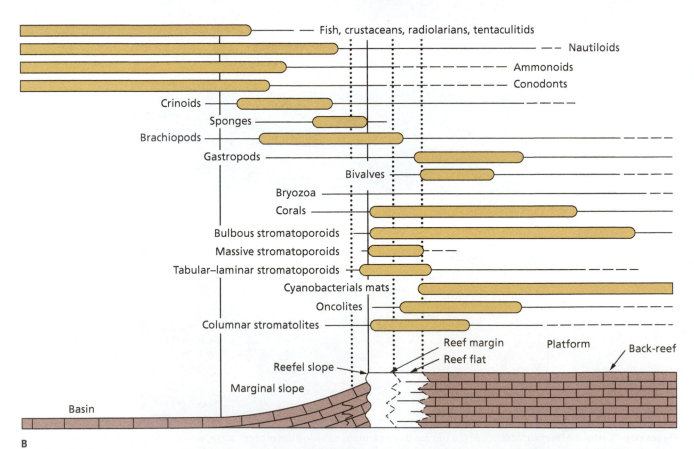

Figure 11.6 Windjana Gorge reef complex in Western Australia. **A.** Cross-section of the reef as exposed in the gorge. **B.** Plot showing the different organisms that dominated each part of the reef community. In the reef basin, the fossils were mostly organisms that swam above them in the surface waters: fish, crustaceans, nautiloids, ammonoids, conodonts, and some other plankton. On the reef slope, the dipping flank beds yielded organisms which could hang on to the slope, including crinoids, sponges, and brachiopods. The reef core was dominated by massive organisms that could tolerate the wave pounding, including corals, massive and tabular stromatoporoids, and even stromatolites. The quiet, flat-lying beds of the back-reef lagoon yielded organisms that liked clear, quiet water: gastropods, bivalves, corals, bulbous stromatoporoids, plus cyanobacterial mats, and stromatolites in the intertidal lagoon.

unusual conditions. These reef-rimmed basins were shallow tropical seas, and in many cases the reefs prevented the free flow of seawater into these **restricted marine basins**. As they filled up, they tended to evaporate their water faster than it could be replaced. Whenever seawater is evaporated without replacement, it becomes a concentrated salty **brine**. As the concentration of salts increases, eventually evaporite minerals such as salt and gypsum will precipitate out.

Such restricted marine basins occur today in the Persian Gulf between the Arabian Peninsula and Iran. There is very little freshwater coming into the Persian Gulf from the deserts of Saudi Arabia, Iraq, and Iran that surround the Gulf (especially since most of the flow of the Tigris and Euphrates Rivers is now diverted to irrigation). Most of the water comes in from the Indian Ocean via the Straits of Hormuz, where American ships have been protecting oil tankers for years. The surrounding area in Saudi Arabia and Iran is one of the hottest deserts in the world, so the rate of evaporation is enormous. Thus, the cold seawater flowing in from the Straits of Hormuz flows north, then evaporates away, leaving large areas of salt and gypsum, especially on the Arabian shore of the Persian Gulf.

You could think of the Michigan Basin or the Appalachian Basin near Syracuse during the Silurian as a restricted basin like the Persian Gulf. The region was situated in the subtropical desert evaporation belt in the middle Paleozoic, and the reefs rimming the basin restricted the flow of normal seawater. In the Michigan Basin alone, the thickness of salt and gypsum (which turns into pure calcium sulfate, or anhydrite, when it is buried) is over 750 m (about 2500 feet) thick. To get that much salt, you would need to evaporate a mass of seawater about 1000 km (about 620 miles) deep! Clearly, the sea was never that deep, and the fossils demonstrate that the basin was extremely shallow, with good light penetration for all the photosynthetic algae. Thus, there must have been continuous replacement of the seawater as it evaporated away over millions of years to pile up this much salt. Salts from the underground salt beds in Michigan have been commercially mined for many years, and the Syracuse Halite used to be one of the biggest sources of salt for New York and adjacent states. Thick evaporites also accumulated in the Silurian and especially Devonian of the Williston Basin in Montana and North Dakota (Fig. 10.7B). Today, salt is mined more cheaply in other parts of the world, but the Silurian salts of North America were an important historical source of these valuable minerals.

11.2 The Kaskaskia Sequence

The Tippecanoe transgression, which began in the Middle Ordovician, lasted through the entire Silurian and Early Devonian. It drowned most of the North American continent except for places in the Appalachian region that had been uplifted during the Taconic orogeny. It produced enormous volumes of limestone from the gigantic shallow-marine shoals that formed when shallow tropical seas flooded most of the continent.

But at the end of the Early Devonian, a major global regression occurred, and the deposition of the Tippecanoe sequence ended. Shortly afterward, the eroded surface in the Lower Devonian limestones was again flooded and buried in sediment. This next epicontinental sequence is known as the **Kaskaskia sequence**, and it persisted from 393 to 300 Ma, or from the Middle Devonian to the Middle Carboniferous (Mississippian–Pennsylvanian boundary). It was named after the battle of Kaskaskia in Illinois on July 4, 1778, where American troops under George Rogers Clark beat the British and their native allies in the westernmost battle of the Revolutionary War.

Like the Sauk and Tippecanoe sequences, the first units of the transgression are shallow nearshore beach sandstones, such as the Oriskany Sandstone and Ridgeley Sandstone of the Appalachian region (Fig. 11.7A), which are extremely clean and made of nearly pure quartz. These represent the first deposits as the sea floods the old eroded land surface. The sandstones are capped by extensive Middle and Upper Devonian limestones as most of the craton again became a giant shallow carbonate shoal, like the Bahamas but ten to a hundred times bigger. Thick successions of Devonian limestones, such as the Helderberg and Onondaga Limestones of New York State (Fig. 11.7B) and many different limestone units in the upper Midwest from Michigan to Iowa, dominated through the rest of the Devonian. Devonian limestones are even found as far west as the Rocky Mountains, where the Jefferson Limestone is a spectacular ridge former in Montana and Wyoming (Fig. 11.7C).

Just as the Sauk sequence did not end at the Cambrian–Ordovician boundary and the Tippecanoe sequence did not end with the Silurian–Devonian boundary, the Kaskaskia sequence did not terminate with the end of the Devonian but persisted into the Early Carboniferous, when the shallow carbonate seas reached their maximum depth and maximum extent across most of North America. We will discuss this further in the next chapter.

11.3 The Acadian Orogeny

Recall that back in the Ordovician, a huge volcanic arc complex we call the Piedmont terrane collided with the Appalachian margin during the Taconic orogeny. By the Late Ordovician, it had produced the Taconic mountain range that was 3000 m (almost 10,000 feet) taller than the Himalayas. That large mountain range was rapidly worn down, forming Upper Ordovician–Lower Silurian molasse deposits, such as the Queenston–Juniata Formations in the Ordovician of New York and Pennsylvania and the Tellico Sandstone and Moccasin Shale in North Carolina and Tennessee. The Taconic mountain range was still not completely eroded away by the Early Silurian, so we find additional Silurian molasse deposits of alluvial fans, river sands and gravels, and nearshore sands that represent the last gasps of

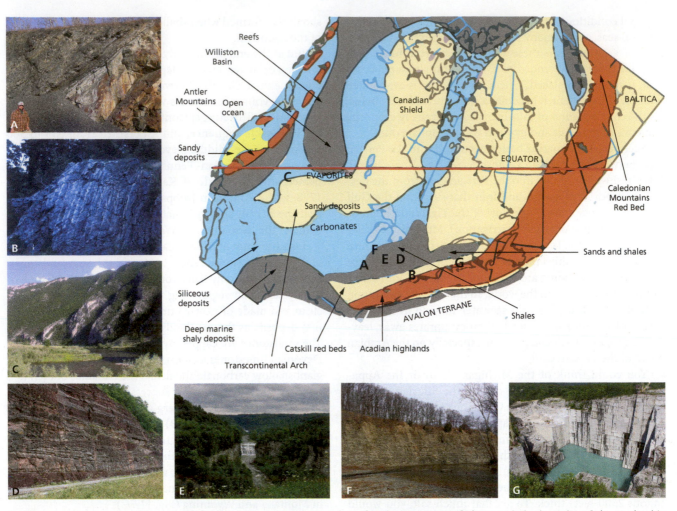

Figure 11.7 Paleogeographic map of the Devonian of North America, showing outcrops of characteristic deposits of the Kaskaskia sequence and the Acadian orogeny. **A.** Middle Devonian Oriskany Sandstone overlain by Needmore Shale, Virginia. **B.** Folded Middle Devonian Onondaga Limestone along the New York State Thruway, deposited in the Kaskakia sequence, then deformed during the Late Ordovician Acadian orogeny. **C.** Outcrop of Jefferson Limestone, Jefferson River Canyon, Montana. **D.** Cross-bedded river and floodplain sandstones and shales of the Catskill Group, formed in the alluvial fans and floodplains shed from the rising Acadian Mountains. **E.** Devonian sandstones and shales shed by the Acadian Mountain belt into central New York, here at Genesee Gorge, western New York. **F.** Cleveland Shale outcrops on the Rocky River, Fort Hill, Ohio. Outcrops like these are full of Devonian fossil fish, including the enormous placoderms like *Dunkleosteus* (see Fig. 11.15B). **G.** Rock of Ages Quarry near Barre, Vermont, a major source of "New Hampshire Granite," intruded during the final phases of the Acadian orogeny.

the eroding Taconic Mountains. These deposits include the Binnewater Sandstone and Shawangunk Conglomerate in the Silurian of eastern New York, the Tuscarora Sandstone in Pennsylvania, and the Grimsby, Whirlpool, and Albion sandstones in western New York and Ontario (forming the base of Niagara Falls). By the Middle Silurian, however, the last remnants of the enormous Himalayan-sized Taconic mountain range had been eroded away.

The erosional surfaces on top of the dying Taconic Mountains were then drowned by the shallow-water limestones of the Tippecanoe sequence during the Middle and Late Silurian and Early Devonian (Fig. 11.7). But the quiet limestone shoals in the Appalachian belt did not last for long. By the Middle Devonian, we find evidence in the rocks

in the Catskill Mountains and Hudson Valley of New York and up and down Appalachia, which suggest that another mountain belt collision is about to occur. The first piece of evidence is that Tippecanoe sequence limestones like the Lower Devonian Helderberg Group and the Onondaga Limestone of upstate New York are folded and highly deformed, especially in the Hudson Valley region to the east (Fig. 11.7B). These rocks are then overlain by pre-orogenic shales and sandstones in the Middle Devonian, such as the Marcellus Shale, which were also deformed by a collisional event (the Marcellus is one of the most important sources of oil and gas in the east today). Overlying them are Middle and Upper Devonian sandstones and shales across the entire east–west transect of New York State (Figs. 11.7 and 11.8)

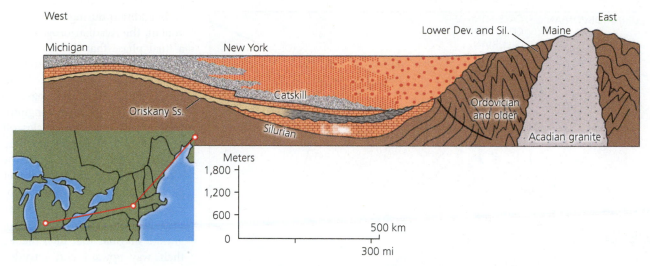

Figure 11.8 Diagram showing the facies changes from the rising Acadian Mountains as they shed sandstones and shales west across New York State.

which grade up into conglomerates and red beds of the Catskill Group (Fig. 11.7D) formed by alluvial fans and rivers at the eastern end (especially in the Catskill Mountains). The red beds and conglomerates are undeformed and represent post-orogenic molasse deposits that eroded off these uplifted mountains as they reached their climax in the Late Devonian. These terrestrial and nearshore rocks gradually change facies to deeper-water sandstones and shales as you go west toward Rochester and Buffalo, with spectacular exposures of these offshore shales in places like Genesee Gorge (Fig. 11.7E) and Taughannock Falls in the Cayuga Valley (Fig. 11.1).

Clearly, something collided with the Appalachian region in the Middle–Late Devonian to fold and deform all the Lower Devonian limestones, then shed huge volumes of sandstones and shales that gradually built westward and flooded the shallow Kaskaskia seas in places like western New York, Ohio, western Pennsylvania, and Michigan. This enormous wedge of clastic sediments is often called the Catskill sequence or Catskill clastic wedge because many of the best exposures are in the Catskill Mountains of eastern New York State (Figs. 11.7D and 11.8).

But New York is not the only region that shows evidence of collision. Similar-looking rocks occur in the east–west transect across Pennsylvania, such as the Mahantango and Harrell Formations of Maryland and Virginia. Middle and Upper Devonian sandstones and shales, such as the Brallier, Scherr, Foreknobs, and Hampshire Formations, can be found in the southern Appalachians, especially in Tennessee and West Virginia. Farther to the west, the Upper Devonian sequence is dominated by shales shed from this mountain-building event, such as the Antrim Shale in Michigan; the New Albany–Chattanooga Shale in Illinois, Indiana, and Tennessee; or the Cleveland Shale in northern Ohio (Fig. 11.7F). To the north, thick Devonian flysch and molasse deposits occur in Maritime Canada, especially the Gaspé Peninsula of Quebec and adjacent areas of New Brunswick.

Up in northeastern Canada, there are thick wedges of pre-orogenic flysch deposits that date back to the Late Silurian. And on the other side of the Atlantic, the Devonian was first recognized in Scotland and northern England by a famous rock unit, the Old Red Sandstone (Fig. 4.3), which is full of fossil fish that once swam in rivers shed on the other side of these rising Devonian mountains. There are also abundant Devonian red beds and shales as far north as Norway. So there must have been a huge collision event that shed red beds and shales and conglomerates on both sides of what is now the Atlantic. The first phase (between Scotland, Norway, Greenland, and northeast Canada) started in the Silurian and is known as the **Caledonian orogeny**. ("Caledonia" is the old Roman name for Scotland.) In the Devonian, it was followed in both England and the Appalachian region by the **Acadian orogeny**. This second mountain-building phase got its name after the early French colony of Acadia in what is now Quebec, New Brunswick, Prince Edward Island, and Nova Scotia, made famous by Henry Wadsworth Longfellow's poem "Evangeline." Many of the rocks of the Acadian orogeny are exposed there.

What crustal blocks collided to form the Caledonian and Acadian orogenies? From various lines of evidence, we know that the ancient continental core of northern Europe, the Baltic Platform or **Baltica**, had been approaching North America for most of the early Paleozoic (Fig. 10.2). This huge collision between Baltica and North America finally takes place in the Silurian, forming the Caledonian orogeny (Fig. 11.2). It caused most of the rocks of Scotland (on one side) to undergo high-grade metamorphism and deformation (such as the tilted rocks beneath the unconformity at Siccar Point, Fig. 1.7), then capped these deformed and altered rocks with sheets of river deposits of the Old Red Sandstone. The same thing happened to the rocks of Norway, Greenland, and northern Canada.

But what about the Acadian collision? It was not a Silurian event, like the Caledonian orogeny, but a Middle–Late

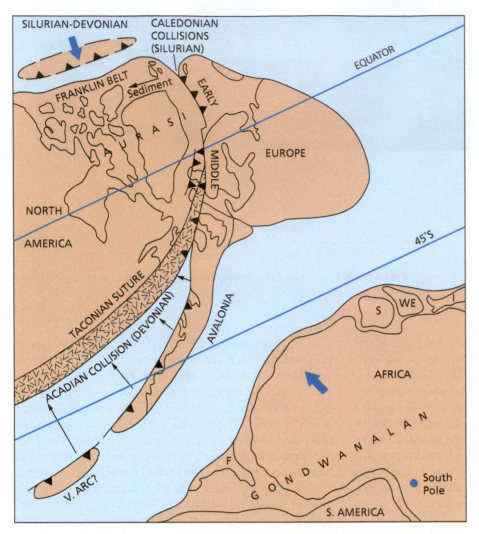

Figure 11.9 Tectonic terrane map of the Caledonian and Acadian orogeny, sowing the Avalon terrane colliding with the future Appalachian mountain belt, and Europe colliding with Greenland to produce the Caledonian orogeny. Notice how the coastal region from the Carolinas to Maritime Canada came from Avalonia, along with the southern part of the British Isles. As the proto-Atlantic Ocean closed, Gondwana was approaching rapidly and would collide in the later Carboniferous.

In addition, during the Late Devonian, the Acadian orogeny had a final phase that did not occur during the Taconic orogeny. As the collisional event finished and the Acadian Mountains began to erode down and shed conglomerates, sandstones, and shales off to the east and west, something else happened deep beneath the mountains. Apparently, the deeper parts of the crust began to undergo partial melting because they produced huge volumes of granite magmas that then melted their way upward and intruded into the collisional belt, especially into the metamorphic rocks of the Piedmont terrane and the Avalon terrane. The **Acadian granites** are true granites, so rich in pink potassium feldspar that they are often dark pink or even red in color. The most famous of these granitic intrusions occur in New Hampshire and Vermont. They have been quarried for years for granite building stone (Fig. 11.7G). They are the reason New Hampshire's nickname is "The Granite State."

The discovery of the Avalon terrane and the plate tectonic history of this region helped solve a long-standing mystery in geology. A century ago, paleontologists noticed that the Cambrian through Devonian trilobites of Scotland, northern England, and northern Wales looked just like those found in most of North America, from Utah and California to western Newfoundland and western Massachusetts (Fig. 11.11). However, the early Paleozoic trilobites of eastern Massachusetts or eastern Newfoundland resembled those of southern Wales and southern England and northern Europe (an example is Cambrian *Paradoxides*, Fig. 10.12D). They could not explain why there was such a sharp boundary between faunal provinces of trilobites that cut Massachusetts and Newfoundland in half and crossed the ocean to cut Great Britain in half as well (Fig. 11.11). Some speculated that there was a deep-water trough that separated "Pacific fauna" trilobites (found in Scotland, in western Newfoundland, in western Massachusetts, and all the way to Utah and California) from "Atlantic fauna" trilobites (found in eastern Massachusetts and eastern Newfoundland and in England and northern Europe). However, the "deep-water trough" explanation made no sense if trilobites from Utah could somehow cross the

Devonian event; and it was restricted to Maritime Canada and regions south. If you look at the Paleozoic predecessor of the Atlantic Ocean (called the Iapetus Ocean) in the early Paleozoic (Fig. 10.11B), you will notice a small continental fragment floating out in the middle of that ocean through most of the Cambrian and through the Silurian. It is known as the **Avalon terrane**, after the Avalon Peninsula in southeastern Newfoundland (Figs. 11.9 and 11.10). The Avalon terrane includes most of southern England and parts of northern Europe south of Baltica, plus parts of Newfoundland, Nova Scotia, eastern New England (especially eastern Maine and Massachusetts), extending all the way down to the Carolina slate belt in the central Carolinas, Virginia, and Georgia (Fig. 10.11). Based on the timing of the deposits shed off these mountains, the Avalon terrane docked first in the north during the Middle Devonian, then the collision "zippered shut" toward the south by the Late Devonian (Fig. 11.9).

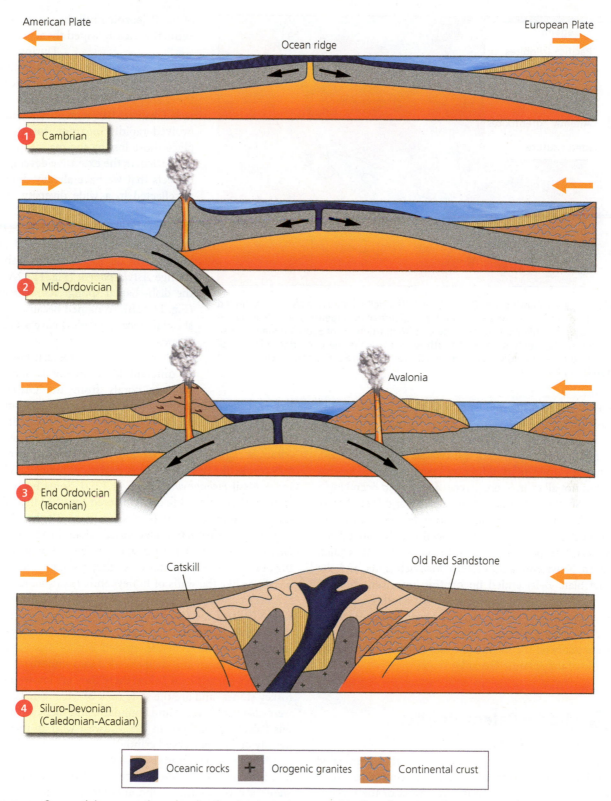

American Plate

European Plate

Ocean ridge

1 Cambrian

2 Mid-Ordovician

Avalonia

3 End Ordovician (Taconian)

Catskill

Old Red Sandstone

4 Siluro-Devonian (Caledonian-Acadian)

Oceanic rocks   + Orogenic granites   Continental crust

Figure 11.10   Sequential cross sections showing the plate tectonic events of the Taconic and Acadian orogenies.

modern Atlantic to Scotland and those from Massachusetts could also swim to England.

When geologists finally put the pieces together and recognized the Avalon terrane, this puzzle was suddenly solved. The trilobites on each half of Great Britain, Newfoundland, and Massachusetts didn't match because they were from the North American terrane on one side and the Avalon terrane on the south and east, with a continental suture (and lots of metamorphic rocks and granites) between them. In fact, that suture between the two trilobite provinces represented

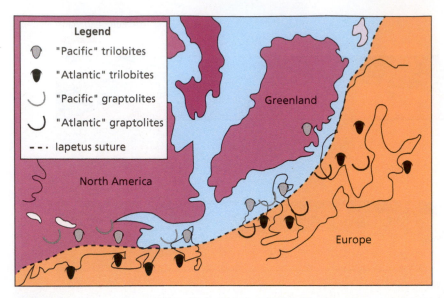

Figure 11.11 Map showing the "Pacific" and "Atlantic" trilobite faunas, now explained by the collision of the Avalon terrane and the closure of the Iapetus Ocean (leaving the "Iapetus suture"). This explains why the Cambrian trilobites of Scotland match those of the central and western parts of North America, while the Cambrian trilobites of the Avalon terrane (including eastern New England, Nova Scotia, and Newfoundland) match those in England.

the line of closure when the Iapetus Ocean, often called the "Proto-Atlantic Ocean" (Figs. 10.2, 11.2), was crushed as Avalonia collided with North America. The modern Atlantic would not begin to open until the Triassic (see Chapter 13), and when it ripped apart, it followed part of the old Iapetus suture but not all of it. Some pieces of what had been North America (Scotland and northern England and Wales) were ripped away from North America and ended up on the other side of the Atlantic. Some pieces of Avalonia that are related to northern Europe and southern England and Wales (such as eastern Newfoundland, eastern Massachusetts, and the Carolina Slate Belt) ended up on the American shore of the Atlantic (Fig. 10.2). Thus, if you visit Scotland, you are actually standing on an old piece of North America, and if you are in eastern Massachusetts or the central Carolinas, you are actually on the Avalon terrane, which came from the other side of the Atlantic Ocean!

11.4 Middle Paleozoic Life

MARINE INVERTEBRATES

The huge areas of shallow carbonate seas that dominated the continents during the Silurian and Devonian harbored an incredible diversity of marine life. Most of the invertebrates we find as fossils are members of the major groups that first took over the oceans during the Great Ordovician Biodiversification Event: brachiopods, bryozoans, corals, crinoids, and cephalopods (sometimes nicknamed the "two B's and three C's"). These five groups dominate the "Paleozoic fauna" that took over from the trilobite-dominated "Cambrian fauna" in the Early Ordovician, and they would continue to be the most important groups through the rest

of the Paleozoic until the great Permian extinction nearly wiped them out.

 See For Yourself: Middle Paleozoic Life   During the Early Silurian, the survivors of the Late Ordovician extinction event (see Chapter 10) soon diversified and evolved rapidly into new forms to replace those lost by the mass extinction. In particular, the extensive development of reefs that we have already discussed is reflected in a wide diversity of reef formers. They include not only the stromatoporoid sponges (Fig. 10.15C, D) but a number of other sponges as well. These included the distinctive golf ball–shaped sponge *Astylospongia* (Fig. 11.12A) and the dish-shaped sponge *Astraeospongia* (Fig. 11.12B), so named because its tiny skeletal pieces (spicules) have a star-like shape.

There was also a remarkable range of different kinds of corals, including tabulate corals from the Ordovician like *Favosites* (Fig. 10.15A) and a distinctive Silurian tabulate called *Halysites*, the "chain coral," whose corallites are arranged in a chain-like pattern (Fig. 11.12C). The horn corals, or rugosids, were even more diverse. Many were solitary corals, like the distinctively rugose, wrinkled Devonian index fossil *Heliophyllum* (Fig. 11.12D). Some horn corals gave up the solitary life and clustered together in a colony that resembles honeycomb or organ pipes. The best known of these is a distinctive Devonian colonial rugosid called *Hexagonaria* (Fig. 11.12E). Its name refers to the close packing of its cylindrical corallites, so they often have a hexagonal shape like the walls of honeycomb. *Hexagonaria* fossils are very common in many Devonian reef localities, such as the Falls of the Ohio near Louisville, Kentucky, and in many Devonian quarries in Iowa (especially near Coralville, Iowa). *Hexagonaria* fossils eroded out of the Devonian rocks of Michigan and then abraded into smooth pieces on the shores of the Great Lakes; there they are known as "Petoskey stones" and are the state rock of Michigan. Crinoids were also very diverse and abundant, and there are numerous Paleozoic localities in the Midwest with an incredible density of well-preserved crinoids.

Living in and around these reefs built by corals, crinoids, and sponges was a wide assortment of brachiopods. Gone were the orthides and strophomenides so typical of the Ordovician (Fig. 10.15F-H). Instead, during the Silurian the most common brachiopods were the pentamerides (Fig. 11.12F). These brachiopods had smooth biconvex shells with narrow hinges, but their insides were more distinctive. The shell volume was divided into five chambers by a number of walls, hence the name *Pentamerus* ("five parts" in Greek). By the Devonian, a new group of brachiopod became incredibly abundant: the spirifers (Fig. 11.12G).

Figure 11.12 Characteristic Silurian–Devonian index fossils. **A.** The Silurian "golf-ball" sponge *Astylospongia*. **B.** The Silurian dish-shaped sponge with star-shaped spicules known as *Astraeospongia*. **C.** The Silurian "chain coral," a tabulate coral known as *Halysites*. **D.** The wrinkled, irregular-shaped horn coral called *Heliophyllum*. **E.** The "Pestoskey stone" coral, a colonial horn coral called *Hexagonaria*. **F.** Internal molds of the typical Silurian reef-building brachiopods known as pentamerides. **G.** A typical Devonian spirifer with a long hinge, known as *Mucrospirifer*. **H.** The goniatite ammonoid Geisenoceras, with the zigzag suture line shown here. **I.** The common Silurian trilobite with the pointy tail, *Dalmanites*. **J.** The common Devonian trilobite with the huge compound eyes and broad cephalon covered with bumps known as *Eldredgeops*. **K.** The large spiny trilobites were particularly common in the Devonian, such as this meter-long *Terataspis*.

These brachiopods got their name because the lophophore inside the shell is arranged in a spiral. Many of them, like *Mucrospirifer* (Fig. 11.12G), have a long, straight hinge and a shallow fold in the middle, so they look almost like the winged lapel pin that pilots and flight attendants wear. In some Devonian localities, the ground is literally paved with spiriferide brachiopods, and it is almost impossible not to step on them. The other group of lophophore-bearing animals, the bryozoans, was still very common and diverse as well.

Among mollusks, snails and clams are found, although they are much less common than the shelled invertebrates like brachiopods. Large predatory mollusks, such as the straight-shelled nautiloids that ruled the Ordovician, were still around in the Silurian and Devonian. But the biggest innovation among mollusks was another branch of cephalopods, the coiled forms known as **ammonoids** (Fig. 11.12H). These were not the direct relatives of nautiloids but instead were more closely related to modern squids. They can easily be distinguished from nautiloids when you see the edge of the walls, or septa, inside their shell. Where these walls meet the outer shell, they form a line known as a **suture** that can be seen on shells with the outer wall worn off. The sutures of nautiloids are simple, flat curves because their dividing septa are also nearly flat. But in ammonoids the edges of the septa start to become complicated and fluted in distinctive ways, giving the suture a complex pattern. Nearly all the ammonoids of the later Paleozoic (Devonian through Permian) have a **goniatite** suture (Fig. 11.12H), which looks like a zigzag on the side of the shell. These ammonoids joined the niche of swimming predators once dominated by the orthocone nautiloids, and through the rest of the Paleozoic and Mesozoic, ammonoids would continue to be abundant and to rapidly evolve, making them excellent index fossils.

While the brachiopods, bryozoans, corals, crinoids, and cephalopods are the dominant groups of the Paleozoic fauna, there were still some elements of the Cambrian fauna hanging on. Most important of these were the trilobites. They had already been severely decimated by the Late Cambrian extinction event and the Late Ordovician extinction event, so their diversity was very low by the middle Paleozoic. A common Silurian trilobite in most localities is

Dalmanites (Fig. 11.12I), with its huge raised eyes and spike on the end of its pygidium. In the Devonian, the most common group of trilobites are the phacopids like *Eldredgeops* (formerly *Phacops*) (Fig. 11.12J); they also had large complex eyes and lots of little bumps on the front of the cephalon, and many were able to roll up in a ball for protection. In some Devonian localities, there were some really remarkable trilobites, such as the enormous spiny fossil *Terataspis* that reached 60 cm (24 inches) in length (Fig. 11.12K). There were many other extremely spiny forms with eyes raised on stalks and many other weird features But this was the last gasp of trilobite diversification. Most of these groups were wiped out by the Late Devonian extinction, and only a few simple lineages survived to the end of the Paleozoic.

▷ See For Yourself: Eurypterids

Trilobites were not the only large arthropods in the Silurian and Devonian seas. The most remarkable arthropods were the **eurypterids** (Fig. 11.13). They are sometimes called "sea scorpions," although they are not true scorpions but are more closely related to horseshoe crabs. Nor did they live exclusively in the sea. Apparently, some of them lived in fresh or brackish waters; some may have even crawled on land. Eurypterids superficially resemble scorpions in the shape of their bodies and their long tail with a spike at the end in some forms, but they were mostly aquatic, with one pair of appendages turned into paddles for swimming and the front appendages shaped like pincers. They first appeared in the Middle Ordovician, where they were already formidable predators, reaching 1.83 m (6 feet) in length. But their heyday in the Silurian produced a wide range of forms (Fig. 11.13A), from the abundant small fossil *Eurypterus* about 20 cm (8 inches) long to the spider-like *Stylonurus* to the huge flat-tailed *Pterygotus*, whose body was up to 1.6 m (5 feet 3 inches) long with long pincers in front, to the enormous *Jaekelopterus* and *Acutiramus* (Fig. 11.13B), which was over 2.5 m (8 feet 2 inches) long! These creatures, as the largest predator of their time, must have been the terror of the Silurian seas since orthocone nautiloids were not as large or as abundant.

MARINE VERTEBRATES

We already saw that during the Cambrian and Ordovician there were just a few fossils of tiny jawless fish (Fig. 10.19), whose skeletons were mostly made of cartilage without bone except for dermal scales. But during the Silurian and especially the Devonian, these early jawless ancestors diversified into a huge range of different kinds of fish, including both extinct groups and most of the fish groups alive today. Indeed, fish fossils are so diverse and abundant in some rocks that the Devonian has often been called "the age of fishes."

The family tree of fishes and their descendants (Fig. 11.14) shows that nearly every group was diverse and abundant in the middle Paleozoic. **Jawless fishes** continued to evolve rapidly (Fig. 11.15A) and developed a variety of forms covered in bony armor, particularly common in Devonian stream deposits like the Old Red Sandstone. Some had flat bodies and upward-facing eyes, suggesting that they sucked up food from the bottom, while others had rounded bodies, simple slit-like mouths, chain mail–like bony armor, and tails which tended to propel their heads

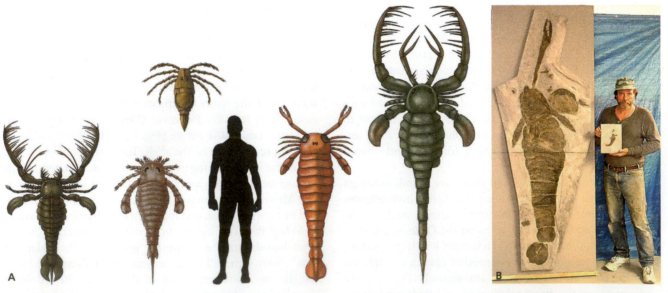

Figure 11.13 Eurypterids (incorrectly called "sea scorpions") were large predatory arthropods, particularly abundant in Silurian seas. A. They came in a wide range of sizes (left to right: from *Megalograptus*, *Eurypterus*, human for scale, *Pterygotus* to huge forms like *Jaekelopterus* and *Megarachne* in the top row. (Drawing by M. P. Williams) B. A nearly complete specimen of *Acutiramus*, showing its immense size compared to the small common species *Eurypterus remipes*.

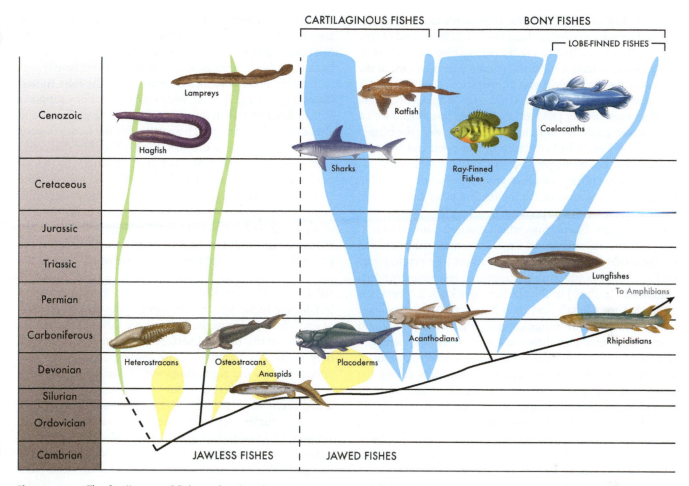

Figure 11.14 The family tree of fishes, showing the early Cambrian–Devonian radiation of jawless fishes (of which only the lamprey and hagfish survive) and the Devonian diversification of jawed fishes, including the extinct placoderms, the sharks and rays, and the bony fish, which eventually branched into the ray-finned fish (most fish alive today) and the lobe-finned fish (including coelacanths, lungfish, and their amphibian relatives).

upward, so they probably swam like tadpoles and filtered out food with their gills. But the major event that allowed fish to diversify so much was the invention of jaws. Without jaws, fish can only use their mouths to suck in food and water and filter out what they need, but jaws give them the power to bite and crush prey, with many other side benefits as well.

See For Yourself: Armored Fish

The largest and most impressive of the fish with jaws was the earliest group to obtain them, known as the **placoderms** (Figs. 11.14, 11.15B, C). Like sharks, these fish had no bone in their skeleton but were supported by a cartilage skeleton instead. They did have bone, however, but it was found only in the dermal shields that covered their heads and thorax, while

Figure 11.15 The Devonian was the "age of fishes." **A.** Painting of a variety of armored jawless fish (left and center) and the lobe-finned fish that gave rise to amphibians (upper right). **B.** The gigantic arthrodire placoderm *Dunkleosteus*, here attacking a primitive shark known as *Cladoselache*. **C.** Head shield, biting plates, and thoracic shield of the giant arthrodire *Dunkleosteus*.

the back half of their body and their shark-like tails generally had no armor covering them (Fig. 11.15B). Instead of true teeth in their jaws, the edges of the dermal plates in the mouth of some placoderms formed sharp cutting devices. One of the largest of the placoderms was *Dunkleosteus* (Fig. 11.15B, C), which reached over 6 m (20 feet) in length, weighed over a ton, and had heavy armor plates on its face and jaw, hinged with a thick armor girdle around the thorax. These were the biggest predators the earth had ever seen to that point and were not surpassed in size by bigger predators until the Mesozoic.

There were dozens of other kinds of placoderms in the Devonian, filling most of the niches that other fish later exploited. A few were huge predators, such as *Dunkleosteus* and the even bigger *Titanichthys*, which reached 8 m (26 feet) in length. Some (the rhenanids) were flat-bodied with broad fins on their side, like rays and skates. Others (the ptyctodonts) had heavy jaws with crushing teeth and resembled modern chimaeras ("ratfish") or Port Jackson sharks. Still others (the antiarchs) were small (less than 30 cm or 12 inches long) with a heavy box of armor covering their entire body, and even hinged armor on their fins that gave them appendages that looked like crab legs. Altogether, there were dozens of species of placoderms in eight separate orders during the Devonian, so they were not only the largest but also the most abundant and diverse group in the "age of fishes."

Not to be overlooked, however, is another important group: the sharks. Although not as abundant as placoderms, by the Late Devonian we find some early sharks like *Cladoselache* (Fig. 11.15B). These sharks reached about 1.8 m (6.5 feet) long and had very primitive cartilaginous skulls with simple teeth composed of numerous tall conical cusps. Although their bodies were long and streamlined like most sharks, they had two dorsal fins with thick bony spines sticking out of the front of the fin, and their front (pectoral) fin was a broad rigid triangle that served as a good stabilizer but was not as maneuverable as the pectoral fins in later sharks.

See For Yourself:
Devonian Fish

In addition to jawless fish, placoderms, and sharks, which had skeletons made of cartilage with bone limited to their teeth, spines, and skin, the second big diversification of fishes are the **bony fish** (Fig. 11.14). They evolved from a Silurian group called the **acanthodians** (misnamed "spiny sharks," although they are unrelated to sharks). These fish had numerous spines along their back and many pairs of fins down both sides of their bodies, as well as a skull and jaw mechanism that is the prototype for the evolution of bony fish. By the Devonian, the bony fish had split into the two groups that are alive today: the **ray-finned fishes** (which make up 99% of living fishes) and the lobe-finned fishes (lungfish, coelacanths, and their amphibian descendants). Ray-finned fishes are the familiar kinds of fish, which support their fins with a number of thin bony rods or rays. Their Devonian ancestors already had evolved a bony skeleton, the fin rays, as well as a skull with a bony braincase and powerful jaws. Although they were relatively rare and overshadowed by placoderms and other fish in the Devonian, they underwent a huge evolutionary radiation and eventually took

over the entire realm of fishes (except for the sharks that are still alive). There are about 30,000 species of ray-finned fish today, more diverse than all other groups of vertebrates (mammals, birds, reptiles, and amphibians) combined.

Finally, the other branch of bony fish is the **lobe-finned fishes** (Figs. 11.14, 11.15A), which have thick bony supports in their fins, rather than thin rays or rods of bone. Those bony elements in their fins match exactly with the bones in your arm and in your leg, so they had the kinds of fins that could evolve into arms and legs. They include lungfishes, which can breathe in air or water and were well represented in the Devonian. A second group of lobe-fins is the coelacanths, common in the Devonian but thought to be extinct until a single living species was found off the deep waters of southern Africa in 1938. The final branch of the lobe-fins is the group that leads to amphibians, which we will discuss next.

INVASION OF THE LAND

In Chapter 10, we saw that for most of the billions of years of earth history (from the Proterozoic through the end of the Ordovician), the land surface was virtually barren. The only organisms found in those ancient landscapes were crusts of fungi and algae, along with lichens. Near the edge of the water, there were algae and mosses growing in the wet areas. There were no plants that could stand up above the surface or survive far from water and no animals on land whatsoever. Finally, during the latest Ordovician, there are signs of the burrows of the first animal to walk on land, the millipedes. We also find the first fossils of spores and plant tissues on land (Fig. 10.21).

See For Yourself:
Land Plants

In the Silurian and Devonian, however, both plants and animals evolved adaptations that made land life possible and opened up whole new realms of ecological habitats. Let's start by looking at the challenges that faced plants that live on dry land. Water plants, like algae, never have to worry about drying out, but in the air this is a serious problem. For this reason, land plants must protect themselves against dry air by covering their outer layer of tissue with a waxy cuticle that reduces evaporation. However, they still must exchange oxygen and carbon dioxide when they perform photosynthesis, so their leaves have pores on the surface (called stomata) that help regulate the exchange of gases and water from the inner tissues. Water plants reproduce by simply releasing their sperm and eggs in the water, but land plants need some sort of waterproof container (spore or seed) to carry their gametes and allow fertilization to happen in the absence of water.

Finally, for plants to live on land and grow taller, they need more complex organ systems to transport fluids uphill against gravity, aid in respiration, remove wastes, and support them. A marine alga such as kelp can have strands many meters long, but because all of it is constantly bathed in seawater, it does not need a system to transport water from one end to the other. The first plants that had water-transport ability are known as **vascular plants** because they had a system of tubes to carry around fluids and nutrients from one part of the plant to the other—just like our own cardiovascular

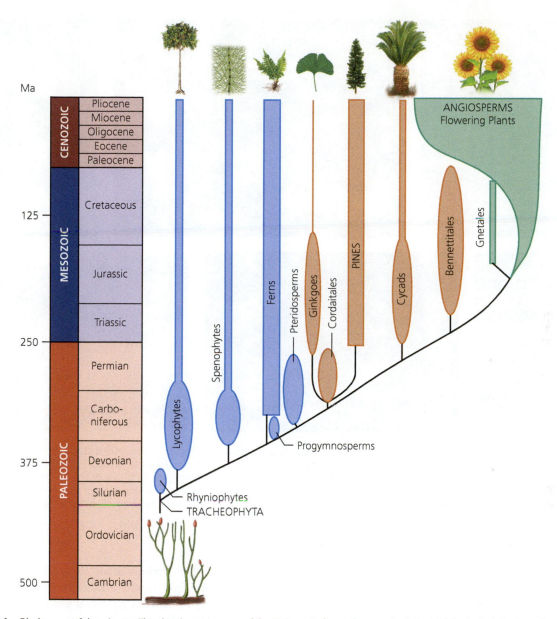

Figure 11.16 Phylogeny of the plants. The dominant groups of the Paleozoic (branches on the left side) included the lycophytes or "club mosses," the sphenophytes or "horsetails," the ferns, and the "tree ferns" or pteridosperms.

system carries a fluid (blood) to all parts of our bodies to supply them with nutrients and take away waste products (Fig. 11.16). In the case of vascular plants, however, they are being "stretched on the rack." The water and nutrients are down in the soil, but the sunlight for photosynthesis comes from above. The root end picks up nutrients and water from the soil and moves them up to the opposite end, where photosynthesis takes place (so carbon dioxide is absorbed and oxygen lost), and a certain amount of water is lost.

Once plants began to grow out of the water, they encountered two problems. First, moisture and nutrients had to be transported to the higher part of the plant. Second, the plant was attempting to stand up against the force of gravity, which kept tugging it down. The solution lay in the evolution of elongate conducting cells, or **tracheids**, lined with a metabolic water product, **lignin.** Lignin is very rigid, thus lending

support. It is also hydrophobic, with a surface that repels water rather than absorbing it (like waxed paper), thus forcing water to bead up rather than be absorbed and speeding water through the tracheids. This conducting tissue occurs as a single central strand within the stem. In more advanced plants, tracheids can become massed to form larger woody trunks. Such vascular plants are formally known as tracheophytes because they have tracheids inside them.

The earliest land plants are tiny Silurian fossils called **rhyniophytes** (Fig. 11.17A, B). These plants are only a few centimeters tall, with simple straight stems and no leaves. The stems were connected beneath the surface by long runners. At the top surface of mature plants are little spore-bearing organs, where their reproductive cells developed and matured. But as we go through the later Silurian, there was a great diversification of simple vascular plants. Then, in the

Figure 11.17   The earliest land plants. **A, B.** Rhyniophytes were simple plants with no leaves but branched stalks with their reproductive structures on top. **A.** A rhyniophyte fossil known as *Cooksonia*. **B.** A reconstruction of *Cooksonia* in life. **C.** Lycophytes are today represented by club mosses, but in the later Paleozoic they became trees that were over 36 m (100 feet) tall. **D.** The sphenophytes are today represented by the ubiquitous water plant *Equisetum*, or the horsetails, also known as the "scouring rushes"; but these primitive plants also reached tree sizes in the later Paleozoic. **E.** Stumps of gigantic primitive gymnosperm trees from the Gilboa Forest in the northern Catskill Mountains of New York. **F.** Restoration of the Gilboa Forest.

Devonian, the plants exploded in diversity. In addition to mosses and liverworts, much more advanced plants, such as ferns, evolved. Two other important groups of living plants also appeared in the Late Silurian or Devonian. One was the lycophytes, or "club mosses," which creep along the ground today. These living descendants are low and unimpressive, but their representatives during the late Paleozoic were gigantic forests made mostly of huge tree-sized "club mosses" over 36 m (118 feet) tall, the largest land plants the world had ever seen up to that point (Fig. 11.17C).

Another important new group was the "horsetails," "scouring rushes," or sphenophytes (Fig. 11.17D). Today, these primitive plants (one genus, a living fossil called *Equisetum*) grow in great abundance in sandy and gravelly soils close to water. Their fibrous stems have tiny particles of abrasive silica in them, so they are hard for animals to eat. Early pioneers called them "scouring rushes" because a crushed handful of them made a good scouring pad for pots and pans. Horsetails are very distinctive because each long hollow stem segment is covered by a series of flutings or ridges along the length. In addition, each stem segment is separated from the next by a distinct joint, from which all the leaves sprout. Each horsetail stem branches from an underground stem called

a rhizome, which sprouts many clones through vegetative reproduction. *Equisetum* is a notoriously tough plant and grows rapidly in the right habitat. It will quickly invade the wet parts of an entire garden if not kept in its own pot, and its underground rhizomes are almost impossible to eliminate, so they will come back no matter what happens to them.

In addition to all these primitive spore-bearing plants, the Late Devonian yields the first plants that reproduced with seeds, which have a hard coating that helps them germinate without being immersed in water. Some of these now extinct "seed ferns" (not true ferns but a more advanced fern-like plant that bore seeds) formed the first large trees, up to 12 m (almost 40 feet) in height (Fig. 11.17E, F). By the Late Devonian, the land was covered by the first true forests, with trees that towered high in the sky (Fig. 11.18). The greening of the landscape that we now take for granted was delayed for a long time, but the earth was no longer barren.

Into this new ecosystem moved a number of different animals that could exploit all these new plant resources or prey on those creatures that did. But just like plants, animals also faced challenges when moving onto dry land. Not surprisingly, the ever-adaptable and rapidly evolving phylum Arthropoda was the first to do it (see Appendix A). Arthropods had advantages

Figure 11.18 The Late Devonian landscape, with the first trees and large land plants in abundance.

over most other invertebrates because their external shell, or exoskeleton, has a hard waterproof cuticle on the outside, so they were naturally protected from drying out. With their flexible joints between body and limb segments and their ability to change their body shape in myriad ways every time they shed their skin, they quickly became masters of the land. As we mentioned in Chapter 10, the millipedes were first on land in the Late Ordovician, exploiting the untapped resources of decaying vegetation (as they still do today). By the Silurian and Devonian, we find fossils of spiders, scorpions, centipedes, millipedes, some of the most primitive wingless insects (such as silverfish), as well as several extinct arthropod groups with no living relatives. Thus, there was a well-established land food web of millipedes, silverfish, and other primitive arthropods that ate plants and decaying matter and predators like spiders, centipedes, and scorpions that fed on them.

 See For Yourself: Early Insects   Almost 100 million years after arthropods had become established on land and dominated the new world of land plants, a second group of invaders crawled out of the water in the Late Devonian: our amphibian ancestors, properly known today as the **tetrapods**, or "four-legged animals." We previously discussed the lobe-finned fish, like the lungfish and coelacanths, which had some features for land life, like lungs and robust bony supports in their fins (Fig. 11.14). But in the Late Devonian, there was a third group of lobe-finned fish (often called the "rhipidistians," although that name is not a natural group) that were even more adapted for land life. One of the best known of these was *Eusthenopteron*, a large (up to 1.8 m, or 6.5 feet long) lobe-finned fish that was much more like an amphibian than either lungfish or coelacanths (Fig. 11.15A). Although *Eusthenopteron* still had a fish-like body, its lobed fins had all the right bones to build the amphibian arm and leg from, and it had the right pattern of bones in the skull to be ancestral to amphibians.

 See For Yourself: When Fish Breathed Air   From *Eusthenopteron*, there is now a tremendous sequence of transitional fossils that shows the steps by which fish became amphibians that first crawled out on land. One of the best-known fossils and the first to be described was

Ichythostega (Fig. 11.19). Like amphibians, *Ichythostega* had four legs with toes, rather than the lobed fins of its ancestors. However, its forelimbs were not strong enough to do much walking, and the most recent analyses suggest that it could only move in short hops, dragging its more flipper-like hindlimbs behind. The forelimbs and especially the hind limbs were much better used in the water, where they propelled the animal along (as newts and salamanders swim today). *Ichthyostega* had robust ribs with flanges that would help support the chest cavity and lungs out of water, but they were not capable of the rib-propelled breathing found in many amphibians. The other amphibian-like feature was the long, flat snout with eyes directed upward and a short braincase; *Eusthenopteron* had a more fish-like cylindrical skull with a short snout and long braincase, eyes facing sideways, and big gill covers. Other than the limbs and bones of the shoulder and hips, however, *Ichthyostega* is very fish-like. It still had a large tail fin, as well as many fishy features of the skull, such as its large gill covers, its hearing (adapted for water), and a lateral line system (canals on the face used in sensing motion and currents in the water). In more recent years, there have been many more fossils that fill in the gap between fish and *Ichthyostega*. One of these fossils was called *Acanthostega*, which was much more fish-like than *Ichthyostega*. The limbs of *Acanthostega* would not have allowed it to crawl on land— it lacked wrists, elbows, and knees. Instead, its limbs were only capable of paddling and pulling it through obstacles underwater. Even more surprising, it had as many as seven or eight fingers on its hands, not the standard five fingers that most vertebrates have! *Acanthostega* had a much larger fin on its tail, and its ribs were too short to support its body on land and allow it to breathe without the support of water. Yet it also had a few advanced amphibian-like features: its ears could hear in air as well as water, and it had strong bones in its shoulder and hip region, four limbs with toes, and a neck joint that allowed it to rotate its head. By contrast, a fish has no "neck" that allows rotation—it must turn the entire front half of its body to change direction or snap at prey.

And in 2004, an even more fish-like amphibian was found in the Canadian Arctic. Called *Tiktaalik*, it is about 10 million years older than *Ichthyostega* or *Acanthostega* and more fish-like in many ways (see Box 11.1).

Thus, by the end of the Devonian, the invasion of the land was complete. Dense forests of club mosses, horsetails, ferns, and seed ferns were inhabited by a wide range of arthropods, both herbivores and predators like spiders, scorpions, and centipedes. And preying on them were the first vertebrates to crawl out on the land and move into this brave new world.

11.5 Devonian Mass Extinctions

Just as the Cambrian and Ordovician ended with mass extinction, there was a major mass extinction near the end of the Devonian. Unlike the previous extinctions, this event occurred in two pulses. The first, known as the Kellwasser Event, occurred at 372 Ma, on the boundary between the last two stages (Frasnian Stage and Famennian Stage) of

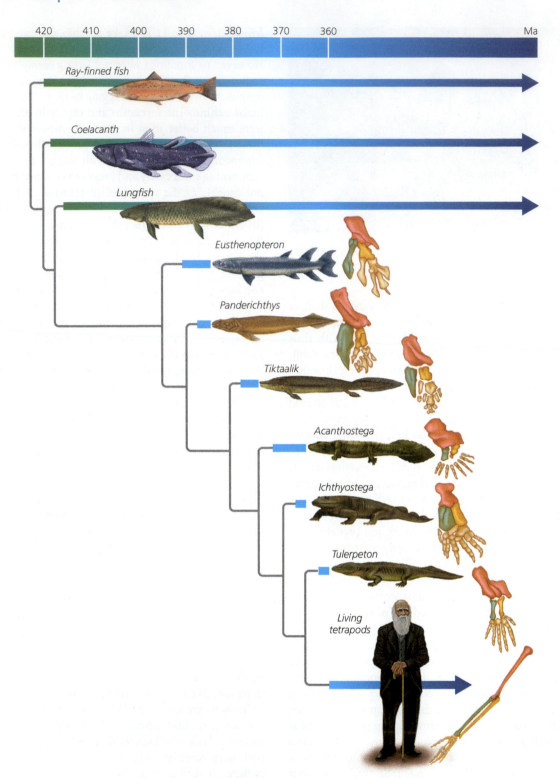

Figure 11.19   Diagram of fish–amphibian transition, showing fossils which go from normal fish to more and more fish-like forms to amphibians.

the Devonian. The second pulse, the Hangenberg Event, occurred at the end of the Devonian (end of the Famennian stage), about 355 Ma. Other scientists have suggested that it was not two distinct events but a series of as many as 8–10 different pulses. Summing up the total of extinctions over the entire interval, it was about the fourth or fifth largest mass extinction in earth history, wiping out 75% of the species, 55% of the marine genera, and about 22% of the marine families in the marine invertebrate community.

 See For Yourself: Devonian Extinction

The waves of extinction struck different groups at different times. The first wave in the late Frasnian (Kellwasser Event) hit the ammonoids, trilobites (most of the typical Devonian groups vanished), brachiopods (especially the diverse spirifers), the armored jawless fish, most of the lobe-finned fish,

BOX 11.1: HOW DO WE KNOW?

How Do We Know About Transitional Fossils?

How do we find important transitional fossils like the specimens that demonstrate the transition from fish to amphibian? A good example of how this is done was described by Neil Shubin in his best-selling book, *Your Inner Fish*. He and his colleagues were looking for a fossil that was intermediate in age between the earliest amphibians that had been collected but not as advanced as the next youngest amphibian fossils known. They knew from previous discoveries that the event occurred in the Late Devonian. So they looked on geologic maps for rocks of a certain age (385–365 Ma) within the Late Devonian Period because there were already good fossils from 385 Ma of very primitive amphibian-like fish, as well as fossils from 365 Ma of more advanced amphibians. After scouring the geologic maps of the world, they found only three areas of the right age and the right sedimentary environment (shallow-marine sandstones and shales, formed in rivers or deltas). Two of them (in Pennsylvania, where there are some very advanced fish, and in Spitsbergen and Greenland in the Arctic, which produced amphibians like *Acanthostega* and *Ichthyostega*) already had been explored and collected. But there was a third place up in Arctic Canada that had never been studied, dated between 365 and 385 Ma.

They raised grant money for a quick visit to the area in 1999, where they found promising bone scraps. So then they had to raise millions of dollars more to mount a full-scale Arctic expedition, which they did for several summers in a row (Fig. 11.20A, B). Only after the third year of very hard, very expensive, very dangerous work dealing with harsh weather and marauding polar bears did they find the fossils of *Tiktaalik*, the critical transitional fossil between fish and amphibians (Fig. 11.20C, D). The name *Tiktaalik* is the Inuit word for a local freshwater fish hunted by the people of the region. As co-discoverer Ted Daeschler said, "We found something that really split the difference right down the middle." Fish–amphibian expert Jenny Clack wrote of their

discovery, "It's one of those things you can point to and say, 'I told you this would exist,' and there it is."

 See For Yourself:
Discovery of
Tiktaalik

The lobed fins of *Tiktaalik* had all the elements ancestral to the amphibian limb but still had the fin rays, rather than toes. It had fish-like scales, a combination of both gills (shown by the gill arch bones) and lungs (shown by the spiracles in its head), and a fish-like lower jaw and palate. Unlike any fish, it had amphibian features too: a shortened, flattened skull with a mobile neck, notches for the eardrums on the back of the skull, and robust ribs and limbs and shoulder and hip bones. Yet like *Acanthostega*, its fins were not strong enough or flexible enough to allow it to crawl on land; instead, they were probably used to paddle in shallow water and push up so that it could see above the surface. Like the other transitional fish–amphibian fossils (and many modern amphibians, especially newts and salamanders), it probably spent most of its time in water, using its limbs to push along and paddle. It could hunt on the margins of the streams in which it lived—but it was not capable of dragging itself across the land very far or walking while holding its belly off the ground.

After this spectacular find, they went back several more times and found many more specimens of *Tiktaalik*, along with a number of other fish and other animals that lived in this ancient river delta around 375 Ma. The discovery of *Tiktaalik* shows the predictive power of geology and evolution. Shubin and his colleagues knew already what time interval to look for (375 Ma), based on the age of the most advanced fish-like transitional fossils at 385 Ma and the most primitive amphibians at 365 Ma. They consulted geologic maps to find rocks of the right age and eliminate those that had already been explored, and the Canadian Arctic proved to be the "missing link" in the chain of rock sequences needed to find the crucial fossils.

---

and the jawed placoderm fish particularly hard, while the snails, clams, and bryozoans escaped with only minor extinctions. The second wave of extinction (Hangenberg Event) at the end of the Devonian wiped out the algal community that produced the acritarch fossils (Fig. 9.10) and finished off the placoderm fish completely. Out of the huge radiation of different groups of fishes in the "age of fishes" (Fig. 11.14), only sharks and bony fish survived into the Carboniferous. In fact, the Devonian extinctions wiped out 97% of the vertebrate species, so the only survivors in the Early Carboniferous were sharks less than a meter long and fish and amphibians less than 10 cm long.

However, the biggest victims of the Devonian extinction were the giant coral–stromatoporoid reefs that had dominated since the early Paleozoic. Stromatoporoid sponges vanished completely, while most of the tabulate and rugosan (horn coral) genera were wiped out as well, with only a few of these corals left to survive into the late Paleozoic. The complete obliteration of the warm-water coral reef

assemblage undoubtedly explains why there is so much extinction in many of the groups that lived in and around the reefs, such as the brachiopods and ammonoids.

There is also a strong temperature signal in the extinction. The major group of brachiopods to vanish was the atrypid spirifers, which were tropical in their distribution. Likewise, the tropical reefs of the world were decimated, whereas corals found in deep, colder water were unaffected and reef communities in the polar latitudes (such as the Parana Basin of South America) also escaped severe extinction. Most striking of all is what organisms replaced the tropical reefs: giant reefs of the glass sponge *Hydnoceras* (Fig. 11.21A), which are particularly common in the Late Devonian of upstate New York. These were previously known from the colder, deeper waters and then spread to the shallows, apparently in response to the dramatic cooling of tropical waters.

The signal of tropical cooling can be seen in many other indicators, such as the geochemistry of oxygen in the oceans,

Figure 11.20 The discovery of *Tiktaalik* in the Canadian Arctic provided another transitional fossil between fish and amphibians. **A.** The outcrops of Upper Devonian red beds on Ellesmere Island that the Shubin–Daeschler team searched for many years before finding a bone bed (here being excavated) rich in fish fossils and *Tiktaalik*. **B.** The camp required tents that could survive hurricane-force winds and the harsh conditions of the polar summer. **C.** The best of the known fossils of *Tiktaalik*, a complete articulated skeleton missing only the tail and hind limbs. **D.** A reconstruction of *Tiktaalik* in life.

Figure 11.21 **A.** The glass sponge *Hydnoceras* dominated the Late Devonian after the extinction of the coral–stromatoporoid tropical reefs that ruled the planet during the Silurian and most of the Devonian. (Photo courtesy Wikimedia Commons) **B.** Upper Devonian glacial till of the Rockwell Formation in the Sideling Hill road cut, Maryland. There are also abundant glacial lake beds and glacial dropstones in the same unit.

which indicates a massive ice buildup. Indeed, there are Late Devonian glacial tills in many places, including the polar regions of Gondwana and even as close to the ancient tropics as Pennsylvania, Maryland, and West Virginia (Rockwell Formation; Fig. 11.21B).

So what caused the Devonian extinctions? Many ideas have been proposed, from asteroid or comet impacts to massive volcanic events to plate tectonic changes to depletion of the oxygen in the ocean. The evidence for impacts has never been confirmed and doesn't seem to match the pattern of extinctions, and the volcanism and tectonics aren't strongly supported as a cause. But the oceanic oxygen depletion is undoubtedly real since the typical Upper Devonian deposits (especially during the Kellwasser Event) are anoxic deep-water black shales, such as the Cleveland Shale in Ohio, the Chattanooga Shale in Tennessee, and the New Albany Shale in Illinois (Fig. 11.7G). However, many geologists think that the anoxia and the clear signal of global cooling are related, so as the ocean basins are dramatically cooled and stratified, their bottoms become starved of oxygen and deadly for organisms. Thus, the only well-supported causes are dramatic cooling in the tropics and oceanic stagnation.

What could have caused this to happen? In recent years, a number of geologists have pointed to the fact that the planet developed large-scale forests for the first time in the Late Devonian. Up until this time, only algae and limited low-growing land plants affected the carbon dioxide balance in the atmosphere. But large trees not only absorb and trap a lot more carbon dioxide but also speed up deep weathering of the soil with their roots, which makes the soils absorb carbon dioxide as well. The earth was not completely done with the greenhouse world yet (as we shall see, it returns for its last hurrah in the Early Carboniferous), but this was the first significant pulse of global cooling; and it affected a planet that was stable and had a huge tropical biomass for millions of years and was unprepared for such a severe cooling. In addition, the extinction was more severe than any previous extinction in important ways. Unlike the Ordovician extinction, which wiped out more genera and species but did not fundamentally change the ecological communities, the Late Devonian extinctions wiped out much of the tropical fauna and changed the world by wiping out the reefs, as well as decimating the huge radiation of fish groups from the "age of fishes." The life of the world's oceans in the Carboniferous would never look the same.

TIMELINE OF MIDDLE PALEOZOIC EVENTS

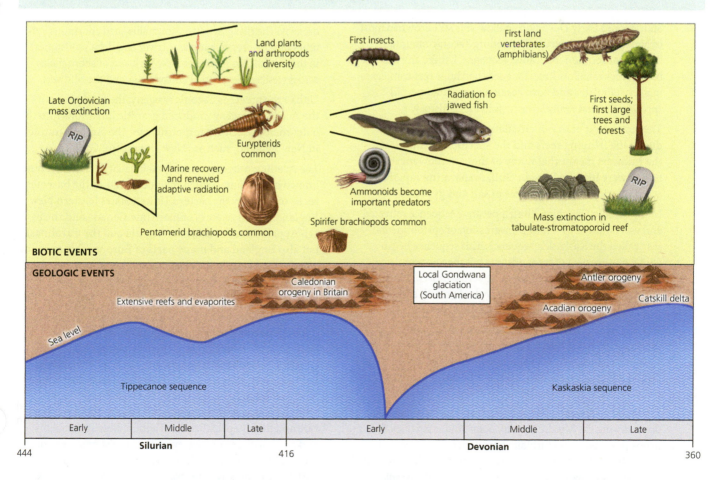

RESOURCES

BOOKS AND ARTICLES

Bird, J. M., and J. F. Dewey. 1970. Lithospheric plate-continental margin tectonics and the evolution of the Appalachian orogeny. *Bulletin of the Geological Society of America* 81:1031–1060.

Blakey, R. C., and W. D. Ranney. 2018. *Ancient Landscapes of Western North America: A Geologic History with Paleogeographic Maps*. Springer, Berlin.

Brannen, Peter. 2017. *The Ends of the World: Volcanic Apocalypses, Lethal Oceans, and Our Quest to Understand Earth's Past Mass Extinctions*. Ecco, New York.

Clack, Jennifer. 2012 (2nd ed.). *Gaining Ground: The Origin and Evolution of Tetrapods*. Indiana University Press, Bloomington.

Dinely, David. 1984. *Aspects of a Stratigraphic System: The Devonian*. Wiley, New York.

Gray, J., and W. Shear. 1992. Early life on land. *American Scientist* 80:444–457.

House, M. R. 1979. *The Devonian System*. Palaeontological Association, London.

Levi-Setti, Ricardo. 2014. *The Trilobite Book: A Visual Journey*. University of Chicago Press, Chicago.

Long, John A. 2010. *The Rise of Fishes*. Johns Hopkins University Press, Baltimore.

MacLeod, Norman. 2015. *The Great Extinctions: What Causes Them and How They Shape Life*. Firefly Books, London.

Maisey, John, and David Miller. 1996. *Discovering Fossil Fishes*. Holt, New York.

McGhee, George. 1996. *The Late Devonian Mass Extinction*. Columbia University Press, New York.

McGhee, George. 2013. *When the Invasion of the Land Failed: The Legacy of the Devonian Extinctions*. Columbia University Press, New York.

Prothero, D. R. 2013. *The Story of Life in 25 Fossils*. Columbia University Press, New York.

Prothero, D. R. 2018. *The Story of the Earth in 25 Rocks*. Columbia University Press, New York.

Schoch, Rainer. 2014. *Amphibian Evolution: The Life of Early Land Vertebrates*. Wiley-Blackwell, New York.

Shubin, Neil. 2008. *Your Inner Fish: A Journey into the 3.5 Billion Year History of the Human Body*. Pantheon, New York.

Williams, H., and R. D. Hatcher, Jr. 1982. Suspect terranes and accretionary history of the Appalachian orogen. *Geology* 10:530–536.

SUMMARY

- Silurian and Early Devonian shallow seas still drowned most of North America (except the eroding Taconic mountain belt) as part of the Tippecanoe sequence. Unlike any time before, these shallow limestone seas had huge reef complexes made of tabulates and horn corals and stromatoporoid sponges, the first really large reefs in earth history.

- The restriction of ocean circulation in the shallow seas due to the fringing reefs and the subtropical latitudes of the Midwest meant that many of these shallow marine basins were like the Persian Gulf, evaporating rapidly and accumulating thick sequences of salt and gypsum.

- In the Middle Devonian, the Tippecanoe sequence retreated and was followed by a new transgression of epicontinental seas called the Kaskaskia sequence. This sequence began with a mature nearshore sandstone but then accumulated enormous thicknesses of limestone across most of North America. The reefs reached their peak of growth at this time.

- In the Middle–Late Devonian, Greenland and Arctic Canada were crushed by the collision with the Baltica protocontinent (modern Scandinavia and northern Europe), causing the Caledonian orogeny, which shed the Old Red Sandstone across Britain and similar-looking rocks in northern Canada.

- In the Appalachian region, another deep-marine trough sank down during the Middle Devonian, accumulating thick flysch deposits of shales and turbidite sandstones. The Devonian and earlier rocks of this region were then intensely deformed by a Late Devonian collisional event known as the Acadian

orogeny. The rising Acadian Mountains shed enormous volumes of post-orogenic molasse (river and delta sandstones) in the Catskill Mountains of New York and offshore shales across western New York, Pennsylvania, and the Midwest.

- Unlike the previous Taconic orogeny, the final phase of the Acadian orogeny produced granitic intrusions into the older rocks, which are responsible for the famous granites in New England, such as "New Hampshire Granite."

- The Acadian orogeny was caused by a large exotic terrane known as Avalonia, which included parts of the basement rocks of not only the eastern United States (eastern New England and large parts of maritime Canada and the eastern Appalachian foothills in Virginia and the Carolinas) but also England and many parts of Europe. This explains why fossils like trilobites have more in common between eastern Massachusetts and England, while fossils of Scotland are like those of the rest of North America.

- Silurian and Devonian sea floors were dominated by many of the same groups that arose in the Ordovician, especially articulate brachiopods (mainly pentamerides and spirifers), bryozoans, crinoids, and the large diversity of sponges and corals that built the great reefs of that time. New predators arose, especially the coiled ammonoids with the zigzag goniatite sutures. Trilobites became more rare and highly specialized, with incredible spiny forms and some with impressive eyes. The largest predators of the Silurian, however, were the eurypterids or "sea scorpions," with their impressive pincers and paddling appendages.

- The Devonian is often called "the age of fish" because it witnessed a huge radiation of different fish groups. Not only were there abundant armored jawless fish, but a new group of jawed predators, the placoderms, ruled the seas, with forms ranging from enormous predators to small armored fish. Sharks arose in the Devonian as well. The most important innovation, however, was the appearance of the first bony fish, including the earliest ray-finned fish (ancestors of most living fish) and the lobe-finned fish (ancestors of land vertebrates).

- By the Late Devonian, the first relatives of amphibians had crawled out on the land, modifying their lobed fins to make legs and arms.

- Long before vertebrates colonized the land, the first vascular plants appeared in the Silurian, and by the Late Devonian there were forests with trees over 30 m (almost 100 feet) tall, composed of giant club mosses, horsetails, and tree ferns. These forests were inhabited by a variety of land arthropods, including spiders, scorpions, centipedes, millipedes, and the first primitive insects.

- The marine realm was severely afflicted by the Late Devonian extinctions, a two-stage event that decimated most of the invertebrate groups (especially trilobites) and the entire coral–stromatoporoid reef complex and wiped out the armored jawless fish, the great diversity of placoderms, and several other archaic fish groups. The causes of the Late Devonian extinctions are unclear, but certainly a global cooling and oceanic stagnation are the major influences, possibly triggered by the decline in greenhouse gases when the first large forests appeared and absorbed enormous amounts of carbon dioxide.

KEY TERMS

Reef complexes (p. 240)
Ecological succession (p. 240)
Pioneer stage (p. 240)
Intermediate stage (p. 240)
Climax stage (p. 240)
Restricted marine basins (p. 245)
Brine (p. 245)
Kaskaskia sequence (p. 245)
Caledonian orogeny (p. 247)
Acadian orogeny (p. 247)
Baltica (p. 247)
Avalon terrane (p. 248)
Acadian granites (p. 248)
Ammonoids (p. 251)
Suture pattern (p. 251)
Goniatite suture (p. 251)
Eurypterids (p. 252)
Jawless fish (p. 252)
Placoderms (p. 253)
Bony fish (p. 254)
Acanthodians (p. 254)
Ray-finned fishes (p. 254)
Lobe-finned fishes (p. 254)
Vascular plants (p. 254)
Tracheids (p. 255)
Lignin (p. 255)
Rhyniophytes (p. 255)
Tetrapods (p. 257)
Devonian mass extinctions (p. 259)

STUDY QUESTIONS

1. How are the reef complexes different from the reefs found in the Cambrian and Ordovician?
2. How does a cross section of a typical Silurian reef show the succession of communities that built it?
3. How does the presence of large barrier reefs and subtropical conditions explain the huge salt and gypsum deposits in the Silurian of Michigan or western New York State?
4. What is the evidence that the Taconic Mountains were eroding down and then completely vanished during the Silurian?
5. What is the evidence that an exotic terrane is about to collide with North America during the Middle Devonian? What kinds of post-orogenic sediments did the Acadian Mountains shed across North America in the Late Devonian?
6. Where is the Avalon terrane basement found today?
7. How does the collision of the Avalon terrane with North America explain the peculiar trans-Atlantic distribution of trilobites?
8. Why is New Hampshire called "the Granite State"?
9. What were the dominant groups of brachiopods, corals, and cephalopods in the Silurian and Devonian? What was the largest marine predator of the Silurian? Of the Devonian?
10. How do jaws allow fish to evolve into many niches that jawless fish could not occupy?
11. What adaptations did plants evolve to allow them to move up and out of the water?
12. Outline the anatomical changes that allowed the aquatic lobe-finned fish to become land tetrapods.
13. What are some of the possible causes of the Devonian extinction?

Carboniferous and Permian 355–250 Ma

Coal, oil and gas are called fossil fuels, because they are mostly made of the fossil remains of beings from long ago. The chemical energy within them is a kind of stored sunlight originally accumulated by ancient plants. Our civilization runs by burning the remains of humble creatures who inhabited the Earth hundreds of millions of years before the first humans came on the scene. Like some ghastly cannibal cult, we subsist on the dead bodies of our ancestors and distant relatives.

—Carl Sagan, Billions and Billions

Diorama of the Pennsylvanian coal swamps, showing the enormous tree-sized lycophytes and sphenophytes and ferns

12.1 The Late Paleozoic: A World of Change

The Late Paleozoic world marked an important transition in earth history. The greenhouse world that had mostly prevailed since the Cambrian ended. The last epicontinental seas depositing huge volumes of limestone across the entire North American continent vanished. By the Late Carboniferous, coal swamps covered the tropics (Fig. 12.1), and the globe was in the grips of polar glaciation in an icehouse world that continued through the Permian. The many smaller continents that had been separate through most of the Paleozoic collided in a series of events that built the supercontinent of Pangea by the Permian (Fig. 12.2). This meant that huge amounts of sediment were shed from these Permian mountains into an icehouse world with extreme climates, producing dunes, red beds, and evaporites in many places. Finally, the marine communities evolved and diversified steadily through the Late Paleozoic until they were decimated by the "mother of all mass extinctions" at the end of the Permian, while land reptiles and amphibians underwent their own remarkable burst of evolution before the Permian extinction.

But at the start of the Carboniferous, you would never know that such changes were on the horizon. The Early and Late Carboniferous look as different as night and day, and the only reason they are combined together in the same period is historical accident. The Carboniferous (meaning "coal-bearing" in Latin) period was first recognized in Great Britain in 1822 as a formal name for what miners had long called the "coal measures," the beds that had produced the coal for the Industrial Revolution. But even in England, they knew that beneath the coal measures was a thick limestone sequence of very different rocks called the Mountain Limestone, which was lumped into the Carboniferous because it was above the Devonian Old Red Sandstone.

In North America, however, the difference is even more striking. Lower Carboniferous rocks (called "Mississippian" in the United States) are normally thick deposits of limestones, rich in crinoids and other fossils (Fig. 12.3A). Yet Upper Carboniferous rocks (called "Pennsylvanian" in the United States) are like the coal measures in Britain and elsewhere, and there is a big unconformity in the middle of the Carboniferous that separates the two (Fig. 12.3B). The Mississippian limestones are products of the last of the greenhouse worlds, while the Carboniferous coal beds are the initial signs of the Late Paleozoic glaciation and mountain-building events. Consequently, American geologists have long used "Mississippian" (named for the extensive limestones in the Mississippi Valley from Davenport, Iowa, to St. Louis, Missouri) and "Pennsylvanian" (from the coal deposits of western Pennsylvania) to emphasize this difference and seldom use "Carboniferous," as the rest of the world does. Today, there is an international compromise, so some versions of the timescale use "Mississippian" and "Pennsylvanian" as the formal early and late global subdivisions of the Carboniferous (even though you rarely hear "Mississippian" anywhere but the United States or "Carboniferous" used in American geological literature).

KASKASKIA CRINOIDAL LIMESTONE SEAS

The Early Carboniferous, or Mississippian, was the last gasp of the continent-wide shallow, clean, mud-free carbonate seas that drowned the mid-continent during the warm greenhouse conditions of most

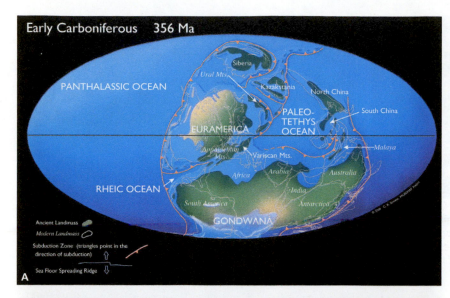

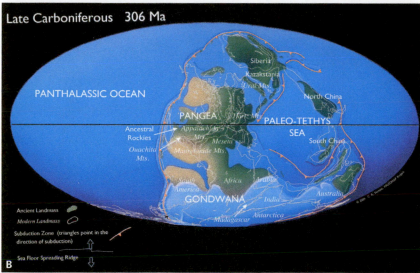

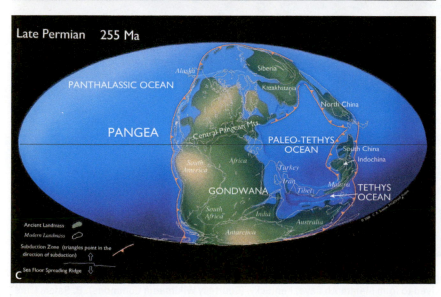

Figure 12.2 Paleogeographic maps of (**A**) Early Carboniferous (**B**) Late Carboniferous, and (**C**) Late Permian.

of the Paleozoic. In North America, the immense volumes of limestones that covered the entire continent (Figs. 12.3, 12.4) were the climax of the great Kaskaskia sequence that began in the Middle Devonian. Like the earlier Kaskaskia seas, the sea floor was home to an immense diversity of marine invertebrates, especially crinoids. In fact, crinoids were so abundant that they make up nearly the entire mass of immense volumes of limestone across several states. When you think "Mississippian," the first words you should associate with it are "crinoidal limestones." For example, in the Mississippi Valley between Iowa and Illinois is the Lower Mississippian Burlington Limestone, a typical unit (Fig. 12.3A). Its volume is 30×10^{10} m³, and it is estimated to contain the remains of 28×10^{16} individual crinoid animals! That's just one formation among many. Everywhere Mississippian rocks crop out, there are thick deposits of crinoidal limestone, with other kinds of limestone and occasionally minor shales represented as well. In central Indiana, it is known as "Indiana Limestone" (made mostly of the microfossils called *Endothyra*). Indiana Limestone is a very popular building stone used in important buildings all over the United States, including many government buildings and even the Empire State Building (Fig. 12.4B). To the south in Kentucky, Mammoth Cave is etched into Mississippian limestone. In fact, the classic sequence of units in the type Mississippian in the Mississippi Valley between Davenport, Iowa, and St. Louis, Missouri, is so distinctive and widespread that many Midwestern geologists know it by heart: Burlington Limestone, Keokuk Limestone, Warsaw Shale, Salem Oolite, Saint Louis Limestone, Sainte Genevieve Limestone, and then the alternating limestone and shale beds of the Chester Group, which foreshadow the deposits of the Pennsylvanian.

 See For Yourself: A Living Crinoid in Motion   In South Dakota, the Mississippian crinoidal limestone is called the Pahasapa Limestone, which is found across most of the central and southern core of the Black Hills. This unit was etched to form Wind Cave and

Figure 12.3   A. A typical sample of Mississippian crinoidal limestone from the Fort Payne Formation, on the shore of Lake Cumberland, southern Kentucky. **B.** Outcrop of the Mississippian–Pennsylvanian unconformity in central Illinois, which is also the boundary between the Kaskaskia (below) and Absaroka sequences. The Pennsylvanian river channel shales and sandstones (top and lower left) cut down through crinoidal Mississippian limestones of the Burlington Formation (resistant ledge in lower right).

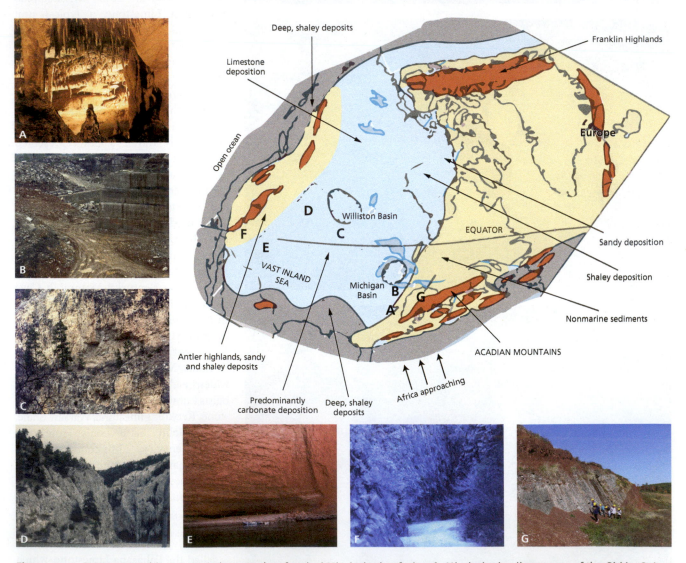

Figure 12.4   Paleogeographic map and photographs of typical Mississippian facies. **A.** Mississippian limestones of the Girkin, Sainte Genevieve, and St. Louis Formations were dissolved to form Mammoth Cave, Kentucky. (Courtesy Wikimedia Commons). **B.** Salem Limestone in Empire State Quarry, Indiana, from which came the "Indiana Limestone" that makes up many important buildings (including the Empire State Building). **C.** The Mississippian Pahasapa Limestone in the Black Hills of South Dakota, from which Wind Cave and Jewel Cave were eroded. **D.** Madison Limestone, Gates of the Mountains, Montana. **E.** Cliffs of Redwall Limestone in the Grand Canyon, as they appear from river level. The actual limestone is gray, but exposed surfaces are colored red from the erosion of the red Supai Group sandstones and shales above it. **F.** Monte Cristo Limestone, Arrow Canyon, Nevada. **G.** Mauch Chunk red beds from Pennsylvania, shed by alluvial fans in the rising Allegheny Mountains.

Jewel Cave and the other caves in the region (Fig. 12.4C). In the northern Rockies, the thick Mississippian limestone is known as the Madison Limestone and forms many of the spectacular cliffs in Montana and Wyoming (Fig. 12.4D). In the Grand Canyon, the spectacular red cliff of the Redwall Limestone is one of the thickest units in the entire region at over 240 m (nearly 800 feet) thick (Fig. 12.4E). In southern Nevada, the Mississippian is represented by a series of limestones of the Monte Cristo Group (Fig. 12.4F).

If we were to take a time machine and scuba dive in the world of the Mississippian, we would see a shallow carbonate shoal setting like in the Bahamas (see Box 10.1) but on an immense scale. These shallow waters would have been packed with extensive meadows of crinoids (and other invertebrates that lived among them), with their umbrella-like filtering arms and heads swaying back and forth with the currents. The earth was still rotating much faster than it does today (400 days in a year, rather than the 365.25 days we now have), and the moon was much closer than it is today, so the tidal currents sweeping across these shoals would have been very powerful. This type of environmental setting extended all the way from Nevada, Arizona, and Montana to the Midwest (especially Michigan, Illinois, Indiana, Kentucky, and south). In fact, the only place that limestones were not being deposited in the Mississippian of the United States was in the foothills of the vanishing Acadian Mountains in the Appalachian belt, where red sandstones and shales representing molasse deposits formed in rivers and deltas eroded off the mountains. These include units such as the brick-red Mauch Chunk Formation (Fig. 12.4G), the Bigstone Gap Shale, Grainger Sandstone, and Pocono Sandstone in Pennsylvania, Virginia, and West Virginia and the New Albany Shale in Indiana and Illinois.

COALS AND CYCLOTHEMS: THE ABSAROKA SEQUENCE

The Kaskaskia sequence came to an end with a major regression that created a deeply incised unconformity carved down into the uppermost limestones of the Mississippian (Fig. 12.3B). We now know that this regression was due to a giant pulse of glaciation in Gondwana, which locked up immense amounts of water in the ice caps and pulled down sea level. Above the unconformity, a new sequence was deposited, and it looks very different from the limestone-dominated Sauk, Tippecanoe, and Kaskaskia sequences that we have seen so far. Called the **Absaroka sequence** by Larry Sloss in 1963 (after the Absaroka tribe in northwestern Wyoming), it is dominated by thick deposits of sandstones and shales across the entire continent—and the valuable coal beds that have generated so much geological study of the Pennsylvanian. Clearly, the world had changed, so the clean, mud-free, fossiliferous limestones that had dominated the craton during most of the Paleozoic were being replaced by something completely different.

One of the signatures of the Pennsylvanian is not only the abundance of shale, sandstone, and coal but the repetitive,

cyclic patterns of deposition. Geologists who worked on the coal geology of Pennsylvania or Illinois first noticed this and called these deposits **cyclothems**. These cyclic patterns of deposition were very predictable in their sequence of deposits, although the individual units vary in thickness and some are incomplete because their upper units might be eroded away by the next cyclothem. In Illinois, the classic cyclothem (Fig. 12.5) began with the unconformity at the base, and a non-marine river channel sandstone filling the basal unconformity. Above this lay a second non-marine unit, a shale formed in a floodplain, often with mud cracks and other indicators of drying and exposure. The third unit is deeply weathered gray clay full of root traces called the **underclay**, which is the soil developed below the coal swamp, represented by the next unit, the coal itself. Above the swampy environment represented by the coal is marine shale, showing that transgression is gradually drowning the non-marine sandstone–shale–underclay–coal sequence. Depending upon the cyclothem, there may be a thin, muddy, marine limestone above the first marine shale or even an alternation of several thin, marine shales and limestones to the top of the cyclothem. Clearly, the lower part of the sequence is non-marine, grading into a marine transgressive deposit in the upper part of the cyclothem (Fig. 12.5A). In some places, there may be dozens or even as many as 100 cyclothems stacked in a row, indicating many cycles of transgression and regression.

This classic Illinois cyclothem (Fig. 12.5A, C) might represent the deposits of a large river delta that built out into the Absaroka seaway across the middle of the continent (Fig. 12.5E). If you travel to the midwest in places like Kansas or Nebraska, you would be in the middle of the seaway, so there would be only a thin river sandstone, floodplain shale, coal, and underclay. Instead, the repetitive cycle is dominated by thick beds of alternating marine shales and limestones that are relatively thin in the Illinois cyclothem. Out in the seas of Kansas or Nebraska, only the very tip of the deltas build out, and only briefly do they do so before being drowned again by a rise in sea level (Fig. 12.5B). On the other hand, if you went east to Pennsylvania or West Virginia or Kentucky, you would be much closer to the mountains that provided all the sediment that built floodplains and deltas westward across the interior seas (Fig. 12.5B). In this region, there is roughly the same sequence of units as seen in Illinois, but the non-marine portion is much thicker, especially the coal seam. The shales and limestones in the upper part of these eastern cyclothems are thought to represent brackish nearshore lagoons with algal limestones, rather than the fully marine sequence of limestones found in Illinois or Kansas. Clearly, as the seas transgressed from west to east and regressed from east to west from the central seaway to the foothills in Pennsylvania, slightly different facies were deposited in each cyclothem—but the pattern of transgression is very regular and repeated and can be correlated over large distances.

12.2 Continental Collision and Mountain-Building

This description of cyclic sedimentation and facies change raises two questions: Where did all that sand and mud building out from the east come from? And why are there nearly 100 cycles in a row during the Late Carboniferous? The answer to the first question is the largest mountain-building event in North America since the Proterozoic: the **Alleghenian orogeny** (sometimes called the Appalachian orogeny) (Fig. 12.6). Across most of Pennsylvania, West Virginia, Ohio, Kentucky, Tennessee, western Virginia, and North Carolina, the modern Appalachian Mountains rose into the sky. Not only are these mountains still with us (unlike the long-gone Taconic and Acadian Mountains), but when we look at the geology of this region, we find evidence of tremendous collision and crustal shortening (Fig. 12.6). The Valley and Ridge province across most of Pennsylvania is characterized by an amazing system of regular folds that remind one of a crumpled carpet in cross section. Some of these folds are incredibly tightly contorted and even flopped over on their side or upside down (recumbent folds). In many places, low-angle thrust faults also break the sequence of folds and stack one package of rocks on top of another. Clearly, some sort of huge tectonic collision must have crushed North America to cause such intensive folding and faulting.

If we look at the ancient tectonic maps of the Paleozoic (Figs. 10.10, 10.11, 11.9, 11.10), we can see that even as the Piedmont terrane collided in the Ordovician and the Avalon terrane in the Devonian, the huge supercontinent of Gondwana was not far offshore from the eastern and southern coast of North America. This long-awaited continent–continent collision finally occurred in the Late Carboniferous, forming the Appalachians (Fig. 12.7). It is analogous to what is happening now with the Indian subcontinent colliding with the belly of Asia, forming the gigantic folded and faulted mountains known as the Himalayas. There is good evidence that the original Appalachians were as high as or higher than the Himalayas in the Pennsylvanian. Today, after over 300 million years of erosion, they have been worn down to their modern, much lower, and less rugged profile.

The Appalachians were the product of the African portion of Gondwana slamming into eastern North America (Fig. 12.7). But the collisional belt was much larger and longer than this. At the same time, the South American portion of Gondwana collided with the southern part of North America, crumpling up mountains such as the Ouachitas in eastern Oklahoma and western Arkansas and the Arbuckle Mountains in south-central Oklahoma. This collisional event is known as the **Ouachita orogeny** (Fig. 12.8). As you visit these ranges, rocks and their folded structure look almost identical to the folded rocks of the Valley and Ridge province in Pennsylvania—except the climate and vegetation are different, and the folks in Arkansas

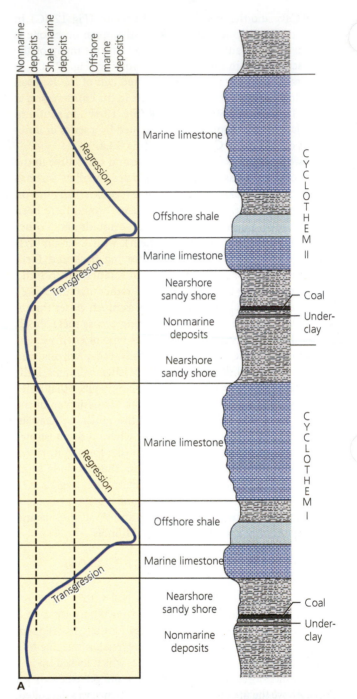

Figure 12.5   A. Classic cyclothem sequence typical of the Illinois Late Carboniferous, showing the repetitive sequence of units. **B.** The cyclothem pattern extends all the way from the Appalachian foothills in Pennsylvania across the Illinois Basin to the open seas of Kansas. **C.** Outcrop photo showing a nearly complete cyclothem from the Caseyville and Abbott formations, on Interstate 24 in southern Illinois. The black coal seam in the middle is clearly visible, as is the gray underclay beneath it. Below those units are river sandstones and shales. Above the coal are alternating layers of marine shales and limestones. **D.** Reconstruction of the sedimentary environments of the transgressing seas that produced most of the rocks of the cyclothem.

(Continued)

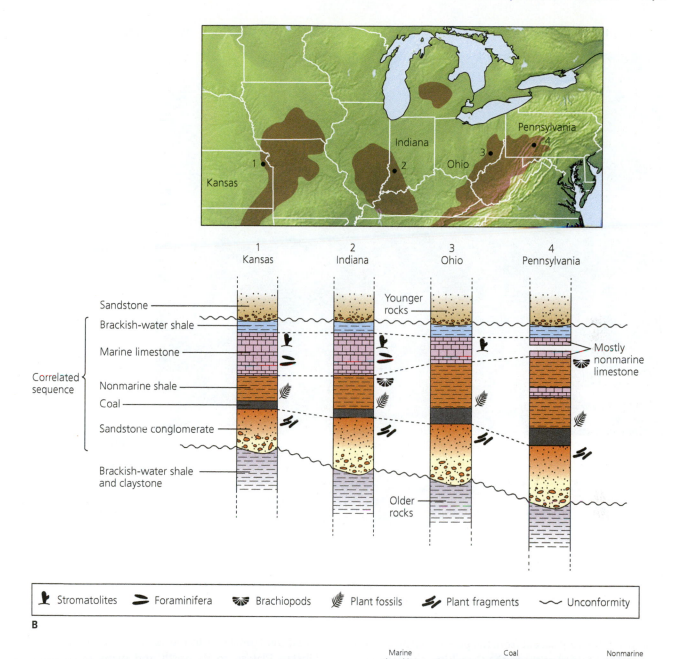

Stromatolites   Foraminifera   Brachiopods   Plant fossils   Plant fragments   Unconformity

B

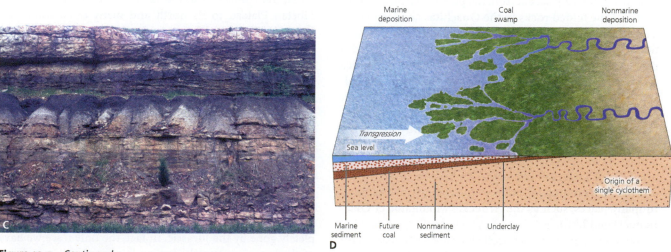

C

Figure 12.5 *Continued*

D

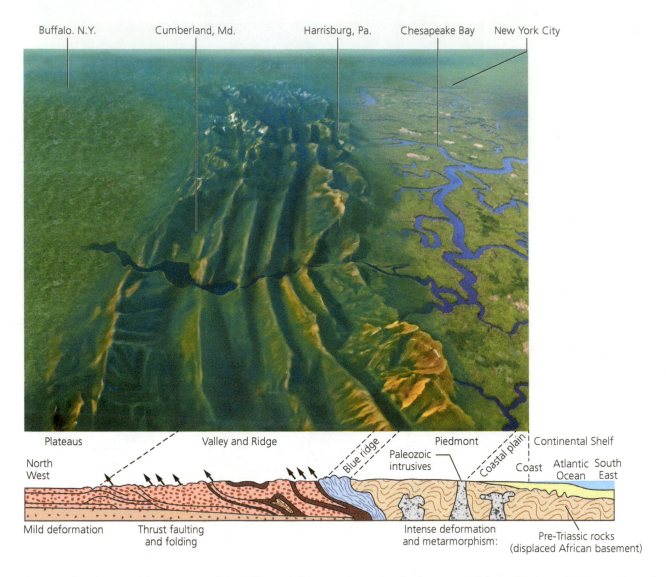

Figure 12.6 Diagram of the structure of the Valley and Ridge province, showing how most of the region is pushed up on thrust faults and tightly folded, forming the valleys and ridges of the terrain.

talk with a different accent from the people in Pennsylvania. Many of the folded rocks in the Ouachitas are deep-water shales and turbidite sandstones (pre-orogenic flysch) of the Stanley and Jack Fork Formations, which filled a deep-water trough in the Mississippian as South America approached and warped a basin downward before the collision. Once the collision occurred, these same deep-water rocks were crumpled up into huge folds and thrust faults (Fig. 12.8D) and rose high into the sky as the original Himalayan-sized Ouachita Mountains. Even as they were still rising in the Late Pennsylvanian, they began to erode away. Thick wedges of conglomerate beds representing the alluvial fans eroding off these young mountains are found in many places, such as the Arbuckle Mountains of Oklahoma (Fig. 12.8E).

But continent–continent collisions are such huge events that the effects are felt far from the edge of the continent. Today, for example, the collision of India with Asia not only raised the Himalayas to the sky but also uplifted the huge Tibetan Plateau to its north and many other mountain ranges in China, southeast Asia, and central-western Asia. Some of the forces were felt far beyond the Appalachian–Ouachita belt and into the cratonic interior of North America, where no mountains had been seen since before the Cambrian. The most striking of these were a series of gigantic mountain ranges in Colorado, Wyoming, Utah, and New Mexico, which rose high into the sky (Fig. 12.9). Deep basins warped down between these ranges and accumulated huge volumes of Pennsylvanian sediment as well. These ranges have been called the **Ancestral Rocky Mountains** because they formed roughly where the modern Rockies lie today, but they eroded away over the rest of the Late Paleozoic and Early Mesozoic and were completely gone by the Jurassic. In fact, the only trace we have of these ancient vanished mountains is the huge volume of sands, shales, and gravels that they shed as they eroded

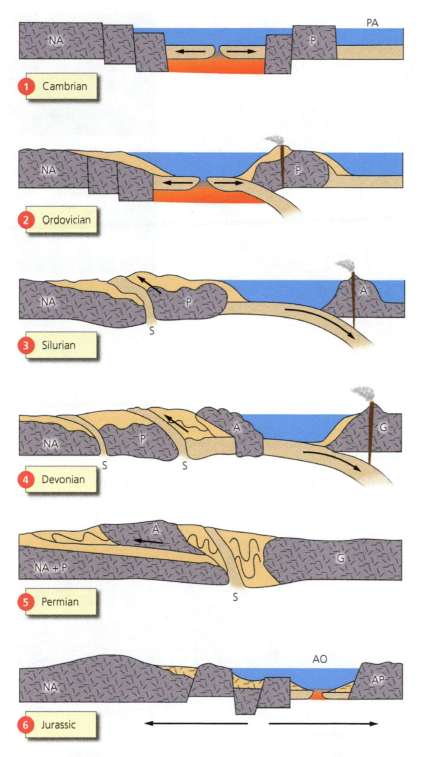

1 Cambrian

2 Ordovician

3 Silurian

4 Devonian

5 Permian

6 Jurassic

Figure 12.7 The sequence of tectonic events in eastern North America (NA). (Top) In the Cambrian, it was a passive margin as the Proterozoic supercontinent broke apart, forming the Proto-Atlantic or Iapetus Ocean. By the Ordovician, the Piedmont terrane (P) was approaching, which collided in the Late Ordovician and Silurian Taconic orogeny, forming a suture zone (S). The Avalone terrane (A) was the next to collide during the Devonian Acadian orogeny. The African portion of Gondwana (G) was still farther offshore, but by the Pennsylvanian and Permian, it collided with North America to form the Appalachians. By the Jurassic (bottom), the continent had rifted apart again, forming the modern Atlantic Ocean (AO).

down. These distinctive red sandstones, shales, and conglomerates are known as the Fountain Formation (Fig. 12.9). This formation forms the remarkable sandstone ridges at Garden of the Gods near Colorado Springs (Fig. 12.9B), the setting for Red Rock Amphitheater west of Denver, and the "Hogback" ridges above Boulder, Colorado, as well as many other distinctive outcrops of red sandstone across the modern Rockies. In the Grand Canyon, the red shales and sandstones of the Supai Group are remnants of an uplift that once lay on the Arizona–New Mexico border.

Finally, there was one more episode of mountain-building in the western United States, but it began in the Late Devonian and continued through the Carboniferous. During this event, a large exotic island arc terrane collided into what was then the Pacific margin of North America. Known as the **Antler orogeny**, the collision was so intense that a huge area of Pacific crust carrying ancient Cambrian and Ordovician sea-floor shales was pushed up over the middle of Nevada along a giant thrust fault called the Roberts Mountains thrust (Fig. 12.10). Thus, the entire western half of Nevada above this thrust fault came from somewhere else, and pieces of Pacific Cambrian–Ordovician sea-floor and deep-water sediments are now stacked on top of Cambrian–Ordovician–Silurian limestones and shales that were deposited in place in Nevada.

In addition to the western half of Nevada, some other blocks accreted to North America during the Late Devonian. Some of them are found in the eastern Klamath Mountains (called the Eastern Klamath Terrane) and today lie just west of Mt. Shasta in northwestern California. Others are found in the foothills of the northern Sierra Nevada Mountains, where they are known as the Shoofly Terrane. Like the rocks above the Roberts Mountain thrust in Nevada, these rocks are deep-water shales that were once far out in the Pacific but have since been stuck on to North America as part of the enormous mountain-building event.

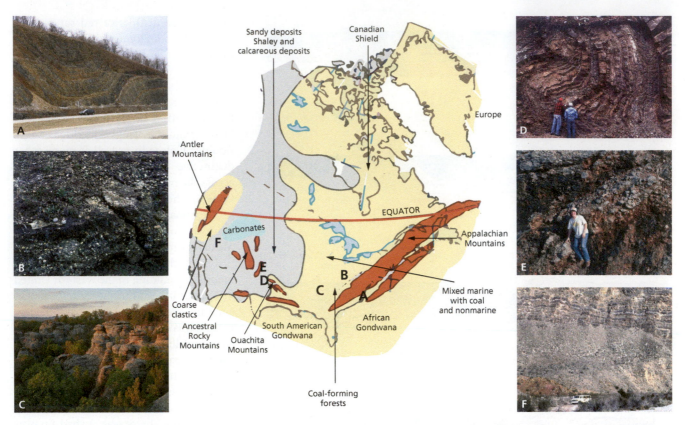

Figure 12.8   Paleogeographic map of the Pennsylvanian, showing the location of the Appalachian–Ouachita–Ancestral Rockies. **A.** Pennsylvanian folds in the Appalachians. **B.** The Pennsylvanian Kyrock Sandstone (here showing a typical pebbly conglomerate layer within the sandstone) overlies the Mississippian limestones that make up Mammoth Cave, Kentucky. **C.** Pennsylvanian cross-bedded sandstones in Garden of the Gods in southern Illinois. **D.** Crumpled Pennsylvanian turbidites in the Ouachita Mountains, Arkansas. **E.** Post-orogenic Collins Ranch Conglomerate, eroded off the newly uplifted Arbuckle Mountains, Oklahoma. **F.** Far west in the Arrow Canyon Range of southern Nevada, the Pennsylvanian is represented by the fusulinid-rich limestones of the Bird Spring Formation.

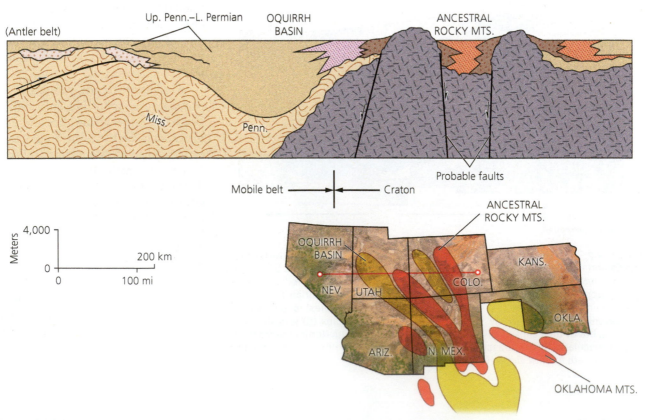

Figure 12.9   A

(Continued)

Figure 12.9 **A.** During the Pennsylvanian, the Ancestral Rockies rose in areas near where the modern Rockies lie. They have been completely eroded away since the Permian, but their sedimentary product, the Fountain Formation, testifies to their original size, elevation, and composition. **B.** Typical outcrops of the Fountain Formation in Garden of the Gods near Colorado Springs. **C.** The Pennsylvanian Supai Group in the Grand Canyon (layered red sandstones and shales between the two limestone cliffs) represents rivers and streams eroded off the Ancestral Rocky uplifts on the border of Arizona and New Mexico.

As the Devonian Antler collision progressed, the continental crust behind this collision began to buckle downward in the Mississippian. This formed an enormous deep-water sedimentary basin in central Utah (Fig. 12.10), known as the Oquirrh (pronounced OAK-er) Basin. During the time this basin sank and flooded with deep water, filling with over 7000 meters (about 23,000 feet) of Mississippian sediment, ranging from shallow-water limestone to deep-water shales and sandstones. The basin continued to subside through the rest of the late Paleozoic, accumulating thousands of meters of Pennsylvanian and Permian limestones and sandstones.

Thus, the Late Carboniferous (Pennsylvanian) was a time of extensive uplift and mountain-building all across North America, from the Appalachians in the east to the Ouachitas and Arbuckles in the south to the Ancestral Rockies and the Antler–Oquirrh belt on the west coast. The Antler belt was a product of collision of an exotic terrane from the Pacific, but the rest of these mountains and their downdropped basins were probably consequences of the enormous continent–continent collision when Gondwana slammed into North America. Such intense uplift and erosion across so much of North America explains why the last of the Mississippian

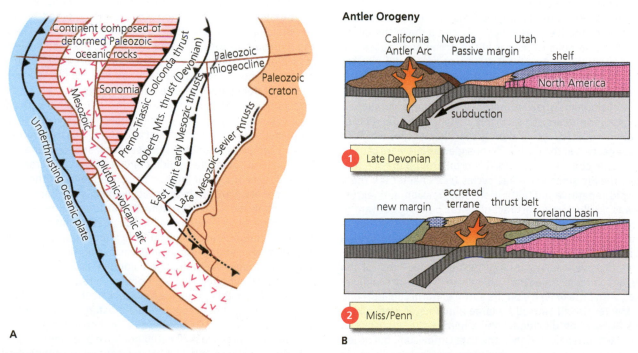

Figure 12.10 Antler orogeny. **A.** Map showing the location of the Antler overthrust belt along the Roberts Mountains thrust in central Nevada. Later collisional belts, such as the Permian-Triassic Golconda Thrust, and the Later Mesozoic Sevier thrusts, are also shown. **B.** Cartoon cross section of the tectonics of the Antler collision, as an arc terrane from the Pacific slammed into northern California and Nevada.

How Do We Interpret the Cyclic Deposition in the Carboniferous?

We have explained why the Absaroka sequence is dominated by shales and sandstones from huge floodplains and deltas that built out from the rising Appalachian–Ouachita uplifts. But what about the second puzzle of the cyclothems: their remarkable repetitive sedimentation patterns? This topic was hotly debated for years. It appeared that sea level went up and down like a yo-yo over a hundred times in order to cause such a repetitious pattern of transgression and regression. For a long time, geologists could not imagine this happening by normal mechanisms. They knew that continents didn't rise and fall that fast to cause so many rapid transgressive and regressive cycles. Some thought that it might be the product of the deltas slowly building out from the mountains, then getting flooded as sea level rose just a little bit faster than the deltas could build out. But that didn't explain the fact that these cycles occur all the way from Pennsylvania to Kansas, not just in the local area of the delta front.

The puzzle was finally solved when geologists working on Late Carboniferous and Permian geology in Gondwana continents (especially South America and southern Africa) documented impressive glacial deposits that formed over the South Pole region of Gondwana at that time (Figs. 5.3, 5.4, 12.11). In many places, they found multiple glacial tills stacked in sequence, proving that there had been many glacial advances and retreats.

As we now know from studying Pleistocene glaciation, the expansion of the ice sheets not only cools down the entire globe, but also pulls water out of the ocean and causes sea level to drop. When the planet warms up and melts glaciers during the interglacial episodes, the glacial meltwater causes global sea level to rise again. Careful analyses of the Carboniferous cyclothems showed that they each had a duration of about 100,000 years. This closely matches the duration of glacial–interglacial cycles during the Pleistocene, which in turn suggests that the Carboniferous glaciations were caused by the same changes in the earth's orbit around the sun that caused the Pleistocene Ice Age cycles (see Chapter 15). Thus, the rapid repetitive fluctuation of transgression and regression was no illusion, no artifact of deltas building outward or local tectonics. It was a reflection of eustatic sea-level change or global rises and falls of sea level in response to the expansion and melting of the Gondwana glaciers.

This discovery raises a larger issue. During most of the early and middle Paleozoic, the earth was a greenhouse planet, with little or no ice on the surface, warm climates, and high sea levels, all caused by large amounts of greenhouse gases like carbon dioxide trapping heat in the atmosphere. There was a brief glaciation in the Late Ordovician, but otherwise we saw almost nothing but warm tropical conditions in North America with huge, shallow, carbonate seas drowning the continents. Yet, as mentioned in Chapter 11, the first signs of Gondwana glaciation appeared back in the Late Devonian, and they were a factor in the Devonian extinctions. The warm tropical conditions of the Mississippian show that there was a return to one final episode of greenhouse conditions. But by the Pennsylvanian, the planet had clearly slipped into an icehouse world, with colder temperatures and large polar icecaps that drew down the oceans from the continents, all caused by the drop in carbon dioxide.

What could have switched the planet from greenhouse to icehouse? Geologists now think that two major factors were involved. The first and most striking is the enormous volume of coal trapped in the earth's crust during the Carboniferous. That coal represents many millions of tons of dead plant matter that removed a lot of carbon dioxide from the atmosphere as well. In the modern world, the processes of decomposition and wood breakdown usually return most of the organic matter in the world's swamp plants back to carbon dioxide fairly rapidly. But in the Carboniferous, there were no termites or any other specialized wood-digesting organisms—they had not yet evolved. So dead trees decayed very slowly. Instead of decaying, most of them dropped into the stagnant swamp water and then slowly turned to coal. Thus, their trapped carbon that they had once pulled out of the atmosphere was locked away in crustal rocks, and took the planet out of its greenhouse state into a full-fledged icehouse.

Another factor is related to the widespread mountain-building of the late Paleozoic. When mountains rise up and their surface rocks weather and erode, the chemical processes of weathering of minerals pull carbon dioxide out of the atmosphere. Major mountain-building events, such as those that dominated the Late Carboniferous, sucked a lot of greenhouse gases out of the atmosphere. Together with the huge volume of carbon locked into coals in the earth's crust, the planet's atmosphere lost a lot of carbon. This transformed it into a complete icehouse, a phase that lasted until the Late Permian (and some effects didn't end until the Late Jurassic).

There was one other interesting consequence of the huge amount of plant growth in the coal swamps of the Carboniferous. Not only did the plants withdraw carbon dioxide from the atmosphere, but they also released a huge amount of oxygen. In fact, numerous geochemical indicators suggest that the level of oxygen in the atmosphere was extraordinarily high at that time. Today, our atmosphere contains 21% oxygen (the rest is mostly nitrogen gas), up from just a few percent in the Early Cambrian. But if the geochemical calculations are right, there was a spike of oxygen up to almost 35%, which would have made breathing Carboniferous air almost like breathing air from an oxygen mask. It could not have been much higher than that because above 35% oxygen content the atmosphere becomes flammable and can combust easily, and we see no

Figure 12.11 Gondwana glaciation left traces all over the southern continents. **A.** The glacial deposit known as the Dwyka tillite from South Africa. (Courtesy Stephanie Scheiber) **B.** A giant boulder or "dropstone," dropped out of a melting iceberg as it floated over finely laminated shales in the open ocean, from the Parana Basin, Brazil.

evidence of widespread wildfires at that time or since then. Thus, the gas mixture we consider normal today was reversed: carbon dioxide dropped to its lowest levels since the Proterozoic snowball earth conditions, while oxygen became so rich that the atmosphere was almost flammable. This had a huge effect on many animals, especially the insects and other arthropods, which grew gigantic when there was so much oxygen available.

shallow carbonate shoals full of clean crinoidal limestones vanished forever. They were flooded with muds and sands eroding down from the Appalachians, the Arbuckles, the Ouachitas, and the Ancestral Rockies, so the waters were too muddy and murky for the classic clear-water carbonate shoal to develop. Instead, the only limestones occur in the more open marine periods when the transgressions across the craton flooded the deltas and rivers building out from the Appalachians or Ouachitas. These are the rocks of the Absaroka Sequence, and they are about as different from the Mississippian Kaskaskia limestones as they could possibly be.

 See For Yourself: The Permian

12.3 The Permian Supercontinent

By the Permian, the Pangea supercontinent was fully assembled (Figs. 5.3, 5.4), with huge ice caps over its southern end on the Gondwana portion (mainly Antarctica, South America, and southern Africa). It also had dense coal swamps across its tropical belt. But it also suffered the effects of being a supercontinent. When continents are smaller and lower in elevation, they have lots of area of coastline, with the cooling, moderating effect of the ocean along most of their edges (called the **maritime climate effect**). But interior regions of gigantic continents do not get any cooling or moderation from nearby oceans, so they experience temperature extremes in what is called **continental climate effect**. Without nearby oceans to buffer the changes in temperature, continental interiors can fluctuate from one extreme to another, often in a single day. For example, in the Great Plains of North America or in the center of Asia, it is not unusual to start out the day with temperatures over 104°F (40°C) and then have a cold front move through and drop the temperature to near freezing by the same afternoon. Likewise, isolation from nearby oceans, and their life-giving moisture in clouds, means that continental centers tend to have drier climates. The great deserts of the central Asian steppe, from the Gobi Desert in Mongolia through the deserts of southern Siberia and western China, as well as the huge Sahara Desert, are all consequences of this isolation. Finally, collision of supercontinents tends to uplift the entire land surface much higher than when the continents are smaller. Thus, you might have the effect of the uplift of the Tibetan Plateau and the Himalayas, which are very high and extremely cold and dry.

Putting this all together, we expect Permian landscapes (except in the tropical coal swamps and polar ice caps) to reflect this drier, more extreme continental climate. Indeed,

that is what we find for North America in particular, which was still tropical to subtropical (Fig. 12.12) during the Permian. The entire eastern half of the continent was still uplifted by the Appalachian–Ouachita orogeny, so there was nothing but erosion; and there are almost no deposits of Permian age in that region (except for minor red beds of the Dunkard Group in fault basins in Ohio and West Virginia; Fig. 12.4G). But in the basins of the Ancestral Rockies and adjacent areas, there are extensive Permian beds. The most famous of these occur in the Colorado Plateau of Arizona, New Mexico, Utah, and Colorado. Most of these sediments are thick cross-bedded sandstones formed in ancient desert dunes (such as the Coconino Sandstone in the Grand Canyon and the De Chelley Sandstone in Monument Valley). These are interbedded with red shales representing arid floodplains, such as the upper Supai and Hermit Formations in the Grand Canyon and the Organ Rock Shale in Monument Valley. In some places, the aridity was so extreme that huge deposits of salt and gypsum were formed. The most famous of these was the Paradox Basin

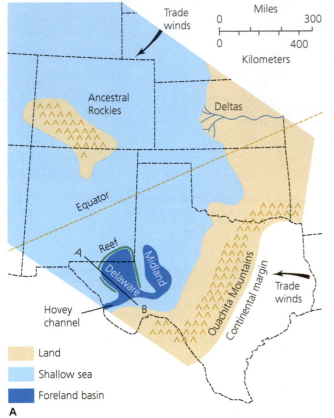

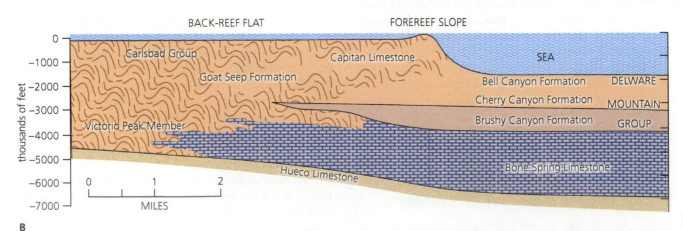

Figure 12.12 Permian reef complex. **A.** Map showing location of basins and reefs in western Texas during the Permian. **B.** Cross section of the Permian reef complex in the Guadalupe Mountains. **C.** El Capitan, the core of the reef complex in Guadalupe Mountains National Park, overlying the older horizontal shales of the basin fill.

of south-central Utah, where over 4600 m (about 15,000 feet) of salt, gypsum, and other deposits accumulated. This area is still an important commercial source of salt even today. Thus, when you think "Permian" in North America, you think of red beds, dune sands, and thick deposits of evaporites.

Most of western North America was in the grip of this harsh desert climate, but there were different conditions on the edge of the continent. In such places, tropical marine limestones were still being formed. We see this effect in the thick sequence of limestones of the Bird Spring Group in southern Nevada and California (Fig 12.8F) and in the Toroweap and Kaibab Limestones that cap the Grand Canyon (Fig 12.9C). One of the most famous of these, however, occurs in southeast New Mexico and adjacent Texas, where large sedimentary basins warped down (Fig. 12.12). Those basins are important for the immense quantity of oil discovered in them, which has powered the economy of western Texas. These basins, such as the Midland Basin beneath Midland, Odessa, and Permian, Texas; the Delaware Basin beneath the Trans-Pecos Texas region and southern New Mexico; and the Marfa Basin on the Mexico–Texas border, were once very deep and filled by thousands of meters of deep-water shales that were the source of all the region's oil.

But on the rim of the Delaware Basin was a shallow carbonate shoal, and on the edge of the rim grew huge buildups of carbonate, which have come to be known as the "**Permian reef complex**" (Fig. 12.12B). Technically speaking, they are not classic reefs held up by corals, like the reefs of the Silurian and Devonian, or the reefs we have today. Instead they are a series of mounds of small bead-like sponges, algae, bryozoans, and other less massive organisms that provides a hard wave-resistant core and a back-reef lagoon. This Permian reef complex is beautifully exposed in Guadalupe Mountains National Park in southeastern New Mexico and across the Texas border (Fig. 12.12). The landmark rock known as El Capitan (Fig. 12.12C) represents the massive core of the reef complex. To the north in the Guadalupe Mountains, you can walk through outcrops that preserved the back-barrier, quiet-water lagoons full of fossils; in other places (like McKittrick Canyon), you can see the original flank beds of the reef that gently dip down from the reef crest. Because of its beautiful accessible exposures, rich fossils, and connection to the oil wealth of west Texas, the Permian reef complex is one of the best-studied reef-like features on earth.

See For Yourself:
Permian Basin and
Reefs

12.4 Life in the Late Paleozoic

MARINE INVERTEBRATES

 Thanks to the Devonian extinctions, the life of the seas of the later Paleozoic was very different from that of the Early and Middle Paleozoic. Gone were the huge reefs made of corals and stromatoporoids. Gone was the huge diversity of trilobites that had dominated the Cambrian and held on through the Devonian. Only one rather simple lineage of trilobites

See For Yourself:
Carboniferous Life

was left, barely hanging on in low numbers and low diversity and rarely found in most Late Paleozoic localities. The only remaining corals were smaller and mostly solitary, with only a few examples of reefs around the world. Gone was the diversity of forms during the huge radiation of the "Age of Fishes," especially the armored jawless fish, the diversity of jawed placoderms, and most of the lobe-finned fish (Figs. 11.14, 11.15). Only small sharks and bony fish remained at the top of the marine food chain. In fact, the age of giant fish with heavy bony armor was gone forever, and most of the fish opted for less armor and greater mobility, perhaps driven by the presence of larger mobile predators like more advanced sharks. Even the smaller predators, like the goniatite ammonoids (Fig. 11.12H) that dominated the Devonian and were nearly wiped out in the Devonian extinction, recovered and underwent another huge evolutionary radiation in the Carboniferous. The thick-shelled nautiloids, on the other hand, did not recover, so once again we see a trend toward more lightly built, mobile, agile predators.

As we have already seen, the extensive shallow seas of the Mississippian were dominated by crinoids, so much so that some people call the Early Carboniferous the "age of crinoids" (Fig. 12.3A). Crinoids reached their heyday in numbers and diversity as they spread across the shallow carbonate seas that had once harbored large coral reefs. Living alongside the crinoids was another important group of echinoderms related to crinoids and sea stars, known as **blastoids** (Fig. 12.13A, B). Like crinoids, they had a head with long delicate arms for catching food particles and a long stalk that allowed them to feed high above the sea floor. However, their detailed anatomy is very different from that of the crinoids, so they are a different order. Most of their fossils are incomplete, with just the heads being preserved. However, these blastoid fossils are very distinctive, especially the five-sided, flowerbud-shaped *Pentremites* (Fig. 12.13A, B), which is extraordinarily abundant in certain localities and a good index fossil of the Mississippian.

In the slightly deeper, muddier habitats next to the shallow banks dominated by crinoids and blastoids are found lots of bryozoans (Fig. 12.13C–E). Bryozoans occurred in nearly every fauna of the Paleozoic after the Cambrian, but they reached the pinnacle of their abundance in the Mississippian. The typical Mississippian bryozoans were a group that had a very "lacy" appearance, with thousands of tiny individual bryozoan animals inhabiting the pin-prick-sized holes in the lacework (Fig. 12.13C). The most distinctive of these was the genus *Archimedes* (Fig. 12.13D, E), whose preservation usually consists only of its corkscrew-shaped central supporting column. However, the complete colony had a helically spiraled fan of lacy bryozoan latticework that was usually broken off when they were fossilized. The genus got its name from the famous ancient Greek mathematician and scientist Archimedes, who invented a simple pumping device of a screw-like blade inside a cylinder, which is still used to pump water uphill in many parts of the world. In some units, such as the Warsaw Shale in Illinois and Iowa, *Archimedes* are so abundant that the

Figure 12.13   Typical Mississippian invertebrate fossils. **A, B.** The blastoid *Pentremites*, an index fossil of the Mississippian, showing both the flowerbud-shaped heads and a econstruction of the animal showing the arms. **C.** The lacy fenestellid bryozoans were extremely common in the Mississippian. **D, E.** The most striking of the bryozoans was the corkscrew-shaped *Archimedes*, another Mississippian index fossil. It wrapped its lacy fenestellid bryozoan structure into a spiral, with the fossilized "corkscrew" structure forming the central axis. **F.** The genus *Spirifer* was a common Mississippian brachiopod. **G.** Typical brachiopods of the later Paleozoic were the cup-shaped productids, with the deep cup-like shell sitting in the mud held up by stilt-like spines and a flat lid-like shell on top. **H.** Diorama of a typical Permian reef assemblage, dominated by spiny productids along with a few lacy bryozoans and crinoids and trilobites. **I.** The most peculiar of the productids were the soap dish–shaped fossils known as *Leptodus*.

formation used to be called the "*Archimedes* beds." *Archimedes* is an index fossil of the Mississippian, along with the lacy bryozoans, most of which vanished by the Pennsylvanian.

The Devonian coral reefs vanished, and were replaced by thick stands of stalked echinoderms (crinoids and blastoids) and various forms of lacy bryozoans, all feeding in the higher waters above the sea floor. The floor of the ocean itself still had the usual snails and clams and especially a diversification of brachiopods. The spirifers (Fig. 11.12G) went through a crash in the Devonian extinction but recovered in the Mississippian, so the genus *Spirifer* itself is a common index fossil of that period (Fig. 12.13F). But a new group of brachiopods evolved that soon would dominate the shelly faunas of the later Proterozoic: the productids

(Fig. 12.13G, H). These were peculiar brachiopods that didn't have a stalk or attachment to the sediment or to rocks. Instead, they were adapted to living unattached and sitting on muddy sea bottoms, with lots of tiny spines holding them up out of the mud like stilts (Fig. 12.13G, H). The lower shell was shaped like a deep cup and rested on the sea floor, while the other was like a hinged flat lid on the top, which opened to admit water currents bearing food and oxygen. Their larvae, however, still had threads in the hinge area so that they could grow attached to the stalk of a crinoid or some other raised surface; once they reached a certain size, they broke loose and settled on the loose mud of the sea bottom.

Thus, a typical assemblage you might collect in the Mississippian is dominated by crinoids, blastoids, bryozoans, and brachiopods, with minor clams, snails, solitary corals, and extremely rare trilobites. This was the last time that clean, shallow, carbonate seas dominated the continents of the earth; and most of this assemblage changed dramatically in the Pennsylvanian and Permian. The clean, mud-free limestones were replaced by sandy and muddy habitats in the Pennsylvanian and Permian. Such changes meant that crinoids, blastoids, and bryozoans were much less abundant. Instead, the marine rocks of the typical cyclothem (Fig. 12.5) are loaded with the productids, which are well adapted to muddy bottoms with their spiny stilts. The muddy habitats also favored clams and snails, which prefer these habitats where they could burrow. Crinoids, blastoids, bryozoans, corals, and trilobites were still found in some places; but they are very rare compared to their dominance in the Early and Middle Paleozoic.

One of the major new groups to take over during the Late Carboniferous was a group of single-celled amoeba-like creatures known as **foraminiferans**. These amoebas are still abundant in modern oceans, both in the plankton and in the marine sediment; but in the Late Paleozoic they were mostly sea floor dwellers. The most abundant group was the foraminifers called **fusulinids** (Fig. 12.14). Shaped like a large grain of rice, fusulinids grew in a tight spiral around a central axis, so they ended up somewhat spindle-shaped. Even more incredible is the fact that, though the shell of the fusulinid can be up to 1 cm or more long (about half an inch), it was produced by a single-celled organism! (Bottom-dwelling foraminiferans have gotten this large several times in the history of the oceans.) Most importantly, they were hugely abundant in late Paleozoic seas (especially during the Permian), when they made up huge, thick units of limestone with trillions of fusulinids in some places. Their other great value is that they evolved very rapidly (Fig. 12.14), so they are the best time indicator for the Late Carboniferous and Permian.

By the end of the Permian, the sea floor was inhabited by a very strange and specialized fauna. The inhabitants of the Permian reef complex in Texas and New Mexico are typical. As discussed already, the "reef" is actually a densely overgrown mound of algae, sponges (especially small sponges shaped like a string of beads), bryozoans, and dense masses of fusulinids (Fig. 12.15). Living in the quiet waters behind the reef were dense clusters of productid brachiopods. Some of these productids were truly bizarre in their shape. Some of them, known as richthofenids (Fig. 12.16), had modified the simple round cup of a typical productid into a cone-like shape, with the other shell forming a tiny lid inside the cone. But the weirdest of all the brachiopods was a group of productids that had one shell shaped like a shallow soap dish and another shell that was a simple comb-like device that covered part of the soft tissues (Figs. 12.13I, 12.16). How these strange creatures lived is still a mystery, but it is clear that brachiopods were becoming more and more specialized. This incredibly weird and specialized community vanished completely when the greatest mass extinction in earth history occurred. Before we discuss this catastrophe, however, we need to look at life on the land.

LAND LIFE

As we saw in Chapter 11, land life was just getting started by the Late Devonian, with the first large trees and dense forests in the wetter habitats, the first amphibians, and a diverse fauna of insects, spiders, scorpions, millipedes, and centipedes dominating the land surface. From this foothold, land life really began to diversify in the Carboniferous.

The first 15 million years of the Carboniferous, however, have a relatively poor record of land life. This is partly because sea level was so high during the Mississippian Kaskaskia transgression that most continents were drowned. There was very little land habitat except in the molasse deposits around the foothills of the deeply eroded Acadian Mountains in the Appalachian region (see Chapter 11) and around the Caledonian Mountains in Scotland. Lately, new discoveries in Scotland and Pennsylvania have produced fossils that have improved our understanding of land animal evolution. In addition, land life (especially land animals) suffered from the Late Devonian extinction, so there was a lag before they really began to diversify and become common again. There is also geochemical evidence that suggests that oxygen levels were very low in the Early Carboniferous, which would have inhibited the evolution of land animals.

By the Middle and Late Carboniferous, however, there are numerous terrestrial fossil localities, especially in the broad deltas and swamps that formed during the great mountain-building events and cyclothems. In North America, these are best known from the giant tropical swamps that produced the coal that gave the Carboniferous its name. As we have seen, these coal swamps were enormous and extended over areas that ranged from Pennsylvania to Kansas. They were formed by the giant versions of the primitive vascular plants we first saw in the Devonian. The most impressive of all were the giant "club mosses" or lycophytes, some of which formed huge trees, the tallest the earth had seen up to this point. The largest was *Lepidodendron* (Fig. 12.17A, B), which reached about 30 m (about 100 feet) in height and over 1 m (3 feet) in diameter at the base, with a distinctive diamond-shaped pattern of leaf scars on its trunk. Another common lycophyte is *Sigillaria*, with two branches tipped by brushes of leaves (Fig. 12.17C). The next most important group are the giant "horsetails" or "scouring rushes" (sphenophytes), which grew into trees that reached 8 m (about 26 feet) in height (Fig. 12.17D). Dense ferns filled in the undergrowth, as shown by abundant fern fossils found in many coal beds. There were also tree-sized seed ferns, which resembled true ferns but were more advanced in having seeds rather than spores. One of the most famous of these was *Glossopteris*, the Permian seed fern that was found on all the Gondwana continents (Fig. 5.4).

See For Yourself:
Coal Swamp Trees

During the Permian, the drier highlands away from the swamps were the site of a newer, more advanced kind of plant, the **gymnosperms**. These plants invented a method of fertilization that does not require water, a handicap which severely limits ferns, lycopods, sphenopsids, and more primitive plants like mosses and liverworts. Gymnosperms (which means

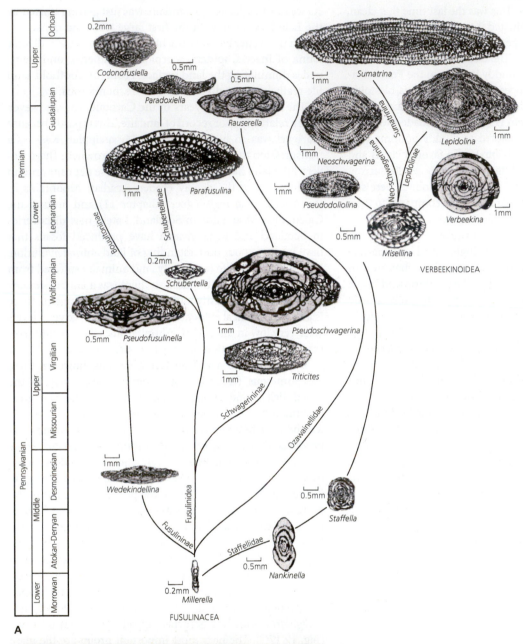

Figure 12.14 **A.** Evolution of late Paleozoic fusulinid foramin-iferans. From simple forms like the Early Pennsylvanian *Millerella* (bottom), fusulinids evolved into larger and more complex shapes, all wrapped like a spindle around a long axis. The de-tails of the cross sections of the growth chambers shown here help the paleontologist identify the species of fusulinid, and these tiny amoeba-like creatures are the best index fossils of the Pennsylvanian and Permian. **B.** Typical Permian limestone full of numerous etched fusulinids.

Figure 12.15 Diorama of Permian reef complex. The surface is covered by numerous bead-like sponges *Girtyocoelia*, tall cylindrical sponges, horn corals, spiriferide and productid brachiopods, as well as predatory goniatite ammonoids swimming above.

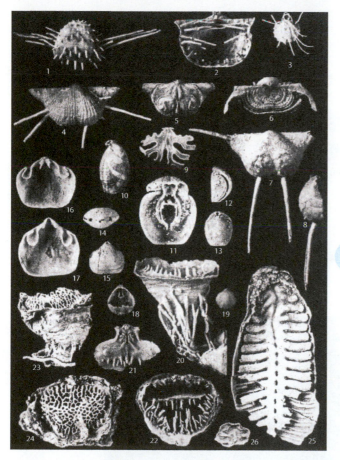

Figure 12.16 This plate shows a spectrum of typical Permian brachiopods. The dominant brachiopods of the later Paleozoic were the cup-shaped productids, which had no pedicle to attach to surfaces as adults but rested on the bottom with spines acting as stilts to keep them from sinking into the sediment. Specimens 1–7 are typical spiny productids. Productids evolved into the bizarre cone-shaped richthofenids (20–22), which had one valve shaped like a cone with its point stuck in the sediment and the upper flat valve as a small lid. The strangest of all productids were genera like *Leptodus* (25), which had a lower valve shaped like a soap dish and a grill-like upper valve (see also Figure 12.13I).

"naked seed" in Greek) usually have seeds placed in cones (as in pine trees and their relatives). The female cone holds the eggs inside, and when it opens, the pollen from the male cone is blown to it and sticks to its sap, fertilizing the egg. This produces a seed adapted for living in dry land far from water. By the Permian, we find the first primitive pine trees, such as *Walchia*, in many upland localities (Fig. 12.17E). There were also huge primitive gymnosperms like *Cordaites* (Fig. 12.17F) that reached 30 m (about 100 feet) in height and formed large coniferous woodlands reminiscent of modern pine forests.

See For Yourself: Carboniferous Coal Swamps

See For Yourself: David Attenborough on the First Forests

Roaming these coal swamps were a wide variety of animals. Among the most spectacular were the arthropods, which grew to enormous size at this time (Fig. 12.18). The dragonflies had wingspans of 75 cm (2.5 feet), the size of an eagle. There were cockroaches over 30 cm (1 foot) long. But the all-time largest terrestrial arthropod that has ever lived was the enormous millipede relative *Arthropleura*, which was almost 2.4 m (almost 8 feet) long (Fig. 12.18B)! If living bugs and creepy-crawlies make you shudder today at their present size, imagine encountering giant dragonflies and cockroaches and having to evade a millipede almost 8 feet long!

Why they got so large has long been a mystery, until the discovery of the high levels of atmospheric oxygen gave us a clue. One of the main things that prevents arthropods like insects, spiders, and scorpions from getting big is their inefficient respiratory system, which does not transfer oxygen to all parts of the body very well. In particular, there are narrow spots and bottlenecks in the trachea, especially in the joints of the limbs that restrict how much oxygen reaches critical parts of the body. But if atmospheric oxygen levels were really 35%, rather than the present-day values of 21%, then it's easier for an arthropod to get big.

See For Yourself: Giant Insects of the Carboniferous

Preying on all this arthropod food was a wide range of amphibians that had evolved from the primitive forms that crawled out of the water in the Late Devonian. The largest and most impressive of these were the flat-headed, flat-bodied amphibians known as **temnospondyls** (Fig. 12.19). In the Carboniferous, these reached up to 6 meters (almost 20 feet) in length and looked vaguely like crocodiles, only with a broad flat skull and snout. In the Permian of Brazil, there was a huge, narrow-snouted, crocodile-shaped temnospondyl called *Prionosuchus*, which reached 9 meters (almost 30 feet) in length. The Lower Permian red beds of north-central Texas are famous for their huge temnospondyls, best known from the sprawling, wide-skulled form *Eryops* (Fig. 12.19B), which at 3 meters (about 10 feet) in length, was the largest animal ever to walk on land at that time. The second great group of amphibians were the **lepospondyls**, which came in a range of different sizes and shapes, from the tiny lizard-shaped microsaurs to the snake-like aistopods to the truly weird *Diplocaulus*, which had broad horns on the side of its head that made its skull look like a boomerang (Fig. 12.19C). The third group, the **anthracosaurs**, had deep skulls with narrow snouts and many features that made

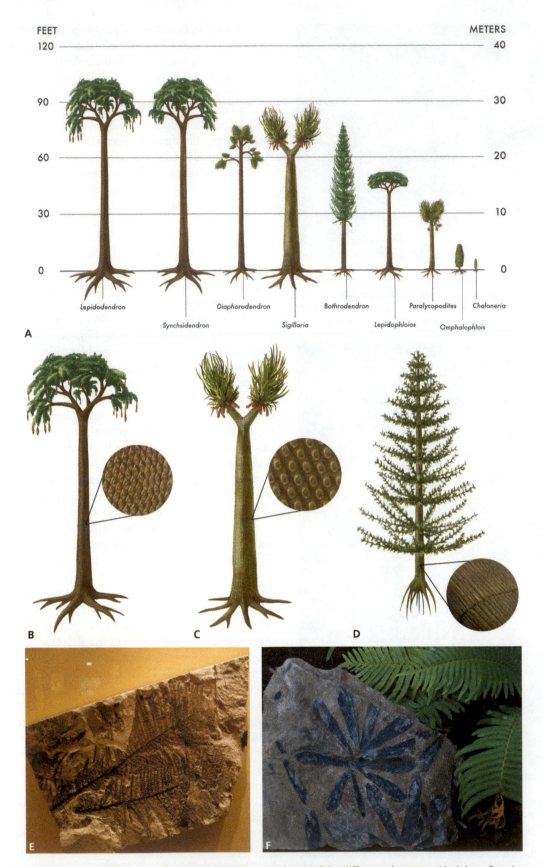

Figure 12.17   Typical plants of the Carboniferous coal swamps. **A.** Sketch of the different shapes and heights of coal swamp trees, mostly lycophytes. **B.** *Lepidodendron*, a giant lycophyte tree. **C.** The bushy lycophyte *Sigillaria*. **D.** Sphenophytes like *Calamites* grew into giant trees. **E.** Primitive conifers like *Walchia* were common in upland, drier habitats in the Pennsylvanian and Permian. **F.** Another large primitive Late Paleozoic gymnosperm was *Cordaites*, with strap-like leaves (here next to a modern fern frond).

Figure 12.18 Giant arthropods of the Carboniferous. **A.** Fossil of the eagle-sized dragonfly *Meganeura*, with the meter-wide wingspan. **B.** Life-sized model of the gigantic millipede relative *Arthropleura*.

them transitional to reptiles (Fig. 12.19C). Finally, by the Early Permian the first transitional fossils linking frogs and salamanders to earlier amphibians appeared. One fossil in particular, *Gerobatrachus hottoni*, has been nicknamed the "frogamander" because it has a salamander-like body with a frog-like head. By the Late Permian and Triassic, primitive frogs had appeared.

Reptiles also originated at this time and flourished during the Late Paleozoic (Fig. 12.20A). The oldest known reptile, *Westlothiana lizziae*, comes from the late Early Carboniferous of Scotland and was built like a small, long-bodied lizard. Similar primitive reptiles called *Hylonomus* are found in old lycopod tree stumps in Nova Scotia. By the Permian, reptiles had truly begun to diversify. They came in a wide variety of shapes and sizes, from the huge hippo-like pareiasaurs (Fig. 12.20B), which weighed up to 600 kg (about 1300 lb), to the swimming *Mesosaurus* (Fig. 5.5), found in Permian lakebeds in Brazil and Africa, to the earliest turtle ancestors, which had a shell on their belly but not on their backs and still had teeth in their beaks. By the Late Permian, the ancestors of the lineage that led to dinosaurs, crocodiles, snakes, and lizards were present, although those groups would not evolve until the Triassic.

But reptiles did not rule the world yet. Far more important was the wide range of synapsids or **protomammals**. These creatures used to be called "mammal-like reptiles," but that term is misleading and has been abandoned because synapsids are a separate lineage from true reptiles, which originated at the same time and evolved side by side with true reptiles. Most protomammals were small lizard-like forms in the Late Carboniferous, but by the Permian they took over the terrestrial realm. In the Lower Permian red beds of north Texas, they are represented by the huge fin-backed predator *Dimetrodon* (Fig. 12.20C), which reached 4.6 meters (15 feet) in length, the largest predators of their world. There were also other primitive protomammals in the

Early Permian, such as the herbivorous finback *Edaphosaurus* and the fish-eating *Ophiacodon*.

 See For Yourself: *Dimetrodon*

This Early Permian radiation of protomammals (sometimes referred to by the outdated name "pelycosaurs") evolved into a much greater variety of protomammals in the Late Permian, often lumped into the wastebasket group "Therapsida." As seen in red beds from South Africa, South America, and Russia, these protomammals included some of the first herbivores known. One group, the dinocephalians ("terrible heads" in Greek) had huge hippo-like bodies up to 2.7 meters (almost 9 feet) long and thick, bony skulls, often with weird flanges and protuberances projecting from around their eyes (Fig. 12.20D). The other main group of herbivorous protomammals was the dicynodonts ("two dog teeth" in Greek), which had a large plant-slicing beak with no teeth except two huge canine tusks (Fig. 12.20E). These ranged from the small dog-sized *Lystrosaurus* (Fig. 5.5) to a variety of forms that were pig-sized and even hippo-sized. In addition to these early herbivores, there was a wide variety of predatory protomammals in the Late Permian, most notably the tiger-sized predators with saber-like teeth known as the gorgonopsians (Fig. 12.20F).

 See For Yourself: Permain protomammals

The Permian landscape was densely forested with primitive swamp plants and dense forests of conifers in the uplands. Gigantic insects and millipedes strolled through the coal swamps. Living in the ponds and rivers was a wide range of amphibians, many of which had the size and shape of large crocodiles. True reptiles of many different sizes and shapes were found on the land and in the lakes and rivers. Ruling over all of them were the synapsids or protomammals, which began with the finbacks of the Early Permian and climaxed with the huge predatory gorgonopsians and a variety of the first herbivorous land animals, the dinocephalians and dicynodonts.

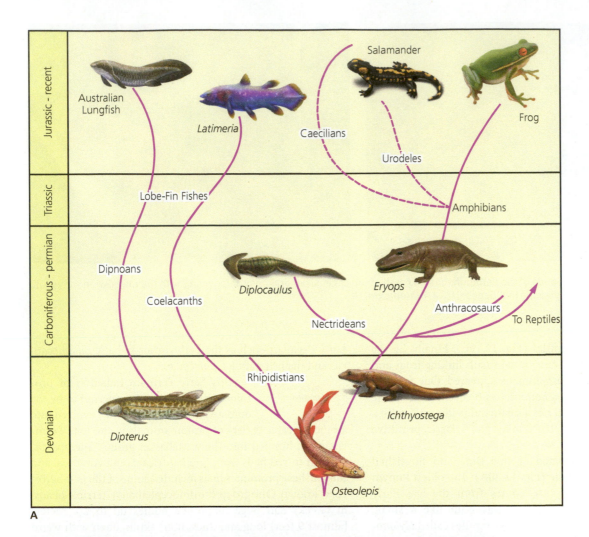

A

B

C

D

Figure 12.19 Amphibians of the Late Paleozoic. **A.** Family tree of the amphibians. **B.** *Eryops*, the largest known complete fossil of a temnospondyl. **C.** The boomerang-headed lepospondyl *Diplocaulus*. **D.** The reptile-like anthracosaur *Seymouria*.

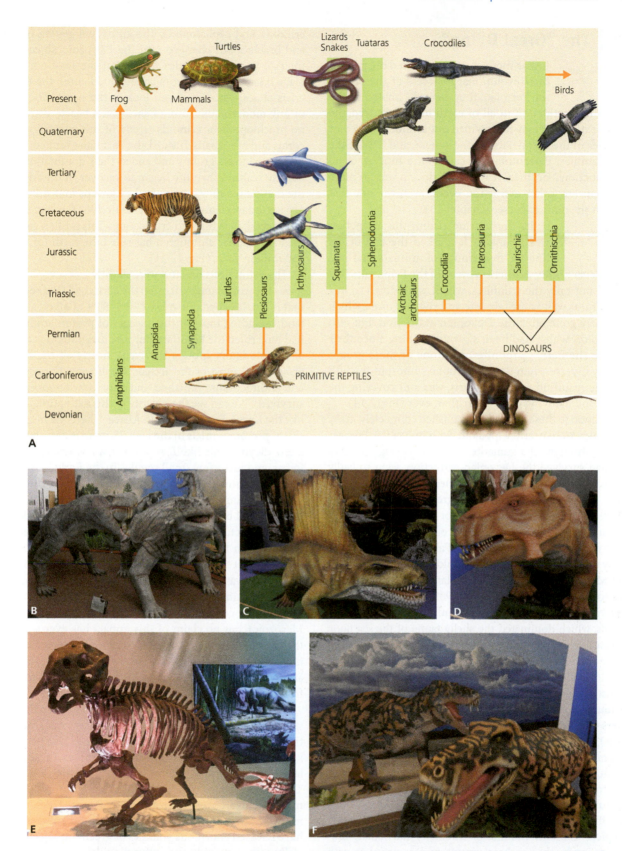

Figure 12.20   A. Family tree of tetrapods. **B.** Pareiasaurs were primitive hippo-sized herbivorous reptiles with huge ugly bumps and ridges on their faces from the Late Permian. This one is being attacked by a top predator, a gorgonopsian. **C.** The top predator of the Early Permian was the fin-backed protomammal *Dimetrodon*. **D.** The huge dinocephalian protomammals of the Late Permian were hippo-like forms with thickened skulls for head butting. Some had bizarre bumps and ridges on their skulls, like this *Estemnosuchus*. **E.** The other main group of Late Permian herbivorous protomammals was the toothless tusked dicynodonts. **F.** The main predators of the Late Permian were the bear-sized gorgonopsians.

12.5 The "Great Dying"

 See For Yourself: Permian Plants

Just as these creatures reached the peak of their evolutionary success at the end of the Permian, they were decimated by the greatest mass extinction in all of earth history. Indeed, the Permian extinction may have wiped out as much as 95% of the species in the ocean, 83% of the genera, and 57% of the marine families. Paleontologists call it "the mother of all mass extinctions" or "the Great Dying" since it was far more severe than any other extinction event, even the one that wiped out the dinosaurs (see Chapter 13).

 See For Yourself: Permian Extinction

In the marine realm, the extinction was dramatic. It marked the end of the line for five groups of animals that had survived every other extinction event of the Paleozoic. This included not only the trilobites, which were down to just two genera, as well as the tabulate and rugose corals, which were nearly gone after the Devonian extinction, but also the blastoids, which had been slowly declining since their heyday in the Early Carboniferous. None of these groups were very diverse when the catastrophe came, but the fifth group, the fusulinid foraminiferans, was still covering the sea floor and evolving even faster than ever (Fig12.14).

In addition to these groups that vanished completely, many other groups were nearly wiped out, with only a few lineages straggling through. The goniatite ammonoids (Fig. 11.12H), which had survived the Devonian extinction, were so badly impacted that only two or three lineages survived, which became the ancestors of another great radiation of ammonoids in the Mesozoic. The major groups of Paleozoic bryozoans vanished, resulting in 80% of the genera going extinct, as did the 98% of the genera of Paleozoic crinoids, leaving only a few minor families that would repopulate the world in the Mesozoic. The extinction of clams was not nearly so severe (only 59% of the genera), but 98% of the marine snail genera vanished. Even more importantly, the ubiquitous brachiopods, which dominated the shelly fauna of the sea floor, were nearly wiped out. About 96% of their genera vanished, marking the end of the orthids (that had been dominant in the Ordovician) and especially the productids, which were the most common group of the Late Paleozoic (Fig. 12.13H). Only four lineages managed to survive into the Mesozoic: the spirifers (which hung on until the extinction at the end of the Triassic), and the three groups that are alive today, the inarticulate brachiopods known as lingulids, the lampshell-shaped terebratulids, and the corrugated rhynchonellids.

All told, this devastating event completely transformed the nature of marine faunas. We saw how the Great Ordovician Biodiversification Event (Chapter 10) pushed aside the typical "Cambrian fauna" of trilobites, inarticulate brachiopods, and other archaic groups and replaced it with the "Paleozoic fauna" (Fig. 12.21). This "Paleozoic fauna" dominated from the Ordovician to the Permian, with such typical fossils you can collect in any Paleozoic locality after the Cambrian, especially brachiopods, bryozoans, corals,

crinoids, and cephalopods. The individual genera of corals and brachiopods changed over the Paleozoic, so each time interval has its own index fossils; but the sea-floor assemblage is largely similar from about 500 Ma until the Permian crisis at 250 Ma. But if you look at the ocean floor today, or any modern collection of shells on a beach, you will see no brachiopods or crinoids or any of the extinct groups of corals and bryozoans. Instead, the Permian event completely decimated these groups and ended their domination of the marine realm. When life began to recover in the Mesozoic, the marine realm was colonized by groups that still rule the oceans today: clams, snails, and sea urchins (among the shell producers), plus fish and crustaceans. This is often called the "Modern fauna," although it has ruled the earth for the past 250 million years (Fig. 12.21).

The extinctions on land were slightly less severe, but still it was the worst land extinction the earth had ever seen up to that point. It wiped out at least 70%–80% of all creatures on land, including most of the lineages of protomammals that had long ruled the Permian (only the dicynodonts and some of the predatory groups survived), nearly all the reptilian groups, and most of the archaic amphibians (lepospondyls and anthracosaurs vanished, and only a few temnospondyls survived). We can see the severity of this extinction in places like the Karoo Desert of South Africa, where the huge diversity of protomammals in the Late Permian is depleted to just a few dicynodonts like *Lystrosaurus*, a few small predatory forms, and just a few small reptiles in the Early Triassic.

At one time, scientists blamed this event on the assembly of the Pangea supercontinent (which wiped the shallow seas crushed between the colliding continental blocks), but that even occurred over 50 million years earlier. Others have blamed it on the growth of the great Gondwana ice sheet (Fig. 5.4), but that too was already in place 100 million years earlier and was actually declining by the Late Permian. A few scientists claimed the earth was hit by a giant impact from a meteor or comet (like what happened at the end of the Mesozoic), but no claim of this impact has ever withstood the scrutiny of other scientists. None of the typical signs of impact (such as droplets of melted crustal rock splattered around the earth or rare elements like iridium which come from space rocks) have even been proven to exist.

Instead, the cause of the "mother of all mass extinctions" seems even more frightening: the biggest volcanic eruptions in all of earth history, which produced lava beds today known as the **Siberian traps**. In many places, these ancient lava flows are still visible in northern Siberia (Fig. 12.22). This event erupted almost 4 million km^3 (almost 1 million miles^3) of lava, covering 2 million km^2 (about 770,000 square miles). Such eruptions from deep mantle sources would have released enormous volumes of greenhouse gases, especially carbon dioxide and sulfur dioxide.

This, in turn, triggered a massive super-greenhouse climate. The oceans then became supersaturated with carbon dioxide, making them too hot and acid and killing nearly everything that lived there. Ocean temperatures are

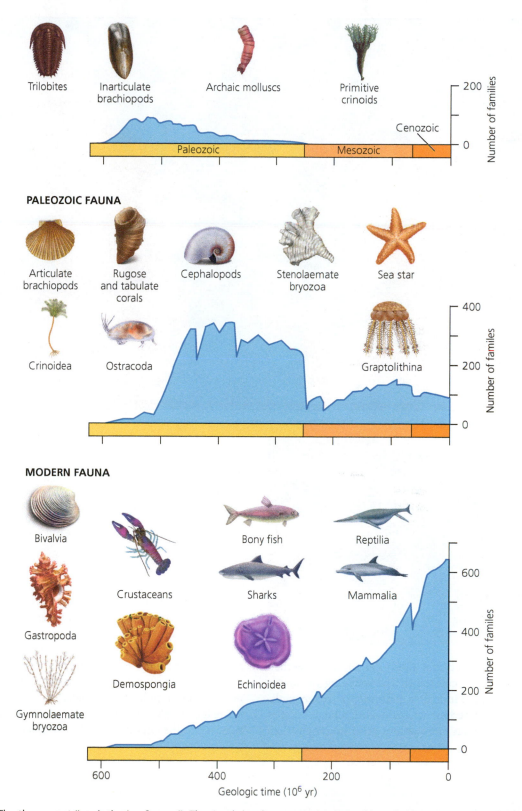

Figure 12.21 The three great "evolutionary faunas". The Cambrian fauna was dominated by trilobites, inarticulate brachiopods, and archaic and experiment echinoderms and mollusks. They declined during the Paleozoic, but were always present in the background. The Paleozoic fauna took over during the Ordovician, and dominated the seafloor until the end of the Permian. It consisted of articulate brachiopods, tabulate and horn corals, crinoids, primitive stenolaemate bryozoans, graptolites, and goniatite ammonoids. These were decimated or went extinct in the Permian event, to replaced by the Modern fauna of clams, snails, sea urchins, crustaceans, and bony fish that have dominated the seafloor ever since. Most of these groups were alive during the Paleozoic, but in the background while the Paleozoic fauna dominated.

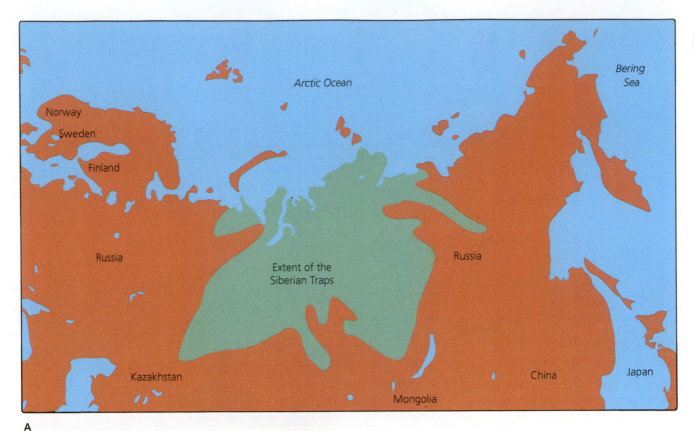

A

B

Figure 12.22 The Siberian trap eruptions were the primary trigger of the Permian catastrophe, sending enormous volumes of greenhouse gases into the atmosphere that triggered a rapid "super-greenhouse" that oversaturated the oceans with carbon dioxide and sulfur. **A.** Map of the areal extent of the Siberian lavas, the largest volcanic event in earth history. **B.** Stacked lava flows in a cliff in Siberia.

estimated to have reached over 40°C (104°F), far hotter than even most tropical life can stand. The warming of the sea floor may have released immense quantities of frozen methane from the bottom sediments, producing a huge burst of methane that is an even more potent greenhouse gas than

carbon dioxide. There are many places where the black shales and geochemical evidence suggest that the waters became depleted in oxygen and maybe even poisoned by hydrogen sulfide. The atmosphere was also low in oxygen and full of excess carbon dioxide, so nearly all land animals above a certain size vanished and only a few smaller lineages of synapsids, reptiles, amphibians, and other land creatures made it through the hellish planet of the latest Permian and survived to the aftermath world of the earliest Triassic.

RESOURCES

BOOKS AND ARTICLES

Benton, Michael J. 2003. *When Life Nearly Died: The Greatest Mass Extinction of All Time*. Thames & Hudson, London.

Bird, J. M., and J. F. Dewey. 1970. Lithospheric plate-continental margin tectonics and the evolution of the Appalachian orogeny. *Bulletin of the Geological Society of America* 81:1031–1060.

Blakey, R. C., and W. D. Ranney. 2018. *Ancient Landscapes of Western North America: A Geologic History with Paleogeographic Maps*. Springer, Berlin.

in Boardman, R. S., A. H. Cheetham, and A. J. Rowell, eds., 1987. *Fossil Invertebrates*. Blackwell Scientific Publishers, Cambridge, Mass

TIMELINE OF LATE PALEOZOIC EVENTS

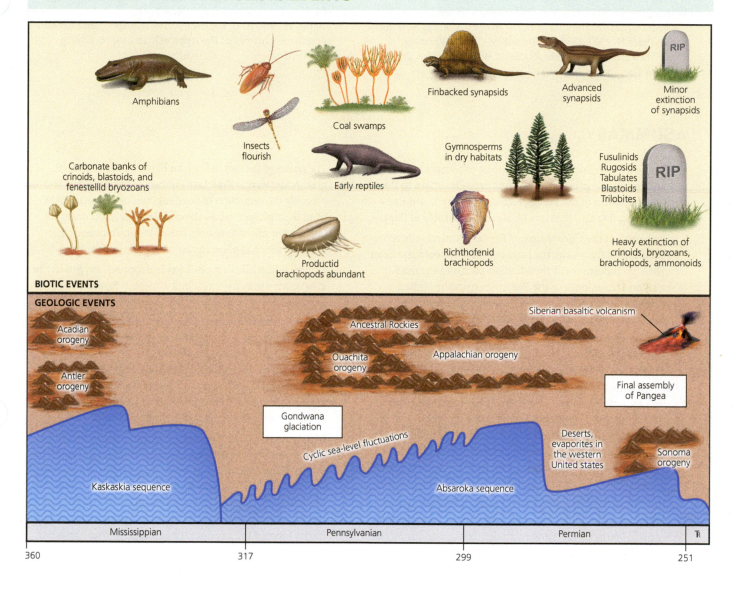

BIOTIC EVENTS

- Amphibians
- Insects flourish
- Coal swamps
- Finbacked synapsids
- Advanced synapsids
- Minor extinction of synapsids
- Carbonate banks of crinoids, blastoids, and fenestellid bryozoans
- Early reptiles
- Gymnosperms in dry habitats
- Fusulinids Rugosids Tabulates Blastoids Trilobites — RIP
- Productid brachiopods abundant
- Richthofenid brachiopods
- Heavy extinction of crinoids, bryozoans, brachiopods, ammonoids

GEOLOGIC EVENTS

- Acadian orogeny
- Antler orogeny
- Ancestral Rockies
- Ouachita orogeny
- Appalachian orogeny
- Siberian basaltic volcanism
- Final assembly of Pangea
- Gondwana glaciation
- Cyclic sea-level fluctuations
- Deserts, evaporites in the western United States
- Sonoma orogeny
- Kaskaskia sequence
- Absaroka sequence

Mississippian	Pennsylvanian	Permian	TR
360	317	299	251

Brannen, Peter. 2017. *The Ends of the World: Volcanic Apocalypses, Lethal Oceans, and Our Quest to Understand Earth's Past Mass Extinctions*. Ecco, New York.

Clack, Jennifer. 2012 (2nd ed.). *Gaining Ground: The Origin and Evolution of Tetrapods*. Indiana University Press, Bloomington.

Erwin, Douglas. 2006. *Extinction: How Life on Earth Nearly Ended 250 Million Years Ago*. Princeton University Press, Princeton, NJ.

Knoll, A. H., R. K. Bambach, R. E. Canfield, and J. P. Grotzinger. 1996. Comparative earth history and Late Permian mass extinctions. *Science* 273:452–457.

MacLeod, Norman. 2015. *The Great Extinctions: What Causes Them and How They Shape Life*. Firefly Books, London.

Maisey, John, and David Miller. 1996. *Discovering Fossil Fishes*. Holt, New York.

McGee, George. 2018. *Carboniferous Giants and Mass Extinction: The Late Paleozoic Ice Age World*. Columbia University Press, New York.

McKee, E. H., S. S. Oriel, et al. 1967. *Paleotectonic Investigations of the Permian System*. US Geological Survey Professional Paper 515. US Government Printing Office, Washington, DC.

McKinzie, Mark, and John Macleod. 2015. *Color Guide to Permian Fossils of North Texas*. McMcPaleo, New York.

Prothero, D. R. 2013. *The Story of Life in 25 Fossils*. Columbia University Press, New York.

Prothero, D. R. 2018. *The Story of the Earth in 25 Rocks*. Columbia University Press, New York.

Sagan, Carl. 1997. *Billions and Billions*. Random House, New York.

Schoch, Rainer. 2014. *Amphibian Evolution: The Life of Early Land Vertebrates*. Wiley-Blackwell, Chichester, UK.

Stewart, W. N., and G. W. Rothwell. 1993. *Paleobotany and the Evolution of Land Plants*. Cambridge University Press, Cambridge.

US Geological Survey. 1979. *The Mississippian and Pennsylvanian carboniferous systems in the United States*. US Geological Survey Professional Paper 1110. US Government Printing Office, Washington, DC.

Ward, Peter D. 2000. *Rivers in Time: The Search for Clues to Earth's Mass Extinctions.* Columbia University Press, New York.

Ward, Peter D. 2004. *Gorgon: Paleontology, Obsession, and the Greatest Mass Extinction in Earth History.* Viking, New York.

Ward, Peter D. 2007. *Under a Green Sky: Global Warming, the Mass Extinctions of the Past, and What They Can Tell Us About Our Future.* HarperCollins, New York.

Ward, Peter, and Joseph Kirschvink. 2015. *A New History of Life: The Radical New Discoveries About the Origin and Evolution of Life on Earth.* Bloomsbury, New York.

Williams, H., and R. D. Hatcher, Jr. 1982. Suspect terranes and accretionary history of the Appalachian orogen. *Geology* 10:530–536.

SUMMARY

- The Late Paleozoic represented a time of transition from the greenhouse planet of the Cambrian–Devonian to the icehouse world of the Pennsylvanian, Permian, and Early Mesozoic; it is also the time of the final assembly of Pangea.

- Even though the Carboniferous Period is universally recognized across the world, the huge contrast between the clean crinoidal limestones of the Mississippian and the muddy coals and limestones of the Pennsylvanian the United States has long made it more convenient to use those terms instead of "Lower" and "Upper Carboniferous."

- Across most of North America in the Mississippian, the Kaskaskia sequence limestone covered enormous areas with thick deposits full of the fossils of crinoids and other shallow-marine life. Only in the foothills of the eroding Acadian Mountains do we see red sandstones and shales that differ from the limestones that prevail across the rest of the continent. This was the last gasp of not only the Kaskaskia sequence but also the giant epicontinental seas making enormous volumes of limestone in a greenhouse world.

- The Kaskaskia seas retreated at the end of the Mississippian, and the Absaroka sequence was deposited on the eroded surface of the ancient limestones. It is dominated by cyclothems of sandstones, shales, coals, and limestones from rivers, deltas, swamps, and shallow-marine deposits. The disappearance of clean, mud-free limestones and the huge volume of sands and muds reflect the uplifts of many mountain ranges in the Pennsylvanian. However, the repetitive cyclicity of these deposits is a reflection of the rapid rise and fall of global sea level due to the advances and retreats of glaciers on Gondwana.

- The Pennsylvanian was dominated by the Himalayan-style collision of North America against Gondwana, producing the Alleghenian or Appalachian orogeny in the east, the Ouachita-Arbuckle orogeny in Oklahoma and Arkansas, and the Ancestral Rocky Mountains in Colorado, Wyoming, and New Mexico.

- In the western United States, there were several smaller exotic terranes colliding in California and Nevada. In the Late Devonian–Mississippian Antler orogeny, a large terrane slammed into North America, emplaced across central Nevada on the Roberts Mountain thrust. It also brought exotic terranes that became part of the northern Sierras and eastern Klamath Mountains in California.

- The causes of the switch from an Early–Middle Paleozoic greenhouse world to the Late Paleozoic icehouse world was probably the extraction of huge volumes of carbon dioxide from the atmosphere into the coal deposits in the crust. Another factor might have been the widespread mountain-building, which can absorb carbon dioxide as it weathers down.

- Conversely, while carbon dioxide dropped, oxygen in the atmosphere may have reached almost 35%, allowing for gigantic insects and millipedes to grow.

- By the Permian, the supercontinent of Pangea was fully assembled, causing the interior of the supercontinent to experience extreme climate patterns. This produced numerous dune deposits, red beds, and evaporites across much of the western United States. On the edges of the continent, thick sequences of limestone still accumulated, especially in the famous "Permian reef" complex of west Texas.

- Mississippian sea life was again dominated by brachiopods (spirifers and productids), lacy bryozoans like *Archimedes*, huge assemblages of crinoids (and their relatives, the blastoids like *Pentremites*), plus goniatite ammonoids and minor corals. In the Pennsylvanian, the loss of clean, mud-free limestone and the big influx of mudstones and shales favored the productid brachiopods over the other groups. However, reefs were absent until the Permian, and trilobites remained very rare after the Devonian extinctions. A new group, the rice-shaped amoebas known as fusulinid foraminiferans, became one of the most abundant and rapidly evolving groups on the sea floor.

- On land, the primitive forests of the Late Devonian were followed by the enormous coal swamps of the Carboniferous, built almost entirely by tree-sized club mosses, horsetails, tree ferns, and seed ferns. These plants were so abundant and had no termites or other animals to quickly decompose them that most of their organic matter got locked up in the rock record as coal. By the Permian, the drier uplands away from the coal swamps were colonized by the first gymnosperms, including primitive conifers.

- Among land vertebrates, amphibians rapidly evolved into the long-bodied, flat, heated temnospondyls; the weird lepospondyls; and the reptile-like anthracosaurs. Reptiles first appeared in the Early Carboniferous and quickly evolved into a number of different forms, from swimming *Mesosaurus* to hippo-like pareiasaurs to early turtles with teeth and only a shell on their bellies, not their backs. Finally, the synapsids or protomammals (formerly called "mammal-like reptiles") dominated the landscape in the Permian, with huge fin-backed predators and bear-sized predators, feeding on a variety of herbivorous forms with strange knobby heads or beaks with tusks.

- The Permian ended with the "mother of all mass extinctions," when 95% of all marine species vanished and a high percentage of land animals did too. The trilobites, blastoids, tabulate and horn corals, and fusulinid foraminiferans vanished completely; and the brachiopods, bryozoans, crinoids, and ammonoids nearly died out, with only a few survivors. This brought the Paleozoic fauna, which had ruled the sea floor since the Ordovician, to an end. The cause of this mass extinction is thought to be the largest volcanic eruption in earth history, the Siberian lavas, which erupted huge amounts of greenhouse gases into the atmosphere and burned up ancient coal seams in the path of the flows. The oceans became supersaturated in carbon dioxide and too acid for life—and probably toxic with sulfur compounds. The atmosphere went through an extreme and rapid greenhouse event that killed most land life as well.

KEY TERMS

Crinoidal limestone (p. 267)
Absaroka sequence (p. 269)
Cyclothem (p. 269)
Underclay (p. 269)
Alleghenian orogeny (p. 270)
Ouachita orogeny (p. 270)

Ancestral Rocky Mountains (p. 272)
Antler orogeny (p. 273)
Maritime climate effect (p. 277)
Continental climate effect (p. 277)
Permian reef complex (p. 279)

Blastoids (p. 279)
Spirifer brachiopods (p. 280)
Productid brachiopods (p. 280)
Fusulinid foraminiferans (p. 281)
Gymnosperms (p. 281)
Temnospondyls (p. 283)

Lepospondyls (p. 283)
Anthracosaurs (p. 283)
Reptiles (p. 285)
Protomammals (p. 285)
Permian extinction (p. 288)
Siberian traps (p. 288)

STUDY QUESTIONS

1. What two big events in plate tectonics and climate dominated the Late Paleozoic?

2. What is the most typical rock found in the Mississippian across most of North America? What kind of environment did it represent?

3. What happened to the Acadian Mountains during the Mississippian?

4. How do the rocks of the Mississippian Kaskaskia sequence differ from the Pennsylvanian rocks of the Absaroka sequence?

5. Describe the pattern of rocks found in a typical cyclothem. What kinds of depositional environments do they represent? What could be the cause of the huge influx of mud and sand? What might explain the numerous repeated cycles of the sequence?

6. What was the cause and what were the effects of the Alleghenian orogeny? The Ouachita orogeny? The Ancestral Rocky Mountains?

7. What was the Antler orogeny? When and where did it occur? How is it different from other Late Paleozoic mountain-building events?

8. What two factors might have absorbed huge amounts of carbon dioxide and pushed the planet from greenhouse to icehouse?

9. Why did the plants of the Carboniferous get buried in the rock record, rather than rotting away as most trees and plants do today?

10. What was the atmospheric carbon dioxide and oxygen content in the Carboniferous? How might it explain the enormous insects and other arthropods of the coal swamps?

11. How did the existence of Pangea in the Permian affect the deposition of typical rocks of the Permian in the continental interior of North America?

12. What kinds of organisms made up the "Permian reef complex"? Why is it not considered a true "reef"?

13. Which groups of fish vanished in the Devonian, and which groups survived in the seas of the Late Paleozoic?

14. You are in a time machine and set the dial to "Mississippian." When the machine stops, you get out and swim away with your scuba gear. Describe the animals you would see on the sea floor. You return to the time machine and do the same for the Pennsylvanian. What different kinds of sea-floor life would you see?

15. What were the typical plants that made up the coal swamps? What were the big changes in the land plants during the later Paleozoic?

16. Why are protomammals or synapsids no longer called "mammal-like reptiles"?

17. Which groups died off at the end of the Permian? Which ones just barely survived? What was the probable cause of the "mother of all mass extinctions"?

Triassic, Jurassic, and Cretaceous 250–66 Ma

The Age of Reptiles ended because it had gone on long enough and it was all a mistake in the first place. A better day was dawning at the close of the Mesozoic Era. There were some little warm-blooded animals around which had been stealing and eating the eggs of the Dinosaurs, and they were gradually learning to steal other things, too. Civilization was just around the corner.

—*Will Cuppy, How to Become Extinct, 1941*

A scene of dinosaurs during the Late Jurassic as recorded in the Morrison Formation. The predator *Allosaurus* attacks a *Stegosaurus*, while the sauropod *Diplodocus* feeds on conifers behind them.

13.1 The Age of Dinosaurs

The Mesozoic Era is often known as the "Age of Dinosaurs" (Fig. 13.1) because those amazing animals evolved in the Late Triassic and ruled the landscape through the rest of the Jurassic and Cretaceous, over 140 million years. However, it was a crucial transition in many aspects of earth and life history. The major event that transformed the planet was the breakup of the supercontinent Pangea (Fig. 13.2). It began with the opening of the rift valley stage of the Atlantic in the Triassic, the opening of the North Atlantic in the Jurassic, and then the rest of the continents breaking apart by the Cretaceous. This event had profound effects on the land and the oceans.

At the same time, the atmosphere made the transition from an icehouse world of the Triassic and Early Jurassic to a greenhouse planet by the Late Jurassic. By the Cretaceous, the seas once again drowned the continents (as they did in the Early and Middle Paleozoic), and polar ice caps disappeared, making the polar regions lush and green. By the end of the Mesozoic, the earth had gone from having a single supercontinent and extreme icehouse climates to having many rapidly separating continents and a global greenhouse world with no ice, high sea levels, and high carbon dioxide levels in the atmosphere.

Through all this, not only did the dinosaurs first evolve and thrive but so did the first mammals, the first birds, the first lizards and snakes and crocodiles, the first frogs and salamanders, and the great radiation of advanced bony fish dominating the seas and freshwaters. By the end of the Mesozoic, life in the sea and on land looked almost modern, with the exception of a few groups like dinosaurs, marine reptiles, and ammonites that vanished in the mass extinction at the end of the Cretaceous.

13.2 Triassic: Beginning of the Breakup

ATLANTIC MARGIN

Back in the Permian, the entire planet had one supercontinent, Pangea, surrounded by a super-ocean, Panthalassa (Fig. 12.2C). The first sign that Pangea was tearing apart was the development of rift valleys, the early stage of continental breakup. As discussed in Chapter 5, similar rift valleys are found in East Africa today and give us a good analogue for the same features of the Triassic. These ancient rift valleys developed along both shores of what eventually became the North Atlantic. Enormous fault blocks pulled apart along normal faults, then sank down and slid along shallow faults, forming a series of valleys and ridges. These occur widely on the margins of the modern North Atlantic, from Spain to western Africa; but one of the best-studied rift valley sequences occurs along the Atlantic coast of North America (Fig. 13.3). Known as the **Newark Supergroup**, it is well exposed and studied in eastern New Jersey just west of New York City underlying Newark and most of the adjacent cities of the Garden State. To the north, a similar-looking Triassic rift valley sequences runs up the Connecticut Valley in Connecticut and Massachusetts, along the shores of the Bay of Fundy in Nova Scotia, and then through Maritime Canada to Spain (which was then to the northeast of Maritime Canada). To the south, the Triassic rifts of the Newark Supergroup run beneath the Gettysburg Battlefield in Pennsylvania, the Culpeper Basin in Virginia, and the Dan River and Deep River Basins in North Carolina (Fig. 13.3A).

The rock sequence of the Newark formations in New Jersey is typical of the entire supergroup (Fig. 13.3B, 13.4F). In most places, there are thick

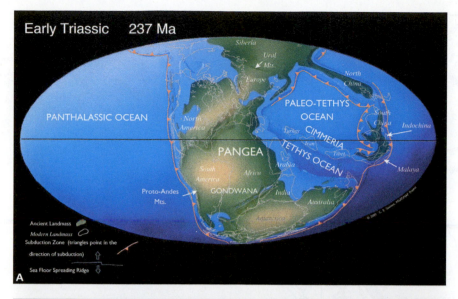

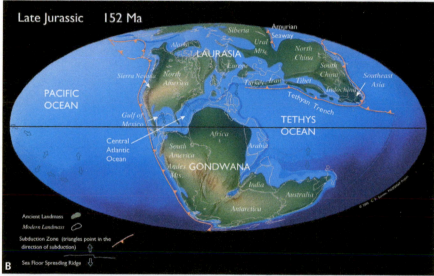

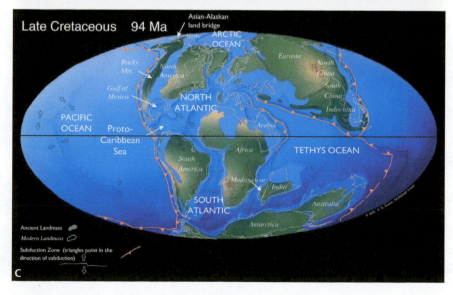

Figure 13.2 Global paleogeographic maps of (**A**) Triassic, (**B**) Late Jurassic, and (**C**) Late Cretaceous.

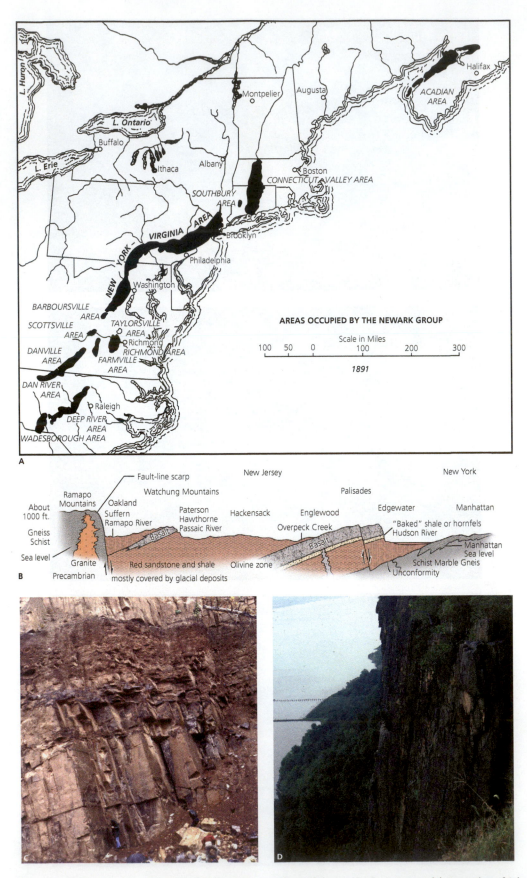

Figure 13.3 The Newark Supergroup was a Triassic rift valley sequence formed during the first stages of the opening of Atlantic as Pangea broke up. **A.** Map showing location of major basins running all the way from the Bay of Fundy in Nova Scotia to the Connecticut Valley in Massachusetts and Connecticut to the Newark, New Jersey, area to Gettysburg Battlefield in Pennsylvania and then down through Virginia and North Carolina. **B.** Geologic cross section of the Newark Basin in New Jersey. **C.** Typical red sandstones and shales of the Newark Group in central New Jersey, formed by ancient lakes and floodplains. **D.** Along the west shore of the Hudson River just west of New York City is the Palisades Sill, a lava intrusion that forced its way between layers of the lower Newark beds.

stacks of red beds formed on muddy floodplains and shallow lakes, much like the sediments accumulating on the floor of the East Africa rifts today. The total thickness in some places is over 6000 m (about 3.7 miles), so the fault-block basin must have been slipping down and sinking for millions of years of the Late Triassic and the earliest Jurassic to accumulate such a thick pile of sediments. In many places, the red sandstones preserve a wide variety of dinosaur tracks (on display in many local museums, such as Dinosaur State Park in New Jersey). An even greater array of dinosaur trackways was found in the Connecticut Valley red beds, now on display in numerous museums. In addition to tracks, there is a good fossil assemblage of primitive Triassic reptiles (including early dinosaurs), amphibians, and a huge assemblage of primitive fish found in the rift valley lakebeds. The red beds are easily eroded and not very resistant, so in most places they are covered by the urban and suburban sprawl of eastern New Jersey.

In addition to the typical red floodplain muds and lake shales, volcanic eruptions are typical of rift valleys. These occur in the East African rift today (exemplified by the many East African volcanoes, such as Kilimanjaro and Ngorongoro) as magmas work their way up through the faults and cracks in the rift valley and reach the surface, where they erupt violently. In the western part of the Newark Supergroup, a series of lava flows that once covered the floor of the rift valley have been eroded to form numerous long, curved ridges called the Watchung Mountains (Fig. 13.3B). Along the west shore of the Hudson River, a thick horizontal lava intrusion known as the Palisades Sill forms the distinctive columnar-jointed black cliff that forms the Hudson shore from Staten Island and Hoboken, New Jersey, north to Haverstraw and Pomona, New York (Fig. 13.3B, D). Ridges formed by lava flows are common in many basins in the Newark Supergroup and have influenced the events that occurred on this landscape. For example, long ridges of basalt formed the high points on Gettysburg Battlefield. The Confederates advanced from Seminary Ridge on the west against the Union positions on Cemetery Ridge and Little Round Top in the east, which gave the Union forces natural fortifications against their attacks, especially during Pickett's charge. Geology (and a few fateful decisions) changed the course of the Civil War.

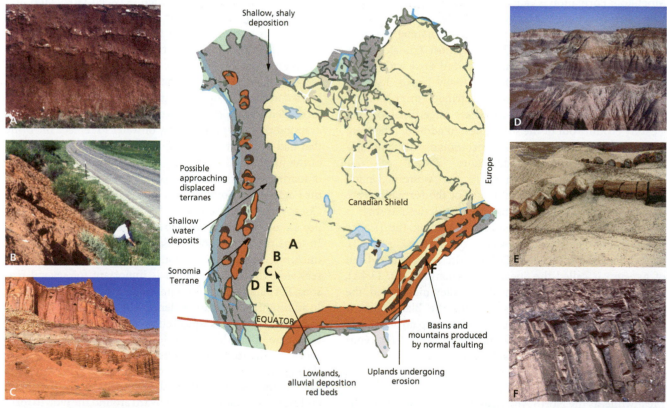

Figure 13.4 Paleogeographic map of North America in the Triassic. Red shales are typical of the Triassic all over the western United States. **A.** In the Black Hills of South Dakota, the Triassic is represented by the Spearfish Shale. Here, it is full of white lenses of gypsum from an ancient dry lakebed. This shot is behind the parking lot of Evans Plunge, Hot Springs, South Dakota. **B.** In Wyoming, Triassic red shales are called the Chugwater Formation, here shown in Shell Canyon on the flank of the Bighorn Mountains, Wyoming. **C.** In the Colorado Plateau and Painted Desert, the Triassic red, gray, purple, and brown shales of the Moenkopi Formation are overlain by the brick-red shales of the Chinle Formation, here at Glen Canyon in Utah. The cliffs are made of Jurassic Navajo, Kayenta, and Wingate Sandstones, while the slope below is Chinle Formation, and the slope in the foreground with the reddish ledges is Moenkopi Formation. This exposure is at Capitol Reef National Monument. **D, E.** Petrified Forest National Park is famous for its huge fossil logs of *Araucaria*, eroding out of the colorfully banded river red and brown floodplain deposits of the Chinle Formation. **F.** Exposure in a quarry in central New Jersey, showing the red lake shales alternating with sandstones from Triassic rivers.

WESTERN UNITED STATES

Outside the Newark Supergroup, there are almost no Triassic deposits in eastern North America because the entire region was still high in elevation from the slowly eroding mountains formed during the Appalachian collision (Fig. 13.4). Most of the best exposures of Triassic rocks are found in places like western North America, as well as South America, Russia, and southern Africa. These Triassic rocks resemble the rocks of the Permian: red shales and sandstones formed on hot, sunny floodplains, with dune sands and occasional evaporites (Fig. 13.4A). This is not surprising because the extreme continental interior climates that dominated the Permian in much of the world continued into the Triassic since Pangea was still together. If you travel around the Colorado Plateau in northern Arizona, southern Nevada, eastern and southern Utah, western New Mexico, and western Colorado, there is a consistent Triassic sequence above the gray Permian rocks such as the Kaibab Limestone that forms the rim of the Grand Canyon (Fig. 13.4C). The first unit above the Kaibab is the Moenkopi Formation, a floodplain shale which comes in many hues, from brick red to gray to brown to lavender to green, giving the colors to the region known as the "Painted Desert." In many places, the Moenkopi Formation is capped by a resistant ledge of sandstone and conglomerate known as the Shinarump Formation, which represents a widespread river channel sequence that once covered the area. The uppermost Triassic unit is the famous Chinle Formation, a brick-red shale with minor sandstones that forms red cliffs in many places. Its most famous exposures, however, are in Petrified Forest National Park in Arizona (Fig. 13.4C–E), where huge fossil logs are steadily eroding out of the soft shales. In addition, there are fossils of many different reptiles, protomammals, amphibians, and some of the earliest dinosaurs.

Elsewhere in the western United States, the names of the formations are different, but the distinctive red shale is still the signature of the Triassic. In Wyoming, the Triassic red shale is called the Chugwater Formation (Fig. 13.4B). In the Black Hills of South Dakota, the red shale bed full of gypsum is called the Spearfish Formation (Fig. 13.4A). It forms an easily eroded oval shale valley nicknamed "the Racetrack" that completely surrounds the Black Hills. It lies between resistant layers of Mississippian Pahasapa Limestone on the inside track and the resistant Cretaceous Dakota Sandstone on the outside rim.

SONOMA OROGENY

Meanwhile, as the North American plate tore away from the rest of Pangea, it began to move rapidly to the west and to override the oceanic plates in the Pacific region. The first sign of this rapid collisional movement was the arrival of another huge exotic terrane from out in the Pacific that accreted to northern California and Nevada during the Late Permian and Early Triassic. Known as the **Sonomia terrane**, it makes up the entire northwestern corner of Nevada (Figs. 12.10A,

13.4) north of Reno, including the Black Rock Desert where Burning Man is held. This tectonic block was pushed up over the rest of Nevada on a huge thrust fault known as the Golconda thrust (Fig. 12.10A). The arrival of this exotic block in the Permo-Triassic is called the **Sonoma orogeny**. Other pieces of the Sonomia terrane were accreted into the northern Sierra Nevada Mountains, where the Triassic Calaveras terrane was plastered up against the old Shoofly terrane of the Devonian–Mississippian Antler orogeny. Yet another Triassic block was accreted into the Klamath Mountains of California, once again accreted to the older Paleozoic rocks that arrived during the Antler orogeny.

13.3 Jurassic Tectonics

THE NORTH ATLANTIC OPENS

The breakup of Pangea continued through the Jurassic, but the rift valleys of the Triassic were replaced by the next stage of the breakup, the proto-oceanic gulf stage (see Fig. 5.17). First the North Atlantic rift valley got wider and wider, until the rift valleys were no longer floored by stretched-out continental crust but instead by oceanic crust. This Jurassic sea floor is some of the oldest still found on the floor of the modern ocean. Rifts also opened between Greenland and Scandinavia to form the first stages of the North Sea, and another rift opened to form the beginning of the Gulf of Mexico (Fig. 13.2B). This last rift evolved into a shallow proto-oceanic gulf in the Jurassic and was closed at one end, so seawater flowed only from the other end. In addition, it was located in the subtropical latitudes where the world's deserts form, so it was exposed to intensely dry climates and lots of evaporation but limited inflow of water to replace the brine that had evaporated. Consequently, the proto-oceanic gulf of the future Gulf of Mexico accumulated thick sequences of marine evaporites. Best known from the Louann Salt and the Eagles Mills Formation in Texas and the rest of the Gulf Coast, this immense body of Middle Jurassic salt underlies almost the entire Gulf from Florida down to Mexico. In some places, the Louann Salt is several thousand meters thick, representing the evaporation of enormous volumes of seawater. This salt formation is important for several reasons. For one thing, it is commercially mined for salt today. In addition, the pressure of denser Cretaceous and Cenozoic rocks on top of it has squeezed up huge blobs and blisters of less dense salt to the surface, forming **salt diapirs**. This process is similar to the way blobs of lower-density magma rise through denser surrounding rock. When these low-density salt diapirs rise and push up the overlying layers, they bend and fault the flanking layers upward around the blob to form a **salt dome** (Fig. 13.5). The upwarped strata on the flank of the impermeable salt rock form a perfect trap for migrating oil and gas, so most of the oil wealth of southeast Texas and Louisiana comes from drilling on the edges of a salt dome.

In some cases, however, drilling on the flanks of salt domes can be risky. On November 20, 1980, a Texaco oil rig drilling on the flank of the Diamond Crystal Salt mine

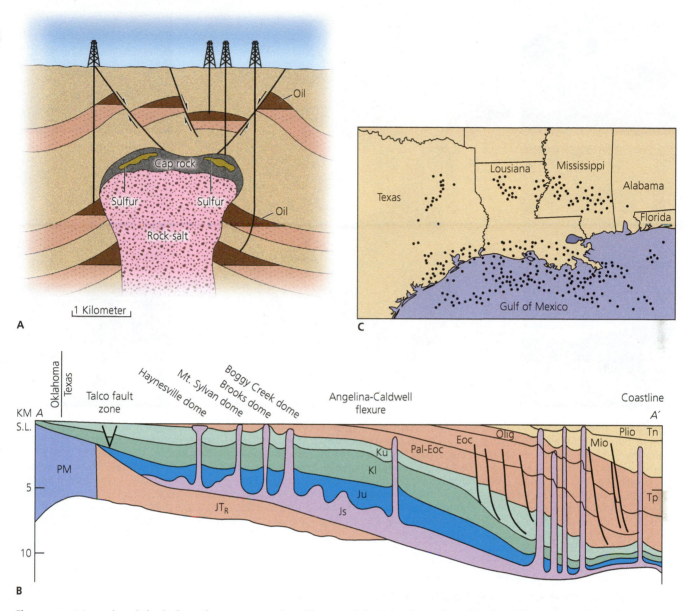

Figure 13.5 Jurassic salt beds from the proto-oceanic gulf stage of the Atlantic are found in the subsurface on the Gulf Coast. **A.** When the lightweight salt rises, it punches up through the dense overlying sandstones and shales and buckles them up, trapping oil against the edge of the dome. **B.** The Jurassic salt (Js) layer comes up in many places, forming multiple domes across the region. (PM = Pennsylvanian-Mississippian bedrock; JT$_R$ = Jurassic-Triassic basement; Ju = Upper Jurassic; Kl and Ku = Lower and Upper Cretaceous; Pal-Eoc = Paleocene-Eocene; Olig = Oligocene; Mio = Miocene; Plio = Pliocene.). **C.** Map showing the major salt dome oil fields on the Gulf Coast of Texas and Louisiana.

under Lake Peigneur, Louisiana, pierced the mine roof. The entire lake rapidly drained down into the chamber of the salt mine at a rate of 0.24 m³/second (8.47 feet³/second), flooding the salt mine and forming a gigantic whirlpool in Lake Peigneur that sucked the drilling platform, 11 barges, a local plant nursery, and 25 hectares (about 62 acres) of the land around the lake down into the hole. The canals that once flowed away from the lake reversed course and flowed backward into the whirlpool, forming the tallest waterfall (50 m, 164 feet) ever seen in Louisiana. Nobody lost their lives, but Texaco ended up paying over $45 million to the mining company and nursery in the aftermath.

JURASSIC WEST

The Jurassic rocks of the western United States (Fig. 13.6) form a distinct sequence that is exposed widely over the entire Colorado Plateau and many other places in the Rockies. In the Colorado Plateau, for example, the most distinctive Lower Jurassic unit is the **Navajo Sandstone**, which forms the striking "White Cliffs" in Zion National Park, Capitol Reef National Park, Glen Canyon National Recreational Area, and many other places in Utah, Colorado, Arizona, and New Mexico (Fig. 4.10C, 13.6A). These white Jurassic sandstones form a blanket of sand across the southwestern United States about 40,000 km³ in volume.

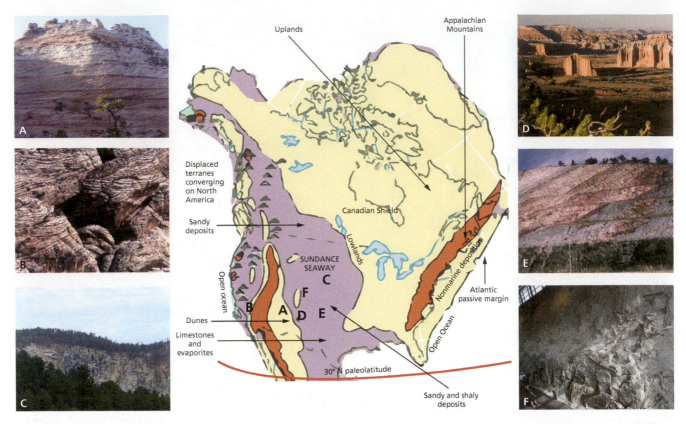

Figure 13.6 Paleogeographic map of the Late Jurassic, showing location of the major mountain ranges and sedimentary basins. **A.** Giant cross-beds of the Navajo Sandstone in Zion National Park, formed in an immense coastal dune field that ran from Las Vegas to central Wyoming. **B.** Similar cross-bedding in the Aztec Sandstone, the Navajo equivalent in Nevada, known as Calico Rocks, in Red Rock Canyon west of Las Vegas. **C.** Outcrop of the Sundance Formation, red and gray shales deposited in the Late Jurassic Sundance Seaway that covered much of the central part of North America. **D.** Upper Cathedral Valley in Capitol Reef National Park, Utah. The red Entrada Formation at the base is capped by Jurassic marine beds of gray Curtis Formation and reddish Summerville Formation of the San Rafael Group. **E.** Typical purple-, gray-, and red-striped Upper Jurassic Morrison Formation (tilted units to the right lower part of the road cut) beneath the tan sandstones of the Lower Cretaceous Dakota Formation, I-70 road cut west of Denver. **F.** Enormous logjam of dinosaur bones in the main quarry, Dinosaur National Monument.

In many places, they show incredibly thick (up to 25 m, or 82 feet) sets of giant cross-beds, which were formed on the back slopes of enormous dunes (Fig. 13.6A). This Early Jurassic dune field stretched from near Las Vegas (Red Rock Canyon State Park), where the white sandstone is called the Aztec Sandstone and is over 760 m (about 2500 feet) thick (Fig. 13.6B), to the Four Corners region, where the Navajo Sandstone and the underlying red river sandstones of the Kayenta Formation are 700 meters (about 2300 feet) thick, and even to central Wyoming, where it is represented by the Nugget Sandstone, which reaches about 655 m (about 2150 feet) in thickness. This enormous Navajo–Aztec–Nugget dune field stretched over a wide area parallel to the Jurassic coastline (Fig. 13.6). Careful measurements of the cross-bedding show that the winds blew the sand from northeast to the southwest, perpendicular to the coastline, which trended northwest to southeast. Although it was a giant dune field, it was not as big as the continent-sized Sahara Desert dune fields of North Africa. Instead, the Navajo dunes are probably more like the huge coastal dune

fields in places like the shores of Namibia today. Dune environments tend not to preserve many fossils, but there are some reptile and pterosaur trackways and a few skeletons of small dinosaurs known.

The Navajo Sandstone is also significant in that it is one of the last indicators of the dry, harsh continental interior climates of the icehouse world that started in the Permian and dominated Triassic landscapes. Immediately above the Navajo in most of the Rocky Mountain region and Colorado Plateau lies the first epicontinental seaway deposit since the greenhouse days of the Devonian and Mississippian, often called the **Sundance Seaway** (Fig. 13.6). It flooded the Great Plains from the Gulf of Mexico to the Arctic Ocean and drowned much of what would become the Rockies and Colorado Plateau. Represented by the rocks of the San Rafael Group (Fig. 13.6D) in the Arizona–Utah borderland around Lake Powell and the Sundance Formation (Fig. 13.6C) over much of the northern Rockies, these thick deposits of shale with rare limestones represent shallow seas drowning most of what would become the Great

Plains and Rocky Mountain region. Major transgressions that flooded the middle of the continent corresponded to the transition of the earth from its former icehouse state to a greenhouse planet once again. These shallow seaways of the Sundance and San Rafael Groups were the home of a wide variety of marine life. These included bizarre oysters on the sea bottom and other kinds of marine snails and clams. Swimming above them were many kinds of ammonites, as well as marine reptiles like ichthyosaurs and plesiosaurs.

See For Yourself: Carmel Formation

See For Yourself: Morrison Formation

The entire Jurassic sequence of the region that would someday become the Rockies and the Colorado Plateau is capped by a distinctive non-marine unit that is the result of a small regression at the end of the Jurassic. Known as the **Morrison Formation**, in most places it consists of a thick stack of purple and gray mudstones (Fig. 13.6E), deposits of ancient floodplains, and weathered soil horizons that covered most of the western United States. But the Morrison represents more than just a Late Jurassic regression event. It is far more famous as "the graveyard of the dinosaurs." Most of the big dinosaur beds occur in the Morrison Formation, especially the outcrops at Dinosaur National Monument on the Utah–Colorado–Wyoming corner (Fig. 13.6F), Dry Mesa Quarry in Colorado, Cleveland-Lloyd Quarry in Utah, and Como Bluff and Bone Cabin Quarry in southeastern Wyoming. Nearly all the huge, long-necked sauropods of America come from the great Morrison quarries in Colorado, Utah, and Wyoming, along with *Stegosaurus*, the predatory allosaurs, and many other famous dinosaurs (Fig. 13.1).

THE SIERRA NEVADA ARC DEVELOPS

Meanwhile, as the seas covered the center of North America in the latter half of the Jurassic, the land was rising out of the sea on the west coast. With the opening of the North Atlantic in the Jurassic, the North American plate began to slide west and override oceanic plates in the Pacific region faster and faster. Starting in the latest Triassic (after the end of the Sonoma orogeny) and then during the Jurassic and especially during the Cretaceous, a huge Andean-style volcanic chain arose across the entire region. The most famous product of this volcanic land is the **Sierra Nevada Mountains** (Fig. 13.7), which are the remnants of ancient magma chambers that once fed Andean-sized Jurassic and Cretaceous volcanoes. Later uplift has since eroded almost all these volcanic rocks away, exposing their deeply buried magma chambers at the surface. Such large masses of long-cold magma chambers that are uplifted and eroded away are known as **batholiths**. The Sierra Nevada is not the only piece of this volcanic arc, however. It once stretched along the entire Pacific coast, from the Coast Range Batholith in British Columbia to the Idaho Batholith to the intrusions in the Klamath Mountains of Oregon and California to the Sierra Nevada (Fig. 13.7E). To the south, the rest of

the volcanic chain is represented by the granitic rocks of the Peninsular Range batholith that runs from southern California down the central axis of Baja California, as well as small fault slices of Cretaceous granitic rock that make up the Salinian block on the central California coast. All of these magma chambers once formed a continuous chain that has since been chopped up and displaced by faulting, so they no longer line up (Fig. 13.7D). Another reason for their displacement is the fact that the Sierra Nevada Mountains moved 200–300 km west of their original position during the later Cenozoic (see Chapter 14).

The early phases of the Sierra Nevada Mountains were much like the modern Andes in many other ways as well. Not only was there an enormous chain of volcanoes (now eroded down to their roots in magma chambers that were once deeply buried), but like the modern Andes, there was a **forearc basin** in front of the volcanic range (Fig. 13.7C). Back in the Jurassic and Cretaceous, it was filled with marine sediments plus volcanic material that erupted out of the arc or eroded from the arc. Today, this ancient Mesozoic forearc basin persists beneath the Central Valley or "Great Valley" of California. The basin is now a giant farm belt, but not long ago it was flooded by the ocean, with volcanoes erupting to its east. Drilling into the basin sediments or found as outcrops on the edges of the Great Valley, geologists have discovered thousands of feet of Cretaceous deep-marine rocks (called the Great Valley Group) that also lie beneath the valley floor (Fig. 13.10B).

Subduction zones usually have a thick **accretionary wedge** between the forearc basin and the top of the downgoing plate (see Fig. 5.20). As described in Chapter 5, accretionary wedges are formed of material sliced or scraped off the subducting plate, so they are predominantly made of sheared oceanic sediments (especially shales, turbidite sandstones, and ribbon cherts), as well as slices of oceanic crust that get plastered onto the overlying plate (ophiolites). Most distinctive are pieces of the unusual high-pressure, low-temperature metamorphic rocks known as blueschist (see Chapter 2), which come from deep in the subduction zone where the pressure is intense but the cold oceanic plate keeps the temperatures down. In the accretionary wedge, all of these components tend to be highly sheared, shredded, and deformed, with no bedding that persists more than a few meters in any direction. For this reason, the accretionary wedge deposit has come to be known as a **mélange**, meaning "mixture" in French. All these typical rocks of the accretionary wedge can be found in the central and northern Coast Ranges of California, where they are known as the Franciscan Complex (Fig. 13.10A). Even though the subduction zone that created the Sierra Nevada Mountains has been gone from central California for at least 30 million years, the remnants persist long after the subduction process that built them ended.

Finally, the Andes have one other feature that also occurred in the Sierran arc. To the east in the Argentinian foothills of the Andes is a huge zone of **back-arc thrusts**

(Fig. 13.7B). These represent the compression of the plate margin as one plate rapidly slides beneath another. The same back-arc thrust belt was developed to the east of the Sierran arc as well. Today, it runs from just west of Las Vegas, where you can see it. If you drive west on Highway 12, the Blue Diamond Highway, you can easily see the Keystone Thrust that brings gray Paleozoic limestones on top of white Jurassic Aztec Sandstone. It is also visible at Redrock Canyon just west of Las Vegas, and in Valley of Fire State Park east of Las Vegas. From there, the back-arc

thrust belt follows the trend of the Sevier Mountains in central Utah (just east of Interstate 15 in Utah), then north to the thrust belt on the Idaho–Wyoming border, and finally to the thrusts that underlie Glacier National Park in Montana. This huge compressional event is sometimes called the "Sevier orogeny," while the eruption of the Sierran arc has traditionally been called the "Sierran orogeny"; but today geologists realize that both events are part of the same process, so it is now called the **Sierra–Sevier orogeny** (Fig. 13.7C).

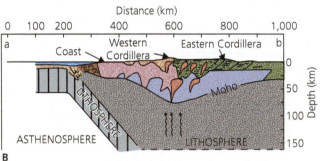

Sevier Orogenic Belt

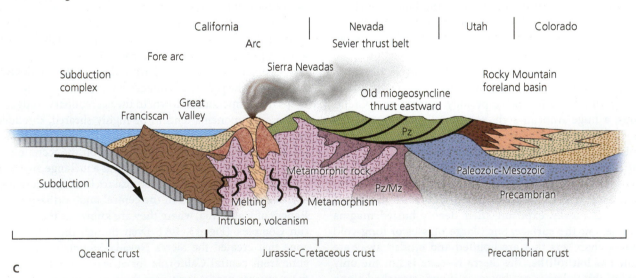

Figure 13.7   **A, B.** Map and cross section of the modern Andes Mountains, a good analogue for the ancient Sierran arc. Cross section in **B** is taken along line a–b in **A. C.** Diagram of the various parts of the Sierran arc complex, including the Great Valley forearc basin (now the Central Valley of California) and the Franciscan accretionary wedge (now the Coast Ranges of California). **D, E.** Position of the batholiths of the later Mesozoic, from the Peninsular Ranges in Baja California to the Sierra Nevada Mountains to the Idaho Batholith to the Coast Ranges Batholith in British Columbia.

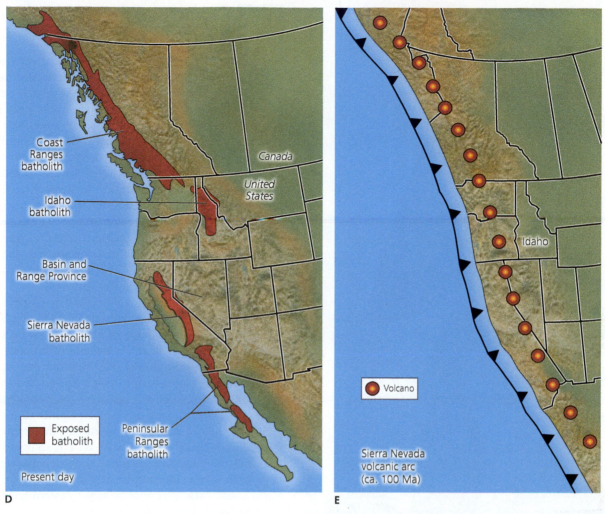

D

E

F

Figure 13.7 **D** shows their present-day position, displaced by faults and the stretching of Nevada during the Cenozoic. **E** shows the original position of the arc volcanoes that made the batholiths before faulting displaced them. **F.** Photograph of typical granitic rocks in the Sierra Nevada Mountains, here in Yosemite National Park.

In addition to the development of the Sierran arc in the Jurassic and Cretaceous, other parts of the Pacific margin of North America were tectonically active. Most important was the continued arrival of more **exotic terranes** from across the Pacific (Fig. 13.8). In section entitled "Sonoma Orogeny" above, we discussed the collision of the terrane called Sonomia in the Late Permian and Triassic. During the Jurassic and Cretaceous, most of the terranes that make up Alaska, British Columbia, and important parts of coastal Washington and Oregon arrived and smashed into North America. By the end of the Cretaceous, nearly all of the pieces of the eastern Pacific that make up Alaska and British Columbia had arrived and docked.

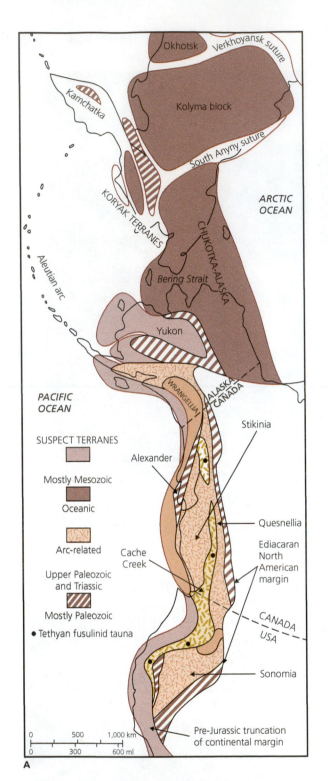

See For Yourself: Cretaceous Tectonics

13.4 Cretaceous World: Greenhouse of the Dinosaurs

The appearance of the great epicontinental Sundance and San Rafael Seaways across the middle of North America in the later Jurassic was just a prelude to the peak greenhouse conditions of the Cretaceous. The prevailing theme of Cretaceous geology around the world is extraordinarily high sea levels and drowning of continents, along with the almost complete lack of polar ice, all caused by extremely high levels of carbon dioxide in the atmosphere. Indeed, the Cretaceous got its name from the Latin word *creta*, which means "chalk." Before the Cretaceous was formally named, European geologists referred to the time interval by the enormous chalk beds that make up not only the White Cliffs in southeastern England (Fig. 13.9A) but similar beds stretching across Belgium and France as well.

What does this chalk represent? Chalk is a form of limestone made entirely of the tiny calcite shells of plankton. The most important plankton making up chalk are a group of submicroscopic algae only a few tens of microns across. Known as **coccolithophores** (Fig. 13.9B), these algae are covered in tiny button-shaped plates known as coccoliths. When the coccolithophore algae died, their coccoliths disaggregated; and trillions of them combined to make the powdery white rock we know as chalk. (The commercially-produced "chalk" used to write on a blackboard is not real chalk but powdered gypsum.) These deposits represent huge blooms of trillions upon trillions of plankton in the warm shallow seas that once drowned western Europe and much of North America (Fig. 13.10).

In addition to submerged Europe, the Great Plains of North America were drowned by these same shallow seas through most of the Cretaceous. This formed a giant marine barrier, the **Western Interior Seaway** (Fig. 13.10), that lasted over 110 million years from the Middle Jurassic (about 180 Ma) until just before the end of the Cretaceous (about 70 Ma). It isolated the land and the dinosaur faunas of the world, so the dinosaurs of Montana have more in common with those from Mongolia than they do with those from New Jersey. The deposits of this great seaway can be found all over the western Plains and Rockies, from the chalk

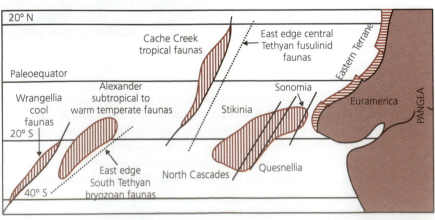

Figure 13.8 During the late Paleozoic and early Mesozoic, exotic terranes rafted from across the Pacific to slam into North America. Today, they make up most of the coastal region of California, Oregon, Washington, and British Columbia and almost all of Alaska. **A.** Map showing the location of some of the major exotic terranes. **B.** Original position of some of the exotic terranes back in the Permian. Both Wrangellia and the Alexander terrane, now largely in Alaska and British Columbia, started in the southwestern Pacific.

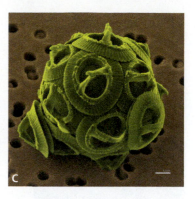

Figure 13.9 During the Cretaceous, most of western Europe was drowned by shallow tropical seas full of plankton, which rained down to the sea floor to build up the deposit known as chalk. **A.** The white cliffs at low tide on the south coast of England near Birling Gap, west of Dover and Beachy Head, Eastbourne, UK. **B.** Scanning electron microscopic image of chalk, showing the button-shaped coccoliths, which armor the planktonic algae known as coccolithophores (shown in **C**) that make up the chalk.

badlands of western Kansas (legendary for marine reptile and fish fossils and flying reptiles like *Pteranodon*) to the Austin Chalk and other marine rocks of the Cretaceous of Texas and Oklahoma to the thick deposits of marine Pierre Shale in South Dakota.

 See For Yourself: The Cretaceous Interior Seaway

Likewise, in the Rocky Mountains, there is a remarkably consistent sequence representing the great transgression of the Cretaceous that drowned the region before the Rockies arose. The base of the sequence is a widespread nearshore river and deltaic sandstone full of fossil leaves known as the Dakota Sandstone (Fig. 13.6E), which represents the first Cretaceous transgression above the regressive deposits of the Upper Jurassic Morrison Formation. In most places, it is capped by a silvery shale full of fish scales, known in different parts of the Rockies as the Mowry Shale, Thermopolis Shale, or basal Mancos Shale. Above these rocks we find an alternation between shallow-marine sandstones and deep-water Cretaceous shales full of ammonites and other fossils, known as the Mancos Shale in Colorado and Utah (Fig. 13.10D), the Cody Shale in Wyoming, and the Pierre Shale in South Dakota (Fig. 13.10E). The stratigraphy of these units is very complex because the deep-water shales alternated with nearshore sandstones such as the Mesa Verde Sandstone in Colorado (home of the Mesa Verde cliff dwellings), the Frontier and Fox Hills sandstones in Wyoming, and many other units (Fig. 4.11, 13.10D). These shorelines built out from the mountains and deltas in the west (some with coal beds; Fig. 4.9B), then retreated again, forming a complex shoreline that zigs and zags from west to east and back each time sea level fluctuated (Fig. 4.11B, 13.10). Even farther west, the nearshore sandstones transition to conglomerates (Fig. 13.10C) and river sandstones that were eroded directly from the mountains even farther west in the Sierran arc and the Sevier thrust belt. Thus, in central Utah are thick conglomeratic sandstones such as the North Horn Formation or Indianola Group, which grade to the east into the river and delta and nearshore deposits of the Mesa Verde Sandstone in central Colorado (Fig. 4.17, 13.10D), which transition to the deep-water Mancos Shale

even farther east. Finally, by the time you are as far east as western Kansas, the shales are replaced by chalk deposits (Fig. 13.10F) from the deepest central part of the Western Interior Seaway.

What could have caused such an extensive flooding of the continents in the Cretaceous? There were several things happening at this time, all of which contributed to the rising sea levels and greenhouse atmosphere, although it is difficult to determine which were most important.

1. *Breakup of Pangea*: During the 80 million years of the Cretaceous, nearly all the remaining pieces of Pangea broke up and separated into their modern continents (Fig. 13.2). The North Atlantic began to open in the Late Jurassic, but most of its widening occurred in the Cretaceous. But the opening of the South Atlantic occurred entirely in the Cretaceous, as did the separation of Africa from Antarctica and the beginning of the split between Australia and Antarctica. Most impressive is how India ripped away from its former position in Gondwana and raced across the Indian Ocean, so that by the end of the Cretaceous it was close to colliding with Asia.

   This breakup of all the Pangea continents into their modern configurations had several important effects. The most important is that after rifting the edges of the continents are wider and more stretched out, so their average continental elevations are lower, allowing the seas to transgress across them. These produce thick sedimentary wedge deposits on the edges of the rifting continents. For example, the bulk of the enormous passive margin wedge over 6100 m (about 20,000 feet) thick on the Atlantic Coast and Gulf Coast of North America (Fig. 5.18) was formed during the 80 million years of the Cretaceous. These sinking continental edges were continuously drowned by shallow seas, producing enormous piles of shallow marine shales and limestones that are today found all over the Atlantic and Gulf coastal plains. Much of the reason for the high Cretaceous sea levels and the drowning of the continents is a product of the sinking of passive margin wedges.

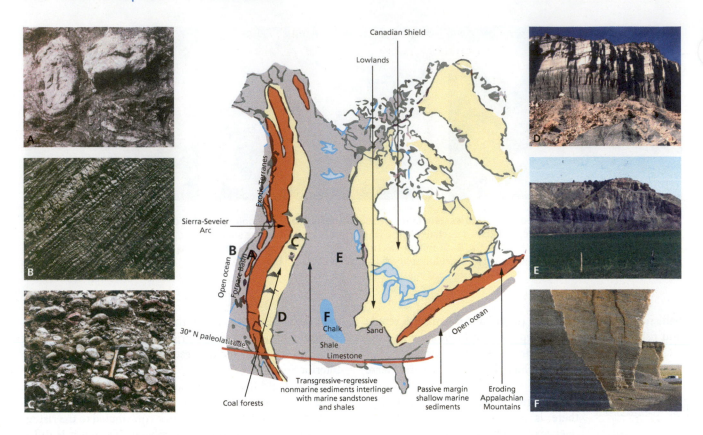

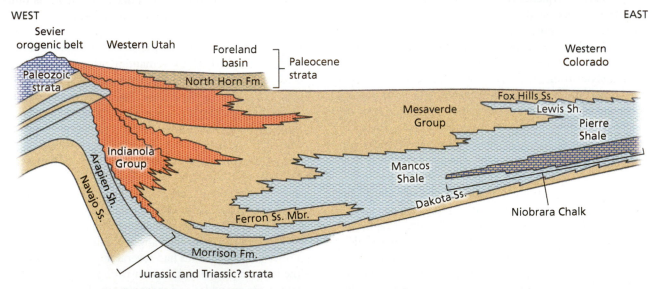

Figure 13.10   Paleogeographic map of North America in the mid-Cretaceous, showing the Western Interior Seaway running down the center of the continent. (Bottom) East–west cross section from the Western Interior Seaway in Colorado (right) to the foothills of the Sevier mountain belt in western Utah (left). Above the non-marine dinosaur-bearing Upper Jurassic Morrison Formation, the Early Cretaceous transgression is indicated by the nearshore and deltaic Dakota Sandstone. The shoreline then zigged and zagged east–west across the region as the sea level rose and fell, so the nearshore sandstones like the Mesa Verde and Fox Hills and Ferron Formations interfinger with offshore units like the Mancos and Pierre Shales and the Niobrara Chalk in Kansas. In western Utah, thick wedges of sandstones and conglomerate from alluvial fans built out from the uplifts and interfingered with the nearshore sandstone of the Mesaverde Group. **A.** Outcrop of typical mélange along the beach near San Simeon and Hearst Castle, California. **B.** Exposure of the typical thickly bedded forearc basin turbidites and shales from the Great Valley Group. **C.** Conglomerates of the Cretaceous Price River Formation in Maple Canyon State Park in central Utah, product of alluvial fans from the Sevier uplifts. (**B** and **C** courtesy of R. H. Dott, Jr.). **D.** This cliff shows the nearshore Mesa Verde Sandstone (on top) capping deeper-marine shales of the Mancos Formation, near Big Water, Utah. This sequence is typical of the transgressing–regressing shoreline deposits of the Interior Seaway. **E.** Typical outcrop of gray Pierre Shale from the center of the seaway, near Hermosa, South Dakota. **F.** Chalk beds in Monument Rocks in western Kansas, formed in the center of the seaway where huge blooms of coccoliths filled the seas.

2. *Rapid sea-floor spreading and ridge volume*: As discussed in Chapter 10, when sea-floor spreading is very rapid, it produces much thicker, taller profiles of rock on the mid-ocean ridge complex (Fig. 10.5). This probably happened when the Pannotia supercontinent broke up in the Cambrian, and the record rates of sea-floor spreading at that time would have produced large mid-ocean ridges that would have displaced much of the ocean's water onto the land, causing the global Sauk transgression. The same would have occurred in the Cretaceous, when all the Pangea continents were pulling apart at a record rate of sea-floor spreading. The increased volume of all those rapidly spreading ridges would have been enormous, and all that extra rock volume made the ocean shallower and pushed the water up onto the land. Numerous studies have shown extraordinarily high rates of spreading starting in the Early Cretaceous (Aptian stage, about 126 Ma) and continuing at lesser rates until nearly the end of the Cretaceous.

3. *Greenhouse gases*: Not only are high rates of sea-floor spreading important in terms of the depth of the ocean, but they also release a lot more greenhouse gases from the mantle during all of the intense submarine eruptions. This would be a major reason that atmospheric carbon dioxide levels may have reached values as high as 2000 ppm (today it's only 415 ppm, even with all the greenhouse gases we are introducing due to global warming). Another contributor was the indirect effect of oceans drowning so much of the land, with limited areas of mountain ranges uplifting at the time. Recall that rapid rates of mountain uplift cause rapid weathering, which is a major absorber of carbon dioxide—but when oceans drown the land surface and there are no large areas of mountains rising around the world, the limited land area also restricts the amount of weathering in soils, and thus how fast they can absorb carbon dioxide. If there had been any significant polar ice caps left in the Jurassic or Early Cretaceous, this extreme warming would have melted them all away, further contributing to the water in the oceans. So those dinosaur movies that show in the mountains in the background are wrong—there was probably little or no ice anywhere in the world during the Cretaceous.

The warming of the "greenhouse of the dinosaurs" was so extreme that the ice-free polar regions were relatively balmy and mild, even if they did experience many months of darkness during the winter. In both the North Slope of Alaska (above the Cretaceous Arctic Circle) and the Antarctic and southern Australia (both near the South Pole then), we find evidence of temperate and even subtropical plants like breadfruit, telling us that temperatures were about 10°C (50°F) warmer on the poles than they are now, along with tree ferns and conifers and other plants adapted to mild climates. Living in these warm, wet, but dark forests were a wide variety of dinosaurs, some of which had large eyes for seeing in dim light. Others had adaptations suggesting that they could live in these polar forests or at least that they migrated up there during the summers. There were also crocodiles and pond turtles and other creatures that cannot stand even a few days of freezing weather.

4. *Mantle superplume eruptions*: Recent research has shown that the Cretaceous was a time of extraordinarily big eruptions of huge hot spots or **mantle superplumes** coming up from the lower mantle and then erupting beneath the oceans. These rocks formed by gigantic mantle eruptions were first discovered in the 1950s and 1960s, when oceanographers mapped huge submarine plateaus, especially in the Pacific Ocean. When scientists drilled and analyzed and dated the rocks, they proved to be the result of huge mantle lava eruptions that spilled enormous volumes of magma into the Cretaceous oceans—and pumped huge volumes of greenhouse gases from the mantle at the same time. The biggest of these is the Ontong-Java Plateau in the southwestern Pacific. About 126 Ma, it was formed by the eruption of 1.5 million km^3 of lava (about 360,000 miles^3) across the ocean floor that lasted less than a million years. This timing coincides with the huge spike in seafloor volcanism in the Aptian stage of the late Early Cretaceous and the rapid rise in sea level around the globe at that time. The Hess Plateaus and the Shatsky Rise also erupted during the Cretaceous, and their huge volumes of lava undoubtedly contributed to the Cretaceous greenhouse gases as well.

Many factors may have contributed to the extraordinarily intense greenhouse world of the Cretaceous. Which was the most important is difficult to determine, but they were all working at the same time and all contributing to greenhouse warming.

13.5 The Laramide Orogeny

Through the last half of the Jurassic and the first 75 million years of the Cretaceous, the entire Pacific Rim of North America was one huge subduction zone. The downgoing oceanic plate plunged into the mantle at a fairly steep angle, producing the enormous batholiths of the Sierran arc and its equivalents from Alaska to Baja California (Fig. 13.7D, E). In this respect, the entire region was much like the modern Andes today, complete with the ancient equivalents of the modern Andean forearc basin, accretionary wedge, and back-arc thrust zone (Fig. 13.7A). To the east of the Sierran volcanic chain and Sevier thrusts, the entire region of the modern Rockies was drowned by the seas of the Western Interior Seaway (Fig. 13.7C). There were no real "Rocky Mountains" yet, and the last time there had been any mountains in regions like Colorado or Wyoming, they were the "Ancestral Rockies" formed during the Pennsylvanian, which had been completely eroded away by the Triassic.

Then about 70 Ma, just four million years before the end of the Cretaceous, everything changed. The Sierran volcanoes stopped erupting, and they would never again erupt in that location. Simultaneously, we see a strange kind of tectonics about 1600 km (about 1000 miles) to the east, in the region of the modern Rockies in Colorado, Wyoming, Utah, and Montana. The Western Interior Cretaceous Seaway abruptly regressed as mountains rose in the region for the first time since 300 Ma in the Pennsylvanian. But these mountains have some unusual features. They were not Andean-style volcanoes with their characteristic back-arc thrusts. Instead, they appear as huge anticlinal folds and steep faults that brought blocks of ancient Precambrian basement up to the surface, while erosion removed all the thousands of meters of Paleozoic and Mesozoic rocks that used to cover them (Fig. 13.11A, B). Suddenly, there were true Rocky Mountains in roughly the same places of the modern Rockies. If the Pennsylvanian Ancestral Rockies can be nicknamed "Rocky I," then this event was "Rocky II." Geologists have known this event for decades as the **Laramide orogeny**, after the Laramie Mountains in southeastern Wyoming where this uplift is very well exposed.

The renewed uplift of the Rocky Mountains not only drained the Interior Seaway from most of the region but raised high mountains and a new landmass with many rivers and deltas flowing out to the sea. These ancient river and delta deposits are the source of our best Cretaceous dinosaur faunas. Up in Alberta, extensive badlands of the Red Deer River expose Upper Cretaceous rocks that are legendary for their dinosaur fossils. The same is true of the Upper Cretaceous rocks of Wyoming (Lance Formation) and Montana (Hell Creek Formation), which are the source of such famous dinosaurs as *Tyrannosaurus rex*, *Triceratops*, and numerous duck-billed dinosaurs and hard-shelled ankylosaurs with the clubs on their tails. To the south, in the San Juan Basin of northwestern New Mexico, a thick Upper Cretaceous sequence of the Kirtland and Fruitland Formations is also rich in dinosaurs. This entire uplifted landmass in the position of the future Rocky Mountains is sometimes called "Laramidia," after the Laramide orogeny.

So what explains this peculiar pattern, where the arc volcanism of the Sierra–Sevier orogeny shuts off and instead there is intense folding and faulting 1600 km (about 1000 miles) farther inland than where normal plate tectonics produce mountains? There are several clues. During the latter part of the Cretaceous, the position of the volcanic arc in the Sierra Nevada Mountains shifted from the west to the east, as did the volcanic eruptions in the Peninsular Ranges of southern California and Baja California. This suggests that the angle of the downgoing plate plunging into the mantle must have gotten shallower since the volcanic chain is shifting farther and farther away from the location of the trench formed by the subducting plate. Following this idea, in the 1970s geologists William Dickinson and Walter Snyder suggested that the peculiar geology of the Laramide orogeny was caused by subduction at an extremely shallow angle, possibly even **horizontal subduction** (Fig. 13.11C). If the downgoing plate slid horizontally beneath the overriding plate, then it would not sink down in the mantle

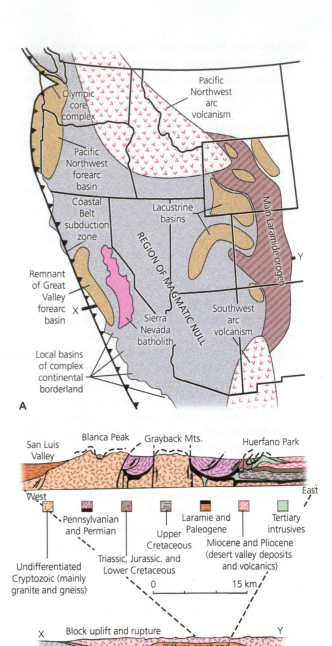

Figure 13.11 At the very end of the Cretaceous and continuing into the Cenozoic (70–40 Ma), the Sierran arc shut off, the Western Interior Seaway vanished, and the Laramide mountain ranges rose in its place. **A.** Map of the location of the Laramide orogeny and other tectonic events of the latest Cretaceous and Early Cenozoic. **B.** The most unusual feature of the Laramide orogeny was the uplift of mountain ranges where the seaway had once been. These ranges are formed by deep basement-cored folds, thrusts, and block faulting that brings Precambrian and Paleozoic rocks to the top of the Rocky Mountains, even today. **C.** The peculiar tectonics of the Laramide orogeny has been explained by nearly horizontal subduction, so the old volcanic arc shuts off since the plate does not plunge into the mantle to melt but instead transfers its stresses to crust about 1600 km (1000 miles) inland in what is now Colorado, Wyoming, and Montana.

deep enough to melt and there would be no volcanism in the Sierras anymore. In addition, if it slid horizontally far enough east, it might transfer stresses inland and possibly push up on the overlying plate, causing the peculiar folding and faulting that brought basement uplifts from tens of kilometers underground into high eroded mountains. Of course, this horizontal subduction is no longer occurring in North America, but there is similar low-angle subduction going on beneath the Pampas of Argentina today. This is considered to be the best model for why both the Sierran volcanic shutoff and the Laramide uplifts occurred at the same time.

The Laramide orogeny did not cease with the end of the Mesozoic but continued into the Cenozoic until about 40 Ma (middle Eocene). It warped down huge basins between the Laramide uplifts that filled with enormous volumes of Paleocene and Eocene sediments and many fossil mammals. But that is a topic for our next chapter.

13.6 Life in the Mesozoic Oceans

THE TRIASSIC RECOVERY

As we saw in Chapter 12, the great Permian extinction nearly wiped out life on the planet, with 95% or more of marine species dying off. Numerous recent studies of the aftermath of this catastrophe have shown that the Early Triassic world was severely depleted in the diversity of marine life. The handful of surviving groups, such as brachiopods, bryozoans, crinoids, and ammonoids, began to slowly evolve to replace some of their lost species; but for more than 5 million years after the Great Dying, the sea floor was still pretty empty.

Into all these empty niches came lots of animals which were considered to be opportunistic or "weedy" species—broad generalists who were extinction-resistant and could repopulate habitats in the absence of specialized competition, just as weeds in our yard take over disturbed habitats that discourage more stable, mature vegetation. Such weedy species included a population boom of the scallop called *Claraia*, which was thought to be able to tolerate the low oxygen levels and harsh conditions of the Early Triassic. Another signature of the weird Early Triassic conditions was the reappearance of abundant stromatolites in many shallow-marine habitats, presumably because most of the snails that kept them cropped since the Cambrian had died out.

Not only did the Permian extinction wipe out most of life on earth, but it completely rearranged the dominant forms of life in the oceans as well. As we saw in Chapter 12 (Fig. 12.21), the "Paleozoic fauna" of brachiopods, bryozoans, corals, crinoids, and cephalopods that had ruled the oceans since the Ordovician was decimated and never again as abundant or diverse. Instead, the sea floor was repopulated by new groups that had long been in the background in the Paleozoic but have ruled the earth ever since the Early Triassic. This assemblage is now called the "**Modern fauna**" because the shells make up most of the fossils of the Mesozoic and Cenozoic and still are the most common seashells today (Fig. 12.21). If you look at any modern beach and pick up seashells, your collection is largely composed of mollusks, especially snails (class Gastropoda) and clams and oysters (class Bivalvia) (see Appendix A), along with sea urchins; these are the groups that dominate the fossil record of marine habitats since the Triassic (Fig. 13.12). Many of these newly dominant groups included the first scallops and oysters, as well as a distinctive triangle-shaped, heavily ribbed bivalve known as *Trigonia* (Fig. 13.12A), which coincidentally happens to be an index fossil of the Triassic. *Trigonia* was once thought to be extinct until a living descendant, *Neotrigonia*, was found in the waters off Tasmania in 1804, making it a true living fossil.

In addition to these bottom-dwelling mollusks, the few surviving lineages of Paleozoic gonitatite ammonoids (Fig. 11.12H) were the ancestors of a new radiation of Triassic ammonoids. These are distinctive and easy to recognize since the suture line between the outer shell and the edge of the septa that divided the chambers has a characteristic "U"-shaped pattern known as a **ceratite suture** (Fig. 13.12B). These ammonoids underwent an explosive evolutionary radiation so that by the Late Triassic there were over 140 species known, all descended from two or three lineages that survived the end of the Permian.

The third most common shell producers in the Modern fauna (as in the Triassic) were the sea urchins, class Echinoidea (see Appendix A). Their primitive relatives were in the background in the Paleozoic but never very diverse until new habitats opened up in the Triassic. The main survivors of the Permian were the slate-pencil urchins, but the more familiar sea urchins evolved in the Triassic, followed by sea biscuits in the Jurassic and heart urchins in the Cretaceous. The most specialized subgroup, the sand dollars, evolved in the Paleocene and Eocene.

Not only were groups like the bryozoans, brachiopods, crinoids, and corals decimated by the Permian extinction but their survivorship patterns are revealing. The crinoids that survived and evolved in the Triassic were characterized by much more flexible stalks and heads, and they were capable of living in unusual habitats, such as floating upside down dangling from driftwood. Eventually, a group of crinoids without stalks evolved which could creep along the sea floor like a brittle star.

The tabulate and rugose corals vanished completely in the Permian, so there were no coral fossils or any true reefs whatsoever in the Early Triassic. By the Middle Triassic, however, a new group of corals had evolved, probably from a soft-bodied anemone-like ancestor. Known as the **scleractinians**, or "hexacorals," these are the corals that have populated the earth ever since the Triassic and make up all living corals (Fig. 13.12C).

Most revealing is the fact that several groups of Paleozoic brachiopods (especially the spirifers) survived into the Triassic but vanished at the end of the period. This has led

Figure 13.12 Typical Triassic marine fossils. **A.** The triangular bivalve known as *Trigonia* is a living fossil typical of the Triassic. It was thought to be extinct until a living relative called *Neotrigonia* was found in the South Pacific. **B.** Triassic ammonoids all have a distinctive ceratitic or "U"-shaped suture pattern. **C.** The modern group of scleractinian corals evolved from anemone-like ancestors once the Paleozoic horn corals and tabulate corals vanished in the Permian extinction. **D.** The largest marine predator was the whale-sized ichthyosaur *Shonisaurus*, found at Berlin-Ichthyosaur State Park near Gabbs, Nevada. It is the official state fossil of Nevada.

several paleontologists to suggest that the new dominance of bivalves, gastropods, and echinoids is more than just selective survivorship of certain groups after the Permian catastrophe. They have pointed out that all the survivors of the Late Triassic have different kinds of anti-predatory adaptations against shell-crushing predators. Many of the snails and clams (especially oysters) developed very thick or very spiny shells that make it hard to be crushed. Still others, like the scallops, developed the ability to swim erratically by clapping their shells together like castanets, which makes it hard for most predators to track them or follow them. Most of the snails and clams evolved the ability to hide by burrowing, so their shells are not exposed on the sea floor. By contrast, most brachiopods (except for the primitive lingulids) sat on the sea floor attached to a surface and could not burrow, so they were easy prey for any shell-crushing predators. And the crinoids that did survive were highly mobile and flexible, so they were also harder for predators to attack.

Taking all these trends together, paleontologists call this trend toward predator-resistant marine life the **Mesozoic Marine Revolution**. Something was driving the prey species to avoid or resist new predators that had not existed in the Paleozoic, and this is another major difference between the more relaxed world of the Paleozoic with few major predators and the **escalation** into a harsher Mesozoic sea floor with a new type of predator. Paleontologists have compared it to an arms race, where a new predator develops a new way of feeding and the prey species have to quickly adapt to resist or escape that predator—or they are eaten.

What were the predators that drove the Mesozoic Marine Revolution? There are actually quite a variety of animals in the Triassic that developed the ability to eat shelled creatures, from the first predatory sea stars that pulled shells open and digested them inside their own shell, to bony fish, sharks, and the newly evolved marine reptiles which had broad pavement stone–like crushing teeth adapted for smashing shells. Finally, the first large-clawed crabs and lobsters also appear, with their powerful claws adapted to crushing shells or peeling them open to get at the soft, tender, tasty flesh inside.

So the Mesozoic seas had new swimming animals to join the bony fish, sharks, and ammonoids that survived the Permian and that became the largest predators. These were the marine reptiles, and they came in many different shapes. Some, known as placodonts, were shaped somewhat like swimming turtles with huge crushing teeth. But the more familiar groups were the first members of the dolphin-like **ichthyosaurs** in the Triassic, which would evolve to become the major marine predators in the Jurassic. The primitive Triassic ichthyosaurs, however, were not all small swimmers. Some, like the gigantic whale-sized *Shonisaurus* of Berlin-Ichthyosaur State Park in Nevada, were over 15 m (about 50 feet) long (Fig. 13.12D). The other group of Mesozoic marine reptiles was the **plesiosaurs**, familiar from the long-necked forms of the later Mesozoic. In the Triassic, though, plesiosaurs are represented by the primitive fossils known as nothosaurs, which were about 3 m (about 10 feet) in length. They had the beginning of the long neck and long fish-catching snout of later plesiosaurs and webbed feet that had not yet evolved into paddles.

TRIASSIC–JURASSIC EXTINCTION EVENT

See for Yourself: Ichthyosaurs

The Triassic Period ended about 201 Ma with another mass extinction, probably the third or fourth largest in earth history. About 76% of all marine species vanished and 34% of marine genera. The most obvious victims were the last of the spirifer brachiopods and a significant percentage of the clams, snails, and especially the diversity of ceratitic ammonoids (Fig. 13.12B), which were reduced from 140 species to just a few lineages that survived into the Jurassic. One group of fossils, the microscopic tooth-like structures from the gill region of primitive jawless eels known as conodonts, vanished completely after having thrived through the entire Paleozoic and Triassic. A number of the primitive marine reptile groups, such as placodonts and the primitive relatives of ichthyosaurs and plesiosaurs, vanished as well. There was a lesser extinction on land, mainly among the primitive groups of reptiles that were replaced by dinosaurs.

The cause of this extinction has been long debated, but one of the key facts is that it appeared to be a very rapid event, possibly happening in less than 10,000 years. This rules out gradual climate change and many of the usual suspects, like tectonic or mountain-building events or sea-level changes. Many scientists have attempted to blame it on an extraterrestrial impact event, but there are no craters the right age. In fact, there is a really large impact that formed Manicougan crater in Quebec, but it triggered no extinctions and is dated at 214 Ma, well before the Triassic–Jurassic extinction. Instead, the geochemical signals from the oceans suggest a super-greenhouse warming event with excess carbon dioxide, just like the main cause of the Permian extinction. Like the Permian event, there was a gigantic eruption of mantle-derived volcanoes at this time. Known as the **Central Atlantic Magmatic**

Province (CAMP), it was a huge flood basalt eruption that poured across the basin of the rapidly opening North Atlantic when it was in transition from the rift valley stage to the proto-oceanic gulf stage (Fig. 13.3). Most of the volcanic rocks like the Palisades Sill (Fig. 13.3D) and the Watchung lava flows (Fig. 13.3B) in the Newark Basin are part of the CAMP.

JURASSIC SEA LIFE

See For Yourself: Late Triassic Extinctions

Marine life eventually recovered after the Triassic–Jurassic extinction, and many of the trends we saw in the Triassic continued as well. Jurassic seas were dominated by an entirely new radiation of ammonoids that survived the mass extinction (Fig. 13.13A, B). These Jurassic and Cretaceous ammonoids had the most complex suture pattern between the edge of the septa and the shell ever seen in this group. Called an **ammonitic suture**, it was incredibly convoluted. The complexity of this pattern makes it possible for ammonite paleontologists to identify almost any species precisely, and thus ammonites are the principal index fossil of the marine Mesozoic since they changed rapidly and are widespread over most of the world's oceans. Some of these Jurassic ammonites, like *Parapuzosia*, were immense, reaching almost 1.8 m (about 6 feet) in diameter (Fig. 13.13B). They must have been built by a huge squid-like creature inside them with tentacles that could reach a long way from their shells.

Swimming with the ammonitic ammonoids was a variety of fish and sharks. One of the biggest was the whale-sized bony fish *Leedsichthys*, which may have been about 16 m (about 53 feet) long. The ichthyosaurs reached their heyday as well, with many that were completely fish-like or dolphin-like and specialized for deep-sea diving (Fig. 13.13C). There were also huge plesiosaurs, including the largest, *Liopleurodon*, which may have reached 6.4 m (21 feet) in length (Fig. 13.13D). Some crocodiles went to the sea, where they evolved into a group known as geosaurs with a tail fin and flippers instead of feet.

Finally, the sea floor was also inhabited by a diverse array of animals, including a variety of snails and clams, many of which have living descendants. Other fossils are typical of the Jurassic, such as the odd-looking oyster *Gryphaea*, known to fossil collectors as the "devil's toenails." These oysters had one thick shell that rested on the sea bottom, shaped like a trough with a coiled end (Fig. 13.13E). The other shell was shaped like a tiny lid that allowed them to close up their shell tightly against predators.

THE CRETACEOUS SEAWAYS

These Jurassic marine animals diversified even further during the 80 million years of the Cretaceous. The abundance and diversity of marine fossils are amplified by the fact that the rise of sea level produced thick marine deposits on the continent, such as the chalk in Europe (Fig. 13.9) or those of the Western Interior Seaway that once covered the center of North America (Fig. 13.10F). Thus, collecting

Figure 13.13 Typical Jurassic marine fossils. **A.** The Jurassic was marked by a big radiation of ammonoids with the complex ammonitic suture. **B.** Some ammonites were gigantic, like this *Parapuzosia* from Germany. **C.** The ichthyosaurs were in their heyday during the Jurassic. This specimen from the Holzmaden Shale of Germany was in the process of giving live birth when it was fossilized. The baby ichthyosaur can be seen emerging from the birth canal. **D.** The huge plesiosaur *Liopleurodon* was the largest marine reptile of the Jurassic. **E.** Among the most common sea-floor fossils were the oysters called *Gryphaea*, or "devil's toenails."

marine fossils from the Cretaceous is much easier and has been done by many paleontologists for a long time.

The marine realm changed from the bottom of the food chain upward. Starting in the Jurassic, there was an explosive radiation of planktonic organisms that produced a rich food source in the oceans that did not exist on this scale before. We have already encountered the enormous diversification of marine algae known as coccolithophores (Fig. 13.9B), which built the enormous chalk deposits of the Cretaceous. Also found in the chalk beds was another group of single-celled amoebas with calcite shells known as foraminiferans. These were already mentioned in the context of the Permian fusulinid foraminiferans (Fig. 12.14), but in the Cretaceous the foraminiferans not only lived on the sea bottom but got small enough to become important in the plankton as well. All of these planktonic foraminiferans are members of the family Globigerinidae (Fig. 13.14A) because their shells are formed like little globes or bubbles that radiate outward in a spiral while getting larger. In addition to these algae and amoeba-like forms that used calcite shells, there was a great evolutionary radiation in plankton that used shells made of silica (SiO_2). These include the amoeba-like creatures known as **radiolarians** (Fig. 13.14B), which diversified into dozens of families in the Jurassic and Cretaceous. Finally, the fourth of the major groups of living marine plankton that leave shells are the marine algae known as **diatoms** (Fig. 13.14C). These, too, underwent a great radiation and diversification during the Jurassic and Cretaceous, providing even more biomass of phytoplankton in the base of the food chain.

Feeding on this rich planktonic food supply was an incredible diversity of marine invertebrates on the sea floor. These included a wide variety of marine snails, many of which are still represented in today's oceans. Even more impressive was the extreme evolution of the bivalves. In addition to the familiar types of clams and scallops, there was a wide range of odd-shaped oysters (Fig. 13.15). Taking things to an extreme was the spiral oyster called *Exogyra* (Fig. 13.15A), whose shell was built like the coiled cup and lid shape of *Gryphaea* but spiraled up an axis like a snail's shell. Other bizarre clams included the "zigzag clam" *Alectryonia* (Fig. 13.15B), with highly crenulated margins on the shell. Even more extreme were the huge flat clams known as **inoceramids** (Fig. 13.15C, D), which were shaped like a dinner plate up to 1.7 m (almost 6 feet) across and rested flat on the sea bottom. Many of them are found with numerous fish fossils inside them, suggesting that they harbored a lot of symbiotic fish in their mantle. But the strangest of all were the weird oysters known as **rudistids** (Fig. 13.15E). These bivalves had one shell shaped like an ice cream cone, and the other was a cup-shaped lid. They were found in densely packed colonies in the tropical seas over the entire Cretaceous and formed gigantic reefs that apparently took the role of coral reefs (which were unimportant during most of the Mesozoic). Some were even weirder in shape, forming twisted shells like ram's horns and many other peculiar forms.

If the clams were strange-looking, the ammonites were even more so. The Jurassic radiation of ammonitic ammonoids continued into the Cretaceous as they swam the extensive seas of the greenhouse world (Fig. 13.15F). But the oddest of all were the ammonites that abandoned the simple spiral pattern in a single plane that had been standard for coiled cephalopods since the Devonian (Fig. 13.14G–K). These were the **heteromorph ammonites** ("heteromorph" means "different shape" in Greek). These varied from forms which were just slightly uncoiled, like the scaphitids, to a variety of ammonites with shells coiled like a paper clip (*Hamites*) (Fig. 13.15I) or a question mark or a corkscrew spiral like a snail (*Turrilites*) (Fig. 13.15II) to the truly weird *Nipponites*, whose coils formed a tight knot (Fig. 13.14J). With these odd arrangements of coils, the squid-like creature inside could not have used jet propulsion with its water nozzle, as a nautilus does, or it would have spun out of control or wobbled around. Instead, they must have floated slowly in the open water or just above the sea bottom, grabbing prey that came within reach of their tentacles. The most common heteromorph of all was the straight-shelled group known as **baculitids**, which had a long conical shell that was completely straight except for a tiny coil at the end (Fig. 13.15K). They are reminiscent of the long, straight-shelled orthocone nautiloids of the Ordovician (Fig. 10.16), except that baculitids evolved from advanced coiled ammonites, so they have complex ammonitic sutures, not the simple curved sutures of nautiloids. Baculitid fossils are common in the deposits of the Western Interior Cretaceous Seaway and evolved so rapidly that they are one of the principal index fossils of the Cretaceous. Their mode of life is curious since they had no dense calcite deposits in the shell to counterweight it and hold it horizontal. Instead, the shell must have floated with the point upward and the squid-like creature inside dangling below, using its tentacles to move along or grab prey as it floated above the sea floor.

In addition to cephalopods with external shells, there was a group of Jurassic and Cretaceous squids known as **belemnites** (Fig. 13.15L), which had an internal shell like living squids do. The belemnite shell is a solid cone of calcite that looks like a large-caliber bullet made of stone. In some places, these fossils are extremely abundant, forming beds made of nothing but belemnites.

Sailing above these bottom-dwellers and ammonites were the marine vertebrates of the Cretaceous seaways. These include some enormous bony fish, such as the gigantic fossil *Xiphactinus*, which is known from many complete skeletons in the chalk beds of western Kansas and reached up to 6 m (about 20 feet) long (Fig. 13.16A). There were huge sharks up to 10 m (33 feet) long as well. But the roles of the biggest sea creatures were taken by the great diversification of marine reptiles. The immense sea turtle *Archelon* was over 4 m (about 13 feet) long, the largest turtle that ever lived (Fig. 13.16B). The dolphin-like ichthyosaurs were declining in the Cretaceous and vanished about 95 Ma in the Early Cretaceous. Instead, two other groups took over the role as dominant marine predators. The paddling plesiosaurs were

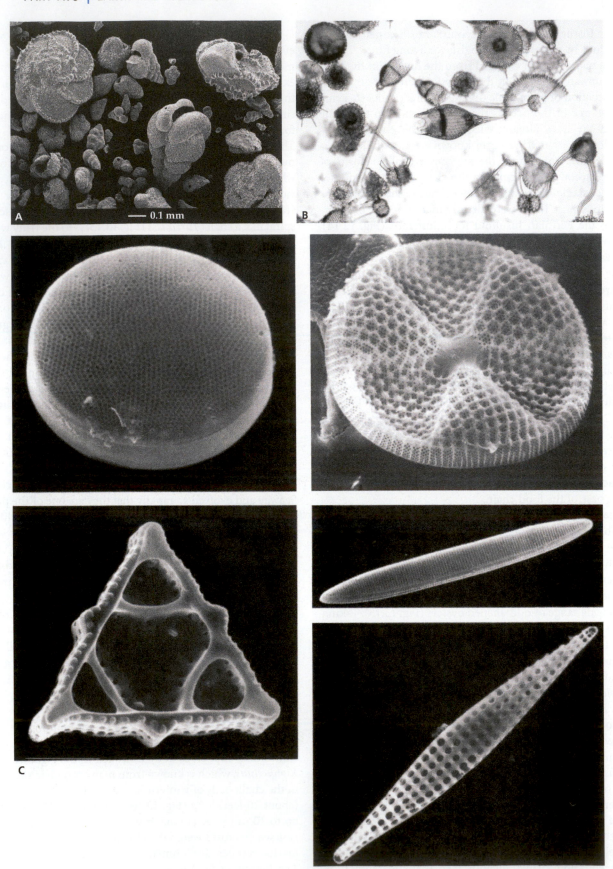

Figure 13.14 The plankton of the Cretaceous seas, which still make up most of the shelled plankton alive today, included not only the submicroscopic algae called coccolithophores (Figure 13.9B) but also the calcite-shelled amoebas called the foraminifera, especially the family Globigerinidae. **A.** These globigerinids have the typical shape of a spiral arrangement of bubble-shaped chambers in their shell or chambers arranged in two or three parallel rows, increasing in size. **B.** Another important group was the amoeba-like radiolarians, which had delicate conical and bubble-shaped lacy shells made out of silica. **C.** In addition to coccolithophores, the phytoplankton included the diatoms, which made their shells out of silica. These diatoms were shaped like circular disks, triangles, or elongate rods; and each algae had two of these shells surrounding its tissue, nested one inside the other like the two parts of a Petri dish.

Figure 13.15   Cretaceous marine life. **A.** The odd asymmetrical oyster called *Exogyra*, with one shell forming a thick spiral and the other a flat lid on top. **B.** The peculiar oyster *Alectryonia*, with its distinctive "zigzag" shell shape. **C, D.** The enormous flat "dinner-plate" clams known as inoceramids, which lived on the sea bottom and often hosted fishes living symbiotically inside them. Some were over 1.7 m (5.6 feet) across. **E.** The peculiar colonial oysters known as rudistids, which had a cone-shaped shell with a hinged cap on top. **F.** A discus-shaped Cretaceous ammonite called *Placenticeras*, with the incredibly complex ammonitic sutures. **G.** Some ammonites were shaped like corkscrews or question marks, like this *Hypantoceras*. **H.** The ammonite *Turrilites* was coiled up in a spiral, like a snail. **I.** The ammonite *Hamites* was curled into a paper-clip shape. **J.** The weirdest heteromorphy ammonites, called *Nipponites*, were coiled up in knots. **K.** *Baculites*, an ammonite with complex sutures that secondarily uncoiled into a nearly straight shell (large tusk-shaped shells in background). It is a common index fossil of Cretaceous seas. **L.** In addition to cephalopods with external shells, there were abundant belemnites. They were squid-like mollusks that had an internal shell that looked a bit like a large-caliber bullet (but made of calcite).

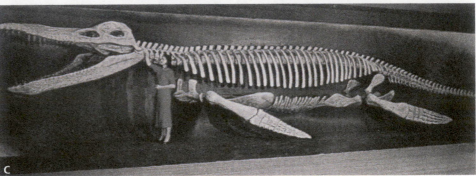

Figure 13.16   Marine vertebrates of the Cretaceous seaways. **A.** The giant 3-m-long (10 feet long) fish *Xiphactinus*, here with another large fish inside its stomach. **B.** The huge sea turtle *Archelon*, largest sea turtle ever known. **C.** The enormous pliosaurine plesiosaur *Kronosaurus* from the Early Cretaceous of Australia. **D.** The mosasaurs were enormous monitor lizards, related to Komodo dragons, which adapted for marine life with hands and feet modified into paddles and a long compressed tail for swimming propulsion. **E.** Artist's conception of a Cretaceous seaway, with the mosasaur *Tylosaurus*, the plesiosaur *Albertanectes*, the fish *Xiphactinus*, and the sea turtle *Nicholssemys*, as well as numerous ammonites.

in their heyday. These included the forms with extremely long necks and tiny heads, known as elasmosaurs, which reached sizes of up to 10.3 m (about 34 feet) in length. But the biggest of all were the short-necked, long-snouted plesiosaurs known as pliosaurs. The largest of these was the immense *Kronosaurus* from the Cretaceous of Australia, which was almost 11 m (about 36 feet) long (Fig. 13.16C).

See For Yourself: Mosasaurs

Finally, the Cretaceous saw a new group of marine reptiles take to the sea and evolve rapidly throughout the period. These were the **mosasaurs**, a group of monitor lizards related to Komodo dragons that became so specialized for swimming that their feet were modified into paddles, their tails were flattened into a tall and narrow fin for propulsion, and their snouts became elongate for catching fast-swimming prey like fish and squids (Fig. 13.16D, E). Although the spectacular mosasaur in the aquarium show of the movie *Jurassic World* is bigger than a whale, in real life they reached only about 17 m (about 56 feet) in length, and most were much smaller. Nonetheless, they were fast, agile, fearsome marine predators, able to outswim and out-maneuver just about any other marine animal in the oceans of Kansas. Most had long snouts with sharp conical teeth for catching fast prey, like fish, squid, and ammonites. Indeed, there are ammonite specimens with a "V"-shaped role of conical bite marks that match the mosasaur tooth pattern. One mosasaur was found with a diving bird, several fish, a shark, and a smaller mosasaur in its stomach, so they were opportunistic feeders. A few of them had globular crushing teeth for eating mollusks, so mosasaurs were flexible in their feeding habits. They diversified into dozens of genera and species, and during the last 20 million years of the Cretaceous, the plesiosaurs vanished, leaving the role of dominant marine predator to the mosasaurs.

13.7 Life on the Mesozoic Landscape

TRIASSIC RECOVERY

Just as we saw in the marine realm, the Early Triassic landscape also took a long time to recover in the aftermath of the Great Dying. Geochemical evidence shows that the oxygen levels were very low, and the carbon dioxide was very high, continuing the trend seen by the extreme super-greenhouse that helped cause the Permian extinction. Most of the survivors in the Triassic appeared to be burrowers, and many had adaptations for surviving in low oxygen levels. The land fauna was very low in diversity for the first few million years of the Triassic. Almost 90% of the fossils were from the pig-sized protomammal *Lystrosaurus* (Fig. 5.5), a few larger protomammal predators, and a handful of reptile lineages. The landscape was apparently stripped of most vegetation because there are no coal deposits in the Early Triassic. In many places, such as South Africa, the surface was drier and less vegetated with few plants to bind the sediment, as shown by the shift to barren, sandy, braided streams instead of the plant-rich floodplains of meandering streams.

 See For Yourself: Triassic Land Life: The plants that dominated the Triassic were mostly the gymnosperms, as we saw in Chapter 12 (Fig. 12.17). These had formerly been abundant in the drier upland habitats of the Permian because their cones and wind-blown pollen allowed them to reproduce without being immersed in water, so they didn't require wet climates. By the later Triassic, a whole range of gymnosperms had evolved. These included not only the familiar pine trees but also the "sego palms" or **cycads**, which look superficially like palm trees but are actually gymnosperms with cone-bearing male and female plants (Fig. 13.17A). Another familiar gymnosperm is the **ginkgo** tree, with its distinctive duck foot–shaped leaves, often used today in herbal medicines (Fig. 13.17B). Most

abundant of all were the trees of the genus *Araucaria*, which today include such plants as the Norfolk Island pine, the monkey puzzle tree, and the kauri pine (Fig. 13.17C, D). Trees very like these were the source of the giant tree trunks found in the Upper Triassic rocks of Petrified Forest National Park (Fig. 13.4E). Besides gymnosperms, there were still ferns in the wetter areas and the ferns plus lycopsids ("club mosses") in the few swampy areas left on the planet. This gymnosperm flora would continue to dominate the landscape until the Early Cretaceous, when flowering plants first diversified. So when you walk in a pine forest or a tree fern forest or any other place with few flowering plants, you are walking in the forests like those of the Triassic or Jurassic. The real "Jurassic Park" had no flowering plants at all (contrary to the movie version, where most of the plants have flowers).

 See For Yourself: Mesozoic Plants   Living in these forests during the later Triassic was a land fauna transitional between the typical Permian and typical Jurassic assemblages. As demonstrated by fossils from places like the Petrified Forest in Arizona, it was a mixture of relics from the Permian plus new forms. These holdovers included the huge flat-bodied temnospondyl amphibians, which enjoyed their last successful period before extinction at the end of the Triassic. Meanwhile, some of the earliest frog fossils come from the Triassic, so the modern groups had already appeared along with their ancient relatives. The main holdover, however, was the diversity of the protomammals (Fig. 13.18A), which ruled the land in the Permian and still dominated in the Early and Middle Triassic. These included the beaked herbivores (dicynodonts), like the pig-sized *Lystrosaurus* and the cow-sized *Kannemeyeria*, and numerous predatory protomammals, from the weasel-like *Thrinaxodon* to the bear-sized killer *Cynognathus*.

 See for Yourself: When Protomammals Ruled the World   Battling for supremacy with these relics of the Permian were several new groups. The most important were the primitive members of the reptilian group **Archosauria**, which includes crocodiles, dinosaurs, pterosaurs, and birds (Fig. 13.18). Archosaurs ("ruling reptiles" in Greek) had more advanced limbs and upright posture than most other sprawling reptiles, and many had additional adaptations for mobility. In addition, they may have had lots of air sacs throughout their bodies as aids for breathing (as their bird descendants have today), which would have given them an advantage in the low-oxygen world of the Triassic and Jurassic. The typical Late Triassic archosaurs (formerly lumped in the wastebasket group "thecodonts") included the crocodile-like phytosaurs (Fig. 13.18A); the long-bodied, deep-snouted predators known as erythrosuchids ("bloody crocodiles" in Greek); the armored aetosaurs (Fig. 13.18B); the beaked, pig-sized herbivores known as rhynchosaurs; and the strange, long-necked reptile *Tanystropheus*, whose 7-m-long (23-feet-long) body had a rigid neck over 3 m (about 10 feet) long. Some of these lineages were related to the crocodile branch of archosaurs,

Figure 13.17   Mesozoic gymnosperms. **A.** Cycads, or "sego palms," are gymnosperms with tough spiky leaves. Male plants (right) bear long, narrow cones, while female plants (left) have short, rounded cones. **B.** *Ginkgo biloba*, the ginkgo tree, a gymnosperm with leaves shaped like duck feet. **C.** *Araucaria*, the Norfolk Island pine, another typical Triassic gymnosperm. **D.** Fossil log of an *Araucaria* relative from the Petrified Forest National Park, showing the detailed tissues and growth lines.

while others were closely related to dinosaurs and pterosaurs. But they all began to decline in the Late Triassic as the two living groups of archosaurs replaced them. One was the true crocodilians, which were delicate long-legged, dog-sized reptiles in the Late Triassic. Only later did they become the big aquatic ambush predators that we think of today, when the phytosaurs died out and vacated that niche.

The most important group of archosaurs, however, was the dinosaurs. They first appeared in the Late Triassic as small, bipedal, fast-running predators like *Eoraptor* and *Herrerasaurus*. Soon they began to evolve into a variety of body forms so that before the Triassic was over there were large ancestors of the long-necked sauropods. Known as **prosauropods**, they were typified by the largest herbivore of the Triassic, *Plateosaurus*, which was about 10 m (about 33 feet) long maximum. Although it was still primarily bipedal, it already showed the lengthening neck and tail that characterized the Jurassic sauropods. It had simple leaf-shaped teeth suitable for chopping vegetation, unlike the sharp pointed teeth of most archosaurs, which were still predators.

While the Triassic–Jurassic extinction hit the marine realm hard, the Triassic–Jurassic transition on land was more gradual. The last of the large protomammals died out, but already by the Late Triassic they had given rise to the first true mammals, which were much like shrews in both their size and their insectivorous diets. The last of the large, flat-bodied temnospondyl amphibians vanished, but both frogs and salamanders had evolved to carry on the amphibian lineages. Most of the archaic archosaur groups also vanished, leaving the world to the main groups of archosaurs that dominated the Jurassic: crocodiles, pterosaurs, and dinosaurs.

THE REAL JURASSIC PARK

When the Jurassic began about 200 Ma, dinosaurs had already replaced the more primitive archosaurs and large protomammals and completely ruled the landscape. Thanks to the high sea levels worldwide during most of the Middle and Late Jurassic, we don't have a lot of land faunas of that age. Our best sample of Jurassic dinosaurs comes from the Late Jurassic, when we have the incredible faunas of the Morrison Formation in western North America (Fig. 13.1), the Tendaguru beds in East Africa, and several important fossil assemblages in Asia and South America.

Figure 13.18   Two scenes of life in the Triassic. **A.** The phytosaur *Smilosuchus* attacks the herbivorous protomammal *Placerias*. **B.** The predatory "bloody croc" *Postosuchus* attacks the armored aetosaur *Desmatosuchus*, while a small early dinosaur waits in the background and an early mammal watches from a branch.

From these faunas, we know that the Late Jurassic was the heyday of the dinosaurs (Fig. 13.1). Huge, long-necked sauropods roamed across floodplains and semi-arid uplands represented by the rocks of the Morrison Formation (Fig. 13.6E, F). These included not only close relatives like the long-faced *Diplodocus* and *Apatosaurus* (formerly called "Brontosaurus") but also short-faced *Camarasaurus* and the towering brachiosaurs like *Brachiosaurus* in North America and *Giraffatitan* in Africa. (Ironically, except for *Brachiosaurus* and *Dilophosaurus*, most the dinosaurs in the *Jurassic Park* book and movie are all actually Cretaceous in age.) Preying on these huge creatures were smaller **theropods** like *Allosaurus, Torvosaurus,* and *Ceratosaurus.* There were also abundant **stegosaurs**, including *Stegosaurus* with the plates on its

back in the Morrison Formation and the spike-backed *Kentrosaurus* in the Tendaguru beds of Africa, as well as primitive **ankylosaurs**. Several other small ornithopod dinosaurs (*Camptosaurus* and *Dryosaurus*) are known, as well as small predators like *Ornitholestes* and *Coelurus*. But there were small animals roaming the Late Jurassic landscape as well. These included a variety of different shrew-sized mammals; diverse frogs, salamanders, and lizards; many types of crocodilians; and some of the oldest bird and pterosaur fossils.

CRETACEOUS LAND LIFE

During the 80 million years of the Cretaceous, the dinosaurs continued to evolve and diversify. The sauropods virtually vanished from most of the continents except for parts

BOX 13.1: HOW DO WE KNOW?

How Do We Know About the Dinosaurs?

Dinosaurs are so incredibly popular today that it's hard for us to imagine when they were not always part of the public imagination. But they were first discovered and identified as giant reptiles as recently as the 1820s, and only *Megalosaurus*, *Iguanodon*, *Hylaeosaurus*, and a few others were known in 1842 when Richard Owen coined the term "Dinosauria" (meaning "terribly great lizards") for them. Most early discoveries were based on just scraps of bone and teeth, so it was not unreasonable for natural historians and their artists to reconstruct them as immense lizards (Fig. 13.19) sprawling on all fours.

 See For Yourself: The Monsters Emerge

It was not until the late 1870s and 1880s that nearly complete articulated skeletons of dinosaurs were discovered in the Rocky Mountain region of Montana, Wyoming, Colorado, and Utah and scientists could see how dinosaurs were actually built for the first time. Even so, these early discoveries remained unknown to the public and barely even illustrated until the early 1900s, when museum curators realized that enormous mounted skeletons and models and painted reconstructions would draw people to visit their museums. Soon they were the hottest item in the media. Pioneering museums like the American Museum of Natural History in New York, the Yale Peabody Museum in Connecticut, and the Carnegie Museum in Pittsburgh set the standard for all later dinosaur reconstructions and made the creatures like "Brontosaurus" (the dinosaur properly known as *Apatosaurus*) and *Tyrannosaurus rex* and *Triceratops* household names.

Dinosaur Relationships

From the 1880s onward, hundreds of different dinosaurs were discovered. Their family tree (Fig. 13.20A) has been well known for a long time, and most of the groups of dinosaurs are well known to just about any kid over 5 years old these days.

First of all, what do we mean when we say that something is a "dinosaur"? Even though the public often uses the word "dinosaur" for any extinct prehistoric animal, being extinct doesn't make something a dinosaur. Some people (and toy dinosaur sets) include extinct mammals such as the saber-toothed cat or woolly mammoth, protomammals like *Dimetrodon*, and marine reptiles like ichthyosaurs and plesiosaurs. But these creatures are *not* dinosaurs. The Dinosauria are a branch of the archosaurs (Fig. 13.20A) that is separate from the crocodile branch, including many of the Triassic creatures that ruled the land before dinosaurs evolved. Nor were all dinosaurs huge. Many of them were tiny, about the size of a chicken, and some even smaller. The dinosaurs are different from other land animal groups in evolving the largest terrestrial creatures the world has ever seen.

The nearest relatives of dinosaurs are the pterosaurs or "flying reptiles." These include such famous fossils as the crow-sized *Pterodactylus* and *Rhamphorhynchus*, the giant crested toothless fossil *Pteranodon* with wingspans of 7 m (about 23 feet) or more, and the huge pterosaurs like *Quetzalcoatlus*, which was the size of a small airplane. How do we

Figure 13.19 When the earliest dinosaur fossils were found, they were so incomplete that they were often reconstructed as gigantic sprawling lizards. Here, what we now know as the *Allosaurus*-like predator *Megalosaurus* (right) and the herbivore *Iguanodon* (with the spike on its nose) look nothing like the swift, slender bipedal dinosaurs we know today. (Drawn by Edouard Riou in 1863)

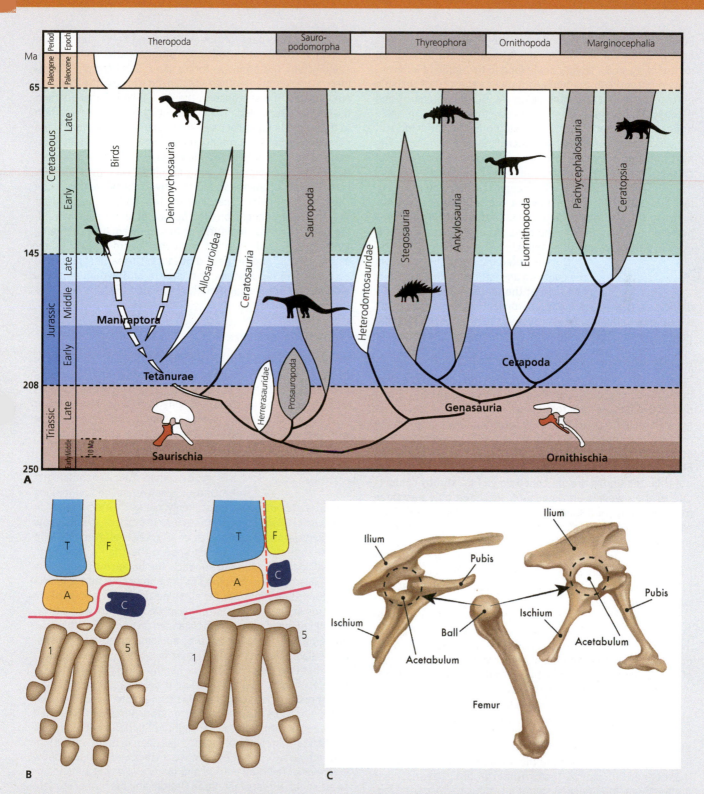

Figure 13.20   A. Dinosaur family tree. The small insets on the bottom show the difference between the "lizard-hipped" or saurischian dinosaurs and the "bird-hipped" or ornithischian dinosaurs. **B.** All dinosaurs, birds, and pterosaurs have a mesotarsal joint (right) in their ankle bone, where the first row of ankle bones (astragalus and calcaneum, A and C) is fused to the end of the shin bone and the hinge is between the first and second rows of ankle bones. By contrast, most other land vertebrates have the joint between the shin bones (tibia and fibula, T and F) and the astragalus and calcaneum (dashed red line), On the left is the ankle of the lineage of crocodilians and their archosaur relatives, with the hinge running between the astragalus and calcaneum. **C.** Comparison of the saurischian hip bones (right) and ornithischian hip bones (right), showing the open hip socket (acetabulum) into which the thigh bone (femur) inserted. The ilium runs along the backbone, and the head of the dinosaur would be to the right.

know they are closely related? Only pterosaurs and dinosaurs have a unique condition of the ankle known as the **mesotarsal joint** (Fig. 13.20B). In your ankle (and in the ankle of all mammals and reptiles), the hinge is between the end of the shin bone and the first row of ankle bones. But in pterosaurs and dinosaurs, the ankle hinges between the first row of ankle bones and the second row of ankle bones, in the middle of the tarsal bones of the ankle (hence, "mesotarsal joint"). The next time you eat your Thanksgiving turkey or get an order of chicken, look at the "handle" end of the "drumstick" (actually the shin bone or tibia). That little cap of cartilage on the meatless end is actually the first row of ankle bones which typically fuses to the end of the shin bone, so the bird ankle hinges between the first and second rows of ankle bones, just as it does in dinosaurs and pterosaurs.

 See For Yourself: Pterosaurs

Once the pterosaur branched out from the common ancestor with dinosaurs, we can look at the features that define the dinosaurs. There are many anatomical characteristics that are unique to dinosaurs, most having to do with the fact that their limbs are fully upright and not sprawling. The easiest to spot is the fact that instead of having a regular enclosed cup-shaped socket in their hip where the thigh bone inserts (as humans and most land animals do), their hip sockets are open holes that go right through the hip bones (Fig. 13.20C). Other distinctive features are found in their feet and hands. All dinosaurs walk on the tips of their fingers and toes, not on the palms of their hands and soles of their feet, as mammals such as bears and apes and humans do. This is true even with the heaviest giant sauropods, which are still walking tippy-toe—but their toe bones are squashed down into stubby columns due to their weight. In dinosaurs, the hand only has three functional fingers (thumb, index finger, middle finger). They have a tiny ring finger, and almost no pinky at all. Finally, in their ankle, there is a distinctive flange of bone that sticks up from the tiny first row of ankle bones and overlaps the lower front end of the shin bone. These and many other unique features define what is and is not a dinosaur—not whether they were gigantic or extinct.

 See for Yourself: Dinosaur Family Tree

The earliest dinosaurs from the Late Triassic, such as *Eoraptor* and *Herrerasaurus*, show that all dinosaurs originated as small, fast-running bipedal creatures about the size of a chicken or a turkey. From these small, primitive ancestors, the huge diversity of dinosaurs evolved by the mid-Jurassic. There are two main groups. The **saurischian** ("lizard-hipped") dinosaurs had the primitive reptilian arrangement of hip bones (Fig. 13.20C), with the ilium running along the spine, the ischium pointed backward, and the pubic bone pointed forward, all arranged around the open hole of the hip socket that defines Dinosauria. The two main groups of saurischians are the huge, long-necked sauropods (like *Apatosaurus*, *Brachiosaurus*, and *Diplodocus*) and the mostly predatory theropods (including birds—see below, "Birds and Dinosaurs").

The second group is the **ornithischian** ("bird-hipped") dinosaurs, which are different in that some portion or all of the

pubic bone runs backward, parallel to and below the ischium (Fig. 13.20C). All the ornithischian dinosaurs were herbivores. They include the primitive forms like *Iguanodon* and *Camptosaurus*, plus many other specialized groups: the armored ankylosaurs, the stegosaurs, the duck-billed dinosaurs, and the horned and frilled ceratopsians (Fig. 13.20A). In addition to their unique hip structure, ornithischians are defined by a number of other evolutionary novelties. They all had an additional bone at the tip of their lower jaw, called a "predentary bone" (an old name for ornithischians was the "predentates"). Most of them had a distinctive "cheek" region in the skull with teeth inset deep into the jaw, so it is thought that they had fleshy cheeks on their faces that aided them in holding vegetation in their mouths as they chewed. (In a confusing twist, birds are descended from the "lizard-hipped" dinosaurs, not the "bird-hipped" dinosaurs. The earliest birds had a saurischian hip system, and later in bird evolution it was modified so that it superficially resembled the hips in ornithischian dinosaurs.)

Dinosaur Renaissance

Since the beginning, the scientific and popular concepts of these dinosaurs were dominated by the idea that they were just stupid, lethargic, sluggish lizards grown gigantic. Most reconstructions had them dragging their tails, eating water plants, and lounging in swamps to support their great weight with the buoyancy of water. These images prevailed for decades, until new discoveries made them obsolete.

 See For Yourself: Dinosaur Biology

The first breakthrough came in the early 1960s, when Yale paleontologist John Ostrom discovered the fast-moving predator *Deinonychus* (mistakenly called "Velociraptor" in the *Jurassic Park* book and movies. The actual fossils show that *Velociraptor* was the size of a turkey.) As everyone who has seen them in the movies knows, they had a long, rigid tail sticking straight out in back for balancing their agile body, a large brain compared to their body size, and enlarged claws on their hind feet for slashing their prey. All of these features suggested to Ostrom that *Deinonychus* was active, intelligent, and fast-moving, balanced on two legs with no tail dragging behind it. Soon, other dinosaurs were reassessed, and paleontologists realized that most dinosaurs had rigid tails that they held straight out behind them. Many dinosaur tails (especially duckbills) are held out straight and rigid by criss-crossing trusses of tendons turned into bone. Thus, bipedal dinosaurs (especially all predatory dinosaurs) balanced on two legs with their tails straight out and did not lean back on their tails in the "kangaroo" pose seen in many early dinosaur mounts. To top it off, dinosaur trackways seldom show evidence of tail drag marks, and many show that some dinosaurs could run very fast. Soon Ostrom and his former student Bob Bakker were promoting the idea that dinosaurs were fast, agile, and intelligent creatures, not slow, sluggish creatures of the swamps. (We now know that the most famous dinosaur beds, such as the Upper Jurassic Morrison Formation, were

Hot and Cold Running Dinosaurs

 See For Yourself: Top Ten Dinosaur Discoveries

The discovery of active dinosaurs led to another controversy: if they were so active, did dinosaurs have a warm-blooded metabolism like mammals and birds? Ostrom and later Bakker made this argument, which reached a peak of controversy in the 1970s and 1980s. But what does it mean to be "warm-blooded"? It's not as simple as most people think. There are actually two independent physiological processes at work. One is the *source* of the heat: if an animal gets its body heat from the environment (as most reptiles, amphibians, fish, and other animals do), it is an **ectotherm**; if it burns its food to generate its own metabolic body heat, it is an **endotherm**. Another factor is how much an animal *regulates* its body temperature. If it allows its body to cool down or heat up with the external ambient temperature, then it has a variable body temperature and it is a **poikilotherm**; if it regulates its body heat so that it remains constant, it is a **homeotherm**.

Most living reptiles, amphibians, fish, and other animals that are ectotherms are also poikilotherms; and most living birds and mammals are both endotherms and homeotherms. But in nature, it's not that simple. A number of ectotherms, such as pythons, can shiver to generate body heat when incubating their eggs, so they can use metabolic energy to make heat like endotherms. Many "cold-blooded" ectotherms, such as desert lizards after they have warmed up in the sun, have a body temperature that is actually warmer than that of a bird or a mammal. Even though most mammals and birds are endotherms and homeotherms, there are exceptions in this generalization, too. For example, large-bodied mammals like camels let their body temperature fluctuate a bit to cope with the extreme heat and cold of deserts. Very small mammals, like shrews and bats, actually slow down their metabolism and let their body temperature drop to cope with periods of cold or starvation. Thus, we can't use the term "warm-blooded" or "cold-blooded" casually because nature is more complex than that.

Ostrom and later Bakker supported their argument for dinosaur endothermy with a number of lines of evidence.

1. *Food pyramids*: In ecology, there is a famous concept called the **food pyramid** (Fig. 13.21). Typically, if you represent the bottom tier of the pyramid as the total biomass of plant matter available for animals to feed on, there is a smaller biomass of herbivores in the next tier up because some of the material they eat must be dedicated to their daily activities other than turning food into body tissue. The next step up the pyramid shows the primary consumers, or the first level of predators on these herbivores. They will have an even smaller biomass because some of what they eat goes to many other functions besides growth and building their bodies. Each step up the pyramid through higher and higher levels of predators means smaller and smaller biomass since the transfer from one level to the next one above it is not 100% efficient.

   In biological systems where the predators are ectothermic, such as crocodiles and their prey, the biomass is only slightly smaller than that of their prey because crocodiles waste almost none of their food on metabolic body heat or other functions that don't build new tissue (Fig. 13.21).

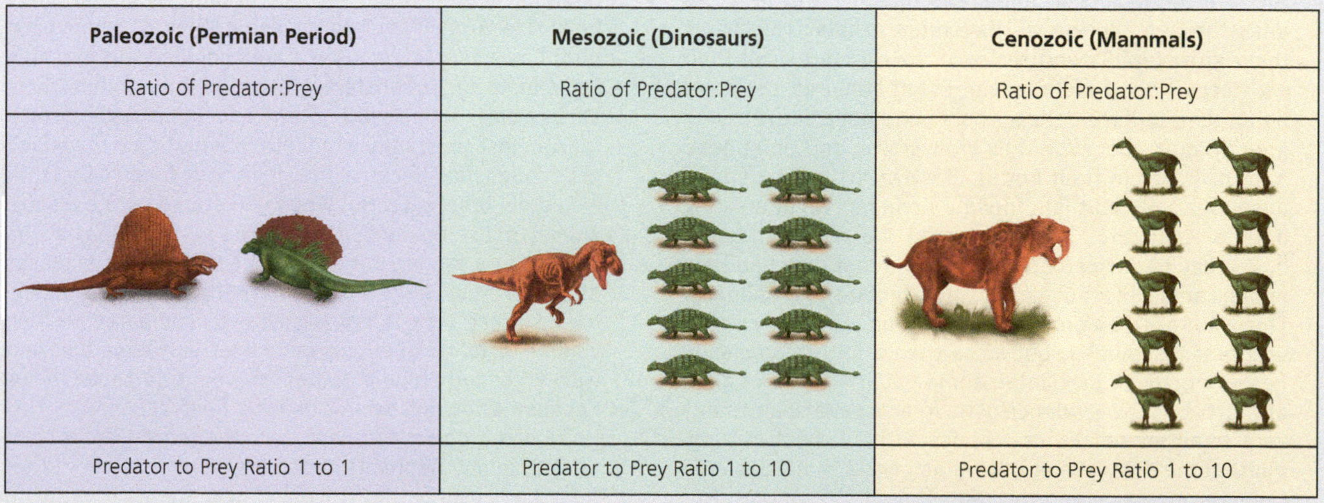

Paleozoic (Permian Period)	Mesozoic (Dinosaurs)	Cenozoic (Mammals)
Ratio of Predator:Prey	Ratio of Predator:Prey	Ratio of Predator:Prey
Predator to Prey Ratio 1 to 1	Predator to Prey Ratio 1 to 10	Predator to Prey Ratio 1 to 10

Figure 13.21 Advocates of dinosaur endothermy pointed out that most ectothermic predators (here, the finbacked protomammal *Dimetrodon*) have a biomass that is roughly equal to the biomass of all possible prey species. However, modern endothermic predators such as the lion require 10 times as much biomass in prey zebras and antelopes to survive because they use up most of their food in maintaining their high body temperature through metabolism. Most dinosaur communities show a similar ratio, about 10 times as much biomass of prey species as there are predators.

But if the predators are endotherms, like lions, then their biomass is considerably less than that of their prey because most of what they eat goes to making metabolic body heat and keeping their body temperature constant. Thus, there can be only few lions compared to the number of prey antelopes and zebra that each needs to support it.

Bakker used this relationship to argue that dinosaurs were endothermic because in many fossil localities only a few predators are found compared to fossils of many prey species. But this is an overly simplistic view of the nature of preservation of the fossil record. In reality, there are lots of factors that determine the number of predators or prey animals that are fossilized so that what you count in a fossil assemblage may have no relationship to what lived there at the time. For example, the famous La Brea tar pits in Los Angeles trapped Ice Age animals by the thousands, but the mammals and birds are overwhelmingly predators or scavengers like the dire wolf and saber-toothed cat, plus vultures, condors, and eagles. This odd ratio only makes sense if the tar pits were a predator "death trap," with one stuck prey animal (horse, bison, camel, sloth) struggling in the tar as bait and luring numerous predators or scavengers to their death. Likewise, the Cleveland-Lloyd Quarry assemblage in central Utah is overwhelmingly predators like *Allosaurus*. What does that tell us about predator/prey ratio? Were *Allosaurus* cannibals since there are few prey species in the quarry? Or was it another predator death trap like La Brea? In short, we cannot take the predator/prey ratio of fossil localities at face value because there are many factors that might distort the ratio and do not reflect the true relative numbers of predators and prey when those organisms were alive.

2. *Bone histology*: For a long time, anatomists have noticed that animals with endothermic metabolism have bones that are full of lots of holes and openings for blood to pass through, known as **Haversian canals**. This helps them grow rapidly and remodel their bone to suit their high activity and metabolism. Ectotherms, on the other hand, tend to have denser, more solid bone with striking growth bands because they grow slower and don't need a lot of blood in their bones. Back in the 1950s, anatomists first noticed that dinosaur bone had an interesting internal structure. They compared the dinosaur specimens they had sliced open and found that they had Haversian canals like those found in mammal and bird bones. This was taken as proof of dinosaur endothermy when it was first discovered, but since then a lot of complexity has emerged. In particular, some ectotherms (like giant sea turtles) show evidence of Haversian canals, and many very small mammals have bones without them. It turns out that the presence of Haversian canals is more related to large body size and the requirement for rapid growth of bones, not metabolic rate. Because dinosaurs clearly grew very fast and reached very large body sizes, they would have had Haversian canals whether they were endotherms or ectotherms

3. *Other arguments*: Early in the debate over dinosaur physiology, scientists pointed to the upright posture of dinosaurs as proof of endothermy since standing upright requires continual expenditure of energy, especially compared to the sprawling posture of a lizard or crocodile when it is resting. However, it turns out that there isn't that much of a difference in energy use to stay upright because it's mostly done by passive mechanisms that do not require muscular effort or much expenditure of energy. Some have pointed to the fact that the isotopic chemistry of the bones shows that their entire bodies were the same temperature, not like reptiles whose limbs tend to stay cooler than the core body temperature. However, this does not prove they were endothermic, only that they were homeothermic—and there are other ways to achieve homeothermy without endothermy. Others pointed to the presence of dinosaurs above the Arctic and Antarctic Circles, but in recent years it became clear that the later Mesozoic was a greenhouse world with mild climates on the poles, so even ectotherms could survive there (as crocodiles did).

4. *Summary*: Since the peak of the debate over hot-blooded dinosaurs in the 1980s, the debate has cooled down considerably. The biggest problem is that dinosaurs span a huge range of body sizes (including much larger than any living land animal), so it is hard to generalize about their metabolism and say that "all dinosaurs were endotherms" or "all dinosaurs were ectotherms." The small-bodied predators, and probably the large predators, were almost certainly endotherms since they required a high metabolism as active hunters. In addition, there is evidence (discussed below, see "Birds and Dinosaurs") that they were covered by feathers, which primarily served as insulation, which only makes sense on an endotherm.

Large dinosaurs, on the other hand, were almost certainly *not* endotherms. The problem comes from their large body size. A small endotherm, like a shrew or a hummingbird, has a very large surface area compared to its internal volume and is constantly losing heat through its large surface area. Animals like shrews and hummingbirds must feed almost constantly or go into hibernation if they don't get enough food because they lose body heat so fast they would die otherwise. But when you start to scale a small animal up to large dinosaur size, the surface area only increases as a square, but the volume of the body increases as a cube. Thus, large animals have the opposite problem: their surface area is not big enough compared to their volume to get rid of excess body heat. Any large dinosaur with an endothermic metabolism would be cooked alive because it wouldn't be able to dump heat fast enough. This issue is a problem for elephants, the largest endotherm on land today. Elephants constantly must find ways to get rid of body heat or they will die of heat stress. They spend much of their time in the shade or in waterholes or mud wallows, and they have huge external ears, whose main function is as a heat radiator, not hearing. Some of the giant sauropods were many times the size of elephants, so if they were endotherms, they would have needed enormous ears or sails or flaps or some other structure to get rid of heat—and there is no evidence of this.

Instead, most paleontologists think that the largest dinosaurs remained ectotherms but employed the advantages of their huge mass compared to surface area by what is known as **inertial homeothermy,** or **gigantothermy**. Since they gain and lose heat very slowly, their body temperature would not have fluctuated much, simply due to their enormous size. In addition, they lived in a greenhouse world with no cold climates and minimal temperature change from poles to the equator, so their environmental temperature was very warm and stable. This is comparable to what desert camels do. Their large body size keeps them thermally stable, and when they build up a lot heat in the daytime, they can let their body temperature run hot for a while until the chilly night cools them down. In the morning, they let their below-normal body temperature slowly rise as the day warms up. This saves them a lot of energy and food if they are not trying to maintain a constant body temperature by metabolic means.

Birds and Dinosaurs

 See For Yourself: How Did Dinosaurs Get So Huge? At the same time as the discovery of *Deinonychus* led to the dinosaur endothermy controversy, Ostrom was also re-examining the handful of specimens of *Archaeopteryx*, the oldest known bird (Figs. 6.1, 13.22). Discovered and described in 1861, the first specimen of *Archaeopteryx* was almost perfectly timed as the "missing link" between birds and reptiles to support Darwin's ideas of evolution, just published in 1859. *Archaeopteryx* was first described by Richard Owen, one of Darwin's critics, who refused to discuss its clear intermediate anatomical position between birds and dinosaurs. But Thomas Henry Huxley, Darwin's chief supporter, not only noticed this but observed something else. Except for its feather impressions (which is why it was first recognized as a bird), the entire rest of the skeleton was completely dinosaurian and not that different from *Compsognathus*, a small predatory dinosaur that Huxley himself described from the same beds in the Solnhofen Limestone in Germany that produced *Archaeopteryx*. But Huxley's ideas were overshadowed by later paleontologists, who could not imagine huge predatory dinosaurs as ancestors for birds and so pushed their ancestry back to some unspecified member of the archosaurs in the Triassic.

This remained the dogma until the early 1970s, when Ostrom's re-examination of the 12 known *Archaeopteryx* specimens caused him to rediscover and add to Huxley's excellent evidence that birds are just modified dinosaurs. He even found that one of the specimens had originally been misidentified as the dinosaur *Compsognathus*, then showed it was really an *Archaeopteryx*. The evidence is clear just by looking at the skeleton of *Archaeopteryx*. It had a long bony tail like a dinosaur, not the short bony "parson's nose" of living birds, which supports their tail with feather shafts. *Archaeopteryx* still had the

teeth of a dinosaur, not the toothless beak of modern birds. Its hand was composed of many separate fingers capable of grasping, not the fused fingers that you find on the end of a chicken wing (modern birds support their wings with feather shafts anchored in the fused hand bones, not their fingers, such as bats do). *Archaeopteryx* and all other birds also have the trademark feature of dinosaurs and pterosaurs, the mesotarsal joint (Fig. 13.21B). Like dinosaurs, this row of ankle bones has a little protuberance sticking up in the front of the shin called the ascending process of the astragalus, a feature found only in dinosaurs and birds (and no other archosaurs).

 See For Yourself: Birds and Dinosaurs The most striking of these features are in the wrist. Birds and some theropod dinosaurs, such as the dromaeosaurs (*Deinonychus* and *Velociraptor* and their kin), all have a wrist bone formed of fusion of multiple wrist bones shaped like a "half moon" (Fig. 13.22B). This bone serves as the main hinge for the movement of the wrist, allowing dromaeosaurs to extend their wrists and grab prey with a rapid protraction and retraction. It so happens that exactly the same motion is part of the downward flight stroke of birds. *Archaeopteryx* had the same three fingers (thumb, index finger, and middle finger) as most other theropod dinosaurs, and the middle digit (the index finger) is by far the longest. In addition, the claws of *Archaeopteryx* are very similar to those of theropod dinosaurs.

The best evidence that birds are dinosaurs comes from embryology. One experiment showed that birds still have the genes for teeth, even though no living bird has teeth. The embryonic mouth tissues of a chick were grafted into the mouth area of a developing mouse. When the mouse grew teeth, they were not normal mouse teeth but conical peg-like teeth similar to those of the earliest toothed birds, or the dinosaurian ancestors of birds. All it took was the removal of the regulatory genes that a chick would normally have (by grafting tissues into a mouse) and the long-suppressed genes for reptilian teeth carried by all birds finally emerged. Other embryonic studies have since managed to change the genes that code for the birds' short, stumpy, bony tail, resulting in the development of a long bony tail like a dinosaur. Another genetic modification experiment in developing chickens gave rise to chickens with dinosaur-like feet, not bird feet. Yet another experiment produced a bird with a dinosaurian snout with teeth instead of a normal toothless beak. Birds have nearly all their old dinosaurian genes residing in their genome; they are just not expressed.

Finally, *Archaeopteryx* has been joined by hundreds of new fossils, especially those found in the Lower Cretaceous beds in Liaoning Province, China. These fine lake deposits have extraordinary preservation, so we can see fossilized feathers in birds, fur in mammals, stomach contents, and occasionally even the pigments that tell us what color some animals were. These Chinese fossils have produced not only an enormous array of Cretaceous toothed birds (including whole orders and families that are now extinct) but quite a few feathered dinosaurs that clearly were not able to fly and were not closely related to birds. It turns out that feathers first evolved for

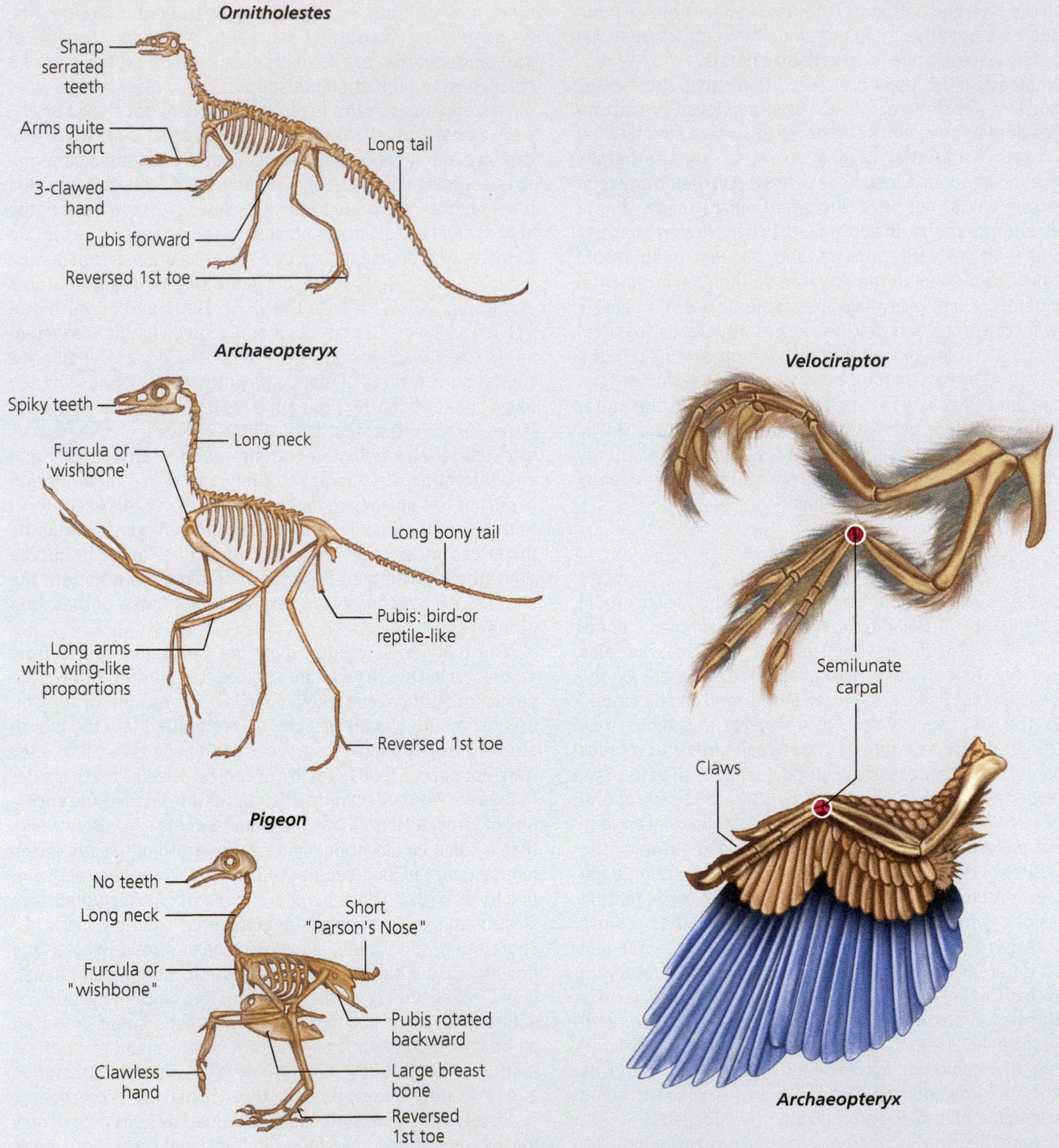

Ornitholestes

Sharp serrated teeth

Arms quite short

3-clawed hand

Pubis forward

Reversed 1st toe

Long tail

Archaeopteryx

Spiky teeth

Furcula or 'wishbone'

Long neck

Long bony tail

Pubis: bird-or reptile-like

Long arms with wing-like proportions

Reversed 1st toe

Velociraptor

Semilunate carpal

Claws

Archaeopteryx

Pigeon

No teeth

Long neck

Furcula or "wishbone"

Short "Parson's Nose"

Pubis rotated backward

Clawless hand

Large breast bone

Reversed 1st toe

Figure 13.22   A. Comparison of the skeletons of a bird, a dinosaur, and *Archaeopteryx*, showing the dinosaurian teeth, long bony tail, large finger bones, and diagnostic ankle of *Archaeopteryx*. **B.** Comparison of the wrist bones of *Velociraptor*, *Archaeopteryx*, and a more primitive dinosaur, showing the half moon–shaped wrist bone that is unique to birds and raptors.

insulation, which is their primary function in birds today, if you consider how many down feathers and body feathers they have. Only long after dinosaurs evolved feathers did one lineage modify certain feathers in the hand for flight.

So many feathered dinosaurs have been found in both the saurischians and ornithischians now that it's clear that most dinosaurs probably had feathers of some sort. The really large dinosaurs (many of which preserve skin impressions that are scaly) may have had feathers only as hatchlings and then lost them when they grew to a large body size and no longer needed insulation but instead needed to dump body heat. So any time you see a dinosaur reconstruction (at least for smaller dinosaurs and most predators) that is *not* feathered, it is grossly out of date. For this reason, the "Velociraptors" (actually *Deinonychus* or maybe *Utahraptor*) in the 2015 movie *Jurassic World* were a big disappointment to paleontologists because the filmmakers refused to update their images but instead gave us the outdated scaly appearance of these creatures (which was all we knew when the first movies were made), rather than the fuzzy feathered creatures that we now know they were (and have known since 1996, when the first feathered dinosaurs were discovered).

Birds are dinosaurs. It's a fact. So the next time you go out, watch the dinosaurs feeding at the bird feeder or listen to the dinosaurs singing in the trees or watch them fly across the sky. Dinosaurs are not extinct. Their descendants have been with us for 66 million years.

of Gondwana, such as South America, where they reached their maximum size with titanosaurs like *Argentinosaurus* and the newly discovered giant *Patagotitan*, currently the largest dinosaur ever found. The stegosaurs declined and eventually vanished in the Early Cretaceous. So too did the primitive theropods like the allosaurs, which were replaced by much bigger predators like the tyrannosaurs.

In their places was a tremendous diversification of three big groups of herbivorous dinosaurs: the duck-billed dinosaurs or **hadrosaurs**, the armored ankylosaurs, and the horned dinosaurs like *Triceratops*. The duckbills and horned dinosaurs, in particular, had a specialized mouth with rows and rows of prism-shaped teeth closely packed together, which form a large grinding surface called a dental battery. These teeth gave them the ability to chew and digest far more plant material than dinosaurs with a single row of peg-like or leaf-shaped teeth.

The success of these two groups may be related to the diversification in the Early Cretaceous of an entirely new form of plants, the flowering plants or **angiosperms**. Today, they make up over 95% of living plants, so that nearly every plant you commonly see, and every plant we eat, from grasses to bushes to trees (except for pine trees, ferns, mosses, and algae), is an angiosperm. There may be some possible angiosperm fossils in the Late Jurassic; but by the Early Cretaceous, angiosperms were clearly established and diversifying rapidly, and they soon took over virtually all land habitats. Some of those Early Cretaceous plants that were among the first to diversify include living fossils such as the magnolia tree, water lilies, and pepper plants.

See For Yourself: The Flowering of Flowering Plants

Why were the angiosperms so successful? A major advantage they have over more primitive plants is their efficient mode of reproduction—the flower and all of its complex reproductive mechanisms that ensure success. Instead of the inefficient wind-pollinated gymnosperm seed, which wastes a huge amount of pollen and depends on the random breezes, angiosperms evolved flowers specifically as devices to attract pollinators. These are mainly insects (especially moths, butterflies, and bees) but also birds, bats, and other flying creatures. Pollinators ensure that the pollen is carried directly from one flower of the same species to another, which is more efficient than relying on the wind. Not surprisingly, the first bee and butterfly fossils appear in the Cretaceous as well, so the angiosperms coevolved with their pollinators.

The reproductive cycle (Fig. 13.23A) is highly modified: the ovules are fully enclosed within protective covers (modified from leaves) called **carpels,** which form the core of the flower. The carpel protects the ovule from drying up, fungal infection, and the depredations of herbivorous insects. The stamens are surrounded by **petals** (which serve to attract the pollinator and guide it to the ovules in many cases) and an outer covering of **sepals** for protection. Typically, a pollinator gets pollen stuck to it as it climbs into a flower, seeking the nectar that the plant generated to lure it in. The ovule is usually pollinated by sperm carried from a different flower, while the pollinator picks up new pollen on its way out and passes it onto yet another flower, thus minimizing self-fertilization of the flowers. However, angiosperms can also self-fertilize if cross-pollination is not possible.

Once the pollen has been delivered, a pollen tube develops that transports the sperm to the ovules. Here angiosperms have another advantage: **double fertilization**. The pollen carries two sperm nuclei, one of which fuses with the egg nucleus to form the embryo and the other of which fuses with two other nuclei to form a food supply for the embryo. This means that angiosperms don't need to invest a lot of energy creating food stores for each seed until it is fertilized (unlike gymnosperms, which create food even for infertile seeds).

The entire fertilization and embryogenesis process takes place in only a few weeks or days, so angiosperms can sprout, flower, reproduce, and die in a single season if necessary. By contrast, most gymnosperms are slow to grow and reproduce (usually taking at least 18 months between reproductive cycles) and cannot accomplish the entire process in a single season. For gymnosperms to live in highly seasonal, cold-winter climates (like the evergreen conifers), they must be able to survive the cold and shut down much of their physiological systems in the dead of winter. Many

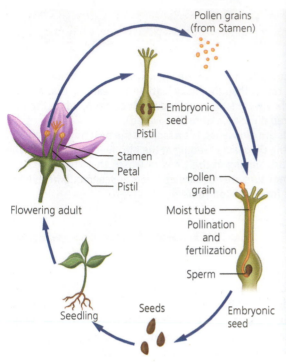

Figure 13.23 **A.** Angiosperm life cycle. **B.** The magnolia tree is a typical Early Cretaceous angiosperm.

angiosperms, on the other hand, are annuals that sprout in the spring, burst into flower, and produce seeds that can survive the next winter while the rest of the plant dies. This rapid reproduction allows them to quickly exploit habitats that other plants cannot.

Finally, angiosperms are known not only for their rapid growth rates but also for their ability to grow back quickly after animals have munched on them. Think of how quickly the grass grows back after you mow it (or after an animal grazes it). By contrast, ferns or cycads cannot grow back so quickly after they have been heavily eaten and often die if the damage is too great (such as when an animal eats the growing tip of the plant). Many angiosperms can be eaten right down to their roots but grow back again. One further advantage of angiosperms that sets them apart from all other seed plants is that they combine the advantages of a seed with the capability for **vegetative reproduction**, with parts of the plant (such as stems or branches) taking root and establishing new plants asexually. Gardeners use this all the time when they plant a branch from an existing plant, or part of a potato, and it grows into a new plant.

Some paleontologists suggest that angiosperms co-evolved these advantages in response to heavy feeding pressure by some sort of efficient herbivore, like the duck-billed dinosaurs of the Cretaceous, which had huge grinding dental batteries for eating large amounts of vegetation. Their highly efficient feeding mechanisms probably put pressure on plants to reproduce and regrow quickly in a sort of coevolutionary "arms race" between plant and herbivore. Angiosperms responded to this pressure and now dominate the earth, while the more primitive plants are mostly on the sidelines.

13.8   The End of the Age of Dinosaurs

The second or third largest mass extinction in earth history is the one at the end of the Cretaceous, which eliminated not only the non-bird dinosaurs but also the ammonites and many other marine groups. Of all the great mass extinctions, this one has garnered the most attention by far and is the only one that most people have heard of, even though it's not the largest mass extinction. The difference, of course, is that it is about popular topics like dinosaurs and impacts of rocks from space, which makes good copy for the media.

Another part of the problem of public misunderstanding of the event is the exclusive and mistaken focus on dinosaurs. Prior to 1980, there were a number of ideas about the extinction of the dinosaurs. Some thought it got too cold for them or too hot. Others blamed it on the spread of an epidemic of disease. For a long time, some paleontologists argued that the dinosaurs vanished because they couldn't digest the new angiosperms—except that angiosperms appeared in the Early Cretaceous, not the end of the Cretaceous. In fact, angiosperms may have actually triggered the diversification of super-herbivores like the duckbills. Still others suggested that the mammals ate their eggs—except that mammals and dinosaurs both appeared in the Late Triassic and lived side by side for 120 million years, so it seems odd that mammals suddenly became voracious egg-eaters at the very end.

But all these ideas missed the crucial point: *the Cretaceous extinctions wiped out not only the dinosaurs but also the ammonites, the marine reptiles, some marine plankton, and many other marine invertebrates.* Any explanation that focuses on only the dinosaurs misses the global nature of this extinction and fails to show why plants and animals from every level on the food chain died out. The extinction of the dinosaurs is really an afterthought. If the proposed cause can wipe out plankton, ammonites, many other marine invertebrates, and most land plants, then it makes sense that the top of the food pyramid—the dinosaurs—would also disappear. Thus, the topic of why the dinosaurs and other victims of the Cretaceous extinctions vanished was mostly the realm of unscientific speculation, largely focused on one group (the dinosaurs) and failing to explain anything else.

The breakthrough came in 1978 when geologist Walter Alvarez and his father, physicist Luis Alvarez, plus geochemists Frank Asaro and Helen Michel, discovered abnormally high levels of the trace element **iridium** in the clay layer that marks the end of the Cretaceous in a section near Gubbio, Italy (Fig. 13.24B). Iridium is extremely rare in crustal rocks but slightly more common in mantle rocks and in extraterrestrial rocks. The Alvarez team wrestled with different explanations for this peculiar geochemical signal until they came up with the idea that an asteroid about 10 km (about 6 miles) in diameter hit the earth at the end of the Cretaceous.

When their paper was published in 1980, it was greeted with a lot of skepticism, as any provocative scientific idea

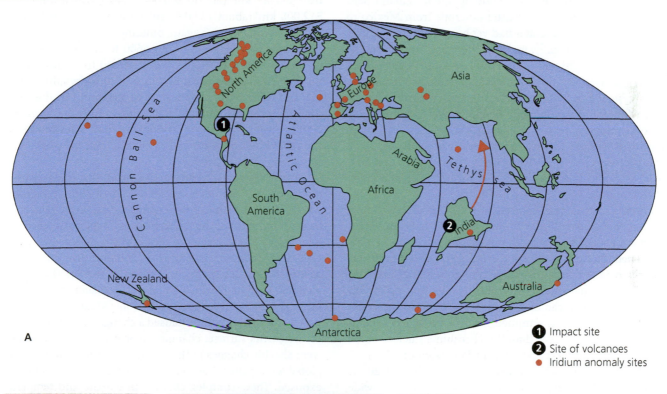

A

1 Impact site
2 Site of volcanoes
● Iridium anomaly sites

B

C

D

Figure 13.24   End-Cretaceous events. **A.** Map showing the location of Chicxulub crater (1), the Deccan volcanic center (2), and dots showing some of the numerous places where the iridium anomaly was detected in boundary clays (both in deep-sea cores and in land sections where the marine Cretaceous–Paleogene rocks have been uplifted and exposed, such as in Gubbio, Italy). The continents have been restored to their latest Cretaceous positions. **B.** Photograph of the clay layer full of iridium at Gubbio, Italy. The gray Cretaceous limestones are below the clay, and the orange Paleocene limestones are above it. **C.** Map showing the location of Chicxulub crater in Yucatán, and the gravity anomaly map that shows its size and ring-like shape, later filled in and covered by the jungles of Mexico. **D.** Stacked lava flows from the Western Ghats in India, part of the enormous Deccan eruptions.

should be. Critics were unconvinced at first because marine clays are notorious for concentrating rare elements. But that reservation began to be doubted when the iridium anomaly was also found in the cliffs at Stevns Klint in Denmark and in deep-sea cores, so it was widespread across the ocean, not a local problem with the Gubbio samples. When iridium was found in the land section in Hell Creek, Montana, the idea that it was a peculiar marine geochemical event was ruled out.

In the ensuing years, many other indicators of impact were found: grains of quartz with shock features, something previously known only from other meteorite impact craters and from nuclear blasts; blobs of melted crustal rock in places like Cuba and Haiti; and apparent tsunami beds in the Gulf of Mexico in coastal outcrops from Mexico to Cuba to Haiti. But no crater had yet been identified, despite many proposals. Finally, in 1991, the crater known by the Mayan name **Chicxulub** (CHIK-zoo-loob) was located under the northern tip of the Yucatán Peninsula in Mexico, deeply buried by younger deposits and covered in jungle (Fig. 13.24C). It had actually been discovered before 1980 by oil geologists, who lost interest when no oil was found (and the asteroid impact model wasn't popular yet), then was rediscovered in 1991 by planetary geologist Glen Penfield, who was looking specifically for evidence of a crater in that region.

According to the Alvarez model, the asteroid traveled at speeds over 100,000 mph. It would have an energy level equivalent to 100 million megatons of TNT. As it approached the earth's atmosphere, it formed a huge fireball in the sky brighter than the sun and temporarily blinded the land animals that witnessed it. Shortly after it lit up the sky, the shock wave ahead of it produced a series of incredible sonic booms. Then it slammed into the earth with energy a billion times stronger than the Hiroshima nuclear bomb. It hit in a spot where the Yucatán Peninsula of Mexico is located today. The asteroid excavated a crater over 20 km (about 12 miles) deep and 170 km (about 106 miles) wide, with a flash brighter than any nuclear blast. The impact caused huge tsunamis to crash around the coastline of the Gulf of Mexico and Caribbean, leaving huge deposits of impact debris and storm waves in places from Mexico to Texas to Cuba and Haiti. Meanwhile, the "mushroom cloud" of dust and debris shot 40,000 feet up into the stratosphere, where it created a "nuclear winter"–like effect. The dust circulated around the global stratospheric layer and blocked the sunlight for years, chilling the planet and cutting off photosynthesis to the plants. In one version of this scenario, the sulfur from the gypsum located in the Yucatán bedrock created acid rain around the globe. Land animals quickly died off, especially the dinosaurs.

In the oceans, darkness caused the food pyramid based on planktonic algae to collapse, wiping out many organisms higher in the food chain. Especially prominent were the ammonites. These had survived all the previous great mass extinctions in the Permian and Triassic, yet succumbed at the end of the Cretaceous to the extreme stresses. Widespread death and destruction occurred throughout the oceanic realm, wiping out marine reptiles, many of the clams and snails, and other marine creatures.

Even as the asteroid extinction model gained popularity in the 1980s, geologists pointed to another huge event, the **Deccan lavas** of western India and Pakistan (Fig. 13.24D). These were already well known, but new research showed that they were the result of the second largest volcanic eruption in all of earth history. They began erupting from the mantle during very end of the Cretaceous, about 68 Ma, then intensified about 67 Ma. Huge floods of red-hot lava poured from cracks in the earth and flooded the landscape over 500,000 km^2 (almost 200,000 miles^2), with a volume of 512,000 km^3 (about 123,000 miles^3) of lava. The volcanoes erupted over and over again, building up a stack of cooled lava flows over 2000 m (about 6600 feet) thick. Immense quantities of volcanic gases were released from these mantle-derived magmas, including carbon dioxide, sulfur dioxide, and other nasty chemicals. The stratospheric dust and gases blocked the sunlight for a time and led to a global cooling event. The global cooling and dark skies produced the same kinds of death and extinction that the postulated impact might have produced.

Finally, geologists had long known that there was a global drop in sea level near the end of the Cretaceous, which dried up the entire Western Interior Cretaceous Seaway and most of the other areas of Cretaceous epicontinental seas. These seas once teemed with marine life, especially huge mollusks, gigantic fish, and marine reptiles such as mosasaurs, plesiosaurs, and giant sea turtles. The loss of so much shallow-marine habitat was devastating to marine life and could trigger mass extinctions. Meanwhile, the exposed continental shelves caused a change in the reflectivity of the earth's surface, changing global temperature. There were also big changes in the ocean circulation patterns and global wind patterns as the seas retreated and the land was exposed. These complex changes in climate and temperature are still poorly understood, but it was bad news for the dinosaurs and much of the plankton that had flourished through most of the Cretaceous. In fact, since all three events (impact, volcanoes, sea-level drop) happened and are well documented, it is clear that the end of the Cretaceous was a bad time on planet earth, and everything was going bad at the same time.

 See For Yourself: Cretaceous Extinctions

Since 1980, these three different explanations have been proposed for the Cretaceous extinctions. The impact model is the most glamorous, so it has gotten the most attention in the media and among some geologists, so most people have only heard the simplistic "an asteroid killed the dinosaurs" explanation. But most paleontologists, who are familiar with the actual fossil record of the end of the Cretaceous, are not convinced that the asteroid is the main killer or even that important. After all, there were three different events happening at the same time, so how do we determine which was most important? The verdict on what really happened, and which cause is the most significant, can only come from a detailed examination of the fossil record. If most or all of the species vanished abruptly at the iridium layer, then the asteroid impact would be the primary culprit. If, however, many were declining long before the impact or died out completely and were already fossils by the time of the impact, then slow protracted climate change driven by the Deccan eruptions is a more important killer.

First, let's look at the marine realm (Fig. 13.25). The most important species are among the plankton (Fig. 13.14), which are the crucial base of the marine food chain and are extremely sensitive to environmental changes. In addition, their tiny shells can be found by the thousands in deep-sea sediment cores, so we can study their extinction in fine detail. The planktonic coccolithophores (the algae that produced chalk) did indeed have a severe crash, as would be expected if the light was dimmed by "nuclear winter" clouds; but two other kinds of algae (diatoms and silicoflagellates) did not

(Fig. 13.14C). Two groups of shelled amoeba-like plankton were also studied. The foraminiferans (Fig. 13.14A) seemed to show a crash (although some micropaleontologists argue differently), but the other group of planktonic amoebas, the radiolarians (Fig.13.14B), did not. Some foraminiferans lived on the sea bottom (benthic habitat) rather than floating in the plankton, and they showed no extinction at all.

Moving up the food chain, we can look at the fossilized shells of the mollusks (clams, snails, and their kin) that once inhabited the Cretaceous sea floor. Some groups, such as the inoceramids (Fig. 13.14C, D) and the rudistids (Fig. 13.14E), were extinct long before the end of the Cretaceous, so the impact is irrelevant to them—they were already fossils. Of the remaining mollusks, 35% of snails and 55% of clams and oysters died out, but every study shows that their extinction was gradual through the end of the Cretaceous.

Critical to this debate were the diverse and rapidly evolving ammonites (Fig. 13.14F–J). They had survived every previous mass extinction since the Devonian. Most studies

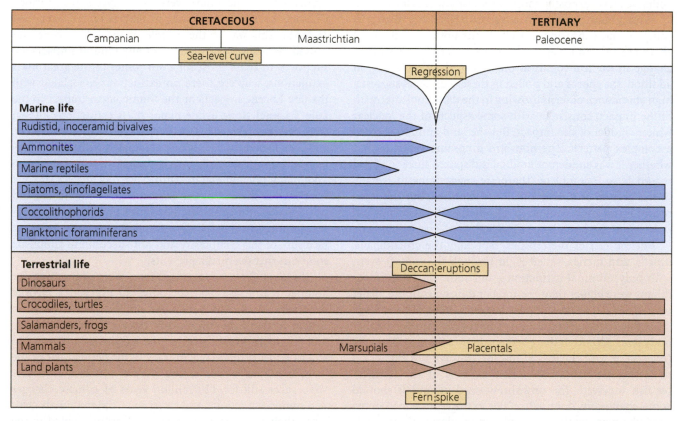

Figure 13.25 During the end of the Cretaceous, a number of marine groups (such as rudistids and inoceramid oysters) died out long before the end of the Maastrichtian, so they clearly were unaffected by the impact but probably victims of volcanic activity–induced climate change. Many of the other marine and terrestrial groups show gradual decline in the latest Maastrichtian, including ammonites and marine reptiles, as well as non-bird dinosaurs on land, so they were affected by environmental changes occurring thousands of years before the impact. Diatoms, dinoflagellates, and radiolarians among the plankton and most mollusks, echinoderms, bryozoans, and other marine invertebrates were unaffected by the extinction. On land, the crocodilians, lizards, and amphibians sailed through the extinction event with minimal losses, while marsupials or pouched mammals were replaced by placentals. Only the change in land plant pollen assemblages (including a "spike" of fern spores) is consistent with the impact model of extinction. Thus, the Chicxulub impact only strongly affected a few groups (non-bird dinosaurs, land plants, some of the plankton, possibly ammonites) but was a coup de grâce for groups already in decline due to the climate changes caused by the Deccan eruptions.

suggested that ammonites died out slowly through the Cretaceous, with only a few species that survived to near the
end of the Cretaceous. In Antarctica, there was a long, protracted extinction in the ammonites through the Upper
Cretaceous rocks. In addition, there are now reliable reports
that a few ammonites survived into the Paleocene. The vast
majority of the marine invertebrates (corals, brachiopods,
bryozoans, crinoids, brittle stars, sea urchins) show either
no effect of the Cretaceous extinction at all or only a gradual
extinction through the interval.

At the top of the food chain are the marine reptiles, especially the giant marine lizards known as mosasaurs (Fig.
13.15D). Their fossil record is not complete enough in the
latest Cretaceous to show whether they were even around to
witness the rock from space, but most of the data suggest that
they were dying out long beforehand. In short, the only sea
creatures that show an obvious abrupt effect consistent with
the asteroid impact model are some (but not all) of the plankton; the rest of the marine realm appears to be deteriorating
slowly throughout the Late Cretaceous, consistent with the
effects of volcanic gases and climate change, as well as the
falling sea level exposing most of the shallow-marine habitat.

What about the terrestrial record? Here, the answer is
complex and confusing as well. At the bottom of the food
chain are the land plants, and there is certainly a striking
change in the leaf fossils at the end of the Cretaceous. In
addition, the spores and pollen in these sediments suggest a
high abundance of ferns growing in the dark, cool aftermath
of the impact, consistent with some aspects of the "nuclear
winter" model of the impact. But the land animals showed
a complex pattern. The non-bird dinosaurs vanished, but
whether it was sudden or gradual is disputed since the fossil
record is so incomplete. There are suggestions that many
of the groups of Cretaceous birds vanished, but the fossil
record is not good enough to tell whether it was abrupt and
precisely at the boundary or gradual.

But the rest of the reptiles, especially the crocodilians,
lizards, and turtles, sailed right through the impact event
with only minimal extinction—and some of those crocodilians were bigger than the smaller dinosaurs. If the planet
was so hellish as the impact suggests, how did the large
crocodiles survive but no dinosaurs of the same or smaller
body size? Although there was a minor extinction in sharks
(consistent with the loss of marine seaways and expansion
of freshwater habitats), more than 90% of the freshwater fish survived. The tiny shrew-sized mammals showed
some changes in dominance. For example, pouched opossum-like marsupials were replaced by placental mammals
in some places like Montana, and groups such as the primitive squirrel-like, egg-laying mammals known as **multituberculates** (distantly related to the platypus and echidna
of Australia) did fine. Thus, there were only a few extinctions within the mammals. In short, the land record does
not strongly support the impact-only model of extinction

(Fig. 13.23). Only the land plants show a clear abrupt effect,
and the rest of the groups other than non-bird dinosaurs
survived through the supposedly extreme climates of the
"nuclear winter" and came right back in the Paleocene no
worse for wear. This casts serious doubt on any model that
is too extreme. In particular, any scenario that argues for
a global bath of acid rain can be rejected immediately, because amphibians sailed right through the end-Cretaceous
catastrophe without any extinction. Yet amphibians are
extremely sensitive to even small amounts of acid in their
habitat, and there would be no frogs or salamanders alive if
the acid rain scenario were true.

In summary, there is a continuing controversy over
which of the causes was more important to the Cretaceous
extinctions. In the 1980s and 1990s, the impact advocates
were dominant, but since then, geological opinion has
shifted in favor of the importance of the Deccan eruptions.
Recent research has shown that no other mass extinction
was caused by an impact (despite premature and overen-
thusiastic claims by impact advocates in the 1980s). Most
revealing of all is the fact that the next biggest impact
events (after Chicxulub) were the impacts that formed
Chesapeake Bay and another in Siberia which happened
35.5 Ma. This was in the middle of the late Eocene and
caused virtually no extinctions of any kind. If impacts are
supposed to cause most (or even some) of the great mass
extinctions, why are there no extinctions associated with
the late Eocene impacts or the Manicougan impact in the
Late Triassic? None of the major mass extinctions known
(the Permian–Triassic, Triassic–Jurassic, Late Ordovician,
and Late Devonian events) are associated with an extraterrestrial impact; the only one is the end-Cretaceous event.
Like many phenomena in nature, the real answer to what
happened at the end of the Cretaceous is complicated and
not amenable to simplistic headlines that the media prefers to present us. As people educated in science, we must
be aware of this complexity and not succumb to simplistic
solutions that the media presents.

The end-Cretaceous extinctions were important for
another reason as well. Not only did everyone's favorite
dinosaurs vanish, but if they had not died out, there is no
reason to think they would not have continued to thrive
and dominate the world for the next 66 million years—and
still rule the earth today. Contrary to the popular use of the
word "dinosaur" as something old and obsolete, dinosaurs
were successful for 120 million years of climate changes
and minor extinctions, and never once did mammals challenge their supremacy. They were superbly adapted to the
world of the normal Mesozoic climates, and only a freakish
coincidence of gigantic volcanic eruptions followed by an
impact ended their dominance. And if they had not vanished, mammals would never have gotten any larger than a
shrew or cat, and humans would not have evolved—and you
would not be reading this book.

TIMELINE SUMMARY OF THE MESOZOIC

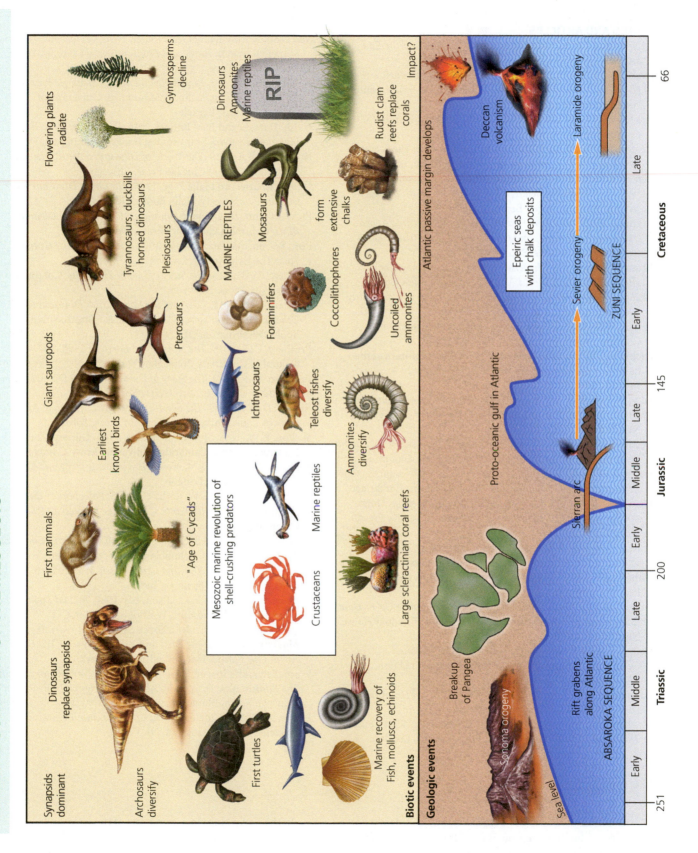

RESOURCES

BOOKS AND ARTICLES

Archibald, J. David. 1996. *Dinosaur Extinction and the End of an Era*. Columbia University Press, New York.

Archibald, J. David. 2011. *Extinction and Radiation: How the Fall of the Dinosaurs Led to the Rise of the Mammals*. Johns Hopkins University Press, Baltimore, MD.

Arkell, W. J. 1956. *Jurassic Geology of the World*. Hafner Publishing, London.

Benton, Michael J. 2019. *Dinosaurs Rediscovered: The Scientific Revolution in Paleontology*. Thames & Hudson, London.

Blakey, R. C., and W. D. Ranney. 2018. *Ancient Landscapes of Western North America: A Geologic History with Paleogeographic Maps*. Springer, Berlin.

Brannen, Peter. 2017. *The Ends of the World: Volcanic Apocalypses, Lethal Oceans, and Our Quest to Understand Earth's Past Mass Extinctions*. Ecco, New York.

Brusatte, Steve. 2018. *The Rise and Fall of the Dinosaurs: A New History of the Lost World*. William Morrow, New York.

Chiappe, Luis. 2007. *Glorified Dinosaurs: The Origin and Early Evolution of Birds*. Wiley-Liss, New York.

Cuppy, Will. 1941. *How to Become Extinct*. University of Chicago Press, Chicago.

Dickinson, W. A., and W. Snyder. 1978. Plate tectonics of the Laramide orogeny. *Geological Society of America Memoir* 141:355–366.

Dingus, Lowell, and T. Rowe. 1998. *The Mistaken Extinction: Dinosaur Evolution and the Origin of Birds*. W. H. Freeman, New York.

Everhart, Mike J. 2017. *Oceans of Kansas: A Natural History of the Western Interior Sea*. Indiana University Press, Bloomington.

Fastovsky, David, and David Weishampel. 2016. *Dinosaurs: A Concise Natural History*. Cambridge University Press, Cambridge.

Foster, John. 2007. *Jurassic West: The Dinosaurs of the Morrison Formation and Their World*. Indiana University Press, Bloomington.

Fraser, Nick, and Douglas Henderson. 2006. *Dawn of the Dinosaurs: Life in the Triassic*. Indiana University Press, Bloomington.

Hallam, A. 1975. *Jurassic Environments*. Cambridge University Press, Cambridge.

Holtz, Thomas. 2007. *Dinosaurs: The Most Complete Up-to-Date Encyclopedia for Dinosaur Lovers of All Ages*. Random House, New York.

Hsü, K. J. (ed.). 1986. *Mesozoic and Cenozoic Oceans*. Geodynamics Series 15. American Geophysical Union, Washington, DC.

Long, John, and Hans Schouten. 2008. *Feathered Dinosaurs: The Origin of Birds*. Oxford University Press, Oxford.

MacLeod, Norman. 2015. *The Great Extinctions: What Causes Them and How They Shape Life*. Firefly Books, London.

MacLeod, Norman, and Gerta Keller. 1996. *Cretaceous–Tertiary Mass Extinctions: Biotic and Environmental Changes*. W. W. Norton, New York.

Meldahl, Keith. 2013. *Rough-Hewn Land: A Geologic Journey from California to the Rocky Mountains*. University of California Press, Berkeley.

Naish, Darren, and Paul Barrett. 2016. *Dinosaurs: How They Lived and Evolved*. Smithsonian Books, Washington, DC.

Norell, Mark A. 2019. *The World of Dinosaurs: An Illustrated Tour*. University of Chicago Press, Chicago.

Officer, Charles, and Jake Page. 1996. *The Great Dinosaur Extinction Controversy*. Helix Books, New York.

Pickrell, John. 2014. *Flying Dinosaurs: How Reptiles Became Birds*. Columbia University Press, New York.

Pim, Kelron, and Jack Horner. 2016. *Dinosaurs—The Grand Tour*. The Experiment, London.

Prothero, D. R. 2013. *The Story of Life in 25 Fossils*. Columbia University Press, New York.

Prothero, D. R. 2016. *Giants of the Lost World: Dinosaurs and Other Extinct Monsters of South America*. Smithsonian Books, Washington, DC.

Prothero, D. R. 2018. *The Story of the Earth in 25 Rocks*. Columbia University Press, New York.

Prothero, D. R. 2019. *The Story of Dinosaurs in 25 Discoveries*. Columbia University Press, New York.

Russell, Dale A. 1989. *The Dinosaurs of North America: An Odyssey in Time*. University of Toronto Press, Toronto.

Skelton, P. W., R. A. Spicer, S. P. Kelley, and I. Gilmour. 2003. *The Cretaceous World*. Cambridge University Press, Cambridge.

Spencer, A. M. (ed.). 1974. *Mesozoic–Cenozoic Orogenic Belts*. Geological Society, London.

Stewart, W. N., and G. W. Rothwell. 1993. *Paleobotany and the Evolution of Land Plants*. Cambridge University Press, Cambridge.

Sues, Hans-Dieter, and Nicholas Fraser. 2010. *Triassic Life on Land: The Great Transition*. Columbia University Press, New York.

Tanner, Lawrence H. 2017. *The Late Triassic World: Earth in Time of Transition*. Springer, Berlin.

Thomas, R. D. K., and E. C. Olson, eds. 1980. *A Cold Look at the Warm-Blooded Dinosaurs*. Westview, Boulder, CO.

Ward, Peter D. 2007. *Under a Green Sky: Global Warming, the Mass Extinctions of the Past, and What They Can Tell Us About Our Future*. HarperCollins, New York.

Ward, Peter, and Joseph Kirschvink. 2015. *A New History of Life: The Radical New Discoveries About the Origin and Evolution of Life on Earth*. Bloomsbury, New York.

White, M. E. 1986. *The Flowering of Gondwana*. Princeton University Press, Princeton, NJ.

SUMMARY

- The Mesozoic world was dominated by the breakup of Pangea, from the opening of the rift valleys in the Triassic to the proto-oceanic gulf stage of the North Atlantic in the Jurassic and finally the breakup of nearly all the Pangea continents in the Cretaceous.

- In the Early Mesozoic, remnants of the icehouse world of the latest Paleozoic persisted, so Triassic and Lower Jurassic rocks are typical red beds, dune deposits, and evaporites formed in harsh continental climates of the interior of Pangea. By the Middle Jurassic, the return of epicontinental seas in North America and elsewhere signaled the transition to a greenhouse planet, with warmer conditions and higher sea levels, which peaked in the mid-Cretaceous with the Western Interior Seaway running from the Arctic to the Gulf of Mexico.

- The Triassic Newark Supergroup of lake shales, alluvial fan sandstones, and lava flows represents the rift valleys that ran from Labrador to Georgia and in western Africa and Spain as the northern continents began to tear apart. Meanwhile, the Triassic beds of the western United States were dominated by red shales and sandstones from floodplains and river deposits, preserving fossils of Triassic land reptiles, amphibians, and primitive gymnosperms.

- The Sonoma terrane arrived in the Late Permian–Triassic and accreted along the Golconda thrust in what is now northwestern Nevada and the northern Sierras and Klamaths in California. It was one of many additional terranes that collided with the west coast over the Jurassic and Cretaceous, making up most of California, Oregon, Washington, British Columbia, and Alaska.

- By the Jurassic, the North Atlantic had opened wide enough to allow seawater in at one end, producing a proto-oceanic gulf. Thick deposits of salt and gypsum formed by evaporation wherever the conditions were hot enough, forming the thick Jurassic salt layers beneath Louisiana and Texas. These have since risen up as salt diapirs, trapping oil deposits on their flanks.

- Lower Jurassic rocks of the western United States are dominated by huge cross-bedded dune deposits of the Aztec–Navajo–Nugget Sandstones or red river sandstones of the Kayenta and Wingate Formations. By the Late Jurassic, however, these rocks were capped by the first marine deposits of the Sundance Seaway that covered much of the mid-continent and signaled the beginning of the greenhouse world. The uppermost Jurassic beds are the terrestrial Morrison Formation, famous for its dinosaurs.

- During the Jurassic and especially the Cretaceous the Sierran arc developed into a chain of Andean-style volcanoes that ran from Baja California to British Columbia. It produced a forearc basin (Central Valley of California) and accretionary wedge (Franciscan rocks of the Coast Ranges), as well as back-arc thrusts in the Sevier thrust system.

- By the Early Cretaceous, the high sea levels of the Western Interior Seaway had flooded nearly all the region that is now the Plains and Rocky Mountains from the Arctic to the Gulf of Mexico. Conglomerates and sandstones eroded from the Sevier uplifts in Utah, while the transgressions and regressions zigzagged back and forth across the region of the future Rocky Mountains. Offshore shales and even chalk deposits formed in what is now the Dakotas, Nebraska, Kansas, Oklahoma, and Texas.

- The Cretaceous greenhouse conditions and global sea-level rise were caused by a combination of a number of factors: the breakup of Pangea and the sinking of the edges of continents into flooded passive margins; extraordinarily high rates of sea-floor spreading, producing thick mid-ocean ridges that displaced water up out of the ocean and onto land; warm greenhouse climates melting all the ice and snow on the planet; and huge mantle plumes erupting beneath the western Pacific, which released large amounts of greenhouse gases and displaced water out of the oceans.

- In western North America, the seas abruptly transgressed in the latest Cretaceous and Early Cenozoic (70–40 Ma) when thrust faults and folds brought up Precambrian and Paleozoic basement rock to form high ranges where the Rocky Mountains now lie. Known as the Laramide orogeny, it coincided with the shutoff of the volcanism in the Sierra Nevada Mountains and was probably due to horizontal subduction, which prevented the plate from melting beneath the Sierras and transferred stresses 1600 km (about 1000 miles) inland in Wyoming and Colorado.

- The marine realm took much of the Early Triassic to recover from the Permian extinctions, leaving the sea floor open to opportunistic "weedy" species that could tolerate the extreme conditions. Eventually, marine life recovered, with the evolution of the first modern scleractinian corals, ceratitic ammonoids, plus the modern fauna of bivalves, gastropods, sea urchins, and other creatures that still dominate the sea floor. Some groups, like the spiriferide brachiopods, that survived from the Paleozoic vanished in the Triassic, probably due to the Mesozoic Marine Revolution of new predators: sea stars, plus shell-crushing fish, crustaceans, and marine reptiles. Only mollusks that could swim, burrow, or build thick shells survived this escalation of new shell-crushing predators. These creatures then suffered another great mass extinction at the end of the Triassic, probably due to climate changes triggered by eruptions of the Central Atlantic Magmatic Province.

- In the Jurassic and Cretaceous, the seas were full of a wide variety of mollusks (especially weird-looking oysters), giant "dinner plate" inoceramid clams, and conical reef-building rudistid bivalves. There was a tremendous radiation of ammonites with complex sutures. Some of these grew not in the simple coil in a plane but uncoiled in many weird shapes. All of these marine animals were supported by a food chain of many new kinds of plankton, including phytoplankton (coccolithophores, diatoms) and amoeba-like plankton (foraminiferans, radiolarians). Swimming above them was a great diversity of bony fish, sharks, and marine reptiles, including enormous sea turtles, the dolphin-like ichthyosaurs, the paddling plesiosaurs, marine crocodiles, and the mosasaurs, which evolved from Komodo dragon relatives.

- On land, the Triassic was dominated by holdovers from the Permian such as giant flat-skulled amphibians, protomammals, and some archaic reptiles; but it was mostly dominated by archosaurs related to crocodiles. By the Late Triassic, the protomammals had vanished, replaced by tiny mammals; the frogs and salamanders replaced the archaic amphibians; archaic reptiles were replaced by the first land turtles; and most of the primitive archosaurs were being replaced by the first dinosaurs, pterosaurs, and the first crocodiles.

- During the Triassic and Jurassic, the landscape was dominated by gymnosperms, including primitive relatives of the conifers, ginkgoes, cycads, and trees like *Araucaria* (Norfolk Island pine). Ferns were abundant in the undergrowth, and Paleozoic holdovers like club mosses and horsetails still lived in the wetlands; but flowering plants had not yet evolved.

- The Jurassic and Cretaceous was the time of the great dinosaurs, which occupied almost all large body size niches. Huge sauropods and stegosaurs were prominent in the Jurassic on most continents, and by the Cretaceous duck-billed dinosaurs, horned dinosaurs, and armored ankylosaurs were dominant. However, some lineages evolved into birds by the Late Jurassic, which had their own evolutionary radiation in the Cretaceous. Some crocodilians were as big as the dinosaurs and preyed upon them. Mammals remained mostly shrew-sized and hiding in the undergrowth, along with frogs, salamanders, lizards, and the first snakes by the Cretaceous. By the Early Cretaceous, the first flowering plants (angiosperms) took over the planet and soon came to dominate the world's flora.

- Our understanding of dinosaurs has undergone a tremendous "renaissance" since the 1970s. No longer are dinosaurs thought to be slow, sluggish, tail-dragging lizards wallowing in swamps but active creatures with upright posture, tails held out straight behind them, and a covering of feathers on many of them. The smaller ones were almost all endothermic, generating their own body heat from metabolizing food (like birds and mammals); but the largest sauropods probably achieved constant body temperature by their sheer size without endothermy.

- The Cretaceous ended with a great mass extinction that wiped out some of the marine plankton, the last of the ammonites, and the marine reptiles in the oceans, as well as the non-bird dinosaurs and many kinds of plants on land. However, there were almost no effects on many plankton, most marine mollusks and other invertebrates, or most smaller land animals, including crocodilians, turtles, lizards, frogs, salamanders, and mammals. Although the media prefers the simplistic model of an asteroid hitting Yucatán as the only explanation for the mass extinction, the evidence shows that the gigantic eruptions of the Deccan lava flows in India and Pakistan were also important, especially in explaining the gradual decline and disappearance of many groups of animals well before the impact occurred.

KEY TERMS

Newark Supergroup (p. 296)
Sonoma orogeny (p. 300)
Louann Salt (p. 300)
Salt diapirs (p. 300)
Salt domes (p. 300)
Navajo Sandstone (p. 301)
Sundance Seaway (p. 302)
Morrison Formation (p. 303)
Sierra Nevada arc (p. 303)
Batholith (p. 303)
Great Valley forearc basin (p. 303)
Accretionary wedge (p. 303)
Mélange (p. 303)
Sevier back-arc thrusts (p. 303)
Exotic terranes (p. 305)
Coccolithophores (p. 306)
Chalk (p. 306)

Western Interior Seaway (p. 306)
Mantle superplumes (p. 309)
Laramide orogeny (p. 310)
Horizontal subduction (p. 310)
Modern fauna (p. 311)
Ceratite ammonoids (p. 311)
Scleractinian corals (p. 311)
Mesozoic Marine Revolution (p. 312)
Ichthyosaurs (p. 313)
Plesiosaurs (p. 313)
Triassic–Jurassic extinction (p. 313)
Central Atlantic Magmatic Province (CAMP) (p. 313)
Ammonitic ammonoids (p. 313)
Globigerinid foraminiferans (p. 315)

Radiolarians (p. 315)
Diatoms (p. 315)
Inoceramids (p. 315)
Rudistids (p. 315)
Heteromorph ammonites (p. 315)
Baculitids (p. 315)
Belemnites (p. 315)
Mosasaurs (p. 318)
Cycads (p. 319)
Ginkgo (p. 319)
Archosauria (p. 319)
Sauropod (p. 321)
Theropod (p. 321)
Stegosaurs (p. 321)
Mesotarsal joint (p. 324)
Saurischian (p. 324)
Ornithischian (p. 324)
Ectotherm (p. 325)

Endotherm (p. 325)
Poikilotherm (p. 325)
Homeotherm (p. 325)
Food pyramid (p. 325)
Haversian canals (p. 326)
Inertial homeothermy (p. 327)
Gigantothermy (p. 327)
Angiosperms (p. 329)
Carpels (p. 329)
Petals (p. 329)
Sepals (p. 329)
Double fertilization (p. 329)
Vegetative reproduction (p. 330)
Iridium (p. 331)
Chicxulub crater (p. 332)
Deccan lavas (p. 332)
Multituberculates (p. 334)

STUDY QUESTIONS

1. What is the major global tectonic event that began in the Triassic and dominates the geology of the Mesozoic world?
2. How did the global climate change from the Triassic to the Late Jurassic–Cretaceous?
3. What are the typical rocks found in the Newark Supergroup? Explain how each of these is formed in a typical rift valley setting (think of the East African rift as an example).
4. How did Triassic geology influence U.S. history?

5. What do the typical Triassic rocks of the western United States look like? What environments did they form in?

6. What was the Sonoma orogeny? Where are the remnants of Sonomia found today?

7. What happened in the Gulf of Mexico region when the Atlantic began to open? How did it influence later geology and lead to economic deposits of salt and oil?

8. What are the risks of drilling for oil near a salt dome?

9. While visiting Zion National Park in Utah, you observe sandstones with huge cross-beds many meters thick. What depositional environment do these beds represent?

10. What does the appearance of Upper Jurassic marine rocks of the Sundance Seaway sitting right above Lower Jurassic dune sands suggest about how the global climate has changed?

11. Describe the features of the Sierran–Sevier arc complex. Where are each of these rock types found today?

12. What is chalk? What is it made of? What depositional conditions does it represent? How does real chalk compare with the stuff used to write on blackboards?

13. Why do the sandstones and shales of the Western Interior Seaway alternate back and forth in a single outcrop exposure?

14. What are some of the factors that might explain the global transgression in the Cretaceous?

15. How were the polar regions in the Cretaceous different from today?

16. Describe the peculiar tectonic style of the Laramide orogeny. What happened in the old Sierran arc? What happened in the future Rocky Mountain region? What plate tectonic model might explain it?

17. What marine life is found in the earliest Triassic after the Great Dying? How is it unusual?

18. What is the modern fauna? What major groups of animals make it up? How is it different from the Paleozoic fauna?

19. What was the Mesozoic Marine Revolution? What kind of anti-predatory adaptations do we see in the prey species? What does this suggest about the predators? What were some of those predators?

20. How was the Triassic–Jurassic extinction like the Permian–Triassic extinction? What volcanic events might have triggered each event?

21. How was the base of the food chain different in the seas of the Jurassic and Cretaceous?

22. How did the ammonites evolve during the Cretaceous?

23. What were the four main groups of marine reptiles during the Jurassic and Cretaceous?

24. What were the dominant groups of land animals during the Early Triassic? How were they replaced by the end of the Triassic?

25. What kinds of plants dominated the landscape in the Triassic and Jurassic? How are they different from the dominant plants today?

26. What kinds of dinosaurs were typical of the Late Jurassic, such as those found in the Morrison Formation? How do they compare with the typical dinosaurs of the Late Cretaceous? What smaller kinds of animals besides dinosaurs were found in the Jurassic and Cretaceous?

27. Most people call any extinct animal a "dinosaur." What does the word "dinosaur" really mean? What anatomical features define what is a dinosaur?

28. How has our concept of dinosaurs changed since the 1970s? How have the movies gotten dinosaurs wrong?

29. Describe the difference between an endotherm and an ectotherm and between a poikilotherm and a homeotherm. What are the exceptions to the rule that most endotherms are homeotherms and ectotherms are poikilotherms?

30. What is the evidence suggesting that dinosaurs were endotherms? What is the counterevidence for each example? Were dinosaurs warm-blooded or cold-blooded—or both?

31. What evidence tells us that birds are descended from dinosaurs?

32. How are angiosperms better equipped to handle dino damage and grow faster?

33. Discuss the major features of the Chicxulub asteroid extinction model versus the Deccan lava model for the Cretaceous extinctions. Which one would leave an abrupt end to land animals, and which one would be more gradual? What does the evidence from the fossil record show?

34. Why does the media constantly portray the extinction of the dinosaurs as simplistic, "the asteroid did it—end of story"?

Paleogene and Neogene Periods 66–2.6 Ma

[The] Age of Mammals [is] an epic evolutionary story that spans 65 million years! But its theme can be distilled into just six words: Continents move. Climates change. Mammals evolve.

—Promotional flyer, Natural History Museum of Los Angeles County

Cenozoic rocks often weather to form spectacular badlands and other erosional features. This is a view of the incredible "hoo-doos" and other rock sculptures at Bryce Canyon National Park, Utah. These features were carved into muddy algal limestone deposits of a great lake that covered much of Utah and Wyoming during the middle Eocene.

14.1 The Transition to Today

The last 66 million years since the extinction of the dinosaurs (other than birds) is called the Cenozoic Era. Since it is the youngest of the three eras of the Phanerozoic and the least likely to be eroded away or buried, Cenozoic rocks are often well exposed at the earth's surface. There are often spectacular fossiliferous outcrops to collect and study (Fig. 14.1). We know a lot more about the Cenozoic, and in a lot more detail, than we do about any older part of the geologic past; and its effects are tremendously important today. In addition, most of the animals we are familiar with evolved in the Cenozoic. Thus, we will devote more space to the Cenozoic than to any other time period in this book.

 See For Yourself: Cenozoic Climate and Tectonics

The Cenozoic is a crucial time in earth history because most of the conditions that exist today developed during the last 66 million years. Two main themes (and many variations) dominate any summary of Cenozoic events. First is the last 66 million years of plate motions as most of the continents that broke away from Pangea in the Mesozoic moved to their present positions. Thus, much of our modern landscape, from the mighty Himalayas to the Rockies to the Andes, is a product of the last 66 million years of tectonics.

Second, there were dramatic changes in climate. The greenhouse world that dominated most of the age of dinosaurs persisted into the Eocene, but by the middle of the Cenozoic (early Oligocene), the first Antarctic glaciers had appeared, and the planet made the transition from a greenhouse world to our modern-day glaciated "icehouse" conditions.

The terminology of the subdivisions of the Cenozoic has undergone a lot of changes over the years. The oldest terms date back to 1759 when early geologists like Giovanni Arduino described the soft horizontal beds in the Apennine Mountains of Italy as "Tertiary" deposits because they laid on top of the hard, folded, and tilted "Secondary" beds, which in turn lay upon granitic and metamorphic rocks called "Primary" or "Primitive." Arduino called the loose deposits that mantled the older rocks the "Quaternary." The Tertiary was subdivided into epochs by none other than Charles Lyell in the third volume (1833) of his revolutionary book *Principles of Geology*. Lyell initially recognized the Eocene ("dawn of the recent"), Miocene ("less recent"), and Pliocene ("more recent"), then later geologists further subdivided the Tertiary by adding the Paleocene ("ancient recent") and Oligocene ("few recent"). Lyell also proposed the term "Pleistocene" for the deposits of what later came to be known as the Ice Ages. In more recent years, geologists have formally rejected the old term "Tertiary," although it is still widely used out of habit. The modern timescales recommend the names "Paleogene" for the Paleocene, Eocene, and Oligocene epochs and "Neogene" for the Miocene and Pliocene epochs (although the Neogene originally included the Miocene to the present). Currently, the timescale uses the Paleocene, Neogene, and Quaternary as the major subdivisions of the Cenozoic; but most geologists just use the names of the epochs since we know these relatively recent events in so much more detail than we do for the older parts of the geologic past.

14.2 Breakup of Pangea

The overarching theme of Cenozoic geology is the continuing gradual breakup of the Pangea continents and their movement to their modern positions. By the end of the Cretaceous, only a few pieces of

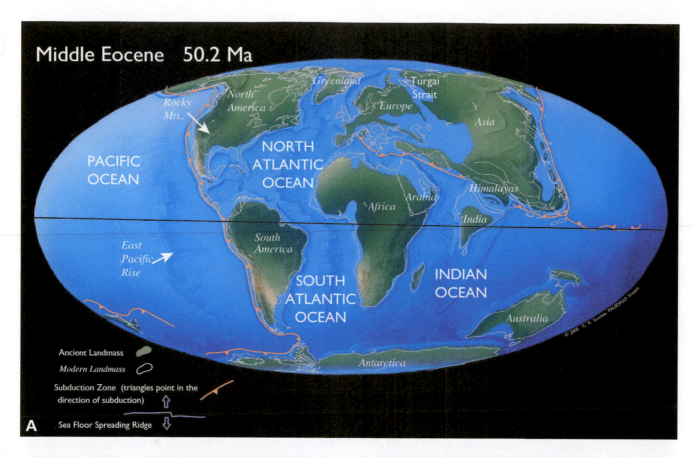

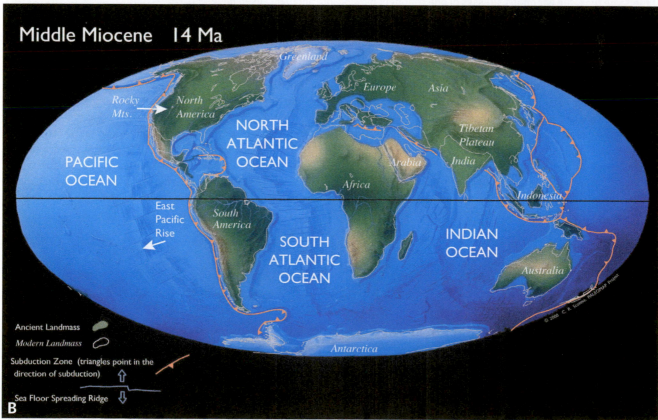

Figure 14.2   Paleogeographic maps of the Cenozoic. **A.** Middle Eocene, about 50 Ma. Australia and South America are still attached to the other Gondwana remnant, Antarctica. Africa is moving north, and India is about to collide with Asia, breaking up the old Tethys Seaway that once stretched from Gibraltar to Indonesia. High sea levels and greenhouse climates drowned much of Europe and the coastal plains of many continents. **B.** By the middle Miocene, about 14 Ma, most of the continents were near their modern-day positions. Australia was still moving north, and South America had just separated from the now ice-covered Antarctica. However, there were high sea levels that drowned much of Europe and the coastal plains around the world.

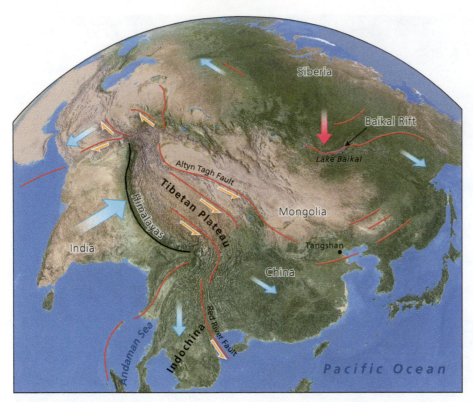

Figure 14.3 Beginning in the Eocene, and continuing today, the collision of India with Asia formed the Himalayas and the Tibetan Plateau, which have risen over 8848 m (29,028 feet) into the sky yet are capped by limestones with fossils from the bottom of the Mesozoic ocean. At the same time, the enormous stresses of one continent plowing into another caused huge faults that have pushed major blocks of China and Indochina to the east and similar stresses on the blocks of the Middle East pushed to the west.

Pangea had separated. The North Atlantic had been opening since the Jurassic and the South Atlantic since the Early Cretaceous (Fig. 13.2), but the separation between Australia and Antarctica had just begun (Fig. 14.2).

The most remarkable tectonic movement was the entire subcontinent of India tearing away from Gondwana and charging across the Indian Ocean (Figs. 13.2C, 14.2). By the Late Cretaceous it was more than halfway to Asia, and

during the Eocene it finally collided (Fig. 14.3). When it did so, there were several remarkable consequences. Whenever there is a collision of one block of buoyant continental crust against another, neither plate can subduct beneath the other (remember, only dense oceanic crust can go down the subduction zone). Instead, a huge collisional mountain belt was born, which eventually became the Himalayas and the Tibetan Plateau to the north. Everywhere you go in the Himalayas, you can find evidence of the enormous stresses that were exerted for more than 40 million years of one crustal block slowly crumpling into another. There are huge folds in many places, as well as giant thrust faults (Fig. 14.4). As these enormous mountains arose (mostly during the Miocene), they shed huge volumes of post-orogenic molasse in the form of Miocene river and floodplain sediments around them on all sides. To the south, the thick fossiliferous deposits of the Siwalik Hills in Pakistan and India record the long period of uplift of the Himalayas in the Miocene. To the north in Tibet and China, there are many other places where thick deposits of Miocene and Pliocene river and floodplain sediments also record this long period of uplift. Even larger are the deep-sea flysch deposits eroded off the Himalayas, which formed huge submarine fans below the Indus River coming from Pakistan and the enormous drainage of the Ganges and Brahmaputra Rivers in eastern India and Myanmar into the Bay of Bengal, which form

Figure 14.4   A. The collision of India with Asia crumpled up rocks all across the region. These enormous folds were caused by the crushing and crumpling of rocks during the collision that formed the Himalayas, here shown in Jammu and Kashmir. **B.** Giant folds in the Zagros Mountains of Iran, crushed when Africa and Arabia collided with Asia.

the Bengal/Nicobar fans—the largest sedimentary deposits on the planet.

The second major trend as the continents began to pull apart was that in other places they began to collide. Back in the Cretaceous (Fig. 13.2C), there was a huge tropical seaway that ran from near Indonesia to the western edge of the Mediterranean. Known as the Tethys Sea, this warm, shallow, tropical seaway teemed with marine life and made huge limestone deposits in many areas. During the later Cretaceous the Tethys Sea was dominated by reefs built of the weird colonial cone-shaped oysters known as rudistids (Fig. 13.15E). In the Eocene, the shallow limestone shoals of the Tethys Sea were the home to the earliest whales (found in Pakistan and Egypt) and the relatives of the manatee and elephants. The most remarkable fossils of the Tethys were giant foraminiferans called **nummulitids** (Fig. 14.5A). Even though they were secreted by single-shelled amoeba-like organisms, their shells reached the size and shape of a quarter, the largest shells ever secreted by a single-celled creature. The shell itself was built in a flat spiral, with many tiny chambers spiraling out from the center. Nummulitids were so abundant in the Tethys that in certain areas the entire bedrock is made of nothing but their shells. In the Giza Plateau west of Cairo, Egypt, most of the limestones that were used to build the pyramids are made of nummulitids (Fig. 14.5B). Back in the fifth century B.C.E., the Greek historian Herodotus thought they were petrified lentils from the meals of the slaves who built the Pyramids. The Tethys Sea would persist well into the Cenozoic until tectonic events destroyed it. First, the collision of India into the belly of Asia cut the Tethys in half. That is why there are shallow-marine Cretaceous limestones on the very top of Mt. Everest.

Meanwhile, as Africa pulled away from Pangea with the widening of the South Atlantic, it moved north and began to collide with the European plate. Just like the Himalayas, another huge collisional mountain range grew in the

Eocene. We see the tectonic effects best in the Alps (Fig. 14.6A), where incredible compressional tectonics has crumpled rocks into tight folds, pushed one slab on top of another with enormous thrust faults, and even faulted folds until they lie on their side as recumbent folds (Fig. 14.6B). These immense compressional fold and fault systems are called the Alpine **nappes** (French for "tablecloth") since they resemble the way a tablecloth crumples and folds up on itself when pushed across a surface (Fig. 14.6). Thick deposits of post-orogenic Cenozoic molasse deposits are found over much of Europe on the fringes of the Alps, especially in southern Germany and northern Switzerland and Austria. This collisional range includes not only the Alps but the folded and faulted rocks that run from the Pyrenees between France and Spain to the uplifts in Italy and Greece. Those areas are the most tectonically active in all of Europe, with numerous earthquakes and many famous volcanoes such as Mt. Vesuvius, which wiped out Pompeii in 79 C.E., and Mt. Etna, which erupted in 2017. The compressional mountain-building trend continues across the mountains of Turkey and then through Iraq, Iran (where the Zagros crush zone forms an intensively folded belt of mountains along the spine of Iran; Fig. 14.4B), Afghanistan, and Pakistan, finally connecting with the Himalayan belt (Fig. 14.2). All of these tectonic events are often combined and called the **Alpine–Himalayan orogeny**.

When the Arabian Peninsula and Africa collided with what is now the Middle East about 20 Ma, another portion of the Tethys vanished, and the Mediterranean Sea was born. Since that time, the Mediterranean has been cut off from the world's oceans except for the narrow passage at the Strait of Gibraltar, which had consequences we will discuss in Box 14.1. But that collision is not over. Over the millions of years to come, Africa will continue to push north against Europe, and the Alps will rise even higher. The Mediterranean will become crumpled up tighter and tighter and become

Figure 14.5 A. Coin-sized fossils of benthic foraminiferans known as nummulitids make up the sediment of much of the Eocene in the Tethys Seaway. The bottom specimens are sliced open, revealing the tight flat spiral of tiny chambers inside the shell. **B.** They are so common in some regions that they make up almost all the limestone bedrock, and in Egypt the nummulitid limestone of the Gizeh Plateau was used to build the pyramids.

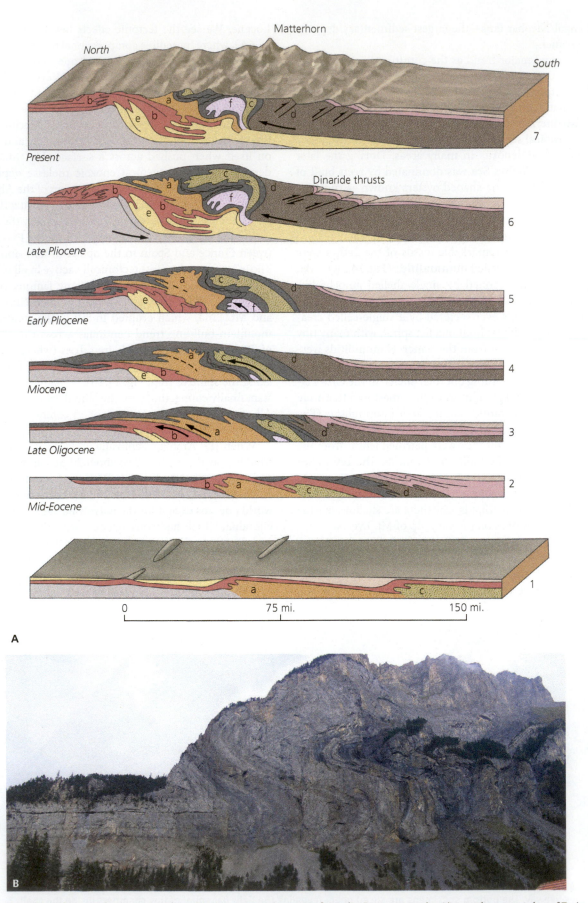

Figure 14.6 When Africa crashed into southern Europe, mountain ranges from the Pyrenees to the Alps to the mountains of Turkey and the Middle East were formed. The Alps underwent extreme compression as numerous folds and faults were formed. Some of the folds are so extreme that they are lying on their sides; these are known as nappes. **A.** Diagram of the complexity of the crustal structure of part of the Alps. **B.** The Dolderhorn nappes in Gasteretal, Switzerland.

BOX 14.1: HOW DO WE KNOW?

How Do We Know that the Mediterranean Was Once a Desert?

The fact that the Mediterranean Sea is surrounded by land, with only the narrow Straits of Gibraltar to bring in Atlantic water, makes it extremely vulnerable to drying up. Indeed, that happened once on a gigantic scale (Fig. 14.7). As early as 1867, geologists noticed immensely thick salt and gypsum deposits over 1500 m (about 5000 feet) thick in the uppermost Miocene beds around the Mediterranean (Fig. 14.7A). They were particularly well exposed in the Straits of Messina, between Italy and Sicily, so the event that caused this thick evaporite deposit was known as the **Messinian salinity crisis**. Then in the late 1960s, geologists looking for hard bedrock to anchor the Aswan High Dam on the Nile River of Egypt discovered an enormous canyon bigger than the Grand Canyon, filled with Pliocene sediments. Below Cairo, it was more than 2500 m (about 8200 feet) deep (a third again deeper than the Grand Canyon), so at one time the Nile River had cut down more than a mile and half from its present level and apparently all the way down to the level of the deepest part of the Mediterranean (Fig. 14.7B).

> See For Yourself:
> The Mediterranean
> Was a Desert

All these mysterious observations suggested that at least parts of the Mediterranean had dried up in the late Miocene, but the clinching evidence came when the Deep Sea Drilling Project drilled the Mediterranean Basin itself in 1970. Deep beneath the Pleistocene muds at the surface of the sea bottom, they found immense layers of salt and gypsum in many different places in the Mediterranean (Fig. 14.7C). This proved that the bottom of the basin had once been a dried-up salt basin, like Death Valley or the Dead Sea, but on an enormous scale. Further drilling found places where there were salty shorelines with stromatolites from bacterial and algal mats that require light (Fig. 9.1). Along with abundant mudcracks, this was proof that the entire basin was exposed to sunlight and had dried up from shallow water, not produced by some deep-sea evaporation process. Other drill sites showed that the Mediterranean must have dried up and partially refilled at least 50 times to accumulate such enormous volumes (over 1 million km³, weighing 4 trillion kg) of salt and gypsum.

How could such an extreme event happen? Remember that the Mediterranean has only the limited flow through the Strait of Gibraltar to bring it Atlantic seawater. Except from the Nile River in Egypt and the Rhône River in France, it gets very little freshwater to fill it. It's in a semi-arid region between dry countries like Spain, Italy, Greece, and the deserts of North Africa so that its rate of evaporation greatly exceeds the rate of freshwater input, and only the huge volume of cold Atlantic water through the Strait of Gibraltar keeps it from vanishing.

Around 5.96 ± 0.02 Ma in the latest Miocene (a time interval known as the Messinian stage), there was a global drop in sea level at the same time the collision between Africa and Spain was causing the Atlas Mountains to rise up. Once sea level dropped below the narrow sill that allowed it to flow in, the Mediterranean was cut off. Between 5.96 and 5.33 Ma, the Mediterranean Basin was alternately dried out completely, then flooded when water leaked through Gibraltar for a while, then flooded again, and again, at least 50 different times. Meanwhile, the basin would have been like an enormous Dead Sea (Fig. 14.7D), and animals that could tolerate the heat and dryness could walk across the Mediterranean or between many areas that are now islands.

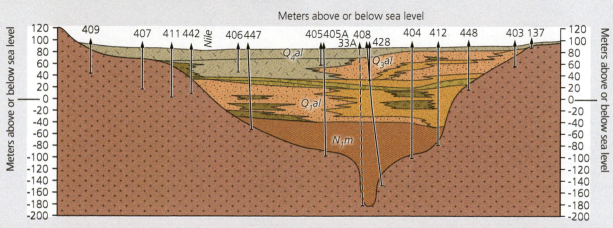

Figure 14.7 (Continued)

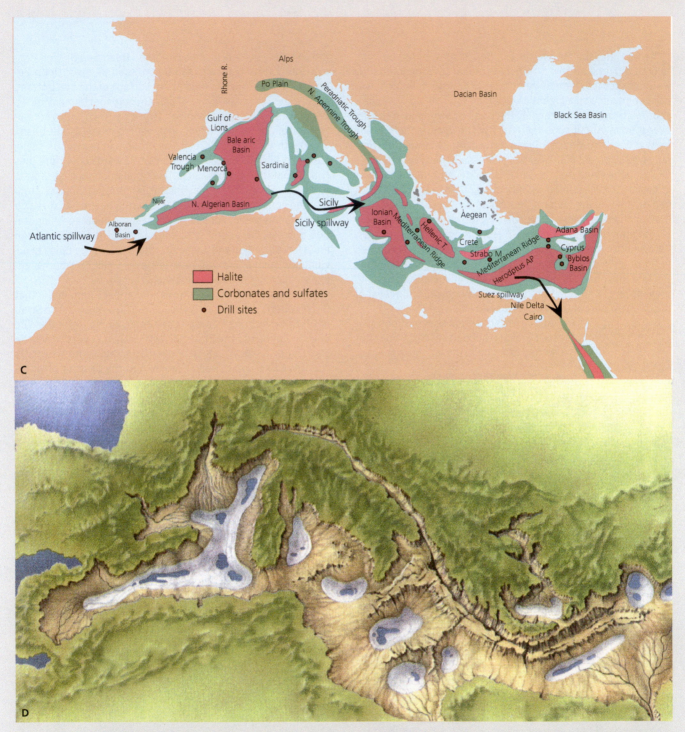

Figure 14.7   At the end of the Miocene, the Mediterranean dried up completely. **A.** Thick beds of gypsum of the Yesares Member, in the Sorba Basin, Spain, since folded by later deformation. The cliff is about 100 m (330 feet) high, requiring the evaporation of a column of water over 10,000 m (33,000 feet) deep to produce it. (Courtesy Wikimedia Commons) **B.** Diagram showing the enormous sediment-filled canyon beneath the Nile Valley. **C.** Map showing the distribution of salt and gypsum deposits formed 5.5–6 Ma in the various parts of the Mediterranean Basin, as determined by deep-sea drilling. (Modified from Ryan, 2008) **D.** Artist's reconstruction of the Mediterranean when it was almost completely dried up, with the deepest parts of the basin forming huge salt lakes lying well below sea level, like the modern Dead Sea.

The realization that the Mediterranean had dried up also solves the final piece of the puzzle: the "grand canyon" beneath the Nile. As the sea level of the Mediterranean began to drop early in the Messinian event, the ancient predecessor of the Nile River would have cut down into its floodplain and canyon to match its gradient. This would continue until the Mediterranean had completely vanished, at which time the Nile Canyon would be cut down all the way to about 2000–2500 m (about 6500-8200 feet) below sea level to flow into the salt pans, then evaporate. Then, when the seas refilled the basin in the Pliocene, the grand canyon of the Nile would have been flooded with seawater and soon filled in with both marine sediments from the ocean as well as sands and muds from the Nile headwaters upriver. This idea was confirmed when studies were made of the Rhône Valley in France, which also has a deep canyon beneath it, now completely filled with Pliocene marine sediments.

Finally, the fact that about 50 smaller episodes of flooding just evaporated away gives us some idea of how much water was needed and how fast it must flow to compensate for the high rate of evaporation. Indeed, an enormous volume of water would be necessary for the great early Pliocene deluge that finally filled the Mediterranean and ended the great cycle of drying and flooding. Scientists calculated that there must have been so much cold Atlantic sea water rushing through the Strait of Gibraltar so fast that it would have made a waterfall about 1000 times as big as Niagara Falls, or 15 times as big as the Zambezi Falls, one of the largest in the world. It turns out that about 34,000 km^3 of water would have to flow through the Gibraltar Falls in a year to fill up the basin in 100 years, and it was probably much more. At those rates, the volume and pressure were so great that the roar might have broken the sound barrier!

shallower and shallower until it vanishes, just like the Tethys did in the Eocene. Africa and Europe will be fused together, just as India is now a part of Asia.

14.3 The Ring of Fire

If the breakup of Pangea means that huge areas of Atlantic and Indian Ocean sea floor are spreading apart and generating new crust, then somewhere else on the planet there must be the same amount of crust consumed by subduction zones. Indeed, since the Jurassic there have been active subduction zones consuming enormous amounts of oceanic crust on the opposite side of the world—the Pacific Basin (Fig. 14.8). The Pacific is the remnant of the giant Panthalassa Ocean that once surrounded Pangea in the Permian and once covered about 75% of the globe. In addition, the East Pacific Rise off the coast of South America is the fastest-spreading ridge on the planet (18 cm/year, three times as fast as the Mid-Atlantic Ridge), so most of the plate consumed in the subduction zones of eastern Asia and the west Pacific is produced in these ridges. Even though much of it has been subducted, the Pacific Ocean is still so huge that it covers about a third of the earth's surface, about half of the entire surface of all the world's oceans; and it is larger than all the area of the continents combined. In other words, you could fit all of the continents inside the Pacific Basin and still have room for more.

For this reason, nearly the entire Pacific is surrounded by huge subduction zones (Fig. 14.8A), source of the biggest earthquakes on the planet and huge volcanic eruptions. It has long been known as the "Ring of Fire" for all the volcanoes and deep earthquakes that surround it. Compared to the relatively quiet tectonics of the Atlantic passive margin,

the Pacific Rim is almost entirely an active margin in every sense of the word. It is an active margin because it is consuming not only the plate produced by the expansion of the Atlantic but also that produced by the East Pacific Rise.

In fact, at one time, there were many major plates subducting around the Pacific Rim (Fig. 14.8C). The East Pacific Rise produces the huge Pacific plate to the west (which subducts beneath Asia and Alaska) and the Nazca plate to the east (which subducts beneath South America). Both plates are still being erupted and subducted, although as the East Pacific Rise has shifted to the east, the Pacific plate has grown larger and the Nazca plate smaller. At the north edge of the Nazca plate, the Galapagos spreading ridge separates it from the Cocos plate to the north, which is currently subducting beneath Central America. But if you look at the old magnetic anomaly patterns and reconstruct the ancient position of the sea floor in the Cenozoic and Mesozoic, the Cocos plate and the smaller Juan de Fuca plate (currently subducting beneath Oregon and Washington to form the Cascade volcanic arc) were once part of a much larger plate known as the **Farallon plate** (Fig. 14.8C). That plate has since broken in two parts (Cocos and Juan de Fuca) as its irregular corner was subducted beneath California and turned into the San Andreas transform fault. Back in the Cretaceous, another plate, called the Izanagi plate, was subducted beneath Japan and Kamchatka; and it is nearly gone. Finally, there also was once a plate to the north of the Farallon and Pacific plates. It no longer exists today (except for a fragment that may be trapped beneath the Bering Sea) but has apparently completely vanished into the subduction zones beneath Alaska and British Columbia. This long-vanished piece of crust was named the **Kula plate**, after the Tlingit word for "all gone."

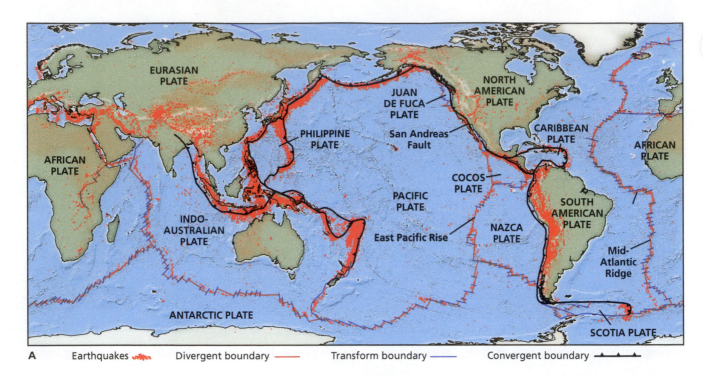

A   Earthquakes ⁓⁓⁓   Divergent boundary ——   Transform boundary ——   Convergent boundary ▲—▲—▲

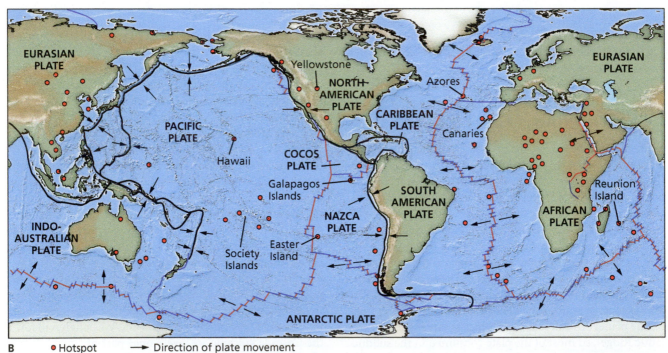

B   ● Hotspot   → Direction of plate movement

Figure 14.8   *(Continued)*

Nearly all the volcanic chains that have formed around the Pacific Rim are products of this subduction during the Cenozoic. These include the island arc chains along the Aleutians, Japan, the Marianas, the Philippines, Indonesia, and New Guinea, along with the Solomon Islands, and the volcanoes from Tonga to New Zealand west of the Tonga–Kermadec trench (Fig. 14.8A). These island arc chains have lots of active volcanoes, which erupted and created the islands mostly during the later part of the Cenozoic. Japan, for example, has volcanoes erupting nearly constantly, especially with the deadly eruption of Mt. Unzen in 1991. In addition, there are dozens of active volcanoes in the Aleutian Islands, as there are in the Philippines, where Mt. Pinatubo erupted catastrophically in 1991. But the chain of volcanoes in Indonesia and Malaysia is perhaps the most active and deadly in the world, with not only hundreds of different volcanoes (several of which are erupting at any given time) but also some of the largest and deadliest eruptions in the Cenozoic. The eruption of Krakatau (between Java and Sumatra) in 1883 killed 36,000 people and made an explosion heard around the world. Even bigger was the 1815 eruption of Mt. Tambora, on the island of Sumbawa

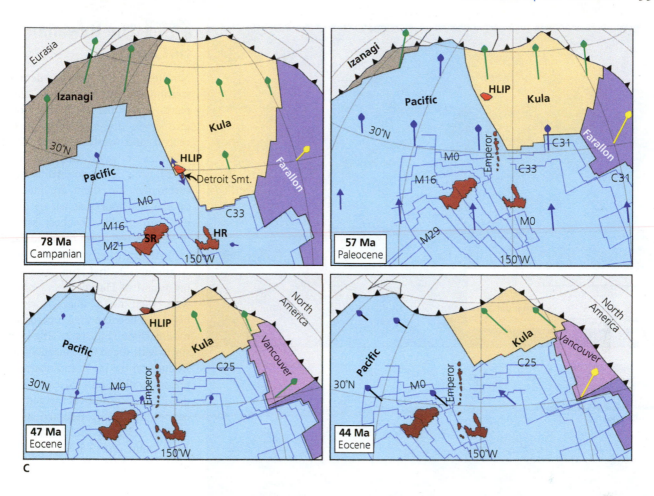

C

Figure 14.8 **A.** Map of the major tectonic plates, shown by the occurrence of earthquakes on each boundary. The huge zone of earthquakes (plus volcanoes) around the rim of the Pacific Ocean is nicknamed the "Ring of Fire" because nearly all the Pacific is bounded by subduction zones that produce arc volcanoes and huge deep-subduction zone earthquakes. **B.** Map of the earth's crustal plates, with their relative motions shown, as well as the position of mantle hot spots whose tracks through the moving plate above them give the direction and velocity of plate motion. **C.** Earlier plate configurations of the Pacific plate, showing plates that have been largely subducted. (HLIP = Hawaiian large igneous province).

east of Java, which killed 71,000 people and scattered so much dust into the stratosphere and darkened and cooled the earth that 1816 was called "the year without a summer," with snow falling in New York and Europe in mid-summer. But the biggest of all was the eruption of Toba in northern Sumatra about 74,000 years ago, which was the largest eruption in the last 28 million years. This enormous event caused a global winter around the world, with dramatically reduced temperatures and lowered snowlines. Even more amazing, there are genetic bottlenecks in the genomes of humans, tigers, pandas, and many other animals suggesting that all life went through a population crash at this time; and only as few as 1000 to 10,000 pairs of humans survived (see Chapter 16). Toba came very close to wiping out land life on earth, including the human race.

In addition to these island arc chains formed by the collision of oceanic plates with each other, there are huge continental volcanic chains formed by subduction of oceanic plates beneath continental crust. The best studied of these are the Andes Mountains that form the western edge

of South America. Like the Sierra Nevada Mountain chain that erupted in the Jurassic and the Cretaceous (Fig. 13.7) from British Columbia, the Andes also began to erupt in the latter half of the Mesozoic as the Nazca plate was subducted beneath South America. Most of the eruptions that make up the highest peaks, however, are Cenozoic in age. The Andes form an incredibly long chain of volcanic mountains, over 7000 km (about 4300 miles) long and 200–700 km (120–430 miles) wide, averaging about 4000 m (about 13,000 feet) in elevation. They include the highest mountains outside of Asia, with the tallest being Mt. Aconcagua at 6961 m (22,838 feet) and the world's highest volcano, Ojos del Salado, which is 6893 m (22,615 feet) high (both are located on the Chile-Argentina border). There are many active volcanoes in this range still, including Nevado del Ruiz in Colombia, which erupted in 1985, killing 30,000 people, and Galeras in Colombia, which killed 9 scientists studying it during a field conference. There are numerous active volcanoes in the rest of the chain, including over a dozen in Chile alone that have erupted since 2000.

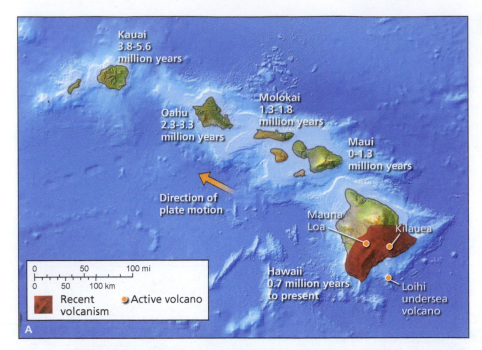

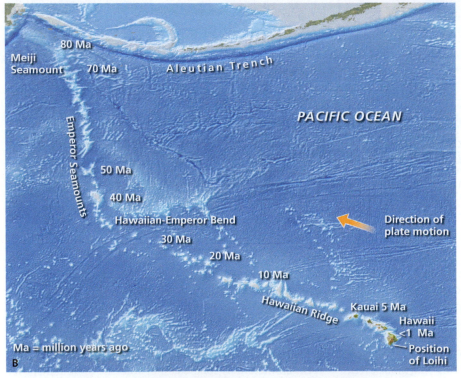

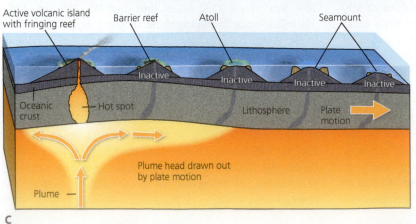

14.4 The Hawaiian Hot Spot

Nowhere is there a better demonstration of the motion of these plates across the earth and down into subduction zones than in the central and western Pacific. Not only can we watch the plates move in real time thanks to satellite measurements, but there are markers on the sea floor that tell us of its past movement. The best known of these is the chain of islands and sunken drowned islands (called "seamounts") that run from the big island of Hawaii to the northwest, then do a sharp clockwise bend and become the Emperor seamount chain (Fig. 14.9). The chain starts with the active volcanoes like Kilauea on the big island, then progresses to older and older and more deeply eroded and submerged volcanoes like Maui (erupted as far back as 1.3 Ma), Molokai (1.3–1.8 Ma), Oahu (2.3–3.3 Ma), and the

Figure 14.9 Evolution of the Hawaiian–Emperor seamount chain. **A.** Map of the Hawaiian Islands, showing their location and the K–Ar dates on their lava flows. Kilauea is currently active and sits over the Hawaiian hot spot, and all the lavas on the Big Island are less than 0.7 million years old. Maui and Molokai are slightly older (1.3 Ma), Oahu is 2.8–3.3 Ma, and the oldest flows on Kauai are 5.6 Ma. Thus, each island stopped erupting as it moved northwest off of the Hawaiian hot spot, while the extinct volcanoes on each island began to erode as the island sank into the sea. **B.** The entire Hawaii–Emperor seamount chain. Beyond Kauai, most of the volcanoes are seamounts below the ocean surface, except for islands like Midway and Laysan. The Hawaiian chain then bent clockwise around 35 Ma, forming the Emperor seamount chain, between 35 and 80 Ma. This suggests that the plate itself changed direction as it slid over the stationary hot spot. **C.** Cartoon showing how seamount chains are formed as they erupt over a hot spot, then sink down and go extinct after they slide past the hot spot.

oldest of the Hawaiian islands, Kauai, which is up to 5.6 million years old. Thus, the entire Hawaiian chain gets older and more deeply eroded as you go from southeast to northwest. But this is just the beginning of the trend. Northwest of Kauai, the volcanoes continue as a chain of extinct volcanic seamounts that sink deeper and deeper beneath the waves as they get older and older (Fig. 14.9B). In some places, the seamounts are tall enough to stick out of the waves and form islands, like Midway Island or Laysan Island; but most are submerged.

The Hawaiian seamount chain continues for almost 3500 km (about 2200 miles). Then it does a sharp bend clockwise about 45° and continues as the north-trending Emperor seamount chain for another 2500 km (about 1550 miles). All of this chain of extinct volcanoes is now seamounts that sink farther and farther into the ocean, so none stand above the waves as islands.

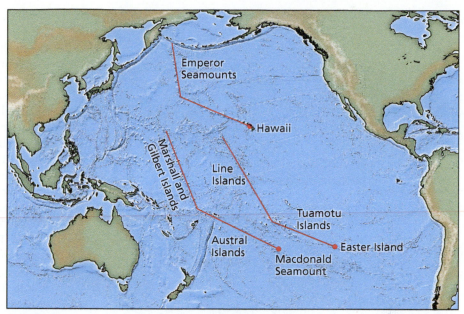

• Hot spot

Figure 14.10 The clockwise bend or "kink" between the Hawaiian and Emperor seamount chains can be found on other hot spot seamount chains in the Pacific, such as the Tuamotu–Line Islands and the Marshall–Gilbert–Austral Islands. They all have kinks dated around 38–40 Ma, suggesting that the entire Pacific Plate changed its direction of movement with respect to the hot spot at that time.

 See For Yourself: Evolution of the Hawaiian Hot Spot

These striking seamount chains were discovered during the early days of marine geology in the 1940s and 1950s. But there was no explanation for them until plate tectonics came along in the 1960s. In 1965, Canadian geologist J. Tuzo Wilson looked at the ages of the dated volcanic rocks in the chain and realized that not only do the Hawaiian Islands get older as you go to the northwest (Fig. 14.9A) but the seamount chain gets even older still. Midway is about 30 million years old, the seamounts at the kink are about 38 million years in age, and those of the Emperor chain range from 40 to 80 Ma, which will be the next part of the chain to be subducted beneath the Aleutian trench (Fig. 14.9B). Wilson proposed that these progressively older volcanoes in the chain were formed when the Pacific plate slid over a stationary hot spot on the mantle (Fig. 14.9C), which currently lies beneath active volcanoes like Kilauea. As the Pacific plate continued northwesterly off the hot spot, it carried the old volcanic island off as well to sink and erode as it became inactive, while a new Hawaiian island was erupted to the southeast after it moved into position above the hot spot. Using the age of islands like Midway, the rate of plate movement can be calculated at about 9 cm (3.5 inches) per year. This process repeated over and over again, burning a complete chain of volcanoes across the western part of the Pacific plate. In fact, the transition may be happening in the near geologic future. Most of the volcanoes on the big island of Hawaii are no longer active; and in a few more million years it will probably move off the hot spot, and Kilauea will become extinct. In its place is the next Hawaiian island, which is the undersea volcano Loihi, just southeast of the big island. It stands about 3000 m (about

10,000 feet) above the sea floor, but it is currently about 975 m about (3100 feet) below sea level. It began erupting beneath the ocean about 400,000 years ago. In about 10,000–100,000 years, it will finally build up enough to rise above the waves and become the next Hawaiian island.

What about the "kink" between the Hawaiian and Emperor seamount chains? It turns out that there are several other seamount chains on the Pacific sea floor with exactly the same trend and kink. One of them runs from Easter Island to the northwest through the Tuamotu Islands and Line Islands (Fig. 14.10). A second starts with the hot spot under Macdonald seamount and continues through the Austral Islands and then the Marshall and Gilbert Islands. Both have kinks that date to about 38–40 Ma, suggesting that the entire Pacific plate changed direction at that time. That timing coincides with a global event of plate reorganization, when India collided with Asia and the entire Indian–African plate system changed its motions, along with readjustments of plate motions in the Caribbean. At the same time, the Laramide orogeny ended and was replaced by a new style of subduction (see "North American Cenozoic Geology").

14.5 North American Cenozoic Geology

COASTAL PLAINS AND CONTINENTAL SHELVES

As we saw in Chapter 13, the entire East Coast and Gulf Coast of North America (and their counterparts in Africa and South America) were born when the North Atlantic ripped open in the Jurassic and Cretaceous. By the Early Cretaceous,

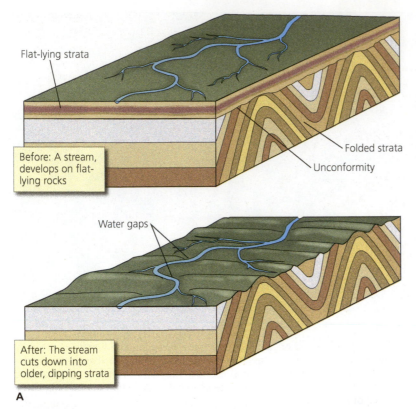

Flat-lying strata

Folded strata

Unconformity

Before: A stream, develops on flat-lying rocks

Water gaps

After: The stream cuts down into older, dipping strata

A

B

Figure 14.11   A. After the Appalachians were folded in the Pennsylvanian, they slowly eroded down in the Mesozoic, then were buried to the top by sediments during the high sea levels of the Cretaceous. This flat-lying blanket of strata covered the ancient erosional surface carving the tops off the Appalachian folds and ridges, and their burial made them invisible to river drainages that developed across the sedimentary blanket during the Cretaceous and early Cenozoic. When the Appalachians were rejuvenated in the late Cenozoic, the rivers stripped away the blanket of soft sediment and exposed the hard ridges of Paleozoic rocks. The existing drainage pattern then carved down through the hard ridges, creating water gaps where the rivers cut through resistant mountains rather than go the easy way around them through the valley of softer sediments. **B.** At the Delaware Water Gap, the Delaware River cuts through the resistant ridges of Silurian Shawangunk Conglomerate from Pennsylvania on its way to New Jersey and the Atlantic Ocean.

there was an extensive passive margin wedge complex of shallow-marine sediments building on the subsiding edge of the continent as it sank below the waves. Not only did the subsiding continental edge invite the drowning of the coastal plain region, but the extraordinarily high sea levels of the Cretaceous covered much of the deeply eroded Appalachian Mountains with sediment, so they nearly vanished entirely (Fig. 14.11A).

Meanwhile, on the subsiding continental shelf, thousands of meters of shallow-marine Cretaceous and Cenozoic sediment accumulated over the past 100 million years (Fig. 5.18B). These marine sediments can still be seen in the outcrop belts that run beneath the modern coastal plain, from New Jersey all the way to Texas. Especially during times of extraordinarily high sea level during the Cretaceous, Eocene, and Miocene, the coastal plain was inundated with shallow shelf sands and muds. You can see these in many places, such as the Calvert Cliffs along Chesapeake Bay in Maryland. During low tide, the Miocene Choptank Formation is exposed and yields thousands of Miocene fossils from dense shell beds that used to cover the region during high sea levels in the Miocene (Fig. 14.12A, B).

On the Gulf coast, the Cretaceous–Cenozoic package of shallow-marine sediments is nearly 1200 m (about 4000 feet) under the coastline of Louisiana (Fig. 14.13). Traveling along the southern part of Alabama and Mississippi and west in Louisiana and Texas, there is an extensive outcrop belt of Cenozoic fossiliferous marine sediments. There is a short sequence of the Paleocene Midway Group, then a very thick package of Eocene deposits (Wilcox, Claiborne, and Jackson Formations, over 1000 m or about 3300 feet thick). These Eocene beds are legendary for their fossil shell beds, rich in many warm-water mollusks that have living relatives today (Fig. 14.12C). The Eocene sequence is capped by a 40-m-thick (about 130 feet) wedge of Oligocene sediments of the Vicksburg Group. It is bounded by unconformities above and below it, due to low sea levels in the Oligocene caused by the first Antarctic glaciers. After the upper Oligocene unconformity, these deposits are capped by Miocene beds from Texas to Georgia, from the lower Miocene Catahoula Formation to the middle Miocene Hattiesburg and Pascagoula clays and finally the Citronelle Formation. This entire Cenozoic sequence wedges seaward and thickens from just a few

Figure 14.12   The Atlantic Coastal Plain is capped by a thick blanket of Cenozoic shallow-marine sediments formed in the recent stages of deposition along the passive margin wedge. **A.** Typical outcrop of the Miocene Choptank Formation along Chesapeake Bay. **B.** The Choptank deposits reflect a shallow shelf setting, as demonstrated by the abundant mollusks found in the sediments. **C.** During the Eocene, thick marine beds full of mollusks were deposited on the Gulf Coast. These shells are from the Lower Eocene Bashi Formation.

tens of meters in the sections exposed on land to almost 3300 m (about 11,000 feet) in the offshore region. Finally, as the weight of all this Cretaceous and Cenozoic sediment squeezed down on the deeply buried Jurassic salt layers, they began to rise upward and flow into salt domes, punching their way through the overlying sediment and deforming them with numerous faulted blocks as well (Figs. 13.5, 14.13). This is the complex geology that petroleum geologists working the Gulf coast must cope with to find pockets of oil and gas.

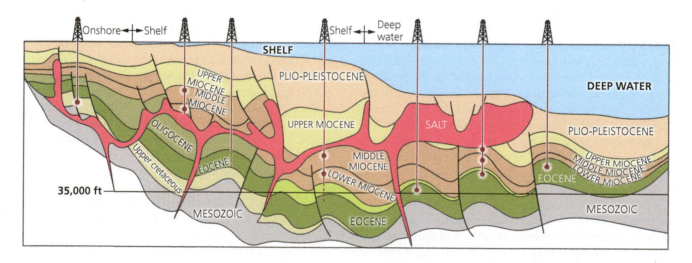

Figure 14.13   Simplified and exaggerated cross section of the rocks of the passive margin wedge on the Texas and Louisiana Gulf Coast. Above the Jurassic salt layer and Cretaceous shallow-marine sediments is a thick layer of Cenozoic shallow marine sediments, some of which is broken up by later faulting and most of which is intruded by salt diapirs rising up from the Jurassic salt beds below.

Throughout this entire time, the ancient Appalachian Mountain chain was deeply eroded after millions of years of uplift that beveled the ranges to flat surfaces, then buried by a blanket of Cretaceous and lower Cenozoic sediments when sea levels were higher. Sometime in the later Cenozoic, the entire Appalachian region began to gently uplift again. First, the blanket of Cretaceous marine sediment was stripped away, exposing the hard ridges of Paleozoic rocks that had been folding into tight anticlines and synclines during the Appalachian orogeny in the Pennsylvanian (Fig. 12.6). But the rivers that eroded away the cover of soft sediment were already established in their courses so that when they exposed and encountered the hard bedrock beneath, they continued to cut down through the ranges, rather than erode their way around them following the softer sediment. In many places in the Appalachians, the rivers mostly run down the centers of the valleys of soft shale and avoid cutting the hard ridges, but in some places they cut through and erode the hard way, forming water gaps. The most famous of these is the Delaware Water Gap on the New Jersey–Pennsylvania border, where the Delaware River cuts through a hard ridge of Silurian Shawangunk Sandstone and Conglomerate, rather than eroding the easy way around this ridge (Fig. 14.11B). This kind of pattern, where a drainage is established on top of soft sedimentary cover, then the river strips away that cover and erodes in the same pattern through the hard bedrock, is called a **superposed** (in Latin, "placed on top of") **drainage** (Fig. 14.11A). The same process caused many of the water gaps in the Rocky Mountains.

THE CORDILLERA

In contrast to the relatively simple geology of the passive margin wedges, the entire western United States from the Pacific coast to the Rocky Mountains (the region known as the **Cordillera**) was very tectonically active and underwent a complex history in the Cenozoic. It can be broken down into roughly three phases.

The First Phase: Laramide Orogeny (70–40 Ma)

As we discussed in Chapter 13, the end of the Cretaceous in the western United States was marked by the peculiarities of the Laramide orogeny (Fig. 13.11). This odd mountain-building event was characterized by two unusual characteristics: the Sierran volcanic arc shut off, so there was no volcanism in the region where it usually occurs in subduction zones. Instead, there were deep crustal uplifts of folded and faulted Precambrian and Paleozoic bedrock in the Rocky Mountain region, about 1600 km (about 1000 miles) east of the normal area of mountain-building. This same region had been covered by the Western Interior Cretaceous Seaway all the way to western Utah during most of the Cretaceous, but during the latest Cretaceous it began to rise up out of the sea to form the "Laramidia" land mass. As we learned in Chapter 13, these uplifts correspond to most of the modern Rocky Mountain ranges, from the Front Range in Colorado to the Laramie and Wind River and Bighorn Mountains of Wyoming and on into Montana. This event was the second largest episode of mountain-building in the region, so if the Pennsylvanian Ancestral Rockies were "Rocky I," then this could be called "Rocky II."

This peculiar pattern only makes sense if the subduction of the plate underneath North America no longer dipped down into the mantle but instead became shallower and shallower until it was horizontal or subhorizontal (Fig. 13.11C). This explains not only why the Sierran volcanism shut off when the downgoing plate no longer plunged down into the mantle and melted but also how the stresses of the subducting plate could be transferred 1600 km (about 1000 miles) inland of where the normal tectonic activity is concentrated. Presumably, the pressure of this horizontal slab compressing on the lower crust beneath the future Rockies transferred stresses into the overlying crust and helped contribute to the uplift and faulting that built the Laramide ranges.

The Laramide tectonism did not stop when the Cretaceous ended but continued through the Paleocene and right into the end of the middle Eocene, from about 70 Ma to about 40 Ma. Most of the uplift of the Laramide ranges occurred during the Paleocene and Eocene, as they continued to rise into the sky and the entire western North America emerged from the Cretaceous seas. Even more impressive, however, were the deep **intermontane basins** that were formed by faulting and downward folding between the Laramide uplifts (Fig. 14.14). As they sank deeper and deeper, these basins were rapidly filled with as much as 3000 m (about 10,000 feet) of Paleocene and Eocene terrestrial sediments, often rich with terrestrial mammal fossils. Most of the deposits were formed in river channels and floodplains that drained between the rising ranges, but in some areas there were lake deposits or swamps or deltas. The same kinds of fossils and environmental settings are found from the Williston Basin on the Montana–North Dakota border to the Wind River Basin south of Yellowstone, the Bighorn Basin east of Yellowstone, the Powder River Basin between the Bighorn Range and the Black Hills, the Green River Basin in southwest Wyoming, the Uinta Basin in northeast Utah, the Piceance (pronounced "PEE-onts") Basin in northwest Colorado, and the San Juan Basin in northwestern New Mexico, along with many smaller basins.

Each of these basins has different formation names, but their sedimentary history is remarkably similar across the region at any given time. For example, in the Paleocene the **Fort Union Group** is found in the basins of Wyoming, Montana, and North Dakota. It is a classic floodplain–deltaic sequence with many swampy areas, legendary not only for its fossil mammals and its petrified logs in Theodore Roosevelt Badlands National Park in North Dakota but especially for its remarkable coal seams (Fig. 14.15A, B). In the Powder River Basin of Wyoming, individual coal seams are over 100 m (over 330 feet) thick in places yet very near the surface, so they can be excavated by strip mining (Fig. 14.15B). This has created enormous mining pits across the entire region, which produces much of the coal that is still mined in the United States. Fort Union coal beds have big advantages in

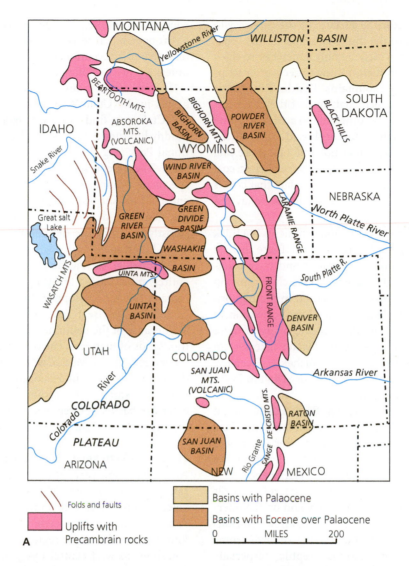

A

Folds and faults

Uplifts with
Precambrain rocks

Basins with Palaocene

Basins with Eocene over Palaocene

0   MILES   200

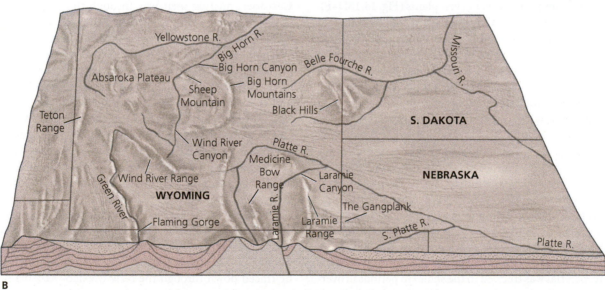

B

Figure 14.14 **A.** Map showing the location of the major ranges uplifted during the Laramide orogeny and the basins which sank down between them accumulating thousands of meters of Lower Cenozoic river, floodplain, lake, and deltaic sediments. **B.** Block diagram of the major northern Laramide basins, showing the position of water gaps allowing the rivers to cut across the heart of mountain ranges to reach basins between them.

that they are thick and near the surface, so they are easy to mine. This is compared to the thin coal seams of the Appalachian region or Illinois, which are only a meter or so thick and require extensive deep shaft mining to recover or stripping off the tops of entire mountains. Powder River coals are also better than the eastern coals in that they have low sulfur content, so they do not contribute to acid rain. However, they were not as heavily metamorphosed as the eastern coals, so they don't tend to burn as hot and a lot more coal is required to get the same energy output. Nevertheless, coal mining is still big in the region, despite the fact that coal is vanishing as a fuel source because natural gas, solar, and wind energy are so much cheaper.

Overlying the Paleocene Fort Union sediments is a thick sequence of Eocene deposits in the region. These include abundantly fossiliferous floodplain muds and river channel sandstones in the early Eocene (Wasatch Formation in most of Wyoming, Willwood Formation in the Bighorn Basin of Wyoming and Montana, San Jose Formation in New Mexico). During the middle Eocene, many of these basins were filled with a gigantic freshwater lake system that filled the Green River and the Washakie, Piceance, and Uinta Basins and extended south across Utah to Bryce Canyon National Park in south-central Utah. Known as the Green River lake system or **Green River Formation**, it consists of over 430 m (about 1400 feet) of laminated lake shales, with minor sandstones and freshwater limestones (Figs. 4.8, 14.15C–F). These finely bedded lake shales are famous worldwide for the extraordinary quality and quantity of fossils they entombed, from the common fish fossils (found in every rock shop in the country and on the internet) to rare fossils like certain kinds of mammals (including the oldest good bat fossil), many types of birds (some preserving their feather impressions), reptiles (especially crocodilians), amphibians, and even plants (Fig. 14.15D–F). Among the many leaves and other plant fossils, abundant palm fronds (Fig. 14.15F) show that the climate was mild and subtropical, as it was for most of the world during the middle Eocene. The bottom waters must have also been very stagnant and quiet because the fossils are undisturbed and apparently didn't even decay very much—a clue that there was very low oxygen and there were no scavengers on the lake bottom.

 See For Yourself: The Green River Shale

The Green River shales are also important because the abundant algae in them (including some freshwater stromatolites) produced a lot of excess organic matter that was later trapped in them. In many places, the shales have millimeter-scale fine laminations (Fig. 14.15C), showing that the lake chemistry was very delicate and rapidly alternated between stagnant periods with dark shale bands (when the organic material was trapped in the shale) to periods of overturn and oxygenation (when the light-colored shale bands were formed). This organic material is rich enough that in some places it has been mined for **oil shale**. To extract this oil, huge volumes of the shale must be mined out, crushed, and then heated until the oil is released. However, since the

energy crisis of the 1970s ended with lower oil prices, there hasn't been much serious effort to pursue this expensive alternative to conventional oil resources, and there won't be unless the price of oil makes it economically feasible.

If you move from the Green River–Uinta–Piceance basins in the Utah–Wyoming–Colorado corner to southern Utah, you find extensive shallow lake limestones and muddy limestones (marls) in places like Bryce Canyon National Park (Fig. 14.1) and Cedar Breaks National Monument. These deposits have long since been eroded to form spectacular pinnacles and hoodoos and many other weird and colorful shapes. But in the middle Eocene, they were one huge lake deposit loaded with algae and invertebrates (mostly crustaceans called ostracodes) that deposited freshwater limestone.

Meanwhile, the Pacific Coast of North America was not entirely quiet during this time. The volcanism in the Sierra Nevada range had stopped, but the old forearc basin (especially in the Great Valley of California) was still subsiding because of the continued subduction and accumulating thousands of meters of Paleocene and Eocene marine sediments (Fig. 14.16). Most of western Oregon and Washington did not yet exist in the Paleocene; but during the Eocene a number of small exotic terranes docked there, and there are extensive basins in Oregon and Washington full of Eocene marine deltas and shallow-marine sands and muds.

The Second Phase: Arc Volcanism Resumes (40–20 Ma)

After 30 million years of no volcanism in the former Sierran arc region, around 40 Ma in the late middle Eocene, arc volcanism returned with a vengeance across western North America. Eocene–Oligocene volcanoes (Fig. 14.16) can be found across the entire region. These include eruptions of volcanoes in west-central Oregon, forming the **Ancestral Cascades**, melting their way through the Eocene terranes that had recently docked there. Although they probably looked much like the modern Cascades of Oregon and Washington, the Ancestral Cascades were farther to the east of the modern volcanic chain. Such extensive volcanic deposits can be seen in the middle Eocene Clarno Formation and the Oligocene John Day Formation of John Day National Monument in central Oregon (Fig. 14.17A, B). The trend continued through Nevada, which formed a high volcanic plateau called the Nevadaplano, over 3000 m (about 10,000 feet) in elevation, that stood higher than the eroded Sierra Nevada Mountains. The volcanic rocks continued through central Utah and then joined with the immense Oligocene volcanic field in the southwestern corner of Colorado known as the San Juan volcanic field, which covered over 25,000 km^2, or 9700 square miles (Fig 14.17C). Additional volcanic eruptions of Eocene and Oligocene age can be traced down through the New Mexico–Arizona border region and then into Mexico.

The resumption of volcanism across the entire Cordillera suggests that the previously horizontal-dipping slab of the Laramide orogeny was once again sinking and plunging into the mantle and beginning to melt in order to form

Figure 14.15 Typical Cenozoic deposits of the Laramide basins of the Rocky Mountains. **A.** The Paleocene Fort Union Group, made mostly of floodplain and deltaic sandstones and mudstones, with thick coal seams from large swamps in the deltas. This exposure in Theodore Roosevelt National Park in western North Dakota is full of petrified logs. **B.** In the Powder River Basin of northeastern Wyoming, enormous thick seams of coal from the Fort Union Group are found just below the surface. They are extracted by gigantic strip mining operations. **C.** Close-up of the fine laminations in the Green River lake shales, showing the dark bands that are organic-rich, making these deposits valuable as oil shales. **D–F.** Typical fossils of the Green River Formation. **D.** A complete articulated skeleton of a crocodilian. **E.** A slab covered with fossils of the tiny herring *Knightia*, the most common fish in the Green River shales. **F.** The shales also trapped leaves of plants from the nearby shore, including this *Sabalites* palm frond, indicating a warm tropical climate in the middle Eocene.

this volcanic belt. But the new volcanic chain was far to the east of the old Sierra Nevada volcanic arc, suggesting that the downgoing slab was plunging into the mantle at a shallow angle and melted much farther inland than where the old Sierran arc once stood. Since the plate was originally undergoing horizontal subduction during the Laramide orogeny, this new subduction suggests that the old horizontal plate had begun to peel away from the overlying crust and to melt again, even if not in the old position. This is often called the "slab rollback" model of subduction.

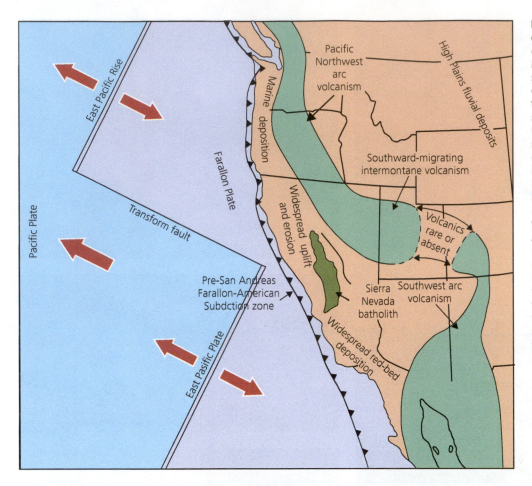

Figure 14.16 Tectonic map of the western United States during the Oligocene and Early Miocene. Once the horizontal subduction of the Laramide orogeny ended around 40 Ma, the slab began to peel off and subduct again to sink into the mantle and melt but at a very shallow angle. Thus, arc volcanism returned to the region but not in the old position beneath the Sierra Nevada Mountains. Instead, it was much farther inland and eastward in a belt that ran from the Ancestral Cascades in eastern Oregon and Washington across skinny Nevada (before it was stretched to twice its original width during the Miocene) through Utah and the San Juan volcanic field of southwestern Colorado and then down through New Mexico and into Mexico. Meanwhile, the old Laramide ranges wore down, and their basins filled up with sediment and volcanic, some of which spilled over to form the Eocene–Oligocene badlands of the High Plains.

Figure 14.17 Photographs of typical deposits of Eocene and Oligocene of the western United States. **A.** The Upper Middle Eocene Clarno Formation in the Painted Hills of central Oregon, which has produced abundant plants and fossil mammals. The deposits are made of deeply weathered volcanic ash, with ancient soil horizons shown by the color bands. **B.** The Upper Oligocene–Lower Miocene John Day Formation, famous for its fossil mammals, composed of volcanic ash that has weathered into a greenish tint. **C.** Wheeler Geologic Area, near South Fork, Colorado, is composed of thick Oligocene volcanic ash deposits of the calderas in the San Juan volcanic field of Colorado, now eroded into these spectacular pinnacles.

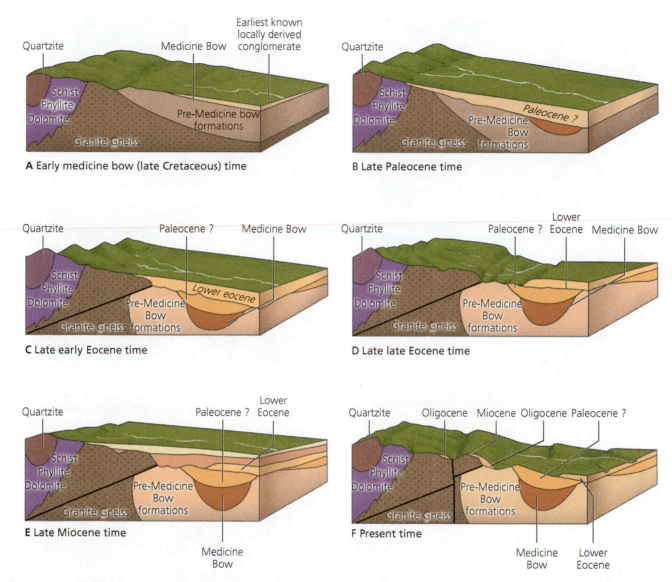

Figure 14.18 By the Oligocene and Miocene, the old Paleogene Laramide basins were filled to the top with sediment eroded from adjacent ranges and from volcanoes to the west. The deeply eroded ranges were buried in their own sediments, so only tiny remnants stuck above the thick Oligocene–Miocene sedimentary blanket that buried most of what became the modern Rockies.

Meanwhile, the old Laramide basins ceased to subside and accumulate sediment, and the old Laramide ranges began to erode away once their uplift ended. Over time, enormous volumes of Oligocene and Miocene sediment from these eroding ranges, along with abundant volcanic ash deposits blown from the newly resurgent arc, filled those ancient basins to the top. By the late Miocene, not only were the basins filled but the sedimentary blanket began to bury the ranges as well (Fig. 14.18). How do we know this? In much of the Rockies, the ranges were eroded down to a nearly flat surface across their tops, with only a few remnants sticking out of the thick mantle of later Cenozoic sediments (Fig. 14.19A). In some places in the Rockies (such as the Granite Mountains in central Wyoming), you can still see these buried mountain ranges with just remnants sticking above the landscape (Fig. 14.19B). On the eastern flank of the Laramie range just west

of Cheyenne, Wyoming, this mantle of Cenozoic sediments formed a gentle ramp up from the Plains to the top of the Laramie range. Nicknamed "the Gangplank" (Fig. 14.14B), this smooth gradient was followed by numerous pioneer trails and then by the transcontinental railroad in the 1860s to give an easy route over the top of the easternmost range of the Rockies (Figs. 14.14B, 14.18).

In most places, however, later uplift of the Laramide ranges in the late Cenozoic has stripped away all this cover and exposed the ancient mountains again (Fig. 14.18), so today they rise above the landscape as they did in the Paleocene or Early Eocene. If the Cretaceous to Eocene uplift of the Laramide orogeny can be nicknamed "Rocky II" (since it is the second incarnation of the Rocky Mountains), then the renewed uplift and exhumation of the ancient Laramide ranges that we see today might be considered "Rocky III."

Figure 14.19 **A.** The tops of the Rockies were beveled flat, as can be seen here in the Beartooth Plateau, Montana, just northeast of Yellowstone National Park. **B.** In other places, the Rocky Mountains are still buried, and just the peaks poke up through the sedimentary blanket, as in the Granite Mountains in central Wyoming. **C.** When the Rockies were uplifted in the late Cenozoic ("Rocky III"), the rivers were forced to carve deep canyons through the mountains that had once been buried and eroded water gaps between the basins. This is Royal Gorge in Colorado, where the Arkansas River cuts through the mountains on its way to the plains. **D.** Another water gap is where the Green and Yampa Rivers cut across the uplifted rocks of the Uinta Arch and Dinosaur National Monument through Flaming Gorge, Split Mountain, and other uplifts. This is a view from a small airplane of the Green River cutting through Split Mountain, near the Split Mountain Campground at Dinosaur National Monument, Utah. **E.** The river channel and floodplain deposits of the White River Group in the Big Badlands of South Dakota, shed from the wearing down of the remnant Laramide uplifts (like the Black Hills of South Dakota, just to the west) and full of volcanic ash from eruptions to the west in Nevada and Utah.

The evidence of this uplift of the ancient buried and flattened surface can be seen in many places. For example, the tops of many of the Rocky Mountain ranges are not steep and jagged but beveled off flat on top (Fig. 14.19A) or have a gently rolling topography. This is true whether you are on the Beartooth Plateau in Montana just northeast of Yellowstone, the Laramie range in southeast Wyoming, or the highest parts of Rocky Mountain National Park in Colorado. In many places, such as Darton's Bluff on top of the Bighorn range in Wyoming, there are remnants of the Oligocene–Miocene blanket of sediment that once covered the entire ranges and buried them, but only tiny portions remain. In a few places, such as the Granite Mountains in central Wyoming, the ancient Laramide uplifts are still buried to the top, with only a few scattered outcrops sticking out of the Late Cenozoic sedimentary blanket (Fig. 14.19B).

The other result of the burial of the Rockies by the late Miocene is that river drainages were established across this flat buried surface without any regard to the hard bedrock of the mountain lying deep beneath the soft sedimentary blanket. Then when the entire region rose into the sky in the latest Cenozoic (Pliocene–Pleistocene) during renewed uplift of the Rocky Mountain tract ("Rocky III"), these established stream drainages began to strip down the soft Oligocene–Miocene sedimentary cover and exposed the hard bedrock beneath. As we saw with the Appalachian drainages in the Cenozoic, the drainages of the Rocky Mountains sometimes cut down the hard way and eroded through the bedrock that was gradually exhumed when the soft sedimentary blanket was stripped away. Rather than eroding around the ranges through softer rocks, the rivers cut down through the hard bedrock right across the mountains, forming water gaps just like those of the Appalachians. This is yet another example of a superposed drainage. This discovery and its explanation were first deciphered by the pioneering geologist John Wesley Powell, who led the first river trip down the Colorado River through the Grand Canyon in 1869.

Water gaps are found throughout the Rockies (Fig. 14.14B). Royal Gorge in Colorado brings the Arkansas River out of the central Rockies and east toward the Plains (Fig. 14.19C). The Wind River flows north out of the Wind River Basin and through Wind River Canyon on the south rim of the Bighorn Basin, whereupon the river changes names and becomes the Bighorn River (Fig. 14.14B). It then flows out of the Bighorn Basin through Bighorn Canyon on the north rim of the same basin and joins the Yellowstone River in Montana. Flaming Gorge and Split Mountain (Fig. 14.19D) on the Utah–Wyoming border bring the Green River down from Wyoming and into the Canyonlands of central Utah, which are also deeply incised by later uplift of the Colorado Plateau. Likewise, the Laramie River cuts through the north end of the Laramie range rather than going a few miles to the north to go around it through softer sediments. There are numerous other examples of Rocky Mountain water gaps.

Finally, not only did the burial of the Rockies during the Oligocene and Miocene fill up ancient basins, change drainages, and make mountains vanish, but the excess sediment spilled over to the east and began to cover the Great Plains (Figs. 14.16, 14.18). In many places across the Plains, there is an enormous blanket of upper Eocene–Oligocene volcanic-rich sediment known as the **White River Group** (Fig. 14.19E). Famous from the exposures in the incredibly fossiliferous Big Badlands of South Dakota, the White River Group extends down into Wyoming and Nebraska and northeastern Colorado, and tiny remnants are even found on top of the Williston Basin in western North Dakota. The White River Group (Fig. 14.20) records a history of dense late Eocene forests near the base (Chadron Formation), then floodplains and river channels in the lower Oligocene (Scenic and Orella members of the Brule Formation), followed by upper Oligocene dune deposits of the Poleslide and Whitney members of the Brule Formation, suggesting a steady drying trend through the late Eocene to the late Oligocene. The White River Group is then capped by another deposit of river channels and floodplain mudstones and some dune deposits, the upper Oligocene–lower Miocene Arikaree Group and the middle Miocene Hemingford Group, capped by the middle Miocene–Pliocene Ogallala Group. The Ogallala forms a widespread blanket of sandstones across the entire High Plains, formed by ancient rivers that eroded down as the renewed uplift of the Rocky Mountains began. In many places from South Dakota to Texas, the resistant cemented beds of the Ogallala form a "caprock" on the bluffs and landforms of the region. Where it plunges into the subsurface, the porous sands of the unit form the Ogallala Aquifer, the most important groundwater resource in the western Plains region, which is now being drained faster than it can recharge.

All of these river deposits are incredibly rich in fossil mammals, producing an enormous number of skeletons of Oligocene and Miocene mammals, from mastodons and rhinos to camels and horses, and many other typical Miocene groups. In some places, like western Nebraska, there is a continuous well-exposed sequence that goes from the late Eocene all the way to the Pliocene, with excellent collections of incredible fossils at every level.

Meanwhile, the Pacific Coast region was no longer experiencing Sierran volcanism, but the subduction continued during the Laramide orogeny and the resumption of inland volcanism from 40 to 20 Ma. Thus, the forearc basins (such as the Great Valley of California and the Willamette Valley of Oregon) continued to subside, accumulating thick deposits of upper Eocene marine sediments. In the Oligocene, there was a global drop in sea level, so most of the Oligocene is represented by a big unconformity. Where there was Oligocene deposition, it formed the reddish brown river and floodplain deposits of the Sespe Formation in California. During the early Miocene, marine deposition returned to the basins of California, Oregon, and Washington, producing thick deposits of marine shale and nearshore sandstones. Some of those units, like the upper Miocene Monterey Formation, are major producers of oil in the region. Others, like the lower Miocene Vaqueros Sandstone and the middle Miocene Round Mountain Silt, are famous for their rich fossil beds full of whales, marine mammals, mollusks, and fish, especially at places like Sharktooth Hill northeast of Bakersfield. The forearc basins of California like the Great Valley, the Ventura Basin, the Los Angeles Basin, and several others remained full of seawater and deposited thousands of meters of deep-water shales and turbidites throughout the late Cenozoic. They only began to emerge and dry out and become exposed to erosion during the Pliocene and Pleistocene, when sea levels dropped due to the Ice Ages and the Sierra Nevada and other ranges rose again after having been quiet through most of the Cenozoic.

Phase Three: Complex Tectonics (20 Ma–Present)

The tectonics of the Cordilleran region during the Laramide orogeny (70–40 Ma) and the later resumption of arc volcanism (40–20 Ma) are relatively easy to understand

Figure 14.20   Stratigraphic section through the sequence of the Big Badlands of South Dakota. The lowest level is the deeply eroded Upper Cretaceous Pierre Shale, full of ammonites and marine reptiles. A large unconformity caps this unit with a deeply weathered ancient soil, on which was deposited the upper Eocene Chadron Formation, (dated 34–37 Ma), full of fossils like the rhino-sized brontotheres and indicators of the last of the warm wet Eocene climate. The next layer above is the lower Oligocene Brule Formation, with the Lower Scenic Member (white layer) full of fossil mammals and land tortoises, representing a drier, brushy scrub habitat, and the Upper Poleslide Member (upper orange layer), representing even drier conditions. The uppermost white layer is the upper Oligocene–lower Miocene Sharps Formation of the Arikaree Group.

tectonically. But about 20 Ma, the tectonic story of western North America became far more complex. The best way to follow all these complex threads is to introduce each region separately, then see how they are all related.

We have already discussed the Ancestral Cascades, which erupted in central Oregon and Washington starting about 40 Ma. By the late Miocene, however, the location of the melting zone in the downgoing Farallon plate produced volcanism in the region where the modern **High Cascades** continue to erupt in recent times (Fig. 14.21A). The present-day Cascades are a continuous chain of volcanoes from Mt. Lassen (last erupted in 1914–1917) to Mt. Shasta (last erupted in 1782) in California, through Crater Lake (which blew its top off about 7700 years ago), the Three Sisters, Bachelor Butte, and Mt. Hood in Oregon (Fig. 14.21A), and culminating with Mt. St. Helens

(erupted in 1980), Mt. Rainier (the highest of the Cascade peaks at 4392 m [14,410feet]), and then on to Mt. Baker in northern Washington, continuing into British Columbia. But at one time (Fig. 14.22) the range extended much farther south through California and Nevada all the way to Las Vegas, running just east of the old extinct Sierra Nevada batholith. This southern extension of the Cascades is clearly no longer erupting. Instead, the active part of the Cascade arc has retreated to the north, and today it begins just east of where the three plates (North American, Pacific, and Juan de Fuca plates) meet in a triple junction just off Cape Mendocino in California (Fig. 14.22). Thus, the arc shutoff trends from south to north during the last 20 Ma, and the southernmost active volcanoes are found just north and east of the Mendocino triple junction, where the subduction zone begins.

Figure 14.21 During the middle Miocene through the Pleistocene, the tectonic style of the western United States changed drastically. **A.** One major event was the eruption of the modern Cascade volcanoes in a chain from northern California to British Columbia, as Andean-style subduction returned to the region. This photo shows the chain in central Oregon, with (left to right, south to north) Mount Bachelor, Broken Top, and the Three Sisters. **B.** About 15–16 Ma, eastern Oregon and Washington were covered by enormous floods of lava from rift eruptions to form the Columbia River basalts. This thick stack of lava flows can be seen in Picture Gorge near John Day, Oregon. **C.** From the air, the parallel mountain ranges and intervening basins of the Basin and Range province are very distinctive.

Behind this plunging Farallon slab was another remarkable event. During the middle Miocene (about 15–16 Ma), huge fissures opened up in eastern Oregon and Washington, and an immense volume of mantle-derived basalts poured across the landscape like a flood of lava (Fig. 14.22). These are known as the **Columbia River flood basalts** because today the Columbia River cuts across them (Fig. 14.21B). But back in the middle Miocene, these enormous eruptions poured across some 40,000 km² (about 15,400 square miles) in a matter of days, covering the landscape with thick basaltic lava flows over and over again. The flows moved about 5 km/hour (3 mph) and each was about 30 m (about 100 feet) thick, and about 100 km (62 miles) wide, and reached temperatures of over 1100°C (over 2000°F). Over about 3.5 million years, flow after flow incinerated the landscape again and again, until they covered 300,000 km² (115,000 square miles) of eastern Oregon and Washington with a sequence of lavas over 4000 m (about 13,100 feet) thick. Between eruptions, the lava flows cooled and became forested landscapes, such as that preserved in Ginkgo Petrified Forest near Vantage, Washington, before the landscape was again incinerated by the next volcanic flood. In many places in Oregon and Washington, you can find cliffs with many stacked lava flows hundreds of feet thick (Fig. 14.21B). In most of the region, however, the Columbia River lavas form

the hard basement rock that covers all the more ancient rocks (except the Blue Mountains in Oregon and the Wallowa Mountains in Washington, which have Paleozoic and Mesozoic remnants still exposed).

One of the most remarkable events to occur in the last 20 million years, however, was the formation of the **Basin and Range province** (Figs. 14.21C, 14.22, 14.23). The entire region from central Utah, all of Nevada, southern Arizona, parts of southern Idaho and Oregon, and the Mojave Desert of eastern California has been broken up into a series of basins and ranges that give this harsh desert region its name. The ranges trend roughly north–south and are parallel to each other and evenly spaced across the entire region (Figs. 14.21C, 14.23). The great geologist C. P. Dutton looked at a hand-drawn physiographic map of the region and said it reminded him of an "army of caterpillars marching to Mexico."

There is a good reason for this remarkable pattern of parallel ranges and basins. The crust of Nevada has undergone tremendous east–west extension in the past 20 million years, stretching to about twice its original dimensions (Fig. 14.23). The brittle crust at the surface fractures into thousands of normal faults to accommodate this extension, with hundreds of **horsts** (upthrown fault blocks) and **grabens** (down-dropped fault blocks).

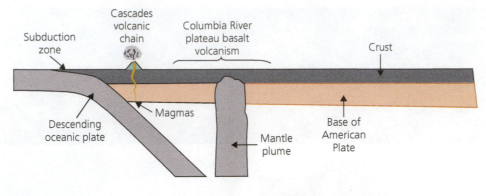

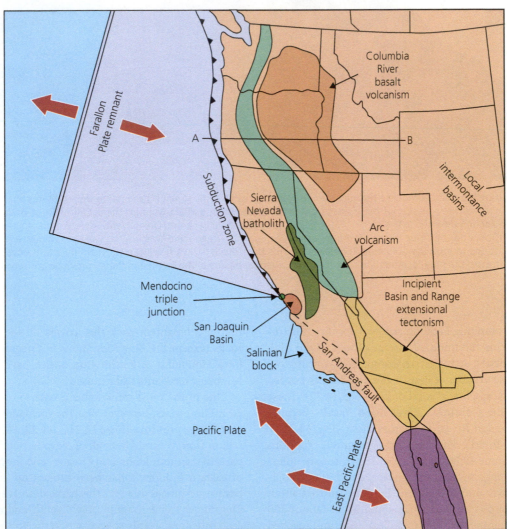

Figure 14.22 Tectonic map of the middle Miocene, showing the beginning of the modern Cascade volcanic arc in western Oregon and Washington, the eruption of the Columbia River flood basalts in eastern Oregon and Washington and down into Nevada and California, and the beginning of Basin and Range extension down in Arizona and eastern California. (Modified from Dickinson and Snyder, 1979)

As the crust pulls apart, the grabens drop down between the horsts, and sediment eroded from the ranges fills the grabens, often with thousands of meters of sands and gravels and lake deposits. The basin sediment varies depending upon where you encounter it. In the foothills of the ranges and against the fault scarps, these sediments are coarse conglomerates, arkoses, and other deposits of flash floods and debris flows. These roar out of confinement of the mountain canyons and then spread out once they reach the valley floor, where they lose energy and drop most of their coarse boulders, gravel, and sand to form alluvial fan conglomerates. In the middle of the basin, however, the fine-grained muds accumulate to form a **playa lake**. In many basins, the water floods during the rains, then evaporates completely, so many playa lakes are covered with extensive deposits of salt, gypsum, and other chemical sediments.

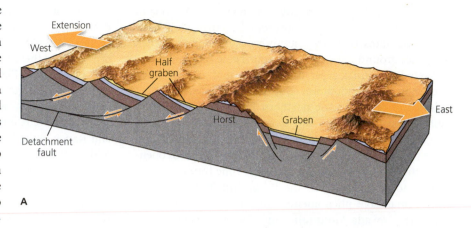

Not all the ranges in the province are simple horsts and grabens. There is a lot of strike–slip faulting in certain areas that may have contributed to the tension and shearing that produced the normal faulting. Recent research shows that most of the ranges are tilted blocks sliding down a ramp known as a **detachment fault** (Fig. 14.23). These detachment surfaces tend to form steep fault scarps near the surface but then flatten out to horizontal at depth, so the crustal blocks are sliding down a ramp like roller-coaster cars, one after another. The tilted blocks are bounded by a normal fault on one side of the valley, but the other side is formed by the dip slope of the top of the tilted block.

The region is famous for all the rising mantle heat that can be measured in many places, lots of hot springs, as well as volcanism all around (especially in the Long Valley Caldera, near the Mammoth Mountain Ski Resort and Bishop, California). The Basin and Range province is very seismically active, with active faults running from the Wasatch fault that runs right through Salt Lake City and Provo to the many faults that separate the ranges and basins to the active faults along the Sierra Nevada front in the Owens Valley of California (which last generated the enormous 1872 Lone Pine earthquake, with a magnitude of 7.9 on the Richter scale). Putting all the geophysical data together, geologists have discovered that the Basin and Range crust is extremely thin, stretched out to only 20–30 km (about 12-18 miles) thick, while most continental crust is 50–150 km (about 30-100 miles) thick. Thus, the mantle is very close to the surface in the region, which explains the high heat flow, the volcanism, the hot springs, and the continuous earthquake activity.

In addition to being tectonically active, the Basin and Range apparently opened in the south first about 30 Ma, in what is now the region around central Arizona, then spread up into Nevada by the middle Miocene (Fig. 14.22). The most actively spreading part of the Basin and Range is now in northern Nevada and southern Oregon and Idaho. This can be seen in the

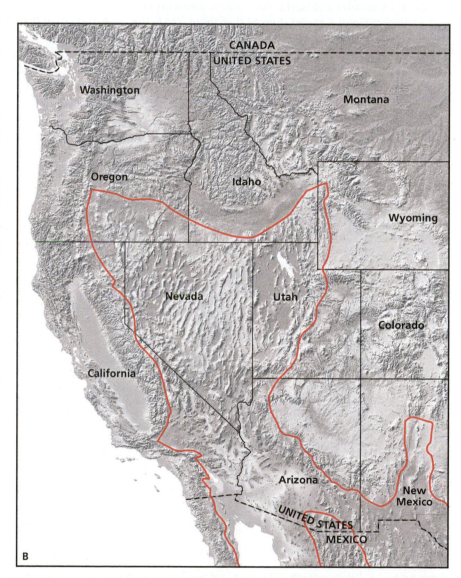

Figure 14.23 The Basin and Range province of Nevada, western Utah, and parts of California and Arizona formed during the middle and late Miocene as the crust of Nevada was stretched to twice its original dimensions. **A.** Block diagram of how the stretching was accommodated by normal faults, with the downdropped fault grabens alternating with the upthrown blocks of the ranges. Some ranges are actually half grabens that have slid down huge detachment faults. **B.** The topography of the Basin and Range province, with the distinctive trend of north–south basins and ranges.

topography of the desert region. The region in Arizona of the oldest activity is now deeply eroded, so instead of fault-block mountains, there are just tiny erosional remnants such as you see around Phoenix or Tucson. These remnants were blanketed in a thick layer of sediment that eroded down from those same mountains. But the more recently faulted valleys and ranges in the northern Basin and Range have fresh, sharp fault scarps and steep mountain fronts, with very little valley fill. Just as we saw a south-to-north shutoff of the Cascade arc, so too we can see a south-to-north opening of the Basin and Range in the same time frame.

Tied to the opening of the Basin and Range was another remarkable phenomenon: the **tectonic rotation of the Sierra Nevada Mountains and Cascades** (Fig. 14.24). In the 1970s and 1980s, paleomagnetic directions from the rocks of the Cascades and Sierra Nevada Mountains (as well as the adjacent Coast Ranges) showed that about 30 Ma the Sierra Nevada Mountains were about 210–340 km (130-210 miles) farther east than they are now. If you restore them to their original position, then Nevada would have been only half as wide as it is today, which we already knew from

the data on Basin and Range extension. So the forces that caused the Sierra Nevada Mountains and Cascades to swing to the southwest, like a door on a hinge, also explain why the crust behind these mountains got stretched out so thin. It is like a paper fan (Fig. 14.24C), with the hinge up in north-western Washington. If you swing the fan open, one side of it (the Sierra–Cascade range) is analogous to the rigid frame that holds one side of the fan steady, and the Basin and Range faulted region is similar to the way the pleated folds of the fan open up as they are stretched when the fan opens. Notice how the opening is earliest and fastest at the tip of the fan (the southern end of the Basin and Range), and this is consistent with the opening of the fan first in the south (central Arizona), then gradually to the north in the hinge of the fan (Fig. 14.24).

By the end of the Miocene and the early Pliocene (about 5 Ma), all of these trends were happening at the same time (Fig. 14.25). The Basin and Range started to open across southern Arizona and New Mexico around 30 Ma and was working its way north into central Nevada. The Cascade volcanic arc was still erupting as far south as the northern

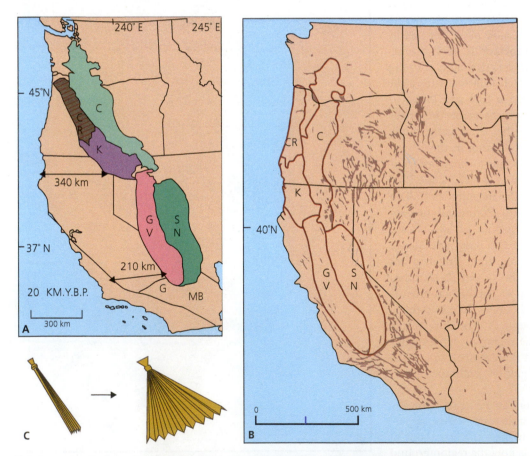

Figure 14.24 **A.** Paleomagnetic data show that the Sierra Nevada Mountains (SN) and Great Valley (GV) of California were originally much farther east about 30 Ma and have swung to the west 210–340 km (170-210 miles) around a hinge in western Washington (triangle) over the last 30 million years. **B.** Modern position of the Sierra Nevada Mountains, Great Valley, Klamath Mountains (K), Coast Ranges (CR), and Cascades (C) compared to their original positions in (**A**). The network of parallel lines shows the many faults in the Basin and Range province that have occurred as a result of this stretching of Nevada and western Utah. **C.** An analogy for the Basin and Range would be a paper fan. As one side of the fan swings open, the pleats of the fan pull apart from the tip to the hinge, so the faults at one end of the Basin and Range (the southern end in Arizona) opened first and have stretched the most, while the region near the hinge (the northern end in Nevada, Oregon, and Idaho) is just beginning to open up and stretch apart.

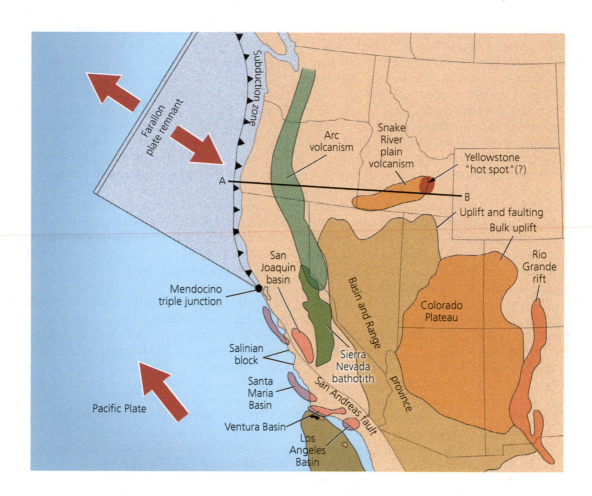

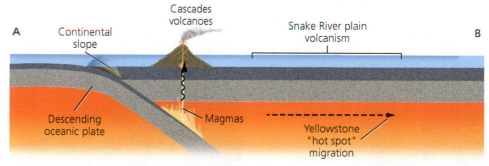

Figure 14.25 By the early Pliocene, about 5 Ma, the tectonic trends that began in the middle Miocene were in full swing. The Cascade arc was still erupting from central California to the north. The Basin and Range province continued to open to the north, stretching and ripping central Nevada apart. In addition, the crust of the Snake River Plain in southern Idaho slid over a stationary hot spot (probably the future Yellowstone hot spot), erupting the numerous lava flows of that region. East of the Basin and Range, the Colorado Plateau rose into the sky as a single block with minimal faulting and folding. In central New Mexico, the Rio Grande Valley rifted open along numerous block faults. (Modified from Dickinson and Snyder, 1979)

Sierra Nevada Mountains because the Mendocino triple junction was near San Francisco at the time. The Sierra–Cascade block had rotated farther west, although these mountains were not yet at their present position.

In addition, there were some new events that occurred about 5 Ma. A hot spot (possibly the one that created the Columbia River flood basalts at 16–15 Ma) was burning a trail beneath southern Idaho, forming the **Snake River plain**, which is floored by huge volumes of Pliocene and younger lavas. Some of those more recent eruptions can be

seen at Craters of the Moon National Monument, near Arco, Idaho. This hot spot now underlies Yellowstone National Park and its huge active caldera.

Even more impressive is the uplift of a huge crustal block known as the **Colorado Plateau** (Fig. 14.25), so named because it is drained by the Colorado River and its tributaries. This region runs from the plateaus around the Grand Canyon in Arizona to the red rock canyons of Monument Valley on the Utah–Arizona border to the spectacular cliffs of Zion, Bryce, Canyonlands, Capitol Reef, Arches, and many other national parks and monuments in Utah. The Colorado Plateau is remarkable not only as a high-elevation plateau (some areas are over 3350 m, or about 11,000 feet, above sea level) but also because its rocks (mostly flat-lying Paleozoic and Mesozoic sedimentary rocks) are relatively undeformed and continuously exposed in the desert climate of that region. But the most remarkable feature of the region is how the rivers have carved down their canyons as the region rose into the sky over the past 5 million years. As is apparent from many places (such as Goosenecks of the San Juan [Fig. 14.26] or Horseshoe Bend on the Colorado River just south of Page, Arizona), the river drainages were originally established at low elevations close to sea level because they have wide, meandering river channels just like the Mississippi or any other river close to the sea. Yet those **entrenched meanders** are today cut down into steep-sided canyons, meaning that the older meander pattern is overprinted by the "V"-shaped canyons of a region undergoing rapid uplift. This is stark proof of how quickly the plateau has arisen from being near sea level to its present elevation.

Meanwhile, as the Colorado Plateau began to rise about 5 Ma, so too did the long-buried Rocky Mountains, which stripped off their soft Miocene sedimentary blanket, elevated their flat surfaces into the sky, and entrenched their water gaps with their superposed drainages (already discussed in the section "Rocky Mountains"). Whatever was causing the Colorado Plateau to rise was having an effect as far east as the Rocky Mountains as well.

By now, most readers see all these clues to the puzzle and wonder how they fit together. What explains all these phenomena happening over this Cordilleran region at the same time? Let's look at one last piece of the puzzle: the San Andreas fault. Look at Fig. 14.8C. You will notice that the boundary between the Farallon and Pacific plates has a distinctive irregular shape, with lots of jogs back and forth caused by offset spreading ridges and transform faults. Around 30 to 25 Ma, the corner of the Pacific plate first reached the subduction zone that had been consuming the Farallon plate (Figs. 14.16, 14.22, 14.27A). At that point, the North American and Pacific plates came into direct contact, and the Farallon plate between them vanished into the subduction zone. But the Pacific plate cannot subduct beneath the North American plate because it is moving northwesterly from the East Pacific Rise, not east like the Farallon plate, as would be required for it to subduct underneath North America. Instead, since it is moving parallel to North America, it forms a transform plate boundary known as the **San Andreas fault**. An active margin that had been a subduction zone since the Triassic was now a transform boundary.

As more and more of the corner of the Pacific plate meets the North American plate, the subduction zone moves away to the north and south, and the San Andreas transform expands in its place (Figs. 14.22, 14.25, 14.27A). Once the corner began to be subducted, it would cut the Farallon plate into its two modern remnants and form the San Andreas transform boundary between them. The subduction zone would gradually vanish, as would the volcanic arc, and in its place would be a region of strike–slip motion.

But what took place behind the San Andreas transform once subduction ceased in the region? And how is it connected to the extension of the Basin and Range province or the disappearance of the Cascade Arc volcanoes? In the early days of plate tectonics, there were lots of speculative models trying to explain it. The most widely accepted model, however, was proposed by Bill Dickinson and Walter

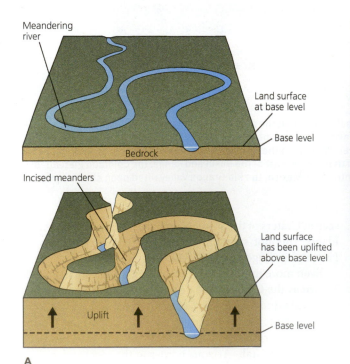

Figure 14.26 **A.** When the Colorado Plateau rose to high elevations, the old river drainages (which had been slow and meandering since they were originally near sea level) were suddenly forced to cut down rapidly through the rising plateau, incising their steep-walled canyons into the old lazy meander pattern. This overprinting of a rapidly downcutting mountain stream pattern on top of the old meandering pattern formed near the mouth of a river indicates rapid uplift of the landscape and rejuvenation of old drainages. **B.** The Goosenecks of the San Juan River in southern Utah show a classic example of entrenched meanders.

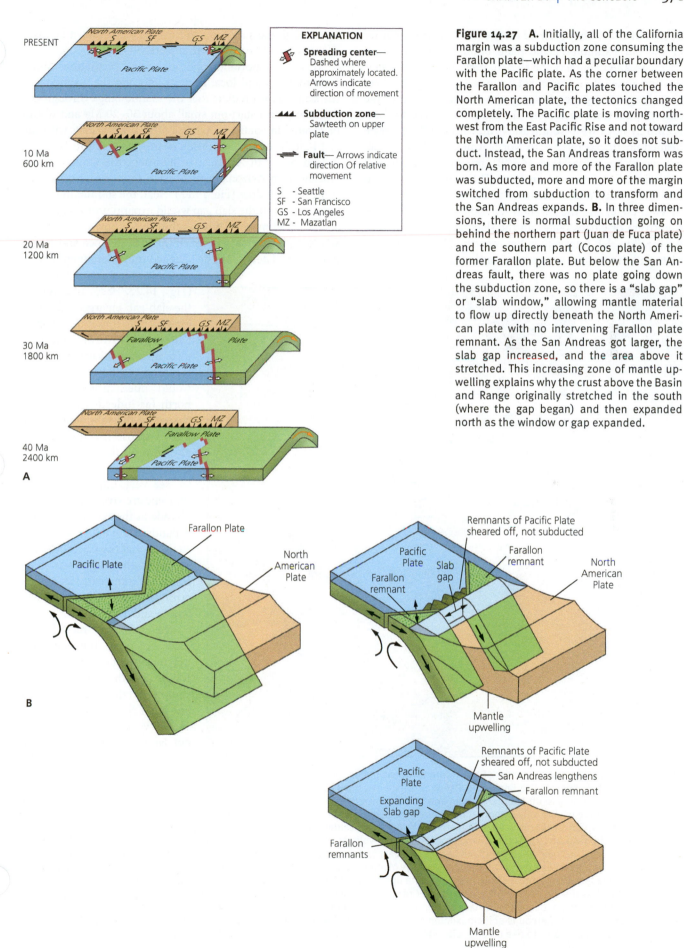

Figure 14.27 A. Initially, all of the California margin was a subduction zone consuming the Farallon plate—which had a peculiar boundary with the Pacific plate. As the corner between the Farallon and Pacific plates touched the North American plate, the tectonics changed completely. The Pacific plate is moving northwest from the East Pacific Rise and not toward the North American plate, so it does not subduct. Instead, the San Andreas transform was born. As more and more of the Farallon plate was subducted, more and more of the margin switched from subduction to transform and the San Andreas expands. **B.** In three dimensions, there is normal subduction going on behind the northern part (Juan de Fuca plate) and the southern part (Cocos plate) of the former Farallon plate. But below the San Andreas fault, there was no plate going down the subduction zone, so there is a "slab gap" or "slab window," allowing mantle material to flow up directly beneath the North American plate with no intervening Farallon plate remnant. As the San Andreas got larger, the slab gap increased, and the area above it stretched. This increasing zone of mantle upwelling explains why the crust above the Basin and Range originally stretched in the south (where the gap began) and then expanded north as the window or gap expanded.

Snyder in 1979. They focused on the fact that as the corner of the old Farallon–Pacific plate boundary (Figs. 14.22, 14.25, 14.27A) contacts the trench, it changes from a subduction zone into a transform boundary. As a result, there is no longer any plate plunging into the mantle east of the San Andreas transform fault. The motion on this transform is entirely northwest–southeast shear, with the Pacific plate headed northwest relative to North America. Thus, a "**slab gap**" or "slab window" (Fig. 14.27B) appeared to the east and behind the San Andreas transform. This region lies between the downgoing Farallon remnant to the north (now the Juan de Fuca plate) and the Farallon remnant to the south (now the Rivera or Cocos plate). In the slab gap, there would be no Farallon slab intervening between the North American plate and the mantle, so the hot mantle material could well up through the slab gap and force its way upward beneath the overlying North American plate.

This has important implications and predictions about plate motions and local geology. Plotting the geometry of the slab gap, it predicts that the mantle bulge through the window would start out small about 30–20 Ma and would be located in southern Arizona. This is consistent with the fact that the earliest extension of the Basin and Range does indeed occur in Arizona. In addition, the Oligocene arc volcanism in Arizona–New Mexico should shut off at this time, which it does (Fig. 14.22). Next, the subducting triangular corner of the slab gap would expand north and south as more and more of it was subducted. Thus, the northern subduction boundary between the Juan de Fuca and North American plates would be changed into a lengthening San Andreas transform fault zone (Figs. 14.22, 14.25, 14.27A). As the slab gap enlarged, it would expand the upwelling mantle beneath the North American plate up into Nevada and Utah, meaning the Basin and Range would open from south to north (as indeed it does). The change of the subduction zone into a transform zone would also shut off the arc volcanism from south to north. Indeed, we see very little Miocene or Pliocene arc volcanic activity in Nevada as the Basin and Range expands. More to the point, this suggests that the Cascade arc is gradually shutting down at its south end (currently its southernmost eruptive center is Mt. Shasta and Mt. Lassen).

The slab gap model predicts not only the south-to-north opening of the Basin and Range and the south-to-north shutoff of the arc volcanism but other features as well (Fig. 14.28). When the mantle bulge got large enough through the slab window to expand to the east of the Basin and Range, it apparently caused the Colorado Plateau to lift over 3500 m (11,500 feet) into the sky in the Pliocene, cutting the Grand Canyon and many other canyons in the region. Indeed, geophysical data show that there is a bulge of mantle material beneath the Colorado Plateau that is causing it to rise into the sky.

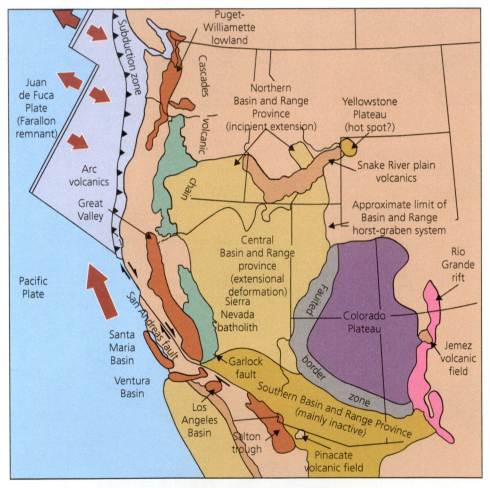

Figure 14.28 Modern tectonic map of the western United States, showing the current status of different tectonic events. The Basin and Range province has opened all the way to northern Nevada and is beginning to rift southern Idaho and southeastern Oregon. Meanwhile, the southern Basin and Range is now fully extended and largely inactive. The hot spot beneath the Snake River plain has moved so that it is now beneath the Yellowstone caldera. The Cascade volcanic arc has shut off from south to north, so there are no longer any volcanoes south of the Mendocino triple junction, and the chain begins with Mt. Lassen just east of the southern edge of the Juan de Fuca plate. The San Andreas extends all the way from the Mendocino triple junction to the Gulf of California, where it switches into a spreading ridge. The uplift of the Colorado Plateau has developed further, as has the Rio Grande rift valley, so the Grand Canyon and many other river canyons were carved to their present depth in the last 5 million years. The mantle bulge beneath the Basin and Range and Colorado Plateau is at its maximum size within the slab gap, and indirectly it has caused the Rocky Mountains to rise up and rejuvenate.

This mantle bulge would explain why the Rocky Mountains renewed their Laramide uplift and incised their canyons after being nearly buried in the Late Miocene (Figs. 14.14B, 14.19A–C).

Today, the complexity of the Cordilleran region (Fig. 14.28) is a reflection of the many different tectonic forces that built it over the past 20 million years. Both the Rockies and the Colorado Plateau continue to rise in elevation as the mantle bulges beneath them. Most of the mantle bulge through the slab gap, however, is focused on the Basin and Range, with its highly stretched crust and abundant heat flow, volcanism, and earthquakes. The Basin and Range continues to be most active at its northern end, where it is just beginning to open. The Sierra Nevada Mountains continue to rise higher in the sky and to swing west. The San Andreas continues to get larger as the Mendocino triple junction slides north and changes its northern boundary from subduction to transform. As this occurs, the volcanoes at the southern end of the Cascade arc (especially Lassen and Shasta) are doomed to go extinct in the distant future as no more subduction occurs beneath them. The Snake River hot spot has now burned its away across the landscape and now lies beneath Yellowstone National Park. In short, the entire western half of the United States is very tectonically active and changing constantly, just as it has done for the entire Cenozoic.

14.6 Cenozoic Life and Climate

LIFE IN THE OCEANS

Paleocene Recovery

At the end of Chapter 13, we discussed the great extinction event at the end of the Cretaceous. Just like the aftermath of the biggest of all mass extinctions in the Permian, the early Paleocene was also a "recovery" interval in the aftermath of an extinction which wiped out many of the dominant marine invertebrate groups (especially ammonites and many of the plankton) as well as nearly all the larger predatory reptiles, such as mosasaurs, which were the top of the food chain in the Cretaceous seas. Although ammonites vanished, the relatives of the chambered nautilus survived and are still in the South Pacific today (and they are common Cenozoic fossils). At first it seems odd that such closely related organisms as ammonites and nautiloids did not suffer similar fates. However, ammonites apparently laid hundreds of tiny, defenseless eggs (like many marine invertebrates), which might be more sensitive to extreme ocean conditions at the end of the Cretaceous, while nautiloids gave birth to a few fairly advanced young with large yolk sacs, which might have made them more resistant to the extreme environmental changes at that time.

However, many of the snails and clams of the Cretaceous survived into the Cenozoic, especially those that burrowed in the sediment and fed on detritus on the sea floor. Although their diversity was low in the earliest Paleocene with "weedy" opportunistic species and low diversity (just as in other post-extinction aftermath periods), most of the modern groups of snails and clams soon became dominant on the sea floor again. Sea urchins and heart urchins were not affected much by the extinction and recovered quickly by the Paleocene. By the late Paleocene, they had evolved to form their most specialized descendants, the sand dollars, which soon spread across the sea floors in the Eocene.

The plankton underwent a huge evolutionary radiation in the Paleocene, so that from only two or three lineages of planktonic foraminiferans that survived the extinction, more than a dozen lineages were established by the early Eocene. Only about 10 lineages of coccolithophores survived, but they also quickly radiated into dozens of genera. The giant reefs of rudistid oysters were gone, so the modern groups of scleractinian corals eventually began to build new reefs in the tropics.

There was also a huge radiation in the advanced bony fish known as the spiny-finned teleosts, which are the most common group of fish in the ocean today. Large sharks also survived, but for a long time in the Paleocene and Early Eocene, there were no large aquatic vertebrate predators to replace the extinct mosasaurs and plesiosaurs until whales appeared in the middle Eocene.

The Blast of Gas from the Past

The end of the Paleocene and the beginning of the Eocene is marked by a dramatic climatic event that affected life both in the oceans and on land. Known as the **Paleocene–Eocene Thermal Maximum (PETM)**, it has been subjected to intensive study since the early 1990s when it was discovered. Paleontologists had long known that the early Eocene was the warmest time of the entire Cenozoic. The region above the Arctic Circle was warm and mild, with temperatures averaging about 20°C (about 68°F) in the summers and fossils showing abundant temperate vegetation. Even the winter temperatures were well above freezing, despite the fact that the area experienced many months of darkness during polar winters. The swamps were warm enough for alligators, turtles, lemur-like primates, tapirs, and many extinct Eocene groups. Likewise, the early Eocene fossils of Antarctica also show dense forests with temperate vegetation and no sign of freezing in the winter. These forests were inhabited by primitive opossum-like creatures that were related to the pouched mammals of Australia and to the many kinds of opossums that live in South America today.

 See For Yourself: Arctic Jungles

But in the 1990s, the geochemical evidence from the ocean showed that a huge amount of very light carbon in the sediments and shells of plankton occurred very suddenly right around the Paleocene–Eocene boundary. Many explanations were suggested, but the volume of carbon needed and the rapidity of the event ruled most of them out. Over the last several decades, the best explanation has been the idea that huge amounts of carbon were released from tiny frozen cages of water ice surrounding methane gas (known as **methane hydrates** or methane clathrates), which even today are trapped in the spaces between subfreezing ocean sediments. If the sea bottom warmed quickly enough, it would have thawed out huge volumes of these ice cages and released enormous amounts of methane, which is an even more powerful greenhouse gas than carbon dioxide. This "blast of gas from the past" explains the peculiar pattern of the PETM: the rapid global warming all over the oceans and continents (warm enough to make the polar regions mild enough for alligators) yet no mass extinction except in the benthic foraminifera that live in and above the sea-floor sediment. After a few thousand years, most of the methane would have oxidized to form carbon dioxide, which continued to warm the planet, so the final pulse of warmth is known as the "Middle Eocene Climate Optimum." Geochemical evidence suggests that the carbon dioxide levels were about 2000 ppm then, comparable to the peak of greenhouse warming in the Cretaceous and 5 times higher than our modern levels of 415 ppm.

See For Yourself: The Paleocene-Eocene Thermal Maximum

We see the effects of this warming in all the world's oceans, with high diversity of tropical and subtropical snails and clams from such famous Eocene beds as the Claiborne and Jackson Groups in the Gulf Coastal Plain (Fig. 14.12C). Even more impressive are the tropical deposits of the remnants of the Tethys Seaway that once ran from Gibraltar to Indonesia in the Mesozoic (Fig. 13.2C). Although the approach of India in the early Eocene was gradually closing the Tethys, there were still areas with abundant tropical sea life. Across the Tethys, from the Mediterranean to southern Asia, were extensive Eocene marine limestones of the Tethys Sea, crammed with their tropical fossils. Typically, they are rich in a group of foraminiferans called nummulitids (Fig. 14.5).

In addition to a great diversification of snails, clams, sea urchins and sand dollars, and bony fish, the middle Eocene Tethys Seaway was home to another important evolutionary story: the origin of whales. Since the early 1990s, numerous fossils have come from middle Eocene Tethyan beds in Pakistan and Egypt that show the amazing transition from land mammal to the familiar creatures of our modern seas (Fig. 14.29). It all starts with a dog-sized creature known as *Indohyus*, which had long legs for leaping but ear bones typical of whales and dense limb bones suggesting that it was a good swimmer. It is a member of an extinct group of mammals that gave rise not only to whales but also to hippopotamuses. Slightly younger is *Pakicetus*, which had a wolf-like body but the skull and teeth of early whales. The most impressive is the 3-m-long (about 10 feet) *Ambulocetus natans*, whose name actually means "walking swimming whale." Its head was completely whale-like, but its limbs were long with webbed feet for swimming. Its spine suggests that it swam with up and down undulations of its body, like a large otter, although it was not a fast swimmer and probably lived on the edges of swamps and ambushed prey that came down to drink, as crocodiles and alligators do today.

See For Yourself: When Whales Walked

Next in the sequence was *Dalanistes*, which had even larger webbed feet and a more whale-like skull, followed by *Takracetus* and *Gaviocetus*, which had more reduced limbs with smaller hands and feet and were built more like

dolphins with legs. Their hind legs were so small that they could no longer walk on land. By the peak of the middle Eocene warming, there were huge whales known as archaeocetes, which were 24 m (about 80 feet) long, with a fully whale-like body weighing about 5400 kg (12,000 pounds), flippers instead of hands, and a long skull with a robust whale-like snout and sharp triangular bladed teeth. Yet they are still primitive in many ways, such as having their nostrils on the tip of their snout rather than on top of their head as a blowhole, as all modern whales have. In addition, complete specimens show that they had tiny hind limbs, about the size of your arm, which are vestiges of the days when whales used to walk. In fact, modern whales still have a tiny remnant of the hip bone and thigh bone, deeply buried and serving no purpose except as a witness of their evolution from four-legged land animals (Fig. 6.9B).

Meanwhile, in yet another part of the tropical Eocene ocean, we find the fossils of a second group of land mammals that returned to the water: manatees and their kin. Fossils of *Pezosiren portelli* from Jamaica show an animal with the same kind of skull and teeth and dense rib bones of living manatees but with four walking limbs and no flippers. By the middle Eocene, manatees and their relatives had spread from the Caribbean to the entire Tethys Seaway region. In the early Miocene, two other groups of marine mammals arose: the seals, sea lions, and walruses (descendants of a common ancestor with bears), and a group of extinct hippo-like mammals called desmostylians. In addition, there was even a ground sloth found in the Miocene beds of Peru that was fully marine.

From Greenhouse to Icehouse

The warm early and middle Eocene greenhouse world, with high carbon dioxide in the atmosphere and temperate forests and alligators on the poles, began to change rapidly at the end of the middle Eocene through the early Oligocene (Fig. 14.31). The records from both plants on land and oceanic geochemistry both show this dramatic change (Box 14.2). The first pulse occurred at the middle–late Eocene transition (37 Ma), when a dramatic cooling of oceanic bottom waters occurred, resulting in 4°–5°C of global cooling in the world's oceans (more than any change during an Ice Age) and a temperature drop of 14°–16°C on some of the continents, as indicated by fossil leaves (Box 14.2). There was a mass extinction in the tropical plankton and significant extinction in the subtropical clams and snails that lived in the shallow waters of the Gulf Coastal Plain. The enormous communities of nummulitids were virtually wiped out (Fig. 14.5).

See For Yourself: Eocene Lethal Lake

The next pulse of cooling occurred in the earliest Oligocene (33 Ma). Deep-sea sediment cores drilled from around the margin of Antarctica show clear evidence of large ice sheets then, proving that polar ice caps had returned for the first time since they'd vanished from Gondwana in the Permian (Fig. 14.31). The world's oceans dropped 5°–6°C in average temperature, a catastrophic cooling event for tropical

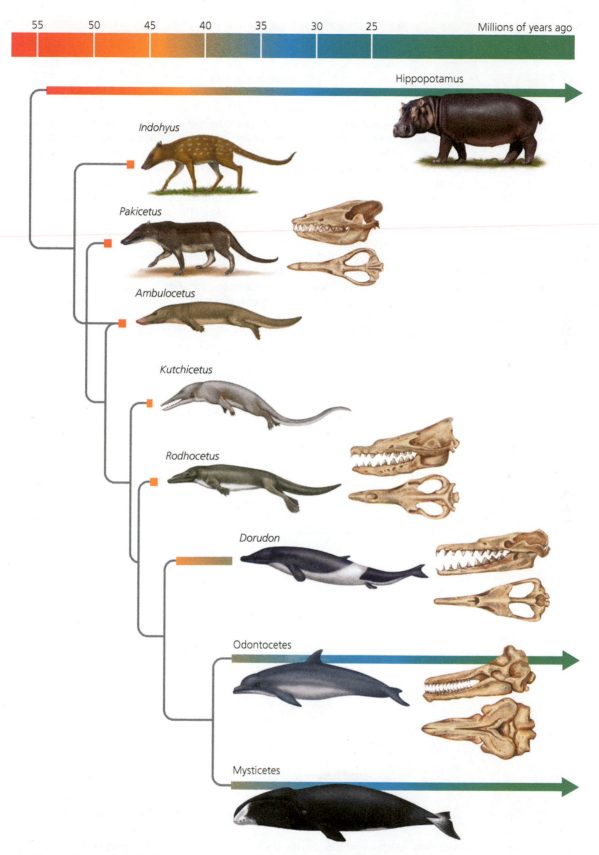

Figure 14.29   The evolution of whales from small terrestrial mammals like *Indohyus* through large land predators like *Pakicetus* to semi-aquatic predators with four webbed feet like *Ambulocetus* and *Dalanistes*, then to forms like *Takracetus*, *Gaviocetus*, and *Rodhocetus* with reduced hind legs. Molecular evidence shows that whales are more closely related to hippos among all living mammals.

BOX 14.2: HOW DO WE KNOW?

How Do We Know Ancient Temperatures?

When geologists speak about ancient temperatures, how can they make such assertions about the past? After all, they weren't there holding a thermometer. Actually, there are lots of ways that nature records temperatures of the distant geologic past, and some of these are crucial to understanding Cenozoic climate changes, as well as those of the more ancient geologic past.

For marine sediments, the most useful method for analyzing paleotemperatures is the **oxygen isotope system**. Normal oxygen atoms (which makes up over 99% of the oxygen in the earth) have 8 protons and 8 neutrons it their nucleui, so their atomic weight is 16 (thus, it's called oxygen-16 or ^{16}O). There is a rare (only 0.2% of the atmosphere and oceans) heavy isotope of oxygen known as oxygen-18 (or ^{18}O), which has 8 protons and 10 neutrons.

Because lighter oxygen is easier to evaporate, it preferentially turns into clouds rich in oxygen-16 and leaving the ocean richer in oxygen-18. When there are higher temperatures, there is more evaporation, and the ratio between oxygen-16 and oxygen-18 changes more than when temperatures are cooler. This effect was particularly true when there were Cenozoic ice caps (Fig. 14.30). Their isotopic composition was extremely rich in oxygen-16 trapped on the land, so when the planet had large ice caps, the ocean was enriched in oxygen-18 by a few parts per thousand (less than 1% but measurable nonetheless). On plots of ancient oxygen isotopes and paleotemperature (Figs. 14.30, 14.31), the more negative the oxygen isotope values (i.e., values of about 0 or 1 on the d^{18}O scale), the warmer it is, and the more positive (i.e., enriched in oxygen-18), the colder it is.

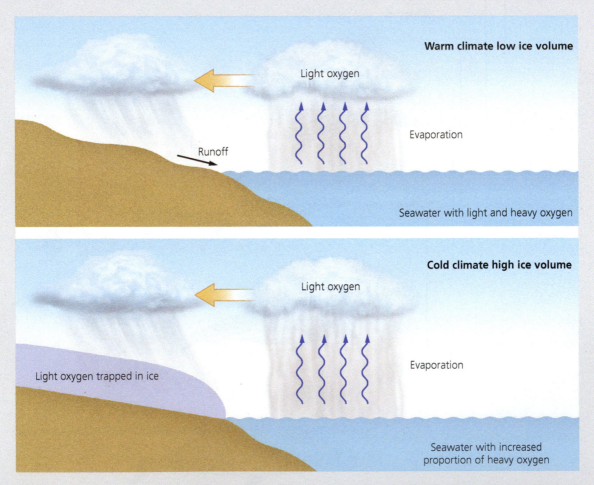

Figure 14.30   Molecules of seawater carry both the light (oxygen-16) and heavy (oxygen-18) atoms of oxygen. Water with lighter oxygen molecules evaporates more easily because it is less dense. The clouds are thus enriched in oxygen-16, and most of the ice and running water on land has similar light oxygen content. Normally, it all flows back into the sea, but when there are large ice caps on the planet, a lot of the oxygen-16-rich water gets locked up in this ice. Not only does this cause sea level to drop as water is trapped in the ice, but the remaining seawater is enriched in oxygen-18. Thus, if we look at the shells of marine organisms that made their calcite out of the chemistry of the ocean, we can tell when there were ice advances from the heavier oxygen isotopes in the water and when the ice was melting from the light oxygen isotopes in their shells.

So how do we know the oxygen isotope chemistry of the ancient oceans? Marine animals make their shells from calcium carbonate ($CaCO_3$) drawn from the carbon dioxide formed from the oxygen in the seawater, so it is the same composition as the seawater from which it formed. The most useful shells in the study of ancient oceans are the microscopic shells of foraminiferans, which are abundant in almost any marine sediment. Just a small number of shells of a key indicator species are needed to be crushed into powder and vaporized and their gases then analyzed in a mass spectrometer, which separates the isotopes by their different masses. These are the best and most reliable indicators of seawater chemistry since they are abundant, and the physiology of the amoeba-like creatures that made the shells is simple and doesn't tend to distort the ratio. You can also use large shells, such as clams or snails, for certain kinds of studies. The best labs in the country have delicate machines which allow you to analyze a scraping less than 1 mm wide and analyze each tiny growth line in a clam shell so that you can determine the seasonal temperature changes in the water from which they grew.

The result of thousands of analyses of both shells of plankton and shells of mollusks and other sea animals is a highly detailed record of ocean temperature for not only the Cenozoic but the entire Phanerozoic (Fig. 14.31). From geochemical constraints, we can precisely determine what temperature the oceans were at any given time and what the ice volume on the poles was like during periods of glaciation. Many other interesting biological and geological problems can be solved with oxygen isotopes (and the very similar system of carbon isotopes), but we will only mention paleotemperature here.

Oxygen isotopes work well with marine organisms, and sometimes they are recorded on land as well (such as in the carbonate formed in ancient soils or the chemistry of fossil bones). But the preferred method of estimating paleotemperature on land is by the shape of fossil leaves. After years of study, a number of paleobotanists have determined that the detailed shape of leaves is strongly correlated to the temperature at which they grew. Tropical leaves tend to be large and thick since they are not shed in the winter, with drip tips if they formed in rainforests (Fig. 14.32). They also tend to have smooth edges (known to paleobotanists as "entire margins"). Temperate leaves, by contrast, tend to be smaller and thinner, since they are often dropped in the fall season, and have jagged margins. When paleobotanists analyzed

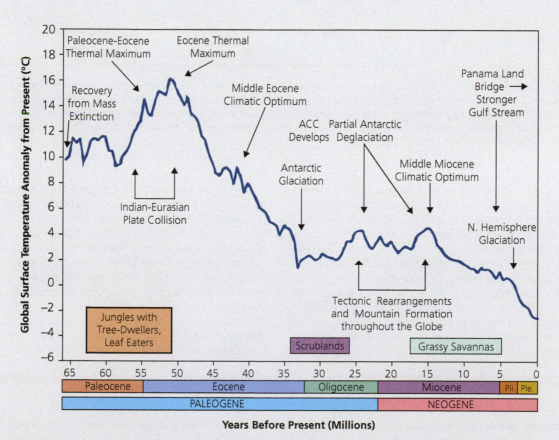

Figure 14.31 History of global Cenozoic temperatures based on oxygen isotope analyses of bottom-dwelling (benthic) foraminifera. The major events, such as the Paleocene–Eocene Thermal Maximum, the Middle Eocene Climatic Optimum, the cooling in the late Eocene, the first Antarctic glaciers in the early Oligocene, the climate changes of the Miocene, and the development of the Arctic ice cap in the Pliocene, are all shown. ACC = Antarctic Circumpolar Current.

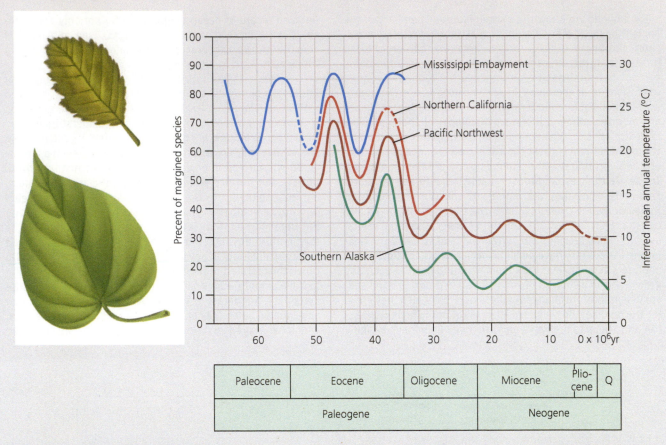

Figure 14.32   On land, the best proxy of ancient temperature is the shape of leaves. Leaves with smooth "entire" margins tend to occur in warm tropical climates, while those with jagged margins occur in colder climates. The percentage of entire-margined leaves very closely tracks the average temperature that a leaf flora was exposed to, so a plot of percent entire-margined leaves can be converted to paleotemperature, and the ancient temperature curve can be seen. The mean values for different regions from Mississippi to Alaska vary, but the shapes of the curves are very similar.

modern leaf assemblages (whose temperature and climate are known) and plotted the percentage of entire margins in the leaf species (regardless of which plant family the leaves belonged to), they got a remarkably tight correlation with the average temperature. Applying this **leaf margin analysis** to ancient leaves, they found that they could easily convert the percentage of the entire margined leaves in ancient fossil leaf assemblages into the ancient temperature that the leaves formed in. This relationship has been tested over and over again and seems to hold up well. From this, we can create a paleotemperature curve for any one region during the Cenozoic and see how its climate fluctuated (Fig. 14.32).

This observed relationship between leaf margin and temperature was established for decades before anyone knew why it occurred. Then, in a clever series of experiments, paleobotanists showed that the jagged margins are found in leaves that must grow quickly in the spring when the tree leafs out. It turns out that the jagged edge of the young growing leaf greatly expanded its surface area, which helps speed up the exchange of water and gases through its surface, so it could grow faster in the spring when this is crucial. However, many important processes in science are not often fully explained, but if we see that the process is a reliable indicator, it still helps us understand nature whether we know exactly how it works or not.

and subtropical species of plankton. In fact, the diversity of planktonic foraminifera was at an all-time low, confined mostly to species that were adapted to colder waters. Once again, the tropical snails, clams, and sea urchins so typical of the middle Eocene of the Gulf Coast, the Pacific Coast,

and many other shallow seas, were nearly wiped out with a major extinction of most of the tropical species.

In the Oligocene, archaeocete whales were replaced by the two main groups of whales that rule the oceans today: the toothed whales (such as dolphins, porpoises, and sperm whales) and the toothless baleen whales (blue whales, hump-back whales, gray whales, and their kin). These whales have a filter made of tough, fibrous, flexible baleen in their mouth to screen out and trap their tiny planktonic food (mostly crustaceans known as krill). Baleen whales feed by gulping a huge volume of seawater and plankton, then forcing the water out of their mouth cavity while the food is trapped in the baleen screen. Many scientists think that the explosive evolutionary radiation of baleen whales in the Oligocene and Miocene is due to the big increase in plankton in the Southern Ocean as the Antarctic Circumpolar Current developed. Baleen whales include the largest animal that has ever lived, the blue whale. At 30 m (about 100 feet) in length and 190 tonnes (210 tons) in weight, it is larger than any animal that has ever existed, including the biggest dinosaurs and the most monstrous of all the marine reptiles.

The Eocene–Oligocene transition is characterized by two pulses of cooling through the long interval of changing climate, culminating in an "icehouse" world by the early Oligocene, with the first polar ice caps since the mid-Permian at 290 Ma (Fig. 14.31). The effect shows up dramatically in land plants, where a cooling event of about 15°–20°C is suggested by the leaves as well (Box 14.2). What could be the cause of this dramatic shift from a "greenhouse" to an "icehouse" planet? Many ideas have been suggested, but there is no widely accepted mechanism yet.

1. *Himalayan weathering hypothesis*: The collision of India with Asia was underway in the early Eocene, and some scientists have argued that the ensuing uplift of the Himalayas would cause enormous amounts of weathering as the mountains rose into the sky. Increased weathering of soils is one of the mechanisms by which carbon dioxide is drawn out of the atmosphere and incorporated into the crust. The principal problem with this idea is that so far the evidence shows that most of the Himalayan uplift occurred in the Miocene, much too late to explain the cooling at the end of the middle Eocene. More recent analyses have suggested that the northern Tibetan Plateau may have been at high elevations about 40 Ma, but this does not explain why the biggest cooling events occurred at 37 and 33 Ma. Himalayan uplift may have been a major contributor toward the gradual global cooling in the Neogene, but it does not explain the abrupt cooling events at the Eocene–Oligocene transition.

2. *Azolla* hypothesis: Some scientists have found evidence in the cores taken from the Arctic Ocean Basin of spores produced by huge blooms of the fern known as *Azolla*, which can trap huge amounts of carbon and incorporate it into the crust as it decays. However, this event occurred in the middle Eocene, about 49 Ma, and much too early to explain the cooling at 37 Ma and again at 33 Ma. It may have started the process but was too early to contribute to the main phases.

3. *Circum-Antarctic circulation*: In our modern oceans, the enormous **Antarctic Circumpolar Current (ACC)** is the main determiner of ocean temperature and

circulation (Fig. 14.33). It is the largest and fastest of the world's ocean currents, traveling eastward and clockwise around Antarctica (as seen from above the pole) at a rate of 25 cm/second, with a volume of 233 million m³/second moving past Antarctica, more than 1000 times the flow of the world's largest river, the Amazon. It has several effects that control the modern oceans. For one thing, this current brings up huge volumes of nutrients from deeper water, producing enormous plankton blooms in the Antarctic waters that are the food for their large numbers of baleen whales, as well as penguins, fish, seals, and all other Antarctic marine life.

 See For Yourself: Ocean Circulation and Climate

Another effect is that the cold, oxygen-rich Antarctic surface water then sinks to the bottom of the ocean to become the **Antarctic Bottom Water**, the deepest of all the deep ocean water masses. It makes up about 59% of the total deep water in the ocean and flows along the deepest sea floor all the way to the North Atlantic and North Pacific. This cold, oxygen-rich current allows deep-sea life to live

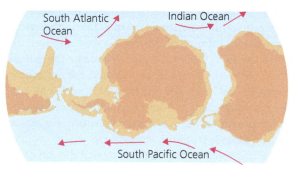

A   Middle Eocene

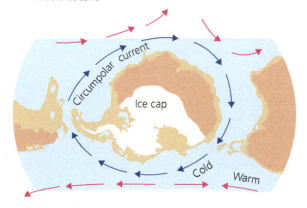

B   Early Oligocene

Figure 14.33 Today, the Arctic Circumpolar Current is the largest and fastest ocean current on earth. It not only forms the Antarctic bottom water that flows across the floor of the Atlantic and Pacific Ocean basins, but it acts as a "refrigerator door" that prevents the exchange of warm water from the tropics with the cold temperatures of the poles. Before the late Eocene, however, Australia and South America were still connected to Antarctica, preventing flow around the Antarctic. As the continents pulled away from each other, a gap between Antarctica and Australia opened, allowing the beginning of deep-water flow between these continents. Naturally, this is the time that permanent ice caps first appeared in Antarctica.

in very deep ocean bottoms, even with extremely high pressures and cold temperatures. It also dictates the flow of other deep-water currents, such as the North Atlantic Deep Water and Antarctic Intermediate Water. The deep-water currents of the modern ocean were established in the early Oligocene when the ACC developed.

But the most striking effect of the ACC is that it acts like a "refrigerator door," isolating the colder waters and air of Antarctica from warmer tropical and subtropical waters. This allowed the cold conditions of the South Pole to be "locked in" and develop permanently frozen in ice caps. During the Paleocene and Eocene, Antarctica and Australia were just beginning to separate, so there was no passage for deep water between them. Likewise, South America was also connected to Antarctica, further blocking any circum-Antarctic flow. Thus, warm tropical and subtropical currents in the Eocene could flow down all the way to the shores of Antarctica and prevent freezing. But sediments from deep-sea cores south of New Zealand clearly show that deep water was passing through the gap between Antarctica and Australia in the early Oligocene, and there is some evidence that it was also passing through the Drake Passage between South America and Antarctica (although the tectonics and timing of this region are complicated). Once the ACC was established, it became the primary reason that Antarctica froze over, and the first ice caps appeared at 33 Ma. The one limitation of this idea is that the rearrangement of ocean currents explains the drop in temperatures, but it is not a mechanism for removing carbon dioxide from the atmosphere.

4. *Impacts without impact*: One thing that did *not* contribute to the extinctions of the Eocene and Oligocene was an asteroid impact, in contrast to the event that happened in the Cretaceous. In fact, there are several well-dated Eocene impact events, one of which formed the crater that became Chesapeake Bay and another called Popigai in Siberia. They are dated at 35.5–36.0 Ma, right in the middle of the late Eocene, about 3 million years too early for the Oligocene extinction and 1.5–2.0 million years too late for the middle–late Eocene extinction. Detailed studies of deep-sea cores show almost *no* extinction at the time of the impact, except for a few species of plankton. Yet the craters these impacts produced are enormous, just slightly smaller than the Chicxulub crater blamed for the Cretaceous extinctions. This shows that impacts do not automatically cause mass extinctions. In fact, we have already pointed out that no other mass extinction horizons except for the Cretaceous extinction are associated with impacts. Back in the 1980s and 1990s, scientists got carried away with the idea that all impacts cause some kind of extinction, but today we know that most impacts have no effect whatsoever and that only the largest collision is potentially capable of triggering a mass extinction—and in the latest Cretaceous, there were also the huge Deccan lava eruptions, which complicates the story (see Chapter 13).

Thus, there are still a lot of unsolved mysteries in the "greenhouse to icehouse" transition of the Eocene and Oligocene.

THE MIOCENE: A BIT WARMER AND THEN BACK INTO ICEHOUSE

The cold, harsh conditions of the Oligocene persisted for almost 10 million years, but in the early Miocene, the planet warmed again (Fig. 14.31). This interval is often called the "Miocene Climatic Optimum" and represents a brief episode of warmer climates (about 4°–5°C warmer in the oceans than today) after the refrigeration of the Oligocene. The causes of this warming are not understood. There is good evidence of another increase in carbon dioxide, although the source of this carbon is not identified. Some geologists believe that the eruption of the huge Columbia River flood basalts 14–16 Ma might be the source of all this excess carbon dioxide, although that only explains the middle Miocene warming and not what happened in the early Miocene between 23 and 16 Ma. Climatic modeling, however, suggests that carbon dioxide levels were about 460–580 ppm, compared to 2000 ppm at the PETM or in the Cretaceous and much higher than the current level of 415 ppm. Nor is there an obvious source for excess carbon dioxide, as there was during the PETM, when the "methane burp" heated the planet into a super-greenhouse.

During the Miocene, planktonic foraminifera speciated rapidly, often evolving into forms that looked much like the tropical forms of the Paleocene and Eocene, which had gone extinct by the Oligocene. The sea floor had a diverse assemblage of clams and snails, which are today represented by the incredible shell beds in the Calvert Cliffs of Chesapeake Bay (Fig. 14.12B) and in the deposits near Sharktooth Hill northeast of Bakersfield, California. These beds, and many others around the world, yield an amazing fauna of marine creatures. At Sharktooth Hill, for example, there are at least 30 different species of shark teeth, including the huge 20-m-long (about 66 feet) giant great white shark, *Carcharocles megalodon* (Fig. 14.34). It is so big that it probably preyed on whales. There were huge sea turtles three times the size of any living sea turtle, dozens of species of primitive seals and sea lions, as well as manatee relatives. In fact, the transitional fossils that demonstrate how seals evolved from bear relatives come from lower Miocene deposits nearby and just slightly older than Sharktooth Hill. Called *Enaliarctos*, this ancestral sea lion superficially resembles a seal with webbed feet and a streamlined body, as well as large eyes, whiskers, and ears suitable for hearing underwater. However, its hands and feet were not yet fully modified into the classic flipper, and its teeth and braincase were still quite bear-like.

Most impressive of the fossils at Sharktooth Hill are the dozens of different types of whales known, from huge baleen whales and primitive relatives of modern gray whales to toothed whales and dolphins of many kinds as well as primitive whale groups that are now extinct. This shows that there was a huge evolutionary radiation of whales in the early and middle Miocene, with archaic Oligocene whales

Figure 14.34 The enormous relative of the great white shark, *Carcharocles megalodon*, roamed the oceans in the middle–late Miocene and was a main predator of whales. **A.** A giant tooth of *C. megalodon* compared to the teeth of Miocene mako sharks. **B.** Restored jaws showing the enormous size of the mouth of this shark. **C.** Life-sized model of *C. megalodon* on display at the San Diego Natural History Museum.

living side by side with new groups that have living descendants. There are even fossils of land mammals such as mastodonts, rhinos, and camels, which must have washed out to sea before being buried in deep water.

But these warm waters and good times in the early and middle Miocene came to an end about 14–15 Ma, when the Miocene Climatic Optimum was followed by an abrupt cooling event worldwide in the late Miocene (Fig. 14.31), known as the Middle Miocene Climate Transition. Over the course of the middle–late Miocene, Pliocene, and Pleistocene, carbon dioxide values dropped from 460–580 ppm to about 280 ppm as the world slipped into Ice Ages. The exact cause of this big drop in carbon is not clear, but it does coincide with the final phase of uplift of the Himalayas and the Tibetan Plateau, so maybe the increased weathering caused by this renewed episode of mountain uplift drew a lot of carbon dioxide out of the atmosphere. Whatever, the cause, it had dramatic effects. The large Antarctic ice sheets of the Oligocene, which were almost gone in the early and middle Miocene, grew to nearly their modern size, so the modern Antarctic ice cap has been on this planet for 14 million years.

See For Yourself:
The Pliocene

By the late Miocene, the cooling trend and drop in carbon dioxide (Fig. 14.31) had reached the point where glacial advances drew huge volumes of seawater out of the oceans and caused big decreases in global sea level. One of these sea-level drops triggered the Messinian salinity crisis (Box. 14.1) at the end of the Miocene, when the entire Mediterranean dried up into a gigantic deep basin full of salts, like an enormous version of the Dead Sea. Apparently, the withdrawal of all this salt from the ocean also had triggering effects on global climate because the Pliocene climate was notably cooler than that of the late Miocene, and it was much drier on land as well (causing extinction of many Miocene land mammal groups). Some of this cooling may also be due to the continued decrease in atmospheric carbon dioxide as the Himalayas

and the Tibetan Plateau uplifted even farther, hastening the weathering of rocks into soils that absorb carbon dioxide.

See For Yourself:
When Rodents
Rafted the Ocean

One crucial event occurred in the late Pliocene, about 2.8–3.0 Ma, which had global effects. Before the late Pliocene, South America had been an island continent, completely isolated from the rest of the world since the Late Cretaceous by a broad Caribbean Sea. In this isolated state, it had developed its own unique endemic community of mammals, including opossum relatives that evolved to look like wolves, hyenas, saber-toothed cats, plus huge ground sloths, gigantic armadillos called glyptodonts, and native hoofed mammals that convergently evolved to resemble camels, horses, rhinos, hippos, mastodonts, and many other animals of other continents. By the middle Eocene, it also acquired early primates and rodents, which rafted over from Africa, where their closest relatives still live today. Once they docked in South America, they evolved into the New World monkeys (spider monkeys, howler monkeys, marmosets, and their kin) and the caviomorph rodents (porcupines, capybaras, guinea pigs, and hundreds of their relatives). By 10 Ma, we find the first fossils of raccoons and coatimundis from North America, along with the first mastodonts, tapirs, peccaries, and several other North American groups. Meanwhile, a few southerners, like capybaras, giant armadillos, and ground sloths, showed up in North America in the early Pliocene. These must have swum or rafted across the short distances of seawater that separated Colombia from Central America.

See For Yourself:
South American
Mammals

Then at 2.8 Ma, the connection between North and South America was finally completed with the formation of the **Panama land bridge**. This allowed a huge exchange of animals from both sides, an event called the "**Great American Biotic Interchange**" (Fig. 14.35A). Although a few South American groups, like anteaters, porcupines, more ground sloths, and terror birds reached Florida and Texas, most of

A

Figure 14.35 A. During the Pliocene, the Panama land bridge finally closed and mammals migrated between the two Americas. Most of South America's native mammals became extinct, although a few groups like ground sloths, armadillos, capybaras, and terror birds successfully reached North America. Meanwhile, a much larger invasion of North American mammals (mastodonts, horses, camels, deer, dogs, cats, bears, raccoons, tapirs, and many others) invaded South America and established most of its modern land mammal community. Today, the typical mammals we associate with Latin America (jaguars, tapirs, llamas and other American camels, peccaries, dogs, bears) are actually late invaders from North America. (Drawing by M.P. Williams). **B.** The world's ocean currents are connected into a huge "global ocean conveyer belt" that carries water from the surface of one ocean to the next and then back via deep-water currents. These currents are so slow that a single water molecule may take 2000 years to complete the loop. One key part of the conveyer belt is the Gulf Stream, which brings warm tropical waters up from the Gulf of Mexico and Caribbean to northern Europe, keeping that region mild and wet. However, before the closure of the Panama land bridge, most of that warm tropical water leaked out into the Pacific. When Panama blocked this flow, it funneled moisture to the Arctic, allowing the Arctic ice cap to grow for the first time.

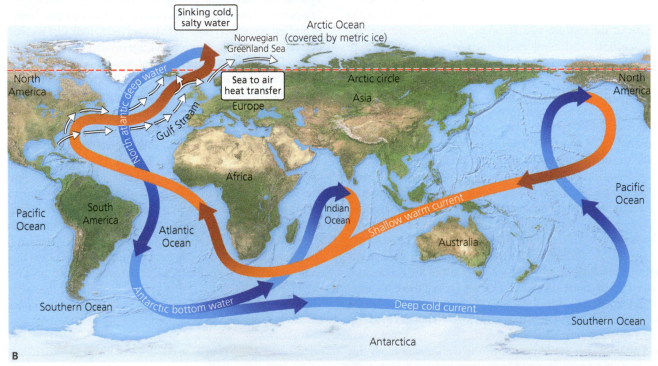

B

the migration went south. Mastodonts plus horses, camels (ancestors of modern llamas and alpacas), deer, dogs, cats (including saber-toothed cats), bears, skunks, weasels, and many others flooded to the south, driving nearly all the native South American mammals to extinction.

▷ See For Yourself: The Terror Birds

Meanwhile, the closure of the Panama land bridge had another effect as well. Before it happened, there was an open marine connection between the Caribbean and the tropical Pacific oceans, and most of the fossil mollusks and sea urchins and foraminifera were the same. But what was a gateway for land animals became a barrier for marine life. After the Panama land bridge closed at 2.8 Ma, there was no longer any marine connection between the marine life of the two oceans. Thus, the marine invertebrate species of the Caribbean began to evolve in a different direction from those in the Pacific, so today there are few species in common.

The closure of Panama and the separation of the Caribbean from the Pacific had another surprising side effect. Today, the flow of the **Gulf Stream** is part of what is called the "**global oceanic conveyer belt,**" which transports deep-ocean water in a big loop around the globe (Fig. 14.35B). The closure of Panama blocked the leakage of warm Caribbean water into the Pacific, and diverted the full force of the warm tropical waters of the Gulf Stream up into the North Atlantic, where it provides not only warmth but moisture for the atmosphere. As it does so, it supplies moisture to the Arctic, which is naturally cold due to the low amount of solar radiation it gets all year. Remember, above the Arctic Circle, it is dark through much of the winter! All the North Pole needs to have an ice cap is a source of moist clouds that can form snow and ice. The geologic evidence from cores in the Arctic Ocean shows that this occurred about 3.5 Ma, coincident with the closure of the Panama gateway and the diversion of Caribbean waters from the Pacific to the Atlantic. Thus, the closure of the Panamanian oceanic gateway apparently produced the Arctic ice cap, just as the development of the ACC produced the Antarctic ice cap. By the latest Pliocene and the beginning of the Pleistocene at 2.6 Ma, the world was in the grip of the Ice Age cycles on both polar ice caps, which is discussed in detail in Chapter 15.

14.7 Cenozoic Land Life

When the Late Cretaceous extinctions removed the non-bird dinosaurs from the landscape, mammals had lived in their shadows for about 130 million years. Suddenly, there was all this ecological space with no occupants, and naturally mammals responded with one of the most remarkable evolutionary explosions ever documented (Fig. 14.36). Since most mammals were cat-sized or smaller, there was a tremendous opportunity for them to evolve into larger forms and exploit the niches left vacant by the dinosaurs. The roots of the mammal radiation were already in place during the Late Cretaceous, with very primitive hoofed mammals called *Protungulatum*, primitive carnivorous mammals such as *Cimolestes*, primitive lemur relatives called *Purgatorius*, and lots of different kinds of insectivorous mammals. From these and just a handful of other lineages, mammals evolved at an astounding pace so that by 50 Ma in the middle Eocene there were dozens of different orders of mammals: rodents, rabbits, primates, sloths and anteaters, carnivores, insectivores, aardvarks, elephants, manatees, even-toed and odd-toed hoofed mammals, and

many others, some of which are now extinct. They not only occupied all the small-, medium-, and large-bodied herbivore, insectivore, and carnivore niches on land but had evolved into the first flying mammals (bats) and aquatic forms like the primitive manatees and whales (Fig. 14.29).

 See For Yourself: Cenozoic Land Life

But mammals were not the only creatures that ruled the planet in the early Cenozoic. In fact, it might be said that dinosaurs still ruled the earth because among the largest predators in the Eocene were huge "terror birds" known as *Gastornis* or *Diatryma*, found in North America, Europe, and Asia, with a different family of terror birds in South America. These huge birds were up to 2 m (6.6 feet) tall and flightless, but they had powerful hind legs for kicking and grabbing their prey and thick beaks for crushing and disemboweling them. The absence of large mammalian predators in the Paleocene and early Eocene gave other reptiles a chance to take over that role. In the Paleocene swamps of Colombia, there were gigantic anacondas known as *Titanoboa* that reached the length of a school bus, and turtles over 3.3 m (about 11 feet) long with jaws powerful enough to eat crocodilians as well as fish and mammals. Large crocodilians and smaller pond turtles are common in nearly all the Paleocene and Eocene beds from the tropics to the poles, another sign of warm swampy conditions and the absence of other large semi-aquatic predators.

PALEOCENE–EOCENE JUNGLES

The world of the Paleocene and Eocene was a "super-greenhouse" climate, especially during the PETM. For that reason, we find abundant plant, mammal, and reptile fossils from this time in river and floodplain deposits in the Laramide basins (Fig. 14.14), such as the Bighorn and Wind River Basins of Wyoming, Williston Basin of North Dakota and Montana, and the San Juan Basin of New Mexico. These fossils demonstrate that nearly the entire globe was covered by dense jungles like those of Central America today (Fig. 14.37). Places like Wyoming and Montana, which today are dry grasslands in the summer and deeply frozen in the winter, were subtropical rainforests in the Paleocene and early Eocene.

These jungles were inhabited by an assortment of primitive mammals, most of which fit into two broad categories: small tree-dwellers, and larger ground-dwelling leaf-eaters (Fig. 14.37). Among the tree-dwellers were the archaic relatives of the lemurs and other **primates**, which were by far the most common mammals of the Paleocene. Like modern lemurs, they had long snouts, forward-facing eyes, and a more primitive body structure than the more specialized monkeys and apes that came later. Many of them had a squirrel-like body form because they occupied the squirrel niche before rodents had even appeared. Another tree-dwelling group were the **insectivorous mammals**, which were the most common mammals of the Cretaceous and continued their diversity in the early Cenozoic. Finally, the jungles of the Paleocene and Eocene were also inhabited by an extremely primitive group of mammals known

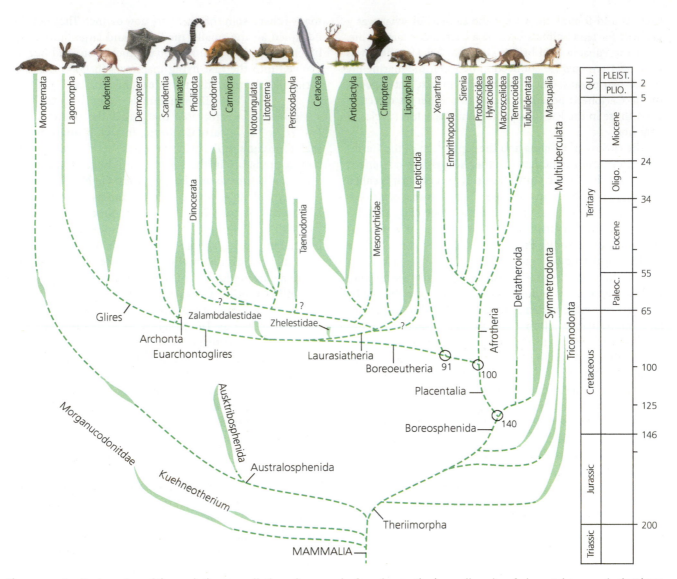

Figure 14.36 Modern view of the evolutionary radiation of mammals, focusing on the huge diversity of placental mammals that have evolved since the beginning of the Cenozoic 65 Ma. In the early Paleocene, there were just a few groups of land mammals, most smaller than a cat and all very primitive and generalized in body form with insectivorous or omnivorous diets. By the middle Eocene, about 50 Ma, nearly all the modern groups of mammals had evolved, including forms as diverse as whales and bats.

as "multituberculates," which were egg-laying primitive mammals related to the platypus but built like squirrels (see Chapter 13).

Moving through the dense jungle undergrowth were many groups of archaic mammals, which had simple teeth for eating leaves and fruits and were not very specialized for running or any other advanced ecology. These were mostly archaic hoofed mammals of a variety of families that ranged from the size of a small dog to the size of a sheep. Some were closely related to the horse–rhino group of odd-toed hoofed mammals, or **perissodactyls**. Others were related to the even-toed hoofed mammals, or **artiodactyls**. But most were extinct groups that vanished in the Eocene without leaving any descendants.

Preying on this wide range of herbivorous mammals was an assortment of mammals, as well as predatory birds

and reptiles. Since they lived in dense jungles (Fig. 14.37), nearly all the predators relied on stealth and ambush to catch their prey and did not have to be fast runners or agile tree-climbers. Most of the predatory mammals were members of an archaic extinct group known as **creodonts**, which vaguely resembled lions and wolverines and weasels but had few specializations for running and tended to be small-brained (Fig. 14.37). Eventually, they were outcompeted by the mammals of the order Carnivora, or **carnivorans**, which includes almost all the flesh-eating mammals alive today (dogs, cats, weasels, bears, skunks, raccoons, hyenas, mongoose, seals, sea lions, walruses, and all their extinct relatives). But the earliest Carnivora were quite weasel-like or mongoose-like, and none of these more specialized descendants had appeared yet. There were even two groups of archaic hoofed mammals which switched to a

Figure 14.37 Diorama of typical mammals and vegetation of the middle Eocene of North America. The largest animals were the bizarre uintatheres, with six knobby horns on their heads and big tusks in their mouth. The forest floor is dominated by a variety of primitive leaf-eating, hoofed mammals, while lemur-like primates climb the trees. The predators are members of archaic extinct groups like the creodonts, along with crocodilians that lived in the marshy areas.

meat-eating diet, so a much more diverse array of carnivorous creatures was around in the Paleocene, compared to the single order Carnivora today which includes all meat-eating placental mammals.

Most of these groups, such as the creodonts, small Carnivora, archaic hoofed mammals on the ground, plus lemur-like primates, insectivores, and multituberculates in the trees, continued to dominate in the early Eocene. But a number of new groups appeared which gradually pushed out the archaic groups. These included the odd-toed perissodactyls such as the earliest relatives of horses, rhinos, and tapirs. These animals are called "odd-toed" because they have one or three fingers or toes on their hands and feet, and the axis of symmetry of their foot runs through the middle finger/middle toe. Their relatives first appeared in east Asia in the late Paleocene, then spread across the northern continents in the early Eocene when the warm Arctic region connected all the landmasses.

Also appearing in the early Eocene were the even-toed artiodactyls, including some of the first relatives of pigs, camels, and deer-like creatures, although most were archaic extinct groups with no living descendants. They are called "even-toed" because they have either two or four toes on their feet and hands, and the axis of symmetry of the hand/foot runs through the middle finger/toe and the fourth finger/toe.

Finally, a major addition to Eocene mammal faunas was the first **rodents**. These first appeared in east Asia in the late Paleocene, then spread across the entire northern hemisphere in the early Eocene. By the end of the Eocene, they may have helped drive the archaic egg-laying multituberculates (which were also squirrel-like in body form) to extinction, along with the rodent-like primates.

The Eocene jungles also had some extraordinarily large mammals. Perhaps the most remarkable were the rhino-sized **uintatheres**, which had huge hoofed bodies and huge heads with three pairs of knobs on the top of the skull and enormous canine tusks protruding below the jaw (Fig. 14.37). Despite their size, they had remarkably small and simple leaf-chopping teeth, and they vanished at the end of the middle Eocene.

PARADISE LOST: THE EOCENE–OLIGOCENE TRANSITION

The dense Eocene tropical "greenhouse" world cooled dramatically (Fig. 14.31) at the end of the middle Eocene (37 Ma) and again in the earliest Oligocene (33 Ma). In the previous section we saw the effect that the transition from Eocene greenhouse to Oligocene icehouse climates had on the marine realm, especially with the first appearance of the Antarctic ice sheet in the early Oligocene.

The effects on land were more subtle. The loss of the great jungles and forests of the Eocene decimated the tree-dwelling groups like lemur-like primates, which vanished completely from North America and Asia and barely held on in Europe. Other tree-dwellers, such as the squirrel-like multituberculates, vanished completely. So did a number of archaic hoofed mammals that were adapted to be leaf-eaters and jungle-dwellers. The most dramatic effects, however, occurred not in the mammals but in everything else. Ancient soil horizons from the Big Badlands of South Dakota show that the late Eocene forests gave way to an open scrubland, with sand dunes in places (Figs. 14.20, 14.38). The land snails of the late Eocene were large-shelled forms typical of tropical jungles, while in the early Oligocene they were small, thin-shelled forms adapted for drought. The late Eocene reptile assemblage was dominated by pond turtles and alligators, which vanished from North America in the Oligocene, only to be replaced by dry-land tortoises (Fig. 14.20).

In Europe, the high Eocene sea levels meant that most of the region was an archipelago of small islands, inhabited by bizarre and unique species of rodents, perissodactyls, and artiodactyls that were found nowhere else. But the big drop in sea level when the Antarctic ice cap grew in the early Oligocene and pulled water out of the ocean basins suddenly exposed land between the European islands and created a land connection after millions of years of isolation. Invading groups from Asia, such as rhinos, the ancestors of hippos, primitive pigs, primitive deer-like ruminants, plus more advanced carnivorans and a variety of rodents including primitive beavers and hamsters, pushed most of the European natives to extinction. Since 1910, European paleontologists have called this the *Grande Coupure* ("great break" in French) since the late Eocene and early Oligocene mammals are so different. The changes, however, were due to a massive immigration event of Asian mammals replacing European natives, rather than directly due to climate change.

In Asia, most of the archaic groups that dominated the Eocene vanished, and in their place was a world dominated by very large mammals (including gigantic rhinos that weighed 20 tons, larger than the largest elephant or mammoth; Fig. 14.39) or by lots of rodents and other small mammals. Ancient plant and pollen fossils suggests that Asia became a semi-arid scrubland, so there were only habitats for small burrowing mammals and very large mammals that could eat off tall treetops.

See For Yourself:
The Big Cats

By the late Oligocene–early Miocene, the surviving mammal groups include those that we still find today. In North America, the odd-toed hoofed mammals included horses, rhinos, and tapirs, while the most important even-toed hoofed mammals were camels, primitive deer-like ruminants, pig-like peccaries, and a few archaic groups that are now extinct. The first true dogs appeared in the late Eocene, along with a group of "false cats" or nimravids, which convergently evolved to mimic the cat-like and saber-toothed body form, even if they were not closely related to cats. These nimravids vanished in the early Miocene, leaving a "cat gap" with no cat-like creatures until the middle Miocene, when the first true cats appeared. Other predators, like bears, raccoons, and weasels, were also evolving at this time.

MIOCENE SAVANNAS

By the middle and late Miocene, the modernization of the mammalian assemblages was nearly completed, and most of the familiar families of land mammals were in place. The vegetation, meanwhile, had adapted for the colder and drier

Figure 14.38 During the late Eocene and early Oligocene, the climate got dramatically cooler and drier in places like the Big Badlands of South Dakota. Scrublands with patchy forests replaced the dense Eocene jungles, and many of the typical Eocene groups vanished. In their place were some of the earliest relatives of horses (extreme right center), camels (white and brown animals in dead center), hornless rhinos (in wash on left and running across the center and right), tapirs (lower left), deer-like animals (extreme lower right corner), peccaries (next to the rhinoceros lying in the wash), and true dogs, along with the last members of archaic groups such as the huge rhino-like brontotheres (big beasts in the upper right), "false cats" (in the bush in the lower right), and the archaic creodont predators (center right). The huge pig-like beasts battling in the center are the enetelodont *Archaeotherium*, which is only distantly related to pigs and other pig-like mammals.

climates of the Miocene, so enormous areas of North America, Africa, and Eurasia were covered in savanna grassland (Fig. 14.40), much like the Serengeti plain in East Africa today. In Africa and Eurasia, this Miocene grassland was dominated by the distant relatives of most of the modern groups that still live there, including mastodonts, rhinos,

hippos, primitive short-necked giraffes, a variety of primitive antelopes and cattle, warthog relatives, horses related to zebras, cats of many sizes, dogs, hyenas, and other relations of the modern Serengeti mammal assemblage.

North America was also covered by grasslands, especially in the Great Plains from the Dakotas to Texas. However,

Figure 14.39 The largest mammals of the Oligocene were the enormous horn-less rhinoceroses known as *Paraceratherium*, which dwarfed the modern elephant. They were descended from the small long-legged rhinoceros standing below this reconstruction.

only distant relatives of pigs that occupied the pig-like niche in the New World. The list goes on and on, but it shows the power of nature to mold animals to fit preexisting niches by convergent evolution, even as they originated from entirely unrelated stems. And the predators were very different as well. Until 11 Ma, there were no true cats in North America, so the role was played by a group of "pseudocats" related to dogs. Nor did North America ever have hyenas, so their role was played by a group of specialized bone-crushing dogs that could scavenge carcasses like hyenas do. Then true cats finally appeared in the late Miocene, and the predatory landscape was complete.

 See For Yourself: Bone-Crushing Dogs   At the end of the Miocene, the abrupt climate change caused by the pulse of cooling and glaciation plus the global side effects of the drying of the Mediterranean (Box 14.1) caused a significant extinction in many groups of land mammals, especially in North America. Rhinos vanished completely after being the largest land mammals in North America through most of the interval between 5 and 50 Ma, as did musk deer and a number of archaic groups of deer-like hoofed mammals and certain groups of rodents. Most of the huge diversity of horses and camels and peccaries vanished, to be replaced by just a few lineages that survived through the colder, drier steppe climates of the Pliocene until the Pleistocene. This was the fauna that survived to inhabit North America during the Ice Ages (Chapter 15).

FAMILY TREES

The effects of evolution and climate change can be seen not only in the overall community of mammals but also in the individual histories of some of the major families of mammals. Since mammals have an incredibly complete fossil record compared to any other vertebrate group (such as dinosaurs or birds), they are good examples of how evolution works in the real world.

Horses are the most familiar example, having been used to demonstrate the evolutionary history of groups since the early fossils were found in the 1880s. Most of the versions that are published in textbooks, however, are grossly out of date. They show horses as a single lineage evolving through time, from the tiny beagle-sized early Eocene horse (once called *Hyracotherium* but now *Protorohippus* and several other genera) with four fingers on its hand and three toes on its foot to Oligocene horses like *Mesohippus* and *Miohippus*, which were three-toed with reduced side toes, to more advanced horses with only a single toe and no side toes, like modern horses (Fig. 14.41A). There is also a trend from the simple low-crowned, low-cusped

instead of the familiar groups found in Africa today, their niches were taken by "mimics" from native North American families that convergently evolved to fill the ecological niches of different groups of animals in Africa. Thus, instead of elephants, North America had primitive mastodonts (which arrived from the Old World about 18 Ma). Instead of hippos, there was a rhinoceros built like a hippo, complete with the stumpy legs and barrel-shaped body, living in the rivers and ponds of the Americas. There were also rhinoceroses that resembled African rhinos except they were hornless. Instead of giraffes, one group of American camels evolved to have extremely long legs and necks and performed the giraffe role. American camels also evolved to fill the roles of antelopes and gazelles. Instead of African antelopes (which are related to cattle), America had many different kinds of pronghorns, which are not related to true antelopes at all but are a uniquely American family. At one time in the Miocene there were many different lineages of horses that performed the roles not only of zebras but of many other African grazers (Fig 14.41). Instead of pigs (which were native to the Old World), North America had peccaries or "javelinas," which are not pigs but

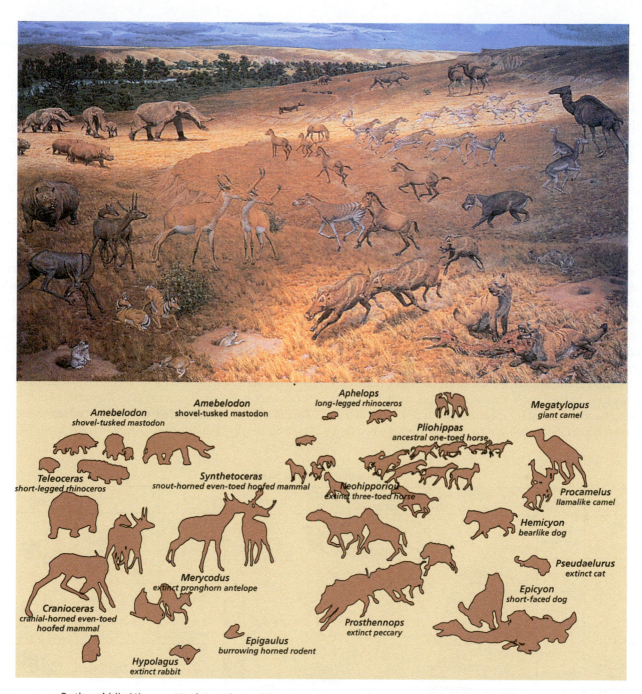

Figure 14.40 By the middle Miocene, North America and Eurasia were covered by savanna grasslands similar in many ways to the modern East African savanna. Many of the prey species had adapted for life in the open grassland by developing long legs for running, and some of their predators were also adapted for running. In North America, the native families replaced the roles played by the African savanna mammals. Mastodonts replaced elephant, camels developed into the shapes of giraffes and gazelles, horses were very diverse and occupied the roles of antelopes and zebras, pronghorns replaced true antelopes like those of Africa, peccaries replaced pigs, rhinos played not only their own roles but also sometimes mimicked hippos, and bone-crushing dogs took the roles of hyenas. This diorama shows many of these mammals roaming the grasslands of the middle Miocene.

teeth of early Eocene horses through horses with higher- and higher-crowed teeth that could wear down through a horse's lifetime as it ate gritty grasses that ground down its teeth but never run out of crown. However, since the 1920s when these diagrams were published, we have found dozens of more horse genera and species, and it turns out that horse evolution is not a single lineage through time

but bushy and branching with multiple lineages that overlapped in time (Fig. 14.41B). For example, in the Oligocene badlands of Wyoming, there are three species of *Mesohippus* and two species of *Miohippus* found in the same beds. One late Miocene fossil quarry in Nebraska yields the remains of 12 different species of horses, all of which must have lived in the same place at the same time.

Some of this Miocene diversity is caused by a split between a group of horses called anchitherines, which maintained the primitive low-crowned teeth for browsing leaves, even as some of them got as large as living horses. They lived alongside multiple lineages of horses that evolved high-crowned teeth for eating gritty grasses (Fig. 14.41B). The extinction at the end of the Miocene wiped out nearly all the three-toed horse lineages, leaving only the lineage that evolved from *Dinohippus* to the modern one-toed horse genus *Equus*. That genus contains not only our domestic horse and its wild ancestor, but all the species of zebras, donkeys, and asses as well.

▷ See For Yourself: Horse Evolution
When early Eocene horses began to evolve from the Paleocene common ancestor of perissodactyls, they were primitive creatures that looked nothing like a modern horse. Only a few key anatomical features and an excellent fossil record of intermediate forms between modern horses and their ancestors (Fig. 14.41) demonstrate this amazing transformation of their body form from a small, three-toed creature the size of a beagle to the modern horse. Even more impressive is the fact that the earliest Eocene tapirs and rhinos, which are found in the same beds with the earliest horses, look almost identical in their skulls, skeletons, and most of their teeth. Only a few small details allow an expert paleontologist to tell them apart when they

began to diverge and evolve in different directions. Yet even a toddler knows the difference between a modern horse and a rhinoceros.

From these roots in extremely similar forms in the early Eocene, all three lineages evolved in amazing directions. Tapirs showed a wide variety of body forms, but already by the middle Eocene they had developed their short proboscis and their characteristic leaf-chopping teeth. Even more remarkable was the evolution of rhinoceroses (Fig. 14.42). All the early rhinoceroses lacked horns but still had the distinctive teeth and a few other features that are unique to rhinoceroses. One Eocene family of rhino relatives evolved into huge hippo-like creatures in the Oligocene with large flaring tusks for combat. A second family developed very long, slender limbs and toes early in their evolution and became specialized for running. Even when they evolved into the enormous indricothere rhinos of the Oligocene of Asia (Fig. 14.39), they retained their long toes as a legacy of their running ancestry, rather than the short, stubby toes of other huge animals like elephants and dinosaurs. The true rhinoceroses (family Rhinocerotidae) include not only the five living species but most of the fossil species as well. The early ones were hornless, but in the late Oligocene two different lineages first experimented with a pair of horns side by side on the tip of their noses. (Rhino horn is made of hairs

The evolution of the horse.

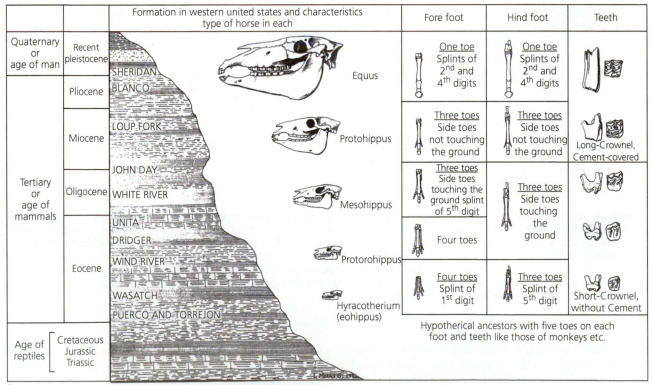

Figure 14.41 The evolution of horses. **A.** The old 1925 simplified version of horse evolution, showing a single lineage changing through time. The early horse fossils showed how they gradually lost their side toes and developed one big central toe in the foot, longer toes and limbs for running, and higher-crowned cheek teeth for eating gritty grasses.

(Continued)

glued together, not bone, so we can tell the size and placement of horns in rhinos by a roughened area on the skull where they once attached.) Later groups of rhinos developed a single horn on their nose (like living Asian rhinos and woolly rhinos), a huge single horn over 2 m (about 6.6 feet) long on their forehead, a tandem set of horns on nose and forehead (like living African rhinos). Many were hornless, others were dwarfed, and some evolved to look like hippos with stumpy legs and barrel chests. Once they vanished from North America at the end of the Miocene,

only the Eurasian and African lineages were left; and these are nearly all extinct in the wild thanks to poaching for their horns (which is more valuable ounce for ounce in China and Vietnam than gold or cocaine, even though it has no medicinal value).

▷ See For Yourself: Are Rhinos Dinos?

▷ See For Yourself: Rhino Pompeii

In addition to the amazing evolution of these perissodactyls, there are many excellent examples of evolution in artiodactyls. Perhaps one of the least

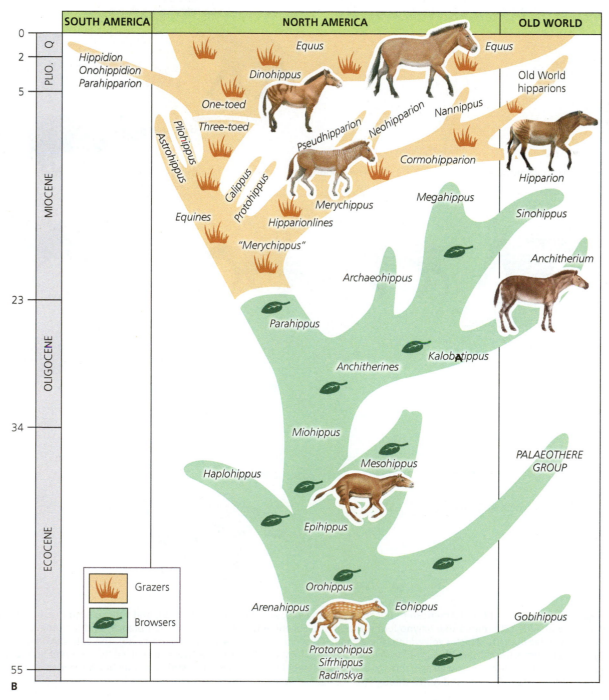

Figure 14.41   B. The modern version of horse evolution, with the improved number of new specimens and new species showing that horse evolution is branching and bushy, with many lineages coexisting.

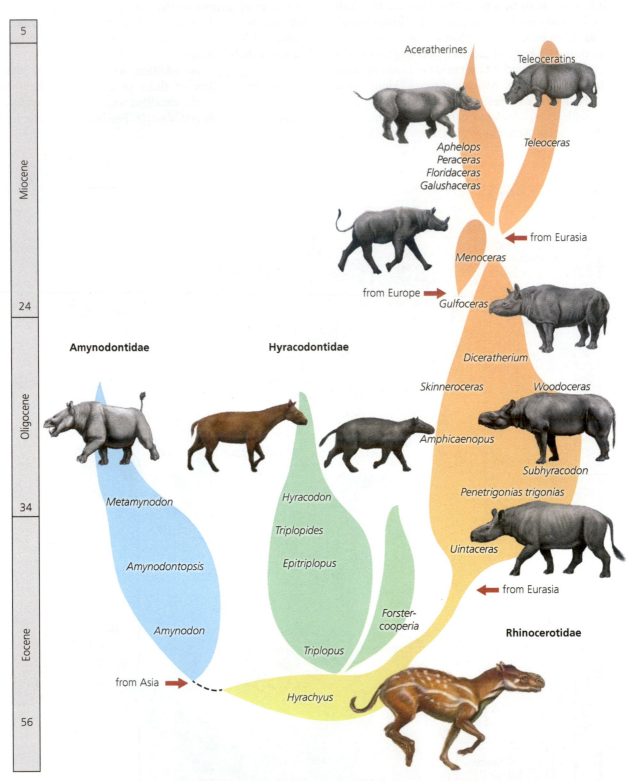

Figure 14.42   The evolution of North American rhinoceroses. Nearly all fossil rhinos were hornless, and many played the roles of other mammals. These included the hippo-like amynodont rhinos of the Eocene and early Oligocene (left branch) and the long-legged hyracodont rhinos (center branch), which evolved into the gigantic *Paraceratherium* in Asia (see Figure 14.39). The only surviving lineage, the family Rhinocerotidae (right branch), started with small hornless forms and later in the Miocene evolved two different lineages with paired horns on the tip of the nose as well as a lineage of rhinos shaped like hippos. Rhinos vanished from North America at the end of the Miocene but survived in Eurasia and Africa through the Ice Ages to today.

appreciated or publicized stories is the evolution of camels (Fig. 14.43). Camels started in North America and lived only on that continent until they finally spread to Eurasia about 8 Ma (where they evolved into the one-humped dromedary and two-humped Bactrian species). They also crossed into South America on the Panama land bridge around 2.8 Ma, giving rise to four humpless species, the llama, alpaca, vicuña, and guanaco. Thus, two-thirds of all living camels are without a hump, and most fossil camels did not have humps either. Humps are specializations for the desert lifestyle of the African and Asian species. Instead of recognizing camels by their humps, paleontologists use distinctive features of the skull, teeth, and limbs. Camels first appeared in the middle Eocene (about 48 Ma), with

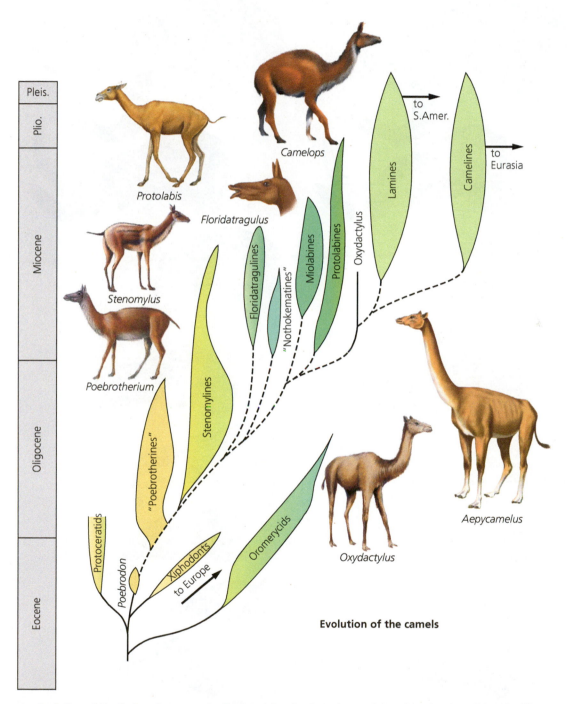

Evolution of the camels

Figure 14.43 Evolution of North American camels. From small animals no larger than a beagle, they diversified into many lineages, some of which were long-legged runners (aepycamelines) and some short-legged forms (protolabines), while others were proportioned like giraffes or gazelles. After evolving in isolation in North America for most of their history, they escaped to Eurasia about 9 Ma, where they gave rise to the living dromedary and Bactrian camels. Another lineage, the lamines, reached South America about 3 Ma, where they evolved into llamas, alpacas, vicuñas, and guanacos. Four of the six living species of camel (all the South American forms) have no hump, and it appears that none of the fossil camels had humps either, which appeared only in the Eurasian and African lineages.

the highest-crowned teeth of any creature of their time and they nearly always had very high-crowned teeth throughout their evolution. However, as they got larger, they also developed longer limbs and elongate toes that fuse into a "cannon bone" and helped them run faster. By the Miocene, they quickly diversified into a wide variety of forms (Fig. 14.43), including some with long necks and legs that performed the roles of giraffes, others with extremely high-crowned teeth and delicate limbs that were like gazelles, and some with relatively short limbs and toes. Like horses and rhinos, they had many different overlapping lineages in the middle and late Miocene, but most died out before the

A

Figure 14.44   *(Continued)*

Pliocene, leaving only the lineages that lead to the llamas and to the African and Asian forms in the Pliocene and Pleistocene. They were still living in their native homeland North America up to 10,000 years ago, until the late Pleistocene extinction wiped them out (as also happened to North American native horses), leaving only forms that immigrated to other continents.

See For Yourself: Camel Evolution

See For Yourself: Evolution of Elephants

Finally, let's consider one of the most amazing evolutionary sequences of all: the elephants, or Proboscidea (named because they have a trunk or proboscis). Their earliest forms are known from many primitive transitional fossils in the Paleocene and Early Eocene of Africa (Fig. 14.44). These fossils do not yet

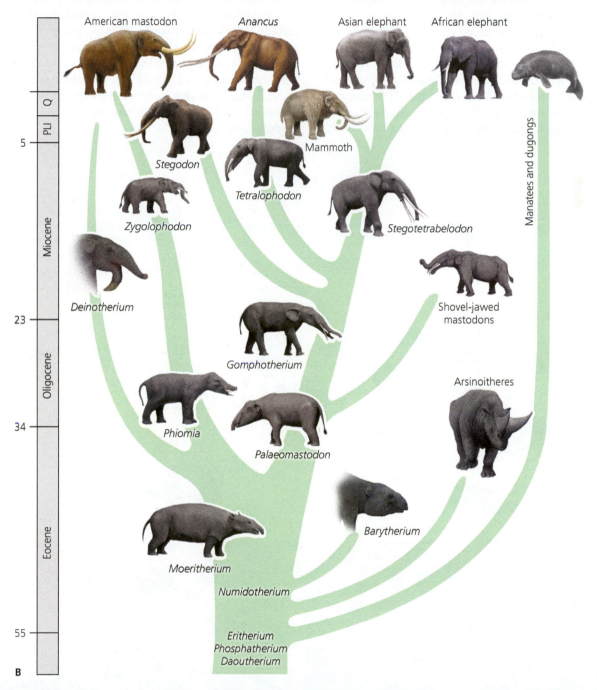

Figure 14.44 Evolution of the Proboscidea and their relatives. **A.** Profiles of skulls, and their reconstructions, from the most primitive proboscideans to the advanced mammoths. During their evolution, their tusks got larger and came in a variety of combinations of both upper and lower tusks, and the nasal bones retracted as they developed a trunk or proboscis. Their teeth also got larger and more complex. At the bottom is *Phosphatherium*, followed (bottom to top) by *Numidotherium*, *Moeritheriun*, *Palaeomastodon*, *Phiomia*, *Gomphotherium*, *Deinotherium*, *Mammut* (American mastodon), and, at the top, *Mammuthus*, the mammoth. **B.** Family tree of the proboscideans and their kin, such as the horned arsinoitheres and the aquatic manatees. The earliest Paleocene forms looked more like a pig or tapir but already had features of the skull and teeth that foreshadowed the elephants. Over the course of the Cenozoic, they diverged into a variety of mastodonts with many different combinations of tusks. In the Ice Ages, only a few lineages were left, of which only the modern Asian and African elephants survive.

show evidence of a trunk, but they have short tusks, the distinctive teeth, and other features unique to the Proboscidea. By the Oligocene of Egypt, they had evolved into forms that resembled pigs or even tapirs, complete with a short trunk. During the Miocene, there were multiple different lineages of mastodonts that escaped from their African homeland and spread across Eurasia and eventually to North America about 18 Ma. These include not only the generalized four-tusked gomphotheres but also the shovel-tuskers, whose lower tusks are shaped like broad shovels, and the gigantic deinotheres, with only a lower pair of tusks pointed sharply downward, and many other forms. Of all these lineages, only the American mastodon survived into the Ice Ages and may have still roamed North America only 4000 years ago. Meanwhile, another lineage developed a shorter face and jaws with a steep forehead and huge, curving tusks. Their teeth turned into huge molars with many convoluted ridges of enamel on a broad surface for grinding many different types of vegetation. These were the mammoths that roamed Eurasia and North America. The living elephant species are survivors of the mammoth lineage, which mostly vanished during the end of the last Ice Age 10,000 years ago. However, dwarf mammoths still survived on some Siberian islands when the Egyptians were building the pyramids.

The Age of Mammals was a truly remarkable time, with incredible evolutionary stories in many groups of animals, all adapting to a global change in climate from a greenhouse world at the beginning to the icehouse that began with the Antarctic ice cap 33 Ma, and the Arctic ice cap about 3 Ma. All of this story was tied to tectonic changes as Pangea broke up and the continents moved to their modern locations.

TIMESCALE AND SUMMARY OF CENOZOIC EVENTS

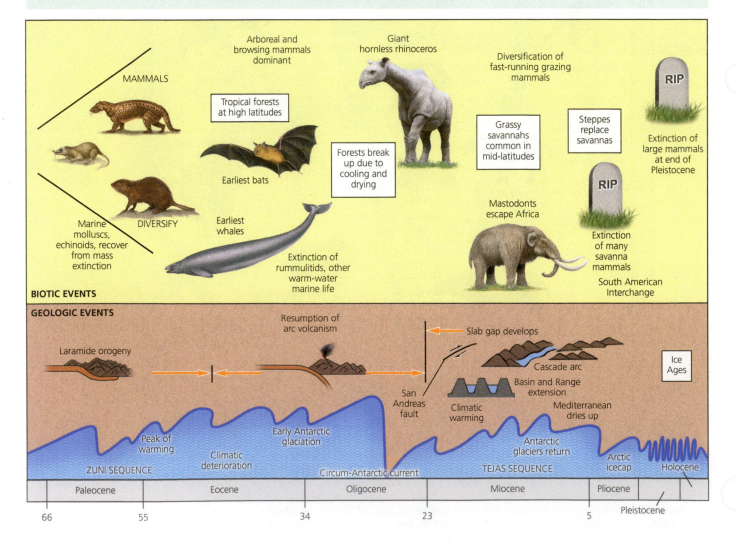

RESOURCES

BOOKS AND ARTICLES

Aubry, M.-P., S. G. Lucas, and W. A. Berggren (eds.). 1998. *Late Paleocene–Early Eocene Climatic and Biotic Events*. Columbia University Press, New York.

Blakey, R. C., and W. D. Ranney. 2018. *Ancient Landscapes of Western North America: A Geologic History with Paleogeographic Maps*. Springer, Berlin.

Dickens, G. R., M. M. Castillo, and J. C. G. Walker. 1997. A blast of gas in the Late Paleocene. *Geology* 25:259–262.

Dickinson, W. A. 2004. Evolution of the North American Cordillera. *Annual Reviews of Earth and Planetary Science* 32:13–45.

Dickinson, W. R., and W. S. Snyder. 1978. Plate tectonics of the Laramide orogeny. *Geological Society of America Memoir* 141:355–366.

Dickinson, W. R., and W. S. Snyder. 1979. Geometry of subducted slabs related to the San Andreas transform. *Journal of Geology* 887:609–627.

Dickinson, W. R., M. A. Klute, M. J. Hayes, S. U. Janecke, E. R. Lundin, M. A. McKittrick, and M. D. Olivares. 1988. Paleogeographic and paleotectonic setting of Laramide sedimentary basins in the central Rocky Mountain region. *Geological Society of America Bulletin* 100:1023–1039.

Hsü, K. J. (ed.). 1986. *Mesozoic and Cenozoic Oceans*. Geodynamics Series 15. American Geophysical Union, Washington, DC.

Hsü, K. J. 1987. *The Mediterranean Was a Desert: A Voyage of the Glomar Challenger*. Princeton University Press, Princeton, NJ.

Kennett, J. P. 1977. Cenozoic evolution of Antarctic glaciation, the Circum-Antarctic ocean, and their impact on global paleoceanography. *Journal of Geophysical Research* 82:3843–3860.

Kennett, J. P. (ed.). 1985. *The Miocene Ocean*. Geological Society of America Memoir 153. Geological Society of America, Boulder, CO.

Magill, J., and A. V. Cox. 1980. Tectonic rotation of the Oregon western Cascades. Oregon Department of Geology and Mineral Industries Special Paper 10. Department of Geology and Mineral Industries, Portland, OR.

McGowran, B. 1990. Fifty million years ago. *American Scientist* 78:30–39.

Meldahl, Keith. 2013. *Rough-Hewn Land: A Geologic Journey from California to the Rocky Mountains*. University of California Press, Berkeley.

Prothero, D. R. 2006. *After the Dinosaurs: The Age of Mammals*. Indiana University Press, Bloomington.

Prothero, D. R. 2009. *Greenhouse of the Dinosaurs: Evolution, Extinction, and the Future of Our Planet*. Columbia University Press, New York.

Prothero, D. R. 2013. *The Story of Life in 25 Fossils*. Columbia University Press, New York.

Prothero, D. R. 2016. *Giants of the Lost World: Dinosaurs and Other Extinct Monsters of South America*. Smithsonian Books, Washington, DC.

Prothero, D. R. 2016. *The Princeton Field Guide to Prehistoric Mammals*. Princeton University Press, Princeton, NJ.

Prothero, D. R. 2018. *The Story of the Earth in 25 Rocks*. Columbia University Press, New York.

Prothero, D. R., and W. A. Berggren (eds.). 1992. *Eocene–Oligocene Climatic and Biotic Evolution*. Princeton University Press, Princeton, NJ.

Prothero, D. R., L. C. Ivany, and E. A. Nesbitt (eds.). 2003. *From Greenhouse to Icehouse: The Marine Eocene–Oligocene Transition*. Columbia University Press, New York.

Rose, Kenneth D. 2006. *The Beginning of the Age of Mammals*. Johns Hopkins University Press, Baltimore, MD.

Ruddiman, W. J. 2013. *Earth's Climate: Past and Future*. W. H. Freeman, New York.

Ryan, W. B. F. 2008. Decoding the Mediterranean salinity crisis. *Sedimentology* 56:95–136.

Severinghaus, J., and T. Atwater. 1990. *Cenozoic Geometry and Thermal State of the Subducting Slabs Beneath Western North America*. Geological Society of America Memoir 176. Geological Society of America, Boulder, CO.

Smith, R. B., and G. P. Eaton (eds.). 1978. *Cenozoic Tectonics and Regional Geophysics of the Western Cordillera*. Geological Society of America Memoir 152. Geological Society of America, Boulder, CO.

Snyder, W. S., W. R. Dickinson, and M. L. Silberman. 1976. Tectonic implications of space–time patterns of Cenozoic magmatism in the western United States. *Earth and Planetary Sciences Letters* 39:91–106.

Spencer, A. M. (ed.). 1974. *Mesozoic–Cenozoic Orogenic Belts*. Geological Society, London.

Turner, A., and M. Anton. 2004. *National Geographic Prehistoric Mammals*. National Geographic Society, Washington, DC.

Wolfe, J. A. 1978. A paleobotanical interpretation of Tertiary climates in the northern hemisphere. *American Scientist* 66:694–703.

SUMMARY

- The Cenozoic Era, or the last 66 million years, was dominated by the continued separation of the Pangea continents until they reached their modern positions. Many of the biggest mountains in the world, such as the Himalayas, the Alps, the Rocky Mountains, and the Andes, are products of Cenozoic tectonism.

- The major climate change in the Cenozoic has been the transition from a greenhouse planet in the early–middle Eocene to the icehouse world we have today, with polar ice caps and lower sea levels. The first Antarctic glaciers appeared in the early Oligocene, and the modern Arctic ice cap appeared in the Pliocene.

- The collision of India with southern Asia in the Eocene not only caused the uplift of the Himalayas and the Tibetan Plateau but also chopped the Tethys Seaway in half. Before that time, the shallow tropical waters of the Tethys hosted a distinctive fauna, dominated by the coin-shaped foraminiferans known as nummulitids.

- At the same time, Africa began to collide with southern Europe, crushing and crumpling up rocks into the Alps, with giant thrust faults and huge recumbent folds called nappes.

- During the latest Miocene (about 6 Ma), the inflow of Atlantic Ocean water through the Strait of Gibraltar was cut off, drying up the Mediterranean Sea and turning it into a deep evaporite basin full of salt and gypsum like the Dead Sea basin. Then, about 5.3 Ma, the Gibraltar dam broke and a gigantic, supersonic waterfall formed to flood the Mediterranean again.

- The Pacific Rim is ringed by subduction zones that consume old crust equivalent to all the new crust formed at the spreading ridge. Several different plates are going down the subduction zone of the Pacific Rim (the "Ring of Fire"), and there are tracks on the sea floor of plates that have completely vanished. Meanwhile, the Pacific plate is moving northwest across a series of hot spots, one of which formed the Hawaiian–Emperor seamount chain.

- In North America, the passive margin wedges on the Gulf and Atlantic coasts continued to subside, adding thousands of meters of shallow-marine sands and muds to the deposits accumulated during the Cretaceous. The Appalachian Mountains were eroded down and buried in their own debris, and only during the Cenozoic have these ranges been uplifted again, exhuming the ancient hard folded bedrock ridges and forcing the drainages to cut across the ranges, forming water gaps.

- The Laramide orogeny, which began in the latest Cretaceous (70 Ma), continued until the middle Eocene (about 40 Ma). It uplifted deep basement rocks into the ranges of the future Rocky Mountains, with deep basins between these ranges subsiding and accumulating thousands of meters of Paleocene and Eocene river, floodplain, delta, and lake sediments. Meanwhile, volcanism ceased in the old Sierran arc region. Horizontal subduction is thought to explain this peculiar pattern of arc shutoff and uplifting of deep basement rocks into folds and faults over 1500 km (about 1000 miles) from the active subduction zone.

- Between 40 and 20 Ma, arc volcanism returned to the Cordillera but in a trend through eastern Washington and Oregon, Nevada, Utah, Colorado, and New Mexico that was far east of the old Sierran arc. This suggests that the downgoing slab had peeled off from horizontal subduction and was again sinking into the mantle but at a much shallower angle than before, producing eruptions farther east than before. As these eruptions continued, the old Laramide ranges eroded away and their basins filled up, so that by the Miocene the sedimentary blanket covered most of the ranges nearly to the top, eroding their summits flat.

- Since 20 Ma, the entire Cordilleran region has experienced complex tectonics. The Cascade arc erupted in its present location due to subduction of the Farallon plate, but its southern end once started in southern California and has progressively shut off from south to north as the Mendocino triple junction moved north. Behind it was a huge Columbia River flood basalt eruption around 15–16 Ma, which covered eastern Washington and Oregon with thick piles of basaltic lava. In the Pliocene, the North American plate moved over a hot spot that burned a trail through the overlying crust to form the Snake River plain of Idaho. This hot spot is currently beneath Yellowstone caldera.

- In the late Miocene, the Basin and Range province began to open in Arizona and then north to Nevada and Utah as crustal extension stretched the region to twice its original width. This produced the characteristic north–south trending ranges bounded by normal faults, with both grabens and blocks sliding down detachment faults. This extension is due to the fact that the Sierra Nevada Mountains have swung about 200–300 km (about120-200 miles) west of their original position, stretching the crust behind them.

- In the latest Miocene and Pliocene, the Colorado Plateau and Rocky Mountains began to uplift to their present elevations. This caused many of the old river drainages to strip off the blanket of sediment that once buried the Rockies and cut through the hard bedrock ridges beneath to form water gaps. In the Colorado Plateau, rivers that once had lazy, low-gradient, meandering patterns were suddenly rejuvenated by the uplift and cut deep vertical-walled canyons.

- The cause of this complex tectonics since 20 Ma is probably the lack of Farallon plate remnant being subducted anywhere east of the San Andreas fault, forming a slab gap or slab window through which mantle can rise up directly below the overlying North American plate without intervening Farallon remnants. As more of the Farallon plate vanished, the San Andreas got longer and the slab window increased in size, expanding the Basin and Range from south to north and transferring stresses east to uplift the Colorado Plateau and Rocky Mountains.

- After the Cretaceous extinctions, life in the oceans slowly recovered, and by the late Paleocene the diversity of marine invertebrates had recovered. Climate reached a peak of warmth in the early Eocene when frozen methane was released from the deep ocean and caused a super-greenhouse global warming event. The warm Eocene seas were home to a big diversity of mollusks, sea urchins and the first sand dollars, the giant foraminiferans known as nummulitids, sharks, fish, and the first whales by the middle Eocene.

- In the late Eocene (37 Ma) and again in the earliest Oligocene (33 Ma), there were two major cooling events that ended the early Cenozoic greenhouse and began the late Cenozoic icehouse world. Glaciers returned to Antarctica about 33 Ma, probably in response to the development of the Antarctic Circumpolar Current and possibly due to some other effects such as the early uplift of the Himalayas and the Tibetan Plateau.

- The early Miocene was a bit warmer after the cold Oligocene, but by 14 Ma the planet had cooled again and the Antarctic ice sheets were permanent. The closure of the Panama land bridge in the middle Pliocene (about 3.5 Ma) not only allowed the Great American Interchange of mammals between the two Americas but also blocked warm tropical water that had poured into the Pacific and forced it to head north and make a powerful Gulf Stream. This, in turn, brought moisture to the Arctic and allowed the Arctic ice cap to form for the first time.

- On land, the entire early Cenozoic greenhouse planet was covered by lush jungles and forests as far north as Montana during the Paleocene and early to middle Eocene. These jungles were inhabited by a diversity of archaic tree-dwelling mammals (early primates, insectivores, multituberculates) in the forest canopy, and primitive leaf-eating hoofed mammals on the ground. By the early Eocene, the first primitive relatives of horses, tapirs, rhinos, and even-toed hoofed mammals were present, competing with the archaic hoofed mammals that had dominated the Paleocene. Most of the tree-dwelling mammals and archaic hoofed mammals vanished in the late Eocene as the climate got colder and drier and the forests were replaced by scrublands in the early Oligocene.

- Through the late Oligocene and Miocene, climate in North America got colder and drier, so during the middle Miocene there were extensive savanna grasslands covering North America, Eurasia, and many other regions. These savannas were inhabited by mammals that resembled those of the modern Serengeti in Africa, although their ecological roles were played by different unrelated groups. In North America, mastodonts performed the elephant role, rhinos took the place of hippos, camels evolved to look like giraffes and gazelles, peccaries replaced pigs, bone-crushing dogs did the jobs of hyenas, and horses and pronghorns took the roles of African zebras and antelopes.

- In the late Miocene, another cooling and drying event, possibly associated with the drying up of the Mediterranean, changed the savannas to a dry grassy steppe habitat during the Pliocene. This may have caused widespread extinction in many of the North American mammal groups and around the world.

KEY TERMS

Tethys Sea (p. 345)
Nummulitids (p. 345)
Nappes (p. 345)
Alpine–Himalayan orogeny (p. 345)
Messinian salinity crisis (p. 347)
Farallon plate (p. 349)
Kula plate (p. 349)
Hawaiian–Emperor seamount chain (p. 352)
Superposed drainage (p. 356)
Cordillera (p. 356)
Intermontane basins (p. 356)
Fort Union Group (p. 356)

Green River Formation (p. 358)
Oil shale (p. 358)
Ancestral Cascades (p. 358)
White River Group (p. 363)
High Cascades (p. 364)
Columbia River flood basalts (p. 365)
Basin and Range province (p. 365)
Horst and graben (p. 365)
Playa lake (p. 366)
Detachment fault (p. 367)
Snake River plain (p. 369)
Colorado Plateau (p. 369)
Entrenched meanders (p. 370)

San Andreas fault (p. 370)
Slab gap (p. 372)
Paleocene–Eocene Thermal Maximum (PETM) (p. 373)
Methane hydrates (p. 373)
Oxygen isotope system (p. 376)
Leaf margin analysis (p. 378)
Himalayan weathering hypothesis (p. 379)
Circum-Antarctic circulation (p. 379)
Antarctic Bottom Water (p. 379)
Panama land bridge (p. 381)

Great American Biotic Interchange (p. 381)
Gulf Stream (p. 383)
Global ocean conveyer belt (p. 383)
Primates (p. 383)
Insectivorous mammals (p. 383)
Perissodactyls (p. 384)
Artiodactyls (p. 384)
Creodonts (p. 384)
Carnivorans (p. 384)
Rodents (p. 386)
Uintatheres (p. 386)

STUDY QUESTIONS

1. Why are geologists abandoning old terms like "Tertiary"? How are divisions of time like the Paleogene and Neogene different from the old divisions of Tertiary and Quaternary?

2. Why is it essential to know the sequence of epochs in the Cenozoic but not necessarily so for older intervals of geologic time?

3. Which parts of Pangea had separated in the Cretaceous? Which ones separated in the Cenozoic?

4. What were the consequences of the collision of India with southern Asia?

5. Where was the Tethys Sea located? What happened to it?

6. What are the Great Pyramids made of?

7. What is the structural geology of the Alps? What caused the Alps to be folded and faulted and uplifted in the Cenozoic?

8. What is the evidence that the Mediterranean once dried up? What caused this event? How did it end?

9. If there is spreading on the Mid-Atlantic Ridge and in the Indian Ocean, where is the crust being consumed to compensate for new crust being formed?

10. Why is the Pacific Rim called the "Ring of Fire"? What is the tectonic cause of these events?

11. Why do the Hawaiian Islands get older and sink down lower as you go to the northwest? What will eventually happen to Kilauea when it moves off the hot spot?

12. Why do all three seamount chains formed by hot spots in the Pacific have a kink dated about 40 Ma?

13. Why are there water gaps cutting through the hard, resistant ridges of the Appalachians? Why don't the rivers just go the easy way around the ridges through the soft valley sediments?

14. What did the Laramide orogeny do to the Cordillera in the Cenozoic?

15. How were the coals of the Powder River Basin in Wyoming formed? How are they economically more valuable compared to the coals mined in the East or Midwest? What are their drawbacks?

16. Why is the Green River Shale rich oil? Why are the fossils so well preserved?

17. What is the evidence that subduction resumed across the Cordillera about 40 Ma? Why didn't the volcanic arc return to the location of the old Sierran arc?

18. Once the Laramide orogeny ended, what happened to the Laramide ranges and basins? What is the evidence for this?

19. How is "Rocky II" (Laramide orogeny) different from "Rocky III" (late Cenozoic uplift of the Rockies)?

20. Why do some of the rivers of the Rocky Mountains cut through water gaps rather than erode the easy way around the ranges through soft valley fill?

21. How did the volcanism and the burial of the old Laramide ranges during the Oligocene produce the typical deposits of the Big Badlands of South Dakota?

22. How are the modern High Cascades different from the Ancestral Cascades in Oregon and Washington?

23. How were the Columbia River flood basalts produced? What did they do to the landscape?

24. Describe the topography and geology of the Basin and Range province. Why are all the ranges oriented north–south? What do the volcanism, heat flow, and frequent earthquakes suggest about the crust under the Basin and Range?

25. How do we know that the Sierras and Cascades have swung about 210–340 km (130-210 miles) west of their original position?

26. What does the sequence of Pliocene volcanic rocks across the Snake River plain in Idaho suggest about the geology there in the Pliocene? Where is that hot spot today?

27. How is the structural geology of the Colorado Plateau different from that of the Basin and Range province?

28. What does the pattern of entrenched meanders tell us about the drainage and uplift history of the Colorado Plateau?

29. How was the San Andreas fault formed? How has it changed through time?

30. Explain the slab gap/slab window model for Cordilleran tectonics. How does it explain such diverse phenomena as the south-to-north opening of the Basin and Range province, the south-to-north shutoff of the Cascade volcanic arc, and the uplift of the Colorado Plateau and Rocky Mountains?

31. How did marine life recover from the Cretaceous extinction?

32. What caused the extreme global warming episode at the Paleocene–Eocene boundary? What effect did it have on life?

33. Describe how whales evolved from land mammals.

34. Explain how oxygen isotopes in seawater are segregated during Ice Ages. What does an increase in heavy oxygen-18 in deep-marine microfossils tell us?

35. How do leaves change shape in response to temperature? Why are jagged margins formed in leaves from colder climates?

36. Describe the steps in the greenhouse-to-icehouse transition, from the early to middle Eocene to the cooling events at 37 and 33 Ma. What happened to Antarctica at 33 Ma?

37. What are some of the explanations for the events that ended the Eocene greenhouse and put the world back into an icehouse state?

38. What is the ACC? How does it affect the conditions over the South Pole? How does it affect deep-ocean circulation?

39. What was the Great American Biotic Interchange? How did affect land mammals in the Americas? How did it affect marine invertebrates in the Caribbean and Pacific?

How did its underlying cause, the closure of Panama, trigger the formation of the Arctic ice cap?

40. What is the global oceanic conveyer belt? How does it transport water among the world's oceans? How long does it take a drop of water to travel the complete loop?

41. What happened to mammals once the non-bird dinosaurs vanished at the end of the Cretaceous?

42. What were climates and vegetation like in Montana and the Arctic Circle in the Early Eocene? What kinds of mammals lived in these conditions?

43. How did the vegetation change in North America during the Eocene–Oligocene transition? What kinds of mammals vanished in response to the loss of the forests and jungles?

44. What kind of vegetation covered North America during the middle Miocene? How did mammals evolve in response to these conditions?

45. How did horses and camels respond to the spread of grasslands in North America during the Miocene?

The Pleistocene
2.6 Ma to 10,000 Years Ago

Macrauchenia 1: Well, why don't they call it The Big Chill? Or The Nippy Era? I'm just sayin', how do we know it's an Ice Age?
Macrauchenia 2: Because . . . of all . . . the ice.
Macrauchenia 1: Well, things just got a little chillier.

—*Ice Age, 2002*

The Aletsch glacier and the peak known as the Monch in Switzerland. Glaciers like these first gave Louis Agassiz and others insights into how ice once covered the landscape.

15.1 The Ice Age Cometh

Today, we take for granted that during the very recent geological past much of our world was once covered by a thick layer of ice (Fig. 15.1) and that there was once a great "Ice Age." Thanks to movies and television, we are familiar with animals such as mammoths and saber-toothed cats that lived during the Ice Ages, along with some of our human ancestors like Neanderthals, who lived near the ice margin in Europe. On the geological timescale, we now refer to this time interval when the ice dominated as the Pleistocene Epoch of the Quaternary Period of the Cenozoic Era. For a long time, the beginning of the Quaternary and Pleistocene was estimated at 1.8 Ma, but recently it has been shifted to 2.6 Ma, which roughly corresponds to the beginning of the major glacial cycles that we consider typical of the "Ice Ages." The most recent 10,000 years is known as the Holocene Epoch (or "recent" epoch) of the Quaternary Period, and it represents the most recent time interval following the last glacial retreat, when all of human history took place.

 See For Yourself: Glaciers and Ice Ages

But the idea of an Ice Age is relatively recent. In the pioneering days of geology, naturalists saw thick deposits of loose sands and gravels (Fig. 15.2A) across much of northern Europe and attributed them to Noah's Flood. These deposits were often called the "**diluvium**" or "**drift,**" as if these sediments had drifted out of the floodwaters in the Bible. However, they were truly puzzled by huge boulders found in many places, often perched precariously on a narrow pedestal (Fig. 15.2B). Close examination showed that the boulders were not made of any local bedrock but came from someplace else. These rocks were called **erratics** because they seemed to have strayed far from their source (from the Latin word *errare*, "to stray or wander"; when we "err" or are "in error" we stray from the truth; if we move erratically, we are straying or wandering). They were also known as "foundlings" or "lost sheep" since naturalists were reminded of lost sheep that have strayed far from their flock. In many cases, their source could be pinpointed as bedrock from many hundreds of miles away, such as examples of rocks in the Netherlands or Germany whose source was in Norway or Sweden. This "Flood" theory was well established in the early days of geology during the 1820s and 1830s.

Not all Europeans saw these strange geologic features as products of floodwaters. Those who lived near the Alps, in particular, had a different perspective, thanks to seeing glaciers in action. As early as 1787, Swiss minister Bernard Friedrich Kuhn argued that erratic boulders were carried by glaciers because he could see Swiss glaciers carrying such boulders even then. When James Hutton visited the Jura Mountains of Switzerland and France a few years later, he reached the same conclusion—something that has been overlooked among all his radical and revolutionary ideas. In 1824, the Norwegian naturalist Jens Esmark argued that glaciers had produced the erratics in Norway, as well as long parallel scratches in the hard bedrock. Influenced by Esmark, in 1832 German naturalist Reinhard Bernhardi published an article arguing that there was once a polar ice cap across all of Europe, even to central Germany.

Meanwhile, in Switzerland the observations of modern Alpine glaciers (Fig. 15.1) and their effects were accumulating and becoming better understood. In 1815, the Swiss mountaineer and chamois-hunter Jean-Pierre Perraudin described the effects of glaciers on the Swiss valleys and inferred that the glaciers had once been larger and extended farther. By 1818, his ideas had impressed a highway engineer by the

Figure 15.2 **A.** Thick deposit of glacial till, showing giant boulders and sand and even fine clays with no sorting or stratification. **B.** A glacial erratic, precariously balanced on a pedestal of totally different bedrock. This image is from Forbes (1843).

name of Ignace Venetz, who had spent much time in the Swiss landscape as part of his work. Venetz gradually became more convinced that glaciers had spread out from the Alps and affected the surrounding area. In presentations given in 1816 and 1821 and finally in 1829 he committed to the idea of a great expansion of glaciers in the past. Meanwhile, the director of the Bex salt mines, Jean de Charpentier, also became convinced after listening to Perraudin and then to Venetz and making his own field observations.

In the audience of de Charpentier's presentation in Luzerne in 1834 was a promising young Swiss paleontologist by the name of Louis Agassiz (Fig. 15.3). He was already famous through much of Europe for his pioneering work on fossil fish. In the summer of 1836, he visited de Charpentier in Bex, with the intention of proving his glacial ideas wrong. Instead, he became a convert to the glacial theory and was eager to spread this idea among geologists outside the handful who knew glaciers well. Unlike earlier advocates of glacial geology who simply gave evidence for glaciation of the Alps and regions nearby, Agassiz was an imaginative thinker, a bold speaker, a provocative writer, and an eager, hardworking, and ambitious man.

In 1837, Agassiz was hosting the annual meeting of the Swiss Society of Natural Sciences in his home town of Neuchâtel. As he got up to give the opening address, the audience expected another presentation on fossil fishes. Instead, he launched into a radical argument in favor of the glaciers as causes of most of the features once attributed to Noah's Flood. After reviewing the evidence and the ideas of Perraudin, Venetz, and de Charpentier, he extended their thinking to most of Europe and argued that it had once been covered by ice during an "Ice Age" (*Eiszeit* in German). His presentation was so unexpected and surprising that the meeting was thrown into confusion, and the long debate completely derailed the program of other talks that were supposed to be given. One of these was by Amanz Gressly, who was planning to introduce his now famous concept of "sedimentary facies" (Chapter 4) but never got up to give his talk during the chaos that followed Agassiz's talk.

Most of the audience was still highly skeptical and critical of Agassiz's ideas, so he responded by throwing together a spur-of-the-moment field trip to the nearest alpine glaciers

Figure 15.3 Portrait of Louis Agassiz, about the time he presented his ideas about the Ice Age.

at the end of the meeting. The greatest geologists in attendance, including Elie de Beaumont and Leopold von Buch, rode in the carriage with him. If Agassiz thought they would be instant converts upon seeing the field evidence, he was unduly optimistic about human nature. They ended the field trip unconvinced, and most of the geological community was still critical of his idea. The great naturalist and explorer Alexander von Humboldt told Agassiz to go back to fossil fishes and "render a greater service to positive geology, than by these general considerations (a little icy besides) on the revolutions of the primitive world, considerations which, as you well know, convince only those who give them birth" (cited in Imbrie and Imbrie, 1986)

If scientists on the Continent gave Agassiz an icy reception, others were more accepting. In England, pioneering naturalist William Buckland heard about some of his ideas and began to have doubts about his own work that explained everything with Noah's Flood. Buckland had hosted Agassiz in 1835 in Oxford while the latter was visiting and studying fossil fish, and they became friends. In 1838, Buckland went to Germany and took the tour of the glaciers with Agassiz but was unconvinced until 1840, when Agassiz addressed the British Association for the Advancement of Science in Glasgow. Finally, Buckland became a convert and was one of the first Britons to become a glacial geologist. Buckland then convinced the most influential critic of all, Charles Lyell, who immediately wrote a paper in support of his ideas. Then Agassiz, Buckland, and Lyell toured the Scottish Highlands, where Agassiz was able to show his colleagues that many of the distinctive features that had so long remained unexplained were due to the fact that Scotland had once been overrun with glaciers. By 1841, Edward Forbes wrote to Agassiz, "You have made all the geologists glacier-mad here, and they are turning Great Britain into an ice house. Some amusing and very absurd attempts at opposition to your views have been made by one or two pseudogeologists" (cited in Imbrie and Imbrie, 1986).

The battle over the Ice Ages raged for many years more, but Agassiz had grown tired of the continuous unresolved conflict. He sailed to the United States in 1846, originally to study fossil fishes, but then received a very generous offer to stay on at Harvard and there he founded the Museum of Comparative Zoology. He remained in the United States for the rest of his life, dying in 1873 after 27 years of teaching and research at Harvard. More importantly, he used the change in venue to take field trips all over the northeastern part of North America, where he found one glacial feature after another, further confirming his idea that it was indeed a global expansion of the Arctic ice cap that had covered all the northern continents (Fig. 15.4).

 See For Yourself: How an Ice Age Is Born

 See For Yourself: The Climate Puzzle

The most significant problem for geologists accepting the Ice Age theory in the 1840s and 1850s was that they could not visualize Europe being covered by an ice cap more than a mile thick. This problem disappeared when an American explorer, Elisha Kent Kane, led an expedition to Greenland (looking for some lost British explorers); he came back in 1857 and published a best-selling account of their harrowing, life-and-death adventures on the Greenland ice sheet. Suddenly, everyone could imagine an entire continent covered with ice, and the last resistance to Agassiz's ideas crumbled.

15.2 A World of Ice

THE ICE AGES BEGIN

As discussed in Chapter 14, the descent into Ice Ages was a steady, stepwise process through the entire Cenozoic. The first step was the development of the Antarctic ice cap about 33 Ma (Fig. 14.31). The size of this ice sheet increased and decreased over the rest of the Cenozoic with climate fluctuations, but by 14 Ma, it was large and permanent. The growth of the Arctic and Greenland ice caps was a much later event, probably triggered by the changes in the Gulf Stream and Atlantic circulation, which brought moisture to the Arctic when it was diverted from flowing out from the Caribbean to the Pacific after Panama closed at 3.5 Ma (Fig. 14.35B).

 See for Yourself: How Do Glaciers Work?

By 2.6 Ma, the beginning of the Pleistocene, we have clear evidence of major ice caps not only over both poles but also flowing down onto the northern continents (Fig. 15.4). These covered nearly all of Canada and Alaska and most of the northern tier of the United States, permanently altering the landscape and burying nearly all the older outcrops of pre-Pleistocene rocks across the region. In many parts of New England and the northern Midwestern and Plains states, the only rock exposures in most places are glacial deposits and bedrock modified by glaciers. Likewise, the higher peaks of the Rocky Mountains, the Sierra Nevada, and the Cascades were heavily carved by glaciers, and in many places there still are glaciers sculpting the landscape.

Figure 15.4 Northern polar view of the world in Pleistocene, showing the great ice caps in North America and Eurasia extending beyond the Arctic ice cap.

Dry land

Sea Ice

Flowing ice sheet

Flowing ice sheet

Dry land

Dry land

MIDWESTERN MORAINES AND LAKES

The most obvious products of the advance of the glaciers in North America are the thick deposits of glacial till (Fig. 15.2A) dumped by glaciers wherever they melted. When the glacier stalls and melts in one place for a long enough time, it piles up a large mound or ridge of glacial till known as a **moraine**. Typically, the farthest advance of a glacier produces a large **terminal moraine** at its greatest extent. As the glacier melts back, it may leave one or more **recessional moraines** if it stalls and melts in one place for a long time.

These moraines may be only a few meters tall, or sometimes over 10 m (about 33 feet) tall; but they are often subtle and hard to see at ground level. In the Midwest and Plains region, the highways cut right across them, and you might not even notice them if you aren't looking for them. They look just like another overgrown roadcut of sand and gravel or just a gentle ridge that you drive over without thinking about it. But geologists using good topographic maps, aerial photos, and satellite images can usually detect them quite easily.

By mapping the distribution of these moraines across North America, we can determine when and where the largest ice advances took place, which areas were glaciated, and which areas were beyond the reach of the glaciers (Fig. 15.5). Even on the flat landscape of the Plains, glaciers flowed out from huge ice sheets because the center of the ice sheet is higher than the edges, so they continually flow from their high point to low point. It is just like pouring honey on a flat table; the honey will always flow away from the highest point in the center no matter how level the surface on which it flows.

From maps such as Fig. 15.5C, geologists have determined that there were at least four or five major ice advances across much of North America. The centers of the North American glaciers were over Labrador and over the Hudson Bay region, and the ice flowed out from there in huge flat sheets. There were also ice sheets flowing down from the Rockies. The ice flowed away from these centers, especially to the south, and left moraines across much of the northern Midwest and Plains. The youngest ice advance is called the **Wisconsinan glaciation**. It began about 115,000 years ago and ended about 18,000 years ago. It is so named because it covered nearly all of the state of Wisconsin as well as most of Minnesota, Michigan, and many other states. Most of the features that you see in those states are products of this most recent sculpting of the landscape.

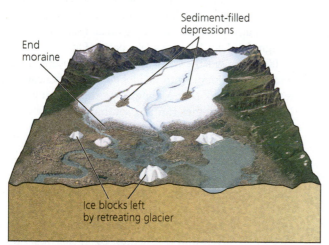

A During glaciation

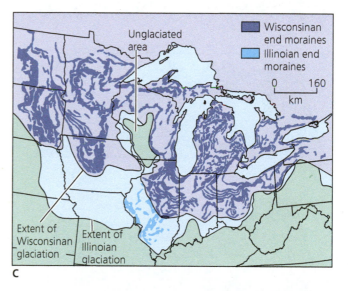

C

Figure 15.5 A–B. As a glacier melts back, it leaves a pile of unsorted till deposits called moraines where the snout of the glacier paused and melted before retreating. Elsewhere it left blocks of ice on the outwash plain, which depressed the ground and left kettle lakes when they melted. **C.** Map of moraines in the Midwest. The youngest moraines from the Wisconsinan stage glaciation (about 115,000–18,000 years old) are in darker purple, with the glacial till sheets covering the rest of the landscape shown in lavender. They form big concentric arcs around the edges of retreating glacial lobes, such as the glaciers that ran down the axis of Lakes Michigan, Erie, and Huron. Slightly older moraines (Illinoian stage, about 130,000–191,000 years old) are shown in dark blue, with the extent of these glaciers shown in light blue. Unglaciated areas are shown in green.

B After glaciation

A About 14,500 years ago.

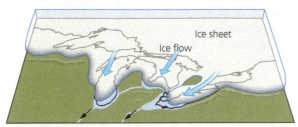

B About 14,000 years ago.

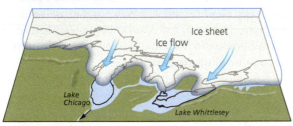

C About 13,000 years ago.

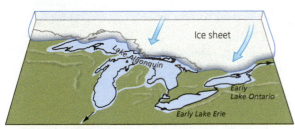

D About 11,000 years ago.

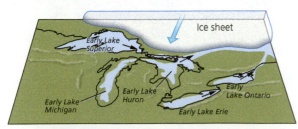

E About 9,500 years ago.

F About 6000 to 4000 years ago.

TIME

To the south of the Wisconsinan ice advance is an older set of moraines from a much larger ice advance, which is buried in some areas by later Wisconsinan deposits. Called the **Illinoian glaciation**, because it bulldozed nearly all the state of Illinois flat and buried it in till deposits, it is dated between 190,000 and 130,000 years ago. If you travel southwest of this glaciated region, you find even older deposits that were not covered by the Illinoian or Wisconsinan ice advances. These older deposits were once called Kansan and Nebraskan, after the states that were touched by the southernmost extent of these glaciers. For a long time, Pleistocene geologists assumed that there were just four major ice advances in North America and gave them the names (from oldest to youngest) Nebraskan, Kansan, Illinoian, and Wisconsinan. However, recent analysis has shown that the Kansan and Nebraskan are actually composites of several different ice advances, not just two discrete events; and these terms have been dropped in favor of the term "Pre-Illinoian."

Not only can you map and date the location of ancient glaciers across the northern United States with moraines, but the glacial ice left many other landmarks as well. The most striking of these is the Great Lakes. Many people have puzzled over the strange shape and arrangement of these lakes, but this became clear when geologists realized that each was sculpted by a major lobe of one of the glaciers advancing across the region and scouring a deep trough that later became a lake basin (Fig. 15.6). About 18,000 years ago, the Great Lakes were all covered in different glacial lobes, with one lobe running beneath the teardrop shape of Lake Michigan, another running down Lake Huron, another running southwest down the axis of Lake Superior, and a major one flowing above what would become Lakes Erie and Ontario. Then they began to melt back in stages during the glacial–interglacial transition about 14,500 years ago. As they melted back, they exposed large lakes on their southern edges that were dammed by moraines. These Late Pleistocene lakes have slightly different shapes from the modern Great Lakes and so are given different names. By 13,000 years ago, the southern part of what would become Lake Michigan had a much larger lake on it called Glacial Lake Chicago, while the huge Lake Whittlesey covered the basin of Lake Erie and much of northern Ohio and Indiana (Fig. 15.6). Through all this time, the meltwater of these glaciers flowed south into what became the Mississippi River system, and there was no St. Lawrence River to drain the Great Lakes, as it does today.

Figure 15.6 The modern Great Lakes were formed by repeated episodes of glaciation. **A.** Initially, ice sheets covered the entire region. **B.** About 14,000 years ago the ice began to melt back, exposing parts of the southern basins of the Great Lakes. All of their meltwater drained south toward the Mississippi River. **C.** About 13,000 years ago, Glacial Lake Chicago (predecessor of Lake Michigan) and Glacial Lake Whittlesey (larger predecessor of Lake Erie) appeared as the ice melted farther back. **D.** About 11,000 years ago, all of the ice had melted back, but Lakes Superior, Huron, and Michigan were connected as a single Lake Algonquin. Meanwhile, the lakes began to drain northeast through the St. Lawrence River for the first time. **E.** As the ice sheet retreated farther about 9500 years ago, the Ottawa River became the major drainage. **F.** Finally, the Ottawa River was abandoned as the St. Lawrence took over the entire drainage about 6000–4000 years ago.

A Before glaciation

B After glaciation

Figure 15.7 A. Before the Pleistocene, northern North America had a huge drainage network that flowed north to rivers that are now beneath Hudson Bay. The Mississippi River drainage, on the other hand, was small and only drained a few states. **B.** After glaciation, the melting ice sheet originally sent most of its water south to the Mississippi, greatly expanding its drainage network (Figure 15.6). Meanwhile, the drainages across Canada that once flowed to Hudson Bay were completely disrupted, so most of the landscape does not have rivers flowing through it but irregular topography caused by till with many tiny unconnected lakes scattered around.

Around 11,000 years ago, a single Lake Algonquin covered all of what would become Lakes Superior, Huron, and Michigan. Meanwhile, about 11,700 years ago, the channel of the St. Lawrence River was exposed when the ice dam melted away, and a huge pulse of fresh water poured out of the Great Lakes and glacial Lake Agassiz, which covered much of Manitoba and parts of Saskatchewan, Ontario, Minnesota, and North Dakota. This catastrophic freshwater flood had a dramatic effect on global climate, as we shall discuss in Chapter 16.

Before these glaciers melted away and formed the Great Lakes and their glacial predecessors, however, the presence of the ice sheet in the region transformed the drainage system as well. Prior to the Ice Ages (Fig. 15.7), a well-developed drainage system flowed north and east across most of Canada and out major rivers across what would later become Hudson Bay (another basin formed by glacial advances and not present before the Pleistocene). Then the ice overran nearly all of Canada and forever buried or rearranged this ancient drainage network (Fig. 15.7). What had been a small network of drainages going out to the Mississippi River enlarged as the ice covered all the old drainages, and all the meltwater was diverted south. The effects of this are remarkable. For example, today the tiny Illinois River is just a small creek (except in flooding seasons) running through a big broad valley in places like Peoria, Illinois. The misfit between the tiny creek and the huge valley is explained when you realize that at one time nearly all the meltwater draining off the Lake Michigan/Glacial Lake Chicago area drained south through the Illinois River on its way to the Mississippi (Figs. 15.6, 15.7).

Today, the Mississippi River system and its tributaries drain nearly all the regions south of the glacial border, while the drainages in Canada that were overrun by the glaciers were completely disrupted. There are almost no branching, connecting river systems in the area surrounding Hudson Bay (Fig. 15.7B) but instead hundreds of lakes and ponds that usually do not connect to one another. This kind of poorly developed, disconnected drainage system is known as a **deranged drainage**. The many ponds were largely created when large blocks of ice depressed the glacial till as the ice sheet melted back. When the ice blocks melted, they left a round depression in the till which filled with water, known as a **kettle lake** (Fig. 15.5A). Kettle lakes are ubiquitous over the entire glaciated region. For example, the nickname of Minnesota is "The Land of 10,000 Lakes," and that's not an exaggeration. Almost the entire state is covered by thousands of lakes of many different sizes, most of which were kettle lakes

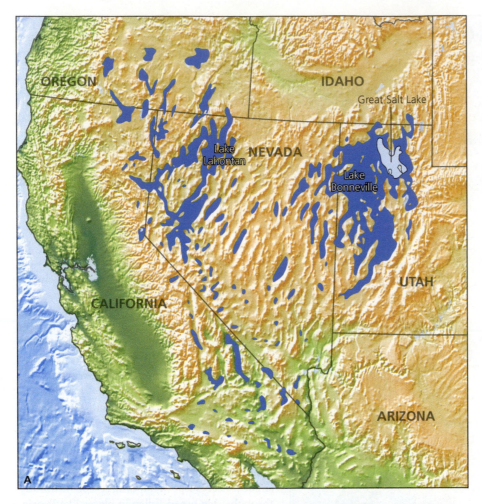

Figure 15.8 A. During the cooler, wetter times of the previous glacial maxima, the dry deserts of the Great Basin were flooded by large lakes, and their climate was cooler and wetter than it is today, with lush vegetation and abundant wildlife like mammoths and bison and horses and camels and saber-toothed cats around the lakes. The "Great" Salt Lake is a tiny remnant of a huge Glacial Lake Bonneville that once covered most of northwestern Utah. Glacial Lake Lahontan covered much of northwestern Nevada, of which only tiny Pyramid Lake remains. **B.** The stair-step lake terraces on Antelope Island in the center of the Great Salt Lake show that at one time the lake level rose and fell in what is now the Salt Lake City area. Similar terraces are found on the Wasatch Mountains to the east of the cities.

formed by ice blocks as the ice sheet vanished.

Meanwhile, nearly the entire Plains and Midwestern region south of the glacial margin was covered by a fine wind-blown dust of ground-up glaciated rocks. These deposits of **loess** (pronounced "luhrss") blew from the edges of the glaciers and piled up in huge thicknesses across most of the region. This fine silt and clay is also rich in organic material, so it formed the rich soils that made the Midwest and Great Plains the "breadbasket" region of North America. Most of our wheat, corn, soy, sorghum, and many other crops grow best in this region; and the United States would not be so productive in agriculture without these loess soils. In fact, loess plains are the basis for most of the world's most productive farming regions. Large areas of loess underlie the farming regions of the plains in Poland and in the Ukraine. Thick deposits of loess mantle much of central China. In short, without this rich soil that is a gift of the glaciers, the world would not be able to feed anywhere near the 7.8 billion people it now feeds.

GLACIERS AND LAKES OF THE WEST

The expansion of lakes across the regions around the edges of the glaciers and across the deglaciated landscape was not restricted to eastern and midwestern North America. During the peak of the last glaciation 20,000 years ago, climate was milder and wetter across nearly all the United States. In particular, the huge area of what are today deserts in Nevada, Utah, Arizona, New Mexico, and eastern California was much moister and more vegetated than it is today (Fig. 15.8A). Nearly every valley in this region once had a glacial lake in the middle of it about 20,000 years ago, and there were huge areas of modern desert covered by giant expanses of water then. What is today called the "Great" Salt Lake is just a tiny remnant of the enormous glacial Lake Bonneville, which covered about 50,000 km² (about 19,000 square miles) of nearly all of northwestern Utah with water at least 335 m (about 1000 feet) deep. In addition to lake deposits, you can see many stair-step shoreline terraces on the flanks of the Wasatch Mountains above Salt Lake and the rest of the cities in the area and on Antelope Island in the middle of the Great Salt Lake (Fig. 15.8B). This lake has completely dried up, leaving just the "Great" Salt Lake and places like the Bonneville Salt Flats, which are so smooth and flat for many miles that they are used to test rocket cars and set land speed records. Much of the northwestern corner of Nevada was flooded by glacial Lake Lahontan. Today only the much smaller Pyramid Lake remains, along with the Black Rock Desert, where the "Burning Man" festival is held each year. Death Valley was once flooded by the huge glacial Lake Manly, which is reduced to salt flats today that only have water during the wettest parts of the year. There are many other former dry lakebeds across most of this desert region, and wherever you go across the area, you find soft tan and pink lakebed silts and muds on the edges of the basin, now eroding away, and salt deposits in the center of the basin, many of which were formed as the glaciers retreated 18,000–10,000 years ago. Many of these ancient glacial lakebeds are full of Ice Age fossils, so mammoths, horses, camels, bison, and other creatures of the glaciated world lived in the lush forests around many of these ancient lakes where there is now only desert.

One of the most remarkable events in the glaciated western states was the draining of glacial Lake Missoula in Montana as the ice dam across northern Idaho broke multiple times, forming the Channeled Scablands (see Chapter 1, Fig. 1.9). Similarly, glacial Lake Bonneville in northwestern Utah drained catastrophically about 15,000 years ago when a natural dam broke in Red Rock Pass, Idaho. A huge wall of water poured through the gap and across the Snake River floodplain in Idaho, incising it deep into the Snake River Gorge on the Idaho–Oregon border and then into the Columbia River drainage. The force of the water was tremendous, exceeding 1.3 km³ (0.3 cubic miles) an hour. Its power was so great that it picked up boulders up to 3 m (about 10 feet) in diameter and deposited them in a unit called the Melon Gravel. In some places in Idaho, the gravel bars are 90 m (about 300 feet) thick and about 2.4 km (1.5 miles) long.

NORTHEASTERN GLACIATION

Although the abundant lakes of the central and western United States are dramatic signs of ancient glaciers, there are many additional relics of glaciation on the landscape of the Northeast, especially New York and New England. Some of the most famous are the Finger Lakes in upstate New York (Fig. 15.9). These form a group of about 15 elongate lakes that lie across the foothills of the Catskill Mountains around Syracuse and Ithaca, New York. From a map view or aerial image, they fan out somewhat like a set of fingers. Once you visit these lakes you find that the valleys all show the classic signs of glaciation, with steep vertical sides and flat bottoms.

Figure 15.9 In upstate New York, glaciers carved a series of parallel valleys filled with lakes that radiate out in a pattern like the fingers on an open hand—hence the name "Finger Lakes."

Figure 15.10 Even in Central Park in the heart of New York City, you can see long parallel scratches carved by glaciers that once flowed across Manhattan and filled the Hudson Valley.

The reason that the Finger Lakes are found only in this belt is that they are underlain by soft Silurian–Devonian shales and sandstones, which the glaciers carved more easily than the bedrock to the north or south of the lakes.

 See for Yourself: Glacial Grooves in New York City

Even in the heart of Manhattan in New York City there are signs of glaciation. In many places in Central Park, you can find big outcrops of the basement rock, a Proterozoic unit known as the Manhattan Schist. These outcrops are smoothly beveled over the top, and in many places there are deep glacial grooves and scratches carved by large rocks at the base of the flowing glacier (Fig. 15.10). If you take the compass bearing

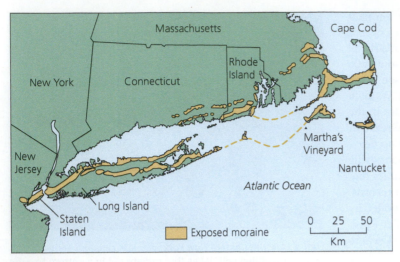

Figure 15.11   Much of Long Island was formed by two separate moraines, with the two points at the east end (Montauk Point on the south, Orient Point and Plum Island on the north) representing the ends of separate moraines. These moraines can then be traced across to Nantucket, Block Island, and Martha's Vineyard, while the southern half of Cape Cod is part of a moraine as well.

buildings are found. The rest of Manhattan is largely underlain by soft glacial beds, and no sky-scrapers can be built on such weak foundations.

Traveling through New England, you can find many more examples of glacially scratched and smoothed bedrock, huge glacial erratics, and many other features. Indeed, one of the reasons that New England has historically not been a major agricultural region is that the glacial soils are very poor and thin, and they are full of large boulders from the unsorted glacial till that discouraged plowing and planting. Most of the stone walls in the New England country-side were built during colonial times out of the hundreds of rocks that the early farmers hit with their plows and had to move.

 See For Yourself: What Is an Ice Age?   Even the shape of the land is the product of gla-ciation. The southern part of Cape Cod is a big glacial moraine (Fig. 15.11), which runs west across coastal Rhode Island and Connecticut. That is why it sticks out away from the coast as it does, while the "hook" of Cape Cod was caused by longshore currents redistributing the glacial sands up the coast. Likewise, the islands offshore (Nantucket, Block Island, and Martha's Vineyard) are all part of a set of moraines formed when the ice sheets were bigger and sea level was lower, expos-ing this whole shelf region about 20,000 years ago. Finally, the funny shape of Long Island is also due to a pair of moraines. The axis of Long Island runs parallel to the two moraines that formed its spine, and the two-pronged eastern end is built from the two moraines. Montauk Point is the eastern end of the southern moraine, while Orient Point and Plum Island form the point to the north (Fig. 15.11).

in the grooves, they all point to the north and the source of the glacial ice up in Labrador. In fact, the distribution of this hard basement rocks determines where there are skyscrap-ers in Manhattan. The hard bedrock only comes near the surface beneath Lower Manhattan and Wall Street and near Midtown and Central Park, and that's where all the tallest

15.3 What Caused the Ice Ages?

THE LAND RECORD

When Louis Agassiz first pro-posed the idea of ice ages in 1837, it was enough of a stretch for his fellow geologists to imagine just one thick ice sheet across northern Europe, let alone more than one. But by the late 1800s, careful mapping of the moraines and places where more than one glacial

Figure 15.12   Timescale for the last 650,000 years of the Pleistocene. At one time, there were only four glacial stages recognized in North America (from oldest to youngest: Nebraskan, Kansan, Illi-noian, and Wisconsinan) and five in the Alpine region (Donau, Gunz, Mindel, Riss, and Würm). For decades, it was impossible to decide how they precisely matched up. Then in the 1960s and after-ward, deep-sea cores and eventually ice cores provided a much more detailed record of the climate changes, and it turns out there were about 25 glacials in the past 2.6 million years. The Würm and Wisconsinan matched, as did the Riss and Illinoian; but the Kansan and Mindel were a composite of at least three glacial cycles, and the Nebraskan, Gunz, and Donau were composites of multiple glaciations and even more poorly defined.

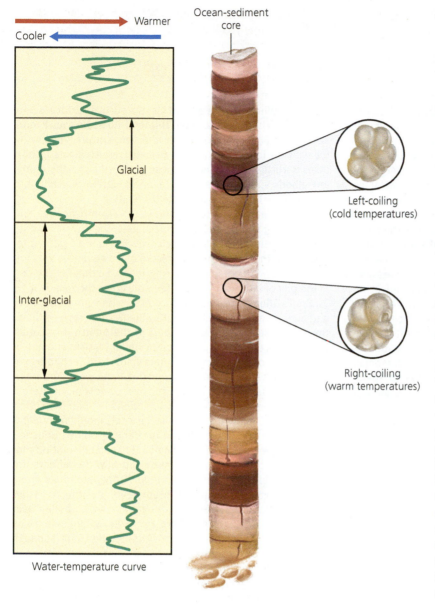

Warmer

Cooler

Ocean-sediment core

Glacial

Inter-glacial

Left-coiling
(cold temperatures)

Right-coiling
(warm temperatures)

Water-temperature curve

Figure 15.13 In some deep-sea cores, it is possible to track the temperature through time by recording the ratios of left-coiling to right-coiling shells of certain foraminifera. Apparently, as these species begin their development, their coiling is influenced by a critical temperature threshold.

ice advances documented by glacial deposits on land. There was no way to get a more detailed record out of the land record because the terrestrial sequence was known to be incomplete, with many unconformities. In fact, the four or five largest glaciations were known to have bulldozed away most of the smaller glacial advances between them, so the record had to be more complex than the land data allowed us to see.

SOLUTION IN THE OCEAN

The answer to the problem came from the deep sea—in particular, long cores taken out of deep-sea muds full of the shells of plankton from all over the world's oceans. Starting in the late 1940s and 1950s, oceanographic vessels routinely took cores in many parts of the ocean as they gathered data of many different kinds. Many of those data from the oceans led to the discovery of plate tectonics (Chapter 5). But by the late 1960s and 1970s, enough cores had been collected that there was an excellent record of the Ice Ages in the tops of many of them. Unlike the land, which erodes more sediment away than it preserves, the deep-sea floor is covered by a steady gentle "rain" of fine mud washed from land rivers plus the shells of plankton which lived at the surface, then died and drifted to the bottom by the trillions. The record is very complete in many places, with some cores recovering several million years of continuous record, while others have faster sedimentation rates and record a shorter interval of time with higher resolution, in some cases almost every year in detail.

Not only are the cores complete, but they record many different kinds of information in them as well. Micropaleontologists could estimate the temperature of the seawater just above where the core was taken by the characteristic warm-water and cold-water species of plankton. These still live in the oceans today, and their temperature tolerances are well documented; so as you find them in older core sediments, they give an accurate indication of water temperature. Not only that but some species actually spiral in a different direction depending on water temperature (Fig. 15.13). *Globorotalia menardii* switches from left coiling during cold conditions to right coiling in warmer times (as do *Globorotalia truncatulinoides* and *Neogloboquadrina pachyderma*). Finally, the individual shells of some species could be crushed up and measured in a mass spectrometer, giving the oxygen isotopes of the ocean chemistry in which they lived, and thus the paleotemperature (see Box 14.2). Thus, the deep-sea sediments in cores and especially their plankton were

deposit occurs on top of another made it clear there were multiple ice ages (Fig. 15.12). As we saw, American geologists recognized four events (Nebraskan, Kansan, Illinoian, and, the most recent, the Wisconsinan). Meanwhile, European geologists studied ancient glacial beds in the Alps and found evidence of at least five major ice advances, which they called the Donau, Gunz, Mindel, Riss, and Würm (from oldest to youngest). So were there four glacial advances or five? Was the Wisconsinan equivalent to the Würm, the Riss to the Illinoian, the Mindel to the Kansan, and the Gunz to the Nebraskan? And, if so, what was the Donau? In the twentieth century, the methods of dating got better (especially when radiocarbon dating came in the 1950s to improve the dating of the last 60,000 years of the Pleistocene), but there were still only four or five

BOX 10.1: HOW DO WE KNOW?

How Do We Know What Controls Ice Age Cycles?

Let's step back a century and look at an entirely different set of scientists using a different approach to understanding glaciers and ice ages. One idea that was being discussed among astronomers was that the amount of sunlight received on the earth's surface (known as **solar insolation**) might influence the shift from glacial ice advances to interglacial ice retreats. No one had really worked out the mathematics of the required orbits or calculated the energy differences involved. Into this breach stepped a remarkable Scotsman by the name of James Croll. He was the perfect example of someone who rose from limited means and education yet, based on sheer effort and intelligence, reached the pinnacle of a profession. Born on a farm in Perthshire, Scotland, he had almost no formal schooling and had to work for a living before he was 16 years old. He started as an apprentice wheelwright and mechanic, then became a tea merchant, then failed after trying to run a temperance hotel, and then became an insurance agent. In 1859, he was hired as the janitor at the Andersonian University in Glasgow, where he used his access to spend many hours in the library, teaching himself mathematics, physics, and astronomy.

Building on the work of astronomer Urbain Le Verrier, who had first shown that the earth's orbit around the sun and its axial tilt were constantly changing, Croll showed his work to Sir Charles Lyell and Sir Archibald Geike, who were impressed. Geike hired him as keeper of the geologic maps and correspondence of the Geological Survey of Scotland in Edinburgh, where he had plenty of spare time to read and do his research with all the necessary documents around him. In 1875, he wrote a book, *Climate and Time, in their Geologic Relations*, which laid out the basics of how the earth's orbital motions change the amount of solar radiation received and, thus, trigger ice ages. This earned him a university research post, an honorary degree by the University of St. Andrews, and eventually election to the Royal Society.

Croll did calculations for the known cycles of the earth's orbit around the sun, to see how much they might explain the ice ages. He pointed out that astronomers as early as Johannes Kepler in 1609 knew the earth's orbit around the sun was not a circle but an ellipse. That ellipse changed shape from nearly circular to slightly more egg-shaped very slowly (we now know it takes about 110,000 years). This is the cycle of the **eccentricity** of the earth's orbit (Fig. 15.14). Another cycle, known since the days of Hipparchos in 130 B.C.E., was the **precession** or "wobble" cycle. As the ancient Greeks knew, the earth's axis wobbles like a top, with its spin axis pointing in different directions. Today, for example, it points to Polaris, the North Star; but 10,000 years ago it pointed to a completely different star, Vega. As it points in different directions, it affects the amount of sunlight the poles receive. This cycle we now know takes about 21,000–23,000 years to complete, the fastest of the three cycles (Fig. 15.14).

Croll also pointed out that the ice can grow and melt back very rapidly due to the albedo feedback loop (see Chapter 8, Fig. 8.14). When there is a lot of ice, the earth's surface is very reflective, or has a high albedo. This tends to bounce back more solar energy to space and cool the temperature, further increasing the ice. But if you have an ice-covered landscape, all it takes is a little melting and it exposes dark, sunlight-absorbing, low-albedo surfaces like seawater or vegetation, which in turn absorb more heat and accelerate the melting.

Croll's ideas in his 1875 book were very provocative and worth taking seriously, but unfortunately, there were no data to test it at that time. No geological events could be dated reliably back then, and the land record of ice advances was too incomplete and poor to evaluate his ideas. So they languished in the pile of interesting but untestable ideas for decades.

Croll's ideas were nearly forgotten when an astronomer and mathematician named Milutin Milankovitch revived them. Born in what is now Croatia (part of the Austro–Hungarian Empire back then) in 1879, he was a stellar student and got an engineering degree from the Vienna Institute of Technology in 1904. Despite his day job as an engineer, he was more interested in fundamental research, and by 1912 he took an interest in solving the problems of how variations in solar insolation might affect climate (insolation is the total amount of sunlight received at a point on earth). In 1912 and 1913, he published several papers calculating the amount of solar insolation that the earth receives at each latitude and how this affects the position of climatic belts. Then in July 1914, Archduke Franz Ferdinand was assassinated in Sarajevo, and the crisis between Serbia and the Austro–Hungarian Empire boiled over to become World War I. As a Serbian citizen, Milankovitch was arrested during his honeymoon in Austria, then imprisoned in Esseg Fortress. Fortunately, he had powerful connections in Vienna, so he was then interned in Budapest, with access to materials so he could continue his research. Even though he was technically a prisoner, he used his undisturbed time in Budapest, and his university library access, to make huge advances in mathematical meteorology, which he published in a series of papers from 1914 to 1920. Finally, the war ended, and Milankovitch and his family returned to Belgrade in March 1919, where he resumed his professorship at the University of Belgrade.

Milankovitch built on this research to calculate exact models for how solar insolation cycles might cause ice ages. Crucially, he realized that the key factor is how much sunlight the earth's surface receives in the summertime, which determines how much of the ice melts back—or remains. Milankovitch built on previous understanding of Croll's eccentricity cycle and the precession cycle and added a third that had been discovered by Ludwig Pilgrim in 1904: the obliquity or "tilt" cycle. The rotational axis of the earth is not straight up and down with respect to the plane of its motion around the sun but tilted at 23.5° (Fig. 15.14). That angle is not always constant but fluctuates between about 21.5° and over 24.5°. When the angle is as steeply tilted as 24.5°, the polar regions get a lot more sunlight and the ice melts; when it is as vertical

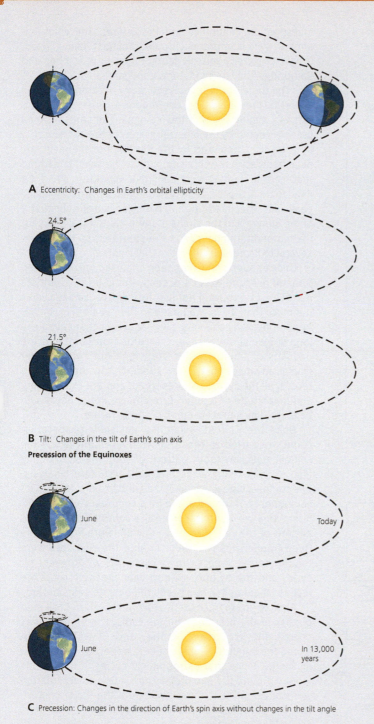

A Eccentricity: Changes in Earth's orbital ellipticity

24.5°

21.5°

B Tilt: Changes in the tilt of Earth's spin axis

Precession of the Equinoxes

June   Today

June   In 13,000 years

C Precession: Changes in the direction of Earth's spin axis without changes in the tilt angle

as 21.5°, the poles get much less sunlight and ice forms. This is known as the "tilt" or **obliquity** cycle of the earth's orbit and takes about 41,000 years to run completely from 21.5° to 24.5° and back again. Milankovitch had all the pieces needed to do the painstaking calculations and plots by hand on many reams of paper, with no computers or electronic calculators. After dozens of scientific papers and short books on the topic of the earth's solar radiation and climate, by the late 1930s Milankovitch focused on putting it all together in one book, *Canon of Insolation of the Earth and Its Application to the Problem of the Ice Ages*.

▷ See For Yourself: Milankovitch Cycles

Once again, world events reached a crisis and interfered with Milankovitch's life and work. Four days after he sent the book to the printer in 1941, the Germans invaded Yugoslavia, and the printing house was destroyed during the bombing of Belgrade. Luckily, the printed pages were in another warehouse and undamaged, and eventually they were bound and published. As the Nazis invaded Serbia in May 1941, two German officers and some geology students came to his home to help him; and he gave them his only bound copy of the book for safekeeping in case something should happen to him or his work. Milankovitch died in 1958 at the age of 79, without ever knowing if his ideas would be supported by geologic data.

But even though Milankovitch had brought the problem of the astronomical cycles and the causes of the ice ages as far as any astronomer or mathematician could do by sheer calculation, there was still no clear evidence from geology to support it. The deposits on land still only showed four or five major glacial advances, and even in the late 1950s their dates were still problematic. The entire idea of astronomical cycles and ice ages was still an unconfirmed speculation. This problem could never be solved with the record of glaciations on land because most of the land record is prone to being eroded away. It is full of gaps that make it incomplete.

▷ See For Yourself: Climate Cycles and Milankovitch Orbital Variations

The solution finally came when long cores of deep-sea sediment were analyzed in the early 1970s. Leading the research on this problem was a group of scientists funded by the National Science Foundation Project CLIMAP (an acronym for Climate: Long-Range

Figure 15.14 The three cycles of orbital variation described by Croll and Milankovitch which control how much solar energy the earth receives. **A.** The longest is the cycle of eccentricity, or how long it takes for the earth's orbit around the sun to go from nearly circular to slightly more elliptical and back. This cycle takes about 110,000 years to complete. **B.** The second fastest is the cycle of tilt or obliquity. Today, the earth's axis is tilted 23.5° from the plane of the earth's orbit around the sun, but in the past it has been as steep at 24.5° or as shallow as 21.5°. When the earth is tilted as steep as 24.5°, the poles receive more sunlight and the ice melts back; when it is as shallow as 21.5°, the poles receive less sunlight and their ice advances. This cycle takes about 41,000 years to complete. **C.** The third cycle, the precession or "wobble" cycle, describes how the earth's spin axis is tilted toward a different part of the sky as it "wobbles" like a top. This cycle takes about 21,000–23,000 years to complete. Together, all three cycles influence how much solar radiation is received at the poles, which changes ice volume and causes glacial–interglacial cycles.

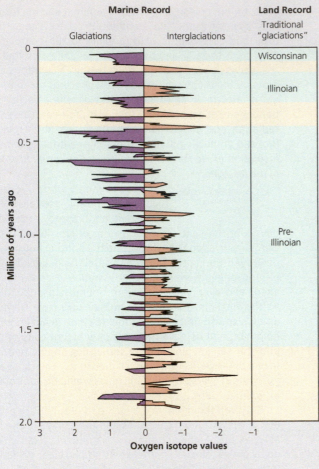

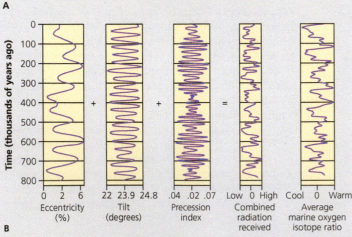

Figure 15.15 **A.** The detailed record of temperature (as determined by oxygen isotopes in deep-sea cores; see Box 14.2) for the last 2 million years. Instead of only four or five Ice Ages, as suggested by land glaciations, deep-sea cores indicated at least 20 glacial–interglacial cycles in this time span. The shape of the temperature curve has a jagged sawtooth pattern. **B.** The three Croll-Milankovitch cycles (left three curves) interact in complex ways to form a pattern of solar radiation received on earth (second curve from right). This signal matches the one measured in deep-sea cores (right curve).

Investigation, Mapping, and Prediction). The chief scientists were Lamont micropaleontologist James Hays, Brown University micropaleontologist John Imbrie, and Cambridge University isotope geochemist Nick Shackleton. The three of them analyzed many different deep-sea cores that preserved long, continuous records of the entire last 2–3 million years of the ice ages. The cores were all precisely dated by the biostratigraphy of the microfossils, by volcanic ashes, and by the flip-flops of the magnetic field recorded in the core sediments. The scientists found that certain temperature-sensitive plankton could be used to track the temperature changes in the ocean in any particular core. In addition, the oxygen isotope chemistry of the minerals in the shells of the plankton is a proxy for the temperature changes in the seawater, so there were several indicators of climate in these cores.

Once enough cores had been analyzed and correlated from all over the world's oceans, the CLIMAP scientists discovered there were not just four or five ice age cycles but more than 20 in the past 2 million years (Fig. 15.12)! Apparently, only the very biggest cycles that cause the largest glacial advances are recorded on land, and the deposits of smaller cycles are wiped out by the larger glacial advances that follow them. But in the deep-sea record, all the cycles are preserved, and the exact timing and magnitude of temperature change in the ocean can be plotted as a precise curve of temperature.

Once the CLIMAP scientists got their temperature curve, they tried to tease apart what might cause the complicated, sawtooth pattern of warming and cooling (Fig. 15.15A). Using a method called spectral analysis, this complex signal was analyzed and broken down to its components. It turned out to be a composite of three different sine waves that formed the complex interference pattern of the real data (Fig. 15.15B). Sure enough, the three frequencies that were behind the real data were the 110,000-year eccentricity cycle, the 41,000-year tilt cycle, and the 21,000–23,000-year precession cycle—just as Milankovitch had predicted over 30 years before.

And so in 1975, exactly 100 years after Croll had published the first book on the topic, the problem was solved. In 1976, Hays, Imbrie, and Shackleton published the legendary "pacemaker" paper, which laid out all the evidence that astronomical cycles of the earth's motion around the sun dictate how much solar radiation we receive and that those cycles are the major controllers or "pacemakers" of the ice ages. Since that time, Croll-Milankovitch cycles have been discovered in all sorts of geological records, including the cycles of Carboniferous coal deposits, the chalk seas of the Cretaceous, and many other places. The confirmation of the Croll-Milankovitch hypothesis has been considered one of the landmark discoveries in geology, and the Hays, Imbrie, and Shackleton "pacemaker" paper of 1976 is ranked as one of the most important scientific breakthroughs of the twentieth century.

powerful indicators of ancient ocean temperatures, and thus global temperature changes. But what did the cores show? Read Box 15.1 for the answer.

15.4 Life in the Ice Ages

MARINE LIFE

In the marine realm, most of the species that lived during the Pleistocene are still alive today, so we can use their presence to track the changes in climate and sea level. We have already discussed how the changes in marine plankton are a powerful tool in determining past ocean temperatures and climate changes in the ocean. Similarly, marine mollusk fossils are commonly found on land, showing the rises in sea level in some places or that the land has uplifted out of the sea in others. The most dramatic effect, however, is on the geographic ranges of marine life. Cold-water species moved away from the polar waters and down toward the lower latitudes as the polar waters were covered by ice and the temperate regions got colder. During interglacial warming intervals, these same species migrated poleward again to the regions that they now occupy since we have been in a warmer interglacial for the last 10,000 years.

LAND PLANTS

The shift of temperature belts toward the equator was even more dramatic on land (Fig. 15.16). When ice sheets covered most of Canada and the northern United States, the region to the south of the ice margin was covered by Arctic-style tundra. Thus, all the vegetation belts shifted to the south as well. For example, most of Pennsylvania, Ohio, Indiana,

Iowa, and the Dakotas were in the tundra belt, rather than having the kinds of vegetation that they do today. South of that, the north spruce forest (which grows today in northern Canada) covered nearly all the middle part of North America, from the Carolinas to Kansas. Finally, a pine-hardwood forest (such as we associate with New England and New York today) grew across the southern states and Texas. Only the southern tip of Florida remained subtropical, as it is today.

On a global basis (Fig. 15.16), the shift of the plant communities from the polar region to the lower latitudes is a widespread pattern. During peak glacial intervals, nearly all of Europe was covered either by ice sheets or by tundra. Most of southern Russia and the Middle East were covered by high, dry steppe grasslands. Only the Mediterranean countries like Spain, Italy, and Greece were covered by dense deciduous forests (such as those in Scandinavia and northern Europe and Russia today) and at higher elevations by coniferous forests—rather than the Mediterranean scrublands and oak woodlands that live there today. Even the coast of North Africa was covered by dense woodlands. The Sahara Desert was about the same size as today but shifted farther south. In fact, nearly all the vegetational belts shifted toward lower latitudes. However, the tropical rainforests did not vanish entirely, nor did the tropics become much cooler. Instead, the region of tropical rainforests just shrank in area so that they occupied a very narrow band around the equator, and much of the area that is now tropical rainforest (such as most of the Congo Basin, the Amazon Basin, and much of southeast Asia) became a tropical grassland savanna as in the tropics of East Africa today.

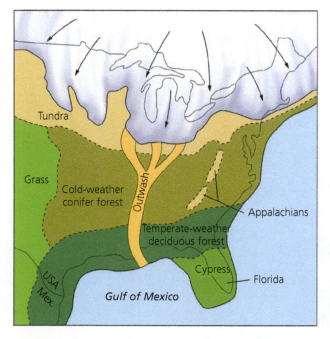

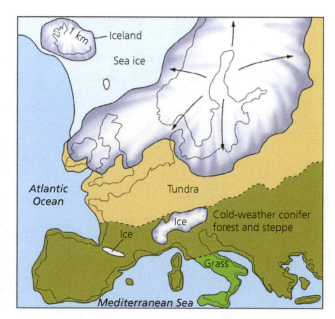

Figure 15.16 Vegetation belts in Europe and North America shifted toward the equator each time the glaciers advanced down from the north. Most of northern Europe and the northwest Midwest of North America not covered by ice was a large belt of tundra. The dense northern forests now found in Canada and northern Siberia moved to regions far south of where they are found today.

ICE AGE ANIMALS

But the best-known examples of Ice Age life are the incredibly diverse mammals, ranging from famous creatures like mammoths and mastodonts and saber-toothed cats and ground sloths to many lesser known animals such as the rodents. Ice Age mammals are now the featured characters in a series of movies and routinely are sold as "dinosaurs" in collections of plastic models of extinct animals.

 See For Yourself: Last of the Mammoths

 See For Yourself: Cloning Mammoths

 See For Yourself: Tar Pits

Ice Age mammals and other land animals (including birds, reptiles, amphibians, fish, and many invertebrates) are found in literally thousands of localities in North America alone, from caves to ancient lakes and ponds to tar pits. But none is richer or better known than Rancho La Brea tar pits in Los Angeles (Fig. 15.17). Over 3 million fossils have been found there, representing 60 species of mammals, 135 species of birds, and 400 other species of reptiles, amphibians, invertebrates, and plants. The largest of these is the Columbian mammoth, which reached 4 m (13 feet) at the shoulder and weighed 10 tonnes (11 short tons), larger than any living elephant. It was also larger than the woolly mammoth, which had longer hair and mostly lived in or near the glaciated areas across Eurasia and northern North America. Next in size was the American mastodon, which had a low, flat skull and longer, less curved tusks and a long shaggy coat. Mastodonts had rounded cusps on their teeth, suitable for browsing leaves (especially in the spruce forests they preferred), while mammoths and elephants had larger molars with nearly flat, ridged grinding surfaces so that they could eat significant amounts of gritty grass. There were also other large herbivores, including horses, llama-like camels, bison (including one bison that had a horn span over 2 m, or more than 6 feet across), and several types of ground sloths.

 See For Yourself: Ice Age Mammals

Preying on these herbivores were a wide variety of predators and scavengers, including the famous saber-toothed cat, the dire wolf, the Ice Age lion (larger than any living cat), the immense short-faced bear (much larger than any living bear), and many others. There were abundant small mammals, too, including many different kinds of rodents, rabbits, and insectivores. Similar mammals roamed much of North America, as well as the armadillo-like glyptodonts, which reached the size of a Smart Car, and the giant beaver, which was the size of a bear.

South America had not only mastodonts, horses, short-faced bears, different species of saber-toothed cats, and rhino-sized capybaras but also 22 kg (about 50-lb) spider monkeys; enormous ground sloths like *Megatherium*, which was 6 m (about 20 feet) tall and weighed over 4 tons, as much as an elephant; and gigantic glyptodonts. It also harbored the last of its strange native mammal groups that dated back to its days as an island continent, including the hippo-like *Toxodon* and the strange *Macrauchenia*, which looked a bit like a cross between a camel and a tapir.

Eurasia had its own distinct assemblage of Ice Age life. In addition to mammoths (some of which are known from freeze-dried specimens with preserved hair and skin and stomach contents), there were woolly rhinoceroses and the enormous elasmothere rhinos, with a straight 2 m (over 6 foot) long horn on the forehead. The predators included different kinds of saber-toothed cats, plus the famous cave bear that was also larger than any living bear. Among the most famous Ice Age mammals is the giant deer, also known as the "Irish elk"—a misnomer since it is related not to the

Figure 15.17 Instead of movie stars, an incredible diversity of mammals roamed the Ice Age Hollywood or any other region south of the glacial margin. This is Los Angeles about 40,000 years ago, when mammoths, mastodons, saber-toothed cats, giant short-faced bears, giant lions, bison, camels, horses, and ground sloths lived and were trapped in the famous Rancho La Brea tar pits.

North American elk or wapiti but to the moose, and it is found in many places besides Ireland. Its enormous moose-like antlers were 3.65 m (12 feet) across and weighed up to 40 kg (88 lb)—and the stag had to shed them every winter and regrow that much bone every year! In Africa there were not only the familiar beasts of the Serengeti today but also huge giraffes with thick, short necks and moose-like horns, which were as tall as living giraffes, and enormous warthogs, almost twice as large as the living species.

Australia also had its own unique assemblage of giant mammals that are now extinct, including the diprotodonts, which looked like wombats but were the size of rhinos; gigantic kangaroos that stood of 2 m (6.6 feet) tall and weighed 230 kg (507 lb); giant koalas; and the "marsupial lion," which looked vaguely lion-like but was a pouched mammal, not a cat. The weird palorchestids had claws like a ground sloth but a proboscis like a tapir or elephant. King of them all, however, was the giant Komodo dragon called *Megalania*, which was at least 4.5 m (about 15 feet) long and possibly up to 7 m (23 feet) long, and weighed 2000 kg (about 4400 lb). There were also enormous flightless birds, just like the emu and cassowary today, only much larger. Islands like Madagascar (elephant bird) and New Zealand (moas) also had flightless birds that reached enormous sizes but were quickly wiped out when humans sailed to these places and hunted them to extinction.

15.5 Ice Age: The Meltdown

See For Yourself:
How Often Do Ice
Ages Happen?

CLIMATE AND OCEAN CURRENTS

The last of the 25 or so Pleistocene glaciations peaked about 18,000–20,000 years ago. The final 8000 years of the Pleistocene, or from about 18,000 years ago to about 10,000 years ago, is called the last **glacial–interglacial transition.**

The peak glaciation at 18,000–20,000 years ago was caused by the Milankovitch cycles of orbital variation (Box 15.1), when all three cycles reached a point of minimal solar radiation to the earth. Then the cycles shifted again, and the earth began to slowly warm out of the peak glacial conditions and toward an interglacial phase by 10,000 years ago, when all three Milankovitch cycles produced the maximum amount of solar input and planetary warming. It was not a simple steady increase in temperature, however (Fig. 15.18). There were many small-scale fluctuations in climate, with both short-term warming events and cooling events, before the climate stabilized into an interglacial mode. Some were due to complex interactions of the Milankovitch cycles, but most were due to events in the earth's oceans and atmospheres.

See For Yourself:
Bill Nye on
Ice Ages

The most famous of these events was a **Heinrich event** about 16,000 years ago, when the Arctic ice sheet began to fall

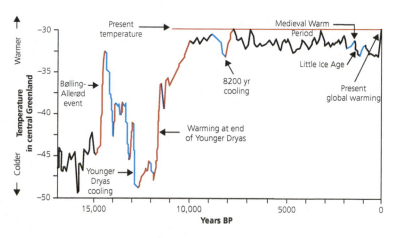

Figure 15.18 The temperature record of the glacial–interglacial transition as recorded in ice cores, mainly from Greenland. The peak of glaciation ended about 18,000 years ago, and about 16,000 years ago was the last of the Heinrich events releasing Arctic icebergs into the ocean. The detailed record shows numerous short-term warming and cooling events known as Dansgaard-Oeschger cycles. About 14,700 years ago was another warming cycle, called the Bølling-Allerød event, followed by the abrupt cooling event from 12,900 to 11,700 years ago known as the Younger Dryas. After the planet warmed up again at the end of the Younger Dryas, climate remained relatively stable for the past 11,000 years of the Holocene.

apart and released huge volumes of floating icebergs and freshwater into the North Atlantic Ocean. Huge Heinrich events release lots of icebergs. These are detected by finding large pebbles and coarse sand grains melted out of icebergs and dropped into the seafloor muds found in deep-sea cores. Heinrich events are known from the last seven glacial periods, dating back 640,000 years. However, a Heinrich event at 16,000 years ago launched the most rapid phase of warming in the glacial–interglacial transition. About 14,680 years ago, an abrupt warming event increased the moisture of Greenland, and the ice sheet almost doubled in thickness with all the extra snow and ice. This warming episode is correlated with another peak in the eccentricity cycle of the Milankovitch orbital variation cycles.

In addition, through this entire interval of climatic fluctuations were a number of **Dansgaard-Oeschger events**, rapid warming episodes lasting a decade or more, followed by gradual cooling. There have been at least 25 of them during the last glacial period and the glacial–interglacial transition, roughly spaced about 1470 years apart. They show up mainly in Greenland ice cores but not in Antarctic ice cores. Based on these records and North Atlantic deep-sea sediment cores, it is thought that they are caused by rapid pulses of freshwater and ice released from the melting of the Arctic ice sheet. What caused these quasi-periodic events is still debated, although they might be due to the cooling and excessive buildup of Arctic ice, followed by a rapid "purge" of ice, or possibly by the changes in the shape and size of northern ice sheets.

Even more dramatic was a rapid cooling event near the end of the Pleistocene, about 11,700–12,900 years ago, known as the **Younger Dryas event**. First spotted in Greenland ice cores with very high-resolution yearly records

(Fig. 15.18) and confirmed in cores of Swiss and Scandinavian glacial lake sediments that record every year precisely, it showed just how quickly the entire climate system can flip from warming interglacial back to full glacial—in about 10 years! The interval got its name after the abundance of pollen of *Dryas octopelta*, a European alpine tundra flower that indicates cold conditions, in cores taken from Swiss lakebeds. The Younger Dryas was long a mystery to paleoclimatologists. It began with an abrupt cooling that lasted a few decades about 12,900 years ago, which plunged the world into another glacial episode until it abruptly warmed in less than a decade about 11,700 years ago.

But then deep-sea cores from the North Atlantic led climate modelers to suggest a startling explanation. In Chapter 14, we discussed the "global ocean conveyer belt" which connects the flow of all the world's oceans in a gigantic looping current that takes a water molecule over 2000 years to complete the circuit. For example, we saw how the Gulf Stream (Fig. 14.35B) was diverted northward to the North Atlantic when the Panama land bridge closed and shut off the flow from the Caribbean to the Pacific. That "conveyer belt" of the Gulf Stream then brought warm, moist air to the Arctic, where it froze and snowed and allowed the Arctic ice sheet to develop. Once the Gulf Stream reaches far enough into the North Atlantic north of Scotland, it runs into cold Arctic waters, where it quickly cools, then sinks, and returns to the South Atlantic in a deep-water current. That deep current then connects with cold Antarctic currents, then runs all the way to the North Pacific, where it warms again. From there it becomes a warm surface current, which flows back into the Indian Ocean, around Africa, and then back into the South Atlantic, eventually to join with the Gulf Stream and complete the conveyer loop (Fig. 14.35B).

That is how the global ocean conveyer works today. But what happened if it were to shut off or slow down? Then all the climate systems that we now take for granted would be thrown into chaos, and there would be an entirely new regime. Apparently, this is what happened during the Younger Dryas event. Deep-sea cores showed evidence of a sudden change in the water chemistry in the North Atlantic when it began. They traced this change to a sudden flood of water from ice-dammed lakes on the fringes of the glaciers melting in North America (especially glacial Lake Agassiz, which covered much of Manitoba and Saskatchewan as well as northern Minnesota and North Dakota). When the ice dam on the St. Lawrence River broke, it released a gigantic flood of freshwater into the North Atlantic. Freshwater is less dense than salt water, so this freshwater "lid" sat on top of the saltier ocean water and caused the "conveyer" to sink much farther south than it does today. This prevented the warmth of the Gulf Stream from reaching the far northern Atlantic, and within a decade, Ice Age conditions had returned to the northern hemisphere. From this, we know that the earth system is very sensitive to slight changes and can switch from warm interglacial to full glacial in just a few years given the right conditions.

SEA-LEVEL CHANGES

As all this ice melted, sea level began a steady and continuous rise from its all-time low during the peak of the last glaciation 18,000–20,000 years ago, when it was more than 137 m (about 450 feet) below present sea level (Fig. 15.19). About 18,000 years ago, during the peak glaciation, huge areas of what is now the submarine continental shelf were exposed as forested floodplains. We know this not only because of the various geological indicators of ancient sea level but also because intact mammoth fossils have been dredged off the continental shelf at 137 m (450 feet) of water depth, and they are too large to have washed out there from modern rivers. When sea level was so low, many areas that are now islands were connected to the mainland. For example, the entire English Channel dried up and was exposed and became a land connection, so mammals like hippos and hyenas migrated from Africa and Eurasia to England (except for the glaciated northern part of Great Britain, especially Scotland). The Adriatic Sea also vanished, so you could walk from Italy to Greece or Croatia across it. Many of the modern-day islands in the Mediterranean were connected to the mainland, allowing mammoths and hippos to move out there. Once they became isolated during episodes of rising sea level, these large mammals became island dwarfs. This is a common effect in large mammals like mammoths, hippos, and rhinos. They no longer need large body sizes for protection when there are no large predators on islands, and they must make do on much smaller food resources on islands as well.

Once the melting began, sea level began to steadily rise from 137 m below modern levels to almost present-day levels by 6000 years ago (Fig. 15.19). The continental shelves were again drowned as the sea poured across them, filling up ancient river valleys to make features like Chesapeake Bay (which is shaped like a river valley that has been flooded). Ancient islands that were previously connected to the mainland were cut off again and became islands once more.

Not every coastline showed the evidence of sea-level rise and transgression, however. The Scandinavian block, which had once been overlain by several miles of ice, was suddenly released of its icy burden in only a few thousand years. But crustal blocks and the mantle below them don't respond that quickly. They are floating like a cork on the mantle of the earth, but they move up and down much more slowly than ice can melt. Thus, the disappearance of many tons of ice allowed the Scandinavian crust to slowly rise higher and higher since nothing is pushing it down into the mantle. Even today, Scandinavia is still rising out of the ocean, as it has been for the last 18,000 years, in a process called **glacial rebound**. The same is happening in much of northern Canada and parts of Siberia. As a result, the shorelines of those regions do not exhibit transgressive seas like those of the rest of the world, but the seas are actually regressing as the land pops up like a floating object riding on liquid pressed down and then suddenly released (Fig. 4.16B). There are even ancient marine shoreline deposits on the

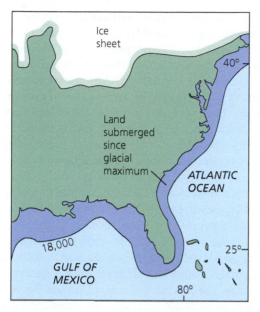
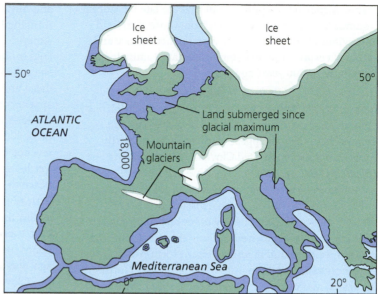

Figure 15.19 When the glaciers were at their maximum about 18,000–20,000 years ago, they withdrew large volumes of water from the ocean and locked it up in ice. This lowered sea level about 150 m (450 feet) so that the entire continental shelves of many of the continents became river floodplains. This not only expanded the land area of North America but created land connections between many islands and mainland Europe. Animals could walk across the English Channel to Great Britain and to many of the Mediterranean islands; other islands were not completely connected to the land, but it was just a short swim or float to reach them. All of these regions were flooded when sea level rose between 18,000 and 11,000 years ago.

Scandinavian coast, showing this rapid rise of the coastline and relative retreat of the ocean.

EFFECTS ON LAND

During the glacial–interglacial transition in the late Pleistocene, the great ice sheets melted back from the continents, leaving numerous terminal and recessional moraines (Fig. 15.5). They also retreated from the great mountain valleys, leaving the vertical-walled, "U"-shaped valleys with waterfalls that are characteristic of glaciated landscapes the world over.

The great glacial lakes, such as Lake Agassiz in Manitoba, Lake Bonneville in western Utah, and Lake Lahontan in northwestern Nevada, also dried up, as did most of the lakes filling the valleys of the Basin and Range province in Nevada and Utah and California (Fig. 15.8). Eventually, the lake basins became giant salt flats surrounded by sagebrush and salt bush. The tops of the ranges in the Great Basin are cloaked with the remnants of ancient forests that used to cover the entire region. As the valleys between the ranges dried up, the forests on the tops of the ranges became isolated. This stranded certain types of rodents and other animals that began to genetically diverge and form new species because they no longer interbred with animals on any other range. Likewise, the tiny permanent springs and ponds that remain in this desert are inhabited by hardy fish that can tolerate extremes of temperature and salinity, like the desert pupfish. Each of these ponds used to be part of a big lake with a common fish population, but now that they are all isolated in a desert, the pupfish are beginning to develop new genetic features and are becoming a different species.

15.6 Where Have All the Mammals Gone?

Almost as soon as people realized in the early 1800s that mammoths and mastodonts were extinct, there were debates about what caused them to vanish. The arguments have been raging for almost 200 years since that time but have boiled down to just a few reasonable ideas.

▷ See For Yourself: Extinction of the Megamammals

- *Climate change:* Most of the extinctions are concentrated at the end of the last Ice Age (especially 12,000–9000 years ago), when the earth rapidly warmed out of the wetter, cooler climates of the last glacial episode into the hotter, drier world we have now. Such climate changes mean dramatic changes in rainfall and vegetation, which many herbivores cannot tolerate. In particular, there were some extreme climate swings, such as the Younger Dryas event about 11,700 years ago, when the Ice Age glaciers returned in just a decade after a long warming period, then melted back almost as quickly—a climate shock that seems to explain some of the extinctions. Other scientists have argued that climates went from being more uniform and mild to more extreme and continental, with rapid shifts in temperature typical of the centers of continents far from the moderating effects of the oceans. Critics of the climate hypothesis have argued that there was no similar mass extinction during previous interglacial periods (such as 125,000 years ago), when climate warmed just as it did 10,000 years ago. But supporters of the climate change scenario have pointed out that the late Pleistocene communities were very different from those at the end of previous glacials, especially with the abundance and extent of certain cold, dry steppe habitats dominated by bison only 12,000 years ago.

- *The "overkill" or "blitzkrieg" model*: Ever since the 1970s, some scientists have argued that humans were the primary agent of mass extinction at the end of the last Ice Age. They point to the first appearance of peoples in North America with sophisticated Clovis-style spear points and arrowheads, right about the time that many of the megamammals vanished. We do have proof that humans killed mammoths, bison, and horses; but we have no direct evidence of them hunting other megamammals. In addition, the bison survived the hunters and became the largest species of land mammal left in North America after the others vanished. It appears that many other North American mammals survived for thousands of years after humans first appeared, so if it was a "blitzkrieg," it wasn't very fast. The "overkill" argument has also been applied to the arrival of Aborigines in Australia, but again it was a complex process because the megamammals mostly died out thousands of years before people appeared. And clearly human hunting is not very important to the animals of Eurasia and Africa, which survived and even evolved for hundreds of thousands of years alongside humans.

- *Giant comet*: In 2007, a trendy new idea made a splash in all the scientific media: a giant comet allegedly impacted over the Carolina Bays region about 12,900 years ago, which might have caused climate changes that affected the North American vegetation and megamammals. However, detailed double-checking and research on the evidence for this impact have pretty well demolished the idea that it ever happened. Even if the evidence for the impact were supported, almost no species vanish at the time of the impact, and it would have no bearing whatsoever on the mass extinctions on other continents.

- *The "keystone" hypothesis*: Some scientists have looked at the megafaunas living in the African savanna today and point out that their diversity is the product of one "keystone" species. That species is the elephant, which breaks down the trees and prevents the forest from becoming too dense. Elephant damage allows for a much more diverse habitat of trees, brush, scrub, and grasses, which sustains the huge populations of antelopes, zebras, giraffes, and other megamammals. If something were to wipe out the mammoths during the last Ice Age (whether climate or humans), it might have a cascading effect on the vegetation and wildlife across the continent, especially in North America. This idea is very interesting but seems to be a secondary effect of whatever major cause killed large mammals such as mammoths in the first place.

Currently, the most popular idea in the media and general public is that humans are mostly to blame for the extinction of the megamammals, especially in the Americas

and Australia (clearly, not in Africa or Eurasia). Conventional thinking has long argued that there were no humans in the Americas until people with Clovis points crossed the Bering land bridge and swept south about 13,000 years ago. But at a site called Monte Verde in southern Chile, the radiocarbon dates are much older. The oldest level bears charcoal that appears to be 33,000 years old, although this is controversial. Higher levels produce good-quality radiocarbon dates of 18,000 years ago, at least 5000 years before Clovis people reached the Americas. Most archeologists believe the Monte Verde people were coastal migrants who traveled from one beach to another in their canoes and did not move far inland with advanced spears and arrows to hunt the megamammals, as did the Clovis people more than 1000 years later. However, there are inland sites like Meadowcroft Rock Shelter in Pennsylvania, which yields radiocarbon dates of 19,000 years ago, and a number of other inland sites in the Americas with pre-Clovis dates. The Bluefish Cave in the Yukon Territory gives reliable dates of 24,000 years ago. The simplistic model of a single big "blitzkrieg" of human hunters arriving only 13,000 years ago has clearly become more complicated.

Not only that but not all the megamammals died out when humans appeared in the New World. Mammoths actually survived on Wrangel Island in the Siberian Arctic until about 3700 years ago (about 1700 B.C.E., later than the building of the Egyptian Pyramids). Mastodonts were still hanging on in North America until about 4000 years ago, and there were ground sloths on some Caribbean islands until 4700 years ago—so not all the megamammals vanished in prehistoric times.

There is another kink as well. The record of fossil bones may not be giving the whole story. In 2009, a group of scientists led by James Haile analyzed samples of soil from permafrost deeply buried in northern Alaska. They found evidence of extinct horse and mammoth DNA in permafrost which was radiocarbon-dated to be 2000 years younger than the youngest bones of these animals! Thus, we must be cautious of overgeneralizing and overinterpreting the record of bones, which we know is incomplete. If these data are correct, horses and mammoths survived for thousands of years after human hunters arrived, which doesn't fit the "blitzkrieg" or "overkill" model at all.

Unfortunately, people (especially reporters, and sometimes including scientists) like clear, simple answers to complex problems. The climate did it! No, humans are to blame! This is the usual way the debate is framed: black-and-white, either/or choices. But most scientists who study the Pleistocene megafaunal extinctions point out that nature is far more complex than such simplistic models that we like to argue about. In most cases, it's clear that there were effects of both climate (weakening and reducing the populations of megamammals as their habitats and diets changed) and possibly human overhunting in some cases. In no case is there an extinction event where only one cause is required. It's not simple or easy to understand, and it can't be reduced to a simple sound bite for reporters and science media; but that's the way most scientific debates are eventually resolved.

RESOURCES

BOOKS AND ARTICLES

Barnosky, A. D., P. L. Koch, R. S. Feranec, S. L. Wing, and A. B. Shabel. 2004. Assessing the cause of Late Pleistocene extinctions on the continents. *Science* 306:70–75.

Blakey, R. C., and W. D. Ranney. 2018. *Ancient Landscapes of Western North America: A Geologic History with Paleogeographic Maps.* Springer, Berlin.

Bolles, Edmund B. 1999. *The Ice Finders: How a Poet, a Professor, and a Politician Discovered the Ice Age.* Counterpoint, New York.

Childs, Craig. 2018. *Atlas of a Lost World: Travels in Ice Age America.* Pantheon, New York.

Cohen, Claudia. 2002. *The Fate of the Mammoth: Fossils, Myth, and History.* University of Chicago Press, Chicago.

Cooper, Alan, C. Turney, K. A. Hughes, B. W. Brook, H. G. Macdonald, and C. J. A. Bradshaw. 2015. Abrupt warming events drove Late Pleistocene Holarctic megafaunal turnover. *Science* 349:602–606.

Fariña, Richard A., and Sergio F. Vizcaíno. 2013. *Megafauna: Giant Beasts of the Pleistocene of South American.* Indiana University Press, Bloomington.

Flint, Richard Foster. 1971. *Glacial and Pleistocene Geology.* John Wiley, New York.

Forbes, James. 1843. *Travels Through the Alps of Savoy.* Longman, Brown, Green, London.

Grayson, Donald. 2016. *Giant Sloths and Sabertooth Cats: Archeology of the Ice Age Great Basin.* University of Utah Press, Salt Lake City.

Gribbin, John, and Mary Gribbin. 2015. *Ice Age: The Theory that Came in from the Cold.* Barnes & Noble, New York.

Guthrie, R. Dale. 1990. *Frozen Fauna of the Mammoth Steppe: The Story of the Blue Babe.* University of Chicago Press, Chicago.

Haile, James, and others. 2009. Ancient DNA reveals late survival of mammoth and horse in interior Alaska. *Proceedings of the National Academy of Sciences USA* 106:22352–22357.

Haynes, Gary. 2009. *American Megafaunal Extinctions at the End of the Pleistocene.* Springer, Berlin.

Imbrie, John, and Katherine Palmer Imbrie. 1986. *Ice Ages: Solving the Mystery.* Harvard University Press, Cambridge, MA.

Koch, Paul, and A. D. Barnosky. 2007. Late Quaternary extinctions: state of the debate. *Annual Reviews of Ecology, Evolution, and Systematics* 37:215–250.

Kurtén, Bjorn. 1988. *Before the Indians.* Columbia University Press, New York.

Kurtén, Bjorn, and Elaine Anderson. 1980. *Pleistocene Mammals of North America.* Columbia University Press, New York.

Lange, Ian. 2002. *Ice Age Mammals of North America.* Mountain Press, Missoula, MT.

Levy, Sharon. 2011. *Once and Future Giants: What Ice Age Extinctions Tell Us About the Fate of the Earth's Largest Animals.* Oxford University Press, Oxford.

Lister, Adrian. 2015. *Mammoths: Giants of the Ice Age.* Chartwell Books, London.

Marshall, L. G. 1984. Who killed Cock Robin? An investigation of the extinction controversy. In *Quaternary Extinctions: A Prehistoric Revolution.* Ed. P. S. Martin and R. G. Klein. University of Arizona Press, Tucson, pp. 785–806.

Martin, P. S. 1984. Prehistoric overkill: the model. In *Quaternary Extinctions: A Prehistoric Revolution.* Ed. P. S. Martin and R. G. Klein. University of Arizona Press, Tucson, pp. 354–403.

Martin, P. S. 2005. *Twilight of the Mammoths: Ice Age Extinctions and the Rewilding of North America.* University of California Press, Berkeley.

Martin, P. S., and R. G. Klein (eds.). 1984. *Quaternary Extinctions: A Prehistoric Revolution.* University of Arizona Press, Tucson.

McDougall, Doug. 2004. *Frozen Earth: The Once and Future Story of the Ice Ages.* University of California Press, Berkeley.

Meldahl, Keith. 2013. *Rough-Hewn Land: A Geologic Journey from California to the Rocky Mountains.* University of California Press, Berkeley.

Pielou, E. C. 1991. *After the Ice Age: The Return of Life to Glaciated North America.* University of Chicago Press, Chicago.

Prothero, D. R. 2006. *After the Dinosaurs: The Age of Mammals.* Indiana University Press, Bloomington.

Prothero, D. R. 2016. *Giants of the Lost World: Dinosaurs and Other Extinct Monsters of South America.* Smithsonian Books, Washington, DC.

Prothero, D. R. 2016. *The Princeton Field Guide to Prehistoric Mammals.* Princeton University Press, Princeton, NJ

Prothero, D. R. 2018. *The Story of the Earth in 25 Rocks.* Columbia University Press, New York.

Pyne, Lydia, and Stephen J. Pyne. 2013. *The Last Lost World: Ice Ages, Human Origins, and the Invention of the Pleistocene.* Penguin, New York.

Ruddiman, W. J. 2013. *Earth's Climate: Past and Future.* W. H. Freeman, New York.

Sutcliffe, A. J. 1985. *On the Track of Ice Age Mammals.* Harvard University Press, Cambridge, MA.

Turner, A., and M. Anton. 2004. *National Geographic Prehistoric Mammals.* National Geographic Society, Washington, DC.

Woodward, Jamie. 2014. *The Ice Age: A Very Short Introduction.* Oxford University Press, Oxford.

SUMMARY

- The Pleistocene epoch, or the ice ages, runs from 2.6 Ma to 10,000 years ago and represents a time when over 20 glacial–interglacial cycles dominated the climate of the planet.

- The existence of the ice ages was first recognized by Louis Agassiz in 1837, and within 20 years it had been established to be a global event that affected the landscape of most of the continents.

- During peak glacial episodes, ice sheets over a mile thick covered not only the poles but the high-latitude regions of the earth as well. Their presence can be detected by the moraines they left when the glacier paused in one place. Much of the landscape of North America is built of ancient glacial moraines.

- Pleistocene glaciers built the Great Lakes and the Finger Lakes and rearranged the drainages of most continents, including North America. They also produced huge lakes and wetter, cooler climates in the dry western deserts of the United States, which were occasionally drained by huge floods when their ice dams broke.

- Mapping and dating of glacial moraines suggested only four big glaciation events in North America and five in Europe because the land record is very incomplete and bigger glacial advances wipe out the record of earlier, smaller advances.

- When deep-sea cores were studied in the 1960s and 1970s, they turned out to have a much longer, more complete record of ice age temperatures preserved in their microfossils. These more complete records showed that there were at least 20 separate glacial advances in the last 2 million years, not the four or five that were recorded on land.

- The answer to the puzzle of what caused the ice age cycles was suggested by mathematicians and astronomers like James Croll in 1875 and Milutin Milankovitch in 1937,

but it wasn't until 1975 that complete enough records from the deep-sea cores and their plankton confirmed that the orbital variation cycles were the cause of ice ages.

- The Croll-Milankovitch cycles consist of three separate cycles: the eccentricity cycle, the tilt cycle, and the precession or "wobble" cycle. Together these three independent cycles interact with each other to produce the distinctive pattern of temperature and climate that has occurred over the past 2.6 million years.

- Land plants had to shift their vegetation belts to lower latitudes when the ice advanced, so northern Europe and northern North America just south of the glacial margin were covered by tundra, and boreal forests were found much closer to the equator.

- The world of the ice ages was ruled by a wide variety of large mammals, including mammoths, mastodonts, saber-toothed cats, giant ground sloths, and many others, most of which are now extinct.

- The cause of the extinction of these megamammals is debated, but it was probably a combination of rapid changes in climate and vegetation, combined with the effects of overhunting of some species.

- The glacial–interglacial transition from 18,000 to 10,000 years ago was not a steady climb out of the peak glacial to the modern interglacial but had many fluctuations, especially the rapid Younger Dryas cooling event that returned the planet to glacial conditions in merely a decade.

KEY TERMS

Pleistocene (p. 404)
Holocene (p. 404)
Drift (p. 404)
Diluvium (p. 404)
Erratics (p. 404)
Moraine (p. 407)
Recessional moraine (p. 407)

Terminal moraine (p. 407)
Wisconsinan glaciation (p. 407)
Illinoian glaciation (p. 408)
Kettle lake (p. 409)
Loess (p. 410)
Würm glaciation (p. 413)

Solar insolation (p. 414)
Croll-Milankovitch cycles (p. 414)
Eccentricity cycle (p. 414)
Precession cycle (p. 414)
Tilt or obliquity cycle (p. 415)
Pacemaker of the Ice Ages (p. 416)

Glacial–interglacial transition (p. 419)
Heinrich events (p. 419)
Dansgaard-Oeschger events (p. 419)
Younger Dryas event (p. 419)
Glacial rebound (p. 420)
Overkill model (p. 422)

STUDY QUESTIONS

1. How did early geologists first explain the odd deposits they called "Diluvium" or the weird rocks out of place they knew as "erratics"?
2. Why was the idea of "ice ages" initiated by Swiss geologists?
3. What was the crucial event that allowed geologists to first visualize Europe covered by an ice sheet over a mile thick?
4. What events led to the first Arctic ice cap about 3.5 Ma? When did the Ice Age cycles begin?

5. Describe the deposits left behind by a glacier after it melts.
6. What were the two youngest glaciations detected by geologic mapping in the Midwest?
7. How were the Great Lakes formed? Why do they have the elongate lobate or elliptical shape that they do?
8. Before the water of the Great Lakes flowed down through the St. Lawrence River, where did its meltwater flow to?
9. Why is Minnesota called "The Land of 10,000 Lakes"? How were most of those lakes formed?

10. What is loess? Why is it crucial to the agricultural productivity of the world?

11. Describe how different places like Salt Lake City, Utah, and Las Vegas, Nevada, would have looked during the peak of the last glaciation 20,000 years ago.

12. Why do the Finger Lakes fan out like fingers in a hand?

13. What explains the odd shape of eastern Long Island, Cape Cod, and islands like Nantucket and Martha's Vineyard?

14. What were the four original glaciations mapped in North America called? What were the five glaciations in the Alps named? Why was it impossible for a long time to determine whether they matched up?

15. Why was the ice age record in the ocean superior to that of glacial advances on land?

16. How do foraminifera respond to changes in water temperature? How does the chemistry of their shells respond?

17. Name and describe the three orbital variation cycles. What are their periods?

18. Why was it impossible for over a century to tell if Croll and Milankovitch were right about orbital variation cycles triggering the ice ages? What breakthrough proved them right?

19. How did the glacial advances affect the vegetation in North America and Eurasia?

20. What was the Younger Dryas event? How did it change our opinions about how quickly an ice age could start? What was the cause of this rapid ice advance?

21. How did sea-level changes during peak glacials affect the ability of mammals to cross between islands?

22. Why is sea level retreating in places like Scandinavia, while it is rising everywhere else?

23. Discuss the pros and cons of the major ideas of what caused the extinction of the megamammals at the end of the Pleistocene.

24. If Haile and others are right about frozen DNA in the Arctic tundra, what does this suggest about inferring the extinction of an animal based on the last hard part fossilized?

We must, however, acknowledge,
as it seems to me,
that man with all his noble qualities,
still bears in his bodily frame the indelible stamp
of his lowly origin.

—*Charles Darwin, 1871, The Descent of Man*

We're all one dysfunctional family
No matter where we nomads roam
Rift Valley Drifters, drifting home
genome by genome
Take a look inside your genes, pardner,
then you will see
We've all got a birth certificate from Kenya

—*Roy Zimmerman, "Rift Valley Drifters"*

A museum display of replicas of the skulls of the many fossil humans brings home the point that there is now an excellent fossil record of human evolution, with many transitional fossils, from those that are very ape-like to those that are only slightly different from modern humans.

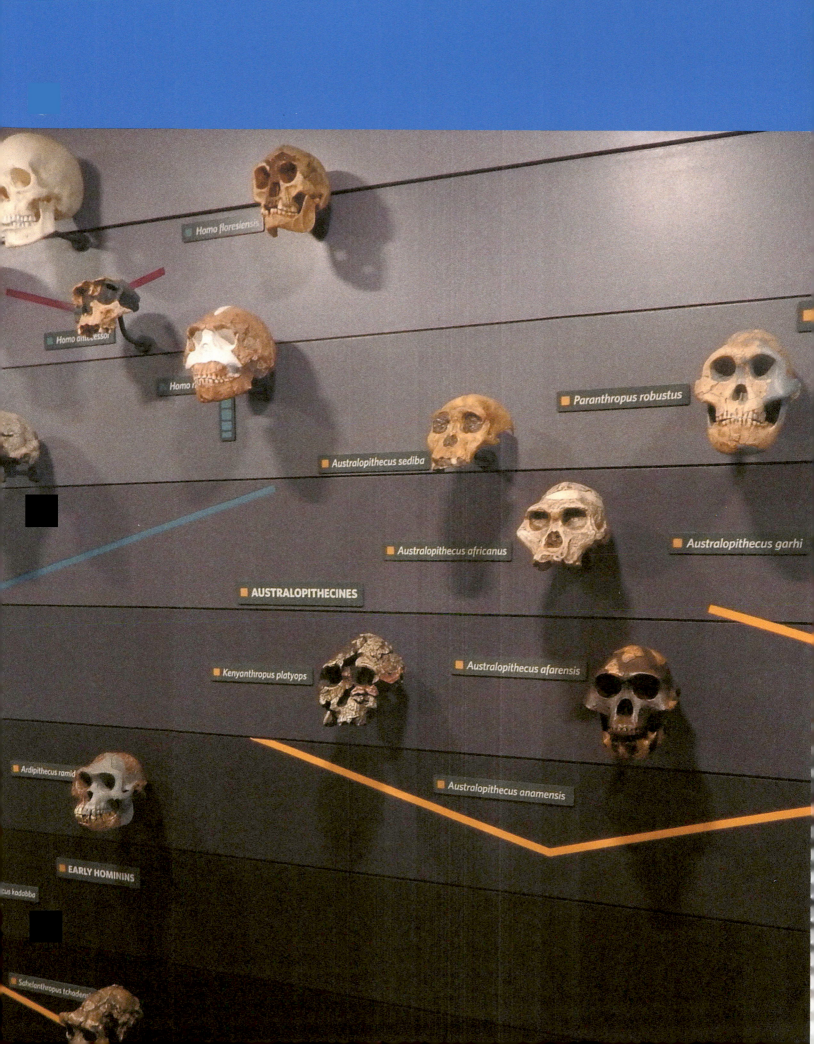

Homo floresiensis

Homo antecessor

Homo r...

Australopithecus sediba

Paranthropus robustus

Australopithecus africanus

Australopithecus garhi

AUSTRALOPITHECINES

Kenyanthropus platyops

Australopithecus afarensis

Ardipithecus ramid...

Australopithecus anamensis

EARLY HOMININS

...cus kadabba

Sahelanthropus tchadens...

16.1 The Descent of Man

As humans, we are always curious about our own origins and our place in nature. Who are we? Where did we come from? What does it mean to be human? Many different cultures have had a variety of creation myths that told the people how they fit in with nature and their deity. But as educated people of the twenty-first century, we are interested in the scientific evidence of how our species evolved and where we came from.

The evidence for human origins has increased exponentially since the 1970s, so the story is now very well known and based on enormous mountains of evidence (Fig. 16.1). But it was not always so. At one time there was almost no fossil record of humans, and the evidence from molecular biology has emerged only since the 1990s. But the bigger problem is the fact that what most people think they "know" about human evolution is just plain wrong or based on outdated or distorted or false notions. These "myths" get in the way of having any meaningful understanding or conversation about the topic. However, just because many people have false, outdated, or mistaken notions in their heads doesn't invalidate the evidence for human evolution.

There are a number of misconceptions about human evolution, but these are some of the major ones.

APES AND HUMANS

Part of the problem with people accepting that humans are part of the kingdom Animalia, and connected to nature, comes from all sorts of myths and misconceptions that are widespread in the public imagination. The idea that we might be related to apes was shocking when it was first proposed. When *On the Origin of Species* was published in 1859, Darwin deliberately avoided the subject in his already controversial book and finally dealt with it in 1871 with the publication of *The Descent of Man*. There was almost no human fossil record for him to point to, and there were still not many human fossils until almost a century later. But Thomas Henry Huxley, Darwin's biggest defender, was not afraid of offending the sensibilities of the Victorian people of England and boldly published *Evidence as to Man's Place in Nature* in 1863, with explicit diagrams showing the detailed skeletal similarities between humans and the great apes (Fig. 16.2). As he showed, humans are bone-for-bone identical with the rest of the apes. The only differences are those of size and proportions, so the arm-swinging apes (gibbons and orangutans) have relatively longer arms than do humans, while humans have relatively longer legs than the other apes.

As the years have passed, however, the gulf between humans and the rest of the apes has narrowed considerably. Instead of the old "screaming, hooting ape" stereotype, we have discovered just how similar the apes are to humans. Decades of field research by pioneering anthropologists like Jane Goodall with the chimpanzees, Birute Galdikas with orangutans, and the late Dian Fossey with the mountain gorilla have demystified these majestic creatures and surprised us with their amazing behavioral similarities to humans. Both chimpanzees and gorillas can learn sign language, communicate in simple sentences, and make and use simple tools. Their societies are very sophisticated compared with those of any other animal and show us many insights into the complexities of human societies. Over a century of research by hundreds of anthropologists has documented more and more connections between apes and humans. The boundary between "human" and "animal" has all but vanished since every time someone

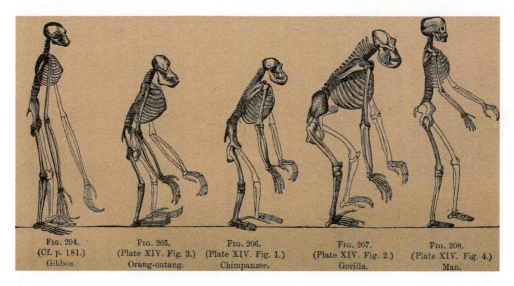

Figure 16.2 Thomas Henry Huxley's famous diagram from *Evidence as to Man's Place in Nature* (1863), which emphasized the complete bone-for-bone similarity in the skeletons of the great apes and humans. We have nearly every bone found in other apes, and the only real differences are size and proportions. For example, apes like gibbons and orangutans have relatively long arms, while we have relatively short arms and long legs.

has tried to prove we were more "special" than any animal, the anatomical or behavioral evidence shows that at least some animals have that feature as well. As a result, the entire scientific community and most educated people now accept that humans are part of nature and connected to other animals in many profound ways.

"MISSING LINKS" AND "MARCH THROUGH TIME"

As we discussed in Chapter 6, *evolution is a bush, not a ladder*. The old idea was the "ladder of life" or *scala naturae*, where the "lower forms" of life like sponges and corals are at the bottom, then mollusks, then fish, amphibians, reptiles, birds, mammals, and then humans at the top. This false notion dates back to the days of Aristotle more than 2500 years ago, but it became obsolete in 1859 when Darwin showed that nature produces a bushy, branching history—a **family tree**, in familiar language. Organisms have branched off the family tree of life at different times in the geological past, and some have survived quite well as simple corals or sponges, while others have evolved more sophisticated ways of living. Corals and sponges, although simple compared to other organisms, are not "lower" organisms, nor are they evolutionary failures for not advancing up the ladder. They are not trying to evolve into a "higher" organism like a worm or a mollusk. They are good at doing what they do (and have been doing for over 500 million years), and they exploit their own niches in nature without any reason to change into something different.

Likewise, the "ladder of life" was often called the "great chain of being" by pre-Darwinian scholars. But life is neither a chain nor a ladder but a branching bush. This is where the ridiculous, outdated idea of a "missing link" in this chain came from. The entire concept of "missing link" is biologically meaningless, and no scientist uses the term—only people who don't understand how evolution works still use

it. We don't discuss missing links in human evolution either because the concept is archaic and useless. However, as we shall outline in this chapter, there are literally thousands of fossils of humans that give us a nearly complete record with only small gaps, more than enough to demonstrate for a fact that humans have evolved.

Nevertheless, this antiquated and long-rejected view of life as a ladder of creation or great chain of being still seems to lurk behind many people's misunderstandings of biology and evolution. For example, it is still common to hear the question, "If humans evolved from apes, why are apes still around?" The first time scientists hear this question, they are puzzled because it makes no sense whatsoever—until they realize this person is still using concepts that were abandoned over 160 years ago. We now know that nature produces a pattern of relationships that is not a ladder but a bush (Figs. 6.2, 14.41B, 14.42, 14.43, 14.44, 16.3). Lineages branch and speciate and form a bushy pattern, with the ancestral lineages sometimes living alongside their descendants. Humans and apes had a common ancestor about 7 Ma (based on evidence from both the fossils and molecular biology), and both lineages have persisted ever since then. Asking "If humans evolved from apes, why are apes still around?" is comparable to asking "If you are descended from your father, why didn't your father die when you were born? Why didn't your grandfather die when your dad was born?" We all understand that we children branch off from our parents, we overlap with them in time, and they do not automatically die when we are born. Similarly, the human family tree branched off from the common ancestors with the rest of the living apes about 7 Ma, but both are still around.

Likewise, the tendency to put things into simple linear order is a common metaphor for evolution—and one of its greatest misrepresentations. The iconic image is the classic

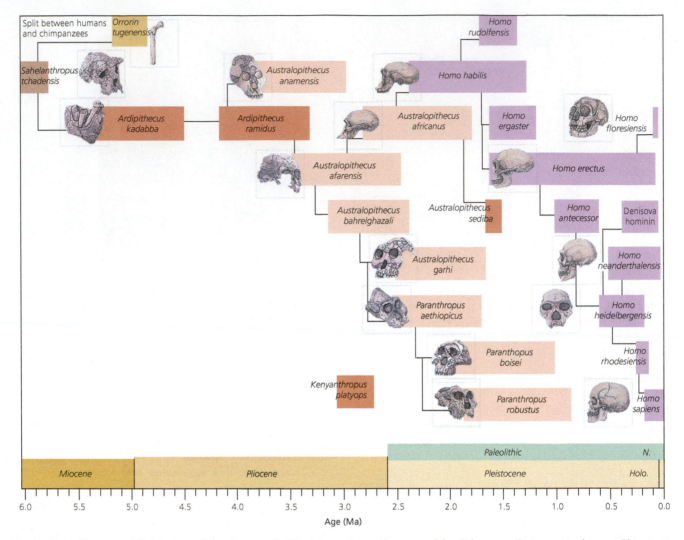

Figure 16.3 The current family tree of the dozens of different species and genera of fossil humans that are now known. The exact relationships between the lineages are still controversial, but the existence of all these different species and their time range is well documented. (Holo. = Holocene; N = Neolithic)

"monkey-to-ape-to-man" "march through time" sequence of organisms walking up the evolutionary ladder. This icon of evolution is so familiar that it is parodied endlessly in political cartoons and advertisements. Most people think that this is an accurate representation of human evolution, but it is not. *Evolution produces a bushy, branching tree of life, not a straight, simple ladder!*

In fact, the main reason that the ladder of life or march through time image for human evolution ever got started is that humans are one of the few groups in nature where only one species still survives today, *Homo sapiens*. Our bushy family tree (Fig. 16.3) has now been pruned down to only one descendant, so it could appear linear if you have only one surviving branch and a few fossil ancestors. (The same is true for horse evolution, which has only one surviving genus, *Equus*; Fig. 14.41). Nearly all the other great evolutionary radiations of animals have multiple living descendants, such as rhinos (Fig. 14.42), camels (Fig. 14.43), and virtually any other family you could name, so it is impossible to render these as a linear trend from ancestor to single descendant marching up a ladder of life or a chain of being.

BRAINS VERSUS BIPEDALISM

In the early days of anthropology, scientists were convinced that the most important thing that made us human was our unusually large brain, and our great capabilities to do complex tasks and form societies and use language that emerge from our large brains. Large brains were supposedly what made humans special and therefore were assumed to have appeared earlier than any other feature, including our bipedal posture and ability to walk and run, rather than walk on all fours like most mammals (including chimps and gorillas).

This dogma was so deeply entrenched that anthropologists rejected fossils that showed that humans were bipedal very early, long before they had large brains. For example, the first truly ancient fossil of humankind to be found was the famous "Taung child" skull of *Australopithecus africanus*, discovered by Raymond Dart in South Africa in 1924 (Fig. 16.4). It showed clear evidence that it once walked on two feet. This was based on the fact that the hole for the spinal column (foramen magnum) was directly beneath the center of the skull, so it had upright posture. Nevertheless,

Figure 16.4 The skull of the Taung child, described by Raymond Dart in 1924 from South Africa. The specimen is a juvenile of *Australopithecus africanus*, with a complete face and a natural cast of its brain. It was the first important early hominin fossil found outside Eurasia and challenged the prevailing notion that humans evolved in Eurasia, not Africa. Most European scientists rejected it at first because it had a small brain yet was clearly bipedal. But decades of further discoveries in Africa proved that Dart was right and the others were wrong—most of early human evolution took place in Africa.

it had a small brain. European scholars rejected this fossil as a human relative for decades because they felt that the brain must enlarge before humans became bipedal or developed any other advanced features. There was also a racist bias among European scholars that the first humans had arisen in Eurasia, not in the "dark continent of Africa," where people have black skins. By contrast, Darwin thought humans arose in Africa, for the simple reason that our closest relatives, chimps and gorillas, live there.

They were also misled by "Piltdown Man," a forgery announced in 1912, which was put together from the skull of a modern human and the jaw of an orangutan cleverly broken in the right places and then stained to make them look ancient. The forger (an amateur archeologist named Charles Dawson and possibly some accomplices) knew exactly what British anthropologists were expecting, so he used a large-brained medieval human skull and a modern orangutan jaw to make it seem plausible to anthropologists of that time. It was the pride of the British anthropological establishment for years, often proudly described as "the first Briton." The Piltdown forgery was exposed in 1953 as more and more discoveries from Africa showed that humans first evolved there, so the Piltdown "fossil" no longer made sense. Then, scientists analyzed it carefully and found the evidence of the forgery.

Since the 1950s all of the fossils of human relatives older than 1.8 Ma have been discovered in Africa, and all of them turned out to be bipedal, as far back as the oldest specimens dated 6–7 Ma. Meanwhile, our large brain capacity only appeared about 100,000–200,000 years ago at the earliest. Large brains were a very late feature in human evolution, whereas bipedalism was one of the first things to evolve in our ancestry.

ONE SPECIES AT A TIME

Probably in part because of the false notion of a "march of human evolution" through time mentioned earlier, for decades anthropologists were convinced that there was only one species of human on the planet at any given time. They looked at how *Homo sapiens* has spread throughout the world as a single species today and could not imagine that our ancestors could have tolerated another species of human in the same place and time. Thus, the linear march through time myth coincided with evolutionary histories of humans where only one species existed at any given time.

By the 1960s and 1970s, there were so many well-preserved fossils from rocks in southern and eastern Africa that clearly looked like different species that this idea no longer held up. In some case, paleoanthropologists resorted to extraordinary lengths to shoehorn all the fossils into a single species. For example, they pointed to the striking differences between male and female gorilla skulls and argued that the robust and gracile fossils of australopithecines were just males and females of the same species. Eventually, the presence of multiple different types of skulls and jaws (not just two, as in males and females) from the same beds, all of the same age and from the same place, made it impossible to hold this notion anymore. In addition, paleontologists were coming to realize that evolution is a branchy, bushing process and that simplistic linear ancestor–descendant sequences of fossils do not represent the complexity of the real story. Consequently, there are now dozens of species of human relatives (**hominins**, members of our subfamily Hominini) in the fossil record, and sometimes there were as many as four or five of them that lived in the same place. The fossil data are so rich now that this can no longer be denied.

16.2 The Human Fossil Record

 See For Yourself: The Incredible Human Journey

 See For Yourself: Great Transitions: The Origin of Humans

 See For Yourself: In Search of Human Origins

 See For Yourself: First Humans

Although most people don't realize it, the fossil record of humans is no longer as poor as it was even in the 1970s. Decades of hard work in the field by hundreds of scientists have turned up thousands of hominin fossils (Figs. 16.1, 16.3), including a few fairly complete skeletons and many well-preserved skulls that show clearly how humans have evolved over 7 million years. This avalanche of discoveries year after year has occurred despite the facts that hominin fossils are delicate and rare and that only one or two are found for every hundred or so fossil pigs or fossil horse specimens in the same beds in eastern Africa. Many museums in Africa, Europe, and Asia now have large

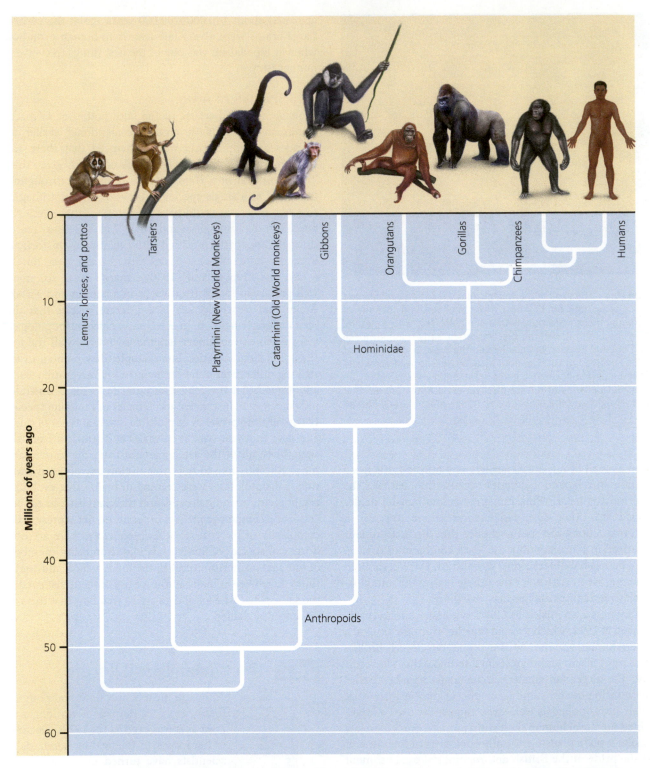

Figure 16.5   The family tree of the primates, showing the branching points between lemurs, lorises, and tarsiers, then the subsequent branching off of the New World monkeys, the Old World monkeys, and the apes (including humans).

collections of our early ancestors, so there are lots of fossils for scientists to work with.

The entire story of human evolution is too long and detailed to be discussed in a single chapter, so we will just cover the highlights. The short version is this: dozens of human species and genera are now known, forming a very bushy family tree that spans almost 7 million years of human evolution, mostly in Africa (Fig. 16.3). The exact details of how all these fossils should be named or how they are interrelated are always being refined and debated because many of the specimens are incomplete and because of the amazing pace of new discoveries.

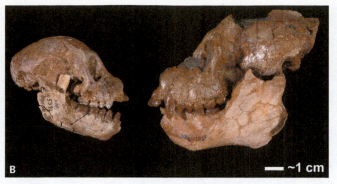

Figure 16.6 An array of early fossil primates. **A.** The early primate *Notharctus*, which had lemur-like proportions and lived in the jungles of the Middle Eocene. **B.** Skulls of a male (right) and a female (left) *Aegyptopithecus*, an early fossil ape from the Oligocene of Egypt. **C.** The skull of *Sivapithecus*, an early fossil orangutan from the Middle Miocene of the Siwalik Hills in Pakistan.

THE PRIMATES

See For Yourself: Ape Man

First, let us place humans in a broader context. Back in 1758, the founder of modern classification, Linnaeus, placed humans in the order Primates (Fig. 16.5), the group that includes not only ourselves and the apes but also the Old World monkeys (Cercopithecidae), New World monkeys (Platyrrhini), and lemurs, lorises, bush babies, pottos, and many other archaic primates still alive today, as well as many extinct lemur-like primates from the Eocene (Fig. 16.6A). We can trace the fossil record of most of these lineages back to the Cretaceous and Paleocene primate *Purgatorius*. In the early Cenozoic, the globe was much warmer and more densely vegetated, and there was a huge diversity of different extinct primates distantly related to lemurs. If you collect fossils in Paleocene or lower Eocene beds in places such as the Bighorn Basin of Wyoming, primates are among the most common fossils. Enormous collections of their jaws and teeth are now stored in museums around the world.

Primates have a number of features in common, but most of them are adaptations to the **arboreal**, or tree-dwelling, lifestyle. Nearly all primates are arboreal except for a few that have later gone back to living on the ground, such as baboons, gorillas, and humans. For a tree-dwelling existence, it is valuable to have good eyesight and good stereovision to accurately estimate the distance to the next branch, and nearly all primates have excellent large eyes and good distance perception. Primates are among the few groups of mammals that all have color vision (most animals don't), presumably to allow them to determine when fruits are ripe or not yet ready. Finally, nearly all primates have thumbs that are opposable so that they can grasp a branch or a food item. This gives them a strong grip for climbing and hanging from trees. Many also have grasping big toes for the same reason.

But as world climate became cooler and drier in the Oligocene and the forests vanished (Chapter 14), primates became scarce, too. They vanished from North America and Europe, where they had once flourished, and became restricted to Southeast Asia and Africa. During the late Eocene, one group of primates, the Platyrrhini, or New World monkeys, managed to cross the South Atlantic, where they radiated into the great diversity of prehensile-tailed monkeys including spider monkeys, colobus monkeys, and howler monkeys, marmosets, and the like (Fig. 16.5). However, most primate evolution occurred in Africa, where the Old World monkeys (baboons, macaques, rhesus monkeys, and their relatives) flourished, alongside the earliest fossils of the ape lineage (*Aegyptopithecus*, *Propliopithecus*, *Apidium*, and many others), which are documented from the Oligocene Fayûm beds of Egypt (Fig. 16.6B).

In the Miocene, apes were actually more diverse and successful than the monkeys and occasionally were found in Europe as well as their African homeland. By the middle Miocene, primitive members of the orangutan lineage (*Sivapithecus*) are known from deposits dated at 12 million years old in Pakistan (Fig. 16.6C), giving us some of the first ape fossils that belong to modern ape lineages. Unfortunately, we have lots of fossil apes but very few fossils of the chimpanzee or gorilla clan just yet, probably because both of those apes have always lived in forests where there is scant chance of fossilization. Most of the fossil apes declined and died out by the end of the Miocene, and Old World monkeys have come to dominate the primate adaptive zones ever since then.

TRIBE HOMININI

The oldest fossil that can be truly described as a member of our own tribe, the Hominini, or "hominins," was discovered and described in 2002. Nicknamed "Toumai" by its discoverers, its formal scientific name is *Sahelanthropus tchadensis*. The best specimen is a nearly complete skull (Figs. 16.3, 16.7A, B) from rocks about 6–7 million years in age from the sub-Saharan Sahel region of Chad (hence the scientific name, which translates to "Sahel man of Chad"). Although the skull is very chimp-like with its small size, small brain, and large brow ridges, it had remarkably human-like features, with a flattened face, reduced canine teeth, enlarged cheek teeth with heavy crown wear, and an upright posture—all of this at the very beginning of human

Figure 16.7   Some of the best fossils of early fossil hominins. **A.** The skull of *Sahelanthropus tchadensis*, nicknamed "Toumai," the earliest known member of our lineage, from beds 6–7 million years old in Chad. **B.** Forensic reconstruction of the head of *Sahelanthropus* by William Munn. **C.** The nearly complete skeleton of *Ardipithecus ramidus*, the oldest hominin fossil known from such a complete skeleton, found in rocks 4.4 million years old in Ethiopia. **D.** The partial skeleton of *Australopithecus afarensis*, nicknamed "Lucy" by its discoverers, Don Johanson and Tim White, from beds about 2.95–3.85 million years old in Ethiopia. **E.** Reconstruction of "Lucy," who was only about 1 m (3 feet) tall but fully bipedal. **F.** Museum display showing a comparison between australopithecines and paranthropines. At the top are the gracile skulls of *Australopithecus africanus*. In the middle row are skulls of three species of *Paranthropus*, showing how much more robust and heavily built they were compared to australopithecines. On the left is the "Black Skull" from beds 2.5 million years old on the shores of Lake Turkana, the oldest member of the genus *Paranthropus*, *P. aethiopicus*. In the center is a replica of Leakey's famous specimen of *Paranthropus boisei*, called the "Nutcracker Man" based on its powerful jaws and huge molars with thick enamel (see the jaws in the bottom row). It was found in beds 1.9–2.3 million years old in Olduvai Gorge, Tanzania. On the right is the first named species of the genus, *Paranthropus robustus*. On the bottom row are the lower jaws of paranthropines, showing their enormous broad molars adapted for crushing food such as nuts, seeds, and roots, very different from those of the australopithecines. **G, H.** Forensic reconstruction of the heads of *Paranthropus aethiopicus* (**G**) and *Paranthropus boisei* (**H**) by William Munn.

evolution. Just slightly younger is *Orrorin tugenensis*, from the upper Miocene Lukeino Formation in the Tugen Hills in Kenya dated between 5.72 and 5.88 Ma (Fig. 16.3). *Orrorin* is known mainly from fragmentary remains, but the teeth have the thick enamel typical of early hominins, and the thigh bones clearly show that it walked upright. Slightly younger still are the remains of *Ardipithecus kadabba*, found in Ethiopian rocks dated between 5.2 and 5.8 Ma. These consist of a number of fragmentary fossils, but the foot bones show that hominins used the "toe off" manner of upright walking as early as 5.2 Ma. Thus, our human lineage was well established by the latest Miocene, and hominins

were fully upright in posture, even though their brains were still small and body size was not much different from that of contemporary apes.

The Pliocene saw an even greater diversity of hominins (Fig. 16.3), with a number of archaic species overlapping in time with the radiation of more advanced hominins. Archaic relicts of the Miocene included *Ardipithecus ramidus*, found in Ethiopia in 1992 from rocks 4.4 million years in age, which had human-like reduced canine teeth and a "U"-shaped lower jaw (instead of the "V"-shaped lower jaw of the apes). *Ardipithecus ramidus* is now known from nearly complete skeletal

See For Yourself: The First Humans

material (Fig. 16.7C), making it the oldest hominin skeleton known. Rocks in Kenya about 3.5 million years in age yield other more primitive forms, like *Kenyanthropus platyops*.

See For Yourself: Paleoworld: Ape Man

By 4.2 Ma, however, the first members of the advanced genus *Australopithecus*, the most diverse genus of our family in the Pliocene, are also found. The oldest of these fossils is *Australopithecus anamensis* from rocks near Lake Turkana in Kenya ranging from 3.9 to 4.2 million years in age. These creatures were fully bipedal, as shown not only by their bones but also by hominin trackways near Laetoli, Tanzania. The most famous of these early australopithecines is *A. afarensis* (from rocks 2.95–3.85 million years old near Hadar, Ethiopia), nicknamed "Lucy" by its discoverers Don Johanson and Tim White. Celebrating by the campfire the night after they made the discovery, they were singing along with a tape of the Beatles' "Lucy in the Sky with Diamonds" and decided to nickname the fossil "Lucy" (Fig. 16.7D, E). When it was discovered in 1974, *Australopithecus afarensis* was the first early hominin to clearly show a bipedal posture (based on the knee joint and pelvic bones) but was not as upright as later hominins. These were still small creatures (about 1 m, or 3 feet, tall) with small brains and very ape-like in having large canine teeth and a large protruding jaw.

See For Yourself: In Search of Human Origins

By the late Pliocene, hominins had become very diverse in Africa (Fig. 16.3). These included not only the primitive forms *Australopithecus garhi* (dated at 2.5 Ma) and *A. bahrelghazali* (dated at 3.5 Ma) but also one of the best-known australopithecines, *Australopithecus africanus* (Fig. 16.4). Originally described by Raymond Dart in 1924 based on a juvenile skull (the "Taung baby"), for decades the Eurocentric anthropology community refused to accept it as ancestral to humans. But as more South African caves yielded better specimens to paleontologists like Robert Broom (especially the adult skull nicknamed "Mrs. Ples"), it became clear that *Australopithecus africanus* was a bipedal, small-brained African hominin, not an ape. *Australopithecus africanus* was a rather small, gracile creature, with a dainty jaw, small cheek teeth, no skull crest, and a brain only 450 cc in volume. On the basis of its gracile and very human-like features, *Australopithecus africanus* is often considered the best candidate for ancestry of our own genus *Homo*.

See For Yourself: Our Earliest Ancestors

In addition to *Australopithecus africanus*, the late Pliocene of Africa yields a number of highly robust hominins. For a long time, they were lumped into a very broad concept of the genus *Australopithecus*, either as distinct species or even as robust males of *Australopithecus africanus*. In recent years, however, anthropologists have come to regard them as a separate robust lineage, now placed in the genus *Paranthropus*. The oldest of these is the curious "Black Skull" (so called because of the black color of the bone), discovered in 1975 on the western shores of Lake Turkana, Kenya, from rocks about 2.5 million years in age (Fig. 16.7F, G). Although it is small in brain size, the skull is robust, with a large bony

ridge along the top midline (called a sagittal crest), massive molars, and a dish-shaped face. Currently, scientific opinion places the "Black Skull" as the earliest member of *Paranthropus*, *P. aethiopicus*. It was followed by the most robust of all hominins, *P. boisei*, from rocks in East Africa ranging from 2.3 to 1.2 million years in age (Fig. 16.7F, H). The first specimen found of this species was nicknamed "Nutcracker Man" for its huge, thick-enameled molars; robust jaws; wide, flaring cheekbones; and strong crest on the top of its head, suggesting a diet of nuts or seeds or even bone cracking. Discovered by Mary Leakey at Olduvai Gorge in 1959, it was originally named *Zinjanthropus boisei* by Louis Leakey, who made his reputation from it. The rocks of South Africa between 1.6 and 1.9 million years in age yield another species of *Paranthropus*, *P. robustus* (Fig. 16.7F). These too had massive jaws, large molars, and large skull crests but were not as robust as *P. boisei*. *Paranthropus robustus* lived side by side in the same South African caves as *Australopithecus africanus*. It is not only more robust but also larger than that species, with some individuals weighing as much as 54 kg (120 pounds).

THE GENUS *HOMO*

See For Yourself: Louis Leakey and the Dawn of Man

See For Yourself: Early Humans in 5 Minutes

Finally, the early Pleistocene produces the first fossils of our own genus, *Homo*, which are easily distinguished from contemporary *Australopithecus* and *Paranthropus* by a larger brain size, flatter face, no sagittal crest on the skull, reduced brow ridges, smaller cheek teeth, and reduced canine teeth. The first of these to be described was *Homo habilis* (whose name literally means "handy man"), discovered in the 1960s by Louis and Mary Leakey in Olduvai Gorge, Tanzania, from beds about 1.75 million years in age (Fig. 16.8A).

See For Yourself: Early *Homo*

Originally, all of the early *Homo* specimens were shoehorned into the species *H. habilis*, but now paleoanthropologists recognize that this material is too diverse to belong to one species, so several are now recognized. These include the more advanced-looking skull (Fig. 16.8A) now known as *Homo rudolfensis* (dated to about 1.9 Ma), which made Richard Leakey's reputation, and the very advanced but short-lived *Homo ergaster* (Fig. 16.8B), from beds 1.6–1.8 million years in age. These species are known not only from bones but also from their primitive stone tools, especially choppers and hand axes of the "Oldowan" technology.

Many of the archaic Pliocene taxa persisted into the early Pleistocene (as recently as 1.6 Ma), including *Paranthropus robustus* and *P. boisei*, *H. ergaster*, and *H. habilis* (Fig. 16.3). The best-known fossil of *H. ergaster* is a nearly complete skeleton of a boy who died when he was about 8–9 years old, found on the western shores of Lake Turkana by Alan Walker and his crew in 1984. Nicknamed "Nariokotome Boy" (Fig. 16.8B), it is estimated that he would have been 2 m (over 6 feet) tall if fully grown.

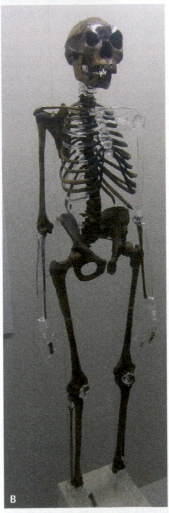

Figure 16.8 Fossils of the earliest members of the genus *Homo*. **A.** Side by side comparison of the skulls of *Homo habilis* (right) and *Homo rudolfensis* (left), the earliest species in our genus. **B.** The nearly complete skeleton of *Homo ergaster*, known as the "Nariokotome boy," found by Alan Walker and crew on the shores of West Turkana in 1984. It dates to about 1.7 Ma.

HOMO ERECTUS

 See For Yourself: *Homo erectus*

 See For Yourself: Peking Man

By 1.9 Ma, however, a new species had appeared: *Homo erectus* (Fig. 16.9). This human not only was bipedal and stood erect (as its species name implies) but also was almost as large in body size as we are. Its brain capacity was about 1 liter (1000 cc),

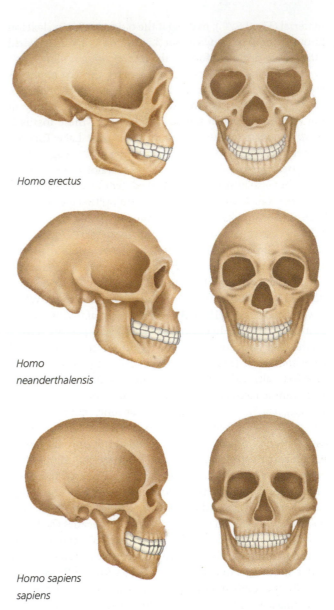

Homo erectus

Homo neanderthalensis

Homo sapiens sapiens

Figure 16.9 Comparisons of the skulls of *Homo erectus* (top), *Homo neanderthalensis* (center), and modern *Homo sapiens* (bottom). All have relatively large brains, with about 1000 cc capacity in *H. erectus*, while both Neanderthals and modern humans have brains in the 1500–1700 cc range. The two extinct species have strong brow ridges, a protruding snout without much of a chin, and broader, heavier cheekbones compared to modern humans. Neanderthals have about the same brain size as modern humans, but their skull is a bit flatter with a point on the back end.

only slightly less than ours. *Homo erectus* made crude choppers and hand axes ("Acheulean culture" tools) and was the first species to make and use fire. Originally, *H. erectus* was confined to Africa, where all of our other ancestors had long lived. By around 1.8 Ma, we have evidence that *H. erectus* migrated outside its African homeland as specimens from Indonesia (originally described as *Pithecanthropus erectus* or "Java man") have been dated at that age. In addition, specimens are known from elsewhere in Eurasia, such as Romania and the Republic of Georgia, that are almost as old. By about 500,000 years ago, we have abundant fossils of *H. erectus* in many parts of Eurasia, including the famous specimens from the Chinese caves at Zhoukoudian, originally called "Peking Man" and dated as old as 460,000 years ago. The latest dating suggests that *H. erectus* may have persisted as late as 143,000 years ago and possibly 74,000 years ago, overlapping with modern *H. sapiens*. *Homo erectus* was not only the first widespread hominin species but also one of the most successful and long-lived species, lasting more than 1.8 million years in duration between 1.9 and 0.143 Ma. During much of that long time, it was the only species of *Homo* on the planet and changed very little in brain size or body proportions. If longevity is a measure of success, then it could be argued that *H. erectus* was even more successful than we are.

NEANDERTHALS

 See For Yourself: Paleoworld: Missing Links   By about 400,000 years ago, another species was established in western Europe and the Near East: the Neanderthals (Figs. 16.9, 16.10). In 1857 these were the first fossil humans ever to be discovered, although their fragmentary fossils were originally dismissed as the remains of diseased Cossacks that had died in caves. The first complete descriptions of skeletons were based on an early specimen from a cave at La Chapelle-aux-Saints in France that suffered from old age and disease, so for decades Neanderthals were thought to be stoop-shouldered, bow-legged, and primitive, the classic stereotypical grunting "cave men."

 See For Yourself: The Secrets of Neanderthals   Modern research has shown that Neanderthals were very different from this outdated image. Although their skulls are distinct from ours in having a protruding face, large brow ridges, no chin, and a flatter skull that sticks out in the back (Fig. 16.9), they had a slightly *larger* brain capacity on average than we do; and they practiced a complex culture that included ceremonial burials, suggesting religious beliefs. Their bones (and presumably bodies) were robust and muscular and slightly shorter than the average modern human because they lived exclusively in the cold climates of the glacial margin of Europe and the Middle East, where their stocky build (like a modern Inuit or Laplander) would have been an advantage. Their toolkits and culture were also more complex, with Mousterian hand axes, spear points, and other complex devices, as well as bone and wooden tools. Some of these tools show complex working and simple carving, so they were artistic as no hominin before had ever been. The famous discoveries at Shanidar Cave in Iraq showed that Neanderthals buried their dead with multiple kinds of colorful flowers, suggesting that they may have had at least some kind of religious beliefs and possibly belief in an afterlife.

For decades, anthropologists treated Neanderthals as a subspecies of *H. sapiens*, but

recent work suggests that they were a distinct species. The best fossil evidence of this comes from Skhul Cave on Mt. Carmel and Qafzeh Cave on Har Oedumim in Israel, where layers bearing Neanderthal remains are interbedded and alternate with layers containing early modern humans. In 1997 their DNA was sequenced, and they are clearly not *H. sapiens* but genetically distinct. However, their DNA shows evidence that all modern humans of non-African descent have a bit of Neanderthal DNA in them, so there must have been some interbreeding between them in Eurasia where they overlapped.

Neanderthals were the only recent extinct species of human known in anthropology until 2010, when molecular biology shocked the world with the announcement that there was yet another species of human in the last 40,000 years. Digging in Denisova Cave in the Altai Mountains of Siberia near the Mongolian–Chinese border, Russian archeologists found a juvenile finger bone, a toe bone, and a few isolated teeth of a hominin mixed with artifacts including a bracelet. The artifacts gave a radiocarbon date of 41,000 years ago, so the age was well established. But when the molecular biology lab of Svante Pääbo and Johannes Krause at the Max Planck Institute in Leipzig, Germany (who first sequenced Neanderthal DNA), analyzed the mitochondrial DNA of the finger bone, they found it had a unique genetic sequence that was distinct from both Neanderthals and modern humans (remember from Chapter 8 that the mitochondrion has its own DNA sequence independent of nuclear DNA since the mitochondrion was once a symbiotic prokaryote). The nuclear DNA was also distinct but suggested that these people were closely related to the Neanderthals. They may also

Figure 16.10   Reconstructed skeleton and life-sized model of a Neanderthal.

have interacted with modern humans because they share about 3%–5% of their DNA with Melanesians and Australian Aborigines. The mitochondrial DNA data suggest that they branched off from the human lineage about 600,000 years ago and represent a separate "out of Africa" migration distinct from the much earlier (1.8 Ma) *H. erectus* exodus or the much younger (300,000 years ago) emigration of *H. rhodesiensis–H. heidelbergensis* from Africa to Eurasia.

These mysterious people whose DNA was so distinctive are now called the **Denisovans**. Since there are so few fossils, we cannot say much about their physical appearance or anything else other than that they have distinctive DNA that is found in no other human species. In fact, most scientists are reluctant to give the Denisovans a formal scientific name because there is not enough fossil material to describe the anatomy of the species in any normal sense. So the Denisovans are mysterious, showing us that the bones don't tell the whole tale but that there may have been numerous other human species on this planet that haven't left a fossil record.

Almost as surprising as the 2010 discovery of the Denisovans was the 2003 announcement of a dwarfed species of humans found only on the island of Flores in Indonesia (Fig. 16.11). Found at a site called Liang Bua Cave, their fossils and artifacts are dated between 1 million and 74,000 years ago. The most striking feature of these people is their tiny size, only about 1.1 m (3 feet 7 inches) tall for a fully grown adult, so they have been nicknamed the "hobbits." Yet these are not modern African pygmies (who are tiny but fully modern *Homo sapiens*) but an entire population of people that appears to have descended from *H. erectus* (or possibly even from *H. habilis*) about 1 Ma, then became dwarfed. Size reduction is a common effect on oceanic islands, with many types of animals (especially elephants, mammoths, and hippos) undergoing dwarfing on islands ranging from Malta to Crete to Cyprus to Madagascar. The reason for this dwarfing is clear: they are living on the smaller food resource base of an island so cannot get the kind of nutrition needed to grow to normal sizes. In addition, on most islands they are typically not under pressure from large predators or competing with the same large herbivores. Although the interpretation of these fossils is controversial, most anthropologists agree that they were a distinct species, which has been formally named *Homo floresiensis*.

HOMO SAPIENS

Finally, we find the first fossil skulls and skeletons that look almost indistinguishable from our own species. Some of these, dubbed "archaic *Homo sapiens*" or, more formally, *Homo heidelbergensis*, are known from deposits in Africa dating as old as 300,000 years. Skulls from Africa (such as those from Klasies Mouth Cave in South Africa, dated to about 90,000 years ago) look almost completely modern and are universally regarded as *H. sapiens* (our species). Like *H. erectus*, early *H. sapiens* spent most of its history in Africa and migrated to Eurasia about 70,000 years ago. There, these people came into contact with Neanderthals, and for about 30,000 years they coexisted. Mysteriously, Neanderthals vanished 40,000 years ago. Whether they were

Figure 16.11 The preserved bones of the skeleton of a "Hobbit," the dwarfed human species from Flores Island in Indonesia, *Homo floresiensis*. The entire skeleton is not known, but based on the limb proportions, these people were only about 1.1 m (3.7 feet) tall as adults.

wiped out by *H. sapiens* or by some other cause is not clear. The subject has been one of endless debate and speculation. Pat Shipman argues that modern humans had an advantage in domesticated dogs, which helped them overcome Neanderthals in hunting and in warfare. Whatever happened, modern *H. sapiens* soon took over the entire Old World, developing complex cultures (the "Cro-Magnon culture") including the famous cave paintings of Europe and many kinds of weapons and tools.

 See For Yourself: Mankind: The Story of Us

This brief review of the hominin fossil record hardly does justice to the richness and quality of the specimens or to the incredible amount of anatomical detail that has been deciphered. If it all seems like too much to absorb, just gaze at the faces of the skulls in Fig. 16.1. They look vaguely like modern human skulls, but they definitely show the change from more primitive hominins that some people see as "mere apes" (even though they were all completely bipedal and had many other human characteristics) up through

BOX 16.1: HOW DO WE KNOW?

What Do Genes Tell Us About Our Relation to Apes?

If the fossil record of human evolution was not evidence enough, the clinching argument is found in every cell in your body. Unbeknown to Darwin or any other biologist before the 1960s, there was another source of data that clearly show our relationships to apes and the rest of the animal kingdom: the structure of our DNA and other biomolecules, such as proteins. Some of the very first molecular techniques demonstrated that our DNA and chimp and gorilla DNA were extremely similar. When you put the serum of antibodies of humans and apes in the same solution, the reactions are much stronger than with humans and any other animal, suggesting that the immunity genes of humans and apes are most similar. This is called the **immunological distance method** for estimating our relatedness to other organisms.

Then in the late 1960s, a technique called **DNA–DNA hybridization** came along. The scientists take solutions of the DNA of apes and humans, then heat the molecules until the two strands of the DNA unzip. When the mixture cools, each strand binds to other nearby strands, creating hybrid DNA with one strand from humans and one from another animal, such as a chimp. When you heat the hybrid DNA again, the more tightly bonded together they are (which reflects how similar they are), the higher the temperature is needed to unzip the two strands again. Doing this with chimps, gorillas, other apes, plus monkeys, lemurs, and other animals gives a rough measure of how similar each is to humans—and, once again, chimp DNA is virtually identical to human DNA.

 See for Yourself: Chimps and Humans Share 99% of their DNA

Since the 1990s, technological leaps like the polymerase chain reaction have made it possible to directly sequence the DNA not only of humans but of many other animals and plants. We first sequenced the entire DNA of humans in 2001 and the DNA of chimps in 2005. When they were compared, we got the same result: humans and chimps share 98%–99% of their DNA. Less than 1%–2% of our DNA differentiates us from chimps and from gorillas. This is because about 60%–80% of our DNA is non-coding DNA (sometimes called "junk" DNA) that is never read or used but is carried around passively generation after generation (see Chapter 6). Some of this junk is made up of endogenous retroviruses, which are remnants of viral DNA inserted into our genes when some distant ancestor was infected, that are still carried around even though they no longer code for anything. One of the pseudogenes we have is an inactive copy of the gene that manufactures vitamin C. Apparently, our primate ancestors ate so much fruit they no longer needed to make their own vitamin C, so this gene got shut off. Now, we humans sometimes don't get enough vitamin C (causing a disease called scurvy). We have to eat more fruit or take vitamin pills to replace it—but our vitamin C gene can no longer be turned back on so that we could make it ourselves. Besides the true junk DNA, a smaller percentage of the genome is made up of structural genes that code for every protein and structure in our body, including genes we no longer use. The 1%–2% that distinguishes us from chimps is mostly made up of regulatory genes, the "on/off switches" that tell the rest of the genome when to be expressed and when not to. They are the reason humans look so different from other apes, even though our genes are nearly identical.

 See For Yourself: Are Humans Just Another Primate?

For example, all apes and humans have the structural genes for a long tail but do not express those genes, except in rare cases where the regulatory genes fail. When such a failure occurs, humans grow a long bony tail (Fig. 16.12). Likewise, living birds have toothless beaks and no longer have teeth like their fossil ancestors, but they still have the genes to make teeth. Experimentally grafting the mouth epithelial tissues of a mouse into chick embryos produced teeth. But the teeth that grow are not mouse teeth but dinosaur teeth! Thus, we all have lots of ancient genes in our DNA that are no longer expressed, but it only takes some sort of modification of gene regulation for these ancient features to return.

The extreme similarity of our genes to those of the two species of chimpanzee (the common chimp, *Pan troglodytes*, and the pygmy chimp or bonobo, *Pan paniscus*) should, all by itself, be overwhelming and convincing evidence of our close relationship. Despite some people's aversion to the idea, we are indeed the ape's reflection. Biologist Jared Diamond puts it this way: imagine that some alien biologists came to earth, and the only samples they could obtain were DNA. They sequenced many different animals, including humans and the two chimp species. Based on these data alone, they would conclude that humans are just a third species of chimpanzee. Our DNA is more similar to that of the two chimp species than the DNA of most species of frog are similar to one another. They are even more similar than lions and tigers are to each other, which share about 95% of their DNA and can interbreed in zoos. The differences in appearance between apes and humans are caused by tiny changes in the regulatory genes, which give huge results. The evidence from the genes, as well as from your anatomy, is overwhelming. The DNA in every cell in your body is a testament and witness to your close relationship to chimps.

In summary, 1%–2% of the genome that differentiates us from chimps must be the regulatory genes that turn on and turn off the structural genes (which make up most of the 98% of the genome that is the same). We still have the genes for most parts of the ape body, and the monkey body too; and every once in a while there is a genetic mistake or atavism, and humans express the long-repressed genes that we still carry to make a tail.

In fact, since the 1920s, many biologists and anthropologists have argued that much of what differentiates us from the chimpanzee is a well-known phenomenon in nature called **neoteny** (literally, "holding on to youth") (see Chapter 6). Many animals in nature find ways to reproduce while their bodies are still in their juvenile form. For example, certain salamanders known as axolotls (genus *Ambystoma*) that live in lakes in Mexico are able to reproduce while they still have their juvenile gills (see Fig. 6.16). But if the lake water goes bad or dries

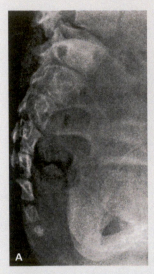
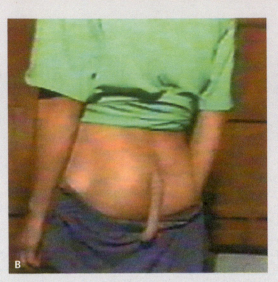

Figure 16.12 Every once in a while, a human is born with an atavistic tail, a throwback to our evolutionary past, when the regulation that normally shuts down our genes for tails fails to operate. The human tail comes complete with fully developed vertebrae, muscles, and other features of animal tails. **A.** X-ray of a human with well-developed tail vertebrae. **B.** Image of two humans with fully developed tails.

up, they complete their development to adult salamanders, lose their gills and develop lungs, and then crawl off to some other pond. Thus, the ability to just change the timing of reproduction with respect to development gives animals great flexibility to take advantage of their existing genetic instructions without having to make a drastic genetic or evolutionary change.

If you look at a juvenile chimpanzee (Fig. 16.13A), its skull is much like that of a human, with a large brain, small brow ridges, short snout, and upright posture. Then during development to an adult (Fig. 16.13B), the chimpanzee develops the larger snout with long canines, big brow ridges, and the forward slouching posture of the head. If regulatory genes tweak our embryonic development a tiny bit, we can make most of

Figure 16.13 **A.** Juvenile chimpanzees have many characteristics of the skull found in adult humans, including upright posture, relatively large brain, small brow ridges, and a less protruding snout. **B.** As they grow into adult chimps, these features all become more ape-like. Since the 1920s, many anthropologists have argued that much of what makes us human is retention of juvenile ape characteristics into adulthood (*neoteny*). (From Naef 1926)

the characteristics that mark us as human just by remaining juvenile apes that reach sexual maturity without ever truly growing up.

You may have read some of the fascinating works of Aldous Huxley, the famous novelist and author of the dystopian classic *Brave New World* (a high school reading list favorite). He was also the brother of the famous evolutionary biologist Julian Huxley, and both were grandsons of Darwin's biggest defender, Thomas Henry Huxley. Aldous knew these ideas about human neoteny very well because of his brother's influence. In 1939, he published a novel entitled *After Many a Summer Dies the Swan*. The theme of the novel is immortality and how humans are always striving to find a way to extend their lives beyond what nature intended. The main character is a millionaire (modeled after legendary press baron William Randolph Hearst, whom Huxley met when he was a

Hollywood screenwriter in the 1920s) named Jo Stoyte, who is attempting to buy eternal life by hiring a classic "mad scientist" character, Dr. Obispo, to do research on delaying aging and prolonging life. Dr. Obispo discovers that the (fictional) Third Earl of Gonister in England had lived several centuries without any signs of aging, apparently by ingesting carp guts. Archival records showed that he had fathered children when he was over 100 years old. In a plot twist, Dr. Obispo seduces the millionaire's mistress (modeled on Hearst's real mistress, actress Marion Davies), and the millionaire accidentally kills Obispo's scientific assistant in a jealous rage. Stoyte and Obispo have to run from the law, so they flee to England and try to find out what happened to the third earl of Gonister. Finally, they break into his castle, and in the basement they find him, still alive and over 200 years old—all grown up to become an adult ape.

forms that everyone would agree look much like "modern humans" (even though they had many distinctive anatomical features, like those found in Neanderthals, that make them a distinct species). Even non-scientists can glance at these fossils and see the hallmarks of their own ancestry.

16.3 Miracles from Molecules

OUR LOW GENETIC DIVERSITY

In addition to discovering how genetically similar we are to the apes (Box 16.1), even more startling breakthroughs have occurred in molecular anthropology. For centuries, the scientific community was deeply racist and treated non-white peoples as inferior or even a different species from white people. But the molecules paint a completely different picture. When you put the DNA of the human "races" in the mix with the DNA of Neanderthals and most of the living populations of chimps and gorillas, a surprising result emerges. As shown in Fig. 16.14, the genetic differences among *all* the human "races" is tiny. More genetic variation among modern humans is found *within* populations than *between* them. All humans are far more genetically similar to each other than the populations of West African chimps are to each other, and the same is true of other populations of chimps and gorillas. As anthropologists have been saying for years, human "races" are genetically meaningless, and their basis is only a tiny part of our genome. In fact, evidence shows that most of the "racial differences" people usually think of, such as skin color and the shape of the eyes, are a very recent change in human evolution, occurring sometime after nearly all modern human lineages emerged from Africa about 70,000 years ago. This is an important thing to think about when issues of race come up in society.

MITOCHONDRIAL "EVE" AND Y-CHROMOSOME "ADAM"

Mitochondrial DNA evolves much quicker than nuclear DNA, so you can see genetic divergence in very recently separated lineages. Using this fact, in 1980 Wesley Brown looked at the DNA of 21 women and found that they all had a common ancestor about 180,000 years ago. Only women could be used because the mitochondria are not carried in the sperm of males but only in the protoplasm of the ovum of females. Allan Wilson at the University of California Berkeley saw the potential of this line of research, and he and his graduate student Rebecca Cann sampled as wide a variety of DNA of different peoples as possible, especially the huge number of distinct populations in Africa, the homeland of humanity. As the sample size increased, the age and divergence estimate got better and better. Although there were still some major uncertainties and problems, their work was finally published in 1987.

 See For Yourself: Human Genetic Origins

What Cann, Wilson, and Mark Stoneking (another graduate student in Wilson's lab) had discovered was startling. By this point they had samples of 147 women from all over the world, representing every genetic grouping known. The majority of the women were from Africa, and the molecular tree they produced showed that not only did all humans came from Africa but numerous lineages had originated in Africa and were still living there. Even more surprising was the fact that all human mitochondrial DNA comes from a common ancestor, a woman who lived in Africa between 140,000 and 200,000 years ago (similar to Brown's original estimate). In other words, most fossils of humans found to be older than 200,000 years were not closely related to anyone living today but members of extinct genetic lineages. All the modern human populations

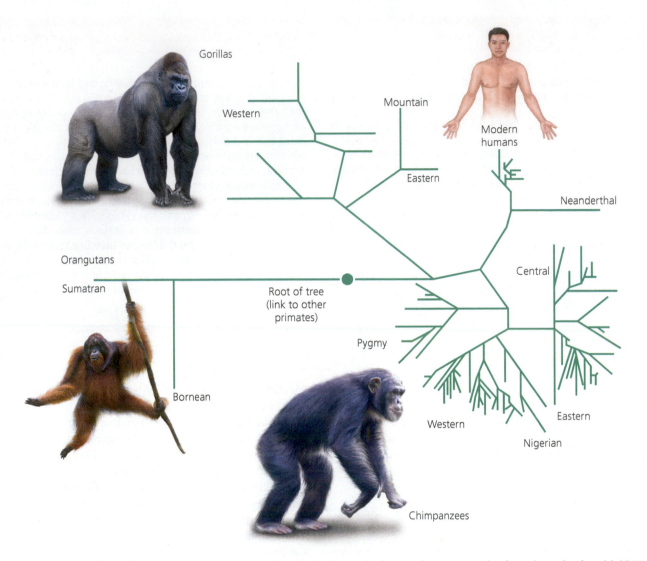

Figure 16.14   Molecular phylogeny of apes and humans, showing their genetic distance from one another based on mitochondrial DNA. All human "races" are much more similar to one another than two populations of gorillas or chimpanzees are to each other. (Redrawn from several sources)

started and evolved in Africa through most of human prehistory, and only late in human evolution did humans who left living descendants migrate to other continents. All modern humans are descended from African ancestors who left Africa relatively recently (less than 100,000 years ago). The media dubbed this discovery "Mitochondrial Eve," although the scientists hated this name and its false implication that the ancestral woman had anything to do with the Eve of the Bible.

 See For Yourself: Genetics of Human Migrations   What does it mean that all humans alive today are descended from a single woman who lived somewhere between 140,000 and 200,000 years ago? All the thousands of other women living in human populations at the same time apparently left no descendants who are alive today. This is simply a result of how exponential growth and reproduction and family trees work. If the "Eve" back then was typical of humans who successfully left fertile living descendants, then one female would leave a bunch of children and grandchildren, dozens of great-grandchildren, hundreds

of great-great-grandchildren, and so on. Over 140,000–200,000 years that successful female and all her descendants would make up a significant portion of all the humans on earth, provided they haven't been pruned back by a population crash that wipes out most of the lineages. All the other females who were alive at the same time may also have left lots of descendants, but for some reason none of them survive today (at least among the 147 women sampled from around the world).

See For Yourself: Human Genetic Evolution   If this study were not provocative enough (and readily misunderstood by the media and the public), another study, by Michael Hammer in 1995, looked at the genetic similarities of the Y chromosome in males from all over the world. The study found that all modern humans are descended from a common African father about 200,000–300,000 years ago, which is consistent with the appearance of fossils of the first anatomically modern humans in Africa at about the same time. Naturally, the media had to mislabel it "Y-chromosome Adam," and the same old misconceptions were spread to the public all over again. When it was possible to get the Y-chromosome sequence of Neanderthals, the divergence time between the Neanderthal and modern human lineages was placed at 588,000 years ago, again consistent with the fossils.

 See For Yourself: Genetic Odyssey

All of these studies confirm abundantly that all modern humans are descended from relatively recent common ancestors who once lived in Africa, and most of the human lineages alive today still live in Africa. For the non-African humans, the next stage is using the genetic divergence of populations to date when human populations left Africa. The latest data suggested that the first migration of modern humans from Africa, nicknamed the "Out of Africa I" event, occurred between 400,000 and 200,000 years ago, resulting in the Neanderthals and Denisovans in Eurasia (Fig. 16.15) and possibly some modern *H. sapiens* in Eurasia. (Remember, *H. erectus* left Africa and spread across Eurasia about 1.8 Ma, so it was actually the second migration of hominins out of Africa.) However, several lines of genetic evidence show that none of these migrating humans have living descendants and that all their lineages died out at one time or another.

HUMAN GENETIC BOTTLENECKS

When a population goes through a crash so that only a small number of individuals survive and become the ancestors of all the later populations, they are said to have gone through a **genetic bottleneck**. The genetic diversity of the surviving population is often very low, with an unusually high frequency of some peculiar genes. This is easily detectable

when you do a genetic analysis, and using molecular clock techniques, you can calculate how long ago this population crash must have occurred. Many animals alive today went through some kind of bottleneck when their populations were near extinction level. Cheetahs today, for example, went through a bottleneck about 10,000 years ago and have a low genetic diversity and are highly inbred, threatening their survival. Their low diversity and inbred populations means that rare genetic diseases can become widespread, and that they don't have a large genetic toolkit to develop useful variants when conditions change. The same is true of both the European and American bison, both of which were hunted until near extinction but have since recovered. The northern elephant seal was hunted down to about 30 individuals in the 1890s but has now recovered and is threatened by its low genetic diversity.

Bottlenecks seem to have occurred in lots of species but most especially in humans. It's shocking to think that the earth today is populated by 7.8 billion people and that the number is rapidly increasing, but that number was much smaller in prehistory. In 2005, a group of scientists found genetic evidence that suggested that only about 70 individuals gave rise to all the Native American populations that spread from Asia to the Americas about 18,000 years ago. This means that all native peoples, from the Inuit of Alaska to the Lakota of the Plains to the Incas and Aztecs and Tolmecs and Mayans and even the Fuegians of Patagonia, are extremely closely related and low in genetic diversity.

In fact, studies of the full diversity of living *H. sapiens* show that they went through a very small bottleneck not that long ago. Estimates vary, but the most recent genetic evidence places the total modern *H. sapiens* population in the bottleneck down to 30,000 people and maybe as few as 4500 people, with some estimates as low as 40 breeding pairs. Most estimates suggest about only 5000 people on earth at that time. This is less than the population of a small town in America today!

So when did humans go through that bottleneck? The latest archeological evidence places it at least 48,000 years ago, which is a bare minimum. There are not many archeological sites over this time interval, so it could be much earlier. We can use the molecular clock to estimate how long ago all the modern humans on the planet diverged from a small survivor population. Some recent genetic studies place it around 74,000 years ago.

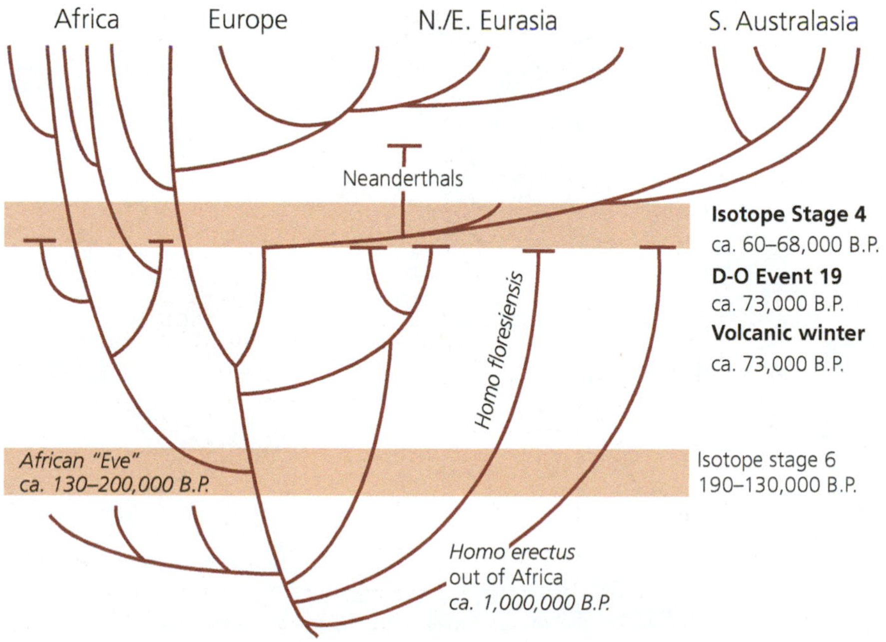

THE TOBA CATASTROPHE

An estimate of 74,000 years ago is an interesting coincidence because that is the date of one of the largest volcanic eruptions in earth history, the explosion of the Toba volcano on

Africa   Europe   N./E. Eurasia   S. Australasia

Neanderthals

Homo floresiensis

African "Eve"
ca. 130–200,000 B.P.

Homo erectus out of Africa ca. 1,000,000 B.P.

Isotope Stage 4 ca. 60–68,000 B.P.
D-O Event 19 ca. 73,000 B.P.
Volcanic winter ca. 73,000 B.P.

Isotope stage 6 190–130,000 B.P.

Figure 16.15 There were many different human lineages on the planet before the Toba catastrophe 74,000 years ago, but most of them (including the Flores people and possibly *Homo erectus*) vanished. Only a few (such as the Neanderthals and the ancestors of all modern non-African humans) survived the event and lived on through the latest Pleistocene. D-O = Dansgaard-Oeschger stage.

the north end of the island of Sumatra in Indonesia. It was the biggest eruption on the planet in the previous 28 million years, so big that it was over 1000 times as large as the catastrophic eruptions of Mount Tambora in Indonesia in 1815 and Krakatoa volcano between Java and Sumatra in Indonesia in 1883, both of which changed the climate for several years after the eruption. In 1816, Tambora caused a "year without summer," but Toba caused global temperature to drop by 3°–5°C (5°–9°F), further amplifying the cold of the ongoing Ice Ages. The tree line and snow line dropped 3000 m (about 10,000 feet) lower than today, making most high elevations uninhabitable. Global average temperatures dropped to only 15°C after 3 years and took a full decade to recover to pre-eruption temperatures. Ice cores from Greenland show evidence of this dramatic cooling in the trapped ash and ancient air bubbles, as well as abundant sulfur dioxide from the volcanic eruption.

 See For Yourself: Mystery of a MegaVolcano

What happened to the people and animals during this terrible time? As we just pointed out, geneticists and archeologists have found evidence that the Toba catastrophe nearly wiped out the human race. Apparently, only about a few thousand humans survived worldwide and passed through the genetic bottleneck. Another study found a similar genetic bottleneck in the genes of human lice and in our gut bacterium *Helicobacter pylori*, which causes human ulcers; both of these date back to the time of Toba, according to their molecular clocks, which show how long since a genetic change took place. The same is true for the genes of a number of other animals, including tigers, orangutans, several monkeys, gorillas, chimpanzees, and pandas—just about every large mammal in southern Asia or Africa that has been sequenced so far. In short, Toba was the biggest eruption ever to occur since humans have been on earth, and it came very close to wiping out humans altogether, along with many other animals. Keep in mind, the Toba catastrophe model is still very controversial among anthropologists, although the fact of the genetic bottleneck is not in doubt and that it happened around the time of the gigantic Toba eruption is also strongly supported by the data.

So the genetic evidence for the "Out of Africa II" model strongly supports the idea that the ancestors of all living people in Eurasia and the Americas did not leave Africa until about 71,000–74,000 years ago, right after the Toba catastrophe (Fig. 16.16). Apparently, they migrated along the southern coastline of Asia and colonized Australia about

Early human migration

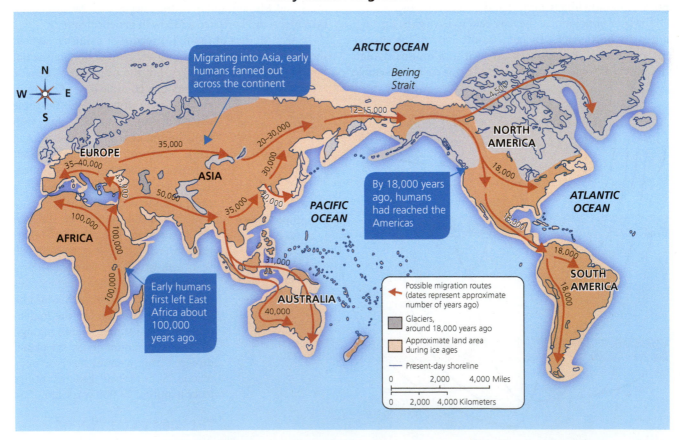

Figure 16.16 According to the "Out of Africa II" model, only the Neanderthals, Denisovans, and possibly some modern *Homo sapiens* lived outside Africa before the Toba catastrophe; and most of them were wiped out by the climate change. The Flores "hobbits" and possibly *Homo erectus* both vanished at that time. After the world recovered and humans survived their genetic bottleneck, modern *Homo sapiens* from Africa spread across southeast Asia and eventually to the Americas, with their population recovering from the bottleneck and exploding about 50,000 years ago.

65,000–50,000 years ago. Another lineage of this migration left the African and Middle Eastern region (which had modern humans some time earlier, possibly 300,000 years ago) and reached southern Europe about 55,000 years ago. Finally, the lineages that crossed the Bering land bridge and reached the Americas were only here about 18,000 years ago, although some controversial dates from the Monte Verde site in Chile suggest that earlier human populations (apparently without modern descendants) got here possibly as soon as 30,000 years ago.

16.4 A Perspective

See for Yourself: Human Population Through Time

Some people may find it disconcerting that we are 98% genetically identical to the chimpanzee, but this is the truth found in every cell in your body. Not only that, but all the human "races" are tiny recent variations in human evolution and represent less difference in the genome than the differences between populations of chimpanzees in West Africa or East Africa. Even more shocking is the idea that the Toba eruption almost wiped humans off the face of the earth 74,000 years ago. If it had, we would not be here to discover and talk about all these things.

Thus, the fundamental characteristics of humanity are very different from what most people think. We are not distinct from nature but a part of it, and genetically we are essentially a third species of chimpanzee. These are the lessons that science has taught us. The sooner we shed our cosmic arrogance, the sooner we can come to understand the real place of humanity in the cosmos.

RESOURCES

BOOKS AND ARTICLES

Beard, K. C. 2004. *The Hunt for the Dawn Monkey: Unearthing the Origin of Monkeys, Apes, and Humans.* University of California Press, Berkeley.

Conroy, G. C. 1990. *Primate Evolution.* W. W. Norton, New York.

Darwin, Charles R. 1871. *The Descent of Man.* John Murray, London.

Delson, E. C. 1985. *Ancestors: The Hard Evidence.* Alan R. Liss, New York.

Diamond, J. 1992. *The Third Chimpanzee: The Evolution and Future of the Human Animal.* HarperCollins, New York.

Foley, R. A., and R. Lewin. 2003. *Principles of Human Evolution.* Wiley-Blackwell, New York.

Gagneux, Pascal, and others. 1999. Mitochondrial sequences show diverse evolutionary histories of African hominoids. *Proceedings of the National Academy of Sciences USA* 96(9):5077–5082.

Harari, Yuval Noah. 2015. *Sapiens: A Brief History of Humankind.* Harper, New York.

Harris, Eugene E. 2015. *Ancestors in Our Genome: The New Science of Human Evolution.* Oxford University Press, Oxford.

Johanson, D., and M. Edey. 1981. *Lucy: The Beginnings of Humankind.* Simon & Schuster, New York.

Johanson, D., and B. Edgar. 1996. *From Lucy to Language.* Simon & Schuster, New York.

Johanson, D., L. Johanson, and B. Edgar. 1994. *Ancestors: In Search of Human Origins.* Villard, New York.

Johanson, D., and K. Wong. 2009. *Lucy's Legacy: The Quest for Human Origins.* Crown, New York.

Jurmain, R., L. Kilgore, D. Trevathan, and R. Ciochon. 2014. *Introduction to Physical Anthropology.* Cengage, New York.

Larsen, C. S. 2014. *Our Origins: Discovering Physical Anthropology.* W. W. Norton, New York.

Lewin, R. 1987. *Bones of Contention: Controversies in the Search for Human Origins.* University of Chicago Press, Chicago.

Lewin, R. 1988. *In the Age of Mankind: A Smithsonian Book on Human Evolution.* Smithsonian Institution Press, Washington, DC.

Marks, J. 2002. *What It Means to Be 98% Chimpanzee.* University of California Press, Berkeley.

Pääbo, Svante. 2014. *Neanderthal Man: In Search of Lost Genomes.* Basic Books, New York.

Potts, R., and C. Sloan. 2010. *What Does It Mean to Be Human?* National Geographic Society, Washington, DC.

Prothero, D. R. 2018. *When Humans Nearly Vanished: The Catastrophic Explosion of the Toba Volcano.* Smithsonian Books, Washington, DC.

Reich, David. 2018. *Who We Are and How We Got Here: Ancient DNA and the New Science of the Human Past.* Pantheon, New York.

Roberts, Alice. 2018. *Evolution: The Human Story.* DK Books, London.

Rutherford, Adam. 2016. *A Brief History of Everyone Who Ever Lived: The Human Story Retold Through Our Genes.* The Experiment, New York.

Sawyer, G. J., V. Deak, E. Sarmiento, and R. Milner. 2007. *The Last Human: A Guide to the Twenty-Two Species of Extinct Humans.* Yale University Press, New Haven, CT.

Shipman, Pat. 2001. *The Man Who Found the Missing Link: Eugene Dubois and His Lifelong Quest to Prove Darwin Right.* Simon & Schuster, New York.

Shipman, Pat. 2011. *The Animal Connection: A New Perspective on What Makes Us Human.* W. W. Norton, New York.

Shipman, Pat. 2015. *The Invaders: How Humans and Their Dogs Drove Neanderthals to Extinction.* Belknap Press, Cambridge, MA.

Sibley, C. G., and J. E. Ahlquist. 1984. The phylogeny of hominoid primates, as indicated by DNA–DNA hybridization. *Journal of Molecular Evolution* 20:2–15.

Stringer, C. 2012. *Lone Survivors: How We Came to Be the Only Humans on Earth.* Times Books, London.

Stringer, C., and P. Andrews. 2005. *The Complete World of Human Evolution.* Thames & Hudson, London.

Stringer, C., and C. Gamble. 1993. *In Search of Neanderthals: Solving the Puzzle of Human Origins.* Thames & Hudson, London.

Swisher, C. C., III, G. H. Curtis, and R. Lewin. 2000. *Java Man.* Scribner, New York.

Tattersall, I. 1993. *The Human Odyssey: Four Million Years of Human Evolution.* Prentice Hall, Upper Saddle River, NJ.

Tattersall, I. 2012. *Masters of the Planet: The Search for our Human Origins.* St. Martin's Griffin, London.

Tattersall, I. 2015. *The Strange Case of the Rickety Cossack and Other Cautionary Tales from Human Evolution.* St. Martin's Press, New York.

Tattersall, I., and J. Schwartz. 2000. *Extinct Humans.* Westview, New York.

Walker, Alan, and Pat Shipman. 1996. *The Wisdom of the Bones: In Search of Human Origins.* Knopf, New York.

Wells, Spencer. 2006. *Deep Ancestry: Inside the Genographic Project.* National Geographic Society, Washington, DC.

SUMMARY

- The fossil record of humans is now incredibly rich and detailed, with dozens of species and genera in the human family tree and thousands of fossils of early humans known.

- Molecular biology has independently confirmed most of the main conclusions about human evolution and given us insights that were not previously possible.

- The public understanding of human evolution is hampered by a number of misconceptions, including that humans are somehow not members of the kingdom Animalia or another species of ape; that humans evolved in a simple linear "ladder of nature" or "chain of being" form, when actually the story of human evolution is a bushy, branching "family tree"; that the brain powered human evolution and bipedal posture came later, when in actuality it was the other way around; that only one species of human could live at any given time, when the fossil record now shows as many as four human species living side by side in the same region of Africa.

- Humans are members of the order Primates, which includes lemurs, tarsiers, New World monkeys, Old World monkeys, and the apes (including us), as well as many extinct lineages.

- The first member of our lineage, the hominins, was *Sahelanthropus*, known from fossils 6–7 Ma in Chad. Many other species of *Orrorin*, *Ardipithecus*, *Kenyanthropus*, *Paranthropus*, and *Australopithecus* are known from beds in eastern and southern Africa over the Pliocene and Early Pleistocene.

- The oldest fossils of our genus *Homo* occur in rocks about 1.75 Ma in East Africa, and several different species emerged during the Early Pleistocene, overlapping with the last members of *Australopithecus* and *Paranthropus*.

- By 1.9 Ma *Homo erectus* appeared in Africa and was the first human species to spread across Eurasia. It lasted 1.8 million years, the longest-lived hominin of all.

- By 400,000 years ago, Neanderthals had evolved and were the dominant humans in the glacial margins of Europe and western Asia; they lasted until about 30,000 years ago. Molecular evidence shows that they may have interbred with early *Homo sapiens*. A mysterious human species, the Denisovans, known from only a few bones and teeth from Siberia, also has a distinctive DNA profile.

- The oldest fossils thought to belong to *Homo sapiens* date back to 300,000 years ago in Africa, and by 100,000 years ago they look almost completely modern.

- Multiple techniques of analyzing human and ape genes independently show that we are 98%–99% identical with chimps and gorillas and that only a small part of the regulatory genome separates us. Genetically, we are virtually a third species of chimpanzee.

- We still have the genes for many of the features of our primate ancestry, including our monkey tails; but those genes are rarely expressed.

- One of the ways in which we changed dramatically from chimps is by neoteny— holding on to juvenile features while becoming reproducing adults. Many features that make humans different from chimps are found in young chimps.

- Genetically speaking, all humans are extremely similar, with a bigger difference between two or more populations of apes and gorillas than all the human "races" put together. We all diverged from a common ancestor about 70,000–50,000 years ago, and all the differences between living human populations are extremely minor and insignificant in genetic terms.

- Studies of the mitochondrial DNA show that the common ancestor of all living humans lived about 140,000–200,000 years ago in Africa, and studies of the Y chromosome DNA yield a similar answer.

- About 74,000 years ago, humans experienced an extreme population crash and almost became extinct. Global human population was reduced to possibly 5000 individuals, causing a genetic bottleneck. It is likely that this crash was due to the climatic effects of the eruption of the Toba supervolcano in Sumatra at that time. All living non-African humans are descendants of that handful of survivors, who then left Africa and spread across southern Asia and Europe and eventually to the Americas.

KEY TERMS

Ladder of life (p. 429)
Family tree (p. 429)
Great chain of being (p. 429)
Bipedalism (p. 430)
Australopithecus (p. 430)
Piltdown forgery (p. 431)
Order Primates (p. 433)
Arboreal (p. 433)
Tribe Hominini (p. 433)

Stereovision (p. 433)
Sahelanthropus (p. 433)
Orrorin (p. 434)
Ardipithecus (p. 434)
Paranthropus (p. 435)
Homo habilis (p. 435)
Homo erectus (p. 436)
Neanderthals (p. 437)
Denisovans (p. 438)

Flores people (p. 438)
Homo sapiens (p. 438)
Immunological distance
(p. 439)
DNA–DNA hybridization
(p. 439)
"Third chimpanzee" (p. 439)
Neoteny (p. 439)
Human genetic diversity
(p. 441)

"Mitochondrial Eve" (p. 442)
"Y-chromosome Adam"
(p. 442)
"Out of Africa I" (p. 443)
Genetic bottleneck (p. 443)
Toba catastrophe (p. 443)
"Out of Africa II" (p. 444)

STUDY QUESTIONS

1. What is the anatomical and behavioral evidence that we are closely related to the other great apes and members of the kingdom Animalia?
2. Why is it false and misleading to sketch human evolution as a "march through time" or talk about a "missing link"?
3. Someone asks you, "If humans are descended from apes, how come apes are still around?" How do you answer?
4. Why did anthropologists think that humans had large brains before they became bipedal? What did the evidence eventually show?
5. Why did anthropologists originally reject the human affinities of the Taung child?
6. What was the Piltdown forgery? Why was it accepted for so long? How was it eventually exposed?
7. Why did anthropologists once think that there was only one human species at a time? What discoveries showed this was wrong?
8. How are humans related to the rest of the Mammalia? How are they related to the rest of the primates?
9. What are some of the defining features of primates?
10. What is the oldest fossil belonging to the human lineage? How old was it, and where was it found? How do we know it was bipedal?
11. How does *Paranthropus* differ from *Australopithecus*?
12. What is the oldest species of the genus *Homo*? How old was it? Where was it found?
13. What was *Homo erectus*? When did it live? Where was it found?
14. Why did early anthropologists think of Neanderthals as brutish, bow-legged, and stoop-shouldered? How do we view their anatomy today?
15. How does the brain capacity of Neanderthals compare to ours? How are their skulls and bones different from modern *Homo sapiens*?
16. What does the DNA of Neanderthals show?
17. Who were the Denisovans? Why can't we give them a formal species name yet?
18. Who were the "hobbits"?
19. How old are the oldest *Homo sapiens*?
20. Over what time range did Neanderthals and *Homo sapiens* overlap? How did modern humans interact with Neanderthals?
21. How does the immunological distance method work? What did it show about humans and apes?
22. How does the DNA–DNA hybridization method work? What did it show about humans and apes?
23. When was the DNA of humans finally sequenced? How does it compare to the DNA of chimps and gorillas?
24. What is the difference between structural and regulatory genes?
25. What does the appearance of humans with tails tell us about our genes?
26. Why did Jared Diamond argue that genetically we are a "third chimpanzee"?
27. What is neoteny? How might it explain how humans have changed from other apes without much genetic change?
28. How does the genetic diversity of human "races" compare to the diversity of different ape populations? What does this suggest about the meaning of "race"?
29. What was the "mitochondrial Eve"? What did this discovery tell us about when and where humans diverged from a common ancestor?
30. What was the "Out of Africa I" event? When did it occur? Are there any living descendants of those first migrants from Africa?
31. What is a genetic bottleneck? What does it do to the genetic diversity of a population?
32. What is the evidence from other animals that a large-scale population crash and bottleneck happened across Eurasia about 74,000 years ago?
33. About how many people were the ancestors of all Native Americans?
34. What was the Toba catastrophe? How might it explain the genetic bottleneck in humans?
35. What was the "Out of Africa II" event? When did modern humans with living descendants reach parts of Asia? When did they reach Australia?

The Holocene—
and the Future:
10,000 Years Ago
to the Future

Civilization exists by geological consent,
subject to change without notice.

—*Will Durant, historian*

A

Almost all of the world's glaciers and ice caps are melting away
at staggering rates. This is visible in most of the world's glaciers,
which are vanishing at rates that can be seen by sequential photographs taken a few years apart. This is Muir Glacier in Glacier
Bay, Alaska, shot from the same point at three different times:
August 13, 1941; August 4, 1950; and August 31, 2004.

B

C

17.1 The Holocene

Full interglacial conditions stabilized about 11,700 years ago after the world warmed out of the final glaciation of the Younger Dryas (Chapter 15). Geologists refer to the last 10,000 years of our current interglacial as the "Recent" Epoch or **Holocene Epoch** (which means "completely recent" in Greek) of the Quaternary Period of the Cenozoic Era. These most recent 10,000 years encompass all of recorded human history and a lot of prehistory.

The Holocene is so short compared to the slow movement of geologic events that there are not a lot of distinctive tectonic events to discuss. Huge masses of ice normally melt slowly, so the great ice sheets that once covered Canada and Scandinavia and Siberia started melting about 18,000 years ago but did not finally vanish until about 6500 years ago, well into the Holocene. Indeed, there are still huge continental ice sheets on Antarctica and Greenland, and large glaciers in the high mountains around the world (although nearly all are vanishing rapidly now).

 See For Yourself: Ice Age Sea Levels

 See For Yourself: Sea Level Rise Isolates Britain

 See For Yourself: Welcome to Doggerland

 See For Yourself: Megaflood: How Britain became an Island

The beginning of the Holocene was marked by the last phase of the postglacial rise of sea level, so between 11,700 and about 9000 years ago, sea level rose about 35 m (about 115 feet) to the steady level that persisted through much of the Holocene. Rising sea levels continued the flooding of old glacial fjords and river valleys like Chesapeake Bay. The English Channel did not form until the early Holocene, when sea level rose to flood it all the way from the Atlantic to the North Sea; and large areas of low-lying northern Europe were also flooded at that time. In particular, the floodplains and river mouths to the north off the coast of the Netherlands, Denmark, and Germany once extended much farther into the North Sea, forming a broad area called **Doggerland**, named after a submarine plateau in the North Sea called the Dogger Banks (Fig. 17.2); today, they are drowned beneath the North Sea. Dredging up artifacts in the North Sea and imaging the sea floor have turned up evidence that large populations of humans once lived there.

17.2 Climate and Human History

HOLOCENE CLIMATIC OPTIMUM AND THE RISE OF CIVILIZATION

The most rapid and important Holocene changes, however, were climatic fluctuations. Even though the earth was in a relatively warm, stable interglacial episode, there were still small fluctuations of warming and cooling over the past 10,000 years (Fig. 17.3). As the planet warmed out of the last Ice Age, the early Holocene is typified by a period of warmer than average temperatures as the peak of solar insolation from the coincidence of the warmest phases of the three Milankovitch cycles occurred. About 9000 years ago, the axial tilt of the earth reached 24°, maximizing the heat on the poles, and the earth was at its closest approach to the sun during the northern hemisphere summer, which substantially increased the amount of solar radiation received. This Early Holocene warming from about 9000 to 5500 years ago has been called the **Hypsithermal** or the **Holocene Climatic Optimum** (Fig. 17.3A). Global average temperatures were about 0.7°C

Figure 17.2   During the low sea-level stands of the peak of the last glaciation 18,000 years ago, most of the floor of the North Sea was an exposed upland called Doggerland, and there was no English Channel cutting Britain off from the continent. As sea level rose in the past 17,000 years, all of this region was eventually flooded.

(1.3°F) warmer than typical for the Holocene, and the polar regions were as much as 4°C (7.4°F) warmer. Temperate North America and Europe were about 2°C (3.6°F) warmer.

See For Yourself: How Climate Influenced History

The most striking effect of the Holocene Climatic Optimum is that the desert regions of central Asia and especially the Sahara Desert were not only a bit warmer but much wetter than they are today. Civilizations began to develop in many parts of Africa and Asia simultaneously. In fact, the Sahara was covered by a huge network of lakes, rivers, and dense forests, dominated by crocodiles and hippos; and a large human population thrived in these rich forests. We can see this by using satellite radar to "look beneath" the Sahara

sand dunes and find the buried river valleys and lakebeds that are now covered completely by sand (Fig. 17.4). It is thought that the extensive overuse of the land and the overgrazing by domesticated animals may have contributed to the destruction of the green Sahara, along with changes in climate around 5500 years ago.

None of this would have been possible without the earlier invention and spread of agriculture, often called the **Neolithic revolution**. It apparently arose somewhere in the Zagros Mountains of Iran about 14,000 years ago during the latest Pleistocene. Archeological sites in that region show the first evidence of domestication of wheat and barley, along with the mortars and grinding stones which were used to grind

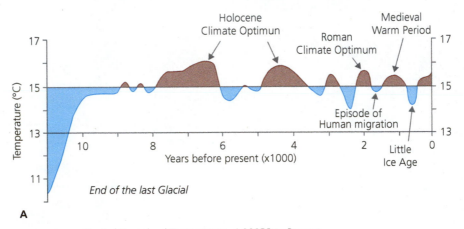

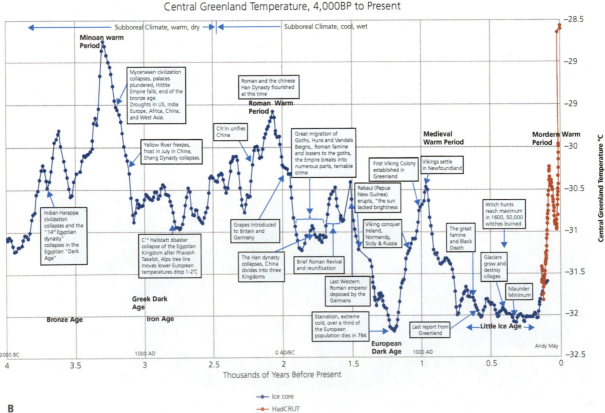

Figure 17.3 **A.** Average surface temperature and climate history of the last 11,000 years in the northern hemisphere, showing the small-scale fluctuations in climate during the relatively stable conditions of the Holocene. **B.** Detail of the last 2000 years from ice cores in Greenland, showing the connection between historic events and climate.

the grain down into flour, as well as bones of domesticated sheep and goats. By 9000 years ago, the earliest cultures in the Tigris–Euphrates valleys below the Zagros Mountains were living in the open in small houses, rather than caves, and raising sheep, goats, pigs, and dogs. By 8000 years ago, sites in Greece show that agriculture and domesticated animals were spreading to Europe, and these practices spread across the recently deglaciated landscapes of Europe, reaching Britain and Scandinavia about 6000 years ago. Meanwhile, around 7000 years ago cattle were domesticated and spread among these early European cultures. All of this expansion of agriculture, domestication of animals, and early civilization was not possible without the warmer, wetter conditions of the early Holocene Climatic Optimum.

Figure 17.4 During the Climatic Optimum about 9000–6000 years ago, the Sahara was a well-watered region with large rivers draining forested plains—all of which are now buried under sand dunes. The left image shows the surface features of Safsaf Oasis in the eastern Sahara, and the right image shows the radar view shot from space shuttle *Columbia* in 1981, which reveals a network of drainages and large rivers that vanished beneath the dunes when the climate dried up.

It is no accident that the most ancient advanced civilizations developed in this same area during the latter part of the Holocene Climatic Optimum. These regions include the Tigris–Euphrates valleys in what is now Iraq (Ubaid Period, about 8500 years ago, followed by the Sumerians, about 6000 years ago), the Nile Valley (Gerzean-Period Egyptians, about 5500 years ago), the Indus River Valley in what is now Pakistan (Harappan civilization, about 4500 years ago), and the earliest Chinese civilization in the Yellow River and Yangtze River valleys about 4200 years ago. The oldest known cities on earth, such as Jericho, were first settled about 10,500 years ago. All of these early civilizations arose during the warm, wet conditions of the Holocene Climatic Optimum, when Asia and northern Africa were much wetter and greener and able to grow many more crops without advanced irrigation techniques. The Bible describes the Middle East as "the land of milk and honey" and "the promised land," and many have described the Middle East as "the Fertile Crescent." This region was certainly fertile before 5500 years ago—but thanks to climate changes of the later Holocene, it is now the harsh desert climate we associate with modern Egypt, Israel, and Iraq.

THE NEOGLACIAL

The last 5500 years of climate history is often called the **Neoglacial Period** because the second half of the Holocene was on average much cooler than the Holocene Climatic Optimum of the early Holocene (Fig. 17.3A). Just as climate changes and warming episodes of wetter climate helped spur the rise of agriculture and eventually civilization, changing climate that made the world cooler and drier

had effects on human history as well. About 1177 B.C.E., there was an abrupt and severe drought along with a cooling event across much of the eastern Mediterranean that wiped out many of the civilizations that existed then, including the archaic Greek cultures of Mycenae and the Minoan civilization of the island of Crete, the Hittites of what is now Turkey, and the New Kingdom of Egypt. This major crisis in world history, called the **Late Bronze Age cultural collapse**, caused the entire civilized world to go into a "dark ages" for over a century (Fig. 17.3B). Its causes are still a matter of debate, but the well-documented extreme drought and rapid cooling are certainly part of the reason.

After this episode, climate began to warm again, and the great civilizations of classical Greece rose and peaked about 500 B.C.E., then were absorbed into the empire of Alexander the Great, followed by the Roman Empire. In fact, the period around 500 B.C.E. to 200 C.E. was another episode of warmer, wetter weather in Europe, called the **Roman Warm Period** (Fig. 17.3B). Many scholars have speculated about the causes of the rise and fall of the Roman Empire, but certainly one of the underlying factors was the dramatic shifts in climate that promoted agriculture and mild conditions. Then as the Roman Empire declined, climate went into the long cooling episode beginning about 500 to 1000 C.E., which coincided with colder, harsher, drier climates of the Middle Ages. At 536 C.E., during the darkest period of the Dark Ages, there was an extreme cooling event that caused a famine and mass death and shocked cultures all over Europe. Its causes are still a matter of debate but probably involved global cooling from the volcanic dust injected into the stratosphere by an Indonesian volcano (possibly an earlier eruption of Krakatau before its 1883 final eruption).

THE MEDIEVAL WARMING AND LITTLE ICE AGE

The cold of the Middle Ages gave way to another warming period between 950 and about 1250 C.E., known as the **Medieval Warm Period** (Fig. 17.3A, B). Temperatures around the North Atlantic region rose by about 0.5°C (1.9°F), enough to melt much of the ice in the North Sea and North Atlantic and bring warmer climates to Europe and North America. In particular, the Vikings from Scandinavia used the opportunity to sail far and wide, invading northern Europe repeatedly. They even sailed to Iceland and then reached Greenland and eventually the Labrador coast of North America. Greenland was warm and ice-free enough on its southern tip to grow a few crops, raise sheep and cattle, and establish small outposts. However, the warming was not so favorable to people in other regions. In the southwestern deserts of the United States, the Medieval Warm Period caused prolonged droughts that wiped out the Anasazi cultures that built the amazing pueblos found throughout that region.

However, no one should mistake the warming of the Medieval Warm Period for being comparable to the current global warming of the planet. For one thing, the warming was localized to the North Atlantic region, and average temperatures actually dropped on a worldwide basis. Also, the warming was tiny—only a fraction of a degree Celsius. By contrast, the warming since the 1960s is almost an order of magnitude bigger, rising more than a full degree Celsius globally in just a few decades. Finally, the causes of the Medieval Warm Period are not fully understood, but there was a maximum of solar radiation at that time and an unusually long interval without volcanic eruptions, which block solar radiation and cool the planet. The deforestation of Europe in the early Middle Ages and burning of so much vegetation released a lot of carbon dioxide into the atmosphere. In addition, changes in oceanic circulation brought unusually warm conditions to the North Atlantic, but that did not mean the rest of the world got warmer. These causes are totally different from the human-caused climate change that we are now experiencing (see Box 17.1).

 See For Yourself: Little Ice Age

About 1500, the Medieval Warm Period gradually began to cool down, and sea ice returned to Greenland, along with expanding ice sheets. These events wiped out the tiny Viking colonies in Greenland and marked the transition to what is known as the **Little Ice Age** (Fig. 17.3A, B). In 1650, 1770, and 1850, there were three particularly cold intervals, which dramatically changed the climates of northern Europe and North America. Regions of Europe that had not frozen in centuries routinely experienced harsh winters, so we see many famous paintings and historical records of the time depicting the freezing of the Thames River in London or the canals of Holland. There were longer and more frequent snowfalls, colder summers, and more severe deep freezes. This led to many crop failures and outright famine numerous times during this period. In North America, the effects were similar. The famous painting of Washington crossing the Delaware through many ice floes is a good example.

Back on Christmas Eve 1776, when Washington made the crossing to surprise the British and Hessian troops in the Battle of Trenton, the Delaware River routinely froze over; today, it rarely gets close to freezing.

The causes of the Little Ice Age are not fully understood. The Milankovitch cycles were reaching a point where the next glacial interval should be starting, so that may be one of the reasons. There was also very weak solar activity at the time, known as the **Maunder minimum**. This meant the earth got less solar warming. Others have suggested that the destruction of Native American cultures in the 1600s and 1700s, which routinely cleared and burned the forests of North America, might have contributed to the cooling as forests grew back and locked up carbon dioxide. Finally, there were far more large explosive volcanic eruptions during this period, blocking the sun with the ash that rose to the stratosphere. The biggest of these were the eruptions of Laki in Iceland in 1783 and 1784 and especially the 1815 eruption of **Tambora** in Indonesia. This last eruption was the largest explosive volcanic event in the past 74,000 years (since the eruption of Toba) and ejected huge amounts of volcanic dust into the stratosphere to cool the planet. The effect was so extreme that 1816 was called the "**year without a summer.**" Summer temperatures were so far below normal that there were deep freezes and snowfalls in June and July in New York and in many European cities known for their summer warmth. Crops froze and farm animals died, causing widespread famine in Europe, just as the world was recovering from the shock of the Napoleonic Wars that ended with the 1815 Battle of Waterloo. In Lord Byron's villa on Lake Geneva, his guests Percy and Mary Shelley were spending their summer holiday huddled by the fireside, instead of enjoying a normal warm summer by the lake. They told ghost stories to pass the time, one of which became the basis for Mary Shelley's *Frankenstein*. The Little Ice Age came to an end not long after the "year without a summer," and the winters of the 1880s were the last gasps of truly cold climates in northern Europe and North America.

17.3 The Anthropocene

In recent years, a number of scientists have proposed a new epoch for the Cenozoic: the **Anthropocene**. The effect of humans on the planet, from its crust to its oceans to its atmospheres, has been as profound as any asteroid impact or massive volcanic eruption or other cosmic catastrophe. Thanks to humans, earth's surface and especially its biosphere, oceans, and atmosphere have been transformed more rapidly and more completely than at any time in the geologic past. Humans have caused a mass extinction that is faster and more extreme than any of the great mass extinctions of the geological past and is still ongoing. We have mined enormous areas of the earth's surface, cut down most of the world's forests, and transformed the landscape, especially with our gigantic network of cities, roads, and other features. We have filled the land and now the oceans with huge quantities of trash and pollutants that will persist, in some cases, for millions of years into the future. Most important of all, we have completely changed the atmosphere that all organisms on earth depend on, at a rate unseen in the geologic past. Without question, a good argument can be made that the world has been completely transformed by humans during the Anthropocene as much as during the transition between any other two geological time intervals.

The Anthropocene has not been officially accepted by the geological committees that make formal changes to the timescale, but it is widely used by many geoscientists now and shows up on the meeting programs and in professional talks at national geological meetings. So far, there has not been an officially agreed-upon definition of the start of the Anthropocene. Some would begin with the mid-1800s, when the Industrial Revolution led to huge increases in mining, land destruction, pollution, and especially the explosion of population and the pollution of the ocean and atmosphere. Others would start with the Neolithic revolution and the beginning of agriculture about 12,000 years ago, which would make the Anthropocene roughly equivalent with the Holocene and virtually redundant. In 2015, a majority of the members of the International Anthropocene Working Group recommended a boundary based on a significant and unambiguous event, the detonation of the first atomic bomb at the Trinity test site in southern New Mexico on July 16, 1945. Not only did this moment spark the beginning of the nuclear age and the Cold War, but the radiation from this test and the many nuclear blasts that followed has left an unambiguous time marker in the rock record, comparable to the iridium layer that marks the end of the Cretaceous. Whatever the definition, there is a lot of merit in thinking about the time interval when humans transformed the planet as a new time unit, one that is profoundly different in many ways from the Holocene or any time that preceded it.

In a book like this, there is not enough space to discuss every way in which humans have changed the planet. Let us focus on just a few examples, starting with the most significant: the transformation of our atmosphere.

CLIMATE CHANGE AND THE ATMOSPHERE

As we discussed in Chapter 8, the earth has a remarkable atmosphere that is unlike that of any other planet we know of, not only in our solar system but across the universe. About 78% of the atmosphere is the inert gas nitrogen, and 21% is made of free oxygen, with trace amounts of other gases. By contrast, Venus has a thick atmosphere made of carbon dioxide and sulfuric acid and is hot enough to melt lead. Mars has a very thin, frozen atmosphere of carbon dioxide.

So far as we can tell, earth is the only planetary body in the universe with abundant free oxygen in its atmosphere, which is entirely due to the presence of life (specifically, photosynthetic bacteria, algae, and more advanced plants). The presence of this oxygen also created the stratospheric ozone layer, which protects life from bombardment of ultraviolet radiation. Without the ozone layer, the intense UV

radiation causes sunburn and in many cases mutations that lead to skin cancer. The earth is in the "Goldilocks zone," the right distance away from the sun so that it is neither too hot (like Venus) nor too cold (like Mars) and its surface is at the temperature for water to remain in its liquid state, not vaporized into steam or frozen into ice. The magnetic field of the earth acts as a shield to protect us from harmful cosmic radiation, especially the charged plasma from the sun called the "solar wind." So far as we know, no other planet in space has all these remarkable and unique conditions, which is one reason many scientists think that it's extremely unlikely that complex life arose on any other planet.

Our unique atmosphere is also very fragile. In the 1970s, measurements from aircraft, balloons, and satellites showed that industrial chemicals used in refrigerants and propellants for spray cans called **chlorofluorocarbons** (CFCs) were destroying the ozone shield that protects us from skin cancer. Each September–November, an **ozone hole** appears over the South Pole as the austral spring warming mixes the CFCs and destroys most of the ozone over the southern hemisphere. But when this was discovered and publicized, people stopped using products with CFCs and the chemical companies soon began to reduce their production of these damaging chemicals. By international agreement in 1987, the nations of the world agreed to stop producing CFCs, and now the ozone layer is gradually healing.

International agreements solved the ozone–CFC problem in the 1970s and 1980s. In the 1970s through 1990s, regulations and agreements also solved the problem of **acid rain**, which was produced in coal-fired power plants and was destroying the world's forests and acidifying lakes and streams. Regulations and new technologies have greatly reduced the pollutants that produce smog and other bad air conditions in many parts of the world. But now we are confronted with an air pollution problem on an even larger scale that is not going to be quite as easy to solve since it involves phasing out and eliminating the use of fossil fuels.

THE GREENHOUSE EFFECT

The problem is the fact that certain chemicals act as "greenhouse gases" that trap heat inside the lower atmosphere and warm up the planet. The story goes back over a century to the 1850 discovery by English physicist John Tyndall that greenhouse gases like water vapor and carbon dioxide trap the heat but let through most solar radiation. Swedish scientist Svante Arrhenius, who received the Nobel Prize in Chemistry for his work, made the next major breakthrough in 1896. Arrhenius discovered that carbon dioxide was the most important greenhouse gas currently in our atmosphere. When the earth gets energy from the sun, the solar radiation arrives in shorter wavelengths (mostly visible light and some ultraviolet) that easily penetrate our atmosphere. After the earth absorbs this energy, it radiates it back out as longer-wavelength infrared radiation (which we call heat), which greenhouse gases prevent from escaping (Fig. 17.5A).

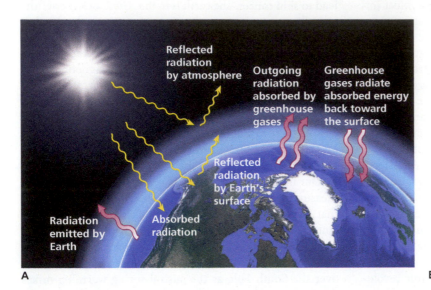

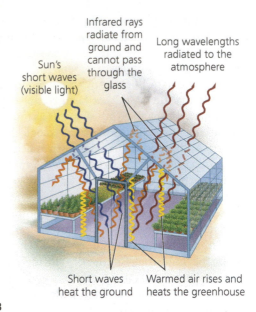

A   B

Figure 17.5   A. The greenhouse effect occurs when the atmosphere lets sunlight (which is mostly in the shorter wavelengths, primarily visible light) pass through, but when that solar energy is absorbed by the earth's surface, it radiates back out into space at a longer wavelength, the infrared. **B.** Greenhouse gases do not let much infrared energy (heat) pass through, just like the glass of a greenhouse lets the short-wave visible light through but traps the heat that tries to escape.

Why do we call them "greenhouse gases"? In a real greenhouse (Fig. 17.5B), the temperatures inside remain very warm because the visible light from the sun easily passes through the glass walls, but the warmth from the heated surfaces inside (infrared radiation) cannot escape through the glass. Another analogy is the inside of a parked car. Just as in a greenhouse, the interior heats up as the sunlight passes through the windows and the solar energy is absorbed by the interior of the car, then radiated back as heat. Thus, it will always be hotter inside the car than outside because the glass windows let light pass through but block heat from escaping.

 See For Yourself:
Greenhouse Effect

Because more heat comes in to the planet than can leave it, the earth's atmosphere warms up. Originally, Arrhenius calculated that doubling the level of atmospheric carbon dioxide would cause global temperatures to rise by 5°–6°C (9–10°F). This is remarkably close to the estimates of scientists in a report done in 2007.

ATMOSPHERIC CHANGES

The next major discovery occurred when Charles Keeling invented one of the first devices for measuring atmospheric carbon dioxide. In 1958 he began to take measurements in two places isolated from the populated areas of the earth (thus minimizing local effects of pollution from cities or industry): Antarctica and the top of Mauna Loa volcano in Hawaii. After a few years of data showing the rapid increase in carbon dioxide, the National Science Foundation ended his Antarctic funding when it foolishly decided that he had proved his point, thus ending the potential for collecting a long-term data set. But the Mauna Loa observatory has

been running continuously for more than 60 years, one of the longest sets of atmospheric data ever collected. By the 1960s, Keeling and his colleague, the oceanographer Roger Revelle, could see the dramatic increase in carbon dioxide in a steady upward trend (Fig. 17.6). Superimposed on the upward trend are the annual cycles of decreasing carbon dioxide in the spring when the northern hemisphere plants take in carbon dioxide and increasing carbon dioxide in the fall when the trees in the north lose their leaves and the carbon dioxide is released as the leaves decay.

From Keeling's initial data to every data set that has been collected since then, the trend is clear: carbon dioxide in our atmosphere has increased at a dramatic rate in the past 150 years. It was barely 315 parts per million (ppm) when the experiment started, and as of 2020, it's over 415 ppm. Not one data set of temperature or carbon dioxide collected over a long enough span of time shows otherwise. A compilation of the past 900 years' worth of temperature data from tree rings, ice cores, corals, and direct measurements of the past few centuries shows the sudden increase of temperature of the past century standing out like a sore thumb (Fig. 17.7A). This famous graph is nicknamed the "hockey stick" because it is mostly long and straight, then bends sharply upward at the end like the blade of a hockey stick. Other graphs show that climate was very stable within a narrow range of variation through the past 1000, 2000, or even 10,000 years since the end of the last Ice Age and even back to the end of the last glacial maximum about 18,000 years ago (Fig. 17.7B). The graph clearly shows the minor warming events during the Climatic Optimum about 7000 years ago, the Medieval Warm Period, and the slight cooling of the Little Ice Age from the 1700s and 1800s. But the magnitude and rapidity

Mauna Loa Observatory, Hawaii
Montly Average Carnon Dioxide Concentration

Data from Scripps CO₂ Program   Last Updated June 2018

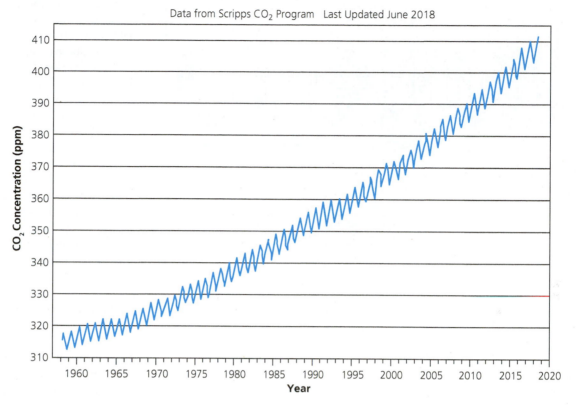

Figure 17.6 In 1958, Charles David Keeling used his newly invented instrument to measure atmospheric carbon dioxide at the top of Mauna Loa Observatory on the Big Island of Hawaii. That experiment has run continuously ever since and clearly shows the steady and rapid increase in carbon dioxide in the atmosphere (verified by many other stations around the world). Since most of the world's plants are on the continents of the Northern Hemisphere, the Keeling curve has an annual cycle of decrease in the northern hemisphere spring (when plants are growing after the winter and absorbing carbon dioxide) and increase in the fall (when those same plants lose their leaves that then decay and release that carbon dioxide back to the atmosphere). In 1958, the concentration of carbon dioxide was only around 315 parts per million by volume (ppmv). In 2018, the upswing of the Keeling curve passed 405 ppmv, and even the downswing was above 400, so the earth's atmosphere has permanently crossed the 400-ppmv threshold. As of 2020, it was over 415 ppmv.

of the warming represented by the last 200 years are simply unmatched in all of geologic history. More revealing, the timing of this warming coincides with the Industrial Revolution, when humans began massive deforestation and the release of carbon dioxide by burning coal, gas, and oil.

See For Yourself: Bill Nye on Climate Change

MELTING POLAR REGIONS

See For Yourself: Chasing Ice

If the data from atmospheric gases were not enough, we are now seeing unprecedented changes in our planet. The polar ice caps are thinning and breaking up at an alarming rate. In 2000, for the first time ever, people flying over the North Pole in summertime saw no ice, just open water. So much for Santa's workshop! The Arctic ice cap has been frozen solid for at least the past 3 million years and maybe longer, but now the entire ice sheet is breaking up so fast that by 2030 (and possibly sooner) less than half of the Arctic will be ice covered in the summer (Fig. 17.8). As you can see from watching the news, this is an ecological disaster for everything that lives up there, from the polar bears to the seals

and walruses and whales to the animals they feed upon. The Antarctic is thawing even faster. In February–March 2002, the Larsen B ice shelf, over 3000 km² (1158 square miles, the size of Rhode Island) and 220 m (about 700 feet) thick, broke up in just a few months, a story typical of nearly all the ice in Antarctica. The Larsen B shelf had survived all the previous Ice Ages and interglacial warming episodes for the past 3 million years and even the warmest periods of the last 10,000 years—yet it and nearly all the other thick ice sheets in the Arctic, Greenland, and the Antarctic are vanishing at a rate never before seen in geologic history. And in 2017, the Larsen C ice shelf began to break apart.

See For Yourself: Melting of the Arctic Ice Cap

SEA-LEVEL RISE

Many people don't care about the polar ice caps, but there is a serious side effect worth considering: all that melted ice eventually ends up as more water in the ocean, causing sea level to rise, as it has many times in the geologic past. At the moment, sea level is rising about 3–4 mm per year, more than 10 times the rate of the 0.1–0.2 mm/year that has occurred over the

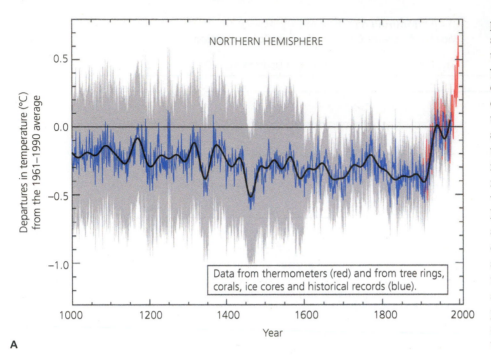

A

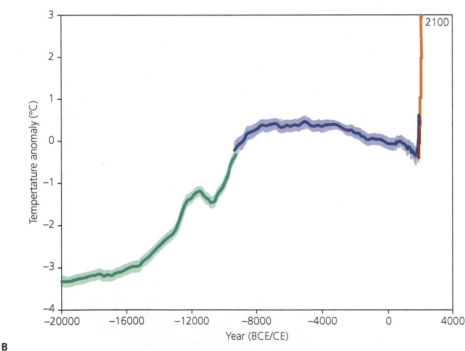

B

Figure 17.7   A. The last 1000 years of temperatures for the northern hemisphere, compiled from tree rings, ice cores, and historical records (blue) and directly measured by thermometers in the last century in red. The trend begins at the end of the Medieval Warm Period around 1000 CE and reaches its lowest point with the Little Ice Age at 1600–1850 CE. Then, the effect of human-caused global warming kicks in, and the temperatures rise dramatically and rapidly to a level not seen since the Eocene. This has been nicknamed the "hockey stick" curve because the rapid inflection at the end after 900 years of stability resembles the way the blade of a hockey stick bends sharply from the handle. Note that the recent rise in global temperature is much faster and more extreme than the warmest part of the Medieval Warm Period. **B.** The last 20,000 years of climate change, from the peak glacial at 20,000 to 18,000 years ago to the warming of the glacial–interglacial transition from 18,000 to 10,000 years ago (green curve) to the Holocene stability of the past 10,000 years (blue curve). Even in this scale, the extreme magnitude and rapidity of the heating of the past 150 years (red curve) stand out as unprecedented, with more and faster warming than in any time in the geologic past.

past 3000 years. Geological data show that sea level was virtually unchanged over the past 10,000 years, since the present interglacial began. A few millimeters here or there doesn't impress people, until you consider that the rate is accelerating and that most scientists predict it will rise 80–130 cm in just the next century. A sea level rise of 130 cm or 1.3 m (over 4 feet) would drown many of the world's low-elevation cities, like Venice and New Orleans, and low-lying countries like the Netherlands and Bangladesh. A number of tiny island nations like Vanuatu and the Maldives, which barely poke out above the ocean now, are already vanishing beneath the waves. Their entire population will have to move someplace else. Already low-lying coastal cities like Miami and Venice are routinely being flooded during really high tides, when they never used to have a problem. If sea level rose by just 6 m (almost 20 feet), nearly all the world's coastal plains and low-lying areas (such as the Louisiana bayous, all of Florida, the low countries like the Netherlands and Belgium, and most of the world's river deltas) would be drowned. Most of the world's population lives in coastal cities like New York, Boston, Philadelphia, Baltimore, Washington, DC, Miami, Shanghai, and London. All of those cities would be partially or completely under water with such a sea-level rise. If all the glacial ice caps melted completely (as they have several times before during past greenhouse worlds in the geologic past), sea level would rise by 65 m (213 feet)! The entire Mississippi Valley would flood, so you could dock your boat in Cairo, Illinois (Fig. 17.9). Such a sea-level rise would drown nearly every coastal region under hundreds of feet of water and inundate New York City, London, and Paris. All that would remain would be the tall landmarks, such as the Empire

Total sea-ice extent*, millions of square kilometers

North Minimum extent measured in September

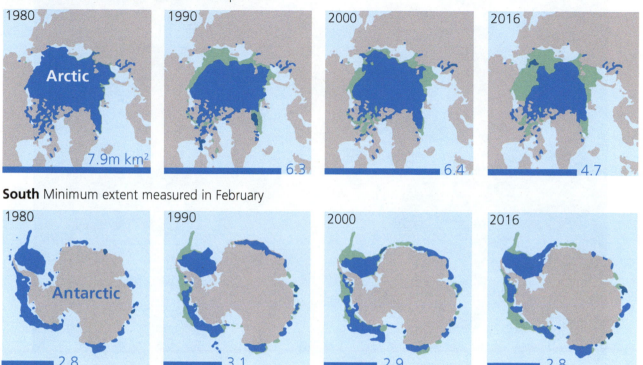

South Minimum extent measured in February

Figure 17.8 Disappearance of the ice caps in both the Arctic and Antarctic over the past decade. Gigantic amounts of ice vanish every summer and fall and are not replaced by the freezing of the following winter.

State Building, Big Ben, and the Eiffel Tower. You could tie your boats to these high spots, but most of the rest of these drowned cities would be deep under water.

OTHER EFFECTS OF CLIMATE CHANGE

The changes are not restricted to polar ice and rising sea level. Climate change has effects around the world. About 95% of the remaining mountain glaciers left over from the Pleistocene are retreating at the highest rates ever documented—or have already vanished. Many of those glaciers, especially in the Himalayas, Andes, Alps, and Sierras, provide most of the freshwater that the populations below the mountains depend upon—yet this freshwater supply is vanishing. The permafrost that once remained solidly frozen even in the summer is now thawing, damaging the Inuit villages on the Arctic coast and threatening all our pipelines to the North Slope of Alaska. A more serious issue with the thawing permafrost is that it stores huge amounts of methane, which is just now being released. Methane is an even more powerful greenhouse gas than carbon dioxide, so if it increases dramatically in the atmosphere, global warming would be even more out of control than it is now.

Not only is the ice vanishing, but we have seen record heatwaves over and over again, killing thousands of people, as each year joins the list of the hottest years on record. In fact, 20 of the hottest years on record have occurred in the last 22 years, with 2016 being by far the hottest year ever measured. 2017 and 2018 were just slightly below that record, and in July 2019 record-breaking heat waves in Europe and North America killed hundreds of people. Natural animal and plant populations are being decimated all over the globe as their environment changes. Many animals respond by moving their ranges to formerly cold climates, so now places that once did not have to worry about disease-bearing mosquitoes are infested as the climate warms and allows them to breed farther north.

 See For Yourself: Al Gore and Optimism About Climate Change

If you have seen any of the recent movies or videos about climate change, the long litany of "things we have never seen before" and "things that have never occurred in the past 3 million years of glacial–interglacial cycles" is staggering. Still, there are many people who are not moved by the dramatic images of vanishing glaciers or by the forlorn polar bears starving to death. Many of these people have been fed misinformation by the powerful lobbies and political organizations and fossil fuel companies who want to cloud or confuse the issue. We examine some of these claims in Box 17.1.

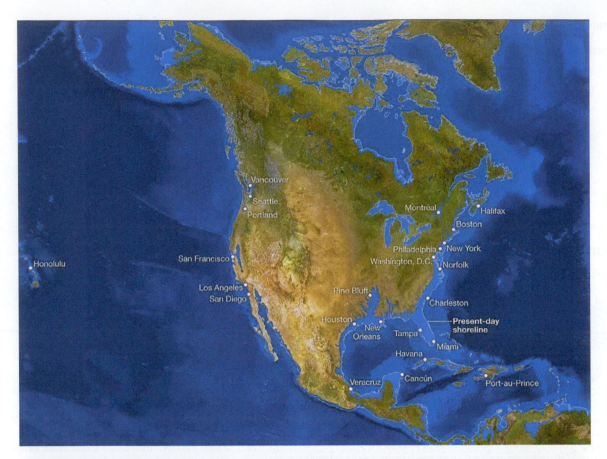

Figure 17.9 The melting of all the world's ice caps and glaciers would cause sea level to rise about 65 m (213 feet). This would completely flood all the coastal cities of North America and make Florida and Louisiana disappear, and have a similar effect on the Low Countries of Belgium and the Netherlands, northern Germany, Poland, Denmark, and much of Great Britain.

BOX 17.1: HOW DO WE KNOW?

How Do We Know That Humans Are Causing Climate Change?

There are people who doubt climate change is happening (despite the overwhelming evidence), and others who admit it's real but don't believe humans are the cause. But the evidence is overwhelming that humans are to blame. How do we know that climate is changing in an unusual manner and not just because of normal "climate fluctuations"?

See For Yourself: How Carbon Dioxide Is Causing Global Warming

• *"It's just natural climatic variability."* As we have seen in this book, geologists and paleoclimatologists know a lot about past greenhouse worlds and the icehouse planet that has existed for the past 33 million years. We have a good understanding of how and why the Antarctic ice sheet appeared at that time and how the Arctic froze over about 3.5 million years ago, beginning the 24 glacial and interglacial episodes of the "Ice Ages" that have occurred since then. We know how variations in the earth's orbit (the Milankovitch cycles) control the amount of solar radiation the earth receives, triggering the shifts between glacial and interglacial periods. Our current warm interglacial has already lasted 10,000 years, the duration of most previous interglacials, so if it were not for global warming, we would be headed into the next glacial any time now. Instead, our pumping greenhouse gases into our atmosphere after they were long trapped in the earth's crust has pushed the planet into a "super-interglacial," already warmer than any previous warming period. We can see the "big picture" of climate variability most clearly in the EPICA-1 core from Antarctica (Fig. 17.10), which shows the details of the last almost 800,000 years of glacial–interglacial cycles. *At no time during any previous interglacial did natural carbon dioxide levels exceed 280 ppm, even at their very warmest.* Our atmospheric carbon dioxide levels are already close to 415 ppm today. The atmosphere is headed to 600 ppm within a few decades, even if we stopped releasing greenhouse gases immediately. This is decidedly *not* within the normal range of "natural climatic variability" but clearly unprecedented in human history. Anyone who says this is "normal variability" has never seen the huge amount of paleoclimatic data that show otherwise.

See For Yourself: Global Change in Temperature Since 1880

• *"It's just the sun or cosmic rays or volcanic activity or methane."* The amount of heat that the sun provides has been cooling down since 1940, just the opposite of what some people claim (Fig. 17.11). If the sun were causing global warming, it should show an increase in activity, not a decrease. Cosmic radiation causes an increase in cloud cover on the earth, so increased cosmic rays would cool the planet, and decreased cosmic radiation would warm it. There are lots of measurements of cosmic radiation, and the result is clear: since the 1960s, cosmic radiation has been increasing (which should cool the planet), while the temperature has been rising, the exact opposite effect expected if cosmic radiation contributed to recent warming. Nor is there any clear evidence that large-scale volcanic events (such as the 1815 eruption of Tambora in Indonesia, which changed global climate for about a year) have any long-term effect that would explain 200 years of warming

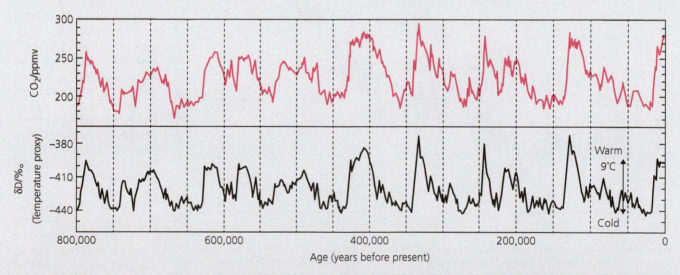

Figure 17.10 The record of carbon dioxide in the EPICA-1 core from Antarctica, the longest and deepest core ever taken. It spans 800,000 years into the past, through at least seven complete glacial–interglacial cycles. Natural climate variability is clearly shown by the core, which proves that the highest level of carbon dioxide (top curve) during the warmest interglacials was only 280 ppm—and today it is over 415 ppm. The bottom curve displays deuterium (hydrogen-2), which is an indirect measure of atmospheric temperature recorded in the core.

and carbon dioxide increase. Volcanoes erupt only 0.3 billion tonnes of carbon dioxide each year, but humans emit over 29 billion tonnes a year, about 100 times as much; clearly, we have a bigger effect. Methane is a more powerful greenhouse gas, but there is 200 times more carbon dioxide than methane, so carbon dioxide is still the most important agent. Every other alternative has been looked at, but the only clear-cut relationship is between human-caused carbon dioxide increase and global warming. We just can't squirm out of the blame on this one.

- *"The climate records since 1998 show cooling"* or *"warming has paused since 1998."* People who make this argument are distorting and misusing the data. Over the short term, there was a slight cooling trend from 1998 to 2000 (Fig. 17.12) because 1998 was a record-breaking El Niño year, so the next few years look cooler by comparison. This is deliberately "cherry-picking" a data point to bias the data to make them show what you want to show. In science, "cherry-picking" data is considered dishonest and unethical—but it is a common tactic in the political sphere and in public relations and propaganda. But since 2002, the overall long-term trend of warming (Fig. 17.12) is unequivocal. Saying that there was a "climate pause since 1998" or "cooling trend since 1998" is a clear-cut case of using out-of-context and outdated information in an attempt to distort and deny the evidence. All of the 20 hottest years ever recorded on a global scale have occurred in the last 22 years. More significantly, the record-breaking temperatures of the last few years (2014, 2015, and 2016, each broke the record of the previous year, and 2017 and 2018 were just below the all-time record hot year of 2016) rocketed upward at a rate never seen in any climatic record, ancient or modern.

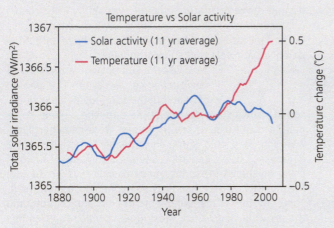

Figure 17.11 The trends of incoming solar radiation and temperature over the past century. If increasing solar activity explained global warming, there should be an upward trend in the solar curve—but solar input has actually decreased since the 1960s.

See For Yourself: Arctic Ice Melt and the Jet Stream

- *"We had heavy snow and freezing temperatures last winter."* So what? This is a classic case of how the scientifically illiterate public cannot tell the difference between *weather* (short-term seasonal changes) and *climate* (the long-term average of weather). Weather is what happens from hour to hour or day to day; climate is the average of weather events over decades to centuries or longer. Climate is what you might read in an almanac or on a website about the expected average temperature for a given day in the future; weather is what you actually see when you step outdoors. Winter doesn't just stop as the globe gets warmer; winters just become *on average* a bit warmer and shorter than usual, but large cold spells, snowstorms, and freezing events will still happen. In addition, our local weather tells us nothing about the next continent, or the global average; it is only a local effect, determined by short-term atmospheric

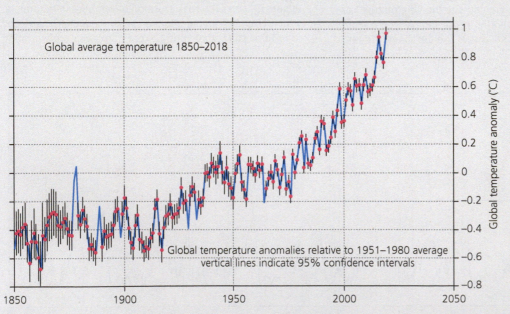

Figure 17.12 The long-term average temperature on earth over the past century. The yearly means are shown with the noisy blue data points, and the 5-year rolling average (which smooths out the noise) is shown in red. Clearly, the planet has been warming dramatically in the past century. Climate deniers people point to a single anomalous data point, the record warm year of 1998 (when it was warmer due to El Niño bringing heat from the ocean), and then claim that "warming has paused since 1998." Not only is this an unethical example of "cherry-picking" one anomalous data point to bias the results in your favor, but it's no longer even remotely true since the average annual global temperatures of the past five years have all broken the previous records and shot way past the warmest part of 1998.

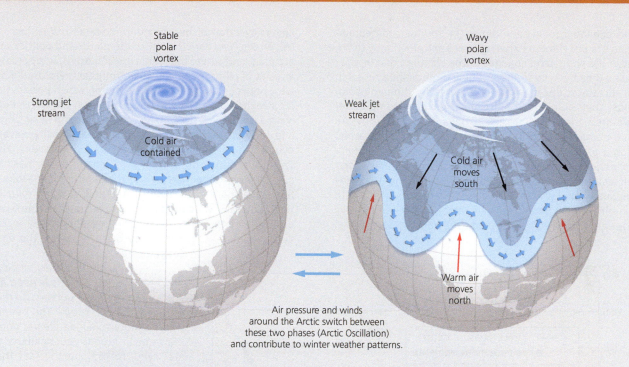

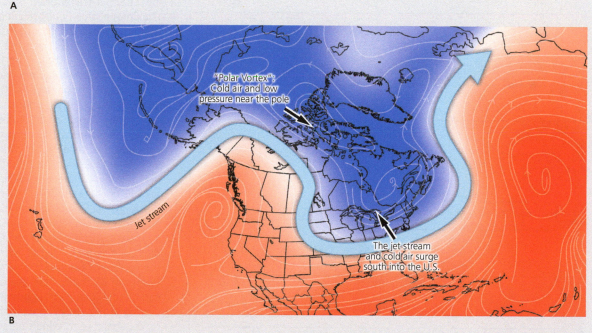

Figure 17.13 When the Arctic warms, the temperature contrast between the poles and equator weakens, which decreases the energy that powers the jet stream. This causes the jet stream to go from a relatively strong, fast, straight flow in the upper atmosphere to a slower, lazier pattern that makes big loops over the northern hemisphere called Rossby waves. If one of these loops gets stuck over northeastern North America, it can pull polar air southward (the "polar vortex"), and it may be stalled in a region for weeks, giving us record long cold spells in the eastern United States, where politicians make our laws. However, a bit of snow in Washington, DC, does not invalidate the fact that overall the globe is warming, so at the same time that it's cold in the east, there may be record warming and drying over the western United States and other regions. What is most important is not weather in one local place but the fact that overall the global temperature is warmer than average. **A.** Cartoon of the irregular looping boundary of the jet stream, with Rossby loops dropping down into the northern continents to bring a low-pressure cell that can stall and cause wintry conditions for weeks. **B.** Weather map of how a cold Rossby loop can stall the polar vortex over one place, while the rest of the country and the world is experiencing record warming and drying. Thus, if it's snowing in Washington, DC or New York, it means nothing on a global scale, when the rest of the planet is warming.

and oceanographic conditions. (In addition, meteorologists are *not* climate experts; they are trained only to analyze the short-term changes in the weather, and in most cases they have no formal training or published research in climate science, which is an entirely different field. Thus, their opinions about climate change are no better informed than anyone else's opinion unless they are also doing research in climate science.) In fact, warmer global temperatures mean *more moisture* in the atmosphere, which increases the intensity of normal winter snowstorms. The past few years have actually had many unusually mild winters in the cold northern part of North America, which people conveniently forget when it finally does get cold. When a long cold spell from the polar vortex lingers over the northeastern United States in Washington, DC, and in New York (where all the policy and politics are dictated, and where the media are centered), it is actually a result of climate change. The strong temperature difference between the pole and the equator, which powers the **jet stream**, gets weaker and weaker when the North Pole warms and melts. This makes the jet stream slow down and develop a lazy, looping path (called **Rossby waves**) over North America, which can get stalled for weeks (Fig. 17.13). If the loop down from the Arctic brings a mass of cold air with it, the freezing will stick around for a while—but somewhere else in North America, a warm loop of the jet stream will get stalled over the northern Midwest or Plains or the west coast, bringing abnormally warm weather to that area while the east coast freezes.

 See For Yourself: Neil deGrasse Tyson on Weather vs. Climate

• *"Carbon dioxide is good for plants, so the world will be better off."* The people who promote this idea clearly don't know much global geochemistry or are trying to play on the fact that most people are ignorant of science. The Competitive Enterprise Institute (funded mostly by money from oil and coal companies) has run a series of misleading ads that insult the intelligence of any educated person, concluding with the tag line "Carbon dioxide: they call it pollution, we call it life." Anyone who knows the basic science of earth's atmosphere can spot the deceptions in this ad. Sure, plants take in carbon dioxide that animals exhale, as they have for billions of years. But the whole point of the global warming evidence (as shown from ice cores) is that the delicate natural balance of carbon dioxide has been disrupted by our production of too much of it, way in excess of what plants or the oceans can handle. Meanwhile, humans are cutting down huge areas of rainforest every year, which means not only that there are fewer plants to absorb the gas but that slash-and-burn practices are releasing more carbon dioxide than plants can keep up with. There is much debate as to whether increased carbon dioxide might help agriculture in some parts of the world, but that has to be measured against the fact that other traditional "breadbasket" regions

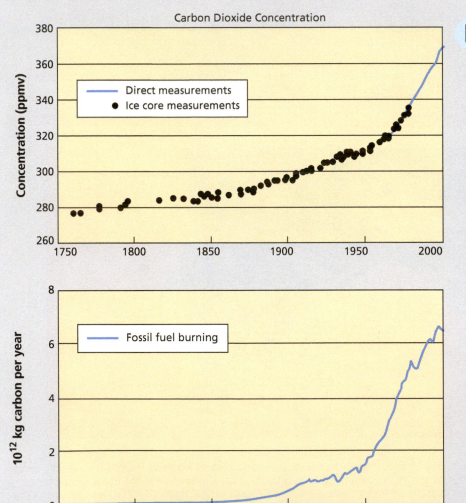

Figure 17.14   The measurements of carbon dioxide in the atmosphere (as determined from ice cores and then direct measurements since the Keeling experiment in 1958) exactly match the amount of carbon we have released into the atmosphere by burning fossil fuels, starting at the same time and showing the same rate of increase. This is one of many lines of evidence that our burning of fossil fuels explains the rise in carbon dioxide since there are no other trends in nature which match the carbon dioxide increase.

(like the American Great Plains) are expected to get too hot to be as productive as they are today. The latest research actually shows that increased carbon dioxide inhibits the absorption of nitrogen into plants, so plants (at least those that we depend upon today) are *not* going to flourish in a greenhouse world. Anyone who tells you otherwise either is ignorant of basic atmospheric science or is trying to fool the public who don't know science from bunk.

But how do we know humans are to blame? The list of lines of evidence for human causes is very long but includes a wide spectrum of different kinds of data.

See For Yourself: Tracking Carbon in the Atmosphere

- We can directly measure the amount of carbon dioxide humans are producing, and it matches exactly with the amount of increase in atmospheric carbon dioxide (Fig. 17.14).
- Through carbon isotope analysis, we can show that this carbon dioxide in the atmosphere is coming directly from our burning of fossil fuels, not from natural sources. The atmosphere has many other chemical fingerprints as well which prove that the greenhouse gases were produced by humans
- We can also measure oxygen levels that drop as we produce more carbon, which then combines with oxygen to produce carbon dioxide.

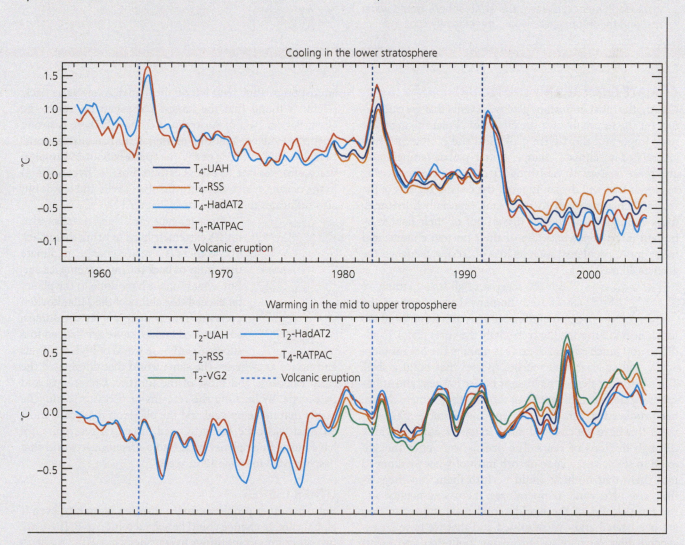

Figure 17.15   The crucial test that warming is coming from sources on the ground and not from solar radiation from space is to measure the relative changes in temperature in the different levels of the atmosphere. If it came from space, then the stratospheric layer high above us would warm and the troposphere (the layer in which we live) would be relatively cool. If it came from below, the troposphere would warm and the stratosphere would remain relatively cool. Satellite measurements show definite stratospheric cooling and tropospheric warming, confirming it comes from ground sources, namely the burning of fossil fuels. This plot also shows spikes in cooling shortly after volcanic eruptions (dashed vertical lines) due to the blocking of sunlight by volcanic ash and dust blowing in the stratosphere for several years after big eruptions.

- We have satellites out in space that are measuring the heat released from the planet and can actually *see and measure* the atmosphere get warmer in real time. They can also see the carbon dioxide and heat emerging from power plants and cities, and it's clear it comes mostly from those sources.
- The most crucial proof emerged only in the past few years: climate models of the greenhouse effect predict that there should be cooling in the stratosphere (the upper layer of the atmosphere above 10 km, or about 6 miles, in elevation) but warming in the troposphere (the bottom layer of the atmosphere below 10 km or 6 miles), where the human-produced gases come from. In contrast, if the warming were due to an increase in solar radiation, the stratosphere would warm first and the troposphere would be relatively cool. Indeed,

our space probes have measured stratospheric cooling and upper tropospheric warming (Fig. 17.15), just as climate scientists had predicted, and proven it's due to greenhouse gases, not the sun.
- Finally, we can rule out any other culprits: solar heat has been decreasing since 1940, not increasing; there are no measurable increases in cosmic radiation, methane, or volcanic gases; nor is there any other potential cause.

 See For Yourself: Answers to Climate Deniers

 See For Yourself: Why People Don't Accept Climate Change

CLIMATE CHANGE AND THE FUTURE

Clearly, the most important geologic events that we can predict for the future involve the effects of climate change. The most important short-term geological and ecological issues are related to climate change: hotter global temperatures and more heatwaves; vanishing ice caps and glaciers; rising sea levels that will drown most of the coastal regions of the world; increasing intensity of storms, especially hurricanes; mass extinctions of species driven out of their habitat by human interference and loss of their normal climate; and many more problems, such as loss of freshwater when the glaciers all melt.

 See For Yourself: Oceans and Climate Change   But the scariest of all is something we land-based humans barely notice: the death of the oceans as overfishing and too much heat kill marine life (especially the coral reefs). In addition, too much carbon dioxide is making the oceans more acidic, killing nearly all marine species that build a shell. The oceans are the earth's biggest buffer for heat and atmospheric gases, absorbing 93% of the heat that climate change has produced so far. They are enormous stable masses of water that have not changed much in temperature or acidity through millions of years. The oceans were once thought to be so stable and able to absorb most of what we dumped into them that nothing could perturb them, but the startling rise of oceanic temperatures, increasing acidity, and the mass deaths of the world's coral reefs show that they are overheated and oversaturated and that we have far exceeded their capacity to absorb our pollution and excess heat. Already we are seeing a shocking die-off in coral reefs ("bleaching") and extinctions in many marine ecosystems that can't handle too much heat and acid. Many of senior biologists and paleontologists are veteran divers who have visited reefs for 30 or 40 years, and they all say that they have watched this destruction happen in their lifetimes. All over the world, the reefs are in bad shape or dying, and there are

almost no pristine reefs left on the planet. A few years back, scientists found that the largest ecosystem on earth, the Great Barrier Reef of Australia, is dying off. There is strong scientific evidence that the "mother of all mass extinctions" (which wiped out 95% of marine species about 250 million years ago) was due to excess carbon dioxide (hypercapnia) in the oceans, which not only dissolves shells and corals but also suffocates marine life (see Chapter 12).

 See For Yourself: Climate Change is Irreversible

 See For Yourself: Bill Nye on Fixing Climate Change

 See For Yourself: David Suzuki on Crunch Time for Climate

Once ice caps vanish and oceans warm up, it's extremely hard to bring them back unless there is a major change in climate that drags us back into an extreme ice age. But what humans have done to the planet far exceeds the ability of the Milankovitch cycles to return us to cooler conditions, so it indeed looks like we are headed to a super-greenhouse world of high sea levels and no ice anywhere—the same world that dominated the early Paleozoic and most of the Jurassic, Cretaceous, and Paleogene. By burning the remains of prehistoric life (the plants that make up coal and the plankton that made most oil) that was once safely trapped in crustal reservoirs, we have pushed our planet back into a greenhouse world that hasn't existed for 55 million years.

OTHER ISSUES

But what other problems might threaten humanity, even if global climate change could be solved tomorrow? There are two other important things to consider.

1. *The world's nuclear arsenals*: Even though the United States and former Soviet Union have dismantled a small portion of the world's nuclear stockpiles since the end of the Cold War, there are still thousands of warheads lying in storage. There are countries like North Korea and Iran that are desperately trying to get the bomb and the

capability of sending it long distances, with leaders whose behavior is unpredictable. So far, the world has avoided "mutually assured destruction" (known during the Cold War as the "MAD" strategy) and the "Dr. Strangelove scenario" by diplomacy and by the fear that nuclear escalation could destroy us all. Unfortunately, that possibility is still real. What is different now in the post–Cold War world is that rogue nations (like North Korea and Iran) or even terrorist cells with nothing to lose might get a device, either by carelessness or by corruption. Experts are particularly concerned about the many warheads still lying around the former Soviet Union. The seriousness of this risk is difficult to assess, but the odds are certainly more likely than most of the scenarios of disaster from outer space. But the scenarios of nuclear winter virtually destroying life on this planet are still realistic, given the size and number of weapons that still exist.

 See For Yourself: World Population Growth

 See For Yourself: The Current Human Population on Earth

2. *Overpopulation*: There is one overriding issue that trumps every other disaster in terms of its impact on our future as a species: overpopulation. As I finish this book in late 2019, the official United Nations estimates place the world's population at 7.8 billion people (up from only 6.8 billion people in 2010). Ever since the end of the Black Death in the 1300s, global population has been increasing, with recent rates at around 2% per year and increasing. For the first 4 million years of human existence, population was fairly stable or gradually increasing, held in check by infant mortality, war, disease, and famine. As we saw in Chapter 16, it may have dropped to only 5000–10,000 during the Toba catastrophe 74,000 years ago. But with the advent of modern medicine, bringing the benefits of lowered infant mortality rates and increasing life spans, the "population bomb" has exploded. This means that while there were 1.6 billion people alive in 1900, world population reached 3 billion by 1960, then 4 billion in 1974, only 14 years later. It hit 7 billion on Halloween 2011 and will pass 8 billion in just a few more years. It will reach 9–11 billion by the year 2050, which is about three decades away. As you are reading this, 14–15 babies are born every 6 seconds around the world. Each hour there are 8800 more mouths to feed, and each year there are at least 80 million more people than the year before.

Even though the attention on human population growth has been diverted elsewhere since the heyday of discussion in the 1970s, the problem has not gone away. Some of the most developed countries in the world, such as Japan and many northern European countries, have reached the point where their populations are declining as women have gained social and financial independence, and medicine has improved their life spans and enabled them to use birth control and avoid infant mortality, so their birth rates drop. This change is called the **demographic transition**. But for the handful of countries that are limiting their population growth, there are far more (especially in Africa, Asia, and South America) that are exploding in population, which is why the problem has not gone away.

Ultimately, overpopulation is at the root of most of the serious problems discussed in this chapter, from the mass extinction of animals and plants due to human influences, and the destruction of all wild lands, to the seriousness of natural disasters in crowded places like China and Southeast Asia. Overpopulation influences how seriously the world suffers from famine and disease. Overpopulation drives the income inequality gap between rich and poor nations. Yet for a variety of political and economic reasons, overpopulation seems to be one of the hardest issues to address. Dealing with the problem realistically is further hampered when pro-growth pundits say that the planet is just fine with 15 or 20 billion people, and certain religious authorities oppose family planning and drive poor, overpopulated countries into even greater poverty.

A PERSPECTIVE

 See For Yourself: Stephen Hawking's 7 Predictions of Earth's Demise

We assume that our civilization is a permanent feature on this earth and that some form of our culture will persist indefinitely. But as any archeologist or historian knows, this is not true. We have many examples of extinct cultures that have vanished, leaving only a few durable artifacts; and often we know very little about how they lived—or why they failed. One only needs to look at the mysterious Etruscans, Minoans, Mycenaeans, Mayans, or Anasazi. The list of failed societies goes on and on. Jared Diamond, in his 2004 book *Collapse: How Societies Choose to Fail or Succeed*, points out that in instances where we do know why a society vanished, it is truly humbling. For example, the Easter Island (off the coast of Chile and Peru) culture vanished completely before the European settlers had much chance to witness it, leaving only the famous huge stone heads, or *moai*, dotting the island. Diamond shows that, to a large extent, the extinction of the Easter Islanders was self-inflicted: too many people, too much overexploitation of their environment when they cut down all the trees on what had been a densely forested island, and finally starvation, disease, and warfare wiped out the survivors. As Diamond shows, such a fate could await our world civilization if we overpopulate this planet or damage our environment or overexploit our resources. After all, 99% of all species that have ever lived are now extinct. There is no

 See For Yourself: David Suzuki on a Sustainable Future

good biological reason to believe that our fate will be different, especially given our accelerated pace of self-destruction.

17.4 The Future of Planet Earth

Setting aside issues like climate change, nuclear holocaust, overpopulation, and whether humanity destroys the planet's ability to sustain our culture, what other events can geologists predict for the far distant future? How can we use the

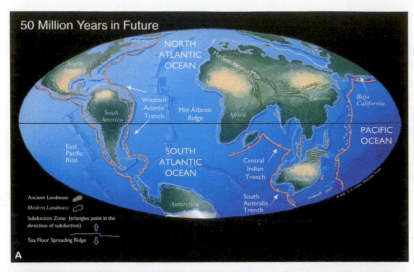

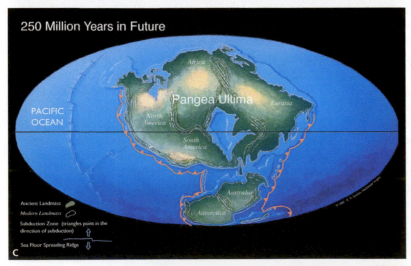

Figure 17.16 We know the speed and direction of the motions of the plates with great precision, so it is fairly easy to predict where they will be in the future. **A.** In 50 million years, Africa will crush the Mediterranean and build the Alps even higher, while the North Atlantic will get wider and California will slide up to Alaska. **B.** In 100 million years, this model predicts that Australia will again attach to Antarctica and Asia will begin to collide with it. Meanwhile, the spreading of the Atlantic could reverse and begin to subduct the sea floor so that it would close up and the Mid-Atlantic Ridge would head down the subduction zone. **C.** Predicting plate motions 250 million years in the future is highly speculative, but it seems likely that another supercontinent will have reassembled, just as we had Pangea 250 million years ago and several supercontinents 500 million years before that.

known trends and trajectories of earth systems and other bodies in space to look at the future?

PLATE TECTONICS

See For Yourself: 10,000 Years in the Future

Thanks to our understanding of plate motions, we have a good idea of where most continents will be for millions of years in the future. Fifty million years into the future, for example, we can expect that Africa will complete its collision with Europe and crush the Mediterranean Sea into a super-alpine mountain belt (Fig. 17.16A). Meanwhile, India's collision with Asia will have slowed down, and the Himalayas and Tibetan Plateau will be eroded down and only a tiny remnant of their former glory. The crustal blocks west of the San Andreas fault, from San Francisco to the southern tip of Baja, will have slid north until they will be in the Gulf of Alaska. The Atlantic Ocean will be almost half again as wide as it is now. Australia will move north as it pulls away from Antarctica and collide with New Guinea and Indonesia. The East African Rift will pull open, forming a proto-oceanic gulf; and eventually the eastern edge of Africa will be a new continent.

Predicting the far future beyond 50 million years from now (Fig. 17.16B) is more difficult, but based on the fact that most passive margin ocean basins that are spreading apart eventually switch to active margins and close up again, it's a good bet that the Atlantic might be closing up again, just as all of its predecessors did (Chapter 5). That means there would be a chain of volcanoes on the east coast of North America as it becomes an Andean-style subduction zone, just as the Cascades formed along the western edge. Likewise, there is a good chance that the spreading between Antarctica and Australia would switch to convergence and subduction so that Australia and Antarctica will be connected again, as they were 100 m.y. ago. The super-alpine range formed by the collision of Africa with Europe would be nearly worn down to nothing.

See For Yourself: The Earth in 250 Million Years

Finally, a map of the earth 250 million years from now (Fig. 17.16C) is less certain. Still, there's a very good chance that some of the world's oceans (especially the Atlantic and Indian Oceans) would close up completely, and there would be a new supercontinent. Its exact shape and the arrangement of the present continents within it are speculative, but since supercontinents seem to form every 500 m.y., then break up for 250 m.y., and reassemble again after another 250 m.y., it is likely that yet another Wilson cycle would be completed.

A CHOICE OF CATASTROPHES

See For Yourself:
Sir Martin Rees
on the Future:

In even the much further future beyond 250 million years, we can look to astronomy to think about the fate of the earth.

See For Yourself:
A Day on Earth 1
Billion Years From
Now

See For Yourself:
A Day on Earth 4
Billion Years From
Now

- *The earth will stop rotating on its axis* and become tidally locked, with one side always facing the moon and one side facing away from it (just as the moon only shows one side toward the earth now). When this happens, the tides will stop moving around the earth's surface, and the sides directly facing the moon and facing away from the moon would have a permanent high tide, while those areas at 90° to the earth–moon axis would have a permanent low tide. The days would get longer as the earth rotates slower and slower, until it takes many "months" to complete a "day." When tidal locking is complete, one complete day–night cycle will take one complete orbit, or 1 year, to go around the sun. That means that we will go from 365 days in a year (365 spins of the earth on its axis for one trip around the sun) to 1 day per year. When this happens, one side of the earth will permanently face the sun and be incinerated, while the opposite will become perpetually dark and frozen. The only places with moderate temperatures will be the transition zones between the scorching day side and the frozen night side of the earth. This isn't likely to happen for 50 billion years in the future, so other things will probably interrupt the process before it is completed.

See For Yourself:
The Sun—Earth's
Fiery Future

See For Yourself:
The Sun as a White
Dwarf

See For Yourself:
How Will the Sun
Die?

See For Yourself:
Death of the Sun

- *The sun becomes a red giant and incinerates the earth.* This scenario is pretty well understood by what we know about stellar behavior from the **Hertzsprung-Russell diagram**, which plots the history and fate of stars of different masses. Contrary to popular myth, the sun will not become a supernova because it is not massive enough to do so. But it is massive enough that when its nuclear fuel begins to run out, it will expand into a red giant that will be 100 times larger than now and 2000 times more luminous. In this scenario, the sun will flare out past the orbits of the inner planets, incinerating them all. The good news is that the timing for this scenario is well constrained based on looking at the histories of other stars in the universe. The best estimates put such an event approximately 5 billion years in the future, which is longer than the 4.6 billion years that the earth has already existed.

See For Yourself:
What Will Happen
to the Earth When
the Sun Dies?

- *The sun's increasing luminosity dries the earth up completely.* We might not have to wait for the sun to become a red giant since as it reaches the end of its life it will become much hotter and brighter with increased luminosity. (Conversely, 4 billion years ago it had a very low luminosity, and it was just barely bright enough to keep the earth from freezing—the "faint young sun paradox" discussed in Chapter 7). In only half a billion to a billion years from now, this event might occur, vaporizing the oceans and turning the earth into a barren desert, as most terrestrial bodies in space (such as the moon and Mars) already are.

See For Yourself:
The Future of the
Sun and Earth

- *Supernovas*: Some other star might become a supernova and destroy the earth with radiation. In 2006, a supernova known as SN2006gy was discovered. It is one of the brightest supernovae ever observed, and it showed just how common these bodies are across the universe. Fortunately for us, most are extremely far away, so the chance of their sending anything lethal our way is very small. For an object to have any effect on the earth at all, it must be much closer than 100 light years away. There are only a handful of such objects in space, so astronomers estimate that one of these might become a supernova every 20 million years on average. However, every search of the rocks representing more than 4.6 billion year history of earth has turned up no reliable evidence that a supernova event struck our planet with enough radiation to leave a trace. Thus, even being generous that there may be another nearby supernova in less than 20 million years, the event is still extremely unlikely to severely impact the earth.

See For Yourself:
Predicting Asteroid
Impacts

- *Extraterrestrial impacts*: Thanks to Hollywood movies like *Armageddon* and *Deep Impact*, people are unduly alarmed that some large object from space will hit the earth and destroy life upon it. For this to occur, the object would have to be much larger than the asteroid that produced the Chicxulub crater at the end of the Cretaceous, which only produced a partial mass extinction (see Chapter 13). If it were as big as the Mars-sized planet Theia that hit us 4.6 billion years ago and blasted part of our mantle into orbit to form the moon, the earth's surface would be completely melted and remodeled and everything—continents, oceans, life, and the rest—would have to start over from scratch. However, all the current research shows that no really large objects are likely to hit us for the foreseeable future. The much smaller objects, which could cause serious problems, are harder to track; and they still present a risk but would not annihilate life on the planet. Nevertheless, a number of scientists are busy keeping track of all the bodies in space that could impact us, yet there is never enough funding or staffing

See For Yourself: 7
Major Events in the
Distant Future of
Earth

to do this research that is much more important than many other things we spend our tax dollars on.

17.5 The Geological Perspective from Earth's History

We have come full circle from the beginning of the formation of the universe and the earth. We have looked at over 4.6 billion years of earth history and projected earth events billions of years into the future. We have come a long way from the viewpoint held only three centuries ago, from the narrow claustrophobic view of a universe only 6000 years old with the earth in the center and humans as the purpose of the universe. Thanks to science, the evidence of nature has shown us a universe that is almost 14 billion years old, with millions of galaxies and solar systems, of which our galaxy is not in the center and in which our solar system is just a small, not very distinguished neighborhood. As Carl Sagan put it, "Who are we? We find that we live on an insignificant planet of a humdrum star lost in a galaxy tucked away in some forgotten corner of a universe in which there are far more galaxies than people." Or as Steven Hawking said (as quoted in the 2014 movie *The Theory of Everything*):

> It is clear that we are just an advanced breed of primates on a minor planet orbiting around a very average star, in the outer suburb of one among a hundred billion galaxies. BUT, ever since the dawn of civilization people have craved for an understanding of the underlying order of the world. There ought to be something very special about the boundary conditions of the universe. And what can be more special than that there is no boundary? And there should be no boundary to human endeavor. We are all different. However bad life may seem, there is always something you can do, and succeed at. While there is life, there is hope.

It has given us a humbling perspective compared to the cosmic arrogance that humans held for centuries. It also reminds us that we are part of nature and cannot ignore its laws. As Dostoevsky put it in his autobiography, "Nature doesn't ask your permission; it doesn't care about your wishes, or whether you like its laws or not. You're obliged to accept it as it is, and consequently all its results as well." Or as the environmental economist Robert K. Watson said, "Mother Nature is just chemistry, biology and physics. That's all she is. You cannot sweet-talk her. You cannot spin her. You cannot tell her that the oil companies say climate change is a hoax. No, Mother Nature is going to do whatever chemistry, biology and physics dictate. Mother Nature always bats last, and she always bats 1,000."

We are just temporary travelers on Spaceship Earth. Our genus *Homo* has been around less than 2 million years, only one-two thousandth of the total history of the earth. Our species *Homo sapiens* first appeared in modern form only about 100,000–200,000 years ago. The consensus of studies of the fossil record is that most species last only a few million years at most, and *Homo sapiens* exhibits all sorts of behaviors that suggest it will not survive much beyond the 100,000–200,000 years it has already existed. It's fun to speculate about the dangers of asteroid impacts, the sun burning up the earth as a red giant, and supernova explosions; but much more prosaic fates are far likelier and more serious. If we can find ways to cope with climate change, contain and eliminate our nuclear arsenals, and mitigate the effects of overpopulation, then we may well survive on this planet. Unfortunately, we act more like bacteria merrily reproducing and expanding in a Petri dish, oblivious to the threats that await us. Whether our intelligence and political will are good enough to save us from a dire fate is the challenge of the future.

RESOURCES

BOOKS

Asimov, Isaac. 1979. *A Choice of Catastrophes: The Disasters that Threaten Our World*. Fawcett Columbine, New York.

Cline, E. 2014. *1177 B.C.: The Year Civilization Collapsed*. Princeton University Press, Princeton, NJ.

Diamond, Jared. 2004. *Collapse: How Society Choose to Fail or Succeed*. Penguin, New York.

Diamond, Jared. 2005. *Guns, Germs, and Steel: The Fates of Human Societies*. W. W. Norton, New York.

Davies, Jeremy. 2016. *The Birth of the Anthropocene*. University of California Press, Berkeley.

Fagan, Brian. 2001. *The Little Ice Age: How Climate Made History 1300–1850*. Basic Books, New York.

Fagan, Brian. 2003. *The Long Summer: How Climate Changed Civilization*. Basic Books, New York.

Fagan, Brian. 2009. *The Great Warming: Climate Change and the Rise and Fall of Civilizations*. Bloomsbury Press, London.

Flannery, Tim. 2006. *The Weather Makers: How Man Is Changing the Climate and What It Means for Life on Earth*. Atlantic Monthly Press, New York.

Gore, Al. 1992. *Earth in the Balance: Ecology and the Human Spirit*. Penguin, New York.

Gore, Al. 2006. *An Inconvenient Truth*. Rodale Press, Emmaus, PA.

Harari, Yuval Noah. 2017. *Homo Deus: A Brief History of Tomorrow*. Harper, New York.

Heinemann, W. 203. *Our Final Century? Will the Human Race Survive the Twenty-First Century?* Arrow, New York.

Henson, Robert. 2014. *The Thinking Person's Guide to Climate Change*. American Meteorological Society, Washington, DC.

Klingaman, W. K., and N. P. Klingaman. 2013. *The Year Without a Summer: 1816 and the Volcano that Darkened the World and Changed History*. St. Martin's Press, New York.

Kolbert, Elizabeth. 2015. *Field Notes from a Catastrophe: Man, Nature, and Climate Change*. Bloomsbury, London.

Lamb, H. H. 1992. *Climate, History, and the Modern World*. Methuen, London.

Mann, Michael. 2012. *The Hockey Stick and the Climate Wars: Dispatches from the Front Lines*. Columbia University Press, New York.

Mann, Michael, and Lee Kump. 2015. *Dire Predictions: Understanding Climate Change*. DK, London.

Mann, Michael, and Tom Toles. 2018. *The Madhouse Effect: How Climate Denial Is Threatening Our Planet, Destroying Our Politics, and Driving Us Crazy*. Columbia University Press, New York.

McKibben, Bill. 1989. *The End of Nature*. Random House, New York.

McKibben, Bill. 2010. *Eaarth: Making a Life on a Tough New Planet*. Times Books, New York.

McNeill, J. R., and P. Engelke. 2016. *The Great Acceleration: An Environmental History of the Anthropocene Since 1945*. Belknap Press, Cambridge, MA.

Oreskes, Naomi, and Erik Conway. 2010. *Merchants of Doubt: How a Handful of Scientists Obscured the Truth on Issues from Tobacco Smoke to Climate Change*. Bloomsbury, London.

Pearce, F. 2007. *With Speed and Violence: Why Scientists Fear Tipping Points in Climate Change*. Beacon Press, Boston, MA.

Prothero, D. R. 2013. *Reality Check: How Science Deniers Threaten Our Future*. Indiana University Press, Bloomington.

Purdy, Jedidiah. 2015. *After Nature: Politics for the Anthropocene*. Harvard University Press, Cambridge, MA.

Rees, M. 2003. *Our Final Hour: A Scientist's Warning: How Terror, Error, and Environmental Disaster Threaten Humankind's Future in This Century—On Earth and Beyond*. Basic Books, New York.

Romm, Joseph. 2015. *Climate Change: What Everyone Needs to Know*. Oxford University Press, Oxford.

Sagan, Carl. 1980. *Cosmos: A Personal Voyage*. Random House, New York.

Scranton, Roy. 2015. *Learning to Die in the Anthropocene: Reflections on the End of Civilization*. City Lights Publishers, New York.

Weiner, Jonathan. 1990. *The Next One Hundred Years: Shaping the Future of Our Lives on Earth*. Bantam, New York.

Wilson, E. O. 2003. *The Future of Life*. Vintage, New York.

SUMMARY

- The Holocene or "Recent" Epoch is the last 10,000 of geologic history.

- The Holocene was marked by a stable, warm interglacial climate after the earth came out of the Younger Dryas glacial event about 11,700 years ago.

- As the planet warmed into the Holocene interglacial, sea level rose from the melting glaciers and drowned the floor of the North Sea (Doggerland), flooded the English Channel and many of the straits and land bridges that once connected islands to the mainland, and caused isolation of populations that had once been connected.

- The early Holocene from 9000 to 5500 years ago was the Climatic Optimum or Hypsithermal, when the planet was slightly warmer due to the Milankovitch cycles. Africa and the Middle East were much wetter and milder in climate than today, so rivers drained the future Sahara Desert; and great civilizations in Egypt, Mesopotamia, the Indus Valley, and China arose at this time.

- At the same time, the Neolithic revolution brought domestication of wheat and barley, sheep and goats to early civilizations across Europe and Asia.

- The planet cooled at the end of the Holocene Climatic Optimum and entered a slightly cooler phase called the Neoglacial Period. Climatic events around 1177 B.C.E. caused the Late Bronze Age cultural collapse, which ended many thriving civilizations. The Roman Empire rose during a period of warmer, milder climates, then fell during a period of colder, harsher climates.

- Between 950 and 1250 C.E. was the Medieval Warm Period, when climates warmed in the North Atlantic, making it possible for the Vikings to settle Greenland and Labrador. However, it was only a local effect in North America and Europe because global temperature actually dropped.

- The Medieval Warm Period ended with the Little Ice Age from 1650 to 1850, when much of the globe was cooler than average, glaciers expanded, and many European and North American cities experienced harsher, colder winters. The Little Ice Age culminated with the "year without a summer" in 1816, due to the ash and dust blocking the sunlight from the 1815 eruption of Tambora volcano in Indonesia.

- Some scholars think that the effects of humans on the planet are so large that the time of our dominance should be called the "Anthropocene." However, there is no official consensus on how to define the Anthropocene or when it officially began.

- The earth has a unique atmosphere that is found nowhere else in the universe. It is the only planet with free oxygen, an ozone layer to protect us from ultraviolet light damage, low levels of carbon dioxide compared to Mars and Venus, the right temperatures for liquid water, and a magnetic field that shields us from the solar wind. It is also a very thin and fragile atmosphere whose composition and temperature have changed many times in the geologic past.

- International agreements and regulation have managed to solve two great environmental risks to the atmosphere, CFCs producing a hole in the ozone layer, and regulation of coal-burning power plants reducing acid rain.

- Since 1850, we have known that the earth behaves like a greenhouse, letting sunlight in (mostly in the shorter wavelengths of ultraviolet and visible light). But when the radiation is absorbed, the earth reradiates the energy back out to space in the long wavelengths of infrared (heat), which greenhouse gases block. This is why the earth is warming up.

- In 1958, Charles Keeling began the first experiment that showed the dramatic increase in carbon dioxide in the atmosphere, from barely 315 ppm in 1958 to over 415 ppm in 2020.

- Global warming is most extreme at the poles, where both the Antarctic and the Arctic are melting at alarming rates, and the Arctic polar ice cap is vanishing faster than anyone thought possible.

- Melting ice means that sea levels are rising worldwide, so already cities like Venice and Miami routinely flood at high tides; and most coastal regions will be under water by the end of the century when sea level rises 6 m (about 20 feet). If all the ice melted, it would rise 65 m (213 feet) and drown most of the world's cities and coastal regions.

- Climate change has meant that mountain glaciers around the world are also vanishing, eliminating an important water supply for most of the world's people.

- Warming is also causing record heatwaves and droughts in much of the world, and warmer climates have enormous effects on life, from the migration of disease-bearing insects to the death of plants and animals from loss of habitat.

- We know that climate change is real and caused by humans and not some normal climate variation because we can determine what natural variation in climate has done thanks to ice cores like EPICA-1 in Antarctica. It shows the record of almost the entire past 800,000 years, and at no time in the warmest interglacial was carbon dioxide higher than 280 ppm.

- We can also rule out effects of increased solar radiation (it's actually been decreasing since the 1960s), cosmic radiation (which is increasing in a way that should cool, not warm, the planet), and volcanic eruptions (which usually cause short-term cooling, not warming).

- Contrary to some people, global warming did not "pause" in 1998; this is cherry-picking the anomalously warm El Niño year of 1998 to bias your result. The past few years have shattered all the heat records, making the planet so much warmer than the 1998 peak that the "1998 pause" is irrelevant and a myth.

- Weather is the short-term (hourly and daily) change in the atmosphere, but climate is a long-term (decade, century, or thousands of years) average of weather. Thus, talking about weather events like snow is irrelevant to the overall climate picture. In fact, climate change can cause more extreme winters in some parts of the world due to the weakening of the jet stream producing a polar vortex that won't move on quickly.

- Carbon dioxide is good for plants, but too much is not a good thing since it throws the entire delicate geochemical cycles of carbon out of balance.

- We know that humans are to blame for global climate change because we can measure how much carbon we have burned and it exactly parallels the increase in carbon dioxide; we can find the chemical fingerprints of human activities in the greenhouse gases; satellites can pinpoint where the heat and carbon are coming from, and it's clearly due to human activity; and the stratosphere is cooling and the troposphere is warming exactly as expected if the heat source comes from the ground, not from space.

- As scary as the effects of climate change on land are, the biggest effects are in the oceans. They were once thought to be so huge and to have such high heat capacity that no one thought we could change them, but the ocean is warming rapidly, its acidity is going up as the carbon dioxide becomes carbonic acid, and this is killing off the coral reefs and much other marine life.

- In addition to climate change, there are other threats to the planet: our nuclear arsenals and their likelihood of causing nuclear winter if they are ever used as well as overpopulation, with a global population that has already topped 7.7 billion people and is not slowing down despite declining birth rates in the developed world.

- Scientists know a lot about the future of geological and astronomical influences on the earth in the distant future. We can predict plate motions for the far future with great confidence. We know the earth is slowing its rotation on its axis, although it will not stop spinning until 50 billion years in the future. We expect the sun to expand into a red giant and destroy the inner planets, although this is about 5 billion years in the future. It may increase luminosity and cook the earth sooner than that. Supernovas are not a realistic problem for earth, and so far there is no evidence of a large asteroid impact in our near future.

- By looking at the past 4.6 billion years of earth history, we have gained a humbling perspective of humanity's place in the cosmos.

KEY TERMS

Holocene epoch (p. 450)
Doggerland (p. 450)
Holocene Climatic Optimum (Hypsithermal) (p. 450)
Neolithic revolution (p. 452)
Neoglacial Period (p. 453)
Late Bronze Age cultural collapse (p. 453)

Roman Warm Period (p. 453)
Medieval Warm Period (p. 454)
Little Ice Age (p. 454)
Maunder minimum (p. 454)
Tambora (p. 454)
"Year without a summer" (p. 454)
Anthropocene (p. 454)

Chlorofluorocarbons (CFCs) (p. 455)
Ozone hole (p. 455)
Acid rain (p. 455)
Greenhouse effect (p. 455)
Keeling curve (p. 456)
"Hockey stick" curve (p. 456)
EPICA-1 ice core (p. 461)

"Cherry-picking" data (p. 462)
Weather versus climate (p. 462)
Jet stream (p. 464)
Rossby waves (p. 464)
Hypercapnia (p. 466)
Hertzsprung-Russell diagram (p. 469)

STUDY QUESTIONS

1. How long ago did the Holocene begin? What major events happened in the Holocene?
2. What was Doggerland? What happened to it?
3. What was the astronomical cause of the Holocene Climatic Optimum? How did it affect human civilization?
4. What was the Neolithic revolution? Where and when did it begin? Where did it spread?
5. What was the Neoglacial Period? How did its climate compare with that of the Holocene Climatic Optimum?
6. How did climate affect the rise and fall of the Roman Empire?
7. What happened in Greenland and Labrador during the Medieval Warm Period?
8. Why is the natural warming of the Medieval Warm Period not comparable to the warming of the past 150 years?
9. What happened to Europe during the Little Ice Age?
10. How did the eruption of Tambora affect the planet?
11. What is the Anthropocene? How is it defined?
12. How are the earth's climate and atmosphere very different from those of any other planet in the universe?
13. Some people claim that there is no political solution to global warming and fossil fuel burning. Give two examples of recent environmental problems that were solved by public pressure and political action.
14. Explain how the greenhouse effect works. In what wavelength does solar radiation arrive on earth? In what wavelength does it reradiate back to space? How do greenhouse gases let one wavelength pass but block the other?
15. What did Keeling's experiment show?
16. What is the "hockey stick" curve? What does it show?
17. What is happening to the polar regions of the planet? What are the consequences of this happening?
18. What is happening to global sea level? What are the consequences of it rising further?
19. What are the consequences of the melting of the world's mountain glaciers?
20. What will happen to the atmosphere if the permafrost melts?
21. How do we know that the current change in climate is not just "normal climate variability"?
22. How do we know that global warming is not caused by increasing solar radiation? By increased cosmic radiation? By volcanoes?
23. A person comes up to you and says "climate change has paused since 1998." What do you say in response?
24. A politician throws a snowball on the floor of Congress and claims that global warming is a myth. How do you respond to this?
25. How can global warming actually cause some regions to get colder winters?
26. Someone tells you that "carbon dioxide is good for plants and for the planet." How do you respond?
27. List the evidence that proves that global warming and the increase in carbon dioxide are caused by humans.
28. What is happening to the oceans thanks to carbon dioxide and heat in the atmosphere?
29. What are the other issues that threaten humanity besides climate change?
30. How will the plates move in the future?
31. What will happen to the earth's rotation on its axis in the future? When will this conclude?
32. What will happen to the sun in the future? When will this happen?
33. What are the risks of supernova explosions or asteroid impacts to life on earth?

BIOLOGICAL CLASSIFICATION

Scientific names may seem long and hard to pronounce for some people, but they are essential to scientific communication. The popular or common name of many living animals and plants differs from culture to culture and language to language. For example, a peccary to English speakers is a *javelina* in Latin America, and a lion to us is *simba* to Swahili speakers. Even within the same language, the common name may not be consistent. If you say "gopher" in some parts of the United States, it means a small burrowing rodent, but in other parts it means a gopher tortoise.

For this reason, every organism (plant, animal, fungus, and even bacterium) has its own scientific name. Scientific names are universal around the world, no matter what language the scientist speaks. For example, you may not be able to read much of a scientific paper written in Mandarin Chinese, but the scientific names are always printed out in Roman script, so anyone can read them and at least guess what animal is the subject of the research. The scientific name for the burrowing rodent some people call a "gopher" is *Thomomys*, but the gopher tortoise is *Gopherus*, so there is no confusion.

For most fossils, knowing their scientific name is essential since most don't even have a common name. You may know the saber-toothed cat by its English name, but it's different in other languages—and to all scientists, it is *Smilodon* (unless it's one of the European saber-toothed cats, like *Homotherium* or *Xenosmilus*). Mammoths are familiar to us by that name, but they are *mamut* in Spanish and *mammouth* in French; they are always *Mammuthus* to a scientist. But nearly all other fossil animals and plants have no common name whatsoever, so there is no choice but to use their scientific names. You actually know quite a few scientific names of prehistoric and living creatures. For example, everyone knows *Tyrannosaurus rex*; but that is its proper scientific name, and no other popular name exists. Nearly all the other dinosaur names you know, from *Brontosaurus* to *Velociraptor* to *Stegosaurus* to *Triceratops*, are scientific names as well.

All organisms on earth actually have a two-part (binomial) scientific name. The first part is the genus name (the plural is "genera," not "genuses"). It is always capitalized and either italicized (in print) or underlined (when handwritten). The names *Tyrannosaurus*, *Brontosaurus*, *Velociraptor*, *Stegosaurus*, and *Triceratops* in the preceding paragraph are all genus-level names or "generic names." But a genus typically includes a number of species. The species name (or "trivial name") is never capitalized (even if it came from a proper noun), but it is always underlined or italicized. Thus, *Tyrannosaurus rex* is a genus and species name; so is *Velociraptor mongoliensis*. Your scientific binomen is *Homo sapiens*, but there are other species of *Homo*, such as *Homo neanderthalensis*, *Homo erectus*, and *Homo habilis*.

Generic names are never used more than once in the animal kingdom, although there are a few cases of the same genus being used for both plants and animals; but there is no likelihood of confusion between a plant and an animal. For example, there is a spider genus called *Erica*, but that genus name means a kind of heath in botany. Species names, however, are used over and over again, so they cannot stand alone in a scientific paper. Thus, you can say *Tyrannosaurus rex* or *Homo sapiens* or *Carcharocles megalodon* but not "rex" or "sapiens" or "megalodon." You can abbreviate the genus name, so *T. rex* (but not "*T-rex*") or *H. sapiens* or *C. megalodon* is correct.

Scientific names were originally based on Latin or Greek words since in the early days of natural history, all scholars read and wrote in Latin or Greek as an international form of communication. Thus, most scientific names can be broken down to their original meaning. *Tyrannosaurus rex* means "kind of the tyrant lizards," and *Homo sapiens* means "thinking human."

The species is the fundamental unit in nature since it is species that evolve due to natural selection on populations within the species (see Chapter 6). The genus is a bit more arbitrary, depending upon the scientists' judgments as to which species cluster together. Genera are clustered into larger groups known as "families." For example, our genus *Homo* belongs to the family Hominidae, along with the great apes *Pan* (the chimpanzee), *Gorilla* (the gorilla), *Pongo* (the orangutan), and the four genera of gibbons. Our immediate human lineage, the subfamily Hominini (hominins), consists of *Homo* and other closely-related extinct genera such as *Sahelanthropus*, *Ardipithecus*, *Paranthropus*, *Australopithecus*, and others. The dogs are all members of the family Canidae, the cats are Felidae, and the rhinoceroses are in the Rhinocerotidae. In the animal kingdom, all family names end with the suffix –idae, which is a clue when you encounter an unfamiliar name. (In the plant kingdom, families end in –aceae, so Rosaceae is the plant family that includes roses.)

Families are clustered into a larger group called an "order." Humans, apes, monkeys, lemurs, and their relatives form the order Primates, while the order Carnivora comprises most of the flesh-eating mammals including cats, dogs, bears, hyenas, weasels, raccoons, seals, and walruses. The rodents are an order (Rodentia), as are the rabbits (Lagomorpha), and most of the larger groups of mammals are orders. Orders are clustered into classes. Among the backboned animals, the orders of mammals are clumped into the class Mammalia, while the reptiles (class Reptilia), the Amphibia, and so on are classes. Classes are clustered into a larger group called a "phylum" (plural is "phyla"). Vertebrates (animals with backbones) are members of the phylum Chordata; other phyla include Mollusca (mollusks, including clams, snails, squids, and their relatives) and Arthropoda (jointed segmented animals, including insects, spiders, scorpions, crustaceans, millipedes, trilobites, and many others), and so on. The highest rank of all is kingdom. We are members of the kingdom Animalia, but there are also kingdoms for the plants, the fungi, and so on (see Fig. 6.7A).

Here is an example of how the hierarchy of groups within groups looks:

KINGDOM	Animalia	Animalia
PHYLUM	Chordata	Mollusca
CLASS	Mammalia	Gastropoda
ORDER	Primates	Neogastropoda
FAMILY	Hominidae	Turritellidae
GENUS	*Homo*	*Turritella*
SPECIES	*sapiens*	*ocoyana*

THE CLASSIFICATION OF LIFE

The following is a simplified classification of the major groups of organisms on the planet, emphasizing those mentioned in the text (Fig. A1). Although it includes bacteria, fungi, and plants, we will focus on the phyla of animals, especially phyla that have a good fossil record.

KINGDOM ARCHAEBACTERIA

Prokaryotic organisms that have the most primitive genetic codes of all life. These include sulfur-reducing and methane-dependent bacteria as well as bacteria that tolerate environments with extreme temperatures or salinities, such as hot springs. Many live in the absence of oxygen.

KINGDOM EUBACTERIA

Advanced prokaryotes, including the more familiar types of digestive and disease-causing bacteria. The cyanobacteria (or cyanophytes or blue-green bacteria) are among the earliest fossils known (Fig. 9.8).

KINGDOM PROTISTA

Single-celled or simply multicellular eukaryotes, including most algae.

Phylum Haptophyta (Calcareous Nannoplankton)

Photosynthetic single-celled planktonic algae (coccolithophores) that are covered by small calcareous, button-shaped plates (Fig. 13.9B, C).

Phylum Sarcodina (Amoebas and Their Relatives)

Class Foraminifera—Single-celled, amoeba-like organisms with a calcareous skeleton surrounded by protoplasm (Figs. 12.14, 13.14A).

Class Radiolaria—Single-celled, amoeba-like planktonic organisms that secrete an intricate internal skeleton of silica (Fig. 13.14B).

KINGDOM PLANTAE

Multicellular, sexually reproducing eukaryotes that obtain their energy by photosynthesis with the aid of chloroplasts. Except for mosses, most plants have vascular tissues for conducting fluids.

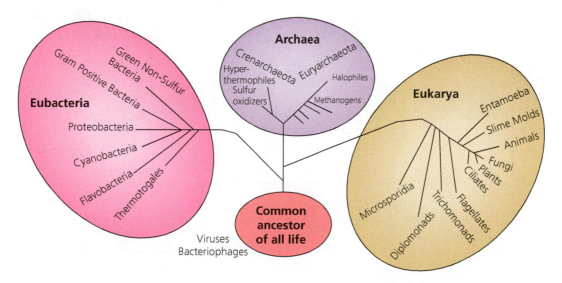

Figure A1 Evolutionary tree of all of life, showing the different branches for archaebacteria or Archaea, the most primitive life on the planet plus the more advanced and familiar Eubacteria and all eukaryotes (including plants, animals, fungi, and many other simpler organisms) in the Eukarya. These relationships are based on the genetic sequences of all these organisms, especially RNA. (Adapted from several sources)

Phylum Bryophyta (Mosses and Liverworts)

Small plants that lack well-developed vascular tissues but have complex reproductive cycles that allow them to live on land.

See For Yourself: The First Forests with David Attenborough

Phylum Psilophyta

Small vascular plants with simple stems but no true roots or leaves. Common in the middle Paleozoic, but only a few survive today.

Phylum Lycophyta (Club Mosses)

Spore-bearing plants with true roots and leaves that are arranged in a spiral pattern around the stem. Most living lycophytes are small and low-growing, but they were the dominant trees of the Carboniferous as represented by *Lepidodendron* and *Sigillaria* (Fig. 12.17A–C).

Phylum Sphenophyta (Horsetails)

Spore-bearing plants whose stems are divided into notes that bear whorls or branches. The living species are mostly small water-loving plants known as "scouring rushes," but some Carboniferous species were tree-sized (Fig. 12.17D).

Phylum Filicinophyta (True Ferns)

Spore-bearing plants with branching leaves, with their spores on the undersides of the leaves.

Phylum Gymnospermae (Seed Ferns, Cycads, Conifers, and Their Relatives)

The most primitive seed-bearing plants. They include a number of fern-like forms such as the extinct seed fern *Glossopteris*; the *Cordaites* trees in the late Paleozoic; the palm-like cycads or "sego palms" and their extinct relatives, the cycadeoids; the ginkgo (or "maidenhair tree"); and many kinds of conifers (*Araucaria*, pines, spruces, firs, and sequoias) (Fig. 13.17).

Phylum Angiospermae (Flowering Plants)

Plants with internal covered seeds that are fertilized with the aid of flowers. They arose in the Early Cretaceous and underwent an explosive radiation so that today they make up the vast majority of plants. They include all hardwood trees and other flowering shrubs as well as all the grasses (Fig. 13.23).

KINGDOM MYCOTA (FUNGI)

Single-celled or multicellular eukaryotes that break down decaying organic matter for their nutrition. The Mycota include molds, slime molds, yeast, true fungi, and mushrooms. Fungi help break down dead tissues and return their nutrients to the food chain. Recent studies of their molecular sequences suggest that fungi are more closely related to animals than they are to plants (Fig. A1).

See For Yourself: Life on Earth: Infinite Variety with David Attenborough

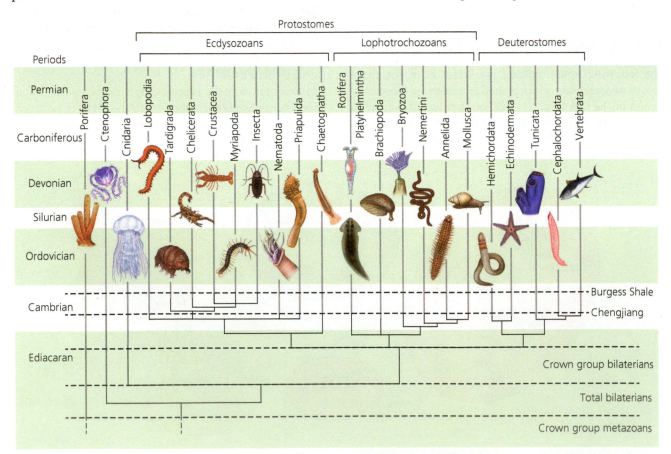

Figure A2 Family tree of the animals, showing their known range in solid lines and inferred range in time and relationships in dashed lines. (Adapted from several sources)

KINGDOM ANIMALIA

Multicellular eukaryotes that obtain their nutrition by consuming other organisms. In recent years, both the evidence of fossils and the evidence from molecular biology have allowed us to draw a family tree showing the relationships of all the animals (Fig. A2).

Superphylum Porifera (Sponges and Their Relatives)

Most people hear the word "sponge" and think of a block of synthetic foam, or of the cartoon character SpongeBob SquarePants. But sponges are actually the simplest and most primitive multicellular animals on the planet. Their fossil record goes back to the earliest Cambrian, and molecular data suggest that they originated even earlier but left no fossils for many millions of years.

Sponges are just one step above single-celled organisms. Each sponge cell is completely independent of the others, and performs all its own biological functions such as feeding, breathing, excreting waste, and reproduction. They have no specialized tissues or organs, as all other animals do. They are loosely connected to each other on a structure (or "skeleton") that is made of tiny, woven, needle-like pieces known as spicules, which the individual cells collaborate to secrete as a support. They are so independent, however, that you can force a sponge through a fine sieve and it will reas-

See For Yourself: The Shape of Life: Sponges

semble itself into a new sponge. I doubt you could do the same if you were forced through a sieve!

The basic sponge is a tube-like or conical structure, shaped roughly like a chimney. The simplest sponges have only a single thin wall (Fig. A3, left) made of interwoven spicules, punctured by many small holes or pores throughout its surface. This is where the phylum gets its name, "Porifera," or "pore-bearing" in Latin. Individual sponge cells line the wall of the tube and especially the canals that connect the outer pores with the inner pores into the central cavity, or spongocoel. These cells have a whip-like flagellum that drives currents past them so that they can trap tiny food particles and oxygen and release their waste products.

All the currents flow from the outside to the **spongocoel** cavity in the middle of the sponge, then out the top of the chimney (Fig. A3), known as the **osculum** ("little mouth" or "kiss" in Latin). The one-way flow of water through the walls and out the top is propelled by the flagella. The flow is also enhanced by the fact that the top of the sponge has a weak suction, drawing currents upward like a chimney. This is because the water flowing over the top of the sponge must flow faster than the water around it, so its increased velocity means less pressure. The decreased pressure over the top of the sponge relative to the rest of the surrounding water forces the water through the pores and out the top. This is the "chimney effect," where the top of a chimney has lower pressure compared to its sides so that it creates an airflow that goes up through the chimney.

Sponges are very efficient at passing water through their canals. The entire internal volume of a typical sponge is replaced with new water nearly every minute. A black loggerhead sponge about 50 cm in diameter and 30 cm tall may draw about 1000 liters of water through its canals in a single day. Some sponges may flush the equivalent of 10,000–20,000 times their internal water volume in a single day. This can be seen if you Google the term "sponge currents." There are several outstanding videos online showing a diver releasing a harmless dye in the water just outside the wall of the sponge. Within a few seconds, you see a strong plume of dye-filled water pouring out the top.

See For Yourself: Sponge Currents

The simplest sponges are just thin-walled tubes or cones or long cylinders. But most sponges have much thicker walls made of many spongy, porous areas perforated by long sinuous canals (Fig. A3, right). These are the kind that formed the bath sponges that were originally collected by divers on the sea floor before synthetic foam replaced them. Most fossil sponges have very simple structures nonetheless, with a shape that can vary a lot if the sponge lived in different environments. Their spicules are also diagnostic as well, and individual tiny spicules are often found in deep-sea sediment. Some make their spicules out of silica, so they are known as "glass sponges" (Fig. 11.21A). Others sponges make spicules out of the common mineral calcite, so they are known as "calcareous sponges." The ones that were once used for bathing make their spicules out of the flexible organic material known as spongin, so this allows their skeleton to compress and be squeezed without destroying it.

Sponge fossils may not be nearly as spectacular as trilobites or dinosaurs, but during certain places and times, they were extremely important to the ecology of the sea floor. Many organisms, including various worms, arthropods, fish, mollusks, and protozoans, seek shelter in sponges because of their large, hollow, protective spongocoel. Some sponge predators may actually eat sponges to get at the sheltered animals inside. A single black loggerhead sponge was reported

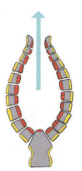

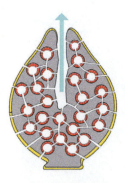

Figure A3 The basic structure of the three types of sponges. The simplest is the ascon type (left), which is a simple thin-walled cylinder that pulls in water from the sides through the pores, then forces it out through the top (blue arrow). The sponge cells (red areas) line the inner surfaces of the sponge and capture their food and oxygen in those currents. The sycon sponge (center) is also built as a simple cylinder but with much thicker walls penetrated by canals lined with sponge cells (red). The most complex is the leucon sponge (right), which has much thicker porous walls and only a small spongocoel in the middle.

to contain over 10,000 organisms within its canals and skeleton. In some regions, sponge fishermen find so many hard-shelled mollusks in them that their catches are worthless as bath sponges.

Phylum Archaeocyatha

During the Paleozoic, there were a number of instances where sponges were the major reef builders or contributed to reefs built with corals. During the Early Cambrian, the first large colonial reef-building organisms on the planet were a group known as the Archaeocyatha (Fig. 10.13A, B). They are organized in a structure similar to a sponge, with a long conical or cylindrical hollow shell that filtered water through it. However, archaeocyathan walls had a distinctive cone-in-cone structure and an "I-beam" wall construction that separates the outer and inner walls. They were perforated by numerous pores, so they almost certainly fed and lived like sponges. However, paleontologists are still debating whether they were truly sponges or only an early extinct independent experiment in sponge-like body form. Archaeocyathans don't have spicules like other sponges, and there are a lot of other fundamental structural differences as well. The answer may never be known because after forming huge reefs in the Early Cambrian in many parts of the world, archaeocyathans vanished completely by the Middle Cambrian, when undoubted sponges replaced them.

Superphylum Radiata

Animals with a radial symmetry and no true head, tail, or one-way digestive tract (Fig. A2).

Phylum Cnidaria ("Coelenterates"; Sea Jellies, Sea Anemones, and Corals)

The next more advanced level of organization beyond sponges is to have cells specialized for certain functions, like an inner layer of tissue or an outer layer of tissue arranged around a central body cavity. In addition, organ systems like the nervous system can be well developed, so there are discrete nerve cells. This is the level of organization you find in the phylum Cnidaria (called "Coelenterata" in older books). The roots of both of these names describe the group well. *Knidos* is the Greek word for "stinger" or "nettles," and all cnidarians have specialized stinger cells in their tentacles that are used to paralyze prey. The obsolete older term "Coelenterata" means "hollow gut" in Greek and describes the fact that this phylum is the first and most primitive group of animals on earth to have an internal body cavity.

See For Yourself:
The Shape of Life:
Cnidaria

You are probably familiar with several kinds of living cnidarians, especially sea jellies (formerly called "jellyfish," but they are not true fish) and sea anemones. All cnidarians are built around the same body plan (Fig. A4). They have a body arranged around an internal cavity, with an opening that serves as both mouth and anus at one end and tentacles with stinger cells arranged around the mouth/anus. The internal wall of the body cavity is lined with **endodermal** cells, which are specialized for digesting prey that they swallow, and the external surface of the body is covered by **ectodermal** cells, which protect it from the outside world. They may also have masses of tissues between the endoderm and ectoderm called mesoderm or **mesoglea**, which provide bulk and thickness to their body walls. In a sea jelly, most of the mass is mesoglea, which makes them "jelly-like." They also have a nervous system and muscular system which reacts to touch and allows them to sting prey and pull it into their mouth using their tentacles. But they do not have eyes or most other sensory systems, nor do they have a respiratory system or excretory system. They take in all their oxygen, and get rid of all their waste gases, through the surfaces of their body.

From this basic body arrangement, cnidarians have two fundamental versions (Figs. A4, A5): the sea jelly-like form, or **medusa** (named after the monster in the Greek myth of Perseus and Andromeda, who had snakes for hair and turned anyone to stone who saw her), and the attached anemone-like body form, or the **polyp**. In the medusa stage, like sea jellies, the tentacles hang down from the floating body, with the mouth below and the rest of the body above. In the polyp stage, the tentacles

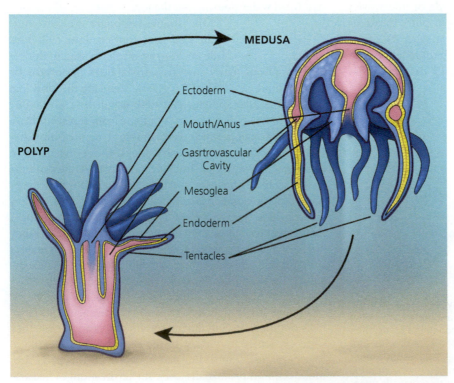

Figure A4 The basic body plan of a medusa or sea jelly (mouth on the bottom surrounded by tentacles) is fundamentally the same as the polyp phase (such as an anemone or hydra) but flipped upside down.

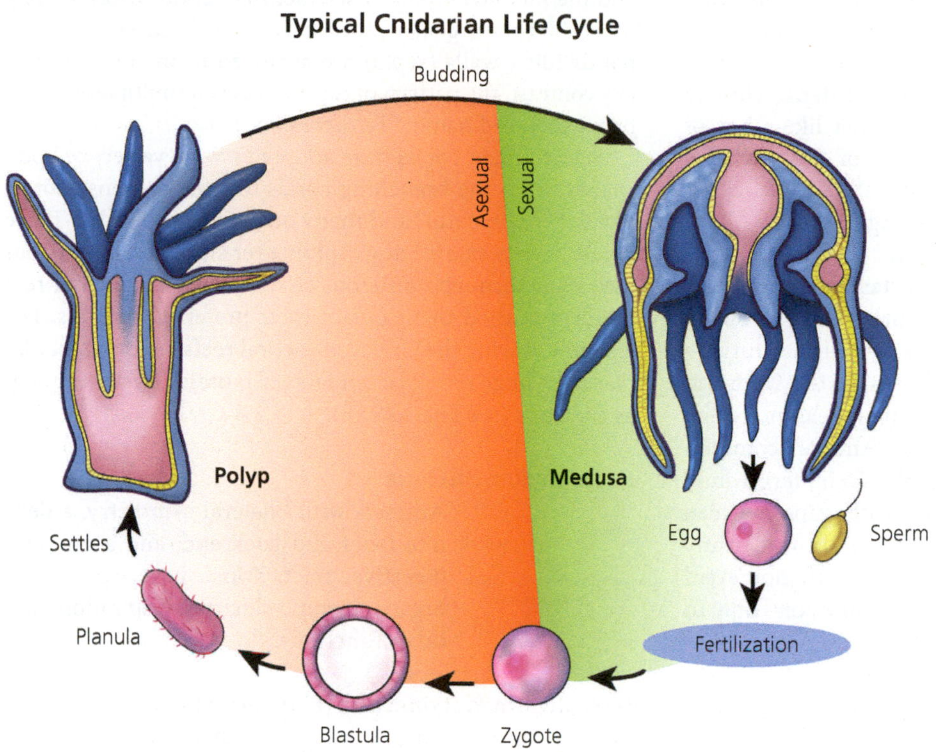

Typical Cnidarian Life Cycle

Figure A5 The polyp and medusa phases are part of the life cycle of all cnidarians, although sea jellies spend most of their lives as medusae, while anemones and corals spend most of it as polyps. Each polyp gives rise to tiny medusa by asexual cloning. The medusae then float freely with ocean currents, and during breeding season they release eggs and sperm, forming the next generation of polyps by sexual reproduction. This alternation of generations between sexual medusae and asexual polyps is found not only in Cnidaria but also in many foraminifera.

combines with their neighboring polyps to form a huge coral "skeleton" or coral reef. When we look at a piece of modern coral or a coral fossil, it's hard to imagine this rock as a living organism; but at one time it was covered with a layer of hundreds of polyps, all busily trapping tiny prey with their tentacles while building their enormous skeletons.

Some corals can live in cold water or in deep water without much light, but they tend to secrete very small skeletons. The vast majority of reef-building corals, however, have symbiotic algae that live in their tissues. The algae are plants, so they provide the coral with oxygen and use up the carbon dioxide they produce; they also help the corals secrete the enormous volume of calcite to make their limestone reefs. But since plants require light for photosynthesis, large reef-building corals can only live in very shallow tropical oceans where there is good light penetration and no sand or mud from land nearby which would make the water dark and murky.

are on the top with the mouth, and the body is attached at the bottom.

See For Yourself:
The Shape of Life:
Anemones Fight

Although they look very different, both of these stages are actually part of the life cycle of the same organism (Fig. A5). Cnidarians reproduced by alternating generations between sexual and asexual reproduction. The polyps are the asexual stage, and they can bud and clone themselves asexually to form huge colonies. Eventually, however, some polyps produce tiny larvae that grow into free-swimming medusae, which have sexual organs. These swarm in the ocean and in a coordinated fashion release sperm and eggs, which fertilize to form a free-swimming larva that eventually settles and grows into a polyp again. Some cnidarians (like *Hydra* in a freshwater pond or corals) spend almost all of their lives as immobile polyps and have only a brief medusa stage to spread to new habitats and colonize new sea floor. Others, like sea jellies, spend most of their lives in the medusa stage and only exist as polyps for a brief time.

Most cnidarians, like freshwater hydras, sea anemones, and sea jellies, are soft-bodied with no hard parts, so they are very rarely fossilized. However, one group of cnidarians leaves an excellent fossil record: corals. Corals are built of hundreds of polyps that look like tiny sea anemones, but the base of their ectoderm secretes a large amount of calcite skeleton that

Thus, coral reefs grow in only a few parts of the world today, mostly tropical or subtropical shallow oceans far from the influence of nearby rivers and their muds. These include the Bahamas, the east coast of Florida (but not most of the Gulf of Mexico side, where the Mississippi mud makes the water murky) and other parts of the Caribbean and Yucatán Peninsula of Mexico, the Persian Gulf, and the South Pacific islands and Great Barrier Reef of Australia, with just a few other places. Paleontologists think that ancient corals had similar restrictions, yet we find fossil corals in places like Iowa. How could this be? At various times in geological history, especially during the early Paleozoic, the entire planet was much warmer, sea level was higher, and shallow, warm tropical seas drowned the continents, so fossil corals are found in many places that are now far from the ocean.

The classification of the Cnidaria recognizes three main groups:

Class Hydrozoa—Hydras and other simple primitive cnidarians.

Class Scyphozoa—Sea jellies.

Class Anthozoa—Corals and sea anemones. These include three major groups of corals, two Paleozoic groups that are now extinct (Tabulata and Rugosa) and the living Scleractinia.

Order Tabulata: Tabulate corals are built of dense clusters of tiny tubes (**corallites**) all packed together like a box of drinking straws. Each corallite had lots of tiny dividing walls inside it, known as **tabulae** ("little table" in Latin), so this feature gives the group its name (Fig. 10.15A). A tiny coral polyp lived in the top of each tube.

Tabulate corals were extremely important as reef builders in the early Paleozoic, especially the Silurian and Devonian. The "honeycomb coral" (*Favosites*) is a typical Silurian–Devonian coral found in many localities in the Midwest, and it gets its name from its "honeycomb" appearance when it is sliced across the top (Fig. 10.15A). Another common Silurian reef builder is the "chain coral" (*Halysites*), whose corallites form little loops or chains with gaps between them in top view (Fig. 11.12C). Tabulate corals were nearly wiped out during the great Late Devonian extinction event but straggled on through the rest of the Paleozoic, only to vanish in the great Permian extinction.

Order Rugosa ("Horn Corals"): The other common group of Paleozoic corals was the horn corals, or rugosids. Most are shaped like the curved horns of a cow, hence their common name. In life, a typical horn coral like *Streptelasma* (Fig. 10.15B) or *Zaphrentis* would have sat on the sea floor with the narrow, pointed end embedded in the sediment, and a large anemone-like animal lived on the bowl-shaped cavity on the top of the cone, or **calyx**. Most rugosids had a wrinkled outer surface on the "horn," hence their name ("rugose" means "wrinkled" in Latin). The most extreme example of this is the very lumpy, irregularly wrinkled Devonian coral known as *Heliophyllum* (Fig. 11.12D). Its name means "sun leaf," possibly because if you look down on the calyx, the septa radiate outward like the rays of the sun.

Although most horn corals were solitary, there were rugosids that clustered together into dense colonies. The most famous of these is *Hexagonaria*, an important reef-building coral of the Devonian (Fig. 11.12E). When you slice across *Hexagonaria*, you can see that the corallites were packed together tightly to form a hexagonal honeycomb pattern, hence their name. Polished pebbles of *Hexagonaria* are common on the shores of the Great Lakes and have come to be known as "Petoskey stones." These are the official state rock of Michigan. Other important late Paleozoic colonial rugosids were *Lithostrotion* and *Syringopora*.

Like tabulates, rugose corals were hit hard by the Late Devonian mass extinction, which wiped out the immense reefs made of tabulate and corals and stromatoporoid sponges. The rugosids straggled on through the rest of the Paleozoic but were finally wiped out by the great Permian extinction.

Order Scleractinia (Living Hexacorals): After both tabulate and rugose corals vanished in the great Permian extinction, there were no corals or reefs of any kind in the Early Triassic. By the mid-Triassic, some group of soft-bodied anemones evolved the ability to build big calcite skeletons again,

and the modern group of scleractinian corals evolved (Fig. 13.12C). They are also called hexacorals because the internal dividing walls (septa) are arranged in multiples of six. (By contrast, the pattern of rugose corals is multiples of four, hence their old name, "Tetracorallia.")

Scleractinian corals have adopted a wide variety of body shapes, from the branching corals to massive dense corals like the "brain coral" and many other shapes besides. These corals have been evolving rapidly ever since the Triassic but have gone through numerous crises as well when other reef builders pushed them out of their preferred habitats. For example, during the Cretaceous coral reefs were almost displaced by huge reefs of colonial reef-building oysters known as rudistids (see Fig. 13.15E).

Superphylum Bilateria

See for yourself: Life on Earth: Building Bodies with David Attenborough

Animals with bilateral symmetry, a definite front and back end, and a one-way digestive tract. Most have specialized organs for reproduction, respiration, and other functions.

Infraphylum Platymorpha (Flatworms)—Bilaterally symmetrical animals with no internal body cavity (= coelom). Flatworms have no known fossil record.

Infraphylum Protostoma—Bilaterally symmetrical animals with or without a coelomic cavity (Fig. A2). They are recognized by a number of embryonic features in addition to their molecular similarities. For example, when the embryo develops into a hollow ball of cells, the opening (blastopore) eventually develops into the mouth. As the first cells cleave, they do so in a spiral fashion; and if broken apart, they form an incomplete animal.

Brachiopods, most other worms except flatworms, mollusks, arthropods, and all the rest of the animals described in this book are much more advanced than cnidarians, which only have tissues and simple nervous systems. The higher animals have not only tissues but a number of specialized **organs** as well. In addition, they are all **coelomates**, built around a fluid-filled body cavity called the **coelom**. In your body, the coelomic cavity includes your thoracic cavity that holds your lungs, and the abdominal cavity which holds your other organs, divided by the muscular wall called the diaphragm. In worms, this internal filling of fluid allows their soft bodies to become rigid like a water balloon, which allows them to burrow into the mud on the sea floor without hard shells. In addition, most advanced animals have a **one-way digestive tract** with a mouth at one end and an anus at the other, not the simple "stomach" with a single opening seen in a sea jelly or anemone. With a mouth at one end and an anus at the other, most higher animals also have **bilateral symmetry**, with a head and tail and right and left sides. By contrast, nearly all cnidarians have simple radial symmetry, with all their body parts arranged around a central axis. There is no true "head" or "tail" or right or left in a sea jelly or sea anemone. Most coelomates are too large to absorb oxygen and excrete waste through their surface area, so they usually have respiratory and excretory systems and a more complex reproductive system, in addition to the nervous and muscular system already present in cnidarians.

Phylum Brachiopoda: Brachiopods or Lamp Shells

Although they are relatively rare today and unfamiliar to most of us, during the Paleozoic the brachiopods, or "lamp shells," were by far the most common of all invertebrate fossils. In many places in the Eastern or the Midwestern United States, the limestones are chock full of them, and they literally pave the ground. They are excellent index fossils, so if you can recognize your basic brachiopod groups, you can easily tell what time it is in the Paleozoic from brachiopods alone. There are only 120 living genera, but over 4500 fossil genera (900 in the Devonian alone). Nearly all the modern brachiopods live in hiding places, such as under rocks or burrowed into the mud, to avoid predators. Consequently, brachiopods have long been intensely studied by paleontologists. However, most marine biologists have never seen one, and brachiopods get only a page in a typical invertebrate zoology textbook.

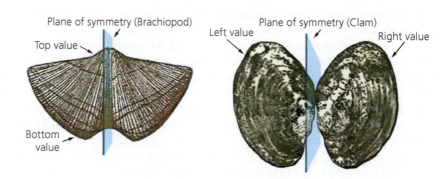

Figure A7 The brachiopod and bivalve shells are similar superficially, but their internal organs and biological affinities are entirely different. The brachiopod shell is typically symmetrical through the valves, so the right half of each shell is the mirror image of the left. Bivalves like clams are symmetrical between the valves, so each shell is the mirror image of the other.

Brachiopods are not familiar to most of us because of their scarcity in modern oceans. The only common name they have is "lamp shells" because one group of brachiopods resembles a biblical oil lamp. Brachiopods look superficially like clams because their bodies are encased in two hard shells, also called **valves**. However, their anatomy inside the shell is completely different (Fig. A6), and they are not closely related to clams or any other kind of mollusk. Even their bivalved shell is very different from the shell of a clam (Fig. A7).

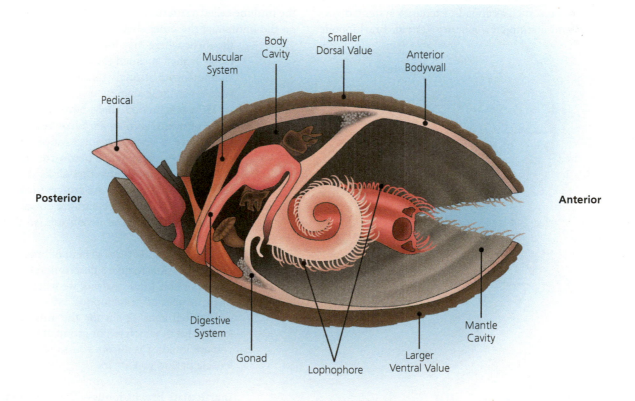

Figure A6 Diagram of the anatomy of a brachiopod. Most of the internal organ systems are crammed into a small area in the back of the shell near the hinge. The bulk of the internal shell volume is occupied by the feathery feeding device called the "lophophore." Many brachiopods have a long fleshy stalk called the "pedicle" that protrudes through the pedicle valve, or ventral valve, that attaches them to the sea floor.

The best rule of thumb to tell them apart is that brachiopods are symmetrical *through* the valves, so the right half of each shell is the mirror image of the left half. By contrast, a clam is a mollusk (an entirely different phylum), and it is symmetrical *between* the valves, so one shell is the mirror image of the other. However, there are a number of oyster-like brachiopods, plus true oysters and other clams that give up on symmetry altogether, so these rules of thumb don't always apply.

Once you have determined the fossil is a brachiopod shell, the rest of the anatomy is straightforward (Fig. A8). One valve is usually larger than the other, and in most brachiopods, there is a small opening on the hinge for the long fleshy stalk known as the **pedicle**, used to attach the shell to the substrate (Figs. A6, A8). This has long been known as the **pedicle valve**, although now it is called the **ventral valve** since it usually sits in the ventral (bottom) position. The pedicle gives the phylum its name, "Brachiopoda," which means "arm foot" in Greek. Hinged to the ventral valve is the other valve that used to be called the **brachial valve** and today is called the **dorsal valve**. It is often slightly smaller than the ventral valve and tends to sit on the top, or dorsal position, in most brachiopods.

Inside their paired shells (bivalved shells), brachiopods have a very simple anatomy (Fig. A6). Their main feature is a large plume-shaped filter-feeding device known as a **lophophore**, which is used to trap microscopic food particles as they pass through a gape in the open shell (Fig. A6). The food then passes down through the mouth and the digestive tract, like in most other coelomates. There are all the other usual organs, such as gonads, excretory system, digestive glands, and a simple circulatory system. Brachiopods

have a nervous system but no eyes. Instead, there are bristles around the margin of the shell called **setae**, which are sensitive to changes outside the shell and warn the animal to close its shell when danger approaches.

Most important are the muscles that close and open the shell. One pair, known as the adductor muscles, pulls the shell closed; a second pair pulls on the lever arm of the internal hinge to open the shell. Because they have paired muscles controlling the shell, brachiopods tend to stay closed when they die and tend to be buried and fossilized as complete shells. By contrast, clams and other bivalve mollusks have adductor muscles to close the shell but a flexible ligament in the hinge that springs-loads the shell so that it opens automatically when the clam is not trying to close the shell. Thus, when clams or scallops die and the adductor muscles relax, they tend to open automatically and often break apart, so it is rare to see both shells preserved together.

There are many details of the shell, especially in the hinge area, the shape of the lophophore, and the detailed microscopic structure, that are used in identifying the brachiopods and creating their classification. For a book of this level, however, we will look at only a few anatomical features that are helpful for identifying the external shape of the shell (Fig. A8). The outside of the shell can have numerous **growth lines**, which form arcs radiating out from the hinge, and fine ribs known as **costae**, which radiate out from the hinge-like spokes. Some shells have large corrugations radiating from the hinge, known as **plications**. In certain groups of brachiopods, there is a large trough-like depression on the midline called a **sulcus**, and a corresponding large ridge on the midline of the other shell called a **fold**. The edge of the shells where they join when the brachiopod closes it shell is called the **commissure**. In brachiopods with strongly plications, the commissure has a zigzag shape.

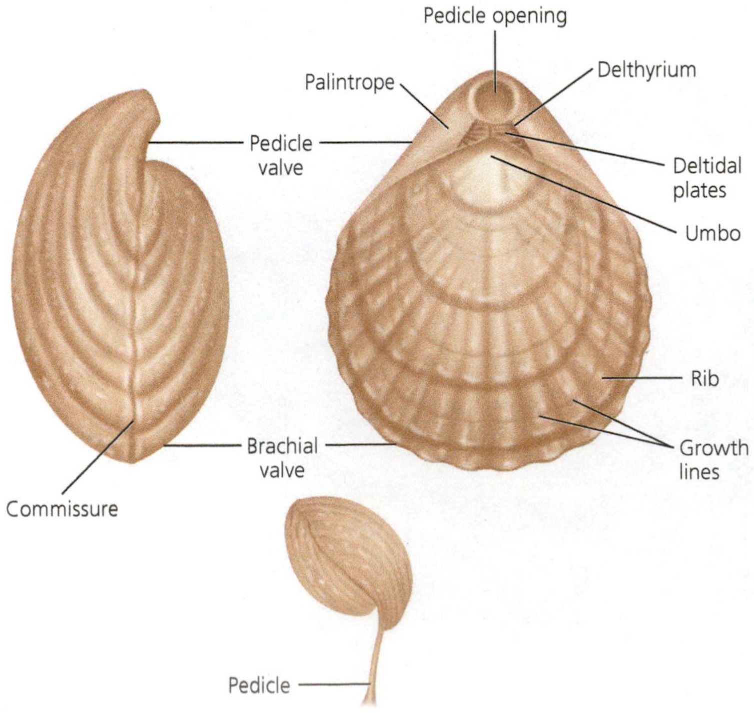

Most brachiopod shells have two shells that bulge outward, so they are **biconvex**. However, some have a convex ventral valve but a flat dorsal valve, so they are **plano-convex**. Some even have a dorsal valve that sinks down into the body cavity, so they are **concavo-convex**. A few have the same shape, but the dorsal valve is convex and the ventral is concave, so they are **convexi-concave**.

With over 4500 fossil genera of brachiopods, it is not possible to give a description to identify every possible kind. Instead, we will look at the major orders of brachiopods, which are fairly easy to tell apart, even for the amateur. Even knowledge of which major group of brachiopod you are looking at allows you to tell which Paleozoic period you are in.

Inarticulate Brachiopods—The most primitive of all the brachiopods are the inarticulates. The name doesn't suggest they can't speak fluently; instead, it refers to the fact that they have no

Figure A8   The features of the exterior of a typical brachiopod shell.

mechanical hinge with teeth and sockets holding the shells together. The hinge is held together only by bands of muscles, with no mechanical articulation. One of the most common brachiopods is alive today and found in mudflats all over the world. It is called *Lingula* ("little tongue" in Latin) since the shell has a tongue-like shape (Fig. 10.13E). *Lingula* live buried deep in mudflats, using a very long, fleshy, stalk-like pedicle to burrow down. This type of brachiopod has been around unchanged since the Cambrian, when inarticulates became the first brachiopods on the planet, so they are a living fossil that has persisted for at least 550 million years. There are a number of other tongue-shaped, coin-shaped, oval, and disk-shaped inarticulates found in the Cambrian; but only *Lingula* survives today. Inarticulates are also interesting in that they make their shells out of calcium phosphate (the mineral apatite), the same material our bones are made of, rather than the calcite that nearly all other marine invertebrates use.

Articulate Brachiopods—All the rest of the brachiopods are articulate, in that their shells have a mechanical hinge that holds together even when the animal dies. All of them make their shells out of calcite, the most common building material among marine invertebrates. They are also peculiar in that they have no anus. When the digested food accumulates to the limit, they expel all their waste products out of their digestive tract.

Order Orthida: The orthides are the most primitive of the articulate brachiopods, originating in the Late Cambrian during the heyday of the inarticulates and flourishing in the Ordovician before crashing during the Late Ordovician mass extinction and straggling on to the end of the Permian extinction. Most are very similar, with a straight hinge and small pedicle opening and lots of fine ribs of costae radiating away from the hinge. During the Ordovician, genera like *Orthis*, *Dinorthis*, *Hebertella*, and *Resserella* (Fig. 10.15H) were very common and now are good index fossils of this period.

Order Strophomenida: The strophomenides are the largest group of articulate brachiopods. They had two great periods of diversification. During the Ordovician, they were by far the most common type of brachiopod, and thus they are an instant indicator of Ordovician rocks. Ordovician strophomenides (Fig. 10.15F, G) tended to look like *Strophomena* or *Rafinesquina*, with long, straight hinges giving them a "D" shape in top view. These strophomenides also tended to have concavo-convex shells, so when the valves were closed, they were like a pair of dishes nested inside one another and had only a tiny internal volume. They apparently lived with their convex side up, so they arched their shell above the substrate; and the lack of a large pedicle opening shows that they did not attach to anything but lived on the open sea floor.

The "D"-shaped strophomenides were nearly wiped out during the Late Ordovician extinction, but the group had another great evolutionary radiation in the late Paleozoic (Carboniferous–Permian). These brachiopods, known as **productids**, had a cup-shaped ventral valve and a tiny flat or concave dorsal valve that formed a lid on top of the shell (Fig. 12.13G, H). They had no pedicle but lived perched on the open sea floor. Most had a dense cluster of spines on their ventral valve that acted as "stilts" or "snowshoes" to prevent their shell from sinking into the sediment. In fossil productids, the delicate spines are usually broken off, but the ventral valve is covered with a dense set of bumps where they once attached.

Productids were by far the most common brachiopods of the Carboniferous and Permian and often lived in dense clusters or colonies. In the Permian they evolved some truly strange forms, known as the leptodids (Fig. 12.13I). These weird creatures had a ventral valve that was shaped like a soap dish, covered by a dorsal valve that looked like a comb or a grill. In contrast to the flattened leptodids, the other extreme were the richthofenids (Fig. 12.13H). These peculiar creatures had a ventral valve that was shaped like an ice cream cone held up by stilt-like spines and a tiny lid-like dorsal valve inside the mouth. This was one of many examples of a group evolving into a conical and colonial coral-like shape or oyster-like habitat. It also happened independently in the rudistid oysters in the Cretaceous (see Fig. 13.15E).

Order Pentamerida: During the Silurian, the most common group of brachiopods was the pentamerides. They had a robust biconvex shell with a narrow hinge, a large pedicle opening, and a relatively smooth shell covered with fine costae. Their name comes from the fact that their internal shell is subdivided by a series of small dividing walls. There are five chambers, hence "pentamerid" (*penta* is "five" and *meros* is "part" in Greek). Pentamerides tended to form dense reef-like clusters during the Silurian, often in areas associated with the huge Silurian coral and sponge reefs. In many places in the Midwest, the Silurian rocks are completely replaced by dolomite, including the filling of the shells. However, the shell itself remained preserved in calcite, so when these rocks weather, the shells dissolve away, leaving a steinkern or internal mold of the more resistant dolomite. Their distinctive internal molds have a large cleft in the middle where one of the dividing walls used to be located.

Order Spiriferida: The spirifers are distinguished by their lophophores, which are arranged in some kind of spiral or corkscrew pattern inside the shell (hence their name, "spirifer," which means "spire-bearing" in Latin). The external shape of *Spirifer*, *Mucrospirifer*, *Neospirifer*, and their kin is also distinctive (Figs. 11.12G, 12.13D), with a long straight hinge, strong plications, and a large fold and sulcus. Their shape resembles a pair of wings. Other spiriferides, like *Atrypa*, were plano-convex with many fine costae and a slightly shorter hinge. Spirifers first appeared in the Ordovician, but during the Devonian they had a huge radiation, so they are a classic index fossil of that period. They were decimated during the Late Devonian extinction; but genera like *Spirifer* flourished in the Mississippian, and the entire group straggled through the rest of the Paleozoic in small numbers. They were nearly wiped out by the great Permian extinction but recovered in the Triassic, only to vanish at the end of the Triassic.

All the previous articulate orders are now extinct. Only two articulate groups survived the Mesozoic and are still in the ocean today.

Order Rhynchonellida: The rhynchonellides first appeared in the Ordovician, but they persisted through the entire Phanerozoic and are still found in great numbers in certain habitats. Nearly all rhychonellides have a short hinge with a pointed beak and strongly corrugated plications that give them a distinctive zigzag commissure. This body form was very conservative and typifies nearly all their order, so they are easy to recognize.

Order Terebratulida: The other surviving group of articulate brachiopods is the terebratulides. These are the most common form alive today and are shaped like the biblical oil lamp, hence the name "lamp shells" (Fig. A6). They are strongly biconvex, with a narrow hinge. They have a large pedicle, and the beak of the ventral valve curves around with a large pedicle opening. Living terebratulides attach to a hard surface and can orient their shells in any position to take advantage of currents.

Phylum Bryozoa: Bryozoans, or "Moss Animals"

The other living phylum of animals which have lophophores are the bryozoans, or "moss animals." In contrast with the relatively large brachiopods, bryozoans were tiny creatures, similar to the polyps of corals (Fig. A9). Like corals, they lived colonially on a massive skeleton of calcite secreted by all of them collectively and filter-fed on small food particles from the seawater which passed through their lophophore. However, individual bryozoan animals are smaller than

even coral polyps. If you are looking at a piece of a colonial fossil and trying to decide if it's coral or bryozoan, the rule of thumb is this: bryozoans leave pinprick-sized holes, while the openings for coral polyps tend to be larger.

Bryozoans are still very common in the world's oceans today, with over 3500 living species, as well as over 15,000 fossil species. They are sometimes known as "moss animals" because a living bryozoan colony looks like a fuzzy coat of moss when all the individual animals are reaching out of their tiny holes and feeding. Despite their incredible living diversity, most beachcombers and marine biologists seldom notice them because they are so tiny and often mistaken for moss or algae.

However, bryozoans are most definitely not built like coral polyps. Instead, they have a large feathery lophophore that protrudes out of their little hole in the colony when they are feeding. At the base of the lophophore is a "U"-shaped digestive tract, which takes food brought down from the lophophore and processes it, then excretes it back out the other end through the anus. The individual animals also have simple gonads, excretory systems, and a set of retractor muscles that pull them back into their holes. Most of them have a small lid that closes behind them when they retract, sealing them in their chambers when they are exposed to danger or drying conditions.

It is hard for nonspecialists to identify most bryozoans because their features are mostly extremely tiny and require microscopes and thin sections to study properly. For a book written at this level, we won't try to go into that much detail. Instead, we will discuss just a few of the more distinctive Paleozoic colonial forms.

Massive and Branching Bryozoans—In the early Paleozoic (especially the Ordovician), there was a great radiation of two groups, the cryptostomes and trepostomes. They are hard to tell apart without a microscope, so we will just note they came in a wide variety of body forms. Some, like *Prasopora*, *Rhombopora*, and *Constellaria* (Fig. 10.15I), formed disk-like massive colonies that occasionally reached 1.5 meters (2 feet) across but were usually only a few inches. Others, like *Dekayella*, *Prismopora*, *Streblotrypa*, *Leioclema*, and *Thamniscus*, grew into distinctive branching forms, with each branch covered with hundreds of tiny pinprick-sized holes.

In the late Paleozoic (especially the Mississippian), there was a great radiation of another group, known as the fenestrate or **fenestrellid** bryozoans, such as *Fenestella* and *Fenestrellina* (Fig. 12.13C). These formed a lacy framework with a grill-like pattern in close-up. Their name refers to the window-like openings in the grillwork

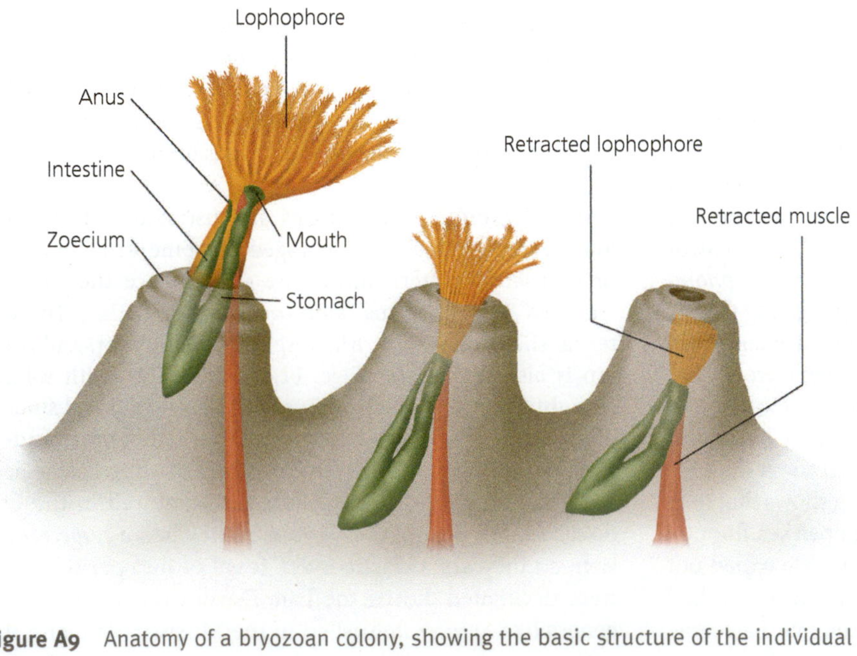

Figure A9 Anatomy of a bryozoan colony, showing the basic structure of the individual animals.

(*fenestra* is "window" in Latin). They are very common in Mississippian limestones and shales.

The most distinctive of all the fenestrellids was a genus named *Archimedes* (Fig. 12.13D, E). This bryozoan colony was shaped like a large corkscrew, with the lacy grillwork fanning around the spiral. Their tiny corkscrew-like central columns are extremely common in some Mississippian shales and a good index fossil of the Mississippian. They get their name from the famous Hellenistic Greek inventor, scientist, and mathematician Archimedes. One of his many famous inventions was a water pump known as "Archimedes screw." It was built of a long tube with a corkscrew-like set of blades inside. If you plunge one end of the tube into water and turn the corkscrew blades, they lift water up the tube and to a different level. The paleontologist who named this fossil was inspired by its corkscrew-like shape to name it after the famous Greek inventor of that device.

The trepostome, cryptostome, and fenestrate bryozoans dominated the Paleozoic; then all of their huge diversity was wiped out by the great Permian mass extinction, along with most of the brachiopods, the tabulate and rugose corals, and most other common Paleozoic groups. Only a few Paleozoic groups, like the cyclostomes, managed to survive. During the Mesozoic, bryozoans slowly recovered, dominated by a Jurassic group known as the cheilostomes, which are the most diverse type in modern oceans. Today, most bryozoans are found encrusting shells of other animals or in small clumps and clusters on the sea floor.

Phylum Annelida (Segmented Worms)

Soft-bodied worms whose body is divided into a number of segments with separation of the internal body cavity into sealed chambers. This gives them hydraulic rigidity for burrowing. They live in marine and freshwaters (polychaetes) and in the soil (earthworms). Although they rarely fossilize, their burrows are common from the late Ediacaran onward.

Phylum Arthropoda: Trilobites and Their Relatives

The "joint-legged" animals, or the phylum Arthropoda, are by far the most numerous and diverse and successful creatures on this planet. The arthropods include not only insects but also spiders, scorpions, millipedes, centipedes, crustaceans, the extinct trilobites, and dozens of other groups. There are over a million different species of arthropods, and some biologists think there may be many million more, most not yet described. Insects alone make up 870,000 species, of which there are 340,000 species of beetles. By contrast, there are only 42,000 species of backboned animals and only about 4000 species of mammals. By nearly any reckoning, arthropods make up about 95%–99% of the animal species on earth.

In addition to their diversity, arthropods are the most numerous animals on the planet. A typical anthill or termite nest might have thousands of individuals. A single pair of cockroaches, if they have unlimited resources and space, could have 164 billion offspring in 7 months. In the tropical rainforests, a single hectare may only contain a handful of birds and mammals but over a billion arthropods, especially mites, bees, ants, termites, wasps, moths, and flies. In the ocean, the plankton in a single cubic meter of seawater may include almost a million tiny crustaceans, especially copepods, krill, ostracodes, and other shrimp relatives.

Arthropods can live in almost any habitat on the planet, from the oceans to freshwater to the land, and can fly through the skies. They can survive in the hottest deserts, the coldest ice, and the saltiest lagoon. Many are parasites that live inside or outside other organisms.

What makes arthropods so successful? Certainly, the fact that they can reproduce rapidly in the right conditions helps, especially when they can lay hundreds of eggs that hatch and develop and then breed again in a matter of days or weeks. Their small body size means they can live in dense numbers on a limited resource and subdivide niches into many small subniches, so they can pack very closely. Most important of all, however, is their modular body plan, with a series of segments that can be increased or decreased. On each segment is at least one pair of jointed appendages, which give the phylum its name (*arthros* means "joint" and *podos* means "foot" in Greek). These appendages can be modified into legs, mouthparts, antennae, claws, pincers, swimming paddles, copulatory structures, and a great variety of other possible functions. This modular construction gives them great evolutionary flexibility, so just by changing their embryonic development, they can end up with more or fewer segments and modifications of the limbs as they grow. In some groups (like the metamorphosis of a caterpillar into a butterfly), they can completely change the fundamental body plan in just a few weeks by rearranging and modifying the modules that control each segment.

Another factor that helps them is their external hard shell or skeleton, known as an **exoskeleton**. In most groups, it is made entirely of the carbohydrate **chitin**, but in others (like trilobites) it is reinforced with the mineral calcite. The exoskeleton provides support for the body, making it a hollow tube or shell, and the muscles run within it to make it move. It also provides some protection against predators since it is harder than the soft bodies of worms or many other invertebrates and protects them against drying out. This allowed arthropods (first millipedes, then centipedes, scorpions, and finally insects like silverfish and springtails) to emerge from the oceans in the Late Ordovician, and they continued to flourish for over 100 million years before the first fish-like amphibians managed to crawl onto land.

About the only ecological niche that arthropods don't dominate is the niche for large body size. As animals get larger, their volume (and thus their mass) increases by a cube of their linear dimensions, so they get heavier much quicker than they get longer. Their length may only increase by a few centimeters, but their volume increases so fast that they soon test the limits of supporting the weight of their bodies under the rapid increase in gravity. Yet arthropods have a key limitation: their exoskeleton cannot grow with them, so they must **molt** by breaking out of it and hardening a new exoskeleton. This is not a problem when they are tiny, but as they get larger and larger, the effects of gravity increase dramatically. Once they pass a certain size, their unsupported soft body after a molt would just collapse without its external shell. Thus, the gigantic ants and praying mantises from old-fashioned science fiction movies are a complete impossibility.

The largest land arthropods (Fig. A10) today are goliath beetles, which are at most about 110 mm (4.3 inches) long; but during the Carboniferous, there were gigantic millipedes like *Arthropleura* that were about 3 m (10 feet) long (Fig. 12.18B), dragonflies with wingspans almost 70 cm (2.3 feet) across (Fig. 12.18A), and cockroaches almost 30 cm (1 foot) long. The Carboniferous, however, was a time when the atmosphere was extremely rich in oxygen, which allowed arthropods (which have relatively inefficient respiratory systems) to get bigger than they ever did at any other time. In the water, the higher density of water supports their bodies better than in air, so there can be larger-bodied arthropods. The largest today is the king crab, whose legs can span about 3 m (10 feet), but there were giant eurypterids ("sea scorpions") that were over 2 m (6.6 feet) in length (Fig. 11.13).

 See for Yourself: The Shape of Life: Arthropods

Even though arthropods are the most successful and diverse and abundant animals on the planet, they have a relatively poor fossil record. Only about 30,000 species have been described from fossils, despite the fact that millions are alive now, and many more millions must have lived in the geologic past. Most arthropods only have a chitinous exoskeleton, which only fossilizes in exceptional conditions. The one group that fossilizes well is the trilobites, which reinforced their chitinous sheath with the mineral calcite, so there are about 2000 genera and thousands of species of trilobites (Fig. A11).

Subphylum Myriapoda (Millipedes and Centipedes)— Multisegmented arthropods with many legs and relatively unspecialized appendages. Millipedes first crawled onto land in the Late Ordovician, making them the earliest land animals. With their relatively thin chitinous exoskeleton, they are rarely fossilized. However, the extinct giant relative of the millipedes known as *Arthropleura* was the largest land arthropod ever known (Figs. 12.18B, A10).

Subphylum Crustacea (Crabs, Lobsters, Shrimp, Barnacles, Sowbugs, Ostracodes, and Many Important Types of Zooplankton)—Mostly aquatic arthropods with specialized antennae and mouth parts. Next to insects, they are the most abundant and successful group. The largest is the Alaskan king crab, also the largest arthropod in the ocean today (Fig. A10).

Subphylum Hexapoda or Insecta (Insects)—Mostly terrestrial arthropods with six legs and a well-defined head, thorax, and abdomen. The most abundant and diverse group of animals on earth, consisting of millions of species that make up about 75% of the animal kingdom. They are the most common organisms in every terrestrial habitat from the tropics to the deserts to the tundra, and a few have become marine as well. The largest insect ever found were the giant dragonfly known as *Meganeura* (Fig. 12.18A).

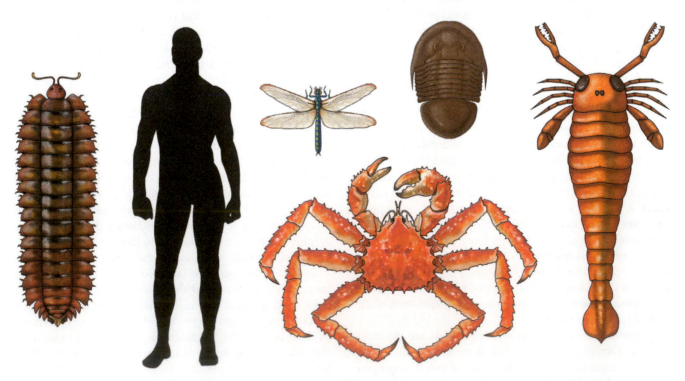

Figure A10 The largest arthropods known include the huge Carboniferous millipede relative *Arthropleura* (left of human silhouette), the largest arthropod on land ever known; the living king crab (bottom), the largest arthropod alive today; the giant dragonfly *Meganeura* and the huge trilobite *Isotelus rex*; and the giant eurypterid *Pterygotus* (right).

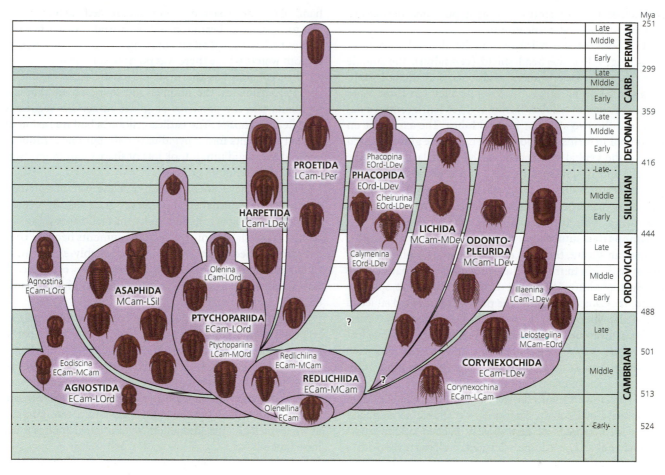

Figure A11   Diagram showing the evolutionary radiation and diversification of the trilobites, with the numerous different families radiating through the Cambrian (Cam) and Ordovician (Ord), then being mostly wiped out by the Late Devonian (LDev) mass extinction. Only one family with a few genera survived until their final extinction at the end of the Permian. E, early; L, late; M, middle; Per, Permian; Sil, Silurian.

Subphylum Trilobitomorpha (Trilobites)—Trilobites are the most popular fossil among collectors, as well as many professional paleontologists. Not only are they abundant and easy to collect in many places, but they have an extraordinary variety of shapes, including some with weird appendages and spines and even some with amazing eyes. Apparently, trilobites have fascinated people for a long time. A Silurian trilobite carved into an amulet was found in a 15,000-year-old Paleolithic rock shelter at Arcy-sur-Cure, France. Australian Aborigines chipped a Cambrian trilobite preserved in chert to form an implement. Both of these specimens were clearly imported from a long distance since they did not come from the local area. The Ute people used to make amulets out of the common Middle Cambrian trilobite *Elrathia kingi*, from the House Range of western Utah (Fig. 10.12B). They called it *timpe khanitza pachavee*, or "little water bug in stone house." This species is so abundant that it can be commercially mined with backhoes; it is found in virtually every rock shop and commercial fossil seller's catalogue.

Trilobites first appeared in the third stage of the Cambrian (Atdabanian stage), and they are by far the most common fossils of the Cambrian, both because they diversified rapidly and because most other Cambrian animals were soft-bodied and did not have hard fossilizable body parts (Fig. A11). By the Late Cambrian, there were 65 families with over 300 genera. In Cambrian shales or limestones, virtually every outcrop yields some trilobites, and they evolved so rapidly that we tell time in the Cambrian with them.

By the Ordovician, trilobites began to decline in importance as other invertebrate groups diversified and came to dominate the sea floor. In addition, trilobites faced their first large predators in the enormous-shelled nautiloids (Fig. 10.16), so the trilobites that did survive became much more specialized and distinctive, with adaptations for rolling up for protection, burrowing, swimming, and hiding from their predators. Trilobites were hit hard during the Late Ordovician extinction event. During the Silurian and Devonian, there were only about 17 families in 60 genera. The Late Devonian extinction nearly wiped trilobites out, and left only a handful of genera in a single family that survived until the end of the Paleozoic, only to finally vanish during the great Permian extinction event that wiped out tabulate and rugose corals, most groups of brachiopods and bryozoans, and many other Paleozoic groups.

The body of the trilobites can be divided into three parts (Fig. A12): the head shield (**cephalon**), the middle part (**thorax**) made of many segments, and the tail shield (**pygidium**). In addition, the trilobite shell has three lobes (hence the name "trilobite") from right to left: a central **axial lobe** and the two sides, or **pleural lobes**. The axial lobe on the cephalon forms a bulbous ridge called the **glabella**. Each side of the cephalon adjacent to the glabella is called the "cheek" and is split by a line called the **facial suture**. The pattern of the suture is very valuable in identifying major groups of trilobites. In primitive forms, the back end of the suture splits the back of the cephalon (known as the **opisthoparian** suture), but in one specialized group with a **proparian** suture, the back end of the suture splits the side of the cephalon in front of the **genal angle** (the back corner of the cephalon). When the trilobite molted, it split the cephalon along this suture. The part that remained attached to the glabella is called the **fixed cheek**, while the two parts that split away are called the **free cheeks**. Many trilobite fossils consist only a partial cephalon: the glabella plus fixed cheeks (known as the **cranidium**). However, for many trilobites, these are often enough to identify the species.

The eyes are usually on each side of the glabella, right along the facial suture. They may be simple semicircular eyes with hundreds of tiny, closely packed calcite lenses (**holochroal** eye) or large, elevated, bulging eyes with a handful of lenses, each built of two of nested parts that correct for spherical aberration (**schizochroal** eyes). A few trilobites had no eyes at all and must have lived in dark muddy bottom waters where there was no need for vision.

The cephalon has the most diagnostic features, but in complete trilobites, the number and shape of the thoracic segments and the presence of spines in certain segments are often diagnostic. The earliest trilobites had no fusion of the tail segments into a pygidium at all, but once trilobites evolved the pygidium, it took lots of shapes and often diagnoses certain trilobite groups.

Trilobites have an exoskeleton made out of calcified chitin, so the external shell preserves quite readily, even if it often breaks apart during molts. In fact, most fossils of trilobites consist of only one or two segments because they are simply shed molts; the animal probably swam away to live another day. Trilobite specialists often only need a few diagnostic parts, like the cranidium, to identify most species.

Collectors usually find only the hard, calcified shells of trilobites; but like all arthropods, trilobites had softer parts: legs, antennae, gills, and other appendages surrounded only by their exoskeleton of chitin but uncalcified. However, under the right circumstances we get preservation of these softer parts, such as in deep-water black shales, where pyrite replaces the organic tissues. This allows us a rare glimpse at what the legs, gills, and antennae of other trilobites must have looked like.

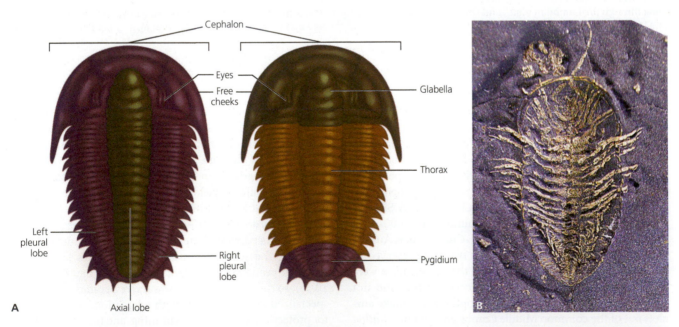

Figure A12   A. The anatomy of the trilobite shell. The three parts of the shell from front to back are: cephalon or "head shield" in green, thorax in the middle, pygidium or "tail shield." The structure in the center of the cephalon that looks like a "nose" is the glabella. On each side of the glabella are the eyes. The two bold lines running through the eyes are the facial sutures, which are lines along which the cephalon splits during molting. The two areas outside the facial sutures are the "free cheeks"; the segment of the cephalon between them including the glabella is called the "fixed cheeks." The three "lobes" of the trilobite from side to side are the axial lobe, and on each side of it, the two pleural lobes. **B.** Occasionally, black shales preserve the soft parts of trilobites, here replaced by pyrite. This specimen shows the legs, gills, and antennae.

With about 2000 genera of trilobites, it is not possible to talk about every group in a book like this. Instead, we summarize some of the most distinctive and commonly collected genera and orders.

Order Redlichiida: The redlichiids are the most primitive group of trilobites, starting in the late Early Cambrian and vanishing at the end of the Middle Cambrian. The most familiar family is the extremely primitive Olenellidae (Fig. 10.12A), which are collected in many places in the western United States. Olenellids were so primitive that they did not fuse their final segments into a pygidium but had a large spike on their tail instead. They had simple half-moon-shaped eyes on each side of a glabella with a round nob at the front tip and large genal spines on the outer corners of the cephalon. They had lots of thoracic segments, often with large spines on the "shoulders" of the first few thoracic segments. Early forms like Olenellidae had no visible suture, but later redlichiids like *Paradoxides* had opisthoparian sutures (Fig. 10.12D).

Order Ptychopariida: The ptychopariids form a large waste-basket group for many different primitive trilobites that dominated the Middle and Late Cambrian. They had a simple glabella that tapered toward the front, a large area in front of the glabella, and many thoracic segments. They include many of the classically well-known Cambrian trilobites, such as *Ptychoparia*, *Olenus*, *Triarthrus*, and the most common trilobite on the commercial market, *Elrathia kingi* (Fig. 10.12B). This last species is mined in huge quantities from the Middle Cambrian Wheeler Shale in the House Range of Utah and can be bought from every commercial market, website, and rock shop in the world.

Order Corynexochida: The corynexochids are another important Early–Middle Cambrian group. They have a box-like glabella and subparallel facial sutures that are opisthoparian. The thorax has only seven or eight segments, and some have a large pygidium that is roughly the same shape and size as the cephalon (**isopygous**).

Order Agnostida: These tiny, peculiar fossils are so unusual that not all scientists agree they are trilobites; they may just be a related group of arthropods (Fig. 10.12C). They have a button-shaped pygidium the same size and shape as the cephalon (isopygous), so only the eyes tell you which is front and which is back. There are only two or three short thoracic segments separating the cephalon and pygidium. They are tiny (only a few millimeters long) and may have lived in the plankton, rather than grubbing in the bottom muds as most trilobites did. They are most common in the Early–Middle Cambrian but straggled on to the Ordovician. Huge numbers of the tiny agnostid *Peronopsis interstrictus* are collected from the Wheeler Shale in the House Range of Utah and are found in collections and rock shops everywhere.

Orders Asaphina and Illaenida: During the Ordovician, most of the unspecialized Cambrian groups declined or vanished, replaced by many highly distinctive and specialized groups. This was probably due to the pressure of new predators, especially the huge new nautiloids that could see and grab any trilobite within reach of their tentacles (Fig. 10.16). Two groups, the asaphids and illaenids, coped with this by becoming "snowplow" burrowers living just beneath the sea-floor surface sediments to hide from predators. In both orders, they have a large, smooth cephalon shaped like a plow, and an isopygous pygidium that mirrored the shape of the cephalon except for the eyes (Fig. 10.15L). They had only about six to nine thoracic segments. Typical asaphids include *Homotelus* and *Isotelus rex* (Figs. 10.15L, A10), the largest trilobite known, which reached almost 1 m (3.3 feet) in length. Illaenids like *Bumastus* and *Illaenus* are also common in collections and on the commercial market.

Order Trinucleida: Another distinctive group of mainly Ordovician trilobites were the trinucleids. Amateur collectors often call them "lace-collar trilobites" because they have a huge cephalon and tiny thorax-pygidum, with a perforated "lacy" edge around the brim of the cephalon. They also have a large knob-like glabella that looks like a big nose. Trinucleids like *Cryptolithus* (Fig. 10.15M) and *Trinucleus* are common Ordovician fossils and probably survived by their tiny body size and burrowed just beneath the surface of the sea floor.

Order Harpida: Harpids are highly distinctive trilobites, recognized by their broad horseshoe-shaped cephalic brim that wraps around the small thorax and pygidium. The broad cephalic brim was also probably suited for plowing through the sea-floor sediment. Unlike trinucleids, harpids had small eyes on tubercles, a glabella that tapers to the front, and lots of thoracic segments.

Order Lichida: The lichids are another peculiar and distinctive group of trilobites. Their glabella stretches to the front edge of the cephalon and is subdivided by long glabellar furrows, and they had relatively small free cheeks. Many of them, like *Arctinurus* and *Radiaspis*, had glabella that tapered sharply forward with a small "prow" on the front. Their most distinctive feature is their relatively large pygidium, often larger than the cephalon, made of three pairs of expanded spiny pleural segments.

Order Odontopleurida: Odontopleurids are easy to recognize because most of them are extremely spiny, with spines sticking up from their cephalic brim, genal angle, top of the glabella, all along the thorax, and even from the back of the pygidium (Fig. 11.12K). They were relatively rare but were most diverse in the Devonian, so they are collected from the Devonian rocks of the Tindouf Basin of Morocco and found in many commercial sources.

Order Phacopida: One of the most interesting and distinctive of the trilobite orders was the phacopids. They are the only group with proparian facial sutures, and most of them were capable of rolling up into a tight ball like sowbugs or pillbugs or "roly-poly bugs" found in leaf litter today (which are not trilobites but actually isopod crustaceans that have moved from the tide pools to the land). In the Ordovician and Silurian, they were represented by the distinctive calymenids like *Calymene* and *Flexicalymene* (Fig. 10.15N). Calymenids had lots of thoracic segments and even segmented furrows on the glabella, with their eyes out on stalks. Another branch includes the Silurian index fossil *Dalmanites* (Fig. 11.12I) and the common Devonian phacopids, including *Phacops, Greenops, Eldredgeops* (Fig. 11.12J), and others. These phacopids have a broad, curved cephalon and a broad glabella covered with tiny bumps that expanded forward. They have large schizochroal eyes that stuck up above the cephalon and gave them almost a spherical view of the world above them, with stereovision between the lenses.

Order Proetida: After the Late Devonian extinction event wiped out the phacopids and odontopleurids and the other dominant Devonian trilobite groups, only one order managed to survive through the rest of the Paleozoic. Proetids were small trilobites with a relatively primitive body form, a large vaulted glabella, and opisthoparian sutures; and most of them were isopygous. They had primitive holochroal eyes and a furrowed pygidium without spines. When these last stragglers of the trilobite radiation finally vanished in the great Permian mass extinction, trilobites were completely extinct.

Subphylum Chelicerata—The chelicerates include the arachnids (spiders, scorpions, whip scorpions, pseudoscorpions, harvestmen or "daddy long legs," mites, ticks, and chiggers) plus the merostomes. With over 70,000 living and extinct species, the chelicerates are the second largest group of arthropods after insects. They all have a distinctive anatomy of head region (prosoma or cephalothorax) and abdomen (or opisthoma). They do not have antennae but instead their first pair of appendages, known as chelicerae, are found in their mouthparts. Their second pair of appendages, the pedipalps, sometimes serve as mouthparts; but in scorpions, some spiders, and eurypterids, they are modified into pincers for capturing prey.

The most familiar and only living merostomes are the horseshoe crabs, which are not closely related to true crabs (subphylum Crustacea) at all. They are often called living fossils since some of the Mesozoic forms look somewhat like modern species, but in the Paleozoic they experimented with many other body shapes.

For paleontologists, however, the most impressive chelicerates in the fossil record are the "sea scorpions," or **eurypterids**. Their nickname is a misnomer since they are not closely related to scorpions (which are arachnids, not merostomes) and apparently lived in freshwater and not just in the sea. However, their body form is reminiscent of scorpions with large pincers out front and spidery legs flaring out

from their prosoma (Fig. 11.13). More impressive is the fact that they were one of the world's largest predators during the Silurian. The largest nearly complete eurypteid fossil is *Pterygotus* (Fig. 11.13), which is about 2.1 m (7 feet) long, and the giant *Jaekelopterus* may have been about 2.5 m (8.2 feet) long. *Pterygotus* had a long body with a flat swimming tail and a pair of appendages modified into paddles, with huge pincers up front. *Stylonurus* was more spider-like, with a spike on its tail and no swimming paddles. *Mixopterus* had long spiny chelicherae and pedipalps, spiny walking legs, and a tail like a scorpion.

Altogether, there were about 70 genera and 250 species of eurypterids, ranging from the Ordovician to the Permian. However, they were the largest marine predators of the Silurian and continued that role in the Devonian. In the United States the famous Silurian Bertie Limestone of upstate New York produces a large majority of the specimens of *Eurypterus remipes* in collections and in the commercial market.

Phylum Mollusca (Clams, Snails, Squids, and Their Relatives)

Next to the arthropods, the mollusks (Fig. A13) are the most abundant and successful phylum of animals on earth, with over 130,000 living species of snails, slugs, clams, scallops, oysters, squid, octopus, and their relatives. Except for slugs, squid, and octopus, most mollusks have hard shells made of calcium carbonate, usually calcite but also the other form, aragonite ("mother of pearl"), which makes them easy to fossilize, so their fossil record is excellent, better than that of any other marine fossil group. Over 60,000 fossil species are known, so more paleontologists work on mollusks than any other invertebrate animal group, most of them specialists in just one class like the snails or clams or ammonoids.

 See For Yourself: The Shape of Life: Molluscs: The Survival Game

Mollusks are also important to human culture. Seashell collecting is not only a popular pastime now but has been for centuries, and in many cultures shells were used as money and currency ("wampum" to native American people). Human cultures eat a wide variety of shellfish, especially clams, oysters, and scallops, as well as abalone, conch, squid (calamari), octopus, and land snails (escargot). Nearly all prehistoric cultures that lived near the coast were big mollusk eaters, as shown by their large garbage heaps known as "shell middens." Over 2 million tons of squid and octopus are harvested each year, along with enormous volumes of clams, oysters, and scallops. In the United States alone, over 220 million pounds of clams, oysters, and scallops are eaten each year. Finally, certain mollusks (especially oysters) produce a valuable gem, the pearl, from the aragonite ("mother of pearl") that their mantle secretes around any irritant that gets inside the mantle cavity.

In addition to being extremely diverse, mollusks have come to inhabit almost as many lifestyles as the arthropods. They live in every marine setting from the deepest ocean to the intertidal zone, and some are also planktonic. Many different freshwater clams and snails are known, and land snails are well adapted to being away from water.

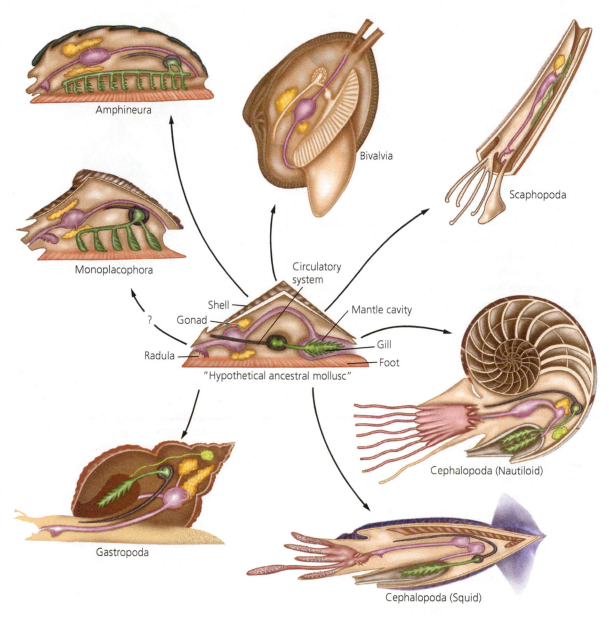

Amphineura

Bivalvia

Scaphopoda

Monoplacophora

Circulatory system

Shell

Gonad

?

Radula

Mantle cavity

Gill

Foot

"Hypothetical ancestral mollusc"

Cephalopoda (Nautiloid)

Gastropoda

Cephalopoda (Squid)

Figure A13 The various groups of mollusks are thought to have all originated from a simple body plan of a limpet-like form nicknamed the "hypothetical ancestral mollusk." It would have had a simple dome-like shell over its back and mantle, a broad foot to help it creep over the bottom, with a head and mouthparts at one end and the gills and anus in the mantle cavity in the back.

They achieve this flexibility despite a rather limited body plan. Nearly all mollusks are built around a softy fleshy body, with a muscular foot that helps propel them and a shell on their back. Their body is covered by an organ called the **mantle**, which secretes the shell. Most mollusks have a simple digestive tract that runs from mouth to anus, plus paired gills for breathing in water and simple versions of most other systems, including circulatory, excretory, reproductive, and other systems. Although most mollusks tend to be small, some are huge. The giant squid allegedly reaches about 18 m (60 feet) in length, and the giant clam is over 1 m (3.3 feet) long, while the giant marine snail *Campanile* has a shell over 1 m (3.3 feet) long.

The simplest mollusks (Fig. A13) are the limpet-like **monoplacophorans**, which have a cap-shaped shell over their back but still have segmented muscles, gills, and other body parts just like their relatives, the segmented worms. Monoplacophorans were the most common mollusks in the Early Cambrian and were thought to have gone extinct by the middle Paleozoic until living examples were found in deep oceanic trenches in 1952.

From this basic body plan of the "hypothetical ancestral mollusk" (Fig. A13), the various classes of mollusks have modified their anatomy to adapt to many different ways of living. Among the least modified are the **chitons** (class Polyplacophora or Amphineura), which are familiar denizens of tide pools around the world. They stick to rocks in the intertidal zone with their broad muscular foot and have modified the primitive limpet-like shell by subdividing it into eight plates, which allow them to bend and flex their body to fit around curves as they cling to rocks. Like limpets, they spent their lives slowly creeping around, scraping algae with their ribbon of tiny teeth made of iron oxide, or magnetite. In some islands in the tropical Pacific, the chitons have scraped so much rock away from the tide pool zone that the islands are on narrow pedestals.

Another minor group of mollusks are the tusk shells (Fig. A13) or **scaphopods** (class Scaphopoda). Most of them have a long conical shell that resembles an elephant tusk with a hole at the tip, hence their name. These burrow the shallow sand in the nearshore zone, with their open tip exposed, allowing them to filter ocean water and trap food while releasing their waste products. In addition, there are a number of extinct molluscan classes, such as the rostroconchs, which are rarely fossilized and mostly known from the Cambrian, when molluscan evolution was undergoing its "experimental" stage.

There are three classes of mollusks which make up the bulk of both the living and fossil species, and deserve detailed discussion.

Class Gastropoda (Snails, Slugs, and Nudibranchs)—The Gastropoda are by far the most diverse of all molluscan classes, with over 100,000 species, making them about 80% of the living mollusks (Fig. A13). The simplest mollusks are the limpets and their relatives, which have not changed that much from the monoplacophorans and other archaic mollusks. They are subdivided into many groups. Most of the marine snails are known as **prosobranchs** ("forward gills" in Greek since they rotate their gills from the rear to over their head during larval development). The bubble shells, plus the snails that have lost their shells (sea hares, sea slugs, nudibranchs) and the planktonic pteropods, are **opisthobranchs** ("backward gills" in Greek). The third group is the land snails and slugs, which have turned their gill cavity into an air-breathing organ; they are known as **pulmonates** ("lunged" in Greek). The opisthobranchs and pulmonates have only a limited fossil record, so we won't discuss them in this book but instead focus on the abundant shell record of prosobranchs.

Among the prosobranchs, the simplest are the limpets and abalones, which are adapted to sticking to hard surfaces and scraping off algae using the ribbon of tiny teeth in their mouth. But the simple cap-shaped limpet shell does not allow for much in the way of evolutionary innovation, so gastropods soon evolved into other forms. If a conical shell gets too large, it is much more stable to carry it if it is coiled into a spiral.

Nearly all the other gastropods have evolved some variation on a coiled shell. The basic terminology of a coiled shell is quite easy to remember (Fig. A14). Each turn of the shell around the axis is called a **whorl**. The number of whorls and the width and rate of turning of the shell are

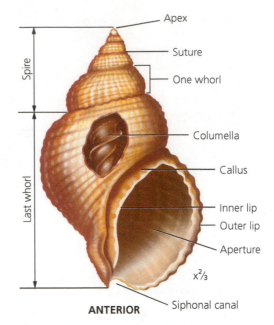

Figure A14   Basic anatomy and terminology of the gastropod shell.

diagnostic of certain groups of snails, and they change as the snail grows larger in its shell and adds more material to the leading edge. The point of the top of the shell is the **apex**, and the shell can be considered **high-spired** if it is long and pointed or **low-spired** if it is shorter and stubbier. If you see the shell broken open, the internal column formed by the junction of the inner part of each whorl is called the **columella**. The opening (**aperture**) of the shell typically has an **inner and outer lip** and an area next to the inner lip called the **callus**, where the shell rested on the back of the snail. Many advanced gastropods have a notch at the bottom of the shell, called the **siphonal notch**. Finally, coiled shells are asymmetric and can be either right-handed (**dextral**) or left-handed (**sinistral**). If you hold the shell with the spire pointed upward, the aperture will be on the right in a dextral shell and on the left in a sinistral shell. Most snails are dextral, although there are some that are sinistral, and a few snails change their coiling direction from one to the other.

However, there are many ways of making a coiled shell. During the early Paleozoic, there were many experiments in shell coiling among the gastropods. Some, like the bellerophontids, coiled their shells in a flat spiral that sat symmetrically over the middle of their backs. *Bellerophon* is a rare but fairly widespread fossil in many Paleozoic localities. Another way to arrange the shell is to hold the spiral diagonally across the back and pointed forward and over the head, an arrangement called **hyperstrophic**. This early experiment in shell evolution is best seen in the common large Ordovician snail known as *Maclurites* (Fig. 10.15K).

These early experiments in shell evolution vanished by the late Paleozoic, and all other prosobranch gastropods have their shell positioned diagonally over their back with the apex pointed backward; this is the **orthostrophic** condition. In addition to the limpets and abalones and other primitive gastropods, there are many coiled gastropods that retain their primitive arrangement of gills, such as the living slit shells, top shells, turban shells, dog whelks, and the sundial shell, *Architectonica*.

The **mesogastropods** are slightly more advanced snails and form a group containing about 30,000 species. They have reduced their paired gills down to a single unbranched gill; they also tend to have a much narrower shell opening. These include the very high-spired shells such as the turritellids, plus periwinkles, moon snails, slipper shells, and many others, including a number of freshwater snail groups.

The most advanced snails are distinguished by having modified their mantle cavity into a long siphon that serves like a snorkel to allow them to take in fresh seawater while burrowing beneath the surface. These snails are known as **neogastropods**, and their shells all have some kind of distinct siphonal notch and often a long flange where the siphon is protruded out of the shell. They originated in the Early Cretaceous and soon became very abundant in the fossil record, with over 16,000 species alive today. Most of the familiar marine snails, such as cone shells, conchs, cowries, whelks, muricids, olive shells, mud snails, volutes, turrids, auger shells, and nutmeg shells, are neogastropods.

Class Bivalvia (= Pelecypoda, = Lamellibranchia) (Clams, Oysters, Scallops, and Their Relatives)—Clams, scallops, and oysters are familiar to all of us and are the second most common group of living mollusks, with about 8000–15,000 living species and a huge number of fossil species (at least 42,000). They are formally known as Bivalvia because the original molluscan cap-shaped shell of a limpet is modified into two shells or valves that hinge with each other. In many books, you may find them called by an obsolete name, "pelecypods," which means "hatchet foot" in Greek. This describes the way their muscular foot is modified into a long probing appendage that can wedge itself into the sand beneath them and help them dig.

Bivalves have a very peculiar body plan (Fig. A15). From the simple limpet-like ancestor (Fig. A14), they have not only enlarged and hinged their shells over their bodies but even lost their heads. Instead, they fill nearly all the internal volume of the shell with their gills. An even older, obsolete name is the "Lamellibranchia" ("layered gills" in Greek), which refers to how their gills are used for not only breathing but also filter-feeding. As they open their shell and let the seawater flow over the gills, they trap not only oxygen but also tiny food particles, which are then captured by mucus and flow toward the mouth at the base of the gills and then through the digestive tract. The rest of the shell volume is taken up by reproductive organs, excretory organs, and muscles used to close the shell (adductor muscles). Finally, most bivalves have a large muscular foot that can dig down into the sand beneath them. Then the foot bulges at its tip to anchor it as it pulls the shell beneath the sand. The clam also rocks the shell back and forth as it knifes through the soft sediment, speeding up its sinking into the substrate. Many clams also pump water out of the mantle cavity to liquefy the sediment into a slurry like quicksand, further accelerating the rate at which they disappear into the sediment and become less vulnerable to predators.

Most of these soft tissue features, however, are not preserved in the shell. Instead, paleontologists use the shape of the shell, details of the hinge with its teeth and sockets, and other aspects of the shell to identify the species. First of all, bivalve shells are symmetrical *between* the valves (the right valve is the mirror of the left valve), in contrast to brachiopods, whose symmetry is *through* the valves (Fig. A7).

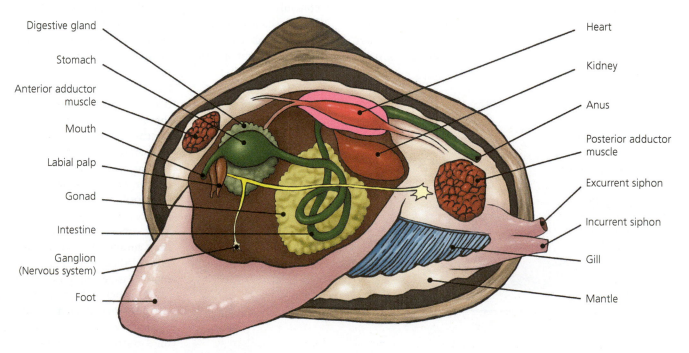

Figure A15 Basic anatomy of a clam, showing the internal organs.

Although bivalves can have many living positions, the standard anatomical orientation used by scientists to describe them is to point the hinge upward and the shorter end of the shell away from your body. When you do this, the right valve will be in your right hand and the left valve will be in your left hand. In this position, the foot and the mouth are on the front edge (anterior) of the shell (pointed away from you), while the long mantle tube called the siphon is at the rear (posterior) of the shell (pointed toward you).

Looking at a single shell on the inside (Fig. A16), there are other useful landmarks that help understand function and aid in identification. Bivalves have stretchy spring-like **ligaments** in the hinge area, which pull the hinge open automatically when their adductor muscles that close the shell relax. This is why clams open automatically when they die and why their shells are usually separated when they fossilize. Where the muscles once pulled on the shell are roughened areas known as adductor muscle scars. Most clams have two adductor muscles, an anterior and a posterior adductor, so there will be two scars on each shell's inside. Scallops are peculiar in that they swim by clapping their valves like castanets, which creates jets of water so that they can swim erratically and escape predators. To do this, they have only one strong column of muscle and only a single muscle scar on each shell. That column of muscle is what you eat when you are served scallops in a restaurant.

Another feature of the inside of the shell is a boundary line dividing the area where the mantle contacted the shell (nearer the hinge) with areas where the mantle did not contact the shell (Fig. A16); this is called the **pallial line**. Typically, on the posterior part of the pallial line there will be an irregular notch called the **pallial sinus**. This is where the siphon protruded through the mantle: the deeper the sinus, the larger the siphon.

On the outside shell, there are also important anatomical features (Fig. A16). Above the hinge is the bulge or hump at the top of the external shell, called the **umbo**, and where the shell spirals to a point is called the **beak**. The **growth** lines spread concentrically from the hinge area, while the **costae** or ribs radiate out from it like spokes in a wheel. Some shells, like scallops, have large corrugations or folds on the shell, known as **plications**. As in brachiopods, the line of the opening of the shell is called the **commissure**. Most clams that burrow must maintain a relatively smooth shell to reduce drag, but oysters and scallops and others that only sit on the bottom and do not burrow may have an irregular outer surface, often with spines (such as the thorny oyster), which makes it harder for a predator to break them in order to feed on them.

The thousands of genera and hundreds of species of bivalves cannot all be discussed in a book like this, but it is easy tell something about the life habit of a bivalve by looking at the shell (Fig. A16). Scallops (subfamily Pectinaceae), for example, have a distinctive shape (familiar as the Shell Oil Company logo). They live on the sea floor with their valves open and a dozen tiny blue eyes on the edge of their mantle. When they detect a threat, they make their escape and jet-propel themselves by flapping their valves together to make them swim in a jerky, irregular movement that confuses most predators.

 See For Yourself:
Scallops Swimming

Oysters (subfamily Ostraceae), on the other hand, sit on the sea floor and give up any attempt to burrow or swim. Their protection is a very thick shell that is hard to open or break, and many oysters live in brackish water that fluctuates between normal marine salinity of 3.5% and freshwater, something that most of their predators can't stand. Mussels (subfamily Mytilaceae) live in rocky tide pools and attach to the rocks with thick, strong fibers called **byssal threads** so that predators have difficulty pulling them off the surface.

Burrowing bivalves come in lots of different shapes and sizes. The most common type are the familiar venus clams (subfamily Veneraceae), some of which are called "quahogs." They are shallow burrowers from the shallow subtidal sea floor, although some species live in the surf zone, and can burrow down out of sight in a matter of seconds. They have relatively short siphons, and this is reflected in the relatively shallow pallial sinus on the shell. Other clams, such as the lucinids (subfamily Lucinaceae) and the tellins (subfamily Tellinaceae), are deeper burrowers, with very long siphons. Some bivalves are very deep burrowers with thick siphons that are so large they cannot close their shell around them but have a permanent gape in the posterior of the shell. These include the soft-shelled clams (subfamily Myaceae), such as the mud-burrowing geoduck (pronounced "gooey-duck")

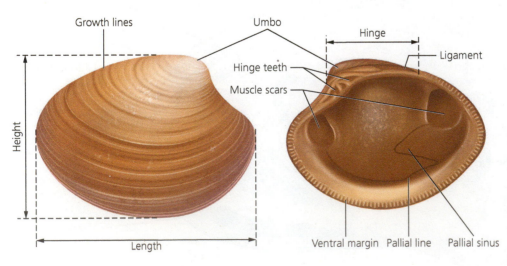

Figure A16 Diagram of the major anatomical landmarks of a bivalve shell.

Panopea, which may have a siphon as long and thick as your arm, and a small shell at the end. Other deep burrowers with long siphons are the razor clams (family Solenidae). A number of clams have adapted for rock boring (family Pholadidae), where their shell is reduced to a pair of cutting blades at the end of a long fleshy body that is protected by their burrow instead of by the shell.

Then there are extreme variations on the basic clam body shape which lose their symmetry when they become oyster-like. For example, the giant clams (family Tridacnidae) have a huge, thick shell that is corrugated but seldom closes completely. Instead, they cover the gape with a thick area of mantle filled with symbiotic algae in their tissues, which help them grow to such giant sizes (just as algae help corals to grow their large reefs of limestone).

There are also a number of interesting extinct oysters that are common in fossil collections. A common Triassic fossil is the triangular clam known as *Trigonia*, whose name means "triangular" in Greek (Fig. 13.12A). Trigoniids survived most of the Mesozoic, then appeared to be extinct until a living descendant, *Neotrigonia*, was discovered in 1802, still living in the South Pacific, a true living fossil. During the Jurassic, one group of oysters grew into a spiral with one valve and a small lid on the other (Fig. 13.13E). Known as *Gryphaea* scientifically, their common name is "Devil's toenails." In the Cretaceous, an even weirder group of oysters evolved with a corkscrew spiral shape that resembles a snail shell in one valve and a lid on the other (Fig. 13.15A). It is known as *Exogyra*, for "spiraling outward" is how its name translates from the Greek.

Another group of Cretaceous clams were the huge, flat inoceramids (Fig. 13.15C, D). Both valves were shaped like shallow plates, and the biggest ones were about 1.5 m (5 feet across). They apparently lay flat on the floor of the shallow waters of the Cretaceous seaways. Many are found with complete fish and other fossils inside them, suggesting they harbored a number of symbiotic organisms that sought shelter inside their huge shells.

But the most bizarre of all bivalves were the Cretaceous reef-building oysters known as rudistids (Fig. 13.15E). One valve was shaped like a large cone, embedded point down into the sea floor. The other valve was shaped like a lid that covered the opening and would open whenever the clams needed to let in currents for food and oxygen. Rudistids formed huge reefs of densely packed shells in the tropical waters of the Cretaceous and even drove corals from the reef region.

Class Cephalopoda (Squid, Octopus, Cuttlefish, Nautiloids, and Ammonoids)—Mollusks have evolved a wide variety of body plans, from the slow-moving snails to the headless clams that burrow beneath the surface and filter-feed to the chitons and tusk shells and the planktonic pteropods. But the champions of mollusks in terms of speed, intelligence, and complex behavior are the cephalopods. Their name "Cephalopoda" means "head foot" in Greek and refers to how the primitive foot of mollusks has

been modified into a ring of tentacles around the head and mouth.

Speed and intelligence are the hallmarks of the group. Squids jet-propel themselves backward through the water faster than most swimming animals. The octopus is legendary not only for intelligence but also for amazing quickness and ability to camouflage with changes in their skin pattern. In fact, many cephalopods can rapidly change their skin patterns and even flash patterns in their skins faster than neon lights, which they use for communication. Most cephalopods also have an amazingly well-developed eyeball that is much like the vertebrate eye, but it evolved independently and converged on the vertebrate eyeball. It has a cornea, lens, and retina; only it is better designed than our eyes. Octopus eyes don't have a blind spot in the retina for the exit of the optic nerve, nor is the sensory layer of the retina covered by blood vessels and other tissues that distort the light, as in vertebrate eyeballs.

Most cephalopods are large, active predators or scavengers, floating above the sea floor in search of prey that they grab with their tentacles armed with suckers. Most of them expel water from the mantle cavity to give them a sort of jet propulsion, especially when they are trying to escape predators. Some, like squid and octopus, even leave a cloud of ink behind them as a smokescreen to cover their escape. In addition to their water jets, squid and cuttlefish can swim slowly forward with the fins along the side of their bodies, and octopus mostly move with their long arms.

Cephalopods range in size from tiny squid only a few inches long to the giant squid, which was about 18 m (60 feet) long. The maximum leg span of an octopus is over 10 m (33 feet). Some ammonite shells were over 1.7 m (5.6 feet) across, suggesting that the soft body was as large as the largest squid today (Fig. 13.13B). Cephalopods live exclusively in marine waters and were unable to colonize the freshwaters or land, as snails and clams did. However, they sometimes live in huge numbers in the ocean, with some squid schools containing over 1000 individuals. Many cephalopods, like the vampire squid, specialize in very deep, dark, cold parts of the ocean from 3000 to 5000 m below the surface.

▷ See For Yourself: The Chambered Nautilus

There are about 1000 living species of squids, octopus, and cuttlefish, which have no external shell, although squids and cuttlefish have a rod of a hard substance (calcite or an organic material) holding their bodies rigid. There is only one group of living cephalopods that still has an external shell, the famous chambered nautilus of the South Pacific (Fig. A17). They are practically the only living analogue for the 17,000 species of extinct shelled cephalopods in the fossil record. There are many different types of extinct nautiloids as well. One extinct group that branched off from the nautiloids is the ammonoids, one of the most successful of all groups in the fossil record. They got their name because their coiled shells resembled the horns of the Egyptian ram god Ammon. In the Middle Ages, some were thought to be petrified snakes ("serpent stones"), and the last part of the shell was sometimes carved with a snake's head to enhance

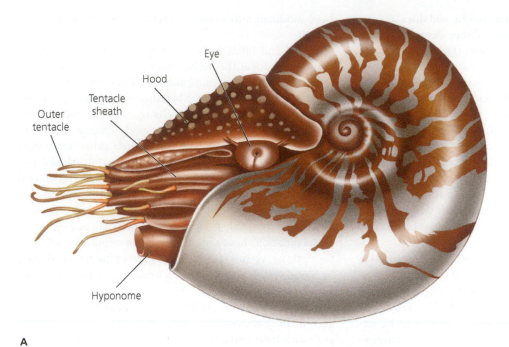

A

B

Figure A17   The chambered nautilus, the only living shelled cephalopod. **A.** Diagram of the external features of the living *Nautilus*. **B.** Cross-sectional diagram of the anatomical features of the interior of the *Nautilus*.

seal off another portion of its shell with a new septum. Once the septum seals the chamber, the nautilus uses a long fleshy stalk from its body called the **siphuncle**, which connects all the chambers, to slowly pull the water out of the new chamber using osmosis, leaving it filled with gases that help the shell float. The blood in the umbilicus is much saltier than the water in the sealed chamber, so the water slowly diffuses into the umbilicus until only gas remains. Contrary to popular myth, nautilus cannot pump water in and out of the old chambers rapidly to change its buoyancy and rise or sink in the depths of the ocean.

Where the edge of the septum meets the outer shell is a line of intersection called the **suture**. These are usually only visible in fossils if the outer layer of shell has been eroded or polished away, exposing the outer edge of the septum (Fig. A17B). The suture pattern is the most diagnostic feature of most cephalopod shells and helps define the different species, genera, and even higher groups. In the primitive nautiloids, it is a simple broad curve. As we shall see when we look at the extinct ammonoids, the suture became more complex and intricate, and this complexity makes it possible to recognize and identify the major groups.

The rest of the body of the nautilus resembles the unshelled squid and octopus in shape (Fig. A17). The large head is covered with a hood and has big eyes that dominate the body. The nautilus has a parrot-like beak in the mouth, surrounded by a ring of tentacles, which it uses to catch prey (mostly crustaceans and carrion). Beneath the head is the mantle cavity, where the gills lie, and the entrance to the mantle cavity is a nozzle called the **hyponome** that funnels and focuses the jet of water to propel them. As the body squeezes down on the mantle cavity, it forces the water out the hyponome and creates a jet of water. Like octopus,

its serpentine appearance and market value. By 1700, the English naturalist Robert Hooke, the father of microscopy, saw the newly discovered chambered nautilus and correctly inferred that ammonites were related to nautilus, not snakes.

The nautilus shell is coiled in a flat spiral, like the snails that coil in a plane. However, unlike snails, the shell interior is divided into chambers by a series of walls called **septa** (Fig. A17B). The mollusk itself lives in only the last chamber. As its shell grows, its body moves forward until it can

however, nautilus uses jet propulsion only for rapid motion, especially when escaping. Most of the time, it uses its tentacles to creep along the sea floor in any direction, especially when hunting.

Subclass Nautiloidea: This group includes not only the living chambered nautilus but also a variety of Paleozoic relatives with long, straight shells. They first appeared in the Cambrian, but by the Ordovician they were very abundant in nearly every marine assemblage. Some got to be huge (Fig. 10.16) and became the largest animals in the ocean. The biggest was *Cameroceras*, with some shells reaching about 11 m (36 feet) in length, and the squid-like creature that once lived in that shell must have been as large as the living giant squid. Slightly smaller is the common fossil *Endoceras*, which was 3.5 m (11.5 feet) in length. Most, however, were much smaller, typically only a few centimeters to a meter long.

Straight-shelled (orthocone) nautiloids were superpredators of the Ordovician and Silurian that ate almost anything they could catch. Their large, heavy shells were counterweighted with calcite deposits inside the shell to serve as ballast, so they floated horizontally above the sea bottom. Otherwise, the gases in the chambers would make the shell float point upward. They could not move fast to chase prey but must have ambushed prey that crept within reach of the tentacles. As the first large marine superpredators in the oceans, they may have influenced how trilobites became more specialized for avoiding their predation in the Ordovician (see Chapter 10).

Different groups of nautiloids are distinguished by where they deposited calcite in their shell as ballast to counterweight their shell and keep it horizontal. The Endoceratoidea filled their shell with cone-shaped deposits of calcite called **endocones**, which were nested one inside the other like cups in the dispenser next to a water cooler. They are most common in Ordovician beds around the world and vanished in the Silurian. Another Ordovician group was the Actinoceratoidea, which are recognized by calcite deposits filling the tube around the siphuncle (**endosiphuncular deposits**). Other groups had deposits of calcite within the chambers (**endocameral deposits**). Usually, the specimen must be sliced open or eroded or broken naturally to reveal these internal features. Straight-shelled nautiloids reached their climax in the Devonian, and specimens from the Silurian and Devonian of the Tindouf Basin of Morocco are so abundant that they are commercially mined in large quantities and sold all over the world.

Nautiloids then declined during the rest of the Paleozoic, and the large, straight-shelled orthocone nautiloids vanished altogether. However, the primitive coiled forms persisted through the Mesozoic in small numbers and actually survived the Cretaceous extinction event much better than their relatives, the ammonites, which died out completely. Rarely, a shell of a coiled nautiloid like *Aturia* is found in Cenozoic beds all around the world, so they continued to survive despite losing the world's oceans to bigger predators, like fish.

Subclass Ammonoidea: During the Devonian, one group of straight-shelled nautiloids (the bactritoids) is thought to have coiled up into a spiral in a flat plane and given rise to the next big radiation of cephalopods, the Ammonoidea. Ammonoids are distinct from nautiloids in that the siphuncle penetrates the septal wall on the outer rim of the shell (the **venter**), rather than in the middle of the septum, as in nautiloids. From the Devonian through the Permian, there was a great evolutionary radiation of the most primitive ammonoids. They are known as **goniatites** and can be recognized by their distinctive zigzag suture pattern on specimens where the outer shell is removed (Fig. 11.12H). They are excellent index fossils for the later Paleozoic, and in some localities (such as the Devonian beds of the Tindouf Basin of Morocco), the goniatite *Geisenoceras* (Fig. 11.12H) occur in huge numbers and are commercially mined on a large scale and sold by shops and dealers all over the world.

During the great Permian extinction event, all but two lineages of goniatitic ammonoids were wiped out. These survivors then were the ancestors of a new diversification of ammonoids during the Triassic. These ammonoids had a distinctive **ceratitic** suture, which typically had a series of "U"-shaped curves in it (Fig. 13.12B). At their peak, there were about 80 families and over 500 genera of ceratitic ammonoids. The ceratitic ammonoids went through another extinction crisis at the Triassic–Jurassic boundary.

The survivors of this extinction underwent a final huge evolutionary radiation in the Jurassic and the Cretaceous. By the Jurassic, there were 90 different families, and even in the Cretaceous there were still 85 families, although they declined to only 11 families near the end of the Cretaceous. These ammonoids had a complex, intricate, florid suture pattern (Fig. 13.15F), known as an **ammonitic suture**. The details of their complex sutures allow the paleontologist to identify any good specimen to species.

During this great Mesozoic diversification, ammonites evolved into many shapes. Some were streamlined into a flat disk with sharp edges (the **oxyconic** shells of *Placenticeras* and related genera—Fig. 13.15F) and must have been relatively fast swimmers (although none were as fast as squids or fish). Others had short, fat **cadicone** shells and must have moved slowly along the bottom to catch their prey. Some had coils that resembled a snake (**serpenticone**), as seen in genera like *Dactylioceras* and *Stephanoceras*.

The most unusual ammonites, however, began to uncoil the simple flat spiral into a number of weird and bizarre forms. They are known as **heteromorph** ("different shape" in Greek) ammonites (Fig. 13.15G–L). Some were just slightly uncoiled, like *Scaphites*. Others uncoiled into huge hairpin shapes (*Hamites* and *Diplomoceras*) (Fig. 13.15I), weird question-mark shapes (*Macroscaphites* and *Didymoceras*—Fig.13.15G), uncoiled flat spirals (*Spiroceras*), and even stranger shapes. The oddest of all were those that coiled in a spiral axis like a snail (*Turrilites*—13.15H) or into a tight knot (*Nipponites*—13.15J). Ammonites like these often had their heads in almost inaccessible positions and could not use jet propulsion, or they would start to spin crazily. So they must have floated passively in the ocean and trapped food nearby with their long tentacles, or maybe they crept slowly along the sea floor using their tentacles. Perhaps the most common and rapidly evolving heteromorphs were the baculitids (Fig. 13.15K), which had secondarily straightened out their originally coiled shell into a nearly straight cone (with just a tiny coil remaining at the tip). Unlike the straight-shelled nautiloids with the counterweight ballast in their shells to hold them horizontal, baculitids had no ballast deposits inside them, so their shell floated point upward, and their heads dangled below. Clearly, they could not swim fast in this position, so they must have either crept along the sea floor using their tentacles or (as recent research suggests) floated in the open waters like plankton, trapping nearby prey (possibly even smaller plankton) with their tentacles.

After surviving the Permian extinction, the Triassic–Jurassic extinction, and several smaller crises, all of this great diversity of ammonites that had long dominated the Mesozoic seas (and make the best fossils for telling time) vanished with the great Cretaceous extinction. Only the coiled nautiloids and the cephalopods without external shells survived.

Subclass Coleoidea: The coleoids include most of the cephalopods without external shells: squids, octopus, cuttlefish, argonauts, and their relatives. Without a hard external shell, they do not fossilize very well, so only a few complete specimens with soft tissues are known. However, there is one type of coleoid that fossilizes well. Known as **belemnites**, they left a conical internal shell (Fig. 13.15L), called a guard or rostrum, which reinforced their squid-like body and counterbalanced the weight of the head and tentacles. (By contrast, living squid have a lightweight, thin, flexible rod that serves to stiffen the back of the body.)

Belemnite fossils look like large stone bullets in shape, and they are particularly common in the Jurassic and Cretaceous, where some beds concentrate hundreds of belemnite fossils. A few belemnites have been found in extraordinary localities that preserve the soft tissue, so we know these fossils came from a squid-like creature that is now extinct.

Subphylum Deuterostomata (Echinoderms, Hemichordates, and Chordates)—Many people are startled to find out that of all the living marine invertebrate groups, echinoderms are closest relatives of backboned animals or

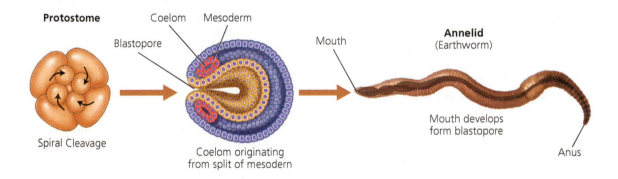

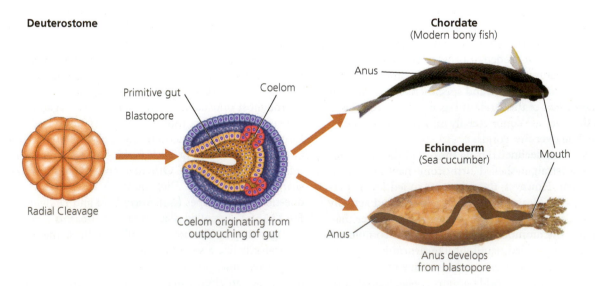

Figure A18   The contrast between the development of deuterostomes and protostomes.

vertebrates, including fish and ourselves. This is shown by many lines of evidence. For example, echinoderms and primitive marine relatives of vertebrates have the same kind of larva, known as a **tornaria** larva. When this larva develops from a fertilized egg into a ball of cells with an opening at one end, that opening becomes the mouth of most invertebrates—but in echinoderms and vertebrates, that opening becomes the anus and another opening develops to become the mouth (Fig. A18). The cells of most invertebrates are determinate, so early in their development they are specialized and cannot change function. However, the cells in an embryo of an echinoderm or vertebrate are indeterminate, and they can be split and develop into a different organ system, or even a complete new animal. The coelomic cavity of most invertebrates comes from the mesodermal wall, but in echinoderms and vertebrates it comes from the endoderm. Finally, analyses of the molecular sequences of echinoderms and other animals have shown over and over again that they are our closest relatives.

Phylum Echinodermata (Sea Stars, Sea Urchins, and Their Relatives)

The third largest phylum in the ocean after mollusks and arthropods is Echinodermata, or the sea stars, sea urchins, brittle stars, sea cucumbers, and crinoids or "sea lilies," and their extinct relatives. Unlike the other two groups, however, echinoderms live only in normal marine waters and have not adapted to freshwater or life on land or in the air. This is because they have no system for respiration or osmotic regulation, so they cannot survive long without normal seawater surrounding them. Nevertheless, they have long been one of the most diverse and successful groups in the ocean. In the right circumstances, they can be incredibly abundant. Sea urchins are usually common in tide pools, along with sea stars. During the late Paleozoic, the sea floor was covered by huge meadows of crinoids and another extinct group called blastoids, representing many trillions of individual animals. The deepest part of the abyssal plain of the oceans, deeper than about1000 m (3300 feet), is dominated by echinoderms as well, especially sea cucumbers and brittle stars, where they make up about 90% of the marine life.

▷ See For Yourself: The Shape of Life: Echinoderms

Even though they are relatively advanced among the invertebrate phyla, echinoderms are truly strange and alien in their body form and construction. First of all, they start as embryos and larvae with bilateral symmetry, but in their adult stages they give up the head and tail and become radially symmetrical (symmetrical around a central axis) like a sea jelly or a sea anemone. A few echinoderms, like heart urchins and sand dollars, even superimpose some bilateral symmetry on top of this radial pattern. Not only are their larvae bilateral in symmetry, but the earliest echinoderms from the Cambrian were bilaterally symmetrical as adults, so the radial symmetry is a later development.

Second, echinoderms are the only group of animals that works by hydraulics, like the fluid control systems in large machinery or the brakes on a truck or car. Hydraulic systems are built around an enclosed tube of fluid; they use this to transmit pressure from one area to another steadily over a long time. Likewise, the echinoderm body has a series of fluid-filled canals that are used to move them across the sea floor using **tube feet**, tiny fluid-filled extensions of the canal system with suckers at the end. The hydraulic system means they move very slowly by our standards, but it also confers advantages. For example, a sea star feeds by wrapping its arms around a clam or mussel and then slowly pulling its shell open. The prey tries to hold its shell closed with its adductor muscles, but these get tired eventually. But the hydraulic system never gets tired, so they can pull steadily and unrelentingly until the prey opens. At that time the sea star turns its stomach inside out, inserts it inside the shell of the prey, and digests its victim in its own shell. The tube feet and hydraulics also give a sea star incredible grabbing power on rocks, so it is almost impossible to pull them off or dislodge them.

With their loss of bilateral symmetry and their hydraulic bodies, echinoderms are truly weird in having lost many other systems we think are normal in higher animals. Echinoderms have a nervous system but no eyes, hearing, taste, or other specialized senses; they sense things by touch. The hydraulic system serves multiple functions, so echinoderms have no circulatory, respiratory, or excretory system. This limits the echinoderms in important ways. Without an excretory system to regulate their water and salt balance, they must they live in normal ocean water where salinity doesn't change much; there are no freshwater echinoderms. In addition, they don't vary in body size as much as do mollusks or arthropods. The smallest echinoderms are at least 1 cm across, so there are no tiny planktonic forms (although echinoderm larvae are planktonic). There are a few sea stars up to 1 m in diameter, and sea cucumbers up to 2 m long but none larger. Some extinct crinoids had arms that flared out over 1 m and stems that were tens of meters in length, but they are nowhere near as bulky or massive as most giant marine animals.

Classes of Echinodermata—The name "Echinodermata" means "spiny skinned" in Greek, and this name describes many of the groups. Most of the groups (especially sea urchins and sea stars) have spines that protrude from their shell or skin for protection. Echinoderms are built of a shell (**test**) composed of plates (**ossicles**) of single crystals of the mineral calcite, so if you break an echinoderm ossicle, it will break along the cleavage planes of the calcite crystal. In most cases, they have a mouth at one end of the body, often surrounded by arms or tube feet, and the anus at the other end.

There are five living classes of echinoderms.

1. Class Asteroidea (sea stars) are the most familiar of echinoderms to most people. They are often called "starfish," but they are not fish in any biological sense, so biologists now call them "sea stars." Sea stars are usually predators on other shelled marine animals and feed

by inserting their stomach inside the victim's shell and digesting it. They have five arms (or arms in multiples of five) radiating around their mouth on the bottom and thousands of tube feet in their arms that give them a powerful grip on the rocks or on their prey. Their ossicles are small and embedded in their soft tissues, so when they die their bodies tend to fall apart. Even though there are about 1500 living species in about 430 genera today, they are rare in the fossil record.

2. Class Ophiuroidea (brittle stars) look much like sea stars, except their five arms are thin and delicate (hence the "brittle" in their name), and their body is a simple disk in the middle. The slender arms are not used for gripping, but instead they allow the brittle star to glide rapidly across the sea floor. Once they are in a good feeding place, they wave their arms in the current to trap tiny food particles. Even though there are about 2000 living species in 325 genera, they are so delicate that they are rarely fossilized, except in extraordinary circumstances.

3. Class Holothuroidea (sea cucumbers) are different from almost all other living echinoderms in that they have a long, fleshy body shaped a bit like a cucumber, with a mouth at one end and an anus at the other, and they have only tiny ossicles embedded in their body. However, tube feet run along the length of the body and help them move along the sea floor. Tube feet around the mouth help them shovel the organic-rich sea-floor sediment into their mouth, where they then digest all the food material out of it and pass it out their anus. They are found in many parts of the ocean, although they are by far the most abundant animals on the deep-sea floor where there is abundant organic detritus and there are few predators.

 See For Yourself: True Facts about the Sea Cucumber However, if a predator does disturb them, they have a truly bizarre defense mechanism. They extrude their internal organs (intestines, respiratory apparatus, and other systems) out their anus at the predator. When they hit the seawater, these organs make a set of sticky threads that cover the intruder in nasty goo. After this happens, the sea cucumber can slowly regenerate an entirely new set of internal organs.

Because they are made entirely of soft tissue, sea cucumbers rarely fossilize, even though they are very successful today, with 1150 species in about 200 genera.

4. Class Echinoidea (sea urchins, heart urchins, and sand dollars) have perhaps the best fossil record of any echinoderm class since they have a hard external shell and many kinds already live in the sea-floor sediment where they can become fossilized.

Of all the echinoids, sea urchins have the most primitive body form (Fig. A19). Their shell (or test) has five rows of defensive spines in its **interambulacral areas**, alternating with five areas with rows of tube feet (in the perforated **ambulacral areas**). Their long tube feet are used to creep along the surface and move detritus off

their shells. The mouth opening is at the bottom, and they have a set of five calcareous plates called **Aristotle's lantern** that serve as jaws to scrape algae off rocks or feed on strands of kelp. The digestive tract then loops around the inside of the shell and out the top, where the anus is located. The rest of the shell is filled with gonads (urchin roe, a delicacy in some places) and other internal organs, including their water vascular system. Sea urchins live in rocky tide pools, where they cruise around during high tide scraping algae. During low tide, they hide in small pockets in the rock that they have carved with their jaws.

By the Jurassic, the radially symmetrical sea urchins gave rise to other branches of echinoids that specialized in burrowing in the sediment, rather than hiding in rocky tide pools. These urchins must burrow in a definite direction, so they have secondarily developed bilateral symmetry overprinting their radial symmetry. They have a front end and a back end and right and left sides and are known as **irregular echinoids**. In most cases, this means that the mouth moves from the center of the bottom of the test to the front of the test and the anus from the top of the test to the rear. With the protection of a burrow, large defensive spines are no longer needed, so their spines have reduced until they resemble a fuzzy coating; the spines are used to help them dig through the sediment. They no longer need jaws since they ingest marine sediment and digest out all the food material from it before passing it through their gut and out the anus.

Irregular echinoids have many different shapes. The first to evolve in the Jurassic were the heart urchins and sea biscuits, which have a tall, rounded shell. These burrow into the sediment and use long tube feet to create a small respiratory canal to the surface of the sediment to bring them fresh seawater. Heart urchins were especially common in the Cretaceous, when they lived in a variety of sedimentary settings. The most extremely modified are the sand dollars, which have gone for a flattened body that helps them burrow just beneath the surface of the sediment. When they are feeding, however, their tests stick out of the seafloor like a set of shingles, trapping currents on their underside and allowing the tube feet to convey the food particles to their mouth. The first sand dollars appeared in the Eocene from highly flattened Paleocene sea biscuits and soon spread to shallow marine sea floors worldwide. They are abundantly fossilized in many parts of the world and indicate a shallow marine setting with strong currents.

5. Class Crinoidea ("sea lilies") was one of the most important groups of echinoderms in the Paleozoic but nearly vanished during the Permian extinction. Only a few groups survived to the present day (Fig. A20). Their bodies have a head with arms on a long stalk rooted into the sea-floor sediment, so they vaguely resemble a flower or a tree (hence the name "sea lily"). However plant-like they may look, they are most definitely

animals, related to sea stars and other echinoderms. If you do a search for a video of "crawling crinoids," there are several videos online that show just how animal-like they really are.

 See For Yourself: A Living Crinoid in Motion

Today, only 25 living genera of stalked crinoids survive (Fig. 10.15J), but there are about 6000 species in 850 genera in the fossil record. These few surviving stalked crinoids are a tiny remnant of their former diversity, but they are all we can study in order to understand the huge diversity of extinct crinoids. Most of them live in highly protected or sheltered habitats where there are few predators, and it is likely that the evolution of new predators in the Triassic explains why stalked crinoids never recovered the diversity or abundance they had in the Paleozoic. Yet enormous volumes of crinoidal limestone made entirely of broken crinoid ossicles demonstrate that during times like the Mississippian they lived by the trillions on the shallow sea floor across the entire continent.

Most of the living crinoids are a specialized form that has lost its stalk. They can crawl around on the sea floor or even swim short distances. They are called "feather stars" since their arms are feathery in appearance, and they vaguely resemble brittle stars and sea stars. About 700 species of these free-living crinoids swim in the oceans today.

The basic body (Fig. 10.15J, A20) of a stalked crinoid consists of a "head," or **calyx** (filled with all the internal organs), topped by the mouth and surrounded by the arms around its perimeter. The arms are covered with smaller branches called pinnules, and they formed a big filter-feeding fan that trapped food particles passing through them in the ocean currents. When the crinoid is feeding, the arms curl backward on themselves into a concave fan, like an umbrella, with the concave side facing the current, so they trap as much food as possible passing through their arms. Trapped food is then passed by sticky mucus down the arms and into the mouth. There a short digestive tract processes the food and passes it to the anus near the mouth.

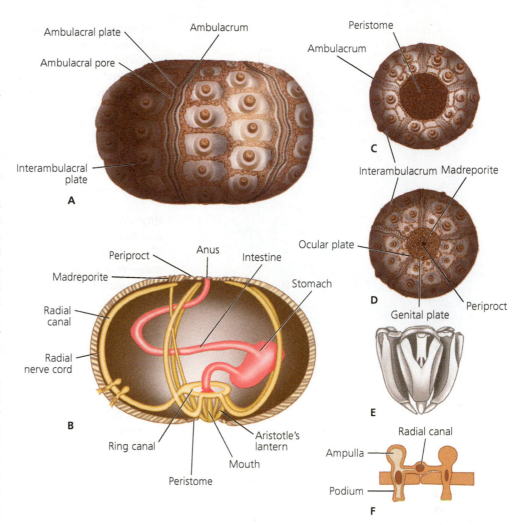

Figure A19 The anatomy of the external features of a sea urchin shell.

The calyx has a distinctive arrangement of plates (Fig. A20), which crinoid specialists use to identify the species and higher groups of crinoids. For the nonspecialist, it is difficult to identify the calyx of any crinoid species without learning the terminology of the plates. At the end of the calyx opposite the mouth is the stalk of the crinoid, made of a set of rings of calcite (**columnals**) that resemble a stack of Lifesavers candies (Fig. 12.3A). The stalk is held together by soft tissue in life, but when the crinoid died, the stalks usually fell apart. Huge volumes of limestone are made almost entirely of millions of broken crinoid columnals, the most easily recognized of any part of a crinoid. At the bottom of the stalk, most crinoids had a root-like set of structures called a **holdfast**, which anchored the crinoid in the sediment (Fig. A20). In fact, all of these parts disarticulated very easily when a crinoid died, so most fossils consist only of a calyx or columnals or a holdfast. Only in a few extraordinary localities is it common to find complete crinoids. Most crinoid specialists need only a decent calyx to identify the species since the stem and holdfast are seldom attached.

Crinoids were enormously successful during the Paleozoic, making up most of the marine biomass in many

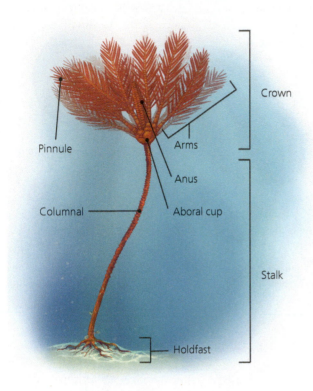

Pinnule

Crown

Arms

Anus

Columnal

Aboral cup

Stalk

Holdfast

Figure A20   The basic anatomical features of a stalked crinoid.

shallow-marine settings, especially during the Mississippian, which has been nicknamed "the age of crinoids." However, as the seas retreated and the planet became colder in the Pennsylvanian and Permian, crinoids began to diminish as well, and they were severely impacted by the great Permian extinction. Only two or three lineages survived into the Triassic, and they never regained their former dominance, probably due to new predators in the seas that could eat defenseless crinoids. Today, crinoids have adapted by being mobile (like the stalkless crinoids) or by living in hidden, protected areas with few predators (like the few remaining stalked crinoids).

Other Extinct Classes: There are as many as 20–25 other extinct classes of echinoderms, most of which are very rare and seldom encountered by most fossil collectors. They will not be discussed in a book like this and are usually only collected and studied by specialists. However, one extinct class is very common (especially in the Mississippian) and deserves mention.

Class Blastoidea (blastoids) looked much like crinoids in that they had a "head" (called a **theca**) with arms, found at the end of a long stalk like Paleozoic crinoids (Fig. 12.13A, B). However, the theca was constructed very differently from the calyx in crinoids. In most cases, the fivefold symmetry of the theca resulted in five ambulacral areas where the tiny, thin tube feet and arms (usually lost in the fossils) once protruded. Inside the theca behind the ambulacral areas was a highly folded pleated structure known as the

hydrospire, which apparently served for respiration. Some of the genera, like the ubiquitous Mississippian index fossil *Pentremites* (Fig. 12.13A), looked a bit like flower buds. Others were spherical in shape, like *Orbitremites*. Blastoids first appeared in the Ordovician but had their greatest diversity in the Mississippian, when they mingled with the immense shoals and meadows of crinoids. They declined again in the later Paleozoic and vanished completely in the great Permian extinction, along with trilobites, tabulate and rugose corals, and many groups of bryozoans, brachiopods, and crinoids that had dominated the Paleozoic.

Phylum Hemichordata (Graptolites and Pterobranchs and Acorn Worms)

The branch of animal life that includes backboned animals (vertebrates, such as fish, amphibians, reptiles, birds, and mammals) has many close relatives. They have some features of vertebrates but do not have key adaptations such as bone tissue or a spinal cord made of bony vertebrae. Nevertheless, they have other key adaptations, such as the presence of a distinct throat region (**pharynx**), and in most of them a flexible rod of cartilage along the backbone, known as the **notochord**. Nearly all vertebrate embryos (including you when you were an embryo or fetus) have a notochord that is later replaced by the bony spinal column. These non-vertebrate animals with a notochord are part of the same phylum, Chordata, to which all vertebrates belong. Some of these chordate relatives include the tiny jawless fish-like group known as lancelets and the tiny filter-feeding creatures known as tunicates (which have a tadpole-like larva with a notochord). None of these non-vertebrate chordates have much of a fossil record, so we will not consider them further here. However, they are important to understanding the transition from invertebrates to vertebrates.

The next closest relative to phylum Chordata is another group of soft-bodied marine creatures known as phylum Hemichordata, whose name means "half-chordates." The living hemichordates include a funny-shaped worm-like creature known as an acorn worm (which has a pharynx and other chordate features, unlike any living worm) and filter-feeding creatures known as pterobranchs. Normally, these would only be the subject of an invertebrate zoology textbook. However, they turn out to be the crucial link in solving a persistent mystery in paleontology.

Since the late 1700s, when geology was first becoming a science, scholars have puzzled over strange-looking fossils found flattened on the surfaces of black deep-water shales. They were preserved only as two-dimensional carbonaceous films on the bedding surface, so it was hard to visualize them in three dimensions (Fig. 10.20). Only a film of graphite remained, without internal structures, so it looked almost as if someone had marked the rocks with a graphite pencil. Hence, they were called "graptolites," a name that means "written on stone" in Greek. Many ideas were proposed as to what animals they might be related to, but there was little evidence to resolve the mystery.

Despite having no idea what kind of biological creature they belonged to, geologists could still use them for biostratigraphy. Graptolites evolved rapidly and were abundant not only in deep-water shales without many other fossils but also in shallow-marine rocks. Soon they became excellent index fossils for the Ordovician, Silurian, and much of the Devonian. Graptolites were so useful for biostratigraphy that they helped solve the 50-year argument over the boundary between the Cambrian and Silurian in the British Isles. The solution came from looking at the evolution of graptolites and led to the proposal of a time period between the two disputed periods named the Ordovician. More importantly, graptolites apparently floated around the world's oceans in the early Paleozoic, so it was possible to date any Ordovician, Silurian, or Devonian rock around the world using them.

The answer to the mystery of what graptolites were did not come until 1948, when paleontologists found uncrushed three-dimensional graptolite fossils that were preserved in limestones and cherts, so they had not been flattened and altered. By carefully slicing these rocks into thousands of tiny slices (or later by etching them in acid), the three-dimensional structure and anatomical details were finally revealed (Fig. A21). Their detailed anatomy with rows of tiny cups (**thecae**) on a long branch (**stipe**) exactly matched the structure of the living pterobranchs still floating in the world's ocean. For this reason, graptolites are now considered to be a group of extinct hemichordates. If the extinct creatures lived anything like their modern relatives,

then each little cup-like theca housed a tiny filter-feeding creature with a fan of tentacles near the mouth for trapping plankton, which then go through their pharynx and digestive tract. Modern pterobranch colonies (called **rhabdosomes**) apparently supported hundreds of individual animals on each stipe, connected to each other down the center by a thread of tissue known as a **stolon**. Likewise, many graptolites had hundreds of thecae and made up huge colonies that once floated across the world's oceans in the early Paleozoic. Some were apparently dangling down from driftwood, while others are preserved with some sort of bubble-like float that kept them hanging down from the ocean surface (Fig. 10.21B).

Their habit of floating on the ocean surface explains why they are found in any kind of marine rock from deep-water shales to shallow-marine limestones. They didn't live on the sea bottom but floated above it everywhere, so when they died and floated to the bottom, they were preserved. Deep-water shales were often formed in quiet, oxygen-poor waters with no other fossils and no scavengers, so only graptolites sinking down from above are preserved there.

Once you look closely at the graptolite fossils, most of them are pretty easy to identify. The earliest forms from the Late Cambrian were bushy branching fossils known as "dendroids" or the genus *Dictyonema* (Fig. 10.21B). This form persisted until the end of the Paleozoic, and some were attached to the hard surfaces on the sea floor. However, by the Ordovician, graptolites had become open-ocean drifters and began to simplify their structures into a few stipes on each rhabdosome. Typical Ordovician fossils include *Diplograptus* and *Didymograptus* (Fig. 10.21A), whose name describes the double rows of thecae on both sides of the stipe (*diplo-* and *didymo-* both mean "double" in Greek). Another Ordovician fossil, *Phyllograptus*, had four leaf-shaped stipes that looked like the vanes on the back end of a dart (*phyllo-* means "leaf" in Greek).

By the Late Ordovician and especially in the Silurian, the most common graptolites are even more simplified. They are called *Monograptus* because they have just a single row of cups on each stipe (*mono* means "single" in Greek). Monograptids diversified into many different shapes in the Silurian, including corkscrew-like rhabdosomes and structures with a large fan of individual stipes flaring out. They are so useful for telling time that they define not only the stages of the Silurian worldwide but the appearance of *Monograptus uniformis*, defines the Silurian–Devonian boundary. By the Middle Devonian, most graptolites had vanished, leaving only the bushy dendroid forms like *Dictyonema* to survive into the Carboniferous, when they vanished forever. We do not know for sure why they went extinct, although the Devonian was a time of a great radiation of jawed fishes, which would have found the tiny branching creatures floating on the surface easy prey.

Graptolites may not be as spectacular as dinosaurs or trilobites, but they are very important in telling time in the early Paleozoic and reconstructing the positions of ancient continents and oceans.

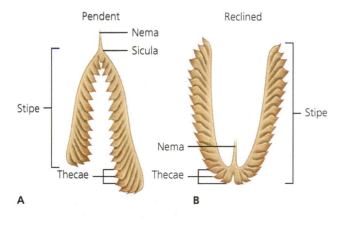

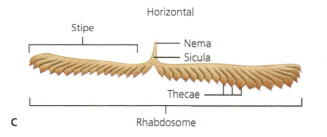

Figure A21   Detailed anatomy of graptolites.

Phylum Chordata (Vertebrates and Their Relatives)

Deuterostomes have a stiffened rod of cartilage, called a notochord, along the back, just above a hollow nerve cord. In some groups, this embryonic notochord is replaced by a bony backbone. Hard parts are made of cartilage, replaced by bone in most groups. Most chordates have gill slits and a pharynx for feeding as embryos, and these are retained in the fish-like groups but lost in the tetrapods (Fig. A21).

Note: Below the phylum level, there are many branching points and subgroups of the chordates. The old ranks of "class Pisces" for fish, "class Amphibia," "class Reptilia," "class Aves" (for birds), and "class Mammalia" as five separate groups of equal rank are no longer used because fish are not a natural group and birds belong within the Reptilia since they are descended from dinosaurs like *Velociraptor*. Instead, the groups are listed below with indications to show how each group is nested within the larger group. The relationships are shown clearly in Fig. A22.

Urochordata (Tunicates, or "Sea Squirts")—Soft-bodied colonial organisms use their large pharynx for filter-feeding. As larvae, they have a tail with segmented muscles, a dorsal

 See For Yourself: Shape of Life: Chordates

hollow nerve cord, and a notochord. These are lost when they attach and metamorphose into sessile adults.

Cephalochordata (Amphioxus, or the Lancelet)—Elongate fish-like animals with a long notochord, segmented muscles, and a dorsal hollow nerve cord. They have tentacles around their mouths for filter-feeding when they lie half buried with their heads protruding from sand.

Craniata—Chordates with a highly specialized head region including a distinct brain with specialized hearing, seeing, and smelling organs and semicircular canals in the ear for balance. They also have paired kidneys and an excretory system, a well-developed heart, and many other unique specializations.

Myxinoidea (Hagfishes): Slimy eel-like marine chordates that burrow in the sea floor and eat polychaete worms or feed on the internal organs of dead fish.

Vertebrata: Chordates that develop phosphatic bone from the embryonic skin layer, or dermis. They have many other

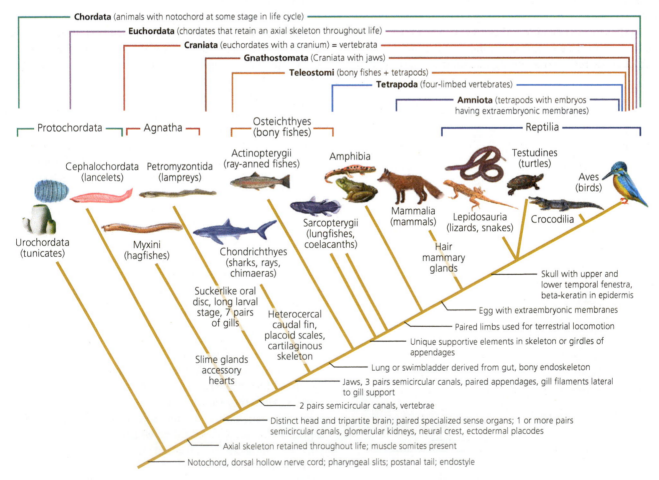

Figure A22 Family tree of the vertebrates, showing the major groups now recognized by their common ancestry and the unique evolutionary specializations that define each group (name given on the top brackets, corresponding to the anatomical specializations shown at the branching points of each group).

unique evolutionary specializations. A great number of jawless fish fall at this point in the diagram, along with the living lamprey, a jawless eel-like parasite that sucks the blood and viscera of its victim.

Gnathostomata (Jawed Vertebrates): Vertebrates with true jaws formed from their cartilaginous gill arches. They have many other unique features, including teeth containing dentin, a specialized gill apparatus, three semicircular canals for balance, pectoral and pelvic girdles in their bodies to support their fins, and dozens more specializations.

Chondrichthyes (Sharks, Skates, Rays, Ratfish, and Their Relatives)—Cartilaginous fish with bony spines in their skins, distinctive teeth, and a complex fin support system.

Osteichthyes (Bony Fish)—Vertebrates with bone replacing their internal cartilaginous skeletons and many other unique specializations.

Actinopterygii (Ray-Finned Fish)—Bony fish with a series of fin rays supporting their fins. They include a number of archaic forms such as the sturgeon, garfish, and bowfin, plus the great diversity of modern bony fish known as teleosts, which make up 90% of living fish.

Sarcopterygii (Lobe-Finned Fish and Their Descendants)—Vertebrates with their muscular fins or limbs supported by simple bony structures, complex enamel in the teeth, and many unique features. Most have the capability to breathe air with fully developed lungs. The lobe-fins include the coelacanth, the lungfishes, and advanced groups related to tetrapods.

Tetrapoda (Four-Legged Vertebrates)—Vertebrates with forelimbs built of a single bone near the body and a pair of bones in the outer limb segment. Their shoulder and hip joints are attached to the spinal column for support, and they have many other advanced features related to life on land.

Amphibia (Frogs, Salamanders, Apodans, and Their Extinct Kin)—The oldest terrestrial vertebrates. These include not only living frogs and salamanders but also the various extinct amphibians known as the temnospondyls, large flat-skulled creatures, common in the late Paleozoic and Triassic. The relationships of other extinct groups, including the lizard-like microsaurs, the eel-like aistopods, and the boomerang-headed nectridians, are less certain. The more advanced anthracosaurs, however, are much closer to tetrapods than they are to amphibians.

Mammalia (Mammals)—Amniotes with mammary glands in females and hair.

Reptilia (Turtles, Snakes, Lizards, Crocodilians, Dinosaurs, Birds)—Amniotes characterized by a number of unique features in their skull bones.

Appendix B

SI and Customary Units and Their Conversions

This appendix provides a table of units and their conversion from older units to Standard International (SI) units.

Length

Metric Measure

1 kilometer (km)	= 1000 meters (m)
1 meter (m)	= 100 centimeters (cm)
1 centimeter (cm)	= 10 millimeters (mm)

Nonmetric Measure

1 mile (mi)	= 280 feet (ft)
	= 1760 yards (yd)
1 yard (yd)	= 3 feet (ft)
1 foot (ft)	= 12 inches (in)
1 fathom (fath)	= 6 feet (ft)

Conversions

1 kilometer (km)	= 0.6214 mile (mi)
1 meter (m)	= 3.281 feet (ft)
	= 1.094 yards (yd)
1 centimeter (cm)	= 0.3937 inch (in)
1 millimeter (mm)	= 0.0394 inch (in)
1 mile (mi)	= 1.609 kilometers (km)
1 foot (ft)	= 0.3048 meter (m)
1 inch (in)	= 2.54 centimeters (cm)
	= 25.4 millimeters (mm)

Area

Metric Measure

1 square kilometer (km^2)	= 1,000,000 square meters (m^2)
	= 100 hectares (ha)
1 square meter (m^2)	= 10,000 square centimeters (cm^2)
1 hectare (ha)	= 10,000 square meters (m^2)

Nonmetric Measure

1 square mile (mi^2)	= 640 acres (ac)
1 acre (ac)	= 4840 square yards (yd^2)
1 square foot (ft^2)	= 144 square inches (in^2)

Conversions

1 square kilometer (km^2)	= 0.386 square mile (mi^2)
1 hectare (ha)	= 2.471 acres (ac)
1 square meter (m^2)	= 10.764 square feet (ft^2)
	= 1.196 square yards (yd^2)
1 square centimeter (cm^2)	= 0.155 square inch (in^2)
1 square mile (mi^2)	= 2.59 square kilometers (km^2)
1 acre (ac)	= 0.4047 hectare (ha)
1 square foot (ft^2)	= 0.0929 square meter (m^2)
1 square inch (in^2)	= 6.4516 square centimeters (cm^2)

Volume

Metric Measure

1 cubic meter (m^3)	= 1,000,000 cubic centimeters (cm^3)
1 liter (l)	= 1000 milliliters (ml)
	= 0.001 cubic meter (m^3)
1 milliliter (ml)	= 1 cubic centimeter (cm^3)

Nonmetric Measure

1 cubic foot (ft^3)	= 1728 cubic inches (in^3)
1 cubic yard (yd^3)	= 27 cubic feet (ft^3)

Conversions

1 cubic meter (m^3)	= 264.2 gallons (US) (gal)
	= 35.314 cubic feet (ft^3)
1 liter (l)	= 1.057 quarts (US) (qt)
	= 33.815 fluid ounces (US) (fl oz)
1 cubic centimeter (cm^3)	= 0.0610 cubic inch (in^3)
1 cubic mile (mi^3)	= 4.168 cubic kilometers (km^3)
1 cubic foot (ft^3)	= 0.0283 cubic meter (m^3)
1 cubic inch (in^3)	= 16.39 cubic centimeters (cm^3)
1 gallon (gal)	= 3.784 liters (l)

Mass

Metric Measure

1000 kilograms (kg)	= 1 metric ton (t)
1 kilogram (kg)	= 1000 grams (g)

Nonmetric Measure

1 short ton (ton)	= 2000 pounds (lb)
1 long ton	= 2240 pounds (lb)
1 pound (lb)	= 16 ounces (oz)

Conversions

1 metric ton (t)	= 2205 pounds (lb)
1 kilogram (kg)	= 2.205 pounds (lb)
1 gram (g)	= 0.03527 ounce (oz)
1 pound (lb)	= 0.4536 kilogram (kg)
1 ounce (oz)	= 28.35 grams (g)

Pressure

standard sea-level air pressure	= 1013.25 millibars (mb)
	= 14.7 lb/in^2

Temperature

To change from Fahrenheit (F) to Celsius (C)

$$°C = °F - 32/1.8$$

To change from Celsius (C) to Fahrenheit (F)

$$°F = °C \times 1.8 + 32$$

Energy and Power

1 calorie (cal)	= the amount of heat that will raise the temperature of 1 g of water 1°C (1.8°F)
1 joule (J)	= 0.239 calorie (cal)
1 watt (W)	= 1 joule per second (J/s)
	= 14.34 calories per minute (cal/min)

Credits

James St. John / Wikimedia Commons. Figure 11.12g: Dwergen-paartje / Wikimedia Commons. Figure 11.12j: Daderot / Wikimedia Commons. Figure 11.13b: A. Langenheim. Figure 11.15a-b, 11.18: Z. Burian. Figure 11.15c: Ghedoghedo / Wikimedia Commons. Figure 11.17a: Hans Steur. Figure 11.17b: Nobu Tamura. Figure 11.17c: Christian Fischer / Wikimedia Commons. Figure 11.17d: Гурьева Светлана / Wikimedia Commons. Figure 11.17f: Frank Mannolini. Figure 11.20a-c: N. Shubin. Figure 11.21a: Ghedoghedo / Wikimedia Commons. Figure 11.21b: J. Merck.

Chapter 12   Figure 12.1: Unknown (photographer), "Pennsylvanian Coal Forest Diorama, Hall of Earth History," AMNH Research Library | Digital Special Collections. Figure 12.2a-c: Chris Scotese. Figure 12.3a-b, 12.4b-f, 12.5c, 12.8b,d-f, 12.13a-b,e,g-i, 12.14a, 12.15, 12.17a-d, 12.19b-c, 12.20b-f, 12.22a: Donald Prothero. Figure 12.4a: NPS / Wikimedia Commons. Figure 12.4g, 12.8a: Callan Bentley. Figure 12.8c: Daniel Schwen / Wikimedia Commons. Figure 12.9b: Robert Corby / Wikimedia Commons. Figure 12.9c: Tobi 87 / Wikimedia Commons. Figure 12.11a: Stephanie Scheiber. Figure 12.11b: J.C. Crowell. Figure 12.13c: Luis Fernández García / Wikimedia Commons. Figure 12.13f: Dwergenpaartje / Wikimedia Commons. Figure 12.13d: Ghedoghedo / Wikimedia Commons. Figure 12.14b: James St. John / Wikimedia Commons. Figure 12.16: Smithsonian. Figure 12.17e: Ghedoghedo / Wikimedia Commons. Figure 12.17f: Jebulon / Wikimedia Commons. Figure 12.18a: Hcrepin / Wikimedia Commons. Figure 12.18b: Joerg Schneider. Figure 12.19d: Daderot / Wikimedia Commons. Figure 12.22b: Timur V. Voronkov / Wikimedia Commons.

Chapter 13   Figure 13.1, 13.18a,b: Julius T. Csotonyi / Houston Museum of Natural Science. Figure 13.2a-c: Chris Scotese. Figure 13.3c,d, 13.4a,b, 13.6a-e, 13.7f-g, 13.9a, 13.10c-f, 13.12a-d, 13.13a-c,e, 13.15a-e,i,k, 13.16a,d, 13.17a-d: Donald Prothero. Figure 13.4c: Famartin / Wikimedia Commons. Figure 13.4d: Adbar / Wikimedia Commons. Figure 13.4e: Nikater / Wikimedia Commons. Figure 13.9b: NEON ja / Wikimedia Commons. Figure 13.9c: Hannes Grobe, Alfred Wegener Institute / Wikimedia Commons. Figure 13.13d: Ghedoghedo / Wikimedia Commons. Figure 13.14a: Courtesy of Smithsonian Institution. Photo by B. Huber. Figure 13.14b: David Lazarus. Figure 13.14c: John Barron/USGS. Figure 13.15h: Hectonichus / Wikimedia Commons. Figure 13.15l: ivtorov / Wikimedia Commons. Figure 13.15j: Masahiro miyasaka / Wikimedia Commons. Figure 13.16b: Frederic A. Lucas / Wikimedia Commons. Figure 13.16c: chensiyuan / Wikimedia Commons. Figure 13.16e: Julius T. Csotonyi / Philip J. Currie Dinosaur Museum.

Figure 13.19: Édouard Riou, 1863. Figure 13.23b: Derek Ramsey, Chanticleer Garden, 2008 / Wikimedia Commons. Figure 13.24b: Lawrence Berkeley Laboratory, U.S. Government / Wikimedia Commons. Figure 13.24d: Nichalp / Wikimedia Commons.

Chapter 14   Figure 14.1, 14.5a,b, 14.15a-f, 14.17a-c, 14.19a,e, 14.21b, 14.34a,c, 14.35a, 14.41a,b, 14.42-44: Donald Prothero. Figure 14.2a,b: Chris Scotese. Figure 14.4a: Prakasam Muthusamy, Wadia Institute of Himalayan Geology, Dehradun, India. Figure 14.4b: Ahmad Aghahosseini, Tehran, Iran. Figure 14.6b: Woudloper / Wikimedia Commons. Figure 14.a: Verisimilus / Wikimedia Commons. Figure 14.11b: Famartin / Wikimedia Commons. Figure 14.12a: Kezee Takoma Park / Flickr / Wikimedia Commons. Figure 14.12b, 14.19d: Mark A. Wilson / Wikimedia Commons. Figure 14.12c: L.C. Ivany. Figure 14.19b: Chris Light / Wikimedia Commons. Figure 14.19c: John Lloyd / Wikimedia Commons. Figure 14.20: G. Retallack. Figure 14.21a: Gabeguss / Wikimedia Commons. Figure 14.21c: Marli Miller. Figure 14.26b, 14.34b: Nikater / Wikimedia Commons. Figure 14.37, 14.38, 14.40: © Jay H. Matternes. Figure 14.39: UNSM.

Chapter 15   Figure 15.1: Dirk Beyer / Wikimedia Commons. Figure 15.2a: Marli Miller. Figure 15.2b: Wellcome Collection Gallery, Photo number: V0025179 / Wikimedia Commons. Figure 15.3: Antoine Sonrel, 1879 / Wikimedia Commons. Figure 15.8b: Mark A. Wilson via Wikimedia Commons. Figure 15.10: Donald Prothero. Figure 16.17: Marc Hallett.

Chapter 16   Figure 16.1, 16.7b,d,e,g, 16.8a,b, 16.11: Donald Prothero. Figure 16.2: Benjamin Waterhouse Hawkins, "Evidence as to Man's Place in Nature" (Huxley, 1863). Figure 16.4, 16.7a: Didier Descouens / Wikimedia Commons. Figure 16.6a: Ghedoghedo / Wikimedia Commons. Figure 16.6b: Proceedings of the National Academy of Sciences. Figure 16.6c: Ryan Somma / Wikimedia Commons. Figure 16.7c: Tim White. Figure 16.7f: B. McGann. Figure 16.10: Photaro / Wikimedia Commons. Figure 16.12a-b: Bar-Maor. Figure 16.13a,b: Adolf Naef, 1926. Figure 16.14: Gagneux et al., Proceedings of the National Academy of Sciences.

Chapter 17   Figure 17.1: U.S. Geological Survey. Figure 17.4: NASA. Figure 17.7a: Michael Mann. Figure 17.8: National Snow and Ice Data Center/The Economist Group Limited. Figure 17.16a-c: Chris Scotese.

Appendix   Figure A.3-7, A.9,10, A15, A20: M.P. Williams. Figure A8, A11, A13, A14, A16-19, A21: Donald Prothero.

Index

Page numbers followed by *b*, *f* and *t* refer to boxes, figures, and tables, respectively.